BIRDS

MAMMALS

REPTILES

AMPHIBIANS

FISH

MOLLUSKS

OTHER MARINE INVERTEBRATES

LIFE LISTS

INDEX

Harper & Row's
Complete Field Guide to
North American Wildlife

Consultants

Birds
Lester L. Short
Curator of Ornithology
American Museum of Natural History

Mammals
Donald Patten
Curator of Mammalogy
Los Angeles County Museum of Natural History
and
Sydney Anderson
Curator of Mammalogy
American Museum of Natural History

Reptiles and Amphibians
Robert L. Bezy
Associate Curator of Herpetology
Los Angeles County Museum of Natural History

Fishes
Saltwater Fishes
John E. Fitch
Research Director
California State Fisheries Laboratory,
Department of Fish and Game, State of California

Freshwater Fishes
Robert R. Miller
Curator of Fishes
Museum of Zoology, University of Michigan

Mollusks
William K. Emerson
Curator of Invertebrates
American Museum of Natural History

Other Marine Invertebrates
James W. Nybakken
Professor and Staff Member
Moss Landing Marine Laboratory

Harper & Row's
Complete Field Guide to North American Wildlife

Western Edition

Covering 1800 species of birds, mammals, reptiles, amphibians, food and game fishes of both fresh and salt waters, mollusks, and the principal marine invertebrates occurring in North America west of the 100th meridian from the 55th parallel to the border of Mexico

Assembled by Jay Ellis Ransom

Illustrations by Biruta Akerbergs, Pamela Carroll, Paul Donahue, William Downey, Jennifer Emry-Perrott, Nancy Lou Gahan, John Hamberger, Walter Hortens, Michel Kleinbaum, Klarie Phipps, Stephen Quinn, Susan Thompson, Guy Tudor, Nina L. Williams, and John Cameron Yrizarry

Harper & Row, Publishers

New York, Hagerstown, Philadelphia, San Francisco
Cambridge, London, Mexico City, São Paulo, Sydney

1817

Library of Congress Cataloging in Publication Data

Ransom, Jay Ellis, 1914–
   Harper & Row's complete field guide to North American wildlife,
Western edition.
   Includes index.
   1. Zoology—North America. 2. Zoology—The West. 3. Animals—
Identification. I. Title.
QL151.R36 1981   591.978   79–2635
ISBN 0-06-181715-5

81 82 83 84 85 10 9 8 7 6 5 4 3 2

Contents

Plates follow page 402

Harper & Row's
Complete Field Guide to North American Wildlife

Sponsoring Editor
Nahum Waxman

Project Editor
Marta Hallett

Production Editor
Bernard Skydell

Project Design Director
Abigail Moseley

Manufacturing Direction
Thomas Malloy

Copy Chief
Dolores Simon

Manufacturing Coordination
Pat Slesarchik

Copy Editing
Carole Berglie
Duane Berry

Design Assistants
Sarah Hartman
Barbara Knight

Bernie Borok
Virginia Ehrlich
Janet Field
Rebecca Finnell

Indexer
Sydney Wolfe Cohen

Composition: TriStar Graphics
Color Separation: Sterling Regal Incorporated
Printing and Binding: R. R. Donnelley & Sons Co.
Jacket Printing: Longacre Press, Inc.

Introduction

Harper & Row's Complete Field Guide to North American Wildlife takes its inspiration from *Complete Field Guide to American Wildlife,* compiled by the late Henry Hill Collins, Jr., and published by Harper & Row in 1959. This important book, covering the Eastern United States and Canada only, was the first single-volume field guide covering all of the major families of larger wildlife—birds, mammals, reptiles, amphibians, fishes, mollusks, and other marine invertebrates.

In the mid-1960s a decision was made at Harper & Row to carry out important revisions in Mr. Collins's field guide and to add a second volume covering the Western United States and Canada. To this end, Jay Ellis Ransom was engaged to prepare a text for the Western Edition, and, in due course, boards of consultants were assembled to review and edit in detail the texts of both volumes.

Working with Mr. Collins's and Mr. Ransom's compilations, the two boards, one for the eastern volume and one for the western, reviewed exhaustively each and every entry, revising, rewriting, and bringing the content into conformity with scientifically attested descriptive information and the latest taxonomic thinking.

In addition, the eastern volume, which did not include species of the southeastern United States, was expanded to cover that region.

Fifteen of the country's outstanding wildlife artists were brought in to work on the projects. All but a dozen of the plates from the original eastern volume were scrapped and replaced with totally new art. The western volume contains only new art, done especially for this work.

Now, after more than ten years in the making, this remarkable collaboration is complete and we are pleased to present it for the use of both professional and amateur field natural scientists.

The purpose of the Field Guides is field identification. Included are more than 3300 species, described by appearance, habitat, life zones, behavior, reproduction, and other important aspects. Covered are all species as follows:

In the East
East of the 100th meridian, north through temperate Canada (approximately the 55th parallel) and south to southern Florida (the 26th parallel, taking in the peninsula but not the Keys). Included are food and game fish species to a depth of 25 fathoms and mollusks and other marine invertebrates to a depth of 10 fathoms.

In the West
West of the 100th meridian, north through temperate Canada (approximately the 55th parallel) and south to the Texas-Mexico border, including all major Texas species but not necessarily specialized coverage of all desert species). Included are all food and game fish species to a depth of 25 fathoms and other marine invertebrates to 10 fathoms.

Names and classification
The common and scientific names used in these books represent the viewpoint of the consultants concerned with each section. In

general the consultants follow the prevailing scientific checklists in their field, and in the instructions to their sections cite the authority they have used. In many individual cases these scientists have developed their own conclusions on certain points, and in these cases the terminology or the order of the materials may not conform to the checklists.

How to Use This Book

In trying to identify any species, note (and jot down, if possible) all the visible characteristics before the animal disappears from sight. If it is a bird, in addition to general color and size, try to determine the color of its beak, legs, feet, and undertail feathers. Notice whether it has wing bars, an eye-ring, white on its tail; stripes on its body; a patch on its cheek or a line over its eyes. Check the shape of its bill. It is important to remember that quick observation is an essential part of field identification, for frequently a species darts in and out of sight very rapidly. Thus, it is necessary for the observer to note visual characteristics quickly, and this can be achieved only if the observer knows what to look for in a particular group of animals.

For mammals, particularly important to note are color patterns—whether they be on the sides, back, or underparts. Color patterns and scales are the key to identification of snakes. So too the number and character of stripes for a lizard. In a fish, the color markings are of great import, but observe them, to the extent possible, while the fish is still alive, for, as with the marine invertebrates, once the animal dies, its colors will change and fade. Naturally, size, habitats, habits, and voice, if any, can be important clues to the identity of any species.

When you turn to the appropriate section of the book, check the ranges and eliminate all species not within your range and season. Check also habitat data, and eliminate those species which appear in a habitat not normally observed. Then turn to the illustrations, which combined with the text, will provide all necessary information for making an identification.

Be sure to note the following points in using these guides:

1. Entries are by common name, with the scientific (Latin) name below. Many marine invertebrate species, however, have no common names and are listed by scientific names only.

2. To the right, opposite each entry, is a set of boldface numbers providing illustration references. Numbers preceded by the designation *Fig.* refer to figures directly in the text. Such figures are generally no more than two or three pages away from the text entry they illustrate. Numbers not preceded by the designation *Fig.* and taking the form of two numerals separated by a colon (e.g., **32:5**) refer to an illustration on one of the plates, with the first half of the number indicating the plate and the second half indicating a specific figure on the plate (thus, **32:5** refers to figure 5 on Plate 32).

3. Plate captions include page references back to the text entries for the species illustrated, so that throughout this book cross-referencing in both directions is complete.

Some species may show color variations because of age, season, sex, environment, or molt—and, as a result, an animal seen in the field may not be colored exactly as the one illustrated. To the extent possible, this factor has been considered, so that in many cases the most prevalent color variations are illustrated. However, if this is not the case with the species sighted, read once again the text description to learn if the species seems to conform in every other respect. It is essential to use both the illustrations and the text for accurate identification. If there is doubt about the identity of a species, don't mark it as having been observed. Scrupulous accuracy in field identification is necessary for any field person.

Life List

This book can be used to keep a life list of species seen and identified. First, there is included in a section, just before the index, an alphabetical list of all the species in the book by category. Second, the reader may make notes in the index or in the margins of the book, right beside the text descriptions of each species identified. Some people may find it useful to specify the place and date of observation. In this way the guide can become a personal lifetime natural history diary of each new sighting.

Acknowledgments

For their aid in providing reference materials for these volumes, thanks go to Mildred Bobrovich, Assistant Librarian for Reference Services at The American Museum of Natural History; Jill Fairchild at the Sea Library in Santa Monica, California; and Mary Ann Nelson at the Duke University Marine Laboratory in Beaufort, North Carolina.

To the consulting editors particular thanks are due, for their exemplary efforts in guiding the plan of each chapter, nurturing and editing the text descriptions, and examining each and every piece of artwork.

And, of course, grateful acknowledgment to the artists, whose fine work and whose enthusiasm for the project have done so much to help make these volumes the accomplishment we believe they are. Special thanks to Guy Tudor, who contributed so much more than his paintings.

The Editors

Birds

Consulting Editor
Lester L. Short
Curator of Ornithology
American Museum of Natural History

Illustrations
Water Birds and Game Birds, Plates 1–7, 10–11, 17, 24–29
Guy Tudor

Flying Ducks, Plates 8–9 Michel Kleinbaum
Birds of Prey, Plates 12–16 Paul Donahue
Owls and Nightjars, Plates 30–31 Stephen Quinn

Shorebirds, Woodpeckers, Perching Birds, Plates 18–23, 32–52
John Cameron Yrizarry

Text Illustrations, Michel Kleinbaum, Guy Tudor,
John Cameron Yrizarry

Birds
Class Aves

A bird is an animal with feathers; most, but not all, birds fly. It is generally presumed that those that do not fly are descended from ancestors that did. For this ability, the bodies of birds are specially adapted. The bones are light but strong, and some contain air sacs connected with the lungs. Wings—adaptations of the vertebrate forelimb—are designed to propel the bird through the air. In flightless forms, they balance the bird as it runs, as in the ostrich, or propel it through the water, as in the penguin.

The tail is made entirely of feathers, with a bony base, in contrast to the mammals in which the bones run down the middle. The tail may be either long or short, and is sometimes forked in strong flying species. What looks like, and is generally called, the leg is more or less the equivalent of the human foot (tarsometatarsus), with toes at the end. What looks like a reversed knee is really a joint somewhat the equivalent of the human ankle, the true knee usually being hidden by feathers next to the bird's body. Most birds have four toes, three forward and one, the hallux, behind. Some birds, like the woodpeckers, parrots, and cuckoos, have two toes forward and two behind. In many birds the toes lock automatically around a perch when a resting birds lowers its body, allowing it to sleep without falling.

Birds have no teeth, so a bill adapted for the bird's preferred diet is used to tear, crush, or seize the flesh, seeds, or insects upon which a particular species subsists. Food is sometimes temporarily stored in a crop, an enlargement of the esophagus. The gizzard, or main stomach, is a tough organ, effectively replacing the teeth and heavy jaw muscles of other animals, and the process of digestion is often aided by stones or gravel, which the bird swallows for that purpose. The products of the digestive, excretory, and reproductive systems are all discharged through a common opening, the cloaca. Birds are warm-blooded, with a body temperature varying between 98°F and 112°F (36.7° and 44.4°C).

Birds show a complex mixture of unmodifiable ("instinctive"), somewhat modifiable, and highly modifiable (or "learned") behavior. Among the most remarkable of the essentially unlearned patterns is the migration flight of some young shorebirds that leave their summer homes independently of their parents and follow, for thousands of miles to their winter homes, a route they have never before traveled.

Evolution

Birds arose from the reptiles; the feather is, in essence, a modified scale such as still occurs on their legs and feet. Birds are seldom preserved as fossils, and their record in the rocks is still scant compared with other groups, such as the mammals. It is known, however, that the gull-like *Ichthyornis* and the flightless, somewhat loonlike *Hesperornis* had already appeared by the Cretaceous period, and that a giant flightless land predator, the seven-foot (2.1-m) tall *Diatryma,* inhabited what is now Wyoming in the Eocene Epoch.

In the Pleistocene, in what is now Los Angeles, the Giant Condor, *Teratornis,* was occasionally engulfed in the La Brea tar pits. Almost into historic times the flightless Elephant Bird, *Aepyornis,* persisted on the island of Madagascar. One of the flightless moas of New Zealand, which reached a height of over ten feet (3.1 m), was contemporaneous with early humans, who exterminated the last of them.

Adaptation

Within the stringent limitations imposed by flight, birds have adapted in diverse ways to varied environments. Birds in their thousandfold varieties are the unchallenged champions of the air. Arctic Terns may fly 22,000 miles (35,405 km) a year in their extended migrations. Some swifts reach speeds of up to 200 miles (321 km) per hour. Hawks, vultures, and albatrosses can soar for long periods, hardly flapping a wing. And, because they are warm-blooded, birds can survive in climates from the Poles to the Equator.

Conservation

Birds have long served humans for game, food, and feathers, as well as in their predatory capacity as destroyers of insects and rodents. Because of the prodigious abundance of wildlife in the early days, the possibility of the permanent disappearance of any species seemed inconceivable. As a result, conservation came too late to save the Great Auk, Labrador Duck, Passenger Pigeon, and Carolina Parakeet, although the Passenger Pigeon was once probably the single most numerous bird species in the entire world.

In the 1890s the public was finally shocked into action at the disappearance of so great a number of birds. The National Audubon Society was formed and set out to save the egrets, which the hunters of plumes for women's hats had almost totally destroyed. Long, hard campaigns resulted in laws that protect virtually all birds. A recent step forward is the Federal Rare and Endangered Species Program.

Urbanization, industrialization, the draining of wetlands, and unwise agricultural practices are continually eliminating the habitats and food supplies of many species of birds, thus causing the greatest modern threat to birds. Also, the widespread use of pesticides has been implicated in the reduction in numbers of certain birds.

Habitat

Birds can be found almost anywhere—even in city parks and backyards. The best places, however, are where different types of habitat meet, such as the edges of woods, shorelines, and marshes. National, state, and county parks, wildlife refuges, and nature sanctuaries are good places to find birds. Gardens, parks, and open suburbs also may have a variety of species, particularly where sufficient undergrowth and cover have been left, and where dead trees and dead limbs (for nesting holes and exposed perches) have not all been cleared away. On the other hand, although deep woods, desert, mountains above timberline, and certain other places of uniform habitat are not rich in birds, the species found there include some not likely to be seen elsewhere.

The best time of day for bird-watching is the early morning, when birds start singing and actively feeding. During the heat of the day most land birds become inactive and silent, but some will start singing and feeding again toward evening. There may be considerable bird activity throughout the day when it is cloudy or raining lightly. Spring, when the migrating birds in breeding plumage are hurrying north, is the best time of the year to see birds; however, the drawn-out fall migration, with the birds in nonbreeding or immature plumage, offers a greater challenge.

Fig. 1
Parts of a Typical Bird

See Glossary for:
axillars
carpal
speculum

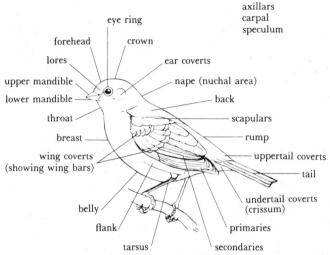

- eye ring
- forehead
- crown
- lores
- ear coverts
- upper mandible
- nape (nuchal area)
- lower mandible
- back
- throat
- scapulars
- breast
- rump
- wing coverts (showing wing bars)
- uppertail coverts
- tail
- belly
- undertail coverts (crissum)
- flank
- primaries
- tarsus
- secondaries

Voice

It is important to be able to identify birds by sound, especially if the bird in question is hidden by foliage or silhouetted against the sky. In some cases, it is the best means of distinguishing between species that visually may appear almost identical.

For each species discussed in the text, such data on voice are given as may be useful for identification. Whenever possible, the song or call has been rendered into English words, or into syllables that can be pronounced. Inadequate as any graphic attempt to portray sound must be, the use of English phrases is the most useful memory aid.

Food

The food a bird eats is correlated closely with the kind of bill, or beak, it has. A sparrow with a stout, conical beak is adapted to a diet of hard-coated seeds; a hawk with a sharp, hooked beak, to the tearing and shearing of flesh. Knowledge of a bird's diet is always of interest and may be of help in identification. Hence, brief descriptions of the principal food of each species are given.

Reproduction

All birds are hatched from eggs laid by the female, usually in a nest, sometimes on the bare ground. The eggs are incubated by the female or male, or by both alternately. In North America, the cowbirds are a notable exception to this practice; they lay their eggs in the nests of other birds.

Incubation takes from eleven to sixty-three days. The young of some species, such as ducks, grouse, and shorebirds, are able to run about almost as soon as they are hatched; these are termed *precocial*. But the *altricial* young of the majority of species must be fed in the nest for several days or weeks until they are fledged. Mating is customarily preceded by a more or less elaborate courtship performance. When their plumages differ, males of the

5

species, except in a few cases, are generally more brightly colored than the females. However, in many groups, such as herons, shorebirds, and sparrows, the sexes are alike in plumage. In the birds of prey the female is often larger than her mate.

Nest and Eggs
With a little practice, the nests and eggs of many species can be identified in the field. The concise descriptions given in the text will be of assistance, but the advanced student will want to use one of the several available guides to birds' nests and eggs. There is often much variation in the color and markings of birds' eggs, so the descriptions given are of typical examples only. When found in the field, a bird's nest should not be disturbed, nor robbed of its eggs.

Plumage
Most birds have more than one plumage; that is, covering of feathers. The variations that occur in the coloration of certain birds usually correspond to specific periods of the year (seasonal), sexual differentiation (male or female), or to stages of development (adult and immature).

There is almost always a difference between the immature, those incapable of breeding, and adult plumages. This is especially prolonged in the water birds and birds of prey. In some groups, gulls for example, there are distinctions between the juvenals and immatures as well. During the juvenal period, usually when the bird is first out of its nest in summer, a bird may exhibit one plumage; this may distinctly change to another plumage when the bird subsequently grows into its immature stage.

Changes that relate to a period of time in the life of the bird may also appear in adults, birds which are capable of reproduction. Generally, the terms *spring, fall, winter,* and *summer,* in relation to plumage, apply to the differences in adult breeding and nonbreeding plumages, rather than to the season of the year. *Spring* and *summer* usually apply to breeding plumages; *fall* and *winter* to nonbreeding. A spring bird, for example, is generally considered to be an adult in its bright breeding plumage. In this context, the term does not apply strictly to the spring season as we think of it (March to June), although it is loosely related to this time of year. The bird may be in this plumage somewhat before, and for a time after, the season as we know it. By the same definition, the "fall" plumage, as used herein, may be acquired by midsummer in a group such as the shorebirds.

The most common plumages of the individual species are discussed within the text when they are a distinct factor in identifying a species.

Range
The area of coverage for this chapter includes species found in eastern North America north of the 25th parallel, and west to the 100th meridian. The central prairie states along the axis of the 100th meridian constitute the "twilight zone," where the ranges of eastern and western birds converge and overlap, to some extent. Ranges apply to the species as a whole only within the geographic coverage of the chapter.

Most birds within the given range that are not permanent residents breed in the north in summer and migrate south for the winter.

Range boundaries are seldom clearly defined in the wild and tend to change slightly as time passes; hence the ranges given should be understood to be approximations only. Species may occasionally be found somewhat outside, or be absent from some space inside, the indicated range. Within a range area, of course, a species will normally occur only in its usual habitat.

Nomenclature

Except in a few cases in which recent studies indicate a change, the English and scientific names and the taxonomy of birds used in this chapter are those set forth in the 1957 *Check-list of North American Birds* by the American Ornithologists' Union, and supplements since published in *The Auk*. The familes, genera, and species are arranged more or less in phylogenetic order; that is, the sequence employed by taxonomists to reflect their understandings of avian evolution, from more primitive to more advanced types. Because of gaps in knowledge and difficulties posed by any linear sequence, the taxonomic arrangement is subject to constant reinterpretation and change. To aid in identifying, the most specific characteristic appears in italics, preceded by a bullet (•), under the description heading for most species.

Illustrations

Virtually every bird that is discussed in this chapter is illustrated and, where appropriate, varying plumages are shown. All the birds are adults unless otherwise specified in the caption. If there is no gender symbol shown, it is because the sexes are identical, or nearly so.

A page of illustrations divided by a rule indicates that there is a change in scale. Otherwise, all the species shown are proportionately sized on the page.

USEFUL REFERENCES

Books

American Ornithologists' Union. 1957. *Check-list of North American Birds*. 5th ed. Baltimore: American Ornithologists' Union.

Austin, O. L., Jr., and Singer, A. 1961. *Birds of the World*. New York: Golden Press.

Bailey, A. M., and Neidrach, R. J. 1965. *Birds of Colorado*. Denver: Museum of Natural History.

Baker, J. H. 1941. *The Audubon Guide to Attracting Birds*. New York: Doubleday & Co.

Barton, R. 1955. *How to Watch Birds*. New York: McGraw-Hill.

Behle, W. H. 1958. *Bird Life of the Great Salt Lake*. Salt Lake City: University of Utah Press. Illus.

Bent, A. C. 1919–68. *Life Histories of North American Birds*. 23 vols. Washington, D.C.: U.S. National Museum.

Berger, A. J. 1971. *Bird Study*. Reprint. New York: Dover Publications.

Burleigh, T. D. 1972. *Birds of Idaho*. Caldwell, Idaho: Caxton.

Collister, A. 1961. *Birds of Rocky Mountain National Park*. Denver: Museum of Natural History.

Darling, L., and Darling, L. 1962. *Bird*. Boston: Houghton Mifflin.

Gilliard, E. T. 1958. *Living Birds of the World.* New York: Doubleday & Co.

Gabrielson, I. A., and Jewett, S. G. 1970. *Birds of the Pacific Northwest.* New York: Dover Publications.

Godfrey, W. E. 1966. *The Birds of Canada.* Bulletin 203. Ottawa: National Museum of Canada.

Gromme, O. W. 1963. *Birds of Wisconsin.* Madison: University of Wisconsin Press.

Harrison, H. H. 1975. *A Field Guide to Birds' Nests.* Boston: Houghton Mifflin.

Hoffmann, Ralph. 1927. *Birds of the Pacific States.* Boston: Houghton Mifflin.

Lanyon, W. E. 1963. *Biology of Birds.* Garden City: Natural History Press.

Oberholser, H. C.; Kincaid, E. B. et al., ed. 1974. *Bird Life of Texas.* Austin: University of Texas Press.

Palmer, R., ed. 1962– . *Handbook of North American Birds.* Vols. 1, 2, 3. New Haven: Yale University Press.

Peterson, R. T. 1947. *A Field Guide to the Birds.* Boston: Houghton Mifflin.

———.1961. *A Field Guide to Western Birds.* Boston: Houghton Mifflin.

———.1963. *The Birds.* New York: Time-Life Books.

———.1953. *A Guide to Bird Finding West of the Mississippi.* New York: Oxford University Press.

Phillips, A., et al. 1964. *Birds of Arizona.* Tucson: University of Arizona Press.

Pough, R. H. 1946. *Audubon Land Bird Guide.* New York: Doubleday & Co.

———.1951. *Audubon Water Bird Guide.* New York: Doubleday & Co.

Reed, C. A. 1965. *North American Birds' Eggs.* New York: Dover Publications. Illus.

Reilly, E. M., Jr. 1968. *The Audubon Illustrated Handbook of American Birds.* New York: McGraw-Hill.

Robbins, C. S.; Brunn, B.; Zim, H. S.; and Singer, A. 1966. *A Guide to Field Identification: Birds of North America.* New York: Golden Press.

Small, A. 1975. *Birds of California.* New York: Macmillan. Illus.

Stefferud, A., ed. 1966. *Birds in Our Lives.* Washington, D.C.: U.S. Department of Interior.

Udvardy, M. 1977. *The Audubon Society Field Guide to North American Birds, Western Region.* New York: Knopf.

Van Tyne, J., and Berger, A. J. 1976. *Fundamentals of Ornithology.* New York: John Wiley.

Welty, J. C. 1975. *The Life of Birds.* New York: W. B. Saunders.

Wing, L. W. 1956. *Natural History of Birds.* New York: Ronald. Lists many state and local books and periodicals.

Periodicals

American Birds. Bimonthly. National Audubon Society, 950 Third Avenue, New York, New York.

Audubon Magazine. Bimonthly. National Audubon Society, 950 Third Avenue, New York, New York.

The Auk. Quarterly. American Ornithologists' Union, Allen Press, Inc., Lawrence, Kansas.

Birding. Bimonthly. American Birding Association, Inc., Box 4335, Austin, Texas.

The Condor. Quarterly. Cooper Ornithological Society, Allen Press, Inc., Lawrence, Kansas.

The Wilson Bulletin. Quarterly. Wilson Ornithological Society, Allen Press, Inc., Lawrence, Kansas.

Records and Tape Cassettes

Gunn, W. W. H., and Kellogg, P. P. 1962. *A Field Guide to Western Bird Songs.* Boston: Houghton Mifflin. More than 500 western species, to accompany Peterson's *A Field Guide to Western Birds,* 2nd ed.

GLOSSARY

♀—Female
♂—Male

Axillar (*pl.* axillars) Feathers of the underwing, between the wing lining and the body ("armpits").

Breeding An adult bird in the plumage in which it reproduces.

Carpal Relating to the wrist (carpus).

Displaying A male bird in breeding display pose.

Form A distinct plumage type, such as of a subspecies, or color phase.

Immature A bird in its first year, usually the first fall and winter following juvenal plumage; occasionally in large birds may extend to second or third year.

Juvenal A bird in its first plumage after natal down during summer; followed by immature plumage.

Molt The loss of one feather plumage and acquiring of another.

Nonbreeding An adult bird that is in a plumage other than that in which it reproduces.

Phase A coloration other than the normal plumage, irrespective of sex, age, or season.

Race A subgroup (or subspecies) within a species.

Resident A species that breeds and winters in the same general latitude.

Sibling Species Closely related, look-alike species that coexist.

Speculum A patch of color, usually iridescent, on the secondaries; mainly waterfowl.

Sub-Adult A plumage between immature and adult; sometimes or regularly breeds in this stage.

Water and Game Birds and Birds of Prey

Loons
Order Gaviiformes

LOONS
Family Gaviidae

Loons are open-water swimming birds midway in size between ducks and geese, with strong, straight, daggerlike bills and thick necks. Both sexes are dark above and white below, with white spots on their backs in summer. Loons can swim long distances under water. On slender, pointed wings they taxi along the surface of the water, finally taking off. They fly with rapid and fairly deep wingbeats that are somewhat slower than those of ducks. Their extended, drooping necks and broad, webbed feet trailing rudderlike behind their stubby tails give them a hunchbacked flight outline. Loons are northern birds that feed on fish and are seen in most of the West area only on migration and in winter.

Loons have more white below than do cormorants and they hold their evenly tapering bills straight. (Cormorants are blacker with a hook at the tip of their bills and long tails; immature birds have thinner necks.) All four of the world's loons are found in western North America.

COMMON LOON
Gavia immer   1:1

Description
Size, 28–36 in. (71.7–91.4 cm). Large; long-bodied, low-swimming; neck thicker than in Arctic or Red-throated; bill heavy and held straight. Summer: dark head, black bill, • *back evenly spotted with white squares* visible only at close range, broken white collar, underparts white. Winter: hindneck and back unspotted gray; throat, cheek, and underparts white.

Similarities
Red-throated is smaller and with lighter back. May also be confused with cormorants.

Habitat
Breeds near large, deep, conifer-bordered lakes; bays, ocean, open lakes; more often in fresh water than other loons.

Habits
As all loons, swims low in water, sometimes with only neck and head showing; patters on surface into wind to take flight; ungainly on land; flies with sagging neck and feet, wingbeats slower than a duck's.

Voice
Wild, quavering (often compared to insane laughter), ringing *ha-oo-oo,* often at night; flight call, a barking *kwuk.* Seldom heard in winter.

Eggs
2; olive-brown, spotted; 3.5 x 2.2 in. (8.9 x 5.6 cm). Nest is debris
pile on grassy lake shore, islet or abandoned muskrat house.
Range
Breeds throughout Canada and in n. U.S.; winters along Pacific
Coast.

ARCTIC LOON
Gavia arctica   1:2

Description
Size, 23–29 in. (58.4–73.7 cm). • *Bill straight* or downcurved, thin.
Summer: • *head gray;* squarish white spots on scapulars, • *2 patches*
on each side. Winter (a difficult identification): • *head and neck
grayish,* back black (often with scaly appearance), body small and
thin, bill slender.
Similarities
Common is larger. Red-throated's bill is not as straight; bill in
winter upturned and back is speckled, whereas Arctic's pale-edged
feathers give a mottled or scaly effect.
Habitat
Large, cold lakes, ponds, tundra waters, ocean; in winter, most
maritime of loons.
Voice
Loud squalls, screams, or deep, barking *kwow;* seldom heard in
winter.
Eggs
2; olive-buff with dark spots; 3.0 x 1.8 in. (7.6 x 4.6 cm).
Range
Breeds in Arctic; winters along Pacific Coast, s. to Gulf of Calif.

RED-THROATED LOON
Gavia stellata   1:3

Description
Size, 24–27 in. (61.0–68.6 cm). Tilts its sharp, • *thin bill* upward.
Summer: head gray, nape striped, rusty-red throat patch,
unpatterned • *back brownish-black speckled with white.* Winter:
head and neck gray above, white below; back brownish-gray, more
conspicuously speckled with white.
Similarities
Other loons are darker colored and heavier. Arctic Loon's bill is
slender, straight. Common Loon is larger, more robust, bill
straight. See also Western Grebe and Red-necked Grebe.
Habitat
Tundra lakes, fresh- or saltwater ponds, bays, estuaries, ocean;
usually salt water other seasons than summer.
Habits
Springs into air without pattering on water; often flies when
disturbed instead of diving; walks, but not easily.
Voice
Usually silent; in Arctic, various notes.
Eggs
2; brownish-olive, sparingly spotted; 2.8 x 1.8 in. (7.1 x 4.6 cm).
Nest is muddy platform at edge of tundra lake or grassy pool, or
on islet.
Range
Breeds along Arctic coastline and on tundra lakes, s. along Pacific
Coast to B.C.; winters along Pacific Coast from Aleutians to
Mexico.

Grebes
Order Podicipediformes

GREBES
Family Podicipedidae

Grebes are highly aquatic, expert swimmers and divers distinguishable from ducks by their thin necks, pointed bills, and no apparent tails. They sit low in the water with lobed-toed feet placed well to the rear and swim with head erect. Labored fliers, they seem to run along the surface prior to takeoff and in flight, fly with neck inclined downward. They are almost helpless on land. They feed on fish and aquatic invertebrates. Their range is nearly worldwide and, of nineteen species, six occur in western North America.

RED-NECKED GREBE
Podiceps grisegena   1:4

Description
Size, 18–22½ in. (45.7–57.2 cm). Large; body short, neck long; • *bill straight, dull yellow.* Summer: dark brown, crown black and slightly tufted, • *cheeks white.* Winter: cap dark, neck red, body gray, • *2 white patches* on each wing and (except in some first-year birds) white crescent on cheek, no contrast between blending—dark upperparts and pale underparts. Neck thicker, head and bill heavier than in other grebes. Most loonlike of all grebes (but no loon has wing patches).
Similarities
Common Loon has larger, longer body; shorter neck; lighter face and neck in winter. Horned and Eared Grebes in winter are smaller, head and back not so heavy, neck paler, bills smaller. Red-breasted Merganser has faster wingbeats, holds neck more horizontally, and shows 1 wing patch in flight. See also Western Grebe.
Habitat
Freshwater ponds, lakes in summer; coastal salt water in winter.
Voice
Usually silent but various loonlike wails and trills; when nesting, a high *keck.*
Eggs
3–5; bluish-white, nest-stained; 2.1 x 1.3 in. (5.3 x 3.4 cm). Nest is reedy floating raft in marshy lake.
Former name
Holboell's Grebe.
Range
Breeds in nw. Canada and n. U.S., from Alaska to Minn., s. to Wash. and e. to s. Wis.; winters along Pacific Coast, from Alaska to cen. Calif.

HORNED GREBE
Podiceps auritus   1:6

Description
Size, 12–15¼ in. (30.5–38.7 cm). Like a small compact duck with a short, pointed bill. Summer: head black, crown and cheeks dark green, ear tufts orange-buff or golden, foreneck rufous, neck and flanks dark chestnut. Winter: gray above; cheeks, foreneck, and underparts clear white. Large white wing patch visible in flight.

Similarities
See Eared Grebe.
Habitat
In summer, freshwater ponds, lakes; in winter, fresh and salt water.
Eggs
4–5; bluish- or olive-white, heavily nest-stained; 1.7 x 1.2 in. (4.3 x 3.0 cm). Nest is floating or anchored to reeds in pond.
Range
Breeds in n. Canada and n. U.S., from Yukon to Man., s. to Wash. and e. to N.Dak.; winters along Pacific Coast, from Alaska to n. Baja Calif.

EARED GREBE
Podiceps nigricollis   **1:7**

Description
Size, 12–14 in. (30.5–35.6 cm). A small diver with a thin neck, dark back, and • *upward-curved bill.* Summer: • *black crest;* neck and back black; golden ear tufts, orange cheeks; flanks dark chestnut. Winter: gray above, light below; lower cheek and spot behind ear are whitish; • *sides of neck grayish.*
Similarities
Horned Grebe has chestnut neck in summer; whiter neck, white face patch, no white spot behind ear in winter; no upturned bill.
Habitat
Lakes, bays, ocean; breeds colonially in shallow, marshy parts of lakes; many winter on salt water.
Voice
Mellow *"poo-eep, poo-eep";* a froglike *"hick-rick-up, hick-rick-up"* (Dawson).
Eggs
4–5; white, stained; 1.7 x 1.2 in. (4.3 x 3.0 cm). Nest is usually a floating mass anchored among freshwater reeds, in colonies.
Range
Breeds from sw. Canada, s. to U.S. desert; winters from Vancouver to Mexico.

LEAST GREBE
Podiceps dominicus   **1:9**

Description
Size, 8–10 in. (20.3–25.4 cm). • *Slate-colored;* very small. • *White wing patches* (often concealed); dark undertail coverts, • *bill black,* slender, pointed; • *eyes golden.* In winter, dark except whitish throat.
Similarities
See larger Pied-billed. Eared Grebe has no wing patches.
Habitat
Ponds, lakes.
Voice
Varied, from ringing *beep* or *peet* to a "trill or chatter" (James); a piercing, reedy *queek;* a rolling *ker-r-r-r-r* (Davis).
Eggs
4–6; buffy; 1.2 x 0.9 in. (3.3 x 2.3 cm). Nest is partially floating reed raft in water or fastened to reeds.
Range
Resident from s. Tex. to S. America.

WESTERN GREBE
Aechmophorus occidentalis   **1:5**

Description
Size, 22–29 in. (55.9–73.7 cm). • *Two-toned, black-and-white.*
Large, long neck; head high above water; dark gray above, white
below; crown and hindneck black; cheek, foreneck, wing patch, and
underparts white; bill light yellow, slightly upturned or straight;
eyes red.

Similarities
Red-necked in winter is smaller, bill duskier yellow, plumage more
dingy gray, especially neck. Loons have shorter neck, no wing
patches.

Habitat
Ponds, lakes; winters on either fresh or salt water; likes bays,
sloughs, ocean.

Habits
Neck droops in middle in flight.

Voice
A "loud, double-toned, whistled *c-r-r-ee-r-r-ee*" (Chapman); also a
rolling croak.

Eggs
3–6; bluish-white to olive-buff, stained; 2.3 x 1.5 in. (5.8 x 3.8
cm). Nesting is colonial; nest usually floating among reeds on
rushy lakes.

Range
Breeds from B.C. and Man., s. to s. Calif. and Mexico and e. to
Plains states; winters along Pacific Coast, from B.C., s. to Mexico
and inland.

PIED-BILLED GREBE
Podilymbus podiceps   **1:8**

Description
Size, 12–15 in. (30.5–38.1 cm). Most likely grebe to be seen in
most areas. Brown above, light below; characteristic "chicken bill,"
short, heavy, whitish; • *black throat and ring around bill* in
summer, black-looking in winter. Swims with tail high, revealing
conspicuous white undertail coverts.

Similarities
Horned, Eared, Least Grebes have slenderer bills, no white
undertail; other grebes have white wing patches.

Habitat
Freshwater streams, ponds, lakes, marshes; saltwater bays in
winter.

Habits
As most grebes, can dive or submerge by sinking whole body until
only head shows. Neck droops in flight.

Voice
Series of harsh *cow-cow-cow-cow-cow-cowm-cowm* notes on 1
pitch, somewhat cuckoolike.

Eggs
5–7; indistinguishable from those of Horned Grebe; 1.7 x 1.2 in.
(4.3 x 3.0 cm). Nest is reedy raft, semifloating.

Range
Breeds from Vancouver and s. Mackenzie, s. to S. America;
winters from cen. B.C., s. throughout w. and s. U.S.

Albatrosses, Fulmars, Shearwaters, and Petrels
Order Procellariiformes

These web-footed, powerfully winged water birds have their
nostrils in a tube on top of the bill. Their home is the open sea,
but they breed on islands.

ALBATROSSES
Family Diomedeidae

These gliding birds of the open seas are much larger than gulls.
They normally range the icy seas of the Southern Hemisphere; of
the world's thirteen species, only one occurs regularly in western
North American waters close to shore.

BLACK-FOOTED ALBATROSS
Diomedea nigripes   **3:6**

Description
Size, 28–36 in. (71.1–91.4 cm). Long, saberlike wings (spread to 7
ft.); • *all-dusky* color; whitish face, pale areas near wing tips; may
show white patch at base of tail; bill and feet dark.
Habitat
Open ocean, occasionally seen from shore.
Habits
Rigid gliding flight.
Range
Breeds on nw. Hawaiian Is.; winters at sea; can be seen in spring
and fall along Pacific Coast of N. America.

FULMARS AND SHEARWATERS
Family Procellariidae

These web-footed, gull-sized birds of the world's open oceans can
glide for long periods on stiff wings close to the surface of the sea.
They are good swimmers, feeding on small fish and squid,
sometimes crustaceans and plankton; they often crowd around ships
for garbage. Wings are narrower than a gull's, tails smaller and
less fanlike; bills (except Northern Fulmar's) are thin with tubelike
external nostrils fused (fulmars) or separate (shearwaters).
Shearwaters are dark above and, except for the Sooty, white below.
Shearwaters are mostly silent at sea, with occasional grunting,
croaking, or gull-like sounds when feeding. They nest on remote
islands or sea cliffs. Of the world's sixty-six species, only six occur
in West Coast waters.

NORTHERN FULMAR
Fulmarus glacialis   **2:7**

Description
Size, 17–20 in. (43.2–50.8 cm). Light phase: back and mantle gray;
head, tail, and underparts white; wings pale gray, darker toward
tips; pale patch at base of primaries; legs bluish. Dark phase: dark
gray all over, bill dusky-brown, wing tips darker. Intermediates
occur frequently. Stubby yellow bill with its tubed nostrils, together
with its manner of flight, distinguish it from gulls.

Similarities
Shearwaters do not flap as much. Dark phase of Sooty is similar to Fulmar, but darker. Gulls do not have the thick bull-neck, stubby tubed bill, or stiff-winged flight.
Habitat
Open ocean.
Habits
Follows ships; a scavenger.
Voice
Grunts, chuckles, cackles.
Eggs
1; white; 2.9 x 2.0 in. (7.4 x 5.1 cm); little or no nest on rock ledge.
Range
Breeds in Arctic; winters along Pacific Coast, s. to Baja Calif.

PINK-FOOTED SHEARWATER
Puffinus creatopus 2:8

Description
Size, 19–20 in. (48.3–50.8 cm). White-bellied, larger than Sooty. Feet pink, bill tipped with black.
Similarities
Manx is similar but smaller, blacker above, more sharply contrasting crown and hindneck, more white on underwing, faster wingbeats, bill pale, flesh-colored. Sooty is faster in flight with brighter bill.
Range
Breeds on islands off S. America; seen in migration off Pacific Coast.

Note: The **NEW ZEALAND SHEARWATER,** *Puffinus bulleri* **(2:5)**, size, 16½ in. (41.9 cm), has a pale gray back, tail coverts and wings. The wings have a prominent, dark M or N pattern. The belly is white, and the feet are yellowish. It is a regular fall visitor (never numerous) to the California coast, especially off Monterey in October; casual to Oregon and Washington.

SOOTY SHEARWATER
Puffinus griseus 2:4

Description
Size, 16–18 in. (40.6–45.7 cm). One of 2 all-dark shearwaters. Gull-like, but dark sooty-brown (black at distance); underwings sometimes lighter but hard to detect at distance; bill thin, black. Looks like a small, black gull that glides (not flaps) on stiff wings.
Similarities
Dark jaegers have base of primaries white, wings angled, hawklike wingbeats. Commonest shearwater, often in great flocks (into millions). Most abundant spring and fall.
Habits
Patters on surface before flight.
Voice
"Low, guttural *wok-wok-wok* when much excited" (Rich). Usually only heard during breeding.
Eggs
1; white; 2.9 x 1.9 in. (7.4 x 4.8 cm).
Range
Breeds in Australia and S. America, winters in N. Pacific at sea; migrates off Pacific Coast, from Alaska, s. to Calif.

MANX SHEARWATER
Puffinus puffinus   **2:12**

Description
Size, 12½–15 in. (31.8–38.1 cm). Only small, white-bellied
shearwater with long, white undertail coverts. Bill, cap,
• *upperparts black;* • *underparts white;* feet pinkish, wing
linings white.
Similarities
Pink-footed is much larger, with less contrast head to throat;
browner above.
Range
Breeds off Baja Calif.; winters at sea, from B.C., southward.

STORM-PETRELS
Family Hydrobatidae

These small dark seabirds appear to run or dance over the surface
of the sea on slender legs dangling webbed feet, often showing a
conspicuously white rump. Their habitat is the open ocean,
sometimes large bays; they feed largely on plankton. They nest in
burrows or rock crevices on offshore islands. Four of twenty-one
species in the world regularly occur in western North American
waters.

FORK-TAILED STORM-PETREL
Oceanodroma furcata   **2:11**

Description
Size, 8 in. (20.3 cm). • *Pearl-gray,* underparts nearly white.
Habits
Readily lands on water; dives.
Voice
High-pitched, twittery.
Range
Breeds from Aleutians to n. Calif. islands; winters at sea; seen
along Pacific Coast.

LEACH'S STORM-PETREL
Oceanodroma leucorhoa   **2:6**

Description
Size, 7½–9 in. (19.1–22.9 cm). Dusky-black; forked tail; wings
long, top of wing with pale band; • *rump conspicuously white;* legs
short.
Similarities
See Black Storm-Petrel.
Habits
Flight erratic, like Common Nighthawk; glides with wings
downcurved; swims in water with wings uplifted, then springs into
air.
Voice
When breeding, 8 low cooing notes; also twitterings, screams, trills
from burrows.
Eggs
1; white; 1.3 x 0.95 in. (3.3 x 2.4 cm). Nest is colonial; burrows on
islands.
Range
Breeds along Pacific Coast, from Aleutians, s. to Baja Calif.;
winters at sea in tropics.

ASHY STORM-PETREL
Oceanodroma homochroa   **2:10**

Description
Size, 7½ in. (19.1 cm). Small, all-black petrel with forked tail.
Head and neck ash-gray, white mottling under wings.
Similarities
Black Storm-Petrel has longer wings, legs; flight less fluttery.
Habits
Flight erratic.
Voice
Most vocal in breeding; twittering notes.
Eggs
1; white; 1.1 x 0.9 in. (3.0 x 2.3 cm). Nest is colonial, under rocks.
Range
Breeds on islands off Pacific Coast, from San Francisco, s. to Baja
Calif.; winters at sea off Pacific Coast.

BLACK STORM-PETREL
Oceanodroma melania   **2:9**

Description
Size, 9 in. (22.9 cm). Largest, most common of all-black storm-
petrels. Wings longer, flight very deep, lazier than other storm-
petrels, not unlike a Black Tern's.
Similarities
Ashy is smaller, with shorter wings and legs and fluttery flight.
Leach's is dusky-black with white rump and faster, more erratic
flight.
Habits
May follow ships.
Voice
A ventriloquial *"puck-apoo, puck-puck-a-poo"* (Fisher), heard
while nesting.
Eggs
1; white; 1.4 x 1.1 in. (3.75 x 2.75 cm); nest is colonial; in burrows
on islands.
Range
Breeds on islands in Gulf of Calif. and in s. Calif.; winters at sea
off Pacific Coast from n. Calif., s. to S. America.

Pelicans and Allies
Order Pelecaniformes

PELICANS
Family Pelecanidae

These large, fish-eating water birds have long, oversized pouched
beaks, long wings, and stout bodies. They fly with necks drawn in
and deflated pouches resting on breast, in single file or orderly V-
shaped lines, alternating several flaps with a glide in a follow-the-
leader pattern. Of the world's eight species, two are western North
American.

AMERICAN WHITE PELICAN
Pelecanus erythrorhynchos   *Fig. 2*

Description
Size, 54–70 in. (137.2–177.8 cm). Very large, • *white*; black
primaries and huge orange-yellow bill.
Similarities
Swans have long, thin necks, small heads, lack black wings. Wood
Stork and Whooping Crane are thin long-necked birds, have long
legs, and fly differently.
Habitat
Breeds inland on lakes, marshes; winters on saltwater bays, also
inland.
Habits
Soars, swims buoyantly and scoops up food with submerged bill.
Voice
Low croaks on breeding grounds.
Eggs
2–3; white; 3.5 x 2.2 in. (8.9 x 5.6 cm). Nest is of vegetation,
sticks, stones, debris on ground or in flattened vegetation; in
colonies on lake islands.
Range
Breeds in Canadian prairies and nw. U.S., from B.C. and
Mackenzie, s. to Nev., Utah, Wyo., N.Dak. and also in marshes w.
of Rocky Mountains; winters along Pacific Coast, from Calif. to
Mexico.

winter

summer

Fig. 2

Brown Pelican
adults

American White Pelican

BROWN PELICAN   **3:4**
Pelecanus occidentalis   *Fig. 2*

Description
Size, 44–55 in. (111.8–139.7 cm). Large, slow-flying. Adults:
• *brown*, with white about head and neck; immatures have
darker head, whitish underparts.
Habitat
Ocean, salt bays; coastal.
Habits
Flies in lines close to ocean surface, almost touching it with wing
tips, characteristically a few flaps and a long glide. Dark color and
spectacular habit of vertically diving bill-first into the sea from a
considerable height distinguish it from American White Pelican.
Often perches on seaside projections, boats.

Voice
Usually silent, rarely croaks.
Eggs
2–3; whitish; 2.8 x 1.8 in. (7.3 x 4.6 cm). Nest in colonial islands, on ground.
Range
Breeds locally along Calif. coast, from Pt. Lobos, southward; winters s. to S. America; seen in migration along s. Calif. coastline.

BOOBIES
Family Sulidae

These are large, cigar-shaped seabirds with big, pointed bills and pointed tails that fly on stiff, long, pointed wings. They are larger than gulls with longer necks, and "pointed" profiles, both sexes alike. They eat fish, which they capture by spectacular plunges from a considerable height.

BLUE-FOOTED BOOBY
Sula nebouxii   3:3
Description
Size, 32–34 in. (81.3–86.4 cm). Bulkier than a gull, resembling a young Brown Pelican at a distance; has longer neck, larger pointed bill, and a • *pointed tail*. White below and on rump, dark back, mottled white head, dark wings and tail, feet blue, face and bill dark. Immatures have brownish head.
Habitat
Coastal waters.
Habits
Dives for food mainly early and late in day.
Voice
Whistling and trumpeting notes.
Eggs
2 or 3; pale blue; 2.4 x 1.5 in. (6.25 x 4 cm).
Range
Breeds s. of U.S., from Gulf of Calif. to S. America; winters from s. Calif. coast or Salton Sea, southward.

CORMORANTS AND ANHINGAS
Family Phalacrocoracidae

These large, blackish, long-necked and slender-billed water birds are about the size of geese, but longer-tailed and darker. They often perch upright on projections showing their S-curved necks, sometimes with wings spread. The sexes are alike. Adults have long, hook-tipped bills, often have colorful faces, gular pouches. Cormorants swim well, sometimes with only the head, neck, and upturned bill showing. They fly in line or wedge formation, with heronlike flaps and intermittent glides. They are often confused with loons, which fly singly or in small flocks, with drooping necks; cormorants maintain the body axis at an upward tilt. They eat fish and crustaceans and nest in colonies on the rocky surface of islands or ledges of sea capes, sometimes in trees. The nest is of seaweed, sticks, twigs, and grasses. Of the world's thirty-seven species, four appear in the West.

DOUBLE-CRESTED CORMORANT
Phalacrocorax auritus   3:7

Description
Size, 30–36 in. (76.2–91.4 cm). Typical; distinguishable along
West Coast by • *orange-yellow throat pouch;* crest seldom evident.
Only cormorant normally occurring inland in western United
States. Adult: black, never any white on flanks or chin. Immature:
brown; breast and forebelly whitish, hindbelly black. Some adults
have bleached throat pouches which look yellowish; some
immatures have unusually light bellies.
Similarities
See Brandt's; also Pelagic, often with distinct double crest.
Habitat
Lakes, rivers, bays, ocean, coast.
Voice
Normally silent; a rare croak when alarmed.
Eggs
3–4; pale blue; 2.4 x 1.5 in. (6.1 x 3.8 cm).
Range
Breeds on islands off Pacific Coast, from Alaska, s. to Baja Calif.
and inland on lakes; winters along Pacific Coast, from Alaska
southward and inland to s. Ariz.

PELAGIC CORMORANT
Phalacrocorax pelagicus   3:8

Description
Size, 15½–30 in. (39.4–76.2 cm). Smallest species; iridescent
plumage, • *thin bill,* slender neck, small head. In spring breeding
season, has a double crest and in flight shows a • *white patch* on
each flank. Part of face (but not forehead) and gular pouch dull
red at close range. Immatures are deep brown, darker on back.
Habitat
Salt water; coastline, bays, sounds.
Habits
Relatively shy; dives into heavy surf.
Voice
Low croaking.
Eggs
3–6; pale blue; 2.2 x 1.4 in. (5.7 x 3.7 cm).
Range
Breeds from Bering Strait s. to c. Calif.; winters along Pacific
Coast, from B.C. s. to Baja Calif.

Note: In Alaska, the **RED-FACED CORMORANT,** *Phalacrocorax
urili,* size, 28–30 in. (71.1–76.2 cm) is similar but a little larger. Its
face and forehead are bright red.

BRANDT'S CORMORANT
Phalacrocorax penicillatus   3:5

Description
Size, 33–35 in. (83.8–88.9 cm). Black with no crest; dark blue
throat patch bordered by • *buffy-yellow band* at rear during
breeding season. Immatures are brown, underparts paler brown; no
white on breast.
Similarities
Very similar to Double-crested; difficult to distinguish, but has
orange pouch, not blue, and no yellow band; immature Double-
crested has whitish breast.

21

Habitat
Coastline, open sea.
Voice
Occasional low grunts.
Eggs
3–6; pale blue; 2.4 x 1.4 in. (6.2 x 2.8 cm).
Range
Resident along Pacific Coast from s. B.C. to Baja Calif.

ANHINGA
Anhinga anhinga   **3:2**

Description
Size, 32–36 in. (81.3–91.4 cm). Snaky-necked bird of swamps and
marshes, with long, pointed bill; thin neck; black or brown above
and below; long fanlike, white-tipped tail.
Habitat
Swamps, lagoons.
Habits
Often swims with head and bill alone above water. Spreads wings
to dry.
Voice
Long chattering call.
Eggs
2–5; chalky-coated, bluish-green; 2.0 x 1.4 in. (5.1 x 3.6 cm). Nest
is singly or in colonies, in trees.
Range
Resident in W., along Gulf Coast in s. Tex., southward.

FRIGATEBIRDS
Family Fregatidae

These are long-winged, gliding and soaring birds of the tropical
seas. They feed at the water's surface on jellyfish, fish, squid, and
sewage items, or by robbing other fish-eating birds. None breed in
the West.

MAGNIFICENT FRIGATEBIRD
Fregata magnificens   **3:1**

Description
Size, 37–41 in. (94.0–104.1 cm); wingspread to 8 ft. (2.4 m). Male:
glossy black. Female: browner with breast and sides of abdomen
white. Immature: white-headed are most often seen. Long, forked
tail and narrow, crooked wings with a long, hooked bill are
diagnostic. Feeds over sea, but in sight of land.
Range
Breeds s. of U.S.; reaches w. U.S., Calif. coast occasionally in
winter.

Long-Legged Waders
Order Ciconiiformes

These usually large wading birds of marshes, mud flats, and shores have long bills, necks, and legs and short tails.

HERONS AND BITTERS
Family Ardeidae

Members of this narrow-bodied family have daggerlike bills; slender necks; large, rounded wings; and long legs. They fly with necks drawn into an S, legs trailing, wingbeats slow and deliberate. Many fly in large flocks to evening roosting sites. They feed on fish and all manner of smaller aquatic vertebrates, insects, crayfish, mollusks, and spiders. The sexes are alike but may show plumes during breeding. They range worldwide, except the Arctic. Of the world's sixty-four species, ten occur in the West.

GREAT BLUE HERON
Ardea herodias   **5:1**

Description
Size, 42–52 in. (106.7–132.1 cm). Largest heron; tall, often miscalled a "crane." Adult: head white with 2 black plumes, neck brownish-gray, shoulder patches black, rest of plumage grayish-blue. Immature: duller, with dark cap and no plumes. Blue coloration and flight with head drawn in distinguish it from the similarly sized Sandhill Crane.

Similarities
Sandhill Crane larger, more uniform gray in color, and flies with neck extended.

Habitat
Shallow water: marshes, swamps, streams, irrigation ditches, rarely mud flats and kelp beds.

Habits
Flies with slow, steady, heavy wingbeats and can soar well; stands motionless for protracted periods in water when fishing.

Voice
Guttural squawks; a flat, harsh *honk*, "*frahnk, frahnk, frawnk*" (Peterson).

Eggs
4; greenish-white; 2.5 x 1.8 in. (6.4 x 4.6 cm). Nest is a platform of sticks in swamp trees, on rocky islets, in marshes; loosely colonial.

Range
Breeds throughout cen. Canada and U.S., from se. Alaska and Sask., s. to Mexico; winters from B.C. southward.

GREEN HERON
Butorides striatus   **4:9**

Description
Size, 16–22 in. (40.6–55.9 cm). Adult: neck chestnut, front white, dark cap, upperparts blue-green (appears dark blue or black at a distance), • *legs orange*. Immature: browner; throat streaked, • *legs greenish*.

Similarities
Least Bittern is only heron that is smaller.

Habitat
Fresh- and saltwater marshes, creeks, sluggish streams.
Habits
Flies crowlike but with wings more arched. Stretches neck, raises crest, and jerks tail when alarmed; frequently alights in trees.
Voice
Squawks, grunts; a loud *skyow* or *skewk;* series of *kuck*'s.
Eggs
3–6; greenish-white or bluish; 1.5 x 1.2 in. (3.8 x 3.0 cm). Nest is a flimsy platform of sticks; in tree, shrub or grass clump; often solitary and not necessarily near water.
Range
Breeds in W., from B.C., s. to Calif. and e. to Ariz., while others from E. range reach Tex.; winters from cen. Calif. e. to Tex.

LITTLE BLUE HERON
Egretta caerulea   **4:6**

Description
Size, 20–24 in. (50.8–61.0 cm). Medium-size, slender; • *bluish, black-tipped bill;* • *feet and legs greenish.* Adult: head and neck deep chestnut (no white in front), other parts slaty-blue; looks black at a distance. Immature: all-white. Changing birds show dark and white patches.
Similarities
Other egrets do not show the blue-chestnut combination. The white egrets lack black-tipped blue bill of the Little Blue.
Habitat
Fresh- and saltwater marshes.
Habits
Gathers at communal roosts at sunset; moves about actively when feeding; in flight, wingbeats more rapid than Great Blue or Great Egret.
Voice
Croaks and screams, *"tell you what, tell you what"* (Chapman).
Eggs
4–5; bluish-green; 1.7 x 1.3 in. (4.3 x 3.3 cm). Nest is of sticks, in bushes and trees near water.
Range
Chiefly E. bird; straggles w. to s. parts of Sw.

Note: The **LOUISIANA HERON,** *Egretta tricolor* **(4:3)**, size, 26 in. (66.0 cm), is a casual straggler west to Oregon and California; it is dark with a white rump, white belly, and white line down the front of the neck.

GREAT EGRET
Casmerodius albus   **4:2, 5:2**

Description
Size, 37–41 in. (94.0–104.1 cm). White; slender with • *yellow or orange bill, black legs and feet.*
Similarities
Snowy Egret smaller; Great does not rush around as much and has slower wingbeats.
Habitat
Streams, ponds, fresh- or saltwater marshes, irrigated fields, mud flats.
Habits
Flocks to evening roosts, waits motionless for prey.

Voice
Low, heavy croak; also *cuk, cuk, cuk.*
Eggs
3–4; bluish-green; 2.2 x 1.6 in. (5.6 x 4.1 cm). Nesting is colonial; nest a platform of sticks in trees or brush near water; in tules.
Former name
American or Common Egret.
Range
Breeds from s. Oreg. and Idaho s. to S. America; winters from sw. states s. to S. America.

SNOWY EGRET
Egretta thula   4:5

Description
Size, 20–27 in. (50.8–68.6 cm). All-white, with many beautiful plumes (aigrettes) on its back in breeding season; • *legs black, feet yellow,* bill narrow and black with yellow or red skin near base.
Similarities
Little Blue Heron immatures are white with greenish feet and legs, two-toned bill.
Habitat
Fresh- and saltwater ponds, marshes, tidal flats, shores, irrigated fields.
Habits
Shuffles legs and dashes about erratically when feeding.
Voice
Harsh hiss; a bubbling *wulla-wulla-wulla.*
Eggs
3–6; pale bluish-green; 1.7 x 1.3 in. (4.3 x 3.3 cm). Nest is colonial; a platform of sticks in trees, shrubs, tules, or marsh grass.
Range
Breeds in Calif., e. to Colo., and throughout sw. states; winters from Calif. southward.

REDDISH EGRET
Egretta rufescens   4:4

Description
Size, 27–32 in. (68.6–81.3 cm). Two color phases, both with rough, shaggy neck; bluish legs; heavy, • *black-tipped, pale pink bill.*
Phases are dark bluish-slate or all-white; dark phase young often all-gray with faint reddish on throat and forewing. In both phases young have uniformly grayish-black bills and greenish-black legs.
Similarities
White phase resembles other white egrets but black-tipped, pink bill and rapid movement of Reddish are diagnostic.
Habitat
Shallow saltwater marshes of coastal bays and open beaches.
Habits
Usually dashes about, actively hunting.
Voice
Guttural croaks in breeding season.
Eggs
3–4; bluish-gray; 2.0 x 1.5 in. (5.1 x 3.8 cm). Nest is on ground or in bush.
Range
Chiefly E. bird, but some along s. Tex. coastline and stragglers to Calif.

CATTLE EGRET
Bubulcus ibis 4:1

Description
Size, 20 in. (50.1 cm). Adult: white, with buff patches on head, breast, and back; buff pale or lacking in nonbreeding; reddish bill and legs. Immature: all-white; bill yellow, legs dark.
Similarities
Stockier than other white-colored herons; bill shorter; eye red. See Snowy and Great Egrets.
Habitat
Wet fields.
Habits
Often associates with grazing cattle, spreading in East.
Voice
Various croaks when breeding.
Eggs
4–5; bluish-white; 1.8 x 1.3 in. (4.6 x 3.3 cm). Nest is of sticks; in trees; in colonies.
Range
Breeds around Salton Sea (California), with some n. to s. Canada; winters s. from Calif.

BLACK-CROWNED NIGHT HERON
Nycticorax nycticorax 4:8

Description
Size, 23–28 in. (58.4–71.1 cm). Squat, stocky; short bill and neck; yellow legs, reddish when breeding. Adult: • *crown and back black,* wings gray, underparts and 2 long plumes from nape white. Immature: brown above with white spot-streaks, white below with brown streaks.
Similarities
See immature American Bittern.
Habitat
Coastlines, larger inland lakes, marshes, shores.
Habits
Active at night, often roosting in groups by day in dense bushes and trees.
Voice
Loud *quawk.*
Eggs
3–6; pale blue-green; 2.0 x 1.5 in. (5.1 x 3.8 cm). Nest is colonial; sticks lined with softer marsh grass in tree of shrub, not always near water.
Range
Breeds from cen. Canada s. throughout U.S. to Mexico; winters from n. U.S. southward.

Note: The **YELLOW-CROWNED NIGHT HERON,** *Nycticorax violacea* **(4:7),** size, 24 in. (61.0 cm), is a rare straggler in the West. It is gray below with a black throat; immatures are very like Black-crowned but brown predominates, white markings more restricted to spotlike form.

LEAST BITTERN
Ixobrychus exilis 4:10

Description
Size, 11–14 in. (27.9–35.6 cm). A tiny heron, rarely seen in marsh vegetation; • *buffy-orange wing patches;* body tiny and thin. Adult:

crown, back, primaries, and tail greenish-black; female brownish; cheeks and neck bright chestnut-buff; neck streaked in browner female; white line down either side of back; underparts buffy. Immature: like a pale female. In very rare dark phase, buff replaced by dark red-brown.

Similarities
Rails have no buff wing patches; see also much larger American Bittern.

Habitat
Cattail or reedy marshes.

Habits
Climbs on plant stems; slips through reeds like a rail; "freezes" like American Bittern when alarmed; flushes with fluttering wings, legs dangling, then quickly drops.

Voice
Dovelike or cuckoolike *coo-coo-coo-coo*.

Eggs
4–5; bluish-white; 1.2 x 0.9 in. (3.0 x 2.3 cm). Nest is solitary; of twigs and grasses; in thick vegetation over water.

Range
Breeds from Oreg., s. to Baja Calif. and e. of Rockies from s. Canada to S. America; winters from s. U.S., southward.

AMERICAN BITTERN
Botaurus lentiginosus   **4:11**

Description
Size, 23–34 in. (58.4–85 cm). Black mark on side of the neck; rather large, stocky. Rich brown above; tan below, heavily marked with brown; bill yellow; throat white; legs green; primaries black, contrasting with general streaky buff-brown of body.

Similarities
Immature Black-crowned Night Heron is grayer, has yellow legs, black bill; no black on wings or neck. Much smaller Least Bittern is more contrastingly patterned in buff and black.

Habitat
Fresh- and saltwater marshes, tules.

Habits
When alarmed, "freezes" with bill pointing straight up. In flight, the wings are less curved, wings beat faster than in other herons. Almost never in trees.

Voice
"Pumping sound, *plum pudd'n*" (Collins), repeated over and over in spring; note as it flies away, a *kok-kok-kok*.

Eggs
3–7; buffy-brown or olive; 1.9 x 1.4 in. (4.8 x 3.6 cm). Nest is solitary; platform of vegetation just above water in marshes.

Range
Breeds from s. Northwest Territories s. to lower Colo. R.; winters from sw. B.C. s. throughout U.S.

STORKS
Family Ciconiidae

The Wood Stork is a large, long-legged heronlike bird found in the West, and one of seventeen world species. Its family is characterized by a long bill, straight, recurved or decurved; several species have naked heads.

WOOD STORK (WOOD IBIS)
Mycteria americana   **5:5**

Description
Size, 34–47 in. (86.4–119.4 cm). Adult: large, white. • *Head dark, naked;* black tail; broad • *black wing areas;* • bill decurved, long, basally thick. Immature: dingier, with lighter head and neck.

Similarities
See White Pelican. Note also that white herons lack black in wings and retract neck in flight. Rare Whooping Crane is very similar, with straight bill.

Habitat
Lakes, ponds, marshes, lagoons.

Habits
Flies with alternate flapping and gliding; often soars in flocks at considerable height, resembling White Pelican, except for outline and visible black; when feeding, usually keeps head down while walking.

Voice
Usually silent, but occasionally a hoarse croak.

Food
Invertebrates and frogs.

Eggs
3–4; white; 2.7 x 1.7 in. (7.0 x 4.4 cm). Nest is of sticks in trees; in colonies.

Range
Breeds from s. Pacific Coast to S. America; winters from Salton Sea, southward.

IBISES AND SPOONBILLS
Family Threskiornithidae

Ibises are long-legged birds with long, decurved bills; slender necks; and short tails. They fly with alternate flaps and glides, head and neck extended, and legs trailing behind. Of thirty-three species, only two are found regularly in the West.

WHITE-FACED IBIS
Plegadis chihi   **5:6**

Description
Size, 19–26 in. (48.3–66.0 cm). Resembles a large, black curlew at a distance. Adult: dark glossy-chestnut (greenish-purple), iridescent, appearing black at a distance; • *long, decurved bill* with a margin of • *white feathers* at base during breeding season; legs, bill tip reddish. Immature: grayer.

Habitat
Fresh- and saltwater marshes, wet fields, mud flats.

Habits
Flies in long, undulating lines or ranks; flocks to roost in evening.

Voice
Piglike grunts with bleats "*ka-onk*, repeated" (Peterson); low-toned *kruk, kruk*.

Eggs
3–5; bluish-white; 2.0 x 1.5 in. (5.1 x 3.8 cm). Nesting is colonial; reed platform among marsh reeds (or sticks among bushes).

Food
Crustaceans, fishes, worms, mollusks, insects.

Range
Breeds from Calif. e. to Idaho; winters in s. Calif. and Mexico.

Note: The **ROSEATE SPOONBILL,** *Ajaia ajaja* (**5:7**), size, 30–34 in. (76.2–86.4 cm), is a casual visitor to the Southwest (California, Utah, Colorado). Its unmistakable long, flat, spatulate bill is used by this long-legged wader to sift small animals from water; the head and bill move side-to-side as it feeds. Adults are • *pinkish* with a dark bill and reddish legs; immatures are white with a yellow bill and legs.

Swans, Geese, and Ducks
Order Anseriformes

SWANS, GEESE, AND DUCKS
Family Anatidae

This is a nearly worldwide family of well-known, medium-sized to large birds of fresh or salt water. They have webbed feet and bills with tiny, toothlike projections along the edges. Their flight is swift and direct. Nearly all make large nests of vegetation on the ground near water. Of the world's 147 species, 40 occur regularly in the West: many of these constitute some of the choicest game birds. Most species breed in the far north, but some individuals may summer well south of their breeding range. The major groups of waterfowl are:

Swans: Very large, long-necked birds usually all-white when adult. They feed on aquatic vegetation from the bottom by extending their necks deep under water. They fly in V-shaped wedges or irregular lines; their necks are held straight out.

Geese: Smaller, shorter-necked than swans and larger than ducks, geese have blunt, triangular bills. The sexes are alike. They fly noisily in a wavering V or line formation, and often feed in stubble and grain fields or grassy marshes, consuming seeds, aquatic plants, and grasses.

Whistling-Ducks: Sometimes called Tree Ducks, these somewhat arboreal, gooselike ducks have long legs and necks and an erect posture. They are inhabitants mainly of warm regions; the sexes are alike, the plumage usually chestnut, varied with black and white. They feed on seeds and grass, and they nest in trees.

Surface-feeding or Pond Ducks (Genus *Anas* and allies): These largely freshwater ducks spring almost vertically into the air when taking flight. They feed on aquatic plants and seeds by "tipping up," with heads below the surface and tails pointing skyward. Males are often highly patterned about the head. Females are predominantly brown and often hard to identify except by the males accompanying them. The iridescent secondaries of many of these "puddle ducks" form a wing patch called the speculum, which is sometimes hard to see when the bird is on the water.

Diving Ducks (Pochards [Genus *Aythya*], Goldeneyes, Eiders, Scoters): The legs of diving ducks are set farther back than those of the surface-feeders, and the hind toe is free and lobed. Confusingly similar females are often best identified by their accompanying males. They dive for aquatic plants, snails, and insects, and patter along the water's surface before they fly.

Mergansers: These fish-eating ducks have crests; long, saw-toothed, slightly hooked bills; long-lined, slender bodies; and some white in the wing secondaries. Rarely do they occur in sizable flocks. They feed on fish, crayfish, and some amphibians.

Stiff-tailed Ducks: These small, chunky ducks, with sexes colored differently, subsist on water plants, insects, and other small aquatic creatures. The stiff tail has eighteen to twenty spike-tipped feathers, often held erect as the duck swims.

WHISTLING SWAN
Cygnus columbianus

6:5
Fig. 3

Description
Size, 47–58 in. (119.4–147.3 cm); 20 lb. (9.1 kg). Large, all-white. Adult: bill black with small yellow spot at base (usually, not always). Immature: bill pinkish, tip dusky, plumage grayish or brownish-white. This is the common wild swan, often heard throughout the West long before the high-flying, ribbonlike flocks become visible. Female is smaller than male.

Similarities
Trumpeter Swan is larger, voice deeper. Snow Goose has black primaries, shorter neck. White Pelican has large bill, black in wings.

Habitat
Tundra lakes, large rivers, reservoirs, bays, estuaries, wet fields.

Habits
Holds neck erect, bill horizontal; feeds in water by tilting up or "dabbling."

Voice
Musical whooping whistle and soft trumpeting, *wow-wow-ou;* cooing, higher-pitched and less harsh than honking of geese.

Eggs
4–5; creamy-white; 4.2 x 2.7 in. (10.7 x 6.9 cm).

Range
Breeds n. of Arctic Circle, from Alaska Peninsula, eastward; winters along Pacific Coast.

Trumpeter Swan

Fig. 3

adults

Whistling Swan

TRUMPETER SWAN
Olor buccinator

Fig. 3

Description
Size, 58½–72 in. (148.6–182.9 cm). Large, white; heavy, • *all-black bill.* At very close range a pinkish spot may be seen on the bill.

Similarities
Whistling Swan is smaller with thinner bill.

Habitat
Lakes, rivers, ponds; saltwater bays in winter.
Habits
May gather in flocks when not breeding.
Voice
Deep bugling notes, louder than Whistling Swan.
Eggs
4–6; whitish; 4.3 x 2.8 in. (4.0 x 7.2 cm). Nest is stick mound on lake shore, island, or beaver house.
Range
Breeds from s. Alaska and Alta. s. to Oreg. and Wyo.; winters from se. Alaska to B.C., and open water in breeding range.

CANADA GOOSE
Branta canadensis   6:6

Description
Size, 22–43 in. (55.9–109.2 cm). Variable, with several different subspecies. • *White cheek patches.* Above, brownish-gray; • *head, neck, tail black;* base of black "stocking" neck clearly defined against whitish underparts; bill, legs black. In flight, black neck stretched out and slightly down-curved. Extremes among individuals of the various subspecies may be identifiable in the field; it is common to see migrant or wintering groups containing individuals of both large subspecies and small subspecies—the latter may be half the size of the former, and have a different, more yelping voice.
Similarities
Cormorants are silent in flight, darker. See also Brant.
Habitat
Tundra, prairies, marshes, lakes, ponds, fields, bays.
Habits
Usually flies high in V formation; very vocal.
Voice
Loud, from a typical 2-syllabled honking to high-pitched yelping of smaller subspecies.
Eggs
4–6; white; 2.3–3.0 x 1.5–2.5 in. (5.8–7.6 x 3.8–6.4 cm). Nest is usually on ground; plant mass on islet, in marsh, or on top of a muskrat house; sometimes in trees, in old nest of a large bird of prey; on cliff; well guarded by adults.
Range
Breeds throughout most of Canada, from Arctic slope s. to Calif. and Kans.; winters from s. Alaska and s. Canada s. to Mexico.

BRANT
Branta bernicla   6:4

Description
Size, 22–26 in. (55.9–66.0 cm). • *Black breast,* either extending onto the belly or clearly set off from the pale posterior underparts. Above dark brown; head, neck, breast, tail black; breast black to below waterline; small white patch on each side of neck, not on throat and cheek; upper- and undertail coverts white, showing white V over tail in flight; • *underparts light brownish-gray;* on water, sides look white.
Similarities
Canada Goose has white cheek patches.
Habitat
Sheltered seacoasts.

Habits
Sits high on water like a gull; flies low over water in irregular flocks with undulating, irregular, "wavy" flight quite rapid for a goose, rarely V formation.
Voice
Deep, loud, grunting honks.
Eggs
3–5; creamy; 2.8 x 1.9 in. (7.1 x 4.8 cm).
Range
Breeds in Arctic, along coastlines; winters along Pacific Coast, from Vancouver s. to Baja Calif.

Note: The western subspecies of the Brant is often called the **BLACK BRANT**. It has black extending over the abdomen. Eastern birds in our area may be distinguished by black restricted to breast and whiter abdomen, but some are dark on abdomen. The **EMPEROR GOOSE**, *Philacte canagica,* size, 26–28 in. (66.0–71.1 cm), is small, bluish-gray, with white head and hindneck, black foreneck, and black-and-white scaled body. It breeds in western Alaska and winters mainly in the Aleutians and southern Alaska, but is casual south to central California.

WHITE-FRONTED GOOSE
Anser albifrons   6:3

Description
Size, 26–34 in. (66.0–86.4 cm). • *Orange or yellow legs.* Adult: gray-brown, • *area around bill* and upper- and undertail coverts *white,* irregular black speckles on belly; bill pink, white crescent on rump. Immature: dusky; bill yellow, no white around bill or black marks on belly.
Similarities
Canada Goose is larger, with black on head and neck, legs black. "Blue" Snow Goose is similar to immature White-fronted, but is less brown and has pink legs.
Habitat
Ponds, rivers, lakes, tundras, bays, prairies.
Habits
Flies rapidly in V formation.
Voice
High-pitched *"wah-wah; kah-lah-a-luck,* uttered 1–3 times" (Moffitt); 2 notes *"low-lyow* or *lyo-lyok"* (Peterson).
Eggs
5–7; white; 3.1 x 2.1 in. (7.9 x 5.3 cm).
Range
Breeds from Alaska tundra to w.-cen. Canada; winters throughout w. U.S., from s. B.C. s. to Mexico.

SNOW GOOSE
Chen caerulescens   6:1

Description
Size, 23–31 in. (58.4–78.1 cm). Usually • *white, with black primaries* in the West; the gray-brown "Blue" phase occurs, especially in the Great Plains; note the long neck and short pink bill. Adult: white or gray-brown, legs pink, head and breast sometimes rust-stained. Immature: dingier, bill darker.
Similarities
Swans have longer necks, no black in wings. Ross' Goose is smaller, neck shorter, head rounder, bill smaller.

Habitat
Tundra, ponds, lakes, rivers, marshes, bays, prairies, grain fields.
Habits
Flies in loose V formation.
Voice
Single, loud nasal honk; "a resonant *whouk* or *houck* given once or twice, rarely thrice" (Moffitt).
Eggs
5–7; white; 3.1 x 2.1 in. (7.9 x 5.3 cm).
Remarks
Blue Goose and Snow Goose formerly were considered separate species, but represent color phases of same species.
Range
Breeds in Arctic; winters along Pacific Coast from Wash., s. to Central and Imperial valleys of Calif.

ROSS' GOOSE
Chen rossii   6:2

Description
Size, 21–25½ in. (53.3–64.8 cm). Adult: • *Bill stubby*, but has • *no black lines along edges*, as does "Blue" Snow Goose. Immature: bill pinker and legs paler than larger Snow's.
Similarities
White phase of Blue-Snow Goose is similar, but smaller, less bulky.
Habitat
Tundra, ponds, lakes, rivers, marshes, bays, prairies, grainfields.
Habits
Gathers in large flocks in fall and winter.
Voice
A "gruntlike *luk-luk*" (Collins); "*kek, kek* or *ke-gak, ke-gak*" (Peterson).
Eggs
4; creamy white; 2.7 x 1.9 in. (7.1 x 5.0 cm).
Range
Breeds in cen. Canadian Arctic; winters in Sacramento Valley of Calif.

FULVOUS WHISTLING-DUCK
Dendrocygna bicolor   7:2

Description
Size, 18–21 in. (45.7–53.3 cm). Long, gangling legs; tawny body; dark back, with a broad creamy stripe down the side; • *white crescent* at base of tail; blackish underwings.
Similarities
Cinnamon Teal is smaller, deeper in color.
Habitat
Freshwater ponds, marshes, wet fields.
Habits
Active by night and day, feeds in fields as well as in water.
Voice
Slurred whistle or squeal *ka-whee-oo* during flight; a "weak, whistled *kill-dee*" (Kincaid).
Eggs
12–17; white; 2.2 x 1.5 in. (5.6 x 3.8 cm). Nest is marsh or grass; occasionally in hollow tree.

Range

Breeds from Imperial and San Joaquin valleys of Calif. s. to Mexico and e. to La.; winters chiefly in Mexico.

Note: The **BLACK-BELLIED WHISTLING-DUCK**, *Dendrocygna autumnalis (Fig. 4)*, size, 20–22 in. (50.8–55.9 cm), is rust-colored with a black belly; long, pink legs; a coral-pink bill; and distinctive broad, white patches along forewings. It frequents freshwater ponds and marshes in southern Texas in the Rio Grande Delta and casually occurs in west Texas to southeastern California along the western border in summer.

Fig. 4

Black-bellied Whistling-Duck

MALLARD
Anas platyrhynchos   7:10, 8:2

Description

Size, 20½–28 in. (52.1–71.1 cm). • *Speculum purple with white borders,* wing linings white. Male: breast chestnut, body gray, tail white, some black tail coverts curl forward, • *head a glossy green* with • *narrow white collar,* bill yellowish, feet orange. Female: brown, bill mottled orangish, tail only with some white.

Similarities

Northern Shoveler male has green head, but white breast, chestnut sides, huge bill. Female Pintail is longer; neck more slender than female Mallard; has only 1 white rear border on speculum; wing linings not white; bill gray, tail more pointed. Female American Wigeon has more white in wing.

Habitat

Almost any water, usually fresh.

Voice

Male, a low, reedy *kwek-kwek-kwek;* female, the familiar, boisterous barnyard *quack-quack-quack.*

Eggs

8–10; olive-green; 2.3 x 1.6 in. (5.8 x 4.1 cm). Nest is soft hollow in grass, not always near water.

Remarks

This duck is the basis for comparisons of other species by the beginner.

Range

Breeds throughout most of Canada and U.S., from Alaska and nw. Canada, s. to cen. Calif. and N.Mex.; winters from s. Alaska and s. Canada, s. to Mexico.

Note: The **MEXICAN DUCK** is a subspecies of Mallard of Mexican border region (southwest New Mexico, southeast Arizona); both sexes resemble the female Mallard, are more heavily streaked and spotted brown below, and have a fully yellow bill. It hybridizes frequently with the Mallard, hybrids being intermediate in coloration. The **BLACK DUCK,** *Anas rubripes* **(8:1)**, a straggler from the East into the Plains states, is 2-toned, very deep brown with a distinctly paler head and neck. The speculum differs from Mallards in lacking front white edge and most or all of rear white border.

GADWALL
Anas strepera   **7:8, 8:3**

Description
Size, 18½–23 in. (47.0–58.4 cm). Surface-feeding duck with a
• *white speculum* (conspicuous in flight) not easily seen on water.
Male: slender; • *gray;* head and neck light brown, shoulder red-brown; • *rump and undertail coverts black;* bill gray; white patch on hind edge of wing; belly white (both sexes). Female: mottled light brown, bill yellow-brown, feet yellow. When swimming, the wing speculum often is hidden.

Similarities
American Wigeon has a conspicuous white wing patch but on forepart, not speculum, of wings; on water the Wigeon shows white on flanks, Gadwall does not. Female Wigeon is gray-billed, ruddy-flanked, gray-headed; female Mallard is larger. Female Pintail has blue-gray bill, pointed tail.

Habitat
Fresh or brackish ponds, lakes, streams, marshes.

Habits
Flight swift, direct; Gadwall's wingbeat is faster than Mallard's or Black's; sits lower on water than Wigeon.

Voice
Male whistles or has a low *bek;* female quacks loudly.

Eggs
7–13; creamy; 2.2 x 1.6 in. (5.6 x 4.1 cm). Nest is hollow in grass, not necessarily near water.

Range
Breeds from Alaska, eastward and s. to Calif. and sw. Arizona.; winters from s. U.S. to Mexico.

NORTHERN PINTAIL
Anas acuta   **7:11, 8:4**

Description
Size, 20–29 in. (50.8–73.7 cm), including long tail of male. Male:
• *long, pointed tail;* long, thin neck; white breasted, with distinctive white line up side of neck; head brown; white patch near black stern. Female: light brown, somewhat darker above; throat and upper breast tan; tail lacks long central feathers of male, but is sharply pointed. Both sexes white below, speculum brown, bill and feet blue-gray. In flight, long-necked and slender ("streamlined"); wings long and pointed; white underparts and white line on trailing edge of wing visible from afar.

Similarities
Female Mallard is longer, heavier-bodied, with shorter neck and tail; orangish bill; speculum with 2 white borders. Female Gadwall has white speculum. Female American Wigeon has gray head, brown breast, paler bill.

Habitat
Ponds, lakes, marshes, rivers; in winter, saltwater bays, prairies, grain fields.

Habits
A fast flier, sits high on water.
Voice
Usually silent; male, a double-toned loud *kwa, kwa* in flight, a
Teal-like whistle; female, a low hoarse quack.
Eggs
6–12; olive-buff; 2.1 x 1.5 in. (5.3 x 3.8 cm). Nest is downy hollow
in marsh or prairie, up to 1 mile from water.
Range
Breeds throughout Canada and U.S., from Alaska eastward, and s.
to Calif. and Colo.; winters along Pacific Coast, from s. Alaska
southward and inland across n. Ariz., s. Utah. and s. Colo.

COMMON TEAL
Anas crecca   **7:5, 8:5**

Description
Size, 12½–15½ in. (31.8–39.4 cm). Male is only duck with
• *chestnut head* and green eye band. Very small. Male gray with
spotted tan breast; underparts and • *vertical white crescent behind
breast;* undertail coverts buffy. Reveals iridescent green speculum
and • *green patch* on side of head in bright sun. Female: brownish-
gray above, pale gray below, speckled; green speculum.
Similarities
Blue-winged has longer neck and body, blue on wing, dark belly in
flight. Cinnamon Teal has light blue wing patches.
Habitat
Marshes, ponds, lakes, streams; in cold weather, brackish and
saltwater bays.
Habits
Sometimes feeds in mud like a shorebird; flight fast, buzzy and
erratic, in compact flocks wheeling like pigeons; wings whistle in
flight.
Voice
Male, piping whistles; female, a high-pitched, crisp quack.
Eggs
10–12; pale buff; 1.8 x 1.4 in. (4.6 x 3.6 cm). Nest is grassy
hollow, often some distance from water.
Range
Breeds from n. Yukon, e. across Canada, and s. to Calif. and Ariz.;
winters from s. Canada to n. Mexico.

Note: The Eurasian form of this species breeds in the Aleutians,
and is accidental in western states; males lack the white vertical
crescent and have instead a horizontal white stripe over the wings.

BLUE-WINGED TEAL
Anas discors   **7:7, 8:6**

Description
Size, 14½–16 in. (36.8–40.6 cm). Small, fast-flying marsh duck;
bill relatively large. Male: grayish above, tan marked with dark
below; white patch on rear of flanks; large • *white crescent* in front
of eye; • *pale blue shoulder patches.* Summer eclipse plumage
maintained into late fall; fall males may resemble females with
little or no white face mask. Female: brownish-gray above, pale
gray with darker marks below, large blue patch on forewing.
Similarities
Female Northern Shoveler is larger, heavier-billed. Female
Common has no blue on wing; bill smaller. Male Cinnamon has

blue wing patches, deep mahogany color; female is virtually indistinguishable from female Blue-winged. Female scaups also have white patch before eye.

Habitat
Freshwater ponds, marshes, rarely salt water.

Habits
Flight erratic; dabbles in mud when feeding.

Voice
Male, a whistling peep; female, a light quack.

Eggs
6–12; white; 1.8 x 1.3 in. (4.6 x 3.3 cm). Nest is a downy depression in grass near water.

Range
Breeds from Canada, s. to n. Calif. and e. to Colo., eastward; winters in C. and S. America, but some winter along Gulf Coast.

CINNAMON TEAL
Anas cyanoptera   **7:9**

Description
Size, 14½–17 in. (36.8–43.2 cm). Both sexes have chalky-blue wing patches. Male: cinnamon-red body and head. Female: body mottled brown; indistinguishable from female Blue-winged.

Similarities
See Blue-winged. Female Northern Shoveler is larger and has much larger bill.

Habitat
Freshwater ponds, rivers, marshes; rarely salt water.

Habits
Erratic flight.

Voice
Relatively silent; male, a low chatter; female, a weak quack.

Eggs
6–12; buff-white; 1.9 x 1.4 in. (4.8 x 3.6 cm). Nest is downy hollow in reeds or grass.

Range
Breeds from s. B.C., s. to Mexico and S. America; winters from cen.-s. Calif., e. to N.Mex., Ariz., and s. Tex., southward.

EURASIAN WIGEON
Anas penelope   **7:3**

Description
Size, 16½–20 in. (41.9–50.8 cm). Male: • *red-brown head, creamy crown;* upperparts and flanks gray; breast pinkish; axillars dusky; white wing coverts, rump, rear flanks, and belly; black primaries and tail coverts; green speculum. Female: resembles American Wigeon, but with • *reddish head.*

Similarities
Male American Wigeon is browner above and on flanks, looks darker and has white crown. Female American is grayer, has gray head and white axillars.

Habitat, Habits, Eggs
Similar to American Wigeon.

Voice
Shrill whistling *whee-you.*

Range
Chiefly a Eurasian duck, but regular winter visitor along Pacific Coast.

AMERICAN WIGEON
Anas americana   7:4, 8:8

Description
Size, 18–23 in. (45.7–58.4 cm). Head and neck gray; speculum green; wing coverts and belly white; bill blue, black-tipped. In flight, shows large white patch on inner forewing. Male: • *shining white crown*; glossy green ear patch, dark at distance; breast and sides pink; rump, rear of flanks, and large patch on front of wing white; primaries and undertail coverts black. Female: wing patches grayish-white, upperparts ruddy-brown, breast and sides tan, undertail coverts white.

Similarities
Eurasian males like a light-bodied American with a dark head; female has browner head. Gadwall lacks white crown and flank patches, sits lower on water. Northern Shoveler has blue wing patches that sometimes appear white, as do Blue-winged's. Female Mallard is larger, darker, with less white on wing. See also female Pintail.

Habitat
Freshwater lakes, streams, rivers, marshes; occasionally saltwater bays.

Habits
Sits high on water, pivots as it feeds; flies in compact, irregular flocks.

Voice
Male: a "wild and musical note, *whew, whew, whew*." Female: "a *qua-awk, qua-awk*" (Kortright).

Eggs
6–12; creamy; 2.1 x 1.5 in. (5.3 x 3.8 cm).

Range
Breeds in Alaska and Northwest Territories, s. to cen. Canada and w. states; winters along Pacific Coast, from Vancouver s. to Mexico; also winters along Gulf Coast.

NORTHERN SHOVELER
Anas clypeata   7:6, 8:7

Description
Size, 17–20 in. (43.2–50.8 cm). Large, flat, spoon-shaped bill. • *Chalky wing patch, green speculum,* orange legs. Male: appears black and white, but head green, looks black at distance; body and tail white; belly and • *sides rufous.* Black primaries and tail coverts and black line down back. Whether swimming or flying, reveals a unique patterning from front to back of dark-light-dark-light-dark. Female: mottled brownish above, paler below.

Similarities
Blue-winged Teal is smaller and smaller-billed, flies faster.

Habitat
Fresh- or saltwater marshes (brackish); estuaries, lakes.

Habits
Flies or swims with big bill pointing down at an angle, wings seem set far back on body; flight slower, more hesitating than teal's; sits low on water, feeds from surface using bill as strainer.

Voice
Male: a low *woh, woh, woh* (Collins); *"took, took, took"* (Peterson); female, a weak quack.

Eggs
6–14; buffy or pale olive-green; 2.1 x 1.5 in. (5.3 x 3.8 cm). Nest is hollow in grass.

Range
Breeds throughout Canada and U.S., but chiefly from w. Alaska to
e. Canada, s. to Ariz. and Tex.; winters from s. B.C. to S.
America.

WOOD DUCK
Aix sponsa   **7:1, 8:9**

Description
Size, 17–20½ in. (43.2–52.1 cm). Crested, surface-feeding. Male:
unmistakable, complex face pattern; body boldly patterned with
iridescent maroon, green, purple, white. Female: head gray,
crested; • *eye-ring, throat, and underparts white;* back gray-brown;
speculum blue. In flight, head held above level of body, bill pointed
down at an angle; short neck and long, square tail conspicuous;
makes a distinctive whistling sound.
Similarities
See American Wigeon.
Habitat
Fresh water, wooded marshes, ponds, swamps, streams.
Habits
Perches and nests in trees; flight swift and direct; sits high on
water.
Voice
Male: *"oo-eek, oo-eek"* (Chapman); female: *"c-r-e-e-k, c-r-e-e-k"*
(Eaton); *"crrek, crrek"* (Peterson).
Eggs
10–15; dull white; 2.0 x 1.6 in. (5.1 x 4.1 cm). Nest is tree cavity,
often far from water.
Range
Breeds from s. B.C., e. to s. Alta., and s. to Wash. and n. Calif.;
winters in s. states and Mexico.

CANVASBACK
Aythya valisineria   **9:2, 11:2**

Description
Size, 19½–24 in. (49.5–61.0 cm). • *Long, sloping profile of bill and
forehead.* Male: • *rusty head;* dark eyes; bill long, sloping, blackish;
neck rusty-red; back very white. Female: head and breast light
brown, back grayish.
Similarities
See Redhead. Female mergansers are red-headed, crested, with
whitish breasts. Scaups have round foreheads.
Habitat
Tundra lakes; freshwater marshes, ponds, lakes, salt bays,
estuaries; rivers.
Habits
Collects in large rafts on water. Flies in lines or V's, with long bill,
head, and neck carried slightly down; pointed wings appear set far
back on body.
Voice
Male makes grunts, low croaks, or growls; female quacks.
Eggs
7–9; grayish-olive; 2.5 x 1.8 in. (6.4 x 4.6 cm). Nest is downy
basket in marsh grass.
Range
Breeds from B.C. e. to s. Man. and s. to Oreg., Utah, and Nebr.;
winters along Pacific Coast, from s. B.C. to cen. Mexico.

REDHEAD
Aythya americana   9:1, 11:1

Description

Size, 18–22 in. (45.7–55.9 cm). Male: • *rounded brownish-red head;* pale eyes, back gray; blue bill with black tip. Female: brown; belly and diffused area around bill white; head rounded. Sexes alike in having broad gray wing stripes.

Similarities

Male Canvasback has blackish bill, long sloping forehead, rustier head, whiter body. Female Common Goldeneye has brown head, white collar. Female scaups have broad white circle around base of bill. Female Ring-necked Duck is smaller, darker, and has white eye-ring and ring around bill. Redhead is shorter and darker than Canvasback in flight, wingbeats more rapid, flight more erratic; wing stripe of Redhead is long and gray, not white as in scaup.

Habitat

Ponds, lakes, rivers, fresh marshes, bays, estuaries.

Habits

Gathers in big flocks ("rafts") on large lakes and bays in winter.

Voice

Male: a catlike *me-ow* (or deep purr). Female: "a growl *r-r-r-rwha, r-r-r-rwha*" (Griscom); "a *squak*" (Peterson).

Eggs

10–15; pale olive-buff; 2.4 x 1.7 in. (6.1 x 4.3 cm).

Range

Breeds from B.C. e. to Man. and s. to Calif., Ariz., and Colo.; winters from s. B.C. eastward, and s. to Mexico.

RING-NECKED DUCK
Aythya collaris   11:3

Description

Size, 14½–18 in. (46.8–45.7 cm). Male: diving duck, on the water shows a • *vertical white crescent* in front of wing; in flight, black-backed with a broad • *gray wing stripe;* head, breast, and back black. In both sexes head is rather triangular in shape, bill with white ring, speculum bluish-gray; head with purple iridescence; sides gray, sometimes seem white; dark chestnut ring on neck seldom visible. Female: brown with white eye-ring and indistinct • *white area near bill.* In flight, shows white margin on trailing edge of wing; head dark crowned, somewhat triangular; eyes dark.

Similarities

Male scaup have lighter backs; in flight, show broad white wing stripe. Female scaup lack ringed bill mark, have more distinct white area around base of bill. Redhead male is larger, paler; female larger with less white on cheeks.

Habitat

Ponds, wooded lakes, streams; in winter, bays, marsh ponds, rivers.

Habits

Travels in small groups, alights without circling.

Voice

Seldom heard; male a low whistle; female, a *scaup,* similar to Lesser Scaup; purring.

Eggs

6–12; buffy-olive; 2.3 x 1.6 in. (5.8 x 4.1 cm).

Range

Breeds throughout w. Canada inland and in n.-cen. states to Great Lakes; winters from Vancouver eastward, and s. to C. America.

GREATER SCAUP
Aythya marila   **9:3, 11:5**

Description
Size, 15½–20 in. (39.4–50.8 cm). Broad white wing stripe
extending almost to wing tip; eyes pale; bill bluish. Male: black
head, breast, primaries, and tail; gray back, white sides; head
glossed with green and quite rounded. Female: dark brown with
sharply defined white patch around bill.

Similarities
Lesser Scaup has thinner neck, shorter wing stripe; purple gloss on
the somewhat angular head of the male; and higher crown. White
of wing in the Greater extends onto the primaries; in the Lesser
only halfway along the rear edge of the wing.

Habitat
Tundra lakes, ponds, rivers; bays, estuaries, ocean.

Habits
Large rafts collect on bays in winter.

Voice
A "loud, discordant *scaup, scaup*" (Kortright).

Eggs
7–10; buffy-olive; 2.5 x 1.7 in. (6.4 x 4.3 cm).

Range
Breeds from Alaska, e. to Great Slave Lake and s. to B.C.; winters
along Pacific Coast from Alaska to cen. Calif.

LESSER SCAUP
Aythya affinis   **9:4, 11:4**

Description
Size, 15–18½ in. (38.1–47.0 cm). Male: general appearance is
white in the middle, black at both ends; • *head iridescent purple,*
varying with angle of light, and somewhat angular; flanks grayish;
flanks and back finely barred; bill blue. Female: dark brown, bill
clearly "masked"with white at base

Similarities
Male Greater Scaup has more rounded, greenish head; thicker
neck; seems larger; female Greater is told with certainty only by
longer wing stripe. Ring-necked Duck has black neck; female has
less distinct white around bill, dark eyes, and white eye-ring.
Redhead and Canvasback females have no white around bill, are
larger.

Habitat
Same as Greater Scaup, but smaller bodies of fresh water and less
often salt water.

Habits
Flight swift, erratic; often in closely bunched large flocks.

Voice
Male: a coarse *scaup;* in flight, a repeated *pppr-pppr* (purring
sound).

Eggs
6–15; buffy-olive; 2.3 x 1.6 in. (5.8 x 4.1 cm). Nest is depression in
grass.

Range
Breeds from cen. Alaska, e. to Man. and s. to Mont. and N.Dak.;
also inland in B.C., s. to Oregon, Idaho, Colo., and Iowa; winters
along Pacific Coast from s. B.C. to Mexico.

COMMON GOLDENEYE
Bucephala clangula   **9:5, 10:3**

Description
Size, 16–20 in. (40.6–50.8 cm). Chunky body, high-domed head, short neck, golden eye, white wing patch; in flight, looks large-headed, short-necked. Male: white breast, large white squares in wings, blackish lining to wings from below, and large, • *round white spot* before the eye. Head, upperparts, and bill black; head with greenish gloss; other parts of body white; legs orange. Female: brown head; gray back and sides; collar, breast, belly, and divided wing patch white; outer third of bill yellow in spring.

Similarities
Barrow's Goldeneye has shorter bill; in winter, females indistinguishable in field; scaups have black breast. Male Common Merganser has longer neck and body, is white-breasted, with rakish rather than stocky look.

Habitat
Lakes, rivers; in winter, salt bays, ocean.

Habits
Wings whistle in flight, hence name of "whistler"; dives frequently when feeding; flock rises all at once as a band, does not string out in flight; female sits lower in water than male.

Voice
Male: "a penetrating *spear, spear*" (Collins); female: a low, harsh quack.

Eggs
8–15; pale green; 2.4 x 1.7 in. (6.1 x 4.3 cm). Nest is a hole in tree near water.

Range
Breeds from Alaska and cen. Canada s. to n. U.S., chiefly e. of Rocky Mountains; winters along Pacific Coast from se. Alaska to Calif. and inland on rivers and deep lakes.

BARROW'S GOLDENEYE
Bucephala islandica   **10:1**

Description
Size, 16½–20 in. (41.9–50.8 cm). Male: black head and a white • *crescent-shaped jowl spot;* • *head has purplish gloss* and bulges fore and aft; black wedge points down at shoulder, separating sides from breast; row of white spots on black scapulars. Female: in spring breeding plumage, bill all-yellow; head darker than female Common; forehead more abrupt; in winter, indistinguishable in field.

Similarities
Common Goldeneye has longer bill, thinner at base; on water, sides and wings show less black.

Habitat
Lakes, rivers; salt bays, ocean in winter.

Habits
In winter, remains farther north than Common.

Voice
Hoarse croaks; when courting makes a "mewing" cry.

Eggs
6–14; pale greenish; 2.4 x 1.7 in. (6.1 x 4.3 cm). Nest is tree cavity or cliff hollow.

Range
Breeds from sw. Alaska to B.C., southward to n. Wyo.; winters along Pacific Coast in Nw. and on deep lakes of Rockies.

BUFFLEHEAD
Bucephala albeola 9:6, 10:2

Description
Size, 13–15½ in. (33.0–39.4 cm). Small (one of smallest ducks, given many local names by hunters); puffy head on chunky body; short-necked; bill blue-gray, stubby; white wing patches conspicuous in flight. Male: • *large white head patch* from eye to rear of crown. Resembles goldeneyes, but flies faster and nearer water. On water one of the whitest ducks, being mostly white, glossy black head feathers puffed out, white patch appearing sunken. Female: dusky; head gray-black, with a slanting white cheek patch, at distance suggests female goldeneye. A dark, compact duck, flight buzzy, fast, large head, small bill, and white wing patch.

Similarities
Hooded Merganser has dark, long head and body; long, thin black bill; black neck; its white head patch has a black border.

Habitat
Lakes, ponds, rivers; winters on salt bays, ocean.

Habits
Occurs in small groups, can dive from wing, and can fly directly into air from under water.

Voice
Usually silent; male whistles or gives a hoarse, rolling note; female quacks.

Eggs
6–14; buffy; 1.9 x 1.4 in. (4.8 x 3.6 cm).

Range
Breeds from Alaska to n. Ont. and s. from B.C. to n. Calif.; winters along Pacific Coast and on open waters throughout U.S.

OLDSQUAW
Clangula hyemalis 9:9, 10:5

Description
Size, male, 19–22½ in. (48.3–57.2 cm); female, 15–17 in. (38.1–43.2 cm). Only saltwater duck with • *all-dark wings* and white on body; male with needle-pointed tail; black and pink bill. Summer male: dusky with • *white belly and face,* white patch around eye. Summer female: largely brown, including crown and ear/cheek patch; white eye patch evident. Winter male: white with dark cheeks, all-white crown, dark back and breast. Winter female: plumage lighter than in summer, white head with dark cheek patch.

Similarities
See Pintail. Head pattern of young female resembles a changing Harlequin.

Habitat
Tundra lakes, bays, ocean.

Habits
In flight, wings low, curved, and pointing to rear; wingbeats rapid; flight erratic, buzzy, low over water, in small flocks veering like shorebirds and flashing black and white; alights with a great splash. Dives for shellfish.

Voice
A "musical, gabbling *south, south-southerly* or *how doodle do*" (Collins).

Eggs
6–10; light grayish-olive; 2.1 x 1.6 in. (5.3 x 4.1 cm).

Range
Breeds along Arctic coastline and s. on tundra in Alaska and n. Canada; winters along Pacific Coast from Aleutians to Wash.

HARLEQUIN DUCK
Histrionicus histrionicus   10:4

Description
Size, 14½–21 in. (36.8–53.3 cm). Both sexes appear dark at a
distance. Small and chunky on water; graceful on land; in flight,
suggests a dark goldeneye; bill small. Male: blue-gray body with
chestnut sides, uniquely patterned with side, wing, neck, and head
markings, even more "patchy" in molt; tail sometimes cocked,
somewhat pointed. Female: brown with 3 white spots on head;
wing patch absent.

Similarities
Female has shape of Bufflehead and pattern of female scoters.
Female Surf and White-winged Scoters are bigger, heavier, show 2
not 3 white head spots. Female Bufflehead has single white head
spot and white wing patch.

Habitat
In winter, the ocean, especially near rocky shores; in summer,
turbulent mountain streams.

Habits
Floats high in water in close formation, often with tail cocked;
feeds around rocks; flies fast, low to water, in compact flocks; can
dive from wing.

Voice
Whistle or squeak; male, *"gua gua gua"* and female *"ek-ek-ek-ek"*
(Peterson).

Eggs
5–10; buffy; 2.3 x 1.6 in. (5.8 x 4.1 cm). Nest is tree cavity, cliff
hole, or on ground near mountain stream.

Range
Breeds in nw. N. America from Alaska to Wyo. and s. to Calif.;
winters along Pacific Coast from Bering Sea to cen. Calif.

KING EIDER
Somateria spectabilis   9:10, 10:9

Description
Size, 18½–25 in. (47.0–63.5 cm). Male: large, orange forehead
shield, with pearly crown; black back, wings, belly; breast, wing
patch, and flank spot near tail white; face white with greenish cast;
cheeks greenish; bill orange. Forehead slopes up abruptly from bill.
In flight shows large white wing patches; at a distance appears
white in front, black to rear; only duck with this effect. Female:
stocky; dusky-brown, barred heavily with black. Immature male:
abrupt forehead; dusky; dark brown head with indication of adult
facial pattern; light breast.

Similarities
Female goldeneyes vaguely resemble young male King.

Habitat
Large coastal lakes, rocky seashores, offshore reefs.

Voice
Male moans or croons, second syllable stressed; female quacks.

Eggs
4-7; dull green; 2.0 x 2.0 in. (5.1 x 5.1 cm).

Range
Breeds in Arctic along coasts and on islands; winters along Pacific
Coast, from Bering Sea s. through Aleutians to B.C.; rarely farther
south.

BLACK SCOTER
Melanitta nigra   **10:6**

Description
Size, 17–20½ in. (43.2–52.1 cm). Smallest-appearing, most
ducklike scoter, primaries with silvery sheen beneath. Male: all-
black; bill with bright orange-yellow knob on it, gives it name
"butternose." A pronounced 2-toned wing effect in flight. Female:
brown with black crown and whitish cheeks.

Similarities
Female White-winged and Surf Scoters have 2 light patches on
cheeks; some immature male Surf Scoters may lack head patches.
Winter Ruddy is smaller, paler, and with a white chest.

Habitat
Coast and tundra, rare inland.

Habits
On water, often cocks pointed tail, rides high with head high and
bill horizontal or uptilted; other scoters carry bill pointed down.
Travels in large flocks.

Voice
Male: a melodious cooing *cour-loo.* Female: a growl.

Eggs
6–10; pale ivory-yellow; 2.5 x 1.7 in. (6.4 x 4.3 cm).

Former name
American or Common Scoter.

Range
Breeds along coast of nw. Alaska and inland to ne. Canada;
winters along Pacific Coast from Aleutians to n. Calif., rare
farther s.

SURF SCOTER
Melanitta perspicillata   **9:12, 10:7**

Description
Size, 17–21 in. (43.2–53.3 cm). No white wing patches; bill forms
an even slanting line with forehead. Male: white forehead and
nape, otherwise black; bill varicolored red, white, orange, and
black. Female: dusky-brown; face has 2 light patches, sometimes
obscure, similar to female White-winged, but without wing patch;
sometimes shows whitish patch on nape.

Similarities
See female White-winged. In flight, resembles Black, but lacks
silvery sheen on under-flight feathers.

Habitat
Oceans, bays, large lakes.

Habits
Unlike White-winged, alights with wings held upward, sits with
bill pointed down; wings hum in flight.

Voice
Seldom heard; a low croak.

Eggs
5–9; creamy; 2.4 x 1.7 in. (6.1 x 4.3 cm). Nest is depression in
brush or marsh grass.

Range
Breeds from w. Alaska eastward; winters along Pacific Coast from
Aleutians to Baja Calif.

WHITE-WINGED SCOTER
Melanitta fusca **9:11, 10:8**

Description
Size, 19–23½ in. (48.3–59.7 cm). • *White wing patches,* often concealed by flank and side feathers when swimming. Male: black with white streak under eye and squarish white patch on trailing edge of wing; bill orange with a black knob; feet pinkish. Female: sooty-brown, with 2 whitish cheek patches prominent in younger birds, sometimes obscure in adults; white wing patch.

Similarities
Surf and Black Scoters are similar but lack wing patches.

Habitat
Bays, ocean, large lakes when nesting.

Habits
Travels in large flocks.

Voice
In flight, a low bell-like whistle (males) or a thinner, reedier whistle (females).

Eggs
9–14; pinkish-buff; 2.6 x 1.8 in. (6.6 x 4.6 cm). Nest is depression in brush.

Range
Breeds from n. Alaska to s. Man. s. to N.Dak.; winters along Pacific Coast from Aleutians to Baja Calif.

RUDDY DUCK
Oxyura jamaicensis **11:9**

Description
Size, 14½–16 in. (36.8–40.6 cm). Small, thick-necked, chunky; bill broad, upturned; • *cheeks white;* wings entirely brown, noticeable in flight; cap dark. Male: in summer, mostly rich red-brown, cap black, bill bright blue, white cheeks. In winter, red is replaced by gray and brown, cap becomes dark brown, bill much duller. Female: similar to winter male, but with black streak on cheek; bill dusky.

Similarities
Female Black Scoter resembles winter male Ruddy but is larger, darker-cheeked, and seen more on ocean.

Habitat
Ponds, lakes, streams; salt bays in winter.

Habits
Swims buoyantly with tail up; can sink slowly under water or dive abruptly; needs long run for takeoff into air.

Voice
Usually silent; weak clucking *cluck-uck-uck-uck-ur-r-r* by courting males.

Eggs
5–15; pale buff; 2.5 x 1.8 in. (6.1 x 4.6 cm). Nest is basket of woven grass attached to reeds above water.

Range
Breeds from nw. Canada and Pacific Coastal valleys eastward through w. U.S. and s. to Calif. and Mexico, then to S. America; winters throughout W. from s. B.C. s. to Mexico.

Note: The **MASKED DUCK,** *Oxyura dominica,* size, 10 in. (25.4 cm), is a casual visitor to southern Texas from Mexico and the West Indies. It shows white in wings when flying; males lack white on black face, females have 2 black face stripes.

HOODED MERGANSER
Mergus cucullatus   9:7, 11:6

Description
Size, 16–19 in. (40.6–48.3 cm). The smallest, slimmest merganser,
male unlike others in pattern; bill small, thin. Male: head,
upperparts, and 2 vertical lower breast stripes black; underparts
and wing patch white; sides brown; unique, vertical, fan-shaped
• *white crest bordered in black.* In flight, crest shows as a white
streak on lowered head. Female: buffy crest without ragged edges;
• *brown head;* gray-brown upperparts, flanks, and breast; white
wing patch and belly. Bill dark, narrow, spikelike; flight profile
long, drawn out.

Similarities
Bufflehead male is smaller, white head patch has no black border;
has white sides. Female Wood Duck has square white wing patch.
Other mergansers have ruddier heads, ragged edges of crests, red
bills.

Habitat
Ponds, swamps, streams, lakes, rivers—usually wooded.

Habits
On water, sometimes cocks tail, can rise into air with great speed;
usually found in pairs or small groups; male frequently raises and
lowers crest.

Voice
Grunting, low *crew, crew.*

Eggs
6–12; white; 2.1 x 1.8 in. (5.3 x 4.6 cm). Nest is hollow tree or
stump.

Range
Breeds from s. Alaska e. across Canada and n. U.S.; winters along
Pacific Coast.

RED-BREASTED MERGANSER
Mergus serrator   9:8, 11:7

Description
Size, 19½–26 in. (49.5–66.0 cm). Male: unruly, often • *double-
pointed crest* and a red bill; head black with greenish gloss; back
black, sides gray; collar, wing patch, and • *underparts white*; breast
red-brown, dark patch with white spots near shoulder. On water,
looks slim, rakish, and dark; in flight, red breastband conspicuous
between white neck and belly; white wing patch appears framed.
Female: see Common Merganser and below.

Similarities
Common Merganser male is uncrested, is mostly white with more
white on wing patch; female Common has brighter head, more of a
crest, and sharp line of demarcation between rufous head and back
and white throat and breast. See also Red-necked Grebe.

Habitat
Lakes, ponds, rivers; salt water in winter.

Habits
Flight is swift, noiseless, and direct, with head, neck, and body
horizontal, flattens crest before diving; flies in string formation.

Eggs
6–12; creamy-buff; 2.5 x 1.8 in. (6.4 x 4.6 cm). Nest is down-filled
hollow under brush or among roots.

Range
Breeds from Alaska and n. Canada eastward and s. to n. B.C.;
winters along Pacific Coast from se. Alaska to Mexico and inland
from s. Calif. to s. Ariz.

COMMON MERGANSER
Mergus merganser   **11:8**

Description
Size, 22–27 in. (55.9–68.6 cm). Male: white sides; • *green-black head;* red bill and feet; breast with rosy blush; rest of bird is white; crest usually not noticeable; back, primaries, and tail black. In flight, shows more • *white on body* and wings than any other duck, and maintains bill, head, neck, and body on a horizontal plane. Female: • *moderate crest,* head bright red-brown, sharply contrasting with white throat and neck; upperparts gray; wing patch large, square, and white; feet and bill also red.

Similarities
Red-breasted has red breast, crest is conspicuous in male; female is more crested and red-brown of head blends gradually into white throat and breast; Female goldeneyes are shorter, stockier, with shorter bill; breast and sides not tinged with buff. Female Redhead has short bill.

Habitat
Wooded lakes, ponds, rivers in summer; open lakes in winter, rarely salt water.

Habits
Submerges by jumping, then diving, or by gradually sinking; flies in string formation, low, loonlike, horizontal; flight shape rakish.

Voice
Usually silent; occasionally "an unmelodious squawk" (Swarth); "a guttural *karrr* (female)" (Peterson).

Eggs
6–17; creamy; 2.5 x 1.8 in. (6.4 x 4.6 cm). Nest is tree cavity or hollow on ground.

Former name
American Merganser.

Range
Breeds from s. Alaska e. to Great Slave Lake and s. to s. Canada and to Sierra Nevada and Rocky Mountain states; winters s. from sw. Canada.

Vultures, Hawks, and Falcons
Order Falconiformes

These birds of prey, often observed soaring in sweeping circles at considerable height, have hooked beaks and, except in the condors and vultures, strong talons and powerful wings. Sexes are alike, but females generally are larger than males. They hunt by day, having keen eyesight; vultures feed on carrion, the others usually on freshly killed small animals.

AMERICAN VULTURES
Family Cathartidae

The vultures are black, naked-headed, carrion-eating, soaring birds of extraordinary visual acuity, highly valued as scavengers. Their voice is a hiss or low grunt. All three species nest in hollow logs, crevices in rocks, or on the ground in a thicket or brush, and feed their young by regurgitation.

TURKEY VULTURE

Cathartes aura   **12:1**

Description
Size, 26–32 in. (66.0–81.3 cm); wingspread, 6 ft. (1.8 m). All-
black; soars on dihedral, 2-toned blackish wings; wing linings
darker than flight feathers, outer flight feathers spread out; tail
long, narrow, rounded.

Similarities
Black Vulture is shorter, with squared tail and white area under
end of wing; it seems "heavier," flaps more. Northern Harrier also
holds wings at an angle, but is smaller, slimmer, white-rumped.
Eagles soar on horizontal wings.

Habitat
Usually seen soaring over unforested land; perches on snags, posts,
carrion, or on ground, shoulders hunched; may roost in groups.
Rocks and tilts in flight, infrequently flaps.

Eggs
1–3; white, blotched with brown and purple; 2.8 x 1.9 in. (7.1 x
4.8 cm).

Range
Breeds from s. Canada throughout U.S. to Mexico and C. and S.
America; winters from cen. Calif., s. and e. to Ariz. and s. Tex.

BLACK VULTURE

Coragyps atratus   **12:2**

Description
Size, 25 in. (63.5 cm); wingspread, 5 ft. (1.5 m). All-black; large;
whitish patches near wing tips; head small, black; tail short,
square; less often protruding in flight; underwings silvery.
Recognizable at a distance by its labored rapid wing flaps and brief
glide.

Similarities
Eagles and dark phases of large hawks have larger heads. Turkey
Vulture is less compact; has longer and slimmer tail, longer,
thinner wings, more "sail area," more graceful; flaps more
deliberately with longer periods of soaring.

Habitat
Open areas, usually flies lower than Turkey Vulture; beaches,
garbage dumps, slaughterhouses.

Habits
Gregarious; in flight holds wings nearly horizontal; tail often
fanned.

Eggs
1–3; bluish-white, blotched with brown, 3.1 x 2.0 in. (7.9 x 5.1
cm).

Range
Resident in the sw. states, s. to S. America.

CALIFORNIA CONDOR

Gymnogyps californianus   *Fig. 5*

Description
Size, 45–55 in. (114.3–139.7 cm); wingspread, 12 ft. (3.1 m). One
of the world's largest birds; nearly extinct, fewer than 50 survive.
Has flat, soaring wing plane. Adult: much larger than Turkey
Vulture; no dihedral wings in flight; fore-edges of wings show
extensive • *white underwing linings;* naked head yellow or orange.
Immature: broader proportioned and about twice the size of
Turkey Vulture; head dusky; no white wing linings.

Habitat
Sparsely inhabited mountains and adjacent range land. Rare.
Eggs
1; white; 4.3 x 2.5 in. (11.0 x 6.6 cm). Nest is cliffside cave or hole.
Remarks
Encroaching Southern California urbanization and other factors little understood are resulting in poor productivity, and heading the species toward extinction. These birds may live 50 to 100 years, and do not mate until 8 years of age. They lay but 1 egg every 2 years, when conditions are favorable, and the young must survive up to several years before becoming independent of the parents.
Range
Resident in mountains of s. Calif.; nearly extinct.

immature

Fig. 5

adult

California Condor

KITES, HAWKS, EAGLES, AND HARRIERS
Family Accipitridae

The predatory habits and seeming love of independence of these diurnal birds of prey have long interested humans. Today they need greater protection for the service they perform in maintaining nature's balance; they are especially valuable for their destruction of rodents, grasshoppers, and other pests. These birds have powerful, hooked talons and fearsome, hooked beaks for tearing flesh. Females usually are larger, sometimes much larger, than the males.

Kites are graceful, falcon-shaped hawks with pointed wings. They often have forked tails, weak bills, and feet adapted for capturing insects and small reptiles and mammals.

The accipiters, or bird hawks (Genus *Accipiter*), have short, rounded wings, finely barred below. Adults are slate-gray above with a dark cap; immatures have streaked breasts and are hard to distinguish. They frequent woods and forest margins, and are low, swift fliers that hunt among trees and brush, feeding mainly on birds. When not in pursuit, they alternately flap, then glide, soaring occasionally.

The buteos (Genus *Buteo*) are medium-sized to large, thick-set hawks with broad wings and short, fan-shaped tails. All are brownish, rufous, or blackish above, but there is considerable variation in individual plumages, particularly underneath, from light to dark. Several species show a distinct dark phase. Buteos hunt by soaring and circling high in the air or flying out after perching motionless on an exposed perch. They often migrate in numbers along ridges. They feed largely on rodents.

Eagles are among the largest birds of prey, distinguishable from very large buteos by their size; big, fully feathered heads; large beaks; and long tails and wings held horizontally in flight. They build huge nests on treetops or ledges.

Harriers are slender birds of prey after small birds or rodents in open rangeland. They have slim wings and long tails, and fly lazily at low altitude on dihedral wings.

WHITE-TAILED KITE
Elanus leucurus   16:4

Description
Size, 15–16 in. (38.1–40.6 cm); wingspread, 3⅓ ft. (1.0 m). Whitish, gull-like, with long pointed wings; sexes alike. Adult: head, • *square tail* and • *underparts white;* rest pale gray except for large • *black patch near fore edge of upper wing,* visible when flying or perched. Immature: back black, rusty streaking above and below, narrow band across tip of tail.

Similarities
White-tailed Hawk also has white tail with narrow black band near tip.

Habitat
Valleys, marshes, grassy foothills.

Habits
Often hovers like American Kestrel, flies like a small gull.

Voice
Abrupt or drawn out, whistled *kee, kee, kee.*

Food
Small mammals.

Eggs
3–5; blotchy; 1.6 x 1.2 in. (4.2 x 3.2 cm). Nest is of twigs in a tree near open country.

Range
Resident in Central Valley and s. coast of Calif.; also found in s. Tex.

MISSISSIPPI KITE
Ictinia mississippiensis   16:3

Description
Size, 14 in. (35.6 cm); wingspread, 3 ft. (0.9 m). Falcon-shaped with • *all-black tail.* Adult: bluish-gray above, • *pale head* with unmarked • *gray below;* long, pointed wings; dark primaries, light secondaries; tail notched or square; eyes and legs reddish. Immature: streaked above with black and white; spotted below with red-brown and buff; 3 gray bands on tail.

Similarities
The White-tailed Kite's tail is white, not all-black.

Habitat
Open country and woods.

Habits
Flight buoyant, gull-like, often tilting to show 2-toned upper surface of gray wings; migrates in flocks.

Voice
"*Phee-phew*" (Sutton).
Food
Insects.
Eggs
2–3; blue-green; 1.6 x 1.3 in. (4.1 x 3.3 cm). Lined nest of twigs in treetop.
Range
Breeds in e. N.Mex. and Tex. and Okla. panhandles; chiefly in se. U.S. but casual to Colo. and Calif.

COOPER'S HAWK
Accipiter cooperii   **13:2**

Description
Size, 14–20 in. (35.6–50.8 cm); wingspread, 3 ft. (0.9 m). Medium-size, with short, rounded wings and • *long, rounded tail;* almost the size of a crow. Adult: breast barred with red-brown. Immature: may show finer streaks below than young Sharp-shinned.
Similarities
Smaller Sharp-shinned is almost identical but has square-tipped tail. Broad-winged Hawk has longer wings, light and unbarred below; tail fan-shaped. Merlin wings are long, pointed. Sharp-shinned has notched or square tip of tail.
Habitat
Mixed woodlands, fields, groves, river canyons.
Habits
Wingbeats slower than Sharp-shinned's; circles and soars more.
Voice
Noisy; "a shrill *quick, quick, quick*" (May); "about nest, a rapid *kek, kek, kek*" (Peterson).
Food
Birds, small mammals.
Eggs
3–5; bluish-white, occasionally spotted with brown; 1.9 x 1.5 in. (4.8 x 3.8 cm).
Range
Breeds through U.S. from s. Canada to n. Mexico; winters from w. B.C. s. to C. America.

NORTHERN GOSHAWK
Accipiter gentilis   **13:1**

Description
Size, 20–26 in. (50.8–66.0 cm); wingspread, 4 ft. (1.2 m). Large, gray, with short rounded wings; long, nearly squared tail. Adult: • *white line over eye,* dark cap and ear patch, finely gray-barred breast. Immature: brown above, streaked below; white eye line less distinct than in adult.
Similarities
Cooper's Hawk is smaller and proportionately less heavy about head and neck; adults reddish below, immatures have only ill-defined eye line. Gyrfalcon has long, pointed wings.
Habitat
Coniferous forests.
Habits
Occasionally soars; usually hunts low, taking prey by surprise.
Voice
Silent except when breeding, then various cackles.

Food
Chiefly large birds; mammals to rabbit-size.
Eggs
2–5; bluish-white; 2.3 x 1.8 in. (5.8 x 4.6 cm). Nest is stick
platform in tree.
Range
Breeds from Alaska through Rockies to N.Mex. and in forest areas
of Wash., Oreg., and Calif.; winters at lower elevations s. to n.
Mexico.

SHARP-SHINNED HAWK
Accipiter striatus   13:3

Description
Size, 10–14 in. (25.4–35.6 cm); wingspread, 2 ft. (0.6 m). • *Short,
rounded wings;* • *long, square tail;* sometimes slightly notched.
Adult: blue-gray back, breast barred with rust. Immature: brown
above, white below with brown streaks. When perched, Sharp-
shinned's wings reach to lower third or half of tail. Tail may look
somewhat rounded when fanned out.
Similarities
Small male Cooper's is often quite similar to a large female Sharp-
shinned. Kestrel and Merlin have wings long, pointed.
Habitat
Coniferous and mixed forests, thickets.
Habits
Rarely soars, except in migration; occasionally ascends in tight
circles with much flapping.
Voice
High *cack, cack, cack.*
Food
Mainly small birds.
Eggs
3–5; bluish-white, splotched with brown and lilac; 1.5 x 1.2 in.
(3.8 x 3.0 cm). Nest is of twigs in a forest tree, usually conifer.
Range
Breeds throughout Canada and U.S. from Alaska, e. to Great Bear
Lake and s. to cen. Calif., N.Mex., and n. Tex.; winters from B.C.
to Mexico.

RED-TAILED HAWK
Buteo jamaicensis   13:7, 14:4

Description
Size, 19–25 in. (48.3–63.5 cm); wingspread, 4½ ft. (1.9 m).
Chunkier than Red-shouldered and with broader, longer wings,
shorter tail. Tail normally red in adults, but whitish if viewed from
below. Adult: from below, whitish breast set off in front by dark
throat, in rear by dark lower breast; often with a black wrist mark.
Veers in soaring, and reveals rufous upper side of tail. Immature:
underparts streaked with brown; no wrist mark; tail brown above
or dark gray, barred below and may or may not show banding;
may show light at base of tail similar to Swainson's or Rough-
legged. Adult of northern Plains race is very pale, with a whitish
head and pinkish tail. Western race typically rufescent below, but
dark phase occurs. Far northwestern race is very dark, tail mottled
without red; winters in small numbers in southern Plains to
Louisiana and Arkansas. All races may mix on southern Plains in
winter.

Similarities
Swainson's has dark forechest band bordered by white throat and white lower breast and abdomen, fine white and black tail bars from below. Red-shouldered is usually finely barred reddish below, without tendency to zoning in as in Red-tailed.
Habitat
Various woodlands, farm country, prairies, marshes, mountains, deserts.
Habits
Soars and circles for long periods, often twisting tail at angle to body, occasionally hovers.
Voice
Squealing *kee-a-a-a-r-r-r* with downward slurring.
Food
Rodents and other small mammals.
Eggs
2–4; white, sparingly spotted with brown; 2.3 x 1.9 in. (5.8 x 4.8 cm). Nest is stick platform in woodland (or isolated) tree, cliff hollow, saguaro cactus.
Former name
The dark northwestern form of this hawk was known as Harlan's Hawk.
Range
Breeds through N. and C. America from Alaska southward; winters from s. B.C. s. through U.S.

RED-SHOULDERED HAWK
Buteo lineatus   **13:5, 14:1**

Description
Size, 17–24 in. (43.2–61.0 cm); wingspread, 4 ft. (1.2 m). Noted for the narrow, white bands on a broad, black tail. Adult: broad wings, longer and slimmer than other buteos; wings with light areas of primaries which appear translucent; shoulders red-brown, duller in immature; upper wings barred with black and white; underparts cross-barred with pale Robin-red. Immature: streaked below, as in other hawks, identifiable mainly by smaller size and banded tail.
Similarities
Immature Red-tailed's streaks tend to be in bands.
Habitat
Mixed woodlands; moister terrain and smaller lower woodlands than Red-tailed; river bottoms.
Habits
Perches less conspicuously, soars less, and has less buoyant flight than Red-tailed.
Voice
Clear whistled *kee-yer*, never wheezy, like Red-tailed, and with descending inflection often imitated by Blue Jay.
Food
Rodents, other small mammals, reptiles, amphibians.
Eggs
2–4; white, blotched with brown and buff; 2.2 x 1.7 in. (5.6 x 4.3 cm). Nest is stick platform in forest tree.
Range
Resident in W. in Central Valley (Calif.) and s. coast to Baja Calif.

Note: The **BROAD-WINGED HAWK**, *Buteo platypterus*, size, 15–17 in. (38.1–43.2 cm), is a buteo east of the Rockies, breeding in southern Alberta and Saskatchewan. It has very broad white-and-black bands in the tail and no rufous in the shoulders.

SWAINSON'S HAWK
Buteo swainsoni   **13:8, 14:2**

Description
Size, 19–22 in. (48.3–55.9 cm); wingspread, 4¾ ft. (1.4 m). Long-winged Plains hawk with • *dark breastband;* • *2-toned underwings;* white in front, dark in rear. Adult: light phase shows throat and belly white; tail dark gray above, lighter at base, with 9–12 indistinct dark bands below; dark phase shows nearly entire plumage sooty-brown, underwings uniformly dark, sometimes with ashy bars on tail. Immature: buff, heavily streaked with dark brown below, usually darker than other young buteos; breast usually darker, wings and tail indistinctly barred with brown. Many variations occur.

Similarities
Red-tailed has white breast, streaks at rear; dark phase of Red-tailed has reddish tail (above). Dark phase of Rough-legged has clear white underwing flight feathers.

Habitat
All open regions and sparsely forested country; alpine meadows.

Habits
Sluggish, tame; wingbeats faster than Red-tailed's; hunts by cruising low over prairie with wings in open V, rather than by high soaring; gregarious, migrates in large flocks.

Voice
Long, "plaintive whistle, *kree-e-e-e*" (Bent), suggesting a Broad-winged.

Food
Grasshoppers, rodents.

Eggs
2–4; dull white with umber spots; 2.3 x 1.8 in. (5.8 x 4.6 cm). Nest is stick platform in isolated tree, bush, tall cactus, yucca, cliff.

Range
Breeds from Alaska, s. to Mexico, winters in S. America.

WHITE-TAILED HAWK
Buteo albicaudatus   **15:4**

Description
Size, 23–24 in. (58.4–61.0 cm); wingspread, 4½ ft. (1.4 m). Large, with • *white tail* narrowly • *banded with black* near tip. Adult: wings long, underparts • *clear white,* tail short, upperparts dark gray, shoulders rusty. Immature: blackish below, may have some white on chest and underparts in front, contrasting blackish to rear.

Similarities
Dark phase of Swainson's has barred grayish tail; more solidly dusky underwing. Ferruginous Hawk has whitish tail lacking black band; melanistic form, rare, resembles young White-tailed but has much cleaner white flight feathers underneath.

Habitat
Desert grasslands, prairie brush.

Food
Rabbits, lizards, rodents.

Eggs
2–3; white or indistinctly spotted; 2.2 x 1.8 in. (5.8 x 4.6 cm). Nest is stick platform on top of scrubby tree or yucca.

Range
Resident in coastal prairies from s. Tex. to S. America.

ZONE-TAILED HAWK
Buteo albonotatus   **15:6**

Description
Size, 18½–21½ in. (47.0–54.6 cm); wingspread, 4 ft. (1.2 m). Dull
• *black,* easily confused with Turkey Vulture (when soaring)
because of long, slender wings with a • *2-toned* effect. Adult: • *tail
bands white,* pale gray on top-side. Immature: black underparts
spotted with white; tail banding narrower.
Similarities
Black Hawk is chunkier; longer legs more "chickenlike"; broader
wings and tail in flight; only 2 white tail bands; white spot
sometimes visible at base of primaries.
Habitat
Desert mountains and rivers.
Habits
Sluggish; soars.
Voice
See Red-tailed; a squealing whistle.
Food
Rodents, birds, lizards.
Eggs
2–3; white; 2.1 x 1.6 in. (5.5 x 4.3 cm). Nest is stick platform, tree
or cliff.
Range
Breeds from cen. Ariz., s. N.Mex., and w. Tex. s. to S. America;
winters chiefly s. of U.S.

ROUGH-LEGGED HAWK
Buteo lagopus   **13:9, 14:6**

Description
Size, 19–24 in. (48.3–61.0 cm); wingspread, 4½ ft. (1.4 m). Large,
occurring mostly in winter and having • *rump and base of tail
white;* otherwise dark, with rather long pointed wings and a
longish tail; feathered legs sometimes visible as feet drop before
swoop. Adult: light phase shows head and upper breast buffy
streaked with brown, black patch at bend of light underwing;
• *belly, broad end of white tail,* wing tips, and rest of plumage
blackish; dark phase shows all-dark except for white on underwing
and base of tail, occasionally even lacks white base of tail.
Immature: similar to light adult, but dark belly band more
pronounced. Birds with intermediate plumages not uncommon.
Similarities
Young Golden Eagle may resemble dark phase of Rough-legged.
Dark phase of Ferruginous usually has some rufous mixed with
the black, and has more white at base of primaries. Dark phase of
Red-tailed usually has some rufous in tail. Northern Harrier is
more slender and has slim wings, slim tail. See also Swainson's
Hawk.
Habitat
Open plains, marshes in winter; tundra, Arctic coast in summer.
Habits
Sluggish, perches on favorite observation post by meadow; hovers
with beating wings like Kestrel or hangs suspended on updraft;
quarters low over meadows like a Northern Harrier, but is larger,
with broader wings, and may light on a tree, which Harrier never
does; hunts especially at dawn, dusk, and on dark days.
Voice
Silent in winter, a squealing *hurry-up* when breeding.

Food
Mice, lemmings.
Eggs
2–5; whitish, blotched with brown; 2.2 x 1.8 in. (5.6 x 4.6 cm).
Range
Breeds from n. Alaska e. to Great Slave Lake and s. to Aleutians
and se. Alaska; winters from s. Canada s. to s. Calif. and e. to
Ariz. and Tex., eastward.

FERRUGINOUS HAWK
Buteo regalis   **13:6, 14:5**

Description
Size, 22½–25 in. (57.2–63.5 cm); wingspread, 2–3 ft. (0.6–0.9 m).
Largest buteo; legs of reddish-brown adult make a distinctive dark
V against light underparts when viewed from below. Light area in
extended primaries in all plumages; head often white. Adult: light
phase shows shoulders, rump, and thighs chestnut; • *head and tail
whitish;* underwings and underparts also whitish, lightly streaked
and barred; dark phase is rare, all-dark with some rufous mixed
in, tail light or with several narrow white bands. Immature:
"underparts and tail whiter than in other hawks" (Bent); less
chestnut above and on thighs than light phase; tail with 4 dark
bars; lacks the dark V formed by legs.
Similarities
Common western subspecies of Red-tailed lacks chestnut
upperparts. Darker Rough-legged has a broad, black band on tail;
Rough-legged is heavier, with shorter tail and smaller bill. Young
Golden Eagle is larger, with wings less pointed and tail shorter.
Habitat
Prairies, badlands.
Habits
Perches on 1 leg on observation post; takeoff is slow and heavy, but
flight is swifter than Rough-legged's; quarters like a Northern
Harrier, alternately flapping and sailing.
Voice
Various squeals; also a gull-like *kaah.*
Food
Rodents, especially ground squirrels.
Eggs
3–5; white, blotched with brown; 2.4 x 1.9 in. (6.1 x 4.8 cm). Nest
is stick platform in tree or cliffside.
Range
Breeds from Canadian prairies, s. to Oreg., Ariz., and Okla.;
winters in Sw. from cen. Calif. to sw. S.Dak.

GRAY HAWK
Buteo nitidus   **13:4, 14:3**

Description
Size, 16–18 in. (40.6–45.7 cm); wingspread short, 3 ft. (0.9 m).
Small; yellow at base of bill. Adult: back gray; • *underparts gray
and white, barred;* rump white, • *tail widely banded;* thighs barred;
underwings gray in front, white with black tips at rear. Immature:
underparts striped buffy, tail narrowly barred.
Similarities
Immature Swainson's is larger, wings longer, reaching almost to
end of tail; reversed underwing pattern; dusky flight feathers
underneath.

Habitat
Lowland streamside woods.
Habits
Flight graceful; soars and feeds from perch.
Voice
"A loud plaintive *cree-eerr*" (Peterson).
Food
Mainly lizards and snakes.
Eggs
2–3; white; 1.9 x 1.6 in. (5.0 x 4.1 cm). Nest is stick platform in cottonwood, mesquite.
Range
Breeds in se. Ariz.; winters from U.S. border southward.

HARRIS' HAWK
Parabuteo unicinctus   15:5

Description
Size, 17½–29 in. (44.5–73.7 cm); wingspread, 3¾ ft. (1.1 m). Chestnut back. Adult: • *shoulders and thighs chestnut* (in good light); • *rump and uppertail a dazzling white;* • *tip of tail broadly banded in black and white.* Immature: shoulders rusty; underparts light, streaked; base of tail conspicuously white.
Similarities
Immature Red-shouldered has no white at base of tail. Black Hawk is much chunkier.
Habitat
River breaks, mesquite, chaparral.
Habits
Frequently perches on a favorite branch.
Voice
Raucous *karrr.*
Food
Rodents, rabbits, birds to size of teals.
Eggs
3–5; whitish; 2.0 x 1.6 in. (5.3 x 4.2 cm). Nest is stick platform in mesquite, yucca, stunted tree.
Range
Resident from Calif.–Mexican border e. to s. Tex. and s. to C. and S. America.

COMMON BLACK HAWK
Buteogallus anthracinus   15:3

Description
Size, 20–23 in. (50.8–58.4 cm); wingspread, 4 ft. (1.2 m). Large, chunky, black, with very broad, long wings and • *long, yellow, chickenlike legs.* Adult: broad • *white band across middle of tail,* broad dark bar with narrow white edge toward tip; occasionally visible during flight may be a whitish spot at base of primaries near wing tip. Immature: back dark; head and • *underparts streaked with bright buff;* tail narrowly banded with 5–6 alternating bands of black and buffy white.
Similarities
See also Zone-tailed, Harris', and melanistic buteos.
Habitat
Stream breaks, bottomlands, usually near stream.
Voice
Weak, high-pitched *quee-quee-quee* (Davis).

Food
Rodents, insects, fish, snakes.
Eggs
1-3; whitish, usually spotted; 2.2 x 1.7 in. (5.8 x 4.5 cm).
Range
Breeds in cen. and s. Ariz.; winters from s. Ariz. and N.Mex.-
Mexico border s. to C. and S. America.

GOLDEN EAGLE
Aquila chrysaetos   **12:6**

Description
Size, 31–40 in. (78.7–101.6 cm); wingspread, 7½ ft. (2.3 m). All-
dark below, including wing linings. Adult: dark brown above, with
golden-brown nape, visible only at close range, and white at base
of tail. Immature: base of tail white both above and below, giving
• *"ring-tailed" effect;* base of primaries white, rest of body dark
brown; amount of white diminishes with age.
Similarities
Bald Eagle has narrower tail and wings; immature Bald has white
on coverts, not flight feathers, and never has sharply ringed tail,
but tail may be mottled with white. Dark Rough-legged Hawk is
smaller, with more white under wings.
Habitat
Open forest mountains, foothills, canyons, badlands (rare).
Voice
Usually silent; a shrill whistled *"kee-kee-kee"* (Bendire); "a
yelping bark, *kya*" (Peterson).
Habits
In flight, tips of flight feathers are outspread and upcurved; soars
high, then dives for prey; in straight flight, alternately flaps and
glides; wings beat faster than Bald Eagle's; occasionally hovers.
Food
Smaller mammals, a few birds.
Eggs
1-3; white, blotched with red-brown; 3.0 x 2.3 in. (7.6 x 5.8 cm).
Nest is bulky stick mass on tree or high cliff.
Range
Resident from Alaska to Mexico, in mountains and rangeland;
northern populations somewhat migratory.

BALD EAGLE
Haliaeetus leucocephalus   **12:3**

Description
Size, 32–40 in. (81.3–101.6 cm); wingspread, 7½ ft. (2.3 m). Adult:
large, brown, with a • *white head and tail;* bill and legs yellow.
Immature: dark brown with mottled white on wing linings, bill
dusky, tail and head gradually assume full white plumage in fourth
year of age.
Similarities
Dark phase of Rough-legged and Ferruginous hawks have shorter
wings, smaller bills. See also Golden Eagle.
Habitat
Near water, rivers, lakes, ocean.
Habits
Wings are horizontal in flight; soars like a Red-tailed, head and
neck stretched far out; feeds chiefly on dead or dying fish and will
often steal a fish from an Osprey.

Voice
Loud, creaking cackles; "a harsh *kleek-kik-ik-ik-ik-ik,* or a lower *kak-kak-kak*" (Peterson).
Food
Dead fish, small animals, rarely birds.
Eggs
2–3; white; 2.8 x 2.3 in. (7.1 x 5.8 cm). Nest is bulky stick mass on tree or high cliff.
Range
Breeds in Alaska, locally in n. and e. Canada, and n. U.S.; winters along rivers inland.

NORTHERN HARRIER
Circus cyaneus   15:7

Description
Size, 18–22 in. (45.7–55.9 cm); wingspread, 4½ ft. (1.4 m). Long, slim wings and tail, white rump. Adult male: • *pale gray above,* white below; black wing tips. Female: dark brown above, lighter below. Immature: like female but underparts rufescent with less streaking.
Similarities
Rough-legged is larger, with broader wings and tail; white mainly is on upper tail, not on rump.
Habitat
Fields, prairies, marshes.
Habits
Flight buoyant, gull-like, tilting, with wings angled upward; flaps and glides low over grass; during courting or migration may soar high and circle on level wings like a buteo.
Voice
Low *chu-chu-chu;* a weak, nasal *pee-pee.*
Food
Small mammals, birds.
Eggs
4–6; whitish; 1.8 x 1.4 in. (4.6 x 3.6 cm). Nest is of reeds or grass on ground in marsh or grass.
Former name
Marsh Hawk.
Range
Breeds throughout most of Canada and U.S. from tundra to desert; winters chiefly in s. U.S.

OSPREYS
Family Pandionidae

OSPREY
Pandion haliaetus   12:5, 15:2

Description
Size, 21–24½ in. (53.3–62.2 cm); wingspread, 6 ft. (1.8 m). Only hawk that dives into water. Wings long, narrow, held "bent." Adult: dark brown above, with narrow blackish bands on tail; • *white crown, throat, and underparts;* • *black eye patch* forming "mask" and wrist mark on underwing; head mainly white. Immature: dark crown.
Habitat
Near water.

Habits
Flaps slowly, sometimes sails, often hovers before plunging; dives
feet first from up to 100 ft. (30.5 m) in the air; in flight, reveals
black carpal patches on whitish undersides.
Voice
Whistled *you-you-you;* "a complaining *shriek, shriek, shriek*"
(Cruickshank); also chickenlike peeps.
Food
Fish.
Eggs
2–4; white to rusty, blotched with deep brown; 2.4 x 1.8 in. (6.1 x
4.6 cm). Nest is massive stick nest high in a tree, cliff ledge, or
rock pinnacle.
Range
Breeds throughout N. America, from Alaska e. to Great Slave
Lake and s. to n. Ariz. and nw. Tex.; winters from s. U.S.
southward.

CARACARAS AND FALCONS
Family Falconidae

This family contains fast-flying, long-legged birds of prey having
large heads; long, narrow, pointed wings; and long tails. The sexes
are alike in color, but females are larger. The various species
primarily eat small animals or carrion. The major group of this
species is the true falcons (Genus *Falco*).

CRESTED CARACARA
Polyborus plancus   **12:4, 15:1**

Description
Size, 22–24 in. (55.9–61.0 cm); wingspread, 4 ft. (1.2 m). Adult:
dark above, long-legged, long-necked; black crest and red face;
throat and breast white; belly black, tail white, tipped with black,
presenting an alternating flight pattern of light and dark viewed
from below; conspicuous whitish patches near wing tips. Immature:
duskier; breast streaked, not barred.
Similarities
Black Hawk is all-black, with white areas near wing tips.
Habitat
Open rangeland, prairies.
Habits
Often observed on fence posts or feeding with vultures.
Voice
Rattling cackle.
Eggs
2–3; whitish, blotched; 2.3 x 1.8 in. (5.9 x 4.6 cm). Nest is reeds or
sticks, atop tree, yucca or saguaro cactus.
Range
Resident of tropics, from s. Ariz. and s. Tex. southward.

GYRFALCON
Falco rusticolus   **16:5**

Description
Size, 20–25 in. (50.8–63.5 cm); wingspread, 4 ft. (1.2 m). Gull-
sized falcon occurring in several color phases: dark, white, gray.
Dark phase (rare in West) is almost solid black with some white
markings; white phase, all-white with some black markings; gray

phase, intermediate and difficult to identify, but shows gray above, paler below, with underparts barred or spotted in adult, streaked in immature.

Similarities
Peregrine is smaller, thinner; solidly dark above, lighter below; throat white; black crown and mustache, relatively shorter tail and faster wingbeats. Prairie Falcon is clay-colored. Northern Goshawk has short, rounded wings, not long or pointed. Dark-phase Rough-legged may lack white at base of tail, sluggish, shows broadly tipped wings.

Habitat
Northern coasts, tundra, barren grounds, mountains.

Habits
Flight swift but heavier than Peregrine's, rapid wingbeats alternate with short glides; may hover before swooping for prey.

Voice
Chattering, screaming.

Eggs
3–4; white to rusty, marked with brown; 2.3 x 1.8 in. (5.8 x 4.6 cm).

Range
Breeds in Arctic, s. to Alaska and B.C.; winters s. from Arctic to n. U.S., especially Oreg., Mont., Wyo.

PRAIRIE FALCON
Falco mexicanus   16:7

Description
Size, 17–20 in. (43.2–50.8 cm); wingspread, 3½ ft. (1.06 m).
• *Black axillars in wing pits;* pale above; clay-colored, pointed wings; tail with narrow, light bands; dark mustache and white collar; below, white marked with brown; legs and feet yellow. Immature: rustier above, more streaked below; legs and feet bluish.

Similarities
Peregrine is much darker above, mustache more prominent. Merlin has smaller, darker tail, broadly gray-banded. Female American Kestrel is smaller, redder.

Habitat
Open rangeland, canyons, ridges, mountains, deserts.

Habits
Far-ranging; pursuit flight is swift, low; cruises with alternate beats and glides; often hovers, perches on conspicuous observation posts.

Voice
"Wert-wert-wert-wert-wert" (May); "a yelping *kik-kik-kik,* etc.; a repeated *kee, kee, kee"* (Peterson).

Eggs
3–6; white, with red-brown spots; 2.1 x 1.6 in. (5.3 x 4.1 cm). Nest is sticks, in cliff niche.

Range
Breeds from prairies of Canada and s. B.C. s. to Mexico–U.S. border; winters from s. U.S. southward.

PEREGRINE FALCON
Falco peregrinus   16:6

Description
Size, 15–21 in. (38.1–53.3 cm); wingspread, 3¾ ft. (1.4 m). Large, dark, with a broad black mustache. Sturdy body, dark above with a black cap, light below; pointed wings; narrow tail; Adult: slaty

above, tail lightly banded, throat and upper breast white, underparts barred. Immature: brown above, streaked below; throat buffy.

Similarities
Gyrfalcon color is more uniform, less contrasting. Prairie Falcon is paler. Merlin is smaller, more streaked below, tail more contrastingly banded, weak or no mustache.

Habitat
Grasslands, meadows, open country, mountains to coast; high buildings; many migrate along seashore.

Habits
Among fastest of birds, wingbeat rapid, flight swift, may glide then beat wings; captures prey by pursuit or by plunging; observation perch a crag or dead branch.

Voice
Loud cackles, *kak, kak, kak,* etc.; wails; a repeated *we-chew* about eyrie.

Eggs
2–4; creamy or reddish, blotched with chocolate; 2.1 x 1.7 in. (5.3 x 4.3 cm). Nest is sticks, in cliff ledge.

Range
Breeds locally throughout w. U.S., but very scarce; winters along Pacific Coast from s. B.C. southward.

MERLIN
Falco columbarius   16:2

Description
Size, 10–13½ in. (25.4–34.3 cm); wingspread, 2 ft. (0.6 m). Medium-sized, with heavily banded tail; suggests a small Peregrine. Male: slaty above, buffy streaked with brown below; broad black bands on gray tail. Female, immature: gray upperparts replaced by dark brown; banded tail. Blacker subspecies occurs along northwest coast.

Similarities
Sharp-shinned has short rounded, not long pointed, wings. American Kestrel has brighter markings; longer, rufous tail.

Habitat
Woodlands mainly, also foothills, rangeland; during migration marshes, as well as seashores.

Habits
Wingbeat rapid, flight swift and low, alternating beats with glides; perching outlook a post, knoll, or dead branch; sometimes hovers and pumps tail like an American Kestrel.

Voice
Seldom heard, but may be various cries and cackles; "at nest a shrill chatter, *ki-ki-ki-ki*" (Peterson).

Eggs
3–6; white to maroon, with dark rusty blotches; 1.6 x 1.2 in. (4.1 x 3.0 cm). Nest is varied, in treetop or cavity, cliff ledge or niche, other birds' (magpie) nests.

Former name
Pigeon Hawk.

Range
Breeds s. of treeline in Canada and s. to mountains in Oreg., Idaho, Mont.; winters s. from s. Canada to S. America.

AMERICAN KESTREL
Falco sparverius   **16:1**

Description
Size, 9–12 in. (22.9–30.5 cm); wingspread, 2 ft. (0.6 m). Back and
tail rufous; head multicolored with black markings; tail with black
band near tip; underparts buffy. Male: wings blue-gray, tail
unbarred, some spots below. Female: duller than male, wings
rusty, rufous tail barred, some streaks below.

Similarities
Merlin is stockier, with shorter wings; back and tail gray or
brown; Sharp-shinned has short rounded, not long pointed wings.
Immature Kestrel may suggest Merlin.

Habitat
Open country, roadsides, cities, prairies, deserts, stream breaks,
farmlands.

Habits
Flight usually unhurried, frequently hovers, pumps tail when
perched; keeps lookout from telephone wires, poles, dead branches.

Voice
High-pitched, loud *killy-killy-killy*.

Eggs
3–5; white or pinkish, spotted with brown; 1.4 x 1.2 in. (3.6 x 3.0
cm). Nest is lined cavity in hole in isolated tree, saguaro, old
building, nest box, cliff, magpie nest, etc.

Former name
Sparrow Hawk.

Range
Breeds throughout Canada and U.S. from Alaska s. to S. America;
winters throughout breeding range, with northern birds migrating
within range.

Fowl-like Birds
Order Galliformes

These small-headed, full-bodied birds with short, rounded wings
and strong legs and feet are usually ground dwellers, good runners
and scratchers; most do not migrate. Their stout wings make a
whirring sound in flight. Their food is largely seeds, fruits, leaves,
buds, occasionally insects. The young run about almost as soon as
hatched; all are considered excellent game birds.

CHACHALACAS
Family Cracidae

PLAIN CHACHALACA
Ortalis vetula   *Fig. 6*

Description
Size, 20–24 in. (50.8–61.0 cm). Pheasant-size, grayish-brown;
blunt-tipped, long tail; small head.

Similarities
Turkey is larger, bare head. Ring-necked Pheasant has pointed
tail, is spotted and barred above.

Habitat
Brushy woods, thickets.

Habits
Secretive, social, readily climbs about trees.

Voice
Cackles, squawks; utters loud *cha-cha-lak,* often by entire group from tree in morning or evening.

Food
Green leaves, seeds, berries, other fruit, some insects.

Eggs
3; dull white; 2.2 x 1.5 in. (5.8 x 4.0 cm). Nest is of twigs, sticks, leaves in a tree.

Range
Resident from lower Rio Grande Valley in Tex., southward.

Fig. 6

Plain Chachalaca

GROUSE, PARTRIDGES, PHEASANTS, QUAIL, AND TURKEYS
Family Phasianidae

Grouse are ground-dwelling birds with feathered legs that feed on seeds, buds, berries, and some insects. Ptarmigan differ from other grouse in having feathered toes as well as feathered legs, an adaptation to their often snowy environment.

Partridges and pheasants are similar in many ways to grouse, differing mainly in that their legs are not fully feathered. They nest in concealed, lined hollows on the ground, and eat insects, grains, and berries.

Quail are small-sized, social, ground-dwelling birds that feed on buds, berries, insects, and seeds.

The Turkey is the largest upland game bird. It is the species from which all domestic turkeys are descended. Its food is acorns and other nuts, berries, plants, seeds, and insects.

BLUE GROUSE
Dendragapus obscurus   **16:10**

Description
Size, 15–21 in. (38.1–53.3 cm). Dusky gray or blackish; black tail, marked with a light band at top. The form in the northern Rockies lacks the tail band. Male: yellow to orange comb above eye. Female: gray-brown barred with black, blackish banded tail.

Similarities
Ruffed is similar to female Blue, but distinguishable by its lighter tail and narrow black tip band. Spruce is smaller, male has black throat, female is rustier.

Habitat
Coniferous forests, logging slash, burned-over timberland.

Habits
Territorial (males) when breeding; may form small groups in winter.

Voice
Hooting or "booming" notes repeated 5–7 times, low and muffled (by male in courtship).

Eggs
5–10; buffy, spotted; 1.9 x 1.3 in. (4.9 x 3.4 cm). Nest is hollow on ground in dense timber.

Range
Resident in forested areas along Pacific Coast from se. Alaska to San Francisco and in Sierra Nevada and Rocky Mountains.

SPRUCE GROUSE

Canachites canadensis   16:8

Description
Size, 15–17 in. (38.1–43.2 cm). Dusky; males with sharply defined black breast, white-spotted at the sides. Male: dusky-brown; bare red comb above eye, chestnut band at tip of tail. Birds from the northern Rockies and West lack chestnut tip of the tail, having a black tail with white spots on either side of its base. Female: rusty-brown, well barred; tail blackish, banded chestnut at top for birds east of Rockies, unbanded to west.

Similarities
Male Blue Grouse is grayer, with less patterning below, larger tail; female is grayer. Ruffed Grouse has paler color; large, black-tipped, black-banded fantail.

Habitat
Spruce and pine forests.

Habits
Solitary; amazingly tame, flushes to small tree and sits.

Voice
Drums with wings; various coos, cackles.

Eggs
6–15; buffy, spotted; 1.7 x 1.2 in. (4.4 x 3.2 cm). Nest is hollow on ground under conifer or brush.

Range
Resident from Alaska and n. Canada s. to n. U.S., including Oreg., Mont., Wyo.

RUFFED GROUSE

Bonasa umbellus   16:9

Description
Size, 16–19 in. (40.6–48.3 cm). Conspicuous black band near tip of tail. Occurs in 2 color phases: • *red brown,* with rufous on tail (typical of brushy woodlands of Pacific states), and • *gray,* with gray on tail (mainly Rocky Mountains). Sexes similar; slight crest; above • *red-* or • *gray-brown;* tail rufous (or gray), white-tipped, fan-shaped; bare red area over eye; black ruff on neck; underparts white, marked with dark brown. Female: duller; lacks crest, red over eye, and ruff.

Similarities
Female Ring-necked Pheasant has longer, pointed tail; wingbeats slower; flight less noisy; found in more open country. See also Sharp-tailed and Blue Grouse.

Habitat
Open woodlands, second growth.

Habits
Seldom seen until they spring abruptly into the air with a startling whir; they are "tame" in the wilds, wary near humans; seek safety

by lying close, flying behind a tree, or running away and flying at a distance.

Voice

Nervous *quit-quit;* various clucks and coos; breeding male makes a characteristic drumming with his wings; "the muffled thumping starts slowly, accelerating into a whir; *bup . . . bup . . . bup . . . bup . . . bup . . . up . . . r-rrr*" (Peterson).

Eggs

6–15; buffy, may or may not be spotted; 1.4 x 1.1 in. (4.3 x 3.2 cm). Nest is hollow in floor of woods, under brush.

Range

Resident from n. Canada s. to ne. Calif., n. Utah, and Colo.; also from Great Lakes southward.

ROCK PTARMIGAN
Lagopus mutus   17:1

Description

Size, 13 in. (33.0 cm). Winter: white, with black patch between bill and eye. Summer: gray-brown. Female looks more like male in summer; in winter, like small Willow.

Similarities

Willow Ptarmigan, which see.

Habitat

Bare, rocky tundra of far north; in summer, mountains above timberline among rocks; lower slopes in winter.

Habits, Voice

Similar to Willow Ptarmigan.

Eggs

6–15; buffy to olive, spotted with brown; 1.7 x 1.2 in. (4.4 x 3.2 cm). Nest is on ground.

Range

Resident mainly in Arctic, s. to Alaska, B.C.

WHITE-TAILED PTARMIGAN
Lagopus leucurus   17:4

Description

Size, 12–13 in. (30.5–33.0 cm). Tail white in all plumages. Summer: brown; belly, wings, tail white. Winter: all-white except black bill and dark eyes.

Similarities

None in its range south of Canada; other species are larger and have black tails, but occur in the Arctic.

Habitat

Tundra, alpine uplands.

Habits

Shifts from alpine tundra to treeline for winter.

Voice

Cackles, clucks, soft hoots.

Eggs

4–10; buffy, spotted; 1.6 x 1.1 in. (4.2 x 2.9 cm). Nest is grassy hollow on open ground.

Range

Resident along Pacific Coast, from Arctic s. to Wash. and along Rocky Mountains from Alaska to N.Mex.

WILLOW PTARMIGAN
Lagopus lagopus   **17:2**

Description
Size, 15–17 in. (38.1–43.2 cm). White, with no black on face in
winter; an Arctic grouse with seasonal change of plumage; tail
black. Spring male: brownish or red-brown on head, neck, breast,
and front of back; white elsewhere. Summer male: brownish,
darkest above; wings white. Spring, summer female: black and
buff; scaly above, barred below. Much variation; individuals in
molt are mottled brown and white. Winter: both sexes pure white,
bill and tail black.

Similarities
Female Rock Ptarmigan is smaller, grayer in summer, but hardly
separable; male is grayer, less rusty in summer; male and some
females have a black patch between bill and eye in winter.

Habitat
In summer, low tundra, upland valleys; in winter, willow bottoms,
muskegs.

Habits
Gregarious after breeding season; roosts in snow; numbers fluctuate
periodically.

Voice
Crows, cackles; "deep, raucous calls, *go-out, go-out*" (Peterson).

Eggs
5–10; blotchy-red; 1.6 x 1.2 in. (4.3 x 3.1 cm). Nest is tundra
depression lined with willow grass, feathers.

Range
Resident in Arctic region, from Alaska s. to limit of forest tundra
and muskeg flats to cen. B.C.

GREATER PRAIRIE CHICKEN
Tympanuchus cupido   **17:3**

Description
Size, 16¾–18 in. (42.5–45.7 cm). Grouse of open country with
short, dark, • *squarish tail (black* in males, barred in females);
brown and buff, with light scales above, • *dark bars* below.

Similarities
Lesser Prairie Chicken is paler; occurs west of Greater. Female
pheasant has long, pointed tail. Sharp-tailed Grouse has white on
sides of pointed tail. Ruffed Grouse has large, fan-shaped tail; in
woods.

Habitat
Prairie grasslands.

Habits
Male, during courtship dance, raises hornlike, blackish neck
feathers and puffs out orange sacs on sides of neck. In flight,
alternately flaps and sails.

Voice
Various clucks, cackles; male in courtship utters a hollow booming,
oo-loo-woo.

Eggs
7–17; olive, spotted; 1.7 x 1.2 in. (4.4 x 3.2 cm). Nest is grassy
depression in prairie.

Range
Resident in prairies of Canada, chiefly Sask. and Man., and U.S.
prairies s. to Okla.

LESSER PRAIRIE CHICKEN
Tympanuchus pallidicinctus

Description
Size, 16 in. (40.6 cm). Very similar to Greater and often considered a subspecies of it; smaller and paler, best identified by range and habitat.
Habitat
Dry grasslands, sand hills, and oak thickets bordering streams.
Habits
Male, in display, shows throat sacs dull violet-red, not orange.
Voice
As Greater, but male in courtship makes a less booming sound.
Eggs
11–13; buffy, spotted; 1.7 x 1.2 in. (4.4 x 3.2 cm). Nest is grassy depression in grass or brush.
Range
Resident in U.S. southern prairie states, from N.Mex., e. to Okla.

SHARP-TAILED GROUSE
Tympanuchus phasianellus   **17:5**

Description
Size, 15–20 in. (38.1–50.8 cm). • *Short, pointed tail;* pale, speckled-brown; brown above, marked with white; white below, with dark V on breast; tail with • *white sides;* neck sacs of male purple; in flight, white spots on wings conspicuous. More rufous and buffier in winter.
Similarities
Female Ring-necked Pheasant has tail pointed but much longer. Prairie chickens have tail short, rounded, dark. Ruffed Grouse has fan-shaped and banded tail.
Habitat
Open grassy woodlands, brushlands, prairies, clearings, forest edges.
Habits
Flight speedy, straight; rapid wingbeats alternate with glides on downcurved wings; in winter, perches in trees; has elaborate courtship.
Voice
Clucking *"whucker, whucker, whucker"* (Bent); during courtship, a low *coo-oot;* also occasional cackling.
Eggs
7–13; brownish-green, spotted; 1.7 x 1.2 in. (4.4 x 3.2 cm).
Range
Resident from Alaska e. to Hudson Bay and s. to Utah and Nebr.

SAGE GROUSE
Centrocercus urophasianus   **17:6**

Description
Size, 22–30 in. (55.9–76.2 cm). Largest grouse (size of small Turkey), in open sagebrush rangeland; only one with • *black belly* and stiff, spikelike tail feathers. Female: tail pointed but shorter, mainly speckled and barred; belly black.
Habitat
Sagebrush. Occurs in mountains, to 10,000 ft. (3,048 m) or timberline; formerly much more plentiful.
Habits
In flight, rises heavily, but attains speed; alternately beats and glides; often occurs in flocks.

Voice
A *kuk-kuk-kuk* when flushed; a "popping" sound during courtship dance.
Eggs
7–13; olive-buff, spotted; 2.1 x 1.4 in. (5.5 x 3.8 cm). Nest is hollow under thick sagebrush.
Former name
Sage Hen.
Range
Resident from s. B.C., Alta., and Sask. s. to e. Calif., Nev., Utah, and w. Colo.

BOBWHITE
Colinus virginianus   29:10
Description
Size, 8½–10½ in. (21.6-26.7 cm). Chickenlike bird, with characteristic head pattern. Body plump, brownish, with much white below; throat and line over eye white in male, buffy in female; tail short and dark. Dark-throated subspecies in Arizona.
Similarities
Ruffed Grouse is larger. Meadowlarks have white outer tail feathers. Other quail are less brown.
Habitat
Brushy areas, roadsides, farms, woodland edges.
Habits
When flushed, explodes into low, fast flight, then glides quickly into cover; covey roosts on ground in circle facing out.
Voice
Clear whistled *bobwhite* or *poor-bob-white;* also "covey call, *ka-loi, lee?* answered by *whoil-kee*" (Roberts).
Eggs
10–20; white; 1.2 x 0.9 in. (3.0 x 2.3 cm). Nest is grassy depression in grass or underbrush.
Range
Resident chiefly in E., but also in foothills of Rockies and Wash., Oreg., and Idaho, and in s. Ariz.

MOUNTAIN QUAIL
Oreortyx pictus   29:6
Description
Size, 10½–11½ in. (26.7–29.2 cm). Small, gray-and-brown mountain quail with a • *pointed, backward-leaning head plume.* Male: • *throat chestnut,* crown gray with white border, forehead white, flanks patterned in white and chestnut. Female: duller.
Similarities
Head plumes of Gambel's and California do not tilt backward.
Habitat
Timbered mountain slopes, brush.
Habits
Flocks when not breeding; moves downslope for winter.
Voice
Rapid, tremulous whistles when alarmed; breeding males, a loud, infrequently repeated cry, "*wook?* or *to-wook?*" (Peterson).
Eggs
5–15; pinkish; 1.3 x 1.0 in. (3.4 x 2.4 cm). Nest is depression on ground, leaf-lined, in brush.
Range
Resident of nw. U.S., from Wash. to Calif.; introduced to B.C.

MONTEZUMA QUAIL
Cyrtonyx montezumae   29:13

Description
Size, 8–9½ in. (20.3–24.1 cm). Somewhat pale, short-tailed; sometimes known as "Fool's Quail" because of its tameness. Male: crest pale, bushy, not always erected; face "harlequin"-marked, black and white; body speckled. Female: brown, face less striped with brown.

Similarities
Male distinctive, female differs from Scaled by cinnamon underparts.

Habitat
Bunchgrass slopes, watered canyons, openly wooded mountains.

Habits
Tends to crouch and hide when approached.

Voice
Ventriloquial; a gentle quaver or whinny somewhat like a Screech Owl's.

Eggs
8–14; dull reddish; 1.2 x 0.9 in. (3 x 2.4 cm). Nest is a grass-lined depression in grass.

Former name
Harlequin Quail.

Range
Resident in Sw. from s. Ariz. to w. Tex., and s. to Mexico.

GAMBEL'S QUAIL
Lophortyx gambelii   29:11

Description
Size, 10–11½ in. (25.4–29.2 cm). Similar to California Quail, including head plume. Male: brown and gray with • *black patch* on a buffy-white belly; black head plume; flanks brown streaked with white; crown reddish-brown, hence local name of "Redhead"; forehead black; throat black with white border; white stripe on sides of head. Female: belly uniformly buffy-white without black patch; gray and brown, with brown crown and plume; both have 1–in. (2.5 cm) crest.

Similarities
Male California Quail has white forehead; both sexes have scaly marked belly. The two have almost exclusive ranges.

Habitat
Desert chaparral or brush, usually near water.

Voice
See California Quail; also "a loud *kway-er* and a querulous 3- or 4-note call, *yuk-kwair' ga-o*" (Peterson).

Eggs
10–16; buffy, blotched; 1.2 x 0.9 in. (3.1 x 2.4 cm). Nest is grass-lined on ground.

Range
Resident of Sw., s. to Mexico.

CALIFORNIA QUAIL
Lophortyx californicus   29:9

Description
Size, 9½–11 in. (24.1–27.9 cm). California's state bird. Male: plump, brown and gray with a • *forward-curving black plume* on

head; flanks brown with white streaks; crown chestnut, forehead white; throat black, with white lines about crown and throat patches. Female: duller, lacks contrasting head markings, but has plume.

Similarities
Male Gambel's Quail has black forehead and black patch on white belly, female has unmarked white belly; inhabits deserts.

Habitat
Coastal brush and timberland edges, parks, estates, farms and ranches, open chaparral.

Habits
In coveys (groups) except when breeding.

Voice
Variously interpreted as *qua-quer'go, where are' you? Chi-ca'go,* etc.

Eggs
10–17; buffy, spotted; 1.2 x 0.9 in. (3.1 x 2.4 cm). Nest is grassy depression on ground.

Range
Resident in Calif., s. to Baja Calif., but introduced in s. B.C., s. Oreg., n. Nev.

SCALED QUAIL
Callipepla squamata   29:12

Description
Size, 10–12 in. (25.4–30.5 cm). Called "cotton top" for • *white crest.* Lacks head plume; has scaly markings on breast and back; head grayish with dark line through eye, but no contrasting white and black marks.

Habitat
Prairies, grasslands, woodland edges, chaparral, brushy desert areas.

Habits
Seeks safety by running with neck outstretched and topknot erect; seldom flushes, but then scales quickly into cover as Bobwhite.

Voice
Nasal *"friendly pe-cos, pecos"* (Simmons); "a guinea-hen-like *chekar"* (Peterson).

Eggs
9–16; speckled; 1.2 x 0.9 in. (3.1 x 2.4 cm). Nest is depression on ground under bush.

Range
Resident chiefly in Mexico, but n. to se. Ariz., Utah and s. Colo., e. to Tex.

CHUKAR
Alectoris chukar   Fig. 7

Description
Size, 13 in. (33.0 cm). Introduced into West from Eurasia; larger than quail; sandy-colored, gray-brown; no crest or head plumes; • *legs and bill bright red;* throat white, bordered by black "necklace"; flanks boldly barred in black and white; tail red-brown.

Similarities
Mountain Quail has dark throat, long head plume, no red in tail.

Habitat
Mountains, canyons, arid reaches, brushy slopes, grasslands.

Voice
Series of *chuck's;* an occasional sharp *wheet-u.*

Eggs
8–15, buffy, spotted; 1.3 x 1.0 in. (3.4 x 2.4 cm). Nest is depression on ground under bush.
Range
Resident chiefly in Great Basin from B.C., s. to Baja Calif. and e. to Colo.

Note: The **GRAY PARTRIDGE**, *Perdix perdix* **(29:8)**, size, 13 in. (33.0 cm), is a native of Eurasia, introduced in parts of the United States. Its bill and feet are gray, and it lacks a "necklace" on throat. It occurs primarily in the Prairie states.

Fig. 7

Chukar

RING-NECKED PHEASANT
Phasianus colchicus 17:7

Description
Size, male, 30–36 in. (76.2–91.4 cm); female, 21–25 in. (53.3–63.5 cm). Originally introduced from Eurasia, a large, chickenlike bird with • *long, sweeping tail.* Male: brilliantly plumaged, usually has • *white ring around neck;* scarlet wattles on face, colors show iridescence, neck ring sometimes absent; tan barred with black. Female: brownish, tail similar to male's but shorter.
Similarities
Ruffed Grouse has shorter, fan-shaped tail. Sharp-tailed Grouse has white on sides of shorter, pointed tail. Sage Grouse is bulkier, with black belly, no bright colors.
Habitat
Farmlands, brushy edges.
Habits
Flushes with loud whir, flies fast, scales quickly into cover; roosts on ground or in trees, flocks in winter, runs swiftly.
Voice
Harsh *c-a-a-a-a;* crowing male, *kork-kok* with wing flapping; roosting gabble, *kutuk, kutuk.*
Eggs
6–14; greenish-brown; 1.6 x 1.3 in. (4.1 x 3.3 cm). Nest is grassy depression in tall grass.
Range
Resident throughout sw. Canada, the U.S. prairies; introduced from China and England.

TURKEY
Meleagris gallopavo *Fig. 8*

Description
Size, male, 48 in. (121.9 cm); female, 36 in. (91.4 cm); wingspread, 5 ft. (1.5 m). Slimmer, better able to fly than domestic turkey. Male: head naked, bluish; wattles red; body iridescent bronze; wings pale; tail erectile, fanlike, bronzy; buff or • *tail tip chestnut.* Female: smaller, duller.

73

Similarities
None.
Habitat
Open timberland, mountain forests, logged-over land.
Habits
Good flier, but prefers to run when escaping; roosts in trees.
Voice
Gobbles like the barnyard turkey; "alarm, *pit!* or *put-put!*"
(Peterson); in flocks, *keow-keow.*
Eggs
8–15; buff, spotted with gray; 2.7 x 1.8 in. (6.9 x 4.6 cm). Nest is
hollow on ground, lined with leaves in brush or woods.
Range
Resident from sw. U.S. to Mexico; introduced widely in woodlands
of w. states.

♂ displaying

Fig. 8

♂

♀

Turkey

Cranes and Rails
Order Gruiformes

These are a diverse group of marsh and open country birds that
nest on the ground. The young can run about almost immediately
after hatching.

CRANES
Family Gruidae

Superficially resembling herons, cranes are the tallest of birds; they
are stockier than herons, with long legs, long necks, and straight
bills, which are shorter and less pointed than the bills of herons.
Greatly lengthened inner secondary feathers curl over the wing tips
in a "tuft." They fly with a quick upstroke and with neck and legs
stretched straight out, often migrating in a line or V-formation.
They are waders and walkers, and never perch in trees. In
courtship they dance. The sexes are alike. Their food is reptiles,
amphibians, insects, grains, aquatic plants. Of the world's fourteen
species, two occur in the West.

WHOOPING CRANE
Grus americana **5:4**

Description
Size, 49–56 in. (124.5–142.2 cm); wingspread, 7½ ft. (2.3 m).
Tallest bird and one of the rarest in North America. Adult: white;
head with naked red face; bill dark with yellow base; legs black.
Immature: tan above, no naked red area, bill all-dark. Adult's
black wing primaries are often concealed by white secondaries
when walking.

Similarities
Snow Goose has shorter bill, neck, legs. White Pelican is heavy,
not tall and slender. Wood Stork has dark head, decurved bill,
inner rear edge of wings black. Egrets have no black in wings.

Habitat
Marshes, breeding muskegs, prairie ponds; very rare.

Habits
Very wary; walks about marsh; in flight, wingbeats slow, flaps in
single file; sometimes spirals to great height and performs aerial
revolutions.

Voice
Shrill, loud, vibrating trumpet call, *ker-loo! ker-lee-oo!*

Eggs
2; buff, blotched with brown; 3.9 x 2.5 in. (9.9 x 6.4 cm). Nest is
flat stick mound in muskeg.

Range
Breeds in nw. Canada (Wood Buffalo Park); winters on Tex. Gulf
Coast (Aransas National Wildlife Refuge).

SANDHILL CRANE
Grus canadensis **5:3**

Description
Size, 34–48 in. (86.4–121.9 cm); wingspread, 7 ft. (2.1 m). Adult:
gray; bill and legs dark; bald red crown and forehead; some
individuals rust-stained. Immature: all-brown, no red; tuft of
feathers over tail. Flies with neck out straight, distinctive wingbeats
slow and with a sharp flick or flap on the upstroke, above body
level.

Similarities
Great Blue Heron is often called "crane"; more varicolored; flies
with neck tucked in; less robust than Sandhill.

Habitat
Marshes and open country, prairies, grainfields, tundra, mountain
meadows.

Habits
Gregarious; walks much; at times spirals high in air; in spring,
groups often leap, hop, flap wings.

Voice
Ringing, trumpeting *"garoo-oo-oo-oo, garoo-oo-oo-oo"* (Laing). Also
"tuk-tuk—tuk-tuk—tuk-tuk; a gooselike *onk"* (Walkinshaw).

Eggs
2; drab olive, sparingly spotted with brown; 3.6 x 2.3 in. (9.1 x 5.8
cm). Nest is grass mound in marsh or wet meadow.

Range
Breeds in Arctic from Alaska coast, e. to cen. Canada and s. to ne.
Calif., Colo., S.Dak., and Mich.; winters from s. Calif., e. to w.
Tex. and s. to Mexico.

RAILS, GALLINULES, AND COOTS
Family Rallidae

Members of this marsh-dwelling family are somewhat chickenlike. Rails seldom fly far, except in migration. They are often heard, especially at dawn, but seldom seen. Their wings are short and rounded, their tail short and usually cocked; they fly short distances with legs dangling. Their nest is a platform in the marsh; young chicks are downy black. There are three major groups: the long-billed rails, the short-billed crakes, and the heavy-billed coots and gallinules. The rails and crakes are skulkers amid marsh reeds and grasses; the more ducklike coots and sometimes the gallinules swim about in open fresh water. Sexes are alike. Rails (Genus *Rallus)* have slightly curved bills and conspicuously barred flanks; crakes are small and short-billed with barred flanks.

CLAPPER RAIL
Rallus longirostris   **18:4**

Description
Size, 14–16 in. (35.6–41.9 cm). Large; body gray-brown, some buff below; tawny breasted; throat and undertail coverts white; flanks heavily barred; legs strong; bill long, slightly decurved.
Similarities
Virginia Rail is much smaller, red-brown, with darker flanks, gray cheeks.
Habitat
Freshwater, brackish, and saltwater marshes.
Habits
Swims well, may dive; often seen walking, occasionally swimming, about salt marsh during high tides; tail often raised to reveal white undertail coverts.
Voice
Frequent, staccato, often repeated *kek-kek-kek-kek.*
Food
Crustaceans, mollusks, worms.
Eggs
5–14; buffy, marked with red-brown; 1.7 x 1.2 in. (4.3 x 3.0 cm). Nest is grassy bowl, arched with grass, in marsh.
Range
Resident along Pacific Coast from cen. Calif., s. to Baja Calif., and S. America.

Note: The **KING RAIL,** *Rallus elegans,* size, 17 in. (43.2 cm), is an eastern rail that reaches the eastern edge of the Great Plains. It is found in freshwater marshes and is very similar to the Clapper Rail. The King Rail has a rust-colored breast and neck. The immature King resembles an immature Clapper.

VIRGINIA RAIL
Rallus limicola   **18:5**

Description
Size, 8½–10½ in. (21.6–26.7 cm). Small, reddish-brown, near size of meadowlark. Adult: above olive streaked with dusky; • *cheeks gray,* eye red; below red-brown with black bars on flanks; forewing reddish, conspicuous in flight; bill reddish, slightly decurved, long and slender. Immature: much black, long bill.
Similarities
Sora has short, yellow bill.

Habitat
Tules, freshwater marshes; in winter, salt marshes (rarely).
Habits
Can climb reeds and vines.
Voice
Harsh *kid-ik, kid-ik, kid-ik;* other "kicking," squealing, clucking, and vaguely chickenlike noises; *wak-wak-wak,* and grunts.
Food
Aquatic animal life, seeds, berries, insects.
Eggs
5–12; buffy, blotched with red-brown; 1.3 x 0.9 in. (3.3 x 2.3 cm). Nest is saucerlike mat of reeds and grass.
Range
Breeds throughout U.S. from s. Canada to s. states; winters from s. U.S. s. to Mexico and in n. marshes that do not freeze.

YELLOW RAIL
Coturnicops noveboracensis   **18:8**

Description
Size, 6–7½ in. (15.2–19.1 cm). • *White wing patch* seen in flight; buffy yellow; back streaking has a "checkerboard" effect; bill very short, yellow. About the size of a week-old chicken.
Similarities
Immature Sora is larger, with no white in wing.
Habitat
Freshwater marsh grass, meadows; may winter around salt marshes.
Habits
Most secretive, almost never flies except at night on migration; can conceal itself in very short, sparse grass. Requires a dog to flush; relatively uncommon.
Food
Little known; includes snails.
Voice
High-pitched, ticking notes in series, *"kik-kik-kik-kik-queeah"* (Ames).
Eggs
7–10; buff, with small red-brown dots; 1.1 x 0.8 in. (2.8 x 2.0 cm). Nest is grassy cup in marsh sedge.
Range
Breeds in Canada, e. of Rocky Mountains and s. to nw. states, rarely to ne. Calif.; winters along Gulf Coast.

SORA
Porzana carolina   **18:6**

Description
Size, 8–9½ in. (20.3–24.1 cm). Black face and throat, chickenlike • *short yellow bill.* Adult: chunky shape, dark brownish above, gray below; black face and throat; tail short and when cocked reveals white under coverts. Immature: buffy below, duller bill; lacks black patch.
Similarities
Buffy Yellow Rail is smaller; rarer; white wing patches. Virginia Rail has long, slender bill.
Habitat
Freshwater marshes, wet meadows, salt marshes in migration or winter.

Habits
Swims well, may dive; migrates at night, often striking obstructions.
Voice
High-pitched, descending horselike whinny; a whistled *cur-wee* in
spring; a single, sharp *keek* when startled; other Rail-like noises.
Food
Insects, wild rice and other seeds.
Eggs
6–15; buffy, marked with dull brown; 1.2 x 0.9 in. (3.0 x 2.2 cm).
Nest is grassy cup in freshwater marsh sedge.
Range
Breeds throughout much of N. America, from B.C. e. across
Canadian prairies and s. to Baja Calif. and the Sw., but not Tex.;
winters from s. U.S. s. to S. America.

BLACK RAIL
Laterallus jamaicensis   18:7

Description
Size, 5–6 in. (12.7–15.2 cm). Tiny, black, with chestnut nape;
about the size of a sparrow. Adult: above black with white dots;
below dark gray with white bars on belly; • *bill black*, nape deep
chestnut. Immature: all-black.
Similarities
Downy young chicks of all rails are glistening black, but they lack
Black's brown nape and white spots as well as barring on flanks;
they cannot fly.
Habitat
Low, grassy salt marshes; grassy edges of freshwater marshes.
Habits
Very hard to flush, runs mouselike with head down and neck out.
Voice
Male: *"kik, kik, kik, kik* or even *kuk, kuk, kuk"* (Wayne). Female:
"croo-croo-croo-o . . . like the commencement of the song of the
Yellow-billed Cuckoo" (Wayne). *"Did-ee-dunk, did-ee-dunk"*
(McMullen).
Food
Isopods (small crustaceans).
Eggs
4–9; white, dotted with red-brown; 1.0 x 0.8 in. (2.5 x 2.0 cm).
Nest is grassy cup under marsh grass.
Range
Breeds along Pacific Coast, from Calif. s. to S. America and locally
in interior; winters s. from Gulf Coast.

COMMON GALLINULE
Gallinula chloropus   18:3

Description
Size, 12–14½ in. (30.5–36.8 cm). Chickenlike marsh bird with
• *red bill* and frontal shield. Adult: head, neck, underparts gray;
back olive, undertail coverts white; conspicuous • *white stripe* along
flanks; bill with yellow tip; legs green. Immature: paler below; bill
dusky.
Similarities
American Coot is plumper, shorter-necked; all slate-gray, larger
headed with white bill. Immature Purple is buffier and white
below.
Habitat
Freshwater marshes, occasionally salt.

Habits
Runs and walks over lily pads jerking tail; tips up like a pond duck to feed in shallow water; also feeds on land; short flights weak with legs dangling; long flights with neck and legs extended; swims stern high, pumping head; can dive.
Voice
Chickenlike, loud, extremely varied; *cac, cac, cac;* "a croaking *kr-r-ruk,* repeated; a froglike *kup*; also *kek, kek, kek*" (Peterson).
Food
Vegetation, snails, insects.
Eggs
6–12; buffy, marked with brown; 1.7 x 1.2 in. (4.3 x 3.0 cm). Nest is reedy shallow saucer semifloating or attached, in marsh.
Other name
Moorhen.
Range
Breeds from n. Calif. e. to Nebr. and s. to S. America; winters from s. Calif. e. to Gulf states, southward.

Note: The **PURPLE GALLINULE,** *Porphyrula martinica* **(18:1)**, size, 13 in. (33.0 cm), rarely occurs in the Southwest and California as a wanderer from the East. Its greenish back and purplish underparts, with yellow legs and a yellow-orange forehead "shield" and white bill make it unmistakable.

AMERICAN COOT
Fulica americana **18:2**

Description
Size, 13–16 in. (33.0–40.6 cm). Slaty-black, ducklike water bird with a conspicuous, short • *white bill.* Adult: head black, bill short, undertail coverts white, legs greenish, toes with gray lobes. Immature: paler below, with duller bill.
Similarities
Common Gallinule is smaller; bill red with yellow tip, less ducklike; immature has white stripe on flanks.
Habitat
Freshwater marshes, ponds, lakes, rivers; salt bays and wet fields in winter.
Habits
Gregarious; swims, dives, as do gallinules; pumps head when swimming, patters along surface before taking flight; in the air shows white on trailing edge of wing, neck and legs are extended, feet protrude behind tail. Where protected, becomes quite tame and often approaches people to be fed.
Voice
"Coughing sounds, froglike plunks, and a rough sawing or filing *kuk-kawk-kuk, kuk-kawk-kuk,* as if the tree saw were dull and stuck . . . [and] a grating *kuk kuk kuk kuk kuk*" (Bailey); various other sounds.
Food
Aquatic plants, grass, grain.
Eggs
8–12; whitish, with pinhead-sized dark brown dots; 1.9 x 1.3 in. (4.8 x 3.3 cm). Nest is reedy shallow basket in marsh sedge or floating vegetation.
Range
Breeds throughout most of North America, e. to Gulf Coast; winters in Pacific Coast states, s. from B.C. and interior wherever water does not freeze.

Shorebirds, Gulls, Auks, and Allies
Order Charadriiformes

OYSTERCATCHERS
Family Haematopodidae

Only one of the world's six or so species of these large shorebirds occurs in the West with regularity.

BLACK OYSTERCATCHER
Haematopus bachmani   **19:17**

Description
Size, 17 in. (43.2 cm). Large, all-black, heavily built; • *bill long, straight, red;* legs pale.
Habitat
Capes, cliffs, rocky coasts.
Habits
Uses laterally flattened bill tip to extract shellfish from shells, rocks.
Voice
Sharp, piercing whistle.
Food
Mollusks, crustaceans, worms.
Eggs
2–3; buffy, spotted; 2.2 x 1.5 in. (5.6 x 3.9 cm). Nest is hollow among rocks.
Range
Resident along Pacific Coast from Alaska to Baja Calif.

PLOVERS
Family Charadriidae

Small to medium-sized, plump, round-headed, short-necked inhabitants of the seashores, plovers are runners rather than waders. They differ from sandpipers in being stouter and having pigeonlike bills that are short and thick, shorter than their heads. They have a stop-and-go manner of feeding, rather than the continuous movement of sandpipers. They eat mostly shellfish, other marine invertebrates, and worms. Plovers nest in an inconspicuous depression in the sand or ground; the eggs are pear-shaped. The young can run about almost as soon as they hatch. The sexes are alike.

SEMIPALMATED PLOVER
Charadrius semipalmatus   **19:2, 23:1**

Description
Size, 6½–8 in. (16.5–20.3 cm). Small, half size of Killdeer, dark-backed, with short bill and • *dark breastband,* not always complete in front. Spring: dark brown above, color of wet sand; black band across upper breast and black line through eye to bill, white forehead, throat, underparts, and sides of tail; yellow-orange bill with black tip; yellow-orange legs. Fall and immature: black areas browner; more white on forehead; legs paler; bill more dusky, all-black in winter.

Similarities
Killdeer is double the size with 2 black rings across chest. Snowy Plover has blackish legs. Piping Plover is lighter, with fewer face markings.
Habitat
Mud flats, beaches, shorelines, wet fields.
Habits
Often flies in compact flocks wheeling in unison; on alighting, birds spread out to feed, stoop intermittently to pick up food.
Voice
A plaintive rising *cheer-wee;* "a plaintive slurred *chi-we,* or *too-li,* 2nd note higher" (Peterson).
Eggs
3–4; buff, marked with blackish-brown; 1.3 x 0.9 in. (3.3 x 2.4 cm). Nest is depression in sand.
Range
Breeds in n. tundra from Alaska s. to cen. B.C.; winters from cen. Calif. coastline to se. Ariz.; migrates along Pacific Coast.

Note: **WILSON'S PLOVER,** *Charadrius wilsonia,* size, 8 in. (20.3 cm), is seen along the Texas coast. It has pinkish legs, a fully black bill, and a broader black chest band.

SNOWY PLOVER
Charadrius alexandrinus   **19:4, 23:2**

Description
Size, 6–7 in. (15.2–17.8 cm). Pale, with slaty legs. Similar to Piping Plover, but slightly paler and "breastband" reduced to 2 black shoulder patches; black ear patch present; • *legs and thin bill dark* at all seasons; loses most of its black markings in fall, but retains dark ear mark.
Similarities
Semipalmated is much darker, legs yellowish, breastband complete. Piping does not have black ear patch.
Habitat
Sandy shores, beaches, salt or alkaline flats, sand flats.
Voice
Low whistled notes, *o-wee-ah;* a trilled *kr-r-r-r* or *pe-e-e-e-et.*
Eggs
2–3; sandy, with small black spots; 1.2 x 0.9 in. (3.0 x 2.3 cm). Nest is hollow on beach or in alkaline flat, lined with bits of shell.
Range
Breeds along Pacific Coast from Wash. s. to Baja Calif. and inland throughout W. along riverbanks, sand dunes, beaches; winters along Pacific Coast from Wash. to Mexico.

PIPING PLOVER
Charadrius melodus   **19:1**

Description
Size, 6–7½ in. (15.2–19.1 cm). Above pale tan, color of dry sand. Spring: bill yellow with black tip, line over forehead, and end of tail black; single black band across upper breast, often broken; base of bill and • *legs yellow-orange;* underparts, rump, and wing stripe white. Fall and immature: black on head and breast lost or replaced by brown, bill darker.
Similarities
Snowy Plover is paler, with black ear patch, slaty legs. Semipalmated is darker, color of mud or wet sand.

Habitat
Sandy beaches and shores; prefers drier sand of higher beach.
Habits
Less social than many plovers, usually alone or in pairs.
Voice
Soft, whistled descending *peep-lo.*
Eggs
3–4; creamy, finely dotted with chocolate; 1.2 x 1.0 in. (3.0 x 2.5 cm). Nest is hollow on dunes or beach.
Range
Breeds in cen. Alta., s. Sask., sw. S. Dak.; winters along Gulf Coast, chiefly in E.

KILLDEER
Charadrius vociferus   **19:3, 23:3**

Description
Size, 9–11 in. (22.9–27.9 cm). Common farmland plover, distinguished by • *2 black breastbands.* Adult: earth-brown above; white forehead, throat, collar which continues around neck, underparts, wing stripes, and edges of tail; • *orange-brown rump* and upper part of rather long tail; black bill; flesh-colored legs. Immature: paler, grayish breastband.
Similarities
Other "ringed" plovers are smaller, with only 1 band.
Habitat
Mud flats, fields, parks, open areas, fill, airports, usually near water.
Habits
Swift runner and flier; often active after dark and on moonlit nights; has spectacular nuptial flight; if nest is approached, circles in air screaming above intruder, or drags wings and a leg on the ground in conspicuous distractive display.
Voice
Loud *kill-dee, kill-dee;* a noisy and persistent trilling; plaintive *dee-dee-dee.*
Eggs
4; buff, blotched with brown; 1.4 x 1.1 in. (3.6 x 2.8 cm). Nest is scrape in field or pasture, gravel, roadway, etc.
Range
Breeds throughout temperate regions of Canada and U.S., from B.C. eastward, and s. to Mexico; winters along Pacific Coast and inland in mild climates.

MOUNTAIN PLOVER
Charadrius montanus   **19:5, 23:7**

Description
Size, 8–9½ in. (20.3–24.1 cm). • *No black below* or on the sides. Adult: in breeding plumage, above dull brown; • *forehead, line over eye,* and underparts white; crown, line from bill to eye, band near end of tail black; tan wash on breast; pale wing stripe visible in flight; bill slender, black; legs tan. In fall plumage, sides of head and breast change to tan.
Similarities
Black-bellied Plover is larger, grayer; back mottled, base of tail white. American Golden Plover in winter has spotted, grayer back; tail dark; no wing stripe.
Habitat
Dry fields, plains, prairies, grassy deserts, plateaus.

Habits
Gregarious; seeks safety in remaining motionless; in flight, wing pits and wing linings show silvery white; bird alternately flaps and sails on downcurved wings.
Voice
Varied low whistles.
Eggs
3; deep olive-buff, spotted and scrawled with black; 1.5 x 1.1 in. (3.8 x 2.8 cm). Nest is bare scrape on ground.
Range
Breeds in w. U.S. prairie, from Mont. e. to N.Dak. and s. to sagebrush area of N.Mex. and Okla.; winters from cen. Calif., e. through Sw. and e. to Mexico.

AMERICAN GOLDEN PLOVER
Pluvialis dominica   **19:8, 21:11, 23:6**

Description
Size, 9½–11 in. (24.1–27.9 cm). No hind toe. Spring: speckled golden above, forehead and sides of neck white, underparts all-black to tail. Fall: brownish or yellowish-brown, darker above; unpatterned viewed in flight. Tail always dark, bill black, legs slate. In flight, wings gray below.
Similarities
Black-bellied Plover is larger; above pale gray, rump and upper tail white; white in wings, and black axillars in wing pits; in winter, grayer, with thicker bill and neck.
Habitat
Marshes, fields, mud flats, prairies, beaches, shores; tundra in summer.
Habits
Gregarious; flight swifter, more buoyant than Black-bellied's; on the ground is more aggressive; raises wings on alighting, often bobs head.
Voice
Quavering whistled *quee-i-ia,* or a harsh *queedle,* descending at end.
Eggs
3-4; buffy olive, spotted with brown and black; 2.0 x 1.3 in. (5.1 x 3.3 cm). Nest is mossy tundra depression; moss-lined beach hollow.
Range
Breeds in Arctic; winters on Pacific islands, also S. America; migrates through Alaska, Great Plains, Pacific States, and Great Basin.

BLACK-BELLIED PLOVER
Pluvialis squatarola   **19:7, 21:12, 23:9**

Description
Size, 10½–13⅓ in. (26.7–34.3 cm). • *Black breast and wing pits* in spring; large, stout head and bill; hind toe present; bill and legs black; wing stripe, rump, black-barred tail, and undertail coverts white. Spring: speckled pale gray above, black below, except white under tail; forehead and sides of neck white, white area larger than in Golden. Fall: mottled gray above, whitish below. Immature: like fall plumage, gray-appearing, stocky, pigeonlike bill.
Similarities
American Golden Plover has brownish back.
Habitat
Beaches, mud flats, salt marsh meadows.

Habits
Sedate, somewhat stolid shoreline figure of erect carriage; white forehead is conspicuous in birds on ground amid grass.
Voice
Wild, plaintive, somewhat Bluebird-like *toor-a-wee;* "slurred whistle, *tlee-oo-eee* or *whee-er-ee*" (Peterson).
Eggs
4; buffy-olive, marked with brown and black; 2.1 x 1.4 in. (5.3 x 3.6 cm). Nest is mossy tundra cup.
Range
Breeds in Arctic; winters chiefly in S. Hemisphere but some stay along Pacific Coast.

SNIPE, SANDPIPERS, AND ALLIES
Family Scolopacidae

This is a varied family of shorebirds that have slender, relatively long bills; long pointed wings; rather long wading legs; and short tails. Most are of small to medium size; they prefer moist areas or shallow water and shorelines. Almost all have a distinctive courtship song, occasionally heard in migration; they are quite gregarious, and sexes are alike in most species. The often diagnostic wing stripes and rump and tail patterns should be noted. Most nest in a depression upon the ground, often lined with grasses, usually near the water. These shorebirds eat crustaceans, mollusks, berries, and insects. The eggs are usually four and pear-shaped; the young can run about almost immediately after hatching.

RUDDY TURNSTONE
Arenaria interpres   **19:9, 21:5**

Description
Size, 8–10 in. (20.3–25.4 cm). Heavy-set, orange-legged, "harlequinned," pied markings are unique and striking, especially in flight. Summer: back russet-red, brown in fall; face and breast most curiously marked black and white; in flight at all seasons, calico pattern with 2 prominent white stripes on each wing, and a white patch on the lower back; bill black, may be slightly upturned; legs short. Immature and winter adult: duller, but recognizable by contrasting black breast pattern.
Similarities
Black Turnstone is similar but lacks ruddy and browns in plumage.
Habitat
Coastal; beaches, shorelines, mud flats, rocks, jetties, islets.
Habits
Pokes bill under pebbles, shells; digs broad, shallow holes in sand; pugnacious; sometimes perches off ground; can swim; large migrating flocks separate into small groups on ground.
Voice
Harsh *chut-chut;* a melodious *quit-tock;* "a staccato *tuk-a-tuk* or *kut-a-kut;* also a single *kewk*" (Peterson).
Eggs
4; cream, splashed with brown; 1.5 x 1.1 in. (3.8 x 2.8 cm). Nest is tundra depression; dune hollow.
Range
Breeds in Alaska, across Canadian Arctic; winters in S. America and s. Pacific; migrates along Pacific Coast.

BLACK TURNSTONE
Arenaria melanocephala   **19:6, 21:8, 23:11**

Description
Size, 9 in. (22.9 cm). Similar to Ruddy Turnstone, especially in
bold flight pattern; heavy-set, blackish above; chest blackish; lower
breast and belly white. Spring: some white speckles on sides; in
front of eye a round white spot.
Similarities
Ruddy Turnstone has rusty or brown back; head more patterned;
legs paler, more orange.
Habitat
Same as Ruddy Turnstone.
Habits
Very active, but with sporadic long pauses; digs hole in sand.
Voice
Similar to Ruddy's but higher; a "rattling" note.
Eggs
4; yellow-olive, blotched with brown and black, 1.5 x 1.1 in. (4.0 x
2.9 cm). Nest is tundra depression along coast.
Range
Breeds in sw. Alaska along coastline; winters along Pacific Coast
from se. Alaska to Baja Calif. and e. to w. Ariz.

COMMON SNIPE
Gallinago gallinago   **20:13, 23:8**

Description
Size, 10½–11½ in. (26.7–29.2 cm). Tight-sitting, chunky; • *long,
slender bill;* striped head. Wings rather pointed; streaked brown
above; breast spotted, belly white; • *tail orange,* tail corners pale;
legs greenish; no seasonal change.
Similarities
Dowitchers have conspicuous white stripe on lower back.
Habitat
Meadows, marshes, bogs, irrigation ditches, fresh or salt water.
Habits
Solitary, secretive; hides by squatting; most active at dawn, dusk,
and on cloudy days; usually rises in zigzags, uttering nasal
"escape" note; can swim and dive.
Voice
A "rasping *escape, escape;* on breeding grounds, a melodious *wheat
wheat*" (Collins); "song, a measured *chip-a, chip-a, chip-a;* in high
flight display, a hollow winnowing *huhuhuhuhuhuhu*" (Peterson).
Eggs
3–4; dull olive, boldly blotched brown; 1.5 x 1.1 in. (3.8 x 2.9 cm).
Nest is grassy cup in marsh, wet meadow, or muskeg.
Range
Breeds throughout much of Canada and U.S. from n. Canada to
Calif., except Rockies; winters from s. U.S. to S. America.

LONG-BILLED CURLEW
Numenius americanus   **19:15**

Description
Size, 20–26 in. (50.8–66.0 cm); bill, 5–7 in. (12.7–17.8 cm).
Magnificent shorebird distinguished by its very long, thin, downcurved bill. Buffy; head unstriped, underwings and • *wing pits
bright cinnamon.*

Similarities
Whimbrel is smaller; less buffy, with striped crown and with no cinnamon underwings.
Habitat
Plains, prairies, open areas near water, rangeland.
Habits
Wary; flies in V-shaped flocks; can swim.
Voice
Loud, whistled, rising *curlew-curlew;* a rapid *"wheety, wheety, wheety,"* and a loud, rattling *"que-he-he-he-he"* (Bent).
Eggs
4; buff, spotted with brown and lavender; 2.6 x 1.9 in. (6.6 x 4.8 cm). Nest is grassy cup in open prairie.
Range
Breeds from s. Canada, s. to Tex. and s. through Great Basin; winters from Sw. to Mexico and along Gulf Coast.

WHIMBREL
Numenius phaeopus

19:14
Fig. 9

Description
Size, 15–18 in. (38.1–45.7 cm); bill, 2¾–4 in. (7.0–10.2 cm). Common; large, brown, with • *prominent stripes on the head;* long decurved bill. Grayish-brown above, barred pinkish-buff underwings, grayish-white below, blue-gray legs. As large as some ducks. Long-billed is larger, longer-billed, buffier.
Habitat
Salt marshes, tidal flats, shores, river bars, tundra, prairies.
Habits
Flight steady, high over land in flocks like ducks, or in long lines low over water; often scales on set wings.

Marbled Godwit, p. 98

Whimbrel

Fig. 9

Willet, p. 90
fall

Hudsonian
Godwit, p. 99
fall

American Avocet, p. 101
spring

Large Shorebirds in Flight

Voice
Soft musical *cur-lew;* a series of harsh *ku-ku-ku-ku* notes on same pitch; "5–7 short rapid whistles, *ti-ti-ti-ti-ti*" (Peterson).
Eggs
4; buff, marked with brown; 2.4 x 1.6 in. (6.1 x 4.1 cm). Nest is grassy tundra depression.
Former name
Hudsonian Curlew.
Range
Breeds from Arctic Alaska eastward; winters from s. U.S. to S. America.

Note: The nearly extinct **ESKIMO CURLEW,** *Numenius borealis,* size, 11 in. (29.2 cm), is a rare migrant in the Great Plains to Texas coast. It is smaller than a Whimbrel, with shorter bill and less conspicuous facial stripes.

UPLAND SANDPIPER
Bartramia longicauda   **19:10, 23:5**

Description
Size, 11–12½ in. (27.9–31.8 cm). Medium-size, in grassland; straight bill; white outer tail feathers. Streaked brownish above, lighter below; small pigeonlike head; slender long neck; rather short bill; long tail; line over eye and outer tips of tail white, underwing black and white; dark rump; yellowish legs.
Similarities
Buff-breasted Sandpiper is smaller, breast unmarked. Pectoral Sandpiper is smaller. Various prairie curlews and godwits are much larger.
Habitat
Grassy inland prairies, plains, fields.
Habits
Flies brief distances with short strokes of downcurved wings like Spotted Sandpiper; expanded flight is swift, buoyant; often perches on fence post.
Voice
Song is an eerie, mournful, mellow whistle *"wh-e-e-e-e-e-e-e-e-o-o-o-o-o-o . . .* [alarm note] *quitty-quit-it-it"* (Knight); "a mellow whistled *kip-ip-ip-ip,* often heard at night . . . weird, windy whistles: *whooooleeeee, wheeeelooooooooo"* (Peterson).
Eggs
4; buff, finely spotted with brown; 1.8 x 1.3 in. (4.6 x 3.3 cm). Nest is hollow in grass clump.
Former name
Upland Plover.
Range
Breeds from Alaska and cen. Canada s. to cen. U.S.; winters in S. America; migrates e. of Rockies.

SPOTTED SANDPIPER
Actitis macularia   **21:18, 22:6**

Description
Size, 7–8 in. (17.8–20.3 cm). Spring: big, • *round, black spots* with dark below; olive-brown above; line over eye, wing stripe and edges and tip of tail white. Fall: spots lacking; dusky sides of neck separated from wing by white mark.
Similarities
Solitary Sandpiper in fall is darker above, streaked breast, no white wing stripe; bobs, does not teeter.

Habitat
Shorelines of lakes and streams; seashores in winter.
Habits
Usually solitary, constantly teeters tail up and down as it walks; flight highly characteristic, somewhat like meadowlark, short rapid beats of quivering downcurved wings alternate with glides low over water; also has a seldom-seen full, free flight like yellowlegs; can swim and dive from water or wing; perches on posts, wires, branches.
Voice
Low, distinct *peet-weet*.
Eggs
4; cream, spotted with gray and chocolate; 1.3 x 0.9 in. (3.3 x 2.3 cm). Nest is grassy scrape near water or in brush.
Range
Breeds throughout most of Canada and U.S., from nw. Alaska eastward and southward; winters from s. B.C. throughout most of U.S., s. to S. America.

SOLITARY SANDPIPER
Tringa solitaria   **21:16, 22:9**

Description
Size 7½–9 in. (19.1–22.9 cm). No seasonal change. Dark back. Dark wings, no wing stripe; black-barred, white-edged tail. Upperparts and underwing blackish; eye-ring white; • *tail appears white* with a dark center; breast streaked, underparts white; bill dark, slender; legs olive. Immature: more dotted with white above.
Similarities
Spotted Sandpiper teeters more, pale above with less white on tail, black spots below in spring and white wing stripe at all seasons. Lesser Yellowlegs has bright yellow legs and whitish rump and tail, which are larger and taller.
Habitat
Freshwater edges, streams, pools in marshes and woods, ditches.
Habits
Rather solitary; flight light, airy, often zigzagging, wings have good upstroke unlike Spotted; engages in extensive aerial maneuvers; short-distance flight is jerky with wings only partially spread; drops abruptly to a landing, raises wings on alighting; bobs head as if hiccuping (does not teeter); can swim, dive.
Voice
One or more *peet* notes, higher-pitched than Spotted's.
Eggs
4–5; pale greenish, with rufous spots and blotches; 1.4 x 1.0 in. (3.6 x 2.5 cm). Appropriates old nests of Robin, Jay, Blackbird, etc., in tree near stream.
Range
Breeds from Alaska e. to Canadian prairie and s. to n. B.C.; winters in C. and S. America; migrates across W.

GREATER YELLOWLEGS
Tringa melanoleuca   **21:21**

Description
Size, 12½–15 in. (31.8–38.1 cm). Both Greater and Lesser Yellowlegs are distinguishable by their • *yellow legs* and • *white rump*, the Greater being the larger. Somewhat slim sandpiper, grayish above, no wing stripe, tail whitish, below white with streaks on breast, long legs, long bill slightly upturned. No seasonal change.

Similarities
Lesser Yellowlegs upperparts may seem lighter; bill is shorter, thinner; legs less orange-yellow. See Voice.

Habitat
Marshes, mud flats, pools, streams, muskeg, bogs.

Habits
Not so gregarious as Lesser; bobs up and down, wades deeply, occasionally swims; very noisy, loud cries warn of intruders; responds to imitations of its note; raises wings over back on alighting.

Voice
Clear whistled *yew,* normally 3 to 4 times, descending scale; sometimes only twice, or a series; "a rolling *toowhee, toowhee*" (Nichols); "a 3-note whistle, *whew-whew-whew,* or *dear! dear! dear!*" (Peterson).

Eggs
4; grayish; splashed with brown and lilac; 1.8 x 1.3 in. (4.6 x 3.3 cm). Nest is a muskeg hollow.

Range
Breeds from Alaska e. through n. Canada; winters from s. Ariz., N.Mex., s. to S. America; migrates throughout W., commonly in wetlands.

LESSER YELLOWLEGS
Tringa flavipes   **21:20, 22:15**

Description
Size, 9½–11 in. (24.1–27.9 cm). Similar to Greater, but smaller, distinguishable by a straight, • *shorter thin bill* and voice. Slight other differences include more delicate build, proportionately shorter neck and legs, softer gray above, more • *lemon-yellow legs;* size alone deceptive, but direct comparison in field is often possible.

Similarities
Stilt Sandpiper in fall is smaller, with greenish legs, bill slightly drooped at tip. Wilson's Phalarope is smaller, with more needlelike bill; swims regularly and looks about in water. See also Solitary Sandpiper.

Habitat
Marshes, mud flats, pools, shores, open boreal forests.

Habits
More gregarious than Greater, less suspicious; flight "more buoyant and hence not so suggestive of momentum" (Brewster); goes north (from winter range in Gulf Coast states) later than Greater and south earlier.

Voice
A *yew* or *you-you* note similar to Greater, but usually in singles or pairs, somewhat flatter and less forceful; also a rolling *toowhee.*

Eggs
4; see Greater; 1.7 x 1.1 in. (4.3 x 2.8 cm). Nest is ground hollow in open, often far from water.

Range
Breeds in n. Canada, from Alaska to James Bay; winters chiefly in S. America, although some stay in Sw.; migrates throughout W., especially through Great Plains.

WANDERING TATTLER
Heteroscelus incanus   **20:16, 23:10**

Description
Size, 10½–11¼ in. (26.7–29.2 cm). Medium-size, coastal;
unpatterned gray above, white line over eye, yellowish legs. Spring:
• *narrow black bars* on underparts. Fall: unbarred.

Similarities
Black Turnstone, Surfbird, Spotted Sandpiper also inhabit rocky
coasts; Wandering Tattler distinguishable easily by complete lack
of pattern in flight.

Habitat
Mountain streams above timberline (breeding); rocky ocean shores,
pebbly beaches.

Habits
Teeters like Spotted Sandpiper.

Voice
Distinct *whee-we-we-we* but less sharp than Greater Yellowlegs;
also a series of 4 to 8 *tweet* notes not unlike Spotted Sandpiper.

Eggs
4; greenish, spotted; 1.6 x 1.2 in. (4.3 x 3.2 cm). Nest is a hollow
of twigs on high mountain, gravel bar.

Range
Breeds in Alaska; winters along Pacific Coast from cen. Calif.,
southward; migrates along coastline.

WILLET
Catoptrophorus semipalmatus   **19:13, 21:17**
                                     Fig. 9

Description
Size, 14–17 in. (35.6–43.2 cm). Plain-colored, with flashy • *broad
white stripe on blackish wings.* On ground rather nondescript;
grayish above, white below with numerous faint black markings in
spring, unmarked in fall; wing pattern occasionally visible; bill
thick, black, and with a bluish base; legs bluish; white rump and
pale tail conspicuous in flight.

Similarities
Greater Yellowlegs is less robust, with plain wings, bright yellow
legs. Smaller than the brown curlews and godwits.

Habitat
Marshes, beaches, sloughs, mud flats; in winter tidal sloughs and
flats.

Habits
Bobs less than yellowlegs; may hold wings aloft for several seconds
after alighting; often perches on fences or posts; wades up to belly.

Voice
Whistling *pill-will-willet* during breeding season; "a loud vehement
wek, wek, wek or *kerwek, kerwek, kerwek*" (Bent).

Eggs
3–4; grayish-olive, spotted with brown; 1.9 x 1.5 in. (4.8 x 3.8 cm).
Nest is grassy cup in grass.

Range
Breeds in Great Basin and n. prairie; winters along Pacific Coast
from s. Calif. to S. America; migrates along Pacific Coast from
Puget Sound southward.

PECTORAL SANDPIPER
Calidris melanotos   **20:8, 22:3**

Description
Size, 8–9½ in. (20.3–24.1 cm). Characterized by • *buffy, streaked breast,* sharply contrasting with white belly. Back streaked rusty-brown; dark rump; greenish legs; wing stripes inconspicuous or lacking; bill greenish, sometimes drooping at tip.

Similarities
See especially Least, White-rumped, Baird's; all peeps are smaller, and all but Baird's have white wing stripe. Baird's has a scaly back pattern.

Habitat
Grassy edges of mud flats, freshwater and saltwater marshes.

Habits
Tame; lies close; if flushed, jumps quickly with a harsh *kriek* and zigzags away, often circling high before pitching abruptly down; suggests Common Snipe in flight actions; occurs singly or in scattered flocks; can swim.

Voice
Reedy *kriek, kriek.*

Eggs
3–4; pale olive, spotted with umber; 1.5 x 1.0 in. (3.8 x 2.5 cm). Nest is grassy tundra hollow.

Range
Breeds in nw. Alaska, e. to Hudson Bay; winters in S. America; migrates chiefly in E., although some along Pacific Coast.

Note: The related **SHARP-TAILED SANDPIPER,** *Calidris acuminata,* size, 8½ in. (21.6 cm), is a rare fall visitor to the Pacific Coast from Siberia. It is very like the Pectoral, but has an evenly colored, grayish-buff breast, with streaks apparent only at the sides, and grading into the belly with no sharp demarcation.

ROCK SANDPIPER
Calidris ptilocnemis   **20:10, 21:7, 22:4**

Description
Size, 8–9 in. (20.3–22.9 cm). Usually seen with Turnstones and Surfbirds on coastal rocks. Spring: above rust-colored, with Dunlin-like black splotch on lower breast (not belly); plump, stocky. Winter: slaty with a white belly, streaked breast, white wing stripes, and a dark rump and tail, legs greenish to dull yellow.

Similarities
Turnstones and Surfbird have banded white-and-black tail, white rump.

Habitat
Rocky coasts, wave-washed jetties.

Habits
Deliberate prober in rocky crevices near surf.

Voice
"A flickerlike *clu-clu-clu*" (Peterson).

Eggs
4; greenish with purple and brown blotches and black lines; 1.4 x 1.0 in. (3.7 x 2.5 cm).

Range
Breeds in Arctic, on Bering Sea islands; winters along n. Pacific Coast, from Alaska to Oreg., casual to Calif.

RED KNOT
Calidris canutus   **20:11, 21:10, 22:11**

Description
Size, 10–11 in. (25.4–27.9 cm). Chunky; with Robin-red breast
(spring) and short bill. Spring: reddish-brown and gray with black
streaks or black and buff mottling above. Fall: gray above, white
below, without much contrast; wing stripe faint; rump whitish, tail
grayish; legs greenish.

Similarities
Spotted Sandpiper is smaller. Dowitchers in spring are also red-
breasted but with long snipe-like bills and white up the back.
Sanderling is smaller, whiter. Dunlin is smaller, darker, with dark
rump.

Habitat
Tundra, shorelines, tidal flats, salt marshes.

Habits
Very gregarious, usually in dense flocks; on ground, sluggish; in
flight, tight flocks twist and turn in unison.

Voice
Rather quiet; a low, harsh *chut;* a soft *wah-quoit;* a low, mellow
tooit-wit; "a soft *whit whit,* like a man whistling for a dog"
(Hoffmann).

Eggs
3–4; light-greenish, marked with brown; 1.8 x 1.2 in. (4.6 x 3.0
cm).

Range
Breeds in Arctic; winters along Pacific Coast from Calif. s. to S.
America; migrates along coastline.

WHITE-RUMPED SANDPIPER
Calidris fuscicollis   **20:4, 21:6, 22:1**

Description
Size, 7–8 in. (17.8–20.3 cm). Straight-billed, with completely
• *white rump.* Above rusty brown in spring, gray in fall; • *breast
streaked,* white wing stripe; bill heavy for a peep, occasionally
slightly drooped at tip; legs dark greenish. Less common than other
peeps. When feeding in mixed flocks, noticeably larger than Least
or Semipalmated.

Similarities
Least is smaller, with dark rump. Baird's has dark rump, scaly
back. Dunlin in fall has dark rump, and longer bill, distinctly
decurved. This species and next 6 are called "peep" sandpipers.

Habitat
Mud flats, rocky beaches, shorelines, salt marshes, fresh marshes.

Habits
Tame, actions deliberate, feeds in water up to its belly.

Voice
"A squeaky mouselike *jeet*" (Nichols); like 2 marbles being struck
together, uttered on wing.

Eggs
3–4; olive, marked with dark brown; 1.3 x 0.9 in. (3.3 x 2.3 cm).

Range
Breeds in Alaska, Yukon, Arctic coastline, s. to cen. Canada;
winters in S. America; migrates through Great Plains, casual to
B.C., Idaho.

BAIRD'S SANDPIPER
Calidris bairdii   **20:5**

Description
Size, 7–7½ in. (17.8–19.1 cm). Buffy-brown, long-winged with distinct, scaly back pattern; no seasonal change. • *Head and breast buffy* (seen in fall), wing stripe weak or missing, rump dark, throat white, legs and bill black.

Similarities
Least and Semipalmated are smaller, slightly darker above, back not scaly; Semipalmated is grayer in fall and always shows paler breast. Pectoral is browner, shorter bill, more slender, greenish legs. White-rumped has white rump, back not scaly. Sanderling has prominent wing stripe; in spring and summer has weak or no white eye stripe. Western has longer bill. Dunlin has droop-tip bill.

Habitat
Grasslands, shorelines, tundras, marsh and pond edges; sometimes tidal flats; high-altitude lakes in migration; rarest peep on coast.

Habits
Allows close approach, less active than other peeps; flies with them but feeds alone.

Voice
Similar to Semipalmated's; a distinctive *kreep* in flight.

Eggs
4; clay-colored, spotted with umber; 1.3 x 0.9 in. (3.3 x 2.3 cm).

Range
Breeds in Arctic; winters in S. America; migrates across plains.

LEAST SANDPIPER
Calidris minutilla   **20:2, 21:2**

Description
Size, 5–6½ in. (12.7–16.5 cm). The smallest sandpiper. • *Yellowish legs;* similar to Semipalmated, but browner in all plumages, with neck and breast more streaked in spring and darker in fall; bill thinner; legs dusky yellow.

Similarities
See Semipalmated. Western is larger, with drooped bill, black legs. White-rumped has white rump, darker legs. Baird's is larger, with darker legs, scaly back.

Habitat
Shorelines, mud flats, wet fields, ponds; most common in muddy or grass-edged areas, not usually on beaches.

Habits
This and Semipalmated among most common shorebirds; quite tame; when flushed, zigzags off like Common Snipe.

Voice
A *peep;* also a short *kreep,* higher-pitched and more squeaky than Semipalmated's.

Eggs
3–4; buff, with rich, dark markings; 1.1 x 0.8 in. (2.8 x 2.0 cm). Nest is mossy depression in bog or tundra marsh.

Range
Breeds in Arctic, s. to cen. Canada; winters along Pacific Coast, although some farther to C. and S. America.

DUNLIN
Calidris alpina   **20:7, 21:9, 22:5**

Description
Size, 8–9 in. (20.3–22.9 cm). • *Bill rather long, stout, with downward droop at tip;* short-legged, rather stocky, legs black. Spring: black patch below on belly, not lower breast as in Rock Sandpiper; bright rusty-red above; white wing stripe. Winter: mouse-gray above, white below with grayish on breast.

Similarities
Sanderling is whiter, with straight bill. Western is smaller, has drooped bill, but not gray on breast. Breeding Black Turnstone, Golden Plover and Black-bellied have black underparts, not in patch below.

Habitat
Beaches, shorelines, tidal flats, summer tundras.

Habits
On ground, tame, sluggish; often feeds in hunched-up position.

Voice
Harsh, rather loud *chee-ur;* "flushing note . . . a fine *chit-l-it*" (Nichols); "a nasal, rasping *cheezp* or *treezp*" (Peterson).

Eggs
4; buff, marked with chestnut brown; 1.5 x 1.0 in. (3.8 x 2.5 cm).

Former name
Red-backed Sandpiper.

Range
Breeds in Arctic; winters along Pacific Coast from cen. B.C. to Baja Calif.

SEMIPALMATED SANDPIPER
Calidris pusilla   **20:3, 21:1**

Description
Size, 5½–6¾ in. (14.0–17.1 cm). The small, common peep with • *dark legs.* Above streaked gray-brown, slightly browner in spring; breast ashy with light streaks, grayer in fall; rump and tail dark in middle, white on sides; bill stout, straight, black; webs between toes scarcely visible in field.

Similarities
Least is smaller, browner, bill thinner, legs yellowish. Western has longer bill, thicker at base with slight droop at tip.

Habitat
Beaches, shorelines, flats; less often in grassy edges than Least.

Habits
Often stands or hops on 1 leg, dashes about feeding with head down; retreats and advances before waves, often in groups or flocks; at high tide rests higher up on beach in groups behind shelter or facing wind.

Voice
Flight call a grating *churp* or *check,* lower than Least's; flushing note, *ki-i-ip*.

Eggs
3–4; variably buffy marked with brown; 1.2 x 0.8 in. (3.0 x 2.0 cm).

Range
Breeds from Alaska, e. across Canada and s. along Pacific Coast to cen. B.C.; winters from cen. Calif., s. and e. to sw. Ariz.

WESTERN SANDPIPER
Calidris mauri

20:1, 21:3, 22:2

Description
Size, 6–7 in. (15.2–17.8 cm). Similar to Semipalmated, but • *bill longer, thicker at base and with slight droop at tip*. Spring: above reddish, breast more streaked than Semipalmated, often with dark V's running down the sides. Fall: almost indistinguishable from Semipalmated, except for drooping bill and, occasionally, some reddish at "shoulder"; legs black.

Similarities
Least is smaller; straight bill; legs yellowish. See also Semipalmated. Western and Least Sandpipers are the 2 common peeps found west of Great Plains.

Habitat
Shorelines, beaches, grassy mud flats, marshes.

Habits
Carries bill pointed down more than does Semipalmated; may feed in slightly deeper water.

Voice
Flight note *chee-rp (ee* as in *kreep* of Least Sandpiper); a thin *jeet.*

Eggs
Similar to Semipalmated's.

Range
Breeds along Alaska coastline; winters along Pacific Coast, from Calif. to S. America; migrates through B.C., Wash., and Oreg., chiefly along coastline.

SANDERLING
Calidris alba

20:6, 21:4, 22:8

Description
Size, 7–8¾ in. (17.8–22.2 cm). Plump, small, active; whitish, with conspicuously flashing • *white wing stripe*. Underparts, sides of rump, and tail white; bill and legs black. Spring: head, upperparts, and breast reddish. Fall: upperparts pale gray, dark at bend of wing, forehead and underparts white.

Similarities
Somewhat larger, whiter, and stockier than other "peep."

Habitat
Sea beaches, tidal flats, sandy edges of bays; lakeshores.

Habits
Closely follows edge of advancing and receding waves, often in small flocks; follows more closely than other sandpipers.

Voice
Sharp and distinct *kit,* singly or in series.

Eggs
4; dull olive marked with brown; 1.4 x 1.0 in. (3.6 x 2.5 cm).

Range
Breeds in Arctic; winters along Pacific Coast from s. B.C. s. to S. America; migrates along Pacific Coast and some through Great Basin and Plains.

STILT SANDPIPER
Micropalama himantopus   **21:13, 22:14**

Description
Size, 7½–9 in. (19.1–22.9 cm). Distinctive in spring with • *rusty ear streak,* dark back and even barring on underparts. In all plumages, has long, greenish-yellow legs; white rump; thin, long bill; "dainty" appearance; lacks wing bars; white on uppertail coverts • *horseshoe-shaped.* Fall: gray above; white line over eye; underparts whitish; wings dark with black trailing edge; bill slightly drooped at tip.

Similarities
Lesser Yellowlegs is larger; legs yellow; indistinct line over eye; no droop to bill. Dowitchers are chunkier with a white wing stripe and back stripe and heavier bill. Spring Wandering Tattler is seen on rocks, with bars below, short legs, dark tail and rump.

Habitat
Fresh- or saltwater shallow pools; mud flats, marshes.

Habits
Flight similar to Lesser Yellowlegs; quiet, usually in loose groups with other shorebirds, especially dowitchers, feeding in pools up to its belly in water; does not bob head as do yellowlegs.

Voice
Low *thu* or *why,* resembling note of Lesser Yellowlegs but hoarser.

Eggs
3–4; dull white, marked with brown; 1.4 x 1.0 in. (3.6 x 2.5 cm).

Range
Breeds in Alaska and n. Canada; winters in S. America; migrates through Rocky Mountain states and Great Plains.

BUFF-BREASTED SANDPIPER
Tryngites subruficollis   **19:11, 23:4**

Description
Size, 7½–8¾ in. (19.1–22.2 cm). No seasonal change. Small, all evenly buff below. Bill short; head rounded; eyes "large"; head, neck and • *underparts plain buff,* a few dark spots on the sides; upperparts blackish-brown tinged with buff; wing stripe obscure; • *wing linings whitish;* tail dark; wings long, pointed, at rest reaching beyond end of tail; legs yellow.

Similarities
Upland Sandpiper is larger; wings at rest do not reach end of tail. Baird's is buff only on breast, legs black.

Habitat
Prairies, plains, fields, pastures.

Habits
Tame; while on ground often raises wing, showing white linings, and extends rather long neck; seeks safety by hiding; flies in compact flocks.

Voice
Rather silent, occasionally a sharp, thin, clicking *tik;* also a low, trilled *pr-r-r-reet* (Wetmore); a *chwup.*

Eggs
4; clay-colored, boldly marked with umber and slate; 1.5 x 1.0 in. (3.8 x 2.5 cm). Nest is hollow on tundra rise.

Range
Breeds in Arctic and n. Canada; winters in S. America; migrates through Great Plains, casual along Pacific Coast.

SHORT-BILLED DOWITCHER
Limnodromus griseus

20:14, 21:14, 22:12

Description
Size, 10½–12 in. (26.7–30.5 cm); bill, male, 2¼ in. (5.7 cm);
female, 2½ in. (6.4 cm). Chunky; wing stripe on trailing edge, not
in middle as in some other shorebirds; tail barred; legs greenish.
Spring: above brown; rusty below. Fall: above gray, with white eye
line.

Similarities
Long-billed has different voice. Red Knot has short bill. Common
Snipe has dark rump and tail, not on beaches or tidal flats.
Yellowlegs have shorter bills, longer legs. Stilt Sandpiper in fall is
slimmer, with shorter bill, longer legs.

Habitat
Marshes and mud flats, shorelines, inner beaches, still waters;
more common near salt water than Long-billed.

Habits
Flies in compact flocks, flight steadier than yellowlegs' and with
bill pointed partly down; feeds in deeper water than most
shorebirds, probing long bill into the mud like a sewing machine;
swims readily.

Voice
Lesser Yellowlegs' flight call but faster, softer, a mellow whistled
too, too, too or *dow-itch* or *dow-itch-er.*

Eggs
4; grayish or greenish, with dark brown spots; 1.6 x 1.1 in. (4.1 x
2.8 cm). Nest is grassy cup in muskeg hillock.

Remarks
The dowitchers have long straight bills, white rumps, continuing
wedges up the back. The bill of the Short-billed averages shorter
than that of the Long-billed, but is not a reliable factor. Normally
dowitchers are the only snipelike birds found on exposed mud flats.

Range
Breeds in s. Alaska e. through cen. Canada; winters along Pacific
Coast, from cen. Calif. s. to S. America.

LONG-BILLED DOWITCHER
Limnodromus scolopaceus

Description
Size, 11–12½ in. (27.9–31.8 cm); bill, male, 2½ in. (6.4 cm);
female, 3 in. (7.6 cm). Similar to Short-billed, but with longer bill;
once considered part of the same species; very difficult to
distinguish from Short-billed except by voice in winter plumage.
Spring: very like Short-billed Dowitcher but ventral red extends to
belly and tail; flanks barred. Fall: indistinguishable from Short-
billed (experienced birders at close range may note its longer bill),
except by voice.

Similarities
Same as Short-billed.

Habitat, Habits
Similar to Short-billed.

Voice
Weak *keek* or less often a series of *keek* notes.

Eggs
4; olive, marked with brown; 1.7 x 1.2 in. (4.3 x 3.0 cm). Nest is
depression in muskeg.

Range
Breeds on coast of w. Alaska, e. to Mackenzie; winters along
Pacific Coast from n. Calif. s. to C. America.

SURFBIRD
Aphriza virgata   **20:9, 23:12**

Description
Size, 10 in. (25.4 cm). Stocky, sandpiperlike, rocky coastal bird, with a conspicuous • *white tail tipped with a broad black triangle.* Above and below dark gray, heavily streaked and spotted with black; below similarly.

Similarities
Black Turnstone is blacker, legs blackish. Wandering Tattler is slimmer, with no wing or tail pattern.

Habitat
Rocky coastlines except when nesting.

Habits
Seeks food at the edges of surf.

Voice
A "sharp *pee-weet* or *key-a-weet*"(Peterson).

Eggs
4; pale buffy, speckled with dark brown; 1.6 x 1.2 in. (4.3 x 3.1 cm).

Range
Breeds in Alaska; winters along Pacific Coast, from s. Alaska to S. America.

MARBLED GODWIT
Limosa fedoa   **19:12**
                               Fig. 9

Description
Size, 16–20 in. (40.6–50.8 cm); bill, 4¼ in. (10.8 cm). Next to Long-billed Curlew in size, with no white in its plumage and an • *upturned or very long and straight bill*; rich buff-brown mottling. Base of bill pink, tip black; legs bluish; buffy-brown with black markings. In flight, patch of cinnamon on upperwing and on entire undersurface; also shows black primaries and black patch near bend.

Similarities
Long-billed Curlew bill turns down. Hudsonian Godwit is smaller, with white rump.

Habitat
Shores, flats, wet grasslands, prairies; salt marshes, beaches, tidal flats in winter.

Habits
Flight swift, strong, direct, head somewhat drawn in, bill straight forward, legs stretched out behind.

Voice
Noisy on breeding grounds; call *god-WIT, god-WIT, god-WIT* or *you're-crazy-crazy-crazy* and *cor-RECT, cor-RECT;* flight note, *queep, queep, queep.*

Eggs
4; buff, spotted with brown; 2.2 x 1.5 in. (5.6 x 3.8 cm). Nest is grassy depression.

Range
Breeds from cen. Canada s. to Mont., N.Dak. and S.Dak.; winters from Calif. to S. America; migrates along Pacific Coast, e. through Great Basin and Great Plains.

HUDSONIAN GODWIT
Limosa haemastica

20:12, 21:15
Fig. 9

Description
Size, 14–16½ in. (35.6–41.9 cm). • *Upturned or long and straight pink bill, black at tip;* wings black with narrow, white wing stripe; underwing and wing pit black, conspicuous in flight; tail black, with base and tip white; legs gray-blue. Spring: above dark gray marked with brown; below rich chestnut marked with black; on ground, looks almost black at a distance. Fall: above dark gray; below whitish.

Similarities
See Marbled Godwit. Willet in flight shows wing whiter, tail whitish, not ringed; standing, sturdier with a thicker, shorter bill. Greater Yellowlegs is smaller; legs yellow not gray-blue; bobs head.

Habitat
See Marbled Godwit.

Habits
Probes in mud and deep water, often up to its belly, and with head submerged; can swim; flies in dark, undulating lines.

Voice
Very silent; a low *qua-qua-qua* when flushed; lower, less harsh than Marbled; a low *ta-it* in flight; "*tawit* (or *godwit*), higher-pitched than Marbled's" (Peterson).

Eggs
2–4; dull olive, marked with brown; 2.2 x 1.4 in. (5.6 x 3.6 cm). Nest is tundra hollow.

Range
Breeds in n. Canada; winters in S. America; migrates through Great Plains.

Note: The **BAR-TAILED GODWIT**, *Limosa lapponica,* size, 15–18 in. (38.1–46.0 cm), breeds in northern Alaska and migrates through the Aleutians. It has a finely barred tail and lacks striking white wing bars.

PHALAROPES
Family Phalaropodidae

The phalaropes are a group of three small sandpiperlike wading or swimming birds with lobed toes. Females are larger and more brilliantly plumaged than the males and take the lead in courtship. Phalaropes fly swiftly, and swim buoyantly, often whirling. The male builds the nest on the ground, incubates the eggs, and cares for the young.

RED PHALAROPE
Phalaropus fulicarius

20:17, 21:23, 22:10

Description
Size, 7½–9 in. (19.1–22.9 cm). Wholly or partly yellow, bill thick. Spring female: • *reddish below* and on neck; black crown; white cheeks and wing stripe; dark-tipped yellow bill; back brownish. Spring male much paler. Fall: both sexes pale, unstreaked gray above; head and underparts white; black patch on side of head behind eye. Bill relatively short, blunt, thicker at base.

Similarities
Fall Northern Phalarope has thinner, darker bill, streaky back, is smaller. Wilson's has thin bill much longer; lacks wing stripe. Fall

Sanderling has paler back, all-black bill, no eye mark; seldom alights on water.

Habitat
Most maritime of the phalaropes; breeds on tundra, winters on open sea in Southern Hemisphere.

Habits
Swims like a tiny gull; buoyant, bobs head and dabs bill for food, or tips up and feeds beneath surface; in feeding, makes twice as many dabs per second as Northern.

Voice
Like Sanderling, but thinner and higher-pitched.

Food
Insects, grit, copepods (small crustaceans), jellyfish, fish.

Eggs
4; buffy or greenish, marked with brown; 1.2 x 0.8 in. (3.1 x 2.2 cm).

Range
Breeds along Arctic coastline s. to n. Canada; winters at sea, chiefly in S. Hemisphere.

WILSON'S PHALAROPE
Phalaropus tricolor   **20:15, 21:19, 22:13**

Description
Size, 8½–10 in. (21.6–25.4 cm). Dark-winged, trim, with a white rump and the thinnest, most needlelike bill of any shorebird. Tail whitish; no wing stripe. Spring female: above gray, underparts and cheeks white; • *black band* from eye turns chestnut on neck and passes to back. Spring male much duller. Fall: both sexes plain brownish-gray above, plain white below, legs turn greenish-yellow; no breast streaks.

Similarities
Other phalaropes have wing stripes, stronger eye mark, and dark rump and tail. Red Phalarope has thicker head and neck and short thick bill. Winter Sanderling has thicker, shorter bill and conspicuous wing stripe. Stilt Sandpiper has similar flight pattern, but is long-legged, darker, and streaked above; bill much heavier. Lesser Yellowlegs has streaks on breast, long lemon-yellow legs.

Habitat
Not oceanic; pools, shallow lakes inland, fresh marshes, shorelines, mud flats; salt marshes in migration.

Voice
Soft, nasal *oit-oit*, unlike other phalaropes; "a low nasal *wurk*. Also *check, check, check*" (Peterson).

Food
Crustaceans, insects, seeds.

Eggs
4; buffy or greenish, marked with brown; 1.2 x 0.9 in. (3.3 x 2.3 cm). Nest is grassy depression in either wet or dry meadow.

Range
Breeds from cen. B.C. e. through Canadian prairie and s. throughout interior of U.S.; winters in S. America.

NORTHERN PHALAROPE
Phalaropus lobatus   **20:18, 21:22, 22:7**

Description
Size, 6½–9 in. (16.5–20.3 cm). The commoner of the 2 seagoing phalaropes; darkest-backed. Small head, thin neck; needlelike black bill; white wing stripe; dark tail and legs. Spring female: gray

above, darkest on head; • *sides of neck chestnut,* throat and belly white; back streaked; spring male much paler. Fall: both sexes blackish, streaked with gray above; white below; black eye stripe ("phalarope mark").

Similarities
Fall Red is larger, shorter, with thicker bill, less sharply defined wing stripe. Fall Wilson's has white rump, no white wing stripe, longer thinner bill. Sanderling has longer wing stripe, no phalarope eye marks, not found at sea or (usually) swimming.

Habitat
Ocean, bays; tundra, inland lakes, ponds.

Habits
Gregarious, tame; flight swift, erratic, often alights on floating seaweed and runs about like a peep on a mud flat; bathes with characteristic jerking motion.

Voice
Similar to Sanderling.

Food
Insects, crustaceans, plankton.

Eggs
4; buffy or greenish, marked with brown; 1.1 x 0.7 in. (2.9 x 2.0 cm).

Range
Breeds in Arctic from Alaska to cen. Canada; winters at sea, chiefly in S. Hemisphere; migrates off Pacific Coast and along w. lakes.

AVOCETS AND STILTS
Family Recurvirostridae

These are slender, very long-legged, thin-billed wading birds spectacularly patterned with black and white. They frequent shallow freshwater, alkaline or brackish ponds and coastal flats, feeding on crustaceans, small aquatic life, and insects.

AMERICAN AVOCET
Recurvirostra americana

19:18
Fig. 9

Description
Size, 15½–20 in. (39.4–50.8 cm). Unique black-and-white pattern, both at rest and in flight, is diagnostic. • *Bill upturned;* legs and feet light blue, feet webbed; head and neck buffy in spring but pale gray in winter.

Similarities
Black-necked Stilt is more slender; has black on head and neck, red legs.

Habitat
Inland shallow lakes and marshy ponds, mud flats; coastal bays in winter.

Habits
Flight direct, strong; feeds by walking quickly, working bill from side to side in the water, or while swimming, by tipping up.

Voice
Loud yelp; "a sharp *wheek* or *kleek,* excitedly repeated" (Peterson).

Eggs
3–4; olive, marked with brown; 2.0 x 1.3 in. (5.1 x 3.3 cm).

Range
Breeds in Great Basin and Canadian prairies, and also prairie lakes from N.Dak. and S.Dak. s. to Okla.; winters in Calif.

BLACK-NECKED STILT
Himantopus mexicanus   19:16

Description
Size, 13–17 in. (33.0–43.2 cm). Very tall; black above, white
below; • *legs long, red;* rump and tail white; bill very slender,
straight, black. In flight, appears to have a white body, black
unpatterned wings. Immatures: browner, legs paler.

Habitat
Fresh or alkaline shallow lakes, grassy marshes, pools, mud flats.

Habits
Aggressive; feeds wading in water; in flight, neck somewhat drawn
in, legs extended; raises wing over back on alighting.

Voice
Noisy; a loud, high-pitched insistent, often irritating, yipping, *kyip,
kyip, kyip.*

Eggs
3–5; buff, marked with dark brown; 1.7 x 1.2 in. (4.3 x 3.0 cm).
Nest is hollow in dry mud flat or marsh.

Range
Breeds from s. Oreg. e. to Nebr. and s. to S. America; winters s. of
U.S.

SKUAS AND JAEGERS
Family Stercorariidae

This family includes a half-dozen species of strong-flying, hook-
billed, predominantly darkish seabirds whose narrow, angled wings
conspicuously flash white wing quills. Various color phases
characterize their plumage. Adult jaegers are generally dark above,
with black caps and whitish collars, and light below, with
protruding central tail feathers. The Northern Skua has no central
tail points. All are predators feeding by harassing other seabirds.

PARASITIC JAEGER
Stercorarius parasiticus   2:1

Description
Size, 16–21 in. (40.6–53.3 cm). Short, • *pointed projecting central
tail feathers* distinguish adults in all color phases. Black legs, pale
breastband. Very difficult to distinguish from other jaegers, unless
adult with fully developed tail.

Similarities
Bill of Pomarine is larger. In light phase, Long-tailed has more
contrast between cap and back, yellow on head is deeper, and
white collar on hind neck wider.

Habitat
Summer tundra, sounds, bays, open sea; rarely on large inland
lakes. Most frequently observed jaeger.

Voice
Various wails, shrieks, "a nasal squealing *eee-air;* alarm, *ya-wow*"
(Peterson).

Food
Small birds, lemmings in summer, fish at other times.

Eggs
2; olive-brown with dark brown spots; 2.2 x 1.6 in. (5.7 x 4.1 cm).
Nest is tundra hollow.

Range
Breeds in Arctic on tundra; winters at sea chiefly in S. Hemisphere
but some also in N. Pacific.

POMARINE JAEGER
Stercorarius pomarinus

2:2
Fig. 10

Description
Size, 20–23 in. (50.8–58.4 cm). Center • *tail feathers blunt and twisted* in adults. Dark and light phases occur, but typically dark-capped, pale below, but also an all-dark phase.
Similarities
Broader winged and larger than other jaegers; young larger, bill heavier, buffier below than others; more aggressive. Other jaegers are smaller, tails differ.
Habitat
Summer tundra, coastal bays, sounds, open sea.
Habits
Very aggressive.
Voice
"A sharp *which-yew*" (Rich).
Food
Mainly fish.
Eggs
2; olive-brown with dark brown spots; 2.4 x 1.7 in. (6.2 x 4.4 cm). Nest is in grass-lined depression on ground.
Range
Breeds in Arctic on tundra, especially in w. and n. Alaska and n. Canada; winters at sea off S. America; often seen along Pacific Coast in spring and fall.

Fig. 10

Northern Skua, p. 104

Pomarine Jaeger
dark phase
immature

LONG-TAILED JAEGER
Stercorarius longicaudus

2:3

Description
Size, 20–23 in. (50.8–58.4 cm), including tail. Small jaeger with very long, pointed, projecting tail feathers, usually 3–6 in. (7.6–15.2 cm); sometimes 9–10 in. (22.9–25.4 cm). Light phase: pale, ashy back contrasts sharply with black cap, white collar on hind neck is broader than Parasitic's and almost surrounds neck, • *legs bluish.*
Similarities
Other jaegers are larger-bodied with shorter central tail feathers. Except when the distinctive tail projection is observable, all jaegers are difficult to identify positively. Parasitic is often darker below, gray breastband, less contrast between cap and collar, black legs, 4–5 white primary shafts instead of 2–3.
Habitat
Summer tundra, open ocean when not breeding.

Habits
Less aggressive, harasses gulls less frequently than larger jaegers.
Voice
Usually silent; "a shrill *pheu-pheu-phey-pheo* . . . followed by a harsh *qua*" (Nelson).
Food
Lemmings, young birds on breeding grounds, fish, invertebrates otherwise.
Eggs
2–3; olive, spotted with sepia; 2.1 x 1.5 in. (5.5 x 3.9 cm). Nest is hollow in rolling tundra.
Range
Breeds in Arctic; winters at sea off Pacific Coast.

NORTHERN SKUA
Catharacta skua   *Fig. 10*

Description
Size, 20–22 in. (50.8–55.9 cm). Jaegerlike, with a conspicuous
• *white wing patch.* Robust, stocky; wings broad, buteolike; above sooty-brown, below rusty; • *tail short,* slightly uptilted, faintly forked in the grayer young.
Similarities
Dark jaegers may lack tail points; smaller; wings narrowed, more pointed.
Habitat
Offshore waters.
Habits
Strong, swift, hawklike flight; attacks other seabirds and forces them to disgorge food.
Voice
A *"skua"* (McGillivray).
Food
As jaegers; carrion.
Eggs
2–3; white, blotched with brown; 2.7 x 1.9 in. (7.0 x 4.9 cm).
Range
Breeds in Iceland and n. of Europe; often seen offshore, sometimes along Pacific Coast from B.C. s. to Calif.

GULLS AND TERNS
Family Laridae

GULLS
Subfamily Larinae

Gulls are fairly large, long-winged, swimming birds, usually white with black wing tips, often with a pearly mantle (i.e., back and tops of wings). They are stouter than terns, with wider wings and longer legs, the tail being usually square or rounded while that of terns is usually forked. Immatures take one or more years to acquire adult plumage, and often are brown or brown and white. Gulls fly with the bill straight forward; they soar, often follow boats, swim buoyantly (rarely diving as terns do), and pick food from the surface. Being scavengers, they often gather at sewer outlets, fish-processing areas, and garbage dumps. They perch on buoys, pilings, and roofs and walk easily. Several species occur inland: California, Ring-billed, and Bonaparte's throughout much of western United States and southwestern Canada; Herring to a lesser extent, and Franklin's concentrated about Great Salt Lake and eastward into the Great Plains. In distinguishing the species, note particularly: color of feet, color or markings of bills, and color or pattern of wing tips. Immature birds are especially difficult to identify.

GLAUCOUS GULL
Larus hyperboreus   **24:1**

Description
Size, 26–32 in. (66.0–81.3 cm). Largest gull, with "frosty" • *white wing tips*. Adult: white with a pale gray mantle, bill heavy, yellow with red spot near tip; eye-ring yellow during breeding; legs pinkish. First-winter immature: pale buffy throughout, • *primaries equally pale*. Second-winter immature: mantle gradually changes to creamy white, then to all-white in third year; bill white, bill spot black.

Similarities
Glaucous-winged has gray primaries, dark eyes; immature has smaller, darker bill.

Habitat
Coastal waters.

Habits
Flight steady, soaring, somewhat hawklike.

Voice
Hoarse *ku-ku-ku, ku-lee-oo*.

Food
Carrion, refuse, birds, marine invertebrates.

Eggs
2–3; olive-buff, marked with dark brown; 3.0 x 2.1 in. (7.6 x 5.3 cm). Nest of seaweed, grasses, etc., up to 3 ft. (0.9 m) high, on cliffs and ledges.

Range
Breeds along Arctic coastline and Alaskan islands; winters along Pacific Coast from Bering Sea s. to Wash. and Oreg., sometimes to Calif.; occasionally visits inland states in W.

GLAUCOUS-WINGED GULL
Larus glaucescens   **24:3**

Description
Size, 24–27 in. (61.0–68.6 cm). Large, • *pink-legged,* resembling the larger Glaucous. Adult: mantle pale gray, • *wing tips patterned in gray*. First-year immature: • *gray-brown* throughout, including primaries. Second-year immature: paler, grayer, primaries becoming grayish-brown.

Similarities
Glaucous is whiter, with yellow eye. Herring has black wing tips, yellow eye; immature is darker and grayer. Western has darker mantle, black wing tips, with small white tips; immature is darker, grayer.

Habitat
Typically coastal; beaches, piers, waterfronts, bays.

Habits
As Herring Gull.

Voice
"A low *kak-kak-kak* or *klook, klook, klook;* a low *wow;* a high-pitched *keer, keer*" (Peterson).

Food
Garbage, carrion, fishes, mollusks, crustaceans.

Eggs
2–3; olive-brown spotted; 2.9 x 2.0 in. (7.3 x 5.1 cm). Nesting is colonial; nest of seaweed, kelp, or grassy cup on cape or offshore island.
Range
Breeds from Aleutians s. to Oreg.; winters along Pacific Coast and tidewater rivers from Alaska s. to Baja Calif.

WESTERN GULL
Larus occidentalis   **24:2**

Description
Size, 24–27 in. (61.0–68.6 cm). Adult: • *mantle very dark,* deep gray to blackish, underparts contrastingly snowy-white; legs pinkish, bill heavy, primaries black. First-year immature: large, gray-brown. Second-year immature: has "saddle-backed" appearance of adult; head and underparts whitish.
Similarities
First-year California Gull is smaller; bill heavier and darker. First year Glaucous-winged has gray-brown primaries. First-year Herring has less contrasting coloration.
Habitat
Coastlines, washes, piers, bays.
Habits
Rarely leaves coastline.
Voice
Raucous calls like other gulls.
Eggs
2–4; buffy brown, spotted; 2.8 x 2.0 in. (7.2 x 5.1 cm). Nest is usually of grass.
Food
Garbage, carrion, crustaceans, eggs, young birds.
Range
Breeds along Pacific Coast from Wash. to Baja Calif. and Gulf of Calif.; winters along Pacific Coast off B.C.

Note: The rare Asiatic **SLATY-BACKED GULL,** *Larus schistisagus,* size, 26–28 in. (66.0–86.4 cm), sometimes visits western Alaska. It is very like the Western Gull, which does not reach Alaska, and has reddish legs.

HERRING GULL
Larus argentatus   **24:4, 26:12**

Description
Size, 22½–26 in. (57.2–66.0 cm). Adult: white with a pearly mantle, black wing tips with white spots, yellow bill with a red spot near tip, • *pinkish legs;* in winter, head and neck streaked with brown. First-year immature: almost uniform sooty. Second-year immature: dusky, broad diffuse dark band near end of tail; body becomes lighter and uppertail coverts whiter with age; bill dark, sometimes with black ring near end.
Similarities
Ring-billed is smaller, with black ring around middle of bill, yellowish legs, more buoyant, dovelike flight; immature with narrow ring of black at tip of tail. Adult California has greenish legs. Adult Western has much darker mantle, dark eye.
Habitat
Shorelines, bays, beaches, lakes, rivers, garbage dumps.

Habits
Gregarious: a ship follower and soarer, often high in air; swims; drops mollusks from height to break shells.
Voice
"Queeeeeah-ah, kak, kak, kak" (Collins); "a loud *hiyah . . . hiyah . . . hiyah-hyah* or *yuk-yuk-yuk-yuk-yuckle-yuckle;* mewing, squeals; anxiety note, *gah-gah-gah"* (Peterson).
Food
Carrion, garbage, refuse, marine animals, eggs, young birds.
Eggs
2–5; variable, often olive-buff, spotted and blotched with dark brown; 2.8 x 1.9 in. (7.1–4.8 cm). Nest is colonial, of seaweed or kelp on offshore island.
Range
Breeds throughout most of Canada and n. parts of U.S., from cen. Alaska e. to n.-cen. Canada and s. to B.C., Great Lakes, and eastward; winters along Pacific Coast from Alaska to Mexico.

Note: The **THAYER'S GULL,** *Larus thayeri,* size, 22 in. (55.9 cm), resembles the Herring Gull closely, but has dark eyes and grayer wing tips; immatures are indistinguishable. It reaches coastal British Columbia and the northwestern United States.

CALIFORNIA GULL
Larus californicus   **24:5, 26:7**

Description
Size, 20–23 in. (50.8–58.4 cm). Like a small Herring with
• *yellowish-green legs.* Adult: mantle gray, wing tips black with white spots; • *bill with red or red-and-black spots.*The commonest of several gray-mantled gulls with black wing tips. First-year immature: mottled dusky-brown, bill pink with black tip. Second-year immature: paler; back medium gray, underparts whiter, base of tail more white.
Similarities
Immature Herring is difficult to distinguish but in first year has all-dark bill. Ring-billed immature has black tail band; adult is smaller, paler backed, complete black ring on bill, less white in wing tips. Western in first year is darker than first-year California and larger, with larger bill. Second year has "saddle-backed" appearance.
Habitat
Shorelines, bays, beaches, farmlands, cities, inland lakes, marshes.
Voice
Similar to Herring; a squealing *kiarr.*
Habits
As other gulls; may gather in very large flocks.
Food
Garbage, carrion, fishes, rodents, birds' eggs, insects, other invertebrates.
Eggs
2–3; variable, often olive-buff, spotted with dark brown, 2.7 x 1.8 in. (6.9 x 4.6 cm). Nest is colonial, in ground hollow on islet in fresh or alkaline lake.
Range
Breeds from n. Canadian prairie sw. to Wyo., Colo., and Calif.; winters along Pacific Coast from Oreg. to Baja Calif., with some farther inland.

RING-BILLED GULL
Larus delawarensis   **24:6, 26:11**

Description
Size, 18–21 in. (45.7–53.3 cm). A small version of the California
Gull. Adult: white with pearly mantle slightly paler than
California's or Herring's; complete black ring around bill; wing
tips black with white spots, but all-black below; • *legs yellowish-
green.* First-winter immature: dusky. Second-winter immature:
light dusky above, whitish below; narrow blackish band near end
of white tail; bill and legs usually flesh-colored, bill with a black
tip.

Similarities
Immature Herring Gull is dark, tail mainly brown; Ring-billed
has dark-tipped white tail, body white. See also California and
Mew Gulls.

Habitat
Shorelines, bays, farmland.

Habits
Gregarious; follows plows or ships; flight is light and buoyant.

Voice
Alarm note, a hawklike *cree-cree;* anxiety, *ka-ka-ka;* a shrill *kyow.*

Food
Garbage, refuse, carrion, aquatic animals, rodents, insects.

Eggs
2–4; variable, often olive-buff, spotted with dark brown; 2.3 x 1.6
in. (5.8 x 4.1 cm). Nest is colonial, on lake islets.

Range
Breeds from Canadian prairie s. to U.S. midwest, with some to n.
Great Basin, s. to ne. Calif.; winters along Pacific Coast and
inland to large bodies of water in Sw.

FRANKLIN'S GULL
Larus pipixcan   **25:2**

Description
Size, 13½–15½ in. (34.3–39.4 cm). Black-headed, with • *white bar*
near the end of its wing. Summer: white bar or "window" between
gray primary bases and white-tipped black wing tips; bill and legs
red; underparts may have faint rosy bloom; bill relatively long.
Winter: head white with dark line behind head from eye to eye.
First-winter immature: small, dark-mantled; tail white with black
band; forehead white; nape black; rump and underparts white; legs
blackish.

Similarities
Bonaparte's has long, white triangle on fore-edge of spread wing.
Ring-billed has white spots in black wing tips, legs, pale, head
white.

Habitat
Prairies, marshes, fields, inland lakes, winter seacoast.

Habits
Gregarious, confiding; follows plow, sometimes soars upward in
spirals, captures insects in air, swims buoyantly; often flies in V-
shaped flocks.

Voice
A "soft *krruk* . . . a louder and more plaintive . . . *pway* or *pwa-ay*"
(Bent); when soaring, a *"weeh-a weeh-a weeh-a po-lee po-lee po-
lee"* (Miller).

Food
Grasshoppers, other insects, fish, frogs, mollusks.

Eggs
2–4; buffy, marked with brown; 2.0 x 1.4 in. (5.1 x 3.6 cm). Nest is colonial, in prairie marsh reeds.
Other name
Prairie Dove.
Range
Breeds throughout Canadian and U.S. prairies from Alta., e. to Man. and s. to Oreg., Mont., and Iowa; winters along Gulf Coast and S. America.

Note: The very similar **LAUGHING GULL,** *Larus atricilla,* size, 17 in. (43.2 cm), of the East nests in the Salton Sea area of southeastern California and is accidental in several western states. Adults are distinguished from Franklin's by dark wing tips, immatures by dark head and gray breast (first year) or lack of black tail band (second year).

MEW GULL
Larus canus   **25:5, 26:8**

Description
Size, 16–18 in. (40.6–45.7 cm). Short, • *unmarked, greenish-yellow bill* and yellow legs. Adult: white head, gray mantle, dark back, black wing tips with larger white spots than either California or Ring-billed. Juvenal: dark gray-brown, bill blackish, legs flesh-brown. First year: uniform grayish-brown. Second year: paler, more like adult.
Similarities
Kittiwake has no white spots on wing tips. Young Ring-billed is larger, but very similar.
Habitat
Marshes, inland summer lakes, tundra, coastal waters, tidal rivers.
Habits
Rests on water, not on roofs or pilings; breaks shellfish by dropping on hard surfaces; flocks often fly in unison; follows plows.
Voice
Low mewing; also *hiyah-hiyah-hiyah* "louder and sharper than . . . Glaucous" (Grinnell).
Food
Garbage, fishes, mollusks, worms, insects, other invertebrates.
Eggs
2–3; olive-buff, marked with brown; 2.2 x 1.6 in. (5.6 x 4.1 cm). Nesting is usually colonial; a depression on ground or in rocks, or twigs and grasses in trees to height of 20 ft. (6.1 m).
Other name
Short-billed Gull.
Range
Breeds from Alaska, e. to cen. Mackenzie and s. to n. Sask., with some along Pacific Coast, s. to B.C.; winters along Pacific Coast.

BONAPARTE'S GULL
Larus philadelphia   **25:3**

Description
Size, 12–14 in. (30.5–35.6 cm). Black-headed, with white wedge down its primaries. Summer: mantle gray, with long, white wedge down outer side of black-tipped primaries; wing tips white from below; bill small, black; legs orange-red. Winter adult: white head with black spot at rear of ear coverts. First-winter immature: similar to winter adult but with narrow black band near tip of tail,

upper wing with grayish diagonal band across it; bill dark with light base, legs pink to red. Bonaparte's Gull is very small, near size of most terns which, however, have forked tails and dive.

Similarities
Laughing adult has much darker mantle; immature, neck and sides of breast dark and broader band on tail. Franklin's has much darker mantle in adult. Kittiwake young is larger; black bar on back of neck, yellowish legs.

Habitat
Coastal and interior waterways, summer muskeg; breeds in coniferous forests.

Habits
Gregarious; flight somewhat ternlike, but seldom dives; points bill down in flight (desultory but rapid); swims buoyantly.

Voice
Shrill, nasal *peer* or *cheer* or *cherr;* "sparrowlike conversational notes" (Jones).

Food
Insects, small fish, crustaceans.

Eggs
2–4; olive-buff, spotted with chocolate; 1.9 x 1.3 in. (4.8 x 3.3 cm). Nest of sticks in conifers, to height of 20 ft. (6.1 m).

Range
Breeds from Alaska e. to Hudson Bay and s. to B.C.; winters along Pacific Coast from B.C. to Mexico; seen in Nw. during spring and fall migration.

HEERMANN'S GULL
Larus heermanni   25:1

Description
Size, 18–21 in. (45.7–53.3 cm). Adult: dark gray above and below with unmarked black wing tips, black tail, • *whitish head* (speckled in winter), • *red bill.* Immature: all-dark, including head; brown or partly red bill.

Similarities
None; darkest of western gulls, easiest to identify.

Habitat
Ocean and coastline.

Habits
Steals from other fish-eating birds, but is not as much a scavenger as most other gulls.

Voice
A "whining *whee-ee;* also a repeated *cow-auk*" (Peterson).

Food
Fishes, mollusks, crustaceans, birds' eggs.

Eggs
2–3; pale gray, cream or blue with olive, brown, or lavender spots; 2.3 x 1.7 in. (5.9 x 4.5 cm).

Range
Breeds on islands in Gulf of Calif. and along Pacific Coast off Baja Calif.; winters along Pacific Coast from Vancouver s. to Calif., with others migrating s. to C. America.

BLACK-LEGGED KITTIWAKE
Rissa tridactyla   25:6, 26:10

Description
Size, 16–18 in. (40.6–45.7 cm). Solid • *black triangular tip* on a gray wing. Adult: wings long, trailing edge white; white tail long,

broad, slightly forked; head white; mantle gray; bill yellow; legs black; nape gray in winter. First-winter immature: dark-tipped wings, dusky bar on nape, bill black, black band at end of slightly notched white tail, black spot behind eye.

Similarities
Young Bonaparte's is smaller; trailing edge of wing black, much white in wing tip, no dark bar on nape. Ring-billed and Mew have black wing tips spotted with white, legs pale.

Habitat
Open seas, occasionally coasts and bays.

Habits
Gregarious; follows ships; flight buoyant, graceful, swallowlike, distinctive at great distance; wingbeats rapid, soars; only gull that dives from wing and swims under water.

Voice
"Kitti-wake, ka-ake; sharp and piercing *ki, ki, ki . . .* harsh rattling *kaa, kaa, kae, kae* and *kaak kaak"* (Townsend).

Food
Fish, crustaceans, mollusks, refuse (but less of a scavenger than larger gulls).

Eggs
1–2; olive-buff, marked with brown; 2.2 x 1.6 in. (5.6 x 4.1 cm). Nest of seaweed on cliffs; in colonies.

Range
Breeds on cliffs of Arctic shorelines; winters from Aleutians s. to Mexico.

Note: The **RED-LEGGED KITTIWAKE,** *Rissa brevirostris,* size, 14–15 in. (35.6–40.6 cm), breeds on Bering Sea islands. It has red legs and is smaller than the Black-legged Kittiwake.

SABINE'S GULL
Xema sabini   **25:4**

Description
Size, 13–14 in. (33.0–35.6 cm). Fully •*forked tail.* Summer: white with black head and gray back; wings with alternating triangles of black at top, white at rear, gray at base; bill black with a yellow tip; legs gray; tail white. Winter: head white, brownish smudge from eye to eye around back of head, darker on nape. First-winter immature: gray above and on head with distinctive pattern of wing triangle as adult; white forked tail with broad black triangle at end; wing pattern like adults.

Habitat
Tundra, coast, ocean.

Habits
Feeds ploverlike over flats at low tide; flight ternlike.

Voice
Similar to Arctic Tern's, but shorter, harsher.

Food
Aquatic life, insects.

Eggs
2–3; olive-buff, faintly spotted with brown; 1.7 x 1.3 in. (4.3 x 3.3 cm).

Range
Breeds in Arctic very far north; winters at sea in tropics; seen in migration off Pacific Coast.

TERNS
Subfamily Sterninae

These graceful water birds, called aptly "sea swallows," are generally smaller, lighter, more streamlined than gulls and have sharply pointed bills; long, pointed wings; and forked tails. Most are whitish, with black caps, changing into fall plumage in midsummer. Terns fly with the bill pointing down, and dive readily for food or at intruders around the nest, rarely swimming and walking little. They feed on fish captured by plunging headfirst into the water, and on large insects. Members of the Genus *Sterna* have thin bills, black caps in breeding plumages, pearl-gray or black mantles, and forked tails.

GULL-BILLED TERN
Gelochelidon nilotica   **26:5, 27:5**

Description
Size, 13–14½ in. (33.0–36.8 cm). Heavy, • *black, gull-like bill;* • *body white, stocky,* feet black. Spring adult, black cap. Fall adult: head white with black ear patch and mottled crown. Immature: similar but with brown band on end of tail, mottling on back, and dark-tipped light bill.

Similarities
Common or Forster's has gray mantle and wings, wings narrowed, tail forked. Immature Gull-billed is particularly gull-like, but gulls don't have notched tail. Bonaparte's is darker, unforked tail.

Habitat
Coasts, marshes, fields.

Habits
Flight characteristic, wingbeats slower than *Sterna* terns, dives less, picks food off surface more; hawks back and forth over marshes for insects; follows plow.

Voice
"*Katydid, katydid;* a gull-like *ka ka ka*" (Collins).

Food
Insects, some crustaceans.

Eggs
2–3; variable, often buff, marked with brown; 1.9 x 1.3 in. (4.8 x 3.3 cm). Nesting is colonial; nest a depression in sand or of vegetation among grass on island.

Range
Breeds in se. Calif. on Salton Sea; winters mostly s. of U.S.

FORSTER'S TERN
Sterna forsteri   **26:4, 27:6**

Description
Size, 14–16½ in. (35.6–41.9 cm). Black-capped orange-red bill, • *white primaries* in breeding plumage. Spring adults: bill orange with black tip; cap black; tail grayish, not contrasting with mantle, and moderately forked with outer edges white, inner edges dark; feet orange-red. Fall adult: narrow black "ear" patch on white head, pale nape, dusky bill, yellowish feet. Bill occasionally lacks black tip.

Similarities
Common has gray primaries, longer wings, shorter tail, and white with gray outer edges; fall Common has a blackish nape.

Habitat
Fresh- and saltwater marshes, coasts, inland waters, bays, beaches.

Habits
Wingbeats are quicker, sharper than Common's.
Voice
Rasping *tsa-a-ap, zreep;* a nasal *kyarr;* also a rapid peeping *pip, pip, pip.*
Food
Insects, fish, frogs.
Eggs
3–4; buff or brown, similar to Common's; 1.7 x 1.2 in. (4.3 x 3.0 cm).
Range
Breeds from Canadian prairies s. to cen. Calif. and s. to Colo.; winters in C. and S. America.

COMMON TERN
Sterna hirundo   26:3, 27:8

Description
Size, 13–15 in. (33.0–38.1 cm). Spring adult: cap black; mantle pearl-gray; tail moderately forked, white, with gray outer edge; legs orange-red; below white; bill orange-red with black tip. Fall adult: head white, black collar around nape from eye to eye, bill blackish, legs paler. At rest, wings extend beyond tip of tail. Immature: like fall adult.
Similarities
Arctic Tern is similar to fall adult and young. Forster's has faster wingbeats, white in wings; fall adult is without black nape.
Habitat
Coast, ocean, shorelines, inland bodies of water, bays, beaches.
Habits
Gregarious; flocks collect over schools of fish.
Voice
Descending *tee-arr,* a rapid *kik-kik-kik;* a rapid *kirri-kirri.*
Food
Fish, aquatic life, insects.
Eggs
2–4; variable, often pale-brown, blotched with gray or lilac; olive, buff, brown, spotted; 1.6 x 1.2 in. (4.1 x 3.0 cm). Nest is colonial; a depression in sand, or saucer of vegetation.
Range
Breeds in Canadian prairies s. to Nebr.; winters s. of U.S.; migrates along Pacific Coast.

ARCTIC TERN
Sterna paradisaea   26:6, 27:7

Description
Size, 13–16 in. (33.0–40.6 cm). A far-northern bird notable for
• *blood-red bill,* long tail, and short legs. Spring: black cap with white border above grayish lower face, tail deeply forked. Fall: bill black, legs blackish; coloration similar to Common; rely on long tail to distinguish.
Similarities
Common's wings reach beyond tail at rest, lower face whiter, longer legged, dark-tipped bill.
Habitat
Coast, ocean, inland lakes, especially tundra in summer.
Habits
Similar to Common. Identifying this species is often difficult; immatures are virtually indistinguishable.

Voice
Rising *key-key-key;* a rising *tee-arr* shriller than Common's.
Food
Fish, aquatic life, insects.
Eggs
2; similar to Common's; 1.6 x 1.2 in. (4.1 x 3.0 cm). Nesting is colonial; sandy depression on beach, tundra, coastal or lake island.
Range
Breeds in Arctic s. to se. Alaska and e. across n. Canada; winters off Antarctica; migrates along Pacific Coast.

Note: The **ALEUTIAN TERN,** *Sterna aleutica,* size, 15 in. (38.1 cm), is seen in western Alaska. This dark, gray-bodied tern has a white forehead and dark bill.

LEAST TERN
Sterna albifrons   26:1, 27:9

Description
Size, 8½–9 in. (21.6–22.9 cm). • *Smallest* tern, with a • *yellow* bill. Spring adult: forehead, underparts, tail white; cap, eye line and outer primaries black; feet yellow; tiny black tip to bill; mantle pale gray. Fall adult: head white, only nape and line to eye black, bill dusky, feet duller. Immature: primaries and front edge of wing dusky, bill darker, back of head dark, fore-edge of wing and tip dark. In fall both adults and young have dark bills and some dull yellow on legs.
Similarities
Black Tern in fall is larger, darker above, dark tail; immature, more evenly gray, including rump and tail, no black at front and little at tip of wings.
Habitat
Sandy coasts, interior islands and shorelines; beaches, rivers, bays, ocean, estuaries.
Habits
Very rapid and distinctive wingbeat.
Voice
A "rasping *zreeeep;* a rapid, high *kik kik kik*" (Collins).
Food
Small fish and aquatic life.
Eggs
2–3; variable, often olive-buff, marked with drab, or speckled; 1.2 x 1.0 in. (3.0 x 2.5 cm). Nesting is loosely colonial; nest a depression in sandy beach or gravel bar.
Range
Breeds along Pacific Coast from n. Calif. to Baja Calif. and along major inland rivers; winters s. of U.S.

ELEGANT TERN
Sterna elegans   26:9, 27:4

Description
Size, 16–17 in. (40.6–43.2 cm). Crest long, black; bill long, slender, deep orange-yellow, no black tip.
Similarities
Royal Tern is heavier, with stronger, more orange bill.
Habitat
Coastline.
Habits
Tends to stay in flocks of its own species.

Voice
Nasal *kareek,* or *ka-zeek,* quite different from call of Royal Tern.
Food
Fish.
Eggs
1; pinkish-buff; 2.0 x 1.4 in. (5.3 x 3.8 cm).
Range
Breeds along Gulf of Calif. and along Pacific Coast near San
Diego; winters n. to n. Calif. coast or s. to S. America; chiefly a
Mexican bird visiting Calif. coastline from Aug. to Oct.

CASPIAN TERN
Sterna caspia   26:14, 27:2

Description
Size, 19–23 in. (48.3–58.4 cm). Big, crested, with • *thick red bill,*
black cap, moderately forked tail; primarily identified by large size.
Spring adult: body heavy; wings broad, at rest extending beyond
end of tail; primaries dusky above, darker beneath; legs black. Fall
adult: crown streaked, duskiness on head covers entire crown down
to forehead below eye. Immature: similar to fall adult, but streaked
and barred blackish above, more orange bill.
Similarities
Royal has tail forked half its length, Caspian one-fourth; bill of
Royal is more slender, orange; clear-white forehead most of year,
more crested look; less black on underside of primaries in adult.
Habitat
Coasts, large interior lakes, rivers, bays.
Habits
Makes powerful boobylike dives; flight heavy, gull-like; flies low
when fishing, at a great height and with bill forward when
migrating; soars occasionally; often in company of Ring-billed
Gulls.
Voice
"A hoarse, croaking *kraaa*" (Bent); also, *karr* and repeated *kak*'s.
Food
Fish.
Eggs
1–3; buff, sparingly spotted with brown; 2.7 x 1.9 in. (6.9 x 4.8
cm). Nesting is colonial; nest a depression in sand.
Range
Breeds locally in w.-cen. Canada and Great Basin, with some in
Calif., winters along Pacific coast from San Francisco Bay, s. to C.
America.

ROYAL TERN
Sterna maxima   26:13, 27:1

Description
Size, 18–21 in. (45.7–53.3 cm). Large, crested, with a • *thick
orange-red bill.* Spring: black cap, sometimes with white forehead;
well-forked tail; underwings light, at rest wings do not extend
beyond edge of tail; feet black. Fall: more white on forehead. Black
head feathers form a bushy crest on back of head. Immature: as
fall adult, sometimes with yellow legs.
Similarities
Caspian is heavier and with thicker red bill. Elegant has deep-
yellow bill, longer, more slender.
Habitat
Coastline.

Habits
Flight heavier than Common's, lighter than Caspian's.
Voice
Shrill *keer,* a squawking *kowk;* a rolling, liquid whistle, *tourrreee;*
also *kaak* or *kak.*
Food
Fish.
Eggs
1–2; whitish, evenly spotted with dark brown; 2.5 x 1.8 in. (6.4 x
4.6 cm). Nest is in depression in sand; in colonies.
Range
Breeds along Pacific Coast from s. Calif. to Mexico; winters mostly
s. of U.S., with visitors common n. to Calif. coast to San Francisco.

BLACK TERN
Chlidonias nigra   **26:2, 27:3**
Description
Size, 9–10¼ in. (22.9–26.0 cm). Spring adult: wings and tail slate-
gray, tail only slightly forked; • *bill, body and head black;* feet dark
red; back and wings gray; undertail coverts white. Fall adult and
immature: head and underparts white; back and wings gray; nape
black with point extending down behind eye; ear mark and patch
on side of breast dark; some brown on mantle of immatures.
Presents a pied appearance when molting; mottled changes begin
by midsummer.
Similarities
Of *Sterna* group, in fall, the others are not as gray, flight less
graceful and airy, larger tail more notched.
Habitat
Prairie sloughs (breeding), marshes, lakes, coast.
Habits
Flight in pursuit of insects like a Nighthawk's; usually picks food
from surface rather than diving. In flight, bill points forward when
not looking for food.
Voice
Short, shrill *crik* or *kik, keek,* or *klea;* a longer screaming *creek.*
Food
Insects, fish, small aquatic life.
Eggs
2–3; variable, often buffy-olive, blotched with brown; 1.3 x 1.0 in.
(3.3 x 2.5 cm). Nest is colonial; on ground or marsh vegetation,
muskrat house.
Range
Breeds throughout much of Canada and U.S., from Canadian
prairies s. to Calif., Nev., and Colo. marshes; winters s. of U.S.;
migrates inland and along Pacific Coast.

AUKS, MURRES, AND PUFFINS
Family Alcidae

The alcids, or auks as these birds are called, with their short necks,
short, pointed wings, short tails, and webbed feet are to the
Northern Hemisphere what the penguins are to the Antarctic.
Their legs are far back on the body but, unlike penguins, they can
fly on rapidly beating, narrow wings, often given to sharp veering.
They land straddle-legged, stand or sit upright, walk poorly, swim
well, dive from the surface, and have to patter along the surface to
get aloft. The plumage is generally black above and white below,

and in flight the buzzy, rolling wingbeats alternately reveal the black upperparts in contrast to the white underparts. These birds frequent rocky coasts and nest in colonies on ledges or in rock crevices or burrows. Larger species feed mainly on fishes, smaller ones on crustaceans, mollusks, and other invertebrates. Murrelets and auklets are notable for their small size. They are compact seabirds with extremely short necks and white throats.

COMMON MURRE
Uria aalge   **28:1**

Description
Size, 16–17 in. (40.6–43.2 cm). Only duck-sized alcid, with a long, thin, slender, pointed bill, other than the Thick-billed Murre; white line on rear of wing. Summer adult: head, throat, and upperparts black; underparts white. Ringed phase: eye-ring and streak behind eye white. Winter adult: throat and side of head white, white line over eye to rear set off by • *black line back from eye.*

Similarities
Winter loons are larger, backs spotted, no white line on rear edge of wing, as in Common Murre.

Habitat
Ocean, coastal rocks, large bays.

Habits
Flies with head and neck extended and drooping; feet project beyond short tail; flies in lines and often raft on water.

Voice
Purring *mur-r-r-r-e;* hoarse moans.

Food
Fish, crustaceans, worms, mollusks.

Eggs
1; light green, marked with brown and lilac; 3.2 x 2.0 in. (8.1 x 5.1 cm). Nest is colonial, on bare cliff ledge.

Range
Breeds along coastlines, from Arctic Alaska to cen. Calif.; winters at sea from far n., s. to cen. Calif.

Note: The **THICK-BILLED MURRE,** *Uria lomvia,* size, 18 in. (45.7 cm), of coastal Alaska is heavier than the Common, has a white streak at the base of its bill, and in winter its face is black to below the eye, without a white ear covert stripe.

PIGEON GUILLEMOT
Cepphus columba   **28:2**

Description
Size, 12–14 in. (30.5–35.6 cm). All-black with white wing patch and bars of black apparent in the white of the wing patch. Summer: • *white shoulder patches,* bill pointed, feet red. Winter: pale; underparts white, wings blackish with summer-type wing patches.

Similarities
White-winged Scoter is larger; smaller wing patches located on rear edge of wing visible only in flight; bill larger.

Habitat
Ocean, rocky shores.

Habits
Dives to ocean bottom to secure food.

Voice
"A feeble wheezy or hissing whistle, *peeeeee*" (Peterson).
Eggs
2; heavily brown-blotched on white; 2.5 x 1.5 in. (6.3 x 3.8 cm).
Range
Resident along rocky coastlines from Alaska to s. Calif.

MARBLED MURRELET
Brachyramphus marmoratus   28:4

Description
Size, 9½–10 in. (24.1–25.4 cm). Only alcid south of Alaska with
• *dark brown* breeding plumage and heavily barred underparts;
dark bill. Fall and winter: • *white stripe* between back and wing.
Similarities
Cassin's Auklet is dark, with dusky throat and sides.
Habitat
Coastal waters, bays, sounds, riptides.
Habits
"Commutes" to and from sea to maintain nesting areas.
Voice
A "sharp *keer, keer,* or a lower *kee*" (Hoffmann).
Eggs
1; yellow-buff, spotted; 2.5 x 1.5 in. (6.3 x 3.8 cm). Nest is among
rocks in high mountains.
Range
Resident from Alaska (Kodiak Is.) s. to cen. Calif.

Note: The **KITTLITZ'S MURRELET**, *Bachyramphus brevirostris*, size,
9 in. (22.9 cm), resembles the Marbled Murrelet, but is speckled
with white above. In winter, the scapulars have a white bar, the
face is all-white to above the eye, and there is a narrow band on
the side of the breast. It frequents the tundra coast, offshore
waters, and glacial bays, in the summer only from coastal Alaska
west and south from Point Barrow to Glacier Bay.

XANTUS' MURRELET
Endomychura hypoleuca   28:6

Description
Size, 9½–10½ in. (24.1–26.7 cm). Resembles small murre,
upperparts unpatterned solid black, crown black to lower edge of
eye, white below, black bill. No seasonal change.
Similarities
Marbled has white wing mark. Ancient has pale bill.
Habitat
Ocean and coastal waters.
Voice
Usually heard only after dark as twittering, finchlike whistles.
Eggs
1–2; variable, spotted; 2.0–1.5 in. (5.0–3.8 cm). Nest is sea island
crevice.
Range
Breeds offshore on islands in Pacific, from s. Calif. to Baja Calif.;
winters at sea from Vancouver to n. Calif.

ANCIENT MURRELET
Synthliboramphus antiquus   28:5

Description
Size, 9½–10½ in. (24.1–26.7 cm). Summer: • *black throat patch*
and sharp, • *white eye stripe;* pale bill; dark slate above, white
below. Winter: similar to Marbled, but lacking white stripe on
scapulars.

Similarities
Xantus' has dark bill; cap black, contrasting with paler gray back;
throat often dusky.

Habitat
Ocean, sounds, bays (rarely).

Voice
Colonial low chirps and whistles, when roosting.

Habits
When nesting, travels to and from nesting site only at night.

Eggs
2; spotted; 2.4 x 1.5 in. (6.1 x 3.9 cm). Nest in island rocks or in
burrow.

Range
Breeds along N. Pacific coastline from Alaska to cen. B.C.; winters
in breeding range, with some moving farther s. to s. Calif.

CASSIN'S AUKLET
Ptychoramphus aleuticus   28:7

Description
Size, 8–9 in. (20.3–22.9 cm). Dull, dark color; dusky throat and
white belly; light spot on dark bill; pale eye.

Similarities
All other similar-sized alcids wintering within the range have
white on throat and sides of neck. Marbled Murrelet in summer is
dusky also but bill more pointed, underparts barred. Immature
Rhinoceros is much larger, with dark eye.

Habitat
Ocean, rocky shores.

Habits
Flies very close to water or ground.

Voice
When roosting in colony, a repeated rasping note.

Eggs
1; white; 1.9 x 1.3 in. (4.9 x 3.5 cm). Nesting is colonial, in island
rock crevice or burrow.

Range
Breeds from Aleutians s. to cen. Baja Calif., with many nesting on
Farallon Is. off San Francisco; winters along Pacific Coast from
Vancouver southward.

PARAKEET AUKLET
Cyclorrhynchus psittacula

Description
Size, 10 in. (25.4 cm). Small; black and white; • *very short, deep,
upturned, red* (in summer) *bill;* throat black, rest of underparts
white. Summer: head all-black with thin white plume back of each
eye, dusky marks on sides and flanks. Winter: lacks plumes, throat
white, few or no flank markings; bill duller.

Habitat
Ocean.

Habits
Erratic flight, feet often dangle.
Voice
A trill, often broken, and rising in pitch.
Food
Crustaceans, mollusks.
Eggs
1; dull white or blue; 2.1 x 1.5 in. (5.5 x 4.0 cm).
Range
Breeds on Bering Sea islands and rocky coastlines of Aleutians; winters along Pacific Coast from Aleutians s. to Oreg., occasionally to cen. Calif.

RHINOCEROS AUKLET
Cerorhinca monocerata   28:8

Description
Size, 14–15½ in. (35.6–39.4 cm). Late winter, spring: above dark, below white; narrow white • *eyebrow plume* above and behind eye, • *white "mustache" plume* back of • *horned,* yellowish bill (horn short, erect, at base). Winter: head plumes present or not but shorter than in breeding plumage, horn absent; swimming or floating bird is rather large, uniformly dark colored and lacks white throat.
Similarities
Cassin's is much smaller, with pale-spotted dark bill, dusky with dark throat. Immature Tufted Puffin is larger, with stouter bill, no facial plumes.
Habitat
Ocean, fjords, riptides.
Habits
More solitary than most auklets, seldom in flocks.
Voice
At night in colonies, growling and shrieking.
Eggs
1; whitish, often spotted; 2.7 x 1.8 in. (6.9 x 4.6 cm). Nest is colonial, sea island burrow.
Range
Breeds from Aleutians s. to cen. Calif.; winters off Pacific Coast from Vancouver to Baja Calif.

HORNED PUFFIN
Fratercula corniculata

Description
Size, 14½ in. (36.8 cm). White-bellied; black collar; black above; feet orange. Summer adult: • *white cheeks;* small, dark horn above eye; deep bill triangular, flat on sides, bright-yellow with red tip. Winter adult: cheeks dusky-gray, bill smaller, blackish with red tip. Immature: same as winter adults but with smaller, all-black bill.
Similarities
Adult Tufted Puffin is darker, pale-eyed; immatures pale-billed.
Habitat
Ocean sounds, bays (rarely).
Habits
As Tufted Puffin.
Voice
Colonial low chirps and whistles when roosting.
Food
Mainly fish.

Eggs
1; spotted; 2.6 x 1.8 in. (6.7 x 4.6 cm). Nest is in island rocks or in burrows.
Range
Breeds on Bering Sea islands and along Alaskan coastline e. to Glacier Bay; winters at sea s. to Calif.

CRESTED AUKLET
Aethia cristatella

Description
Size, 9½ in. (24.1 cm). • *Crested;* slate-gray, with a frontally
• *curved plume* on the forehead and a white stripe diagonally behind eye. Summer adult: very short, • *bright-orange or red bill.* Winter adult: bill duller, crest shorter, plume less distinctive. Immature: sooty.
Similarities
Immature Cassin's is indistinguishable from immature Crested.
Habitat
Ocean, sea cliffs.
Voice
Grunts, honks, chirps, in breeding colonies.
Food
Mainly crustaceans.
Eggs
1; dull white or blue; 2.1 x 1.5 in. (5.4 x 3.9 cm).
Range
Resident about Aleutians and other Bering Sea islands.

Note: The **LEAST AUKLET,** *Aethia pusilla,* size, 6 in. (15 cm), has the same summer range as the Crested but it migrates from Japan to the Aleutians. It is black above, white below; sparrow-sized; and with a tiny bill.

TUFTED PUFFIN
Lunda cirrhata   **28:3**

Description
Size, 14½–15½ in. (36.8-39.4 cm). Body nearly all-black, stocky; head remarkable for the strange bill. Summer adult: • *large triangular brilliant orange-red bill;* contrasting white face; long, • *back-curving ivory ear tufts;* feet bright orange, eyes pale. Winter adult: blackish above, dusky below; eye pale and whitish line over eye observable at close range; ear tufts absent; bill orange-red with black base but smaller than in summer. Immature: dark eye, light grayish underparts; bill smaller, yellowish.
Similarities
Immature Rhinoceros Auklet has slenderer bill.
Habitat
Ocean, fjords.
Food
Mainly fish.
Eggs
1; white, spotted; 2.8 x 1.9 in. (7.2 x 4.9 cm). Nest is on sea island or craggy headland, in burrow or crevice.
Range
Breeds from nw. Alaska s. to Bering Sea islands, Aleutians, and s. to B.C., with a few s. to Calif.; winters at sea in breeding range s. of ice pack.

Land Birds

Pigeons and Doves
Order Columbiformes

PIGEONS AND DOVES
Family Columbidae

Members of this very large family of plump, fast-flying, small-headed birds with slender bills and low, cooing voices have long, pointed wings and short legs. They spend much time on the ground; bob their heads as they walk; and drink without raising their heads, unlike most other birds.

BAND-TAILED PIGEON
Columba fasciata   **31:2**

Description
Size, 14–15½ in. (35.6–39.4 cm). Stout, heavily built, with • *broad, pale band* across the end of its tail, bordered by black transverse band midway up; tail fanlike. Male: upperparts brown, gray below; underparts purplish-pink, abdomen almost white; white collar on back of neck; • *bill yellow, with dark tip*. Female: duller, grayer, often lacks collar of male.

Similarities
Rock Pigeon has white rump, black band on tail tip, red feet; usually not in woods.

Habitat
Mountains, forests, oak foothills, chaparral, canyons.

Habits
Flight strong but not very fast.

Voice
Repeated hollow, owllike *oo-whoo* or *whoo-oo-whoo*.

Eggs
1–2; white, creamy, unmarked; 1.5 x 1.1 in. (3.8 x 2.8 cm). Nest is in trees, usually flat stick platform.

Range
Breeds from s. B.C. s. to Baja Calif. along coastline and forested areas, and in mountains from Utah and Colo. s. to C. America; winters along Pacific Coast from Wash. southward, and from cen. Ariz. and N. Mex., southward.

ROCK PIGEON
Columba livia   **31:1**

Description
Size, 13 in. (33.0 cm). Common town and farmyard pigeon of varied coloration, distinguished by • *white rump,* square tail. Bluish-gray with 2 black bars on wing, 1 black bar on tip of tail; eyes and legs red. Feral birds may be variegated with shades of gray, brown, white, and black, but usually have white rump; populations tend to revert to original wild plumage.

Similarities
Band-tailed lacks white rump; white band is at tip, not middle of tail. Mourning Dove has pointed tail, tan color.

Habitat
Originally cliffs; now also adapted to cities, parks, bridges, freight yards, farmyards, beaches.
Habits
Gregarious, tame; flight strong (to 67 mph, or 107.8 km/hr), has been trained for homing.
Voice
Characteristic cooing, *oo-roo-coo.*
Eggs
2; white; 1.5 x 1.1 in. (3.8 x 2.8 cm). Nest is of trash, on building ledge, cliff.
Other names
Domestic Pigeon, Rock Dove.
Range
Resident throughout N. America, excluding northernmost tundra and taiga.

WHITE-WINGED DOVE
Zenaida asiatica

29:3
Fig. 11

Description
Size, 11–12½ in. (27.9–31.8 cm). Distinctive large • *white patch on wing;* • *tail rounded* with rectangular white corners.
Similarities
Mourning is not as heavy, tail pointed, no white wing patch.
Habitat
Towns, desert brush, breaks and river woods, mesquite, saguaros.
Habits
Gathers in large flocks when not breeding.
Voice
A "harsh cooing, *'who cooks for you?';* also, *ooo-uh-cuck'oo*" (Peterson).
Eggs
2; olive-buff; 1.2 x 0.9 in. (3.1 x 2.1 cm). Nest is crude stick platform in tree or thicket.
Range
Breeds from s. Calif. e. to Ariz. and Tex., and s. to Mexico and S. America; winters s. of breeding range from Mexico southward.

Mourning Dove, p. 124

White-winged Dove

Fig. 11

Common Ground Dove, p. 124

Inca Dove, p. 125

White-fronted Dove, p. 124

MOURNING DOVE
Zenaida macroura

29:2
Fig. 11

Description
Size, 11–13 in. (27.9–33.0 cm). Long, white-edged, pointed tail;
• *wings long, pointed;* head brown, body buffy-gray with a • *bluish cast on wings.*
Similarities
American Kestrel in flight lacks sharp downstrokes of dove, heavier head and shoulders, squared tail. White-winged has rounded tail, wing patches. Rock Pigeon is larger and huskier, has rounded tail.
Habitat
Open woodlands, farmlands, suburbs, roadsides, coastal scrub, grassland, desert, mesquite.
Habits
Feeds on ground, often takes dust baths or picks gravel from roadside; flight direct, wings whir as it rises; in dry areas flies daily to water; forms loose flocks in winter, frequently seen on wire lines; may nest in cities on fire escapes.
Voice
Mournful *coo-ah, coo, coo, coo.*
Eggs
2; white; 1.1 x 0.9 in. (2.8 x 2.3 cm). Nest is usually in tree, shrub, cactus, on ground, loosely built of twigs.
Range
Breeds throughout much of N. America, excluding northern forests, from cen. Canada s. to Mexico; winters in breeding range, plus from n. Calif. s. and e. to Sw. states.

Note: The lower Rio Grande Valley of Texas is the United States range of the **WHITE-FRONTED DOVE**, *Leptotila verreauxi* (**29:4**, *Fig. 11*), size, 11 in. (27.9 cm). It resembles the Mourning and especially the White-winged doves. It is distinguished by a lack of white in wings and nonpointed tail, white on the forehead, and rusty underwings.

COMMON GROUND DOVE
Columbina passerina

29:7
Fig. 11

Description
Size, 6–7 in. (15.2–17.8 cm). Catbird-sized, with wingspread to 11 in. (27.9 cm); short black tail; wings rounded, flashing red-brown spots in flight; feet yellow. Male: body brown; underparts drab pink; breast scaly; black-tipped, reddish bill. Female: duller than male.
Similarities
Inca has longer, white-sided tail. Mourning Dove is larger with pointed tail.
Habitat
Brush, orchards, groves, river bottoms, dirt roads, farmlands, forest fringes.
Voice
Repetitive *woo-oo, woo-oo,* often with rising intonation.
Eggs
2; white; 0.7 x 0.6 in. (1.9 x 1.7 cm). Nest is on ground or in trees to 25 ft. (7.6 m).
Range
Resident in s. U.S. from s. Calif. e. to Gulf Coast.

SPOTTED DOVE
Streptopelia chinensis   **29:1**

Description
Size, 13 in. (33.0 cm). • *Broad collar of black and white spots* on the hind neck; tail blunt to rounded, considerable white in corners.
Similarities
Mourning has faster flight, is smaller, tail pointed.
Habitat
Suburbia, parks, wooded areas.
Voice
Similar to cooing of White-winged; a soft *coo-who-coo*.
Habits
Flight relatively slow; pair displays together in flight.
Eggs
2; white; 1.1 x 0.9 in. (2.8 x 2.3 cm).
Range
Resident in Los Angeles area, along Pacific coast n. to Santa Barbara and inland to Bakersfield; introduced from Asia.

INCA DOVE
Scardafella inca   **29:5**
                                                            Fig. 11

Description
Size, 7½–8 in. (19.1–20.3 cm). Small, with dark-tipped reddish wings; upperparts pale, scaly; • *tail comparatively long,* square-ended, edged with white on sides when spread, but appearing pointed when folded. In flight, shows red-brown in wings.
Similarities
Ground Dove has shorter all-dark tail.
Habitat
Towns, parks, suburbia, farms, chicken yards.
Habits
Allows close approach.
Voice
A monotoned *"coo-coo-hoo* or *hink-a-doo"* (Davis).
Eggs
2; white; 0.8 x 0.6 in. (2.1 x 1.6 cm). Nest is in low tree or shrub or shed, as a saucer of mixed vegetation.
Range
Resident from cen. Calif. e. throughout Sw. to Mexico.

Cuckoos and Allies
Order Cuculiformes

CUCKOOS AND ROADRUNNERS
Family Cuculidae

Members of this family are slender birds with slightly curved bills; long, narrow, rounded tails; and zygodactyl feet (two toes forward, two toes behind). Unlike some Old World cuckoos, these all build their own nest and raise their own young. Sexes are similar, brown above and white below. The family feeds on seeds, grasshoppers, and other insects and small fruits (Roadrunners eat small reptiles).

YELLOW-BILLED CUCKOO
Coccyzus americanus   **36:8**

Description
Size, 11–13½ in. (27.9–34.3 cm). • *Rusty wing areas,* conspicuous in flight; back dull brown; breast white; • *large white tail spots;* eyelids and • *lower bill yellow.*
Similarities
Black-billed has small tail spots, no rufous on wings.
Habitat
Second growth, orchards, stream thickets, willows, mesquite.
Habits
Secretive, moves noiselessly about upper foliage of small trees, sits motionless; flight in open is direct, gliding, horizontal.
Voice
Wooden *kuk-kuk-kuk-kow-kow-kow* (or *kiaow-kiaow-kiaow* or *kowlp-kowlp*), The *"ow"* notes retarded toward the end, some indistinguishable from Black-billed's.
Eggs
2–6; bluish-green; 1.2 x 0.8 in. (3.0 x 2.0 cm). Nest is frail platform of twigs 4–10 ft. (1.2–3.0 m) up.
Range
Breeds locally in West, excluding Rockies and Central Plateau, from s. Canada to C. America; winters in S. America.

BLACK-BILLED CUCKOO
Coccyzus erythropthalmus   **36:10**

Description
Size, 11–12½ in. (2.7.9–31.8 cm). Lacks rufous in wings and has only fine spots in tail; uniform olive-brown above, white below; • *bill black; eye ring red.*
Similarities
Yellow-billed is not as slim.
Habitat
Second growth, forest fringes, thickets.
Habits
Secretive; more active than Yellow-billed at night.
Voice
Evenly spaced 2-syllable notes, *kuk* or *coo* in groups. Yellow-billed may give similar calls.
Eggs
2–6; bluish-green; 1.3 x 0.8 in. (3.3 x 2.0 cm). Nest is platform of twigs, sturdier than Yellow-billed's, 2–10 ft. (0.6–3.0 m) up in shrub.
Range
Breeds chiefly in the e. U.S. w. to se. Alta. and s. to w. Wyo. and n. Colo.; winters in S. America.

ROADRUNNER
Geococcyx californianus   **33:12**

Description
Size, 20–24 in. (50.8–61.0 cm). Unmistakable; the only large, running ground bird with crest and very long tail. Short, rounded wings with curved, white stripe; brownish above, white below heavily streaked with brown; bare red area behind eye; crest shaggy, legs strong.
Habitat
Deserts, sagebrush, chaparral; dry brush and pinyon-juniper country.

Habits
Runs as fast as humans; will fly only if forced.
Voice
A *"coo coo coo ooh ooh ooh ooh"* (Bent); purring sounds; clacks mandibles.
Eggs
3–8; chalky white; 1.6 x 1.4 in. (4.1 x 3.6 cm). Nest is a coarse shallow cup of sticks 1 ft. (0.3 m) wide, in cactus, mesquite, 3–15 ft. (0.9–4.6 m) up.
Range
Resident in desert and scrub of Sw. from s. Calif., s. Utah, s. Kans. southward.

GROOVE-BILLED ANI
Crotophaga sulcirostris

33:13

Description
Size, 13 in. (33.0 cm). Black with short, rounded wings; black, puffinlike • *high-ridged bill;* loose-jointed tail appears as long as head and body.
Habitat
Chaparral, brush.
Habits
Flight weak; alternately flaps and glides.
Voice
Repetitive *whee-o,* or *tee-ho;* "1st note slurring up and thin; 2nd lower" (Peterson); in flight, a low chuckling.
Eggs
3–4; bluish; 1.2 x 0.9 in. (3.1 x 2.3 cm). Nest is leaf-lined, twiggy mass in low tree or thick brush.
Range
Resident from lower Rio Grande Valley, s. to C. America; casual to Big Bend region of w. Tex. and s. Ariz., sw. N.Mex.

Owls
Order Strigiformes

Owls are largely nocturnal birds of prey with big eyes that face forward, broad heads, with facial disks, short "swivel" necks, and soft fluffy plumage that covers the base of their bills and their legs. They fly silently, almost mothlike; have acute hearing. Some species have conspicuous feather tufts, like "horns" or "ears" above and behind the eyes.

BARN OWLS
Family Tytonidae

Members of this family, of which only a single species occurs in the West, are "monkey-faced" and have long legs extending beyond the tail in flight.

BARN OWL
Tyto alba

30:10

Description
Size, 14–20 in. (35.6–50.8 cm). White, • *heart-shaped "monkey face"* and pale breast. Pale overall; legs and wings long; buffy-

brown above; pale below, relatively unmarked; eyes dark; no ear tufts. In flight looks big-headed, slender-bodied, white below, ghostly by dark.

Habitat
Wood edges, farmland, haunts of man; groves, shade trees, barns, cool canyons.

Habits
Nocturnal; frequents old belfries, water towers, deserted buildings; flies with deep wingbeats, buoyantly, legs trailing behind.

Voice
"(1) A discordant scream . . . (2) a snapping of the bill . . . (3) a flight call, resembling *ick-ick-ick-ick*" (Potter and Gillespie); "a shrill rasping hiss or snore: *kschh* or *shiiish*" (Peterson).

Food
Rodents, insects.

Eggs
5–8; white, more oval than most owls; 1.8 x 1.3 in. (4.6 x 3.3 cm). Nest is in hollow tree or other cavity (belfry, cave, hole in stream bank) on litter of disgorged pellets.

Range
Resident in se. Calif., s. Ariz., N.Mex., and Tex. south to C. America.

TYPICAL OWLS
Family Strigidae

These owls have round or squarish faces and short legs that do not reach beyond the tail in flight.

SCREECH OWL
Otus asio   30:4

Description
Size, 7–10 in. (17.8–25.4 cm). Adult: small, with conspicuous • *ear tufts;* eyes yellow. Predominantly gray in West, but some brown birds occur in northern Great Basin and northwestern coastal areas. Immature: lacking ear tufts.

Similarities
Flammulated is smaller, with dark eyes; ear tufts inconspicuous.

Habitat
Orchards, woods, suburbs, small towns.

Habits
Strictly nocturnal; often lives in hollow tree and sits in entry hole by day.

Voice
Series of tremulo whistles on 1 pitch, initial notes separated, picking up speed like a bouncing ball; lacks descending wail of eastern variety.

Food
Rodents, insects, other animal food.

Eggs
3–5; white; 1.4 x 1.3 in. (3.6 x 3.3 cm).

Range
Resident throughout much of N. America from s. Canada s. to Mexico.

Note: The **WHISKERED OWL**, *Otus trichopsis*, size, 7–8 in. (17.7–20.3 cm), is very similar but has "codelike" *boo-boo-boo-boo* call. It is seen in southeastern Arizona and southwestern New Mexico.

FLAMMULATED OWL
Otus flammeolus

Description
Size, 6–7 in. (15.2–17.8 cm). Small, with • *dark eyes,* red-brown or gray with a wash of tawny; short ear tufts quite inconspicuous.
Similarities
Screech and Whiskered Owls have long ear tufts and yellow eyes.
Habitat
Mountain pine woods.
Habits
Perches high in pines; common in ponderosa pine forests.
Voice
Low-pitched, mellow hoot repeated at 2–3 sec. intervals.
Food
Moths, spiders, other insects.
Eggs
3–4; white; 1.1 x 0.9 in. (2.9 x 2.5 cm). Nest is hole in tree, old woodpecker cavity.
Range
Resident from inland B.C. e. to Rockies and s. inland to Mexico.

GREAT HORNED OWL
Bubo virginianus   **30:13**

Description
Size, 18–25 in. (45.7–63.8 cm). Very • *large,* with • *ear tufts;* size of largest hawks; female larger than male. Wings long, broad; dark brown, heavily barred below, and streaked with black; throat white; eyes yellow. In flight, ear tufts flattened; appears neckless and large-headed.
Similarities
Long-eared Owl is smaller; streaked below, not barred; ear tufts closer together.
Habitat
Variable, from dense forests to deserts, grasslands, canyons, stream fringes.
Habits
Most powerful of owls, sometimes hunts by day or by night, often sails on fixed wings; rarely soars.
Voice
Bass, a deep, resonant *hoo-hoo-hoo-hoo, hoo-hoooooo* (Males 4–5 rhythmic hoots, females 6–8, lower-toned); a *waugh-HOO;* less commonly, a blood-curdling shriek.
Food
Small mammals, birds.
Eggs
2–3; white; 2.2 x 1.8 in. (5.6 x 4.6 cm). Nest is variously on ground, in tree, in dense timber, pothole, cliff, river bluff, deserted nest of other birds.
Range
Resident throughout N. America from n. Canada, s. to Mexico.

HAWK OWL
Surnia ulula   **30:6**

Description
Size, 14½–17½ in. (36.8–44.5 cm). Medium-sized, barred across both breast and belly; dark, plump; wings short, pointed; • *tail long,* rounded, graduated, banded. Gray-brown above, head with

white spots and black sideburns and chin. In flight shows diagonal white wing mark.

Habitat
Natural openings in northern coniferous forest, birches.

Habits
Often perches hawklike on tops of trees with body bent forward; pumps, sometimes cocks tail; flight falconlike, direct, swift, close to ground; hunts by day, hovers, drops on prey; tame.

Voice
"A trilling whistle *tu-wita-wit, tuwita-tu-wita, wita, wita*" (Henderson).

Food
Rodents, other small mammals, grouse.

Eggs
3–7; white; 1.5 x 1.3 in. (3.8 x 3.3 cm). Nest is in hollow tree, stump, snag, or deserted hawk's or crow's nest.

Range
Breeds from Alaska and n. B.C. eastward; winters occasionally farther s. to n. U.S.

SNOWY OWL
Nyctea scandiaca   **30:12**

Description
Size, 20–27 in. (50.8–68.6 cm). Head round, eyes yellow, wings rounded. Adult male: • *white* with some dark scaly barring. Female and immature: heavier, dusky barring. Some individuals are much whiter than others.

Similarities
White Gyrfalcon is more slender, smaller head, longer neck, pointed wings. Pale subspecies of Horned Owl in Arctic has ear tufts. Barn Owl is smaller, whitish on underparts only, dark eyes. Immatures of all owl species are covered with whitish down before feathers appear.

Habitat
Rolling tundra, coastal marshes, prairies, farmland, beaches.

Habits
Wary, perches on dune, post, stump; flight strong, direct but jerky, upbeat faster than down; often sails. Often feeds during day.

Voice
Usually silent, except when breeding.

Food
Smaller mammals, especially lemmings.

Eggs
4–10; white; 2.3 x 1.8 in. (5.8 x 4.6 cm). Nest is grassy tundra hollow.

Range
Breeds in far n. tundra; winters from Arctic s. to n. U.S.; occasionally farther s. in winter to n. Calif., Utah, Colo., in search of food.

PYGMY OWL
Glaucidium gnoma   **30:3**

Description
Size, 7–7½ in. (17.8–19.1 cm). Very small, brown, "earless," with • *black "eye" patch* on each side of hind neck; sharply • *black streaked down center of underparts;* tail rather long and barred, often held at a perky angle; head appears small, dotted with white.

Similarities
Saw-whet head is larger, tail stubbier, streaks blotchy brown.
Ferruginous Owl, in desert Southwest, has brownish blurry breast
streakings; fine pale streaks, not dots, on crown.
Habitat
Usually open, coniferous or mixed woods in mountain or canyon.
Habits
Often heard or seen flying shrikelike by day.
Voice
Mellow whistled *hoo,* repeated at 2-second intervals; a rolling
series ending with 2–3 sharp notes.
Food
Various insects, lizards, small birds.
Eggs
3–4; white; 1.0 x 0.9 in. (2.6 x 2.3 cm). Nest is usually in
woodpecker hole.
Range
Resident from se. Alaska throughout W., e. to Rockies, and s. to C.
America.

FERRUGINOUS OWL
Glaucidium brasilianum

Description
Size, 6½–7 in. (16.5–17.8 cm). Distinguishable from Pygmy by
• *brownish breast streakings* and fine pale streaks on crown; also by
lowland habitat.
Similarities
Pygmy Owl has white dots on crown. Elf Owl is smaller, with
short tail.
Habitat
Lowland mesquite, saguaros, river breaks.
Habits
Jerks or flips tail; often heard in daytime.
Voice
Monotonous repetitive *chook* or *took,* 2–3 times a sec.
Eggs
3–4; white; 0.4 x 0.3 in. (1.1 x 0.9 cm). Nest is tree cavity,
woodpecker hole.
Range
Resident from sw. states s. to S. America.

ELF OWL
Micrathene whitneyi   **30:2**

Description
Size, 5–6 in. (12.7–15.2 cm). Sparrow-sized, small-headed, earless,
with white "eyebrows"; tail relatively short, underparts rust-
striped.
Similarities
Pygmy and Ferruginous have longer tails extending beyond wing
tips and "eye" spots on back of neck.
Habitat
Saguaro deserts and watered canyons.
Habits
Often seen peering from its roosting hole during day.
Voice
Puppylike, high-pitched yipping; rapid *whi-whi-whi-whi-whi,* or
chewk-chewk-chewk-chewk.

Food
Chiefly insects.
Eggs
3–4; white; 1.0 x 0.9 in. (2.6 x 2.3 cm). Nest is old woodpecker nest in tree or cactus.
Range
Breeds from se. Calif. e. to Tex. and s. to C. America; winters s. of U.S.

BURROWING OWL
Athene cunicularia   30:1

Description
Size, 9–11 in. (22.9–28.9 cm). Only small owl that lives on ground; • *legs unusually long;* tail short; eyes yellow; brown above, spotted with white; white below, barred with brown.
Habitat
Open country, vacant lots in cities, deserts, farms, prairies, dikes.
Habits
Lives in prairie dog or other holes, often standing near entrance, or perching on eminence above ground or on post, wire, shrub; bobs head and tail up and down in comical fashion on its long legs; may follow moving animals; flies little.
Voice
In flight, a chattering note; alarm, *tsip-tsip;* at night, a mellow *co-hoo.*
Food
Insects, small invertebrates.
Eggs
5–9; white; 1.3 x 1.1 in. (3.3 x 2.8 cm). Nest is grass-lined chamber at end of rodent burrow in open ground.
Range
Breeds throughout w. U.S.; winters from s. breeding range, southward.

SPOTTED OWL
Strix occidentalis   30:9

Description
Size, 16½–19 in. (41.9–48.3 cm). Seldom seen; large, dark brown; • *dark eyes;* round, puffy head; distinguished by eye color and the heavy barring underneath.
Similarities
Barred Owl is similar but different range.
Habitat
Dense forests, wooded slopes, and canyons.
Habits
May allow close approach to roosting area.
Voice
High-pitched hooting, usually grouped by 4's or 3's; a rapid series in crescendo.
Food
Insects, small birds (even small owls), rodents.
Eggs
2–3; white, 1.9 x 1.6 in. (4.9 x 4.1 cm). Nest is hollow tree, cliff cavern, old hawk nest.
Range
Resident along Pacific Coast from s. B.C. to San Francisco, and inland in forests of Rockies and Sierras to s. Calif. and Mexico.

BARRED OWL
Strix varia   **30:11**

Description
Size, 17–24 in. (43.2–61.0 cm). Big, round-headed; brownish
without ear tufts; bill yellow; eyes dark; gray-brown above, pale
below; • *dark crossbars* on breast and collar, dark vertical streaks on
belly; white spots on back.

Similarities
Great Horned and Long-eared have reddish facial disks, ear tufts.
Spotted Owl is browner, lower breast and belly heavily
crossbarred. Great Gray Owl is larger; grayer, eyes yellow.

Habitat
Moist woodlands.

Habits
Sometimes seen by day; flight buoyant, wingbeat slow; inquisitive.

Voice
Higher-pitched hoots than Great Horned, commonly 8 hoots in 2
groups of 4; at a distance sounds like dog barking. Last *hoo* or a
series usually ends in a characteristic *aw*.

Food
Mice, other small animals, birds.

Eggs
2–3; white; 2.0 x 1.6 in. (5.1 x 4.1 cm). Nest is in hollow tree or
deserted nest of hawk, crow, or squirrel.

Range
Resident chiefly in E., but some reach ne. and e. B.C., s. through
Mont. to Colo.

GREAT GRAY OWL
Strix nebulosa   **30:14**

Description
Size, 24–33 in. (61.0–83.8 cm). Large, round-headed; gray, with no
ear tufts (largest owl in North America). Plumage very loose,
fluffy, striped lengthwise below with black; face disks very large,
black chin spot, • *eyes yellow*; no bars on breast; tail long (to 12 in.,
or 30.5 cm) and broad.

Similarities
Barred and Spotted Owls are smaller, browner, bars on breast,
smaller face disk, eyes dark, shorter tail.

Habitat
Northern coniferous forests and adjacent meadows and parks.

Habits
Flies with slow flaps of broad, rounded wings; hunts by daylight in
Arctic summer; tame.

Voice
"Several deep-pitched *whoos*" (Grinnell and Storer); also single
resonant *whoos;* sometimes a wavering cry.

Food
Small mammals.

Eggs
2–5; white; 2.2 x 1.8 in. (5.6 x 4.6 cm). Nest is of sticks from 20
ft. (6.1 m) up in evergreens; in old hawk or crow nest.

Range
Resident in n. regions, from Alaska e. to Man. and s. to cen.
Calif., inland s. to Wyo.

LONG-EARED OWL
Asio otus   **30:8**

Description
Size, 13–16 in. (33.0–40.6 cm). Slim, grayish-brown; long wings
and tail; tall, closely spaced ear tufts, medium-sized. Blackish-
brown above, rusty face, yellow eyes, • *underparts streaked,* brown
spot shows on buff underwing.

Similarities
Great Horned is larger; wide apart ear tufts, white throat,
crossbars below. Short-eared in flight looks lighter, occurs in more
open country. Screech Owl is smaller, shorter ears, no rust on face.

Habitat
Mixed woodlands, preferably coniferous; also other thickets.

Habits
Nocturnal; flight wavering; perches close to trunk of tree, stretches
body to make it very thin, hiding in slight cover; in winter
sometimes in flocks, often in groves of dense evergreens.

Voice
A "dove-like *hoo hoo hoo* . . . a slurred whistle, *WHEE-you*"
(Griscom).

Food
Small mammals.

Eggs
3–8; white; 1.6 x 1.3 in. (4.1 x 3.3 cm). Nest is of sticks in
conifers; in old nest of crow, hawk, magpie.

Range
Breeds from cen. B.C. e. across Canada, and s. to s. Calif., s. Ariz.,
N. Mex., and w. Tex.; winters from n. U.S. s. to Baja Calif. and
eastward.

SHORT-EARED OWL
Asio flammeus   **30:7**

Description
Size, 13–17 in. (33.0–43.2 cm). Day-flying, of open country;
distinguished by lack of ear tufts, streaked tawny breast, irregular
wavering flight. Buffy-brown head and pale breast streaked with
black, eyes yellow, wings with buff patch above and black spot
near bend below; appears neckless in flight.

Similarities
Long-eared is darker brown, with ear tufts, shorter wings.
Northern Harrier has white rump, longer tail, different head
shape. Rough-legged Hawk has smaller head, white rump, more
direct flight; also shows black mark under wing.

Habitat
Open country, grasslands, marshes, tundra.

Habits
Often hunts by day, quarters low over ground like a Northern
Harrier, or sits on observation post; roosts on ground; has
impressive aerial courtship flight.

Voice
Up to 20 *toots,* higher than Great Horned's.

Food
Mice.

Eggs
4–9; white; 1.5 x 1.3 in (3.8 x 3.3 cm). Nest is grassy hollow on
ground near clump of vegetation in marsh or meadow.

Range
Breeds throughout most of n. N. America from Arctic s. to cen.
U.S.; winters from n. U.S. s. throughout U.S.

BOREAL OWL
Aegolius funereus

Description
Size, 8½–12 in. (21.6–30.5 cm). Small forest owl with • *large earless head*; • *yellow bill* and eyes, and • *black rim around face disk*. Adults: brown above with white spots on forehead and back, underparts pale, smudged with rusty brown. Juvenal: dark brown; broad white "eyebrows."

Similarities
Saw-whet is smaller; streaked instead of spotted on forehead, no black rim around disk, black bill. Hawk Owl is larger; grayer, barred below, tail longer. Pygmy has smaller head, longer tail, no facial rim.

Habitat
Coniferous and mixed forests.

Habits
Very tame; hunts by day in Arctic summer.

Voice
Notes like dripping water.

Food
Mice, birds, insects.

Eggs
4–7; white; 1.2 x 1.0 in. (3.0 x 2.5 cm). Nest is tree cavity or abandoned nests of other birds.

Range
Resident from Alaska, e. across Canada and s. to n. B.C., to cen. Alta., and cen. Sask.; accidental to n. U.S.

SAW-WHET OWL
Aegolius acadicus   **30:5**

Description
Size, 7–8½ in. (17.8–21.6 cm). Very small, with black bill and • *no ear tufts*. Adult: brown above with white streaks on head; underparts white, heavily streaked and blotched rufous; wings broad, eyes yellow. Juvenal (in summer): chocolate-brown with dusky face and conspicuous white V "eyebrow" on forehead.

Similarities
Boreal is larger, with yellow bill, black facial rim, white spots on forehead. Pygmy has small head, black "eye" spots on hindneck, longer tail.

Habitat
Forests, swamps, groves.

Habits
Nocturnal, inquisitive, tame; in taking off, drops before flying forward.

Voice
In spring, like filing a saw with notes tapering off at end; also a ventriloquial "soft *co-co-co-co-co-co*" repeated 100–130 times a minute (Eckstrom).

Food
Insects, small mammals.

Eggs
3–7; white, 1.0 x 0.9 in. (2.5 x 2.3 cm). Nest is tree cavity, woodpecker hole, in woods or swamp.

Range
Breeds from se. Alaska and s. Canada e. across Canada, and s. along mountains to Mexico; winters from s. Canada s. to s. U.S.

Goatsuckers
Order Caprimulgiformes

GOATSUCKERS
Family Caprimulgidae

These nocturnal birds have tiny bills belying huge, often bristle-bordered mouths; large flat heads with big dark eyes; long wings; and often long tails. They have very short legs, small weak feet, and fluffy plumage associated with their silent, wavering flight. Their mottled colors provide daytime camouflage when resting motionless, horizontally, on a limb or among ground leaves or gravel. Goatsuckers are most active at night and have distinctive voices.

WHIP-POOR-WILL
Caprimulgus vociferus   **31:10**

Description
Size, 9–10 in. (22.9–25.4 cm). Vigorous night call best identifies this species. Mottled-brown, no white wing spot, tail rounded and longer than wings. Male: throat black, bordered below by white band; white patch on outer tail feathers. Female: buff instead of white on throat, outer tail feathers dark.

Similarities
Nighthawk's wings at rest are pointed and extend beyond tail, Whip-poor-will's tail extends beyond wings; also nighthawks show white wing patches.

Habitat
Wooded areas, mountain slopes.

Habits
Calls mainly near dusk, dawn, and in moonlight. When flushed by day, flits off like moth; feigns broken wing to lure intruders from nest; flight erratic; red eyes may shine in headlights along roadside.

Voice
Cluck often precedes the characteristic *whip-poor-WILL* (accent on 1st and 3rd syllables), or rolling, often repeated *prrrip'puur-rill'*.

Food
Various moths, beetles, flying insects.

Eggs
2; white, mottled with brown and gray; 1.2 x 0.8 in. (3.0 x 2.0 cm).

Range
Breeds from Ariz., N. Mex., s. to Mexico and C. America; winters chiefly s. of U.S.

Note: Largest of the goatsuckers, found only in southern Texas, is the **PAURAQUE**, *Nyctidromus albicollis* **(31:7)**, size, 11 in. (27.9 cm), distinguished in flight by nighthawklike white bands on the tips of its wings and a white stripe down each side of the tail.

POOR-WILL
Phalaenoptilus nuttallii   **31:9**

Description
Size, 7–8½ in. (17.8–21.6 cm). Most easily recognized by night call. Very like Whip-poor-will; mottled gray-brown; breast very dark; wings rounded; tail short, rounded; throat white; only small areas of white in corners of tail, less in female.

Similarities
Nighthawks are larger; pointed wings with conspicuous white bar.
Habitat
Open country, arid or sparsely wooded hills.
Habits
When flushed, flutters up like large gray-brown moth; rests by day on ground, sometimes in low tree; flight low, mothlike; pink eyes often shine in headlights from roadsides.
Voice
At a distance at night sounds like *poor-WILL* or farther off, *p'will;* a loud, repeated *poor-jill* or close up, *poor-jill-ip.*
Food
Nocturnal flying insects.
Eggs
2; white; 1.1 x 0.8 in. (2.8 x 2.0 cm). No nest; eggs laid on bare ground, rock, or in gravel.
Range
Breeds from se. B.C. and Alta. s. throughout w. U.S.; winters in s. U.S. and Mexico.

COMMON NIGHTHAWK
Chordeiles minor   **31:8**

Description
Size, 8½–10 in. (21.6–25.4 cm). Usually seen high in sky with white bar on its long, slim wings. Gray-brown, mottled; sides pale, barred; wings longer, more pointed than Whip-poor-will's, extend beyond end of forked tail. Male: • *broad white band* high on throat and white bar across notched tail. Female: throat bar is buffy.
Similarities
Lesser Nighthawk has different call, white wing bar closer to wing tip.
Habitat
Plains, mountains, pine forests, cities.
Habits
Often seen by day, as well as at dusk and night; flies high in bouncing erratic manner hawking after insects in spectacular aerial maneuvers; also swoops low; seen in migration in loose flocks of 20–100 birds.
Voice
Harsh *peenk,* given in flight with 3 double-speed flips of wings; in courtship when mate dives steeply only to zoom up sharply with a sudden deep whir, it gives a booming sound.
Food
Diverse insects, including mosquitoes.
Eggs
2; grayish-white, speckled with brown; 1.2 x 0.9 in. (3.0 x 2.3 cm). No nest; lays eggs on bare ground or gravel roofs in towns and cities.
Range
Breeds throughout most of N. America from s. Yukon, e. across Canada, and s. to s. U.S., excluding sw. deserts; winters s. of U.S.

LESSER NIGHTHAWK
Chordeiles acutipennis   **31:6**

Description
Size, 8–9 in. (20.3–22.9 cm). Lowland bird, smaller than Common Nighthawk, most easily identified by manner of flight and peculiar calls.

Similarities
Common has similar color and pattern, but white wing bar is farther from the tip.
Habitat
Lowland open scrub, gravelly deserts, prairies, dry range, fields.
Habits
Flies very low; does not dive from altitude.
Voice
Low *chuck chuck,* a whinnying trill, or soft purring.
Food
Insects, including beetles.
Eggs
2; white, spotted; 1.0 x 0.7 in. (2.6 x 1.8 cm). No nest, lays on bare ground.
Range
Breeds from sw. U.S., including s. Calif., Ariz., Nev., Utah., N.Mex., and Tex., s. to S. America; winters s. of U.S.

Swifts and Hummingbirds
Order Apodiformes

SWIFTS
Family Apodidae

The swallowlike swifts have short bills; long, narrow, scythelike wings; short heads and tails; and small, weak feet. They are extremely swift fliers, sailing between spurts, wings rigidly convexed. On the wing constantly, they catch insects in flight, drink and mate on the wing.

BLACK SWIFT
Cypseloides niger   **31:4**

Description
Size, 7–7½ in. (17.8–19.1 cm). Large, • *all-black* except for small splash of white on forehead, visible at close range; tail slightly forked and sometimes fanned.
Similarities
Purple Martin has wider wings, differently shaped and swallow-like flight. Vaux's is smaller, with paler breast, rounded tail.
Habitat
Sky, mountains, coastal bluffs.
Habits
Flight slower than other swifts.
Voice
A "sharp *plik-plik-plik-plik*" (Cogswell).
Eggs
1; white; 1.1 x 0.7 in. (2.8 x 1.9 cm). Nest is of damp moss (or algae) in cliff crevice of mountain or coast, often behind a waterfall.
Range
Resident along Pacific Coast from Alaska s. to Calif. and inland to s. Rocky Mountains.

CHIMNEY SWIFT
Chaetura pelagica

Description
Size, 4¾–5½ in. (12.1–14.0 cm). Small, all-dark, like flying cigar. Wings sicklelike, tail squared, plumage sooty throughout.

Similarities
Vaux's is larger.
Habitat
Sky, usually near habitations.
Habits
Flies and sails alternately in bold sweeps, often high up; 3 birds
often fly together; roosts clinging in chimney, well, cave, supported
by its stiff tail.
Voice
Characteristic repeated twittering in flight.
Eggs
4–6; white; 0.7 x 0.5 in. (1.8 x 1.3 cm). Nest is of twigs, in
chimney or hollow tree.
Range
Breeds chiefly in E., with some locally in Sask. and foothills of
Rockies; winters in S. America.

VAUX'S SWIFT
Chaetura vauxi   31:3

Description
Size, 4–4½ in. (10.2–11.4 cm). Small; dark above, dingy below;
swallowlike or "a cigar with wings"; tail tiny; wings long, slightly
bowed, held stiff; throat pale.
Similarities
Chimney is slightly larger and darker below, east of Rockies.
Habitat
Sky, forest openings and burns.
Voice
Indistinct feeble chipping.
Eggs
3–5; white; 0.7 x 0.4 in. (1.8 x 1.2 cm). Nest is in hollow tree
(rarely chimney), as bracket of twigs glued to side.
Range
Breeds from s. Alaska to cen. Calif. along Pacific Coast; winters s.
of U.S.; migrates along coastline.

WHITE-THROATED SWIFT
Aeronautes saxatalis   31:5

Description
Size, 6–7 in. (15.2–17.8 cm). White-throated, • *black-and-white
patterned;* black above and on flanks; throat, breast, and patches on
wing and sides of rump white; tail slightly forked.
Similarities
Violet-green Swallow is all-white below, with slower wingbeats.
Habitat
Sky over cliffs, foothills, and adjacent valleys; cruises widely.
Habits
Flight very swift (to about 200 mph, or 322 km/hr), and erratic,
usually high in air.
Voice
Shrill laughing *he-he-he-he* or descending *jejejejeje.*
Eggs
3–6; white; 0.8 x 0.5 in. (2.0 x 1.3 cm). Nest is twiggy bracket
glued to side of a crevice or cave in precipitous cliff.
Range
Breeds in interior from B.C., s. to Calif. and Sw.; winters from
cen. Calif., s. Ariz., and w. Tex. s. to C. America.

HUMMINGBIRDS
Family Trochilidae

This family includes the smallest birds—those with the fastest wingbeat and the only ones that can fly backward or vertically. Often called "hummers," they have long, needlelike bills; extensible tongues; partly iridescent plumage; and small weak feet. They frequent flowers, often make a humming noise with their wings, which are so fast as to appear blurred (55–80 complete wingbeats per second, at speeds of forty-five to seventy-five miles per hour, or 72.4–120.7 km/hr); though tiny, they are pugnacious and will attack crows, hawks, and even eagles. They hover when feeding; may perch on a twig, flower stem, or wire. Jewellike gorgets (throat feathers) adorn most males; females are less colorful, usually greenish above, whitish below and lacking gorgets. Although late summer and autumn mountain meadows sparkle with young hummers, they are extremely difficult to identify by species. They feed on nectar and small insects.

LUCIFER HUMMINGBIRD
Calothorax lucifer   **39:13**

Description
Size, 3¾ in. (9.5 cm). • *Bill decurved.* Male: • *throat purple, sides rusty,* crown green, tail deeply forked. Female: underparts uniform buff; recognizable by decurved bill.
Similarities
Costa's resembles Lucifer but does not occur with it.
Habitat
Desert slopes, agaves.
Habits
Much as other hummingbirds.
Voice
Squeaking notes.
Food
Various small insects, probably nectar.
Eggs
2; white; 0.5 x 0.3 in. (1.3 x 1.0 cm).
Range
Breeds from w. Tex. (Chisos Mts.), s. to Mexico; winters s. of U.S.; accidental in summer to se. Ariz.

RUBY-THROATED HUMMINGBIRD
Archilochus colubris   **38:16**

Description
Size, 3–3¾ in. (7.6–9.5 cm). Male: • *ruby throat,* green back, forked tail. Female: throat and tips of outer tail feathers white, tail rounded. Lacks rufous in tail.
Similarities
Broad-tailed male has rose-red throat, tail weakly forked. Male Anna's has red crown.
Habitat
Gardens, woodland edges.
Habits
Flight relatively silent; readily frequents sugar-water feeding stations.
Voice
Shrill squeals, chirps, chippering.

Eggs
2; white; 0.5 x 0.3 in. (1.3 x 0.8 cm). Nest is lichen-covered cup on branch.
Range
Breeds chiefly in E. with some w. to Alta. and Gulf Coast to Tex.; winters s. of U.S.; migrates through Great Plains.

BLACK-CHINNED HUMMINGBIRD
Archilochus alexandri   **39:12**

Description
Size, 3⅓–3¾ in. (8.5–9.5 cm). Male: • *throat black* with conspicuous white collar; band on lower throat blue-violet, visible when light is right. Female: greenish above, whitish below.
Similarities
Costa's and Ruby-throated females are impossible to distinguish in field from female Black-chinned.
Habitat
Foothill suburbs, semiarid regions near water or thinly wooded canyons, river timberlands, chaparral.
Habits
Male's aerial display is a shallow back-and-forth swoop and whir.
Voice
Thin chippering.
Eggs
2; white; 0.4 x 0.3 in. (1.3 x 0.8 cm). Nest is tiny cup in tree or shrub.
Range
Breeds from B.C., s. throughout W. to Mexico, excluding nw. Pacific Coast; winters in Mexico.

ANNA'S HUMMINGBIRD
Calypte anna   **39:17**

Description
Size, 3½–4 in. (8.9–10.2 cm). Male: • *crown and throat red,* metallic bronze-green above, tail dusky black. Female: bronze-green above, white below; sides grayish-green; throat more heavily spotted than in Costa's or Black-chinned, often with central patch of red spots. Immature males of other species east of California may show similar throat spots; Anna's is usually a little larger than other *Calypte* species.
Similarities
Adult male Broad-tailed, east of Sierra Nevada, has no red crown.
Habitat
Gardens, open woods, chaparral. Only winter hummer in California.
Habits
Often common about homes.
Voice
When feeding, a *chip-chip-chip* or *chick;* when perched, a series of squeaks; during aerial display, male reaches bottom of pendulum arc with a sharp *pop.*
Eggs
2; white; 0.5 x 0.3 in. (1.3 x 0.9 cm). Nest is lichen-covered cup in tree or bush.
Range
Resident from s. Oreg. s. to n. Baja Calif.; casual to Vancouver; inland populations migratory.

COSTA'S HUMMINGBIRD
Calypte costae   **39:11**

Description
Size, 3–3½ in. (7.6–8.9 cm). Male: • *throat and crown purple or amethyst,* gorget feathers project greatly at sides. Female: see Black-chinned.

Similarities
Male Anna's is larger; throat and crown rose-red. Male Black-chinned has limited blue-purple on throat, none on crown.

Habitat
Deserts.

Habits
May flycatch for insects.

Voice
Soft *chick;* displaying male may utter a ventriloquial hiss at bottom of U-shaped arc.

Eggs
2; white; 0.4 x 0.3 in. (1.1 x 0.8 cm). Nest is in desert shrub or tree, lichen-thatched.

Range
Breeds in sw. U.S. from cen. Calif. and s. Nev., to Mexico; winters from se. Calif., s. Ariz., southward.

BROAD-TAILED HUMMINGBIRD
Selasphorus platycercus   **39:16**

Description
Size, 4–4½ in. (10–11.2 cm). Principally a hummer of mountain regions, especially Rockies. Male: back green; • *throat metallic reddish-purple or bright rose-red;* most easily recognized by the shrill, trilling sound of its wings in flight. Female: bronze-green above, underparts and tips of outer tail feathers white, some rufous near base of outer tail, but not in center of rump.

Similarities
Male Rufous has bright orange-red throat. Male Allen's has mainly rufous plumage. Male Anna's is different range, red crown. Ruby-throated has redder throat, forked tail. For females, compare Rufous, Allen's, and Calliope.

Habitat
Alpine meadows, glades, open underbrush, willows, foothills, coniferous forests and edges.

Voice
High, squeaky *chip.*

Eggs
2; white; 0.5 x 0.3 in. (1.3 x 0.8 cm). Nest is lichen-covered cup in tree or bush.

Range
Breeds from e. Calif. e. to n. Wyo., Great Basin, and Rockies, s. to Mexico; winters s. of U.S.

RUFOUS HUMMINGBIRD
Selasphorus rufus   **39:18**

Description
Size, 3⅓–4 in. (8.2–10 cm). • *Rufous black,* red throat. Male: upperparts bright reddish-brown (not iridescent), throat scarlet, chest white, underparts pale rufous. Female: back green, sides dull rufous, underparts light rufous, base of tail and rump rufous.

Similarities
Male Allen's is green in middle of back. Female Allen's is indistinguishable in field. Female Broad-tailed is slightly rufous on flanks and in tail base but not rump.
Habitat
Forest edges; flowering areas, habitations.
Habits
Unusually pugnacious; male's aerial display flight a closed ellipse.
Voice
Squeaks, *chip*'s, a low double *chirp,* a high sharp *bzee;* "sound on aerial dive of male, a strident stuttered *v-v-v-v-vrip*" (Cogswell).
Eggs
2; white; 0.5 x 0.3 in. (1.3 x 0.8 cm). Nest is lichen-covered cup in tree or bush.
Range
Breeds from Alaska s. to Oreg.; winters in Mexico; migrates through s. Calif.

ALLEN'S HUMMINGBIRD
Selasphorus sasin   **39:15**

Description
Size, 3⅓ in. (8.5 cm). Male: metallic green above; sides, rump, tail, and cheeks rufous; throat fiery orange-red; breast whitish. Female: variably marked with gray, white, reddish-brown.
Similarities
Male Rufous has rufous back; female Rufous is indistinguishable from Allen's except in hand. (Rufous has wider outer tail feathers.)
Habitat
Parks, gardens, vegetated canyons, mountain meadows in late summer.
Habits
Male's pendulum display is a shallow arc with tail feathers producing a buzzing, then a steep wavering climb (80–150 ft., or 24.4–45.7 m), followed by an abrupt descent with an "air-splitting *vrrip* at the 'focus'; then flies off" (Cogswell).
Voice
Similar to Rufous.
Eggs
2; white; 0.5 x 0.3 in. (1.3 x 0.9 cm). Nest is lichen cup in tree or bush.
Range
Breeds along Pacific Coast from s. Oreg. to s. Calif.; winters in Mexico; migrates through Calif., e. to Ariz.

CALLIOPE HUMMINGBIRD
Stellula calliope   **39:14**

Description
Size, 2¾–3½ in. (7.0–8.9 cm). Smallest U.S. hummingbird, usually in mountains. Male: above golden-green, below white marked with reddish-brown and lavender; • *throat white, streaked with reddish-purple rays to form dark inverted V.* Female: sides buffy, base of tail rufous.
Similarities
Female Broad-tailed is much larger, more rufous on sides. Female Rufous is difficult to separate, but is larger and rustier on sides and has some rusty on central tail feathers.
Habitat
Mountains, canyons, meadows.

Habits
Displaying male plunges in shallow U-shaped arc.
Voice
At bottom of dive, a brief *pfft;* when feeding, *tsip.*
Eggs
2; white; 0.4 x 0.3 in. (1.1 x 0.8 cm). Nest is lichen cup in tree or bush.
Range
Breeds from s. B.C. s. along coast to Baja Calif. and inland from Alta. to Wyo.; winters in Mexico; migrates through mountains of Ariz. and N.Mex.

RIVOLI'S HUMMINGBIRD
Eugenes fulgens   39:7

Description
Size, 4½–5 in. (11.4–12.7 cm). Unusually large. Male: unmistakable, appears all-black at distance; above dull green, • *belly blackish, crown purple, throat bright green.* Female: above greenish; underparts dusky or heavily washed with pale gray corners.
Similarities
Female Blue-throated has blue-black tail spotted with large white corner spots.
Habitat
Mountains, canyons, meadows.
Habits
Wingbeats slow, discernible; may come to sugar-water feeders.
Voice
Thin, sharp *chip.*
Eggs
2; white; 0.6 x 0.3 in. (1.6 x 0.8 cm). Nest is lichen-covered cup in shrub.
Range
Breeds from se. Ariz. e. to w. Tex., and s. to C. America; winters s. of U.S.

BROAD-BILLED HUMMINGBIRD
Cynanthus latirostris   39:10

Description
Size, 3½–4 in. (8.9–10.2 cm). Male: appears all-black at distance; above and below dark green, crown green, • *throat blue, bill bright red* with black tip. Female: throat and underparts • *unmarked pearly gray,* bill like males but duller. Females of most species show some spots on throat.
Similarities
Larger Rivoli's lacks red bill.
Habitat
Agaves, mesquite, desert canyons, and mountains.
Habits
As others.
Voice
Chattering. Male's display hum is high-pitched with "the zing of a rifle bullet" (Willard).
Eggs
2; white; 0.5 x 0.3 in. (1.3 x 0.8 cm). Nest is rough cup on vertical branch near stream.
Range
Breeds from se. Ariz. e. to w. Tex., southward; winters in Mexico.

BLUE-THROATED HUMMINGBIRD
Lampornis clemenciae 39:3

Description
Size, 4½–5¼ in. (11.4–13.3 cm). Very large with • *large white patches* on its dark tail. Male: only U.S. male hummer with white tail spots; throat light blue, black eye stripe bordered by white streaks. Female: underparts uniformly pale gray; face shows white lines above and below eye; large tail, blue-black, with big white corners.

Similarities
Female Rivoli's has mainly green tail, with small pale corners.

Habitat
Mountain canyons containing woods and water.

Habits
Aggressive, drives away other hummingbirds.

Voice
Squeaky *seek*.

Eggs
2; white, 0.5 x 0.3 in. (1.3 x 0.8 cm). Nest is near water, as a feltlike cup on vertical support, often under a bridge.

Range
Breeds from se. Ariz. e. to w. Tex., southward; winters in Mexico.

Note: The **VIOLET-CROWNED HUMMINGBIRD,** *Amazilia verticalis,* size, 3¾–4¼ in. (9.5–10.8 cm), is similar in both sexes. The underparts are white and sharply contrast with the brownish back and violet-blue (male) or greenish-blue (female, immature) crown. The bill is red with a black tip. It frequents streamside woods in extreme southeast Arizona and southwest New Mexico, where it breeds irregularly in summer, and is occasional to the Huachuca and Chiricahua Mountains in southeast Arizona.

Trogons
Order Trogoniformes

TROGONS
Family Trogonidae

This family and order constitute colorful tropical forest birds, usually solitary, with short, stout, dentate bills; short necks; and long, truncated tails.

ELEGANT TROGON
Trogon elegans 33:11

Description
Size, 11–12 in. (27.9–30.5 cm). Male: head, upperparts, chest dark shiny-green; • *belly geranium-red;* narrow white band across breast; tail rather long, square tipped; bill pale. Female: patterned like male but duller, brown above; white in cheeks; white undertail barred and spotted black.

Habitat
Desert mountains, sycamore canyons, pine-oak forests.

Habits
Stolid, perches quietly; gleans some food by hovering in front of leaf clusters.

Voice
Low, "coarse notes . . . *kowm kowm* . . . or *koa koa, koa* . . ."
(Peterson).
Food
Insects, various invertebrates, fruit.
Eggs
3–4; bluish or white; 1.1 x 0.9 in. (2.8 x 2.3 cm). Nest is in hollow
tree.
Range
Breeds from se. Ariz. e. to w. Tex., southward; winters s. of U.S.

Kingfishers
Order Coraciiformes

KINGFISHERS
Family Alcedinidae

These are solitary fishing birds with strong, straight, pointed bills
longer than their large heads; big eyes; short tails; small, weak feet
on small legs. They fish from a perch above water or by hovering
and diving headfirst, feeding on fish, some insects, and lizards. The
scales and bones are ejected in pellets.

BELTED KINGFISHER
Ceryle alcyon   37:10

Description
Size, 11–14½ in. (27.4–36.8 cm). Dives into water; crested, looks
top-heavy. Male: upperparts and breastband gray-blue, underparts
and collar white. Female: second band of chestnut on lower chest
and flanks.
Habitat
Shorelines of rivers, lakes, ponds, coast, bays.
Habits
Each bird has own territory along watercourses with a series of
observation posts; hovers before diving; flight straight, with uneven
wingbeats.
Voice
Loud, high rattle, often heard in flight.
Eggs
5–8; white; 1.3 x 1.1 in. (3.3 x 2.8 cm). Nest is chamber at end of
4–8 ft. (1.2–2.4 m) burrow in riverbank.
Range
Breeds from cen. Alaska e. across Canada, and s. to Mexico and
Gulf Coast; winters from n. Canada s. to C. America.

GREEN KINGFISHER
Chloroceryle americana   37:13

Description
Size, 7–8½ in. (17.8–21.6 cm). Small; typical kingfisher shape,
green coloring. Above deep green spotted with white, collar and
underparts white. Male: • *breastband rusty.* Female:1–2 greenish
breastbands.
Habitat
Watercourses, wet or marshy.

Habits
Flight direct, buzzy.
Voice
Sharp squeak; a sharp *tick, tick, tick.*
Eggs
3–6; white; 0.9 x 0.7 in. (2.3 x 1.8 cm).
Range
Resident from s. Tex. s. to S. America; casual to s. Ariz. and
N.Mex.

Woodpeckers and Allies
Order Piciformes

WOODPECKERS
Family Picidae

Woodpeckers are medium-sized, wood-boring birds with strong
skulls, chisel-like pointed bills, remarkably extensible tongues, stiff
spiny tails which assist in climbing trees, and zygodactyl feet (i.e.,
two toes pointing forward, two back). Males of most species have
red or yellow on the head. Virtually all are arboreal; the bird feeds
largely by clinging to the bark of a tree with its feet, bracing itself
with its tail, and chiseling at the wood with its beak in search of
wood-boring insects. They excavate nests in trees, drum as a form
of communication, and have undulating flights.

COMMON FLICKER
Colaptes auratus   **32:2, 33:1**

Description
Size, 10–14 in. (25.4–35.6 cm). Ground-feeding; barred-backed,
white rumped. All forms are barred brown above with • *white
rump,* black-spotted underparts, and black crescent on breast.
• *Bright yellow or orange-pink underwing* and undertail surfaces;
red nape patch, gray crown, tan cheeks or brown crown and nape,
gray cheeks and throat. Male: "mustache," black or red. Female:
lacks mustache or has brownish mustache. Hybrids are common
and show various mixtures and combinations of these features.
Habitat
Needs open ground to feed; beneath evergreen forest trees, in
woods, streamsides, farms, suburbs, and deserts.
Habits
Feeds largely on ants taken on ground; drums on tree or metal
roofing; has conspicuous displays, bowing, swinging and flashing
wings and tail in 2's or 3's.
Voice
Long series of *wik* notes; alarm, *pee-ah;* a repetitive *wick-a, wick-a*
or *wick-up, wick-up* during displays; many others.
Food
Ants, other insects, fruits, berries.
Eggs
3–8; white; 1.0 x 0.8 in. (2.5 x 2.0 cm). Nest is hole excavated in
tree, stump, post, cactus; rarely in nest box.
Remarks
Several forms hybridize widely and now are treated as a single
species: the eastern and northern former Yellow-shafted Flicker,
the western Red-shafted Flicker, and the southwestern desert
Gilded Flicker.

Range

Resident throughout most of N. America from n. Canada s. to
Mexico; northernmost populations winter s. from B.C.; hybrids of
Yellow-shafted and Red-shafted are widespread, but especially
B.C., e. to sw. Sask., and s. to Mont., e. of Rocky Mountains;
Gilded group are most common in sw. deserts, hybridizing with
Red-shafteds in Ariz. valleys.

PILEATED WOODPECKER

Dryocopus pileatus   **32:1**

Description

Size, 16–19 in. (40.6–49.5 cm). Crow-sized, spectacularly black
with • *red crest.* Male: underwing in flight flashes white; big, heavy
bill; white stripe on thin neck; red "mustache." Female: less red on
crest, no red mustache, forehead blackish.

Habitat

Mixed and conifer forests.

Habits

Flight vigorous with sweeping wingbeats, slow, either straight and
crowlike or in long undulations. Strips quantities of bark off trees
and chisels big rectangular holes in dead trees.

Voice

A *yuk-yuk-yuk,* louder, more hollow, slower than Common
Flicker's; in flight, a slow *puck, puck;* an "irregular *kik—kik—
kikkik—kik-kik*" (Peterson).

Food

Wood-boring beetles and ants, berries, nuts.

Eggs

3–6; white; 1.4 x 1.0 in. (3.6 x 2.5 cm).

Remarks

This big woodpecker seems gradually adapting itself to well-
wooded suburbs, often near large cities.

Range

Resident across n. Canada and along Pacific Coast from n. B.C., s.
to cen. Calif., inland in wooded parts of Prairie provinces;
accidental to Utah and n. Ariz.

GILA WOODPECKER

Melanerpes uropygialis   **33:6**

Description

Size, 8–10 in. (20.3-25.4 cm). Male: barred ("zebra-backed") with
• *round red cap* and white wing patch visible in flight; upperparts
finely barred black and white, underparts and head plain grayish-
brown. Female: lacks red cap.

Similarities

Flicker is brown. Ladder-backed Woodpecker has striped face, no
white wing patch.

Habitat

Desert watercourses and groves, saguaros, suburban areas,
cottonwoods.

Habits

Vocal and common in its habitat.

Voice

Sharp *yip* or *pit;* "a rolling *churr*" (Peterson).

Food

Diverse insects, berries, fruit; will eat honey.

Eggs
3–5; white; 1.0 x 0.7 in. (2.6 x 1.8 cm). Nest is cavity in tree or saguaro.
Range
Resident from se. Calif. to Mexico.

Note: The **RED-BELLIED WOODPECKER**, *Melanerpus carolinus* **(32:5)**, size, 10 in. (25.4 cm), can be seen in central Texas and eastern Colorado. Males have a fully red crown, females a red nape. The **GOLDEN-FRONTED WOODPECKER**, *Melanerpes aurifrons* **(33:5)**, size, 8½–10 in. (21.6–25.4 cm), can be seen in southern Texas. Males of this species have a red forecrown and orange-gold nape; females have only the golden nape patch.

RED-HEADED WOODPECKER
Melanerpes erythrocephalus   **32:10**

Description
Size, 8½–9½ in. (21.6–24.1 cm). Black-and-white, with • *all-red head and neck*. Adult: above, blue-black; white rump and underparts; conspicuous white wing patch; when perched, lower back appears white. Immature: brown barred above with streaky gray-brown head; white wing patch; white rump; white uppertail coverts.
Similarities
None within its range.
Habitat
Orchards, roadsides, farmlands, broken woods.
Habits
Quarrelsome; catches insects in air like a flycatcher; stores nuts in cracks and cavities; tends to drop from tree almost to ground and then fly low.
Voice
Higher-pitched *churr, churr* than Red-bellied; *"ker-r-r-ruck, ker-ruck-ruck-ruck"* (Merriam); "a loud *queer* or *queeoh*" (Peterson).
Food
Beechnuts, acorns, insects; fruit, occasionally eats young birds.
Eggs
4–6; white; 1.0 x 0.8 in. (2.5 x 2.0 cm). Nest is cavity in tree, stump, or pole.
Range
Breeds from Sask. and Man. s. to N.Mex.; winters in S. America.

ACORN WOODPECKER
Melanerpes formicivorus   **33:7**

Description
Size, 8–9½ in. (20.3–24.1 cm). Back black; head distinctively patterned with black, white, and red; rump patch large and white; white wing patch in flight. Sexes alike with red in crown and whitish eyes.
Habitat
Wooded hills, oak-pine slopes and canyons, oak groves; common.
Habits
Stores acorns in rough tree bark. Social, often in groups.
Voice
Easily recognizable *ja-cob, ja-cob* or *whack-up, whack-up, whack-up.*

Food
Omnivorous; insects, other invertebrates, fruit, acorns, other nuts, seeds.

Eggs
4–5; white; 1.0 x 0.7 in. (2.6 x 1.8 cm).

Range
Resident along Pacific Coast from s. Oreg. s. to Baja Calif., and inland from Ariz., N.Mex. and w. Tex. s. to S. America.

LEWIS' WOODPECKER
Melanerpes lewis   32:9

Description
Size, 10½–11½ in. (26.7–29.2 cm). Extensive pink belly. Above black; cheeks red; breast and collar gray; appears all-black at distance; wings wide, black.

Habitat
Mountains, open woods, edges, suburbs, towns, burns, logged-over land.

Habits
Flight direct, strong, seldom undulates; glides, soars, catches insects in air; often perches crosswise; stores acorns; gregarious in fall.

Voice
"A harsh *chirr* and a high-pitched squalling *chee-up*" (Hoffmann).

Food
Diverse insects, berries, other fruit, acorns.

Eggs
6–8; white; 1.0 x 0.8 in. (2.5 x 2.0 cm). Nest is cavity in tree or snag.

Range
Breeds from s. B.C. and Alta. s. to s.-cen. Calif. and n. Ariz.; winters from Oreg. to Colo. and s. to Mexico, occasionally along coast.

RED-BREASTED SAPSUCKER
Sphyrapicus ruber   33:8

Description
Size, 8½–9½ in. (21.6–24.1 cm). Both sexes have bright red head and breast, with other markings similar to Red-naped.

Habitat
Forests, groves, wood lots, orchards.

Habits, Voice, Food, Eggs
Same as Red-naped.

Remarks
Formerly a subspecies of Yellow-bellied Sapsucker *(Sphyrapicus varius)*.

Range
Breeds from se. Alaska s. along Pacific Coast to n. Calif. and in Cascade–Sierra mountains s. to high mountains of s. Calif; winters along Pacific Coast to Baja Calif.; casual to Ariz.

RED-NAPED SAPSUCKER
Sphyrapicus nuchalis   32:8

Description
Size, 8–9 in. (20.3–22.4 cm). • *Red forehead*, black upper breast; • *long white wing patch*, visible at rest. Male adult: red forehead and throat. Female adult: throat fully to partly red. Immature: barred and streaked brownish; identifiable by wing patch.

Similarities
Red-breasted has red head and breast.
Habitat
Forests, groves, wood lots, orchards.
Habits
Perforates bark of trees with even rows of small holes to get sap, perches against bark at 45° angle when feeding.
Voice
Noisy in spring, with nesting drumming distinctive—several rapid thumps followed by several slow, rhythmic beats; discordant and varied calls and squawks; a ringing *cleur* 5–6 times.
Food
Insects, sap.
Eggs
4–7; white; 0.9 x 0.7 in. (2.3 x 1.8 cm). Nest is tree cavity.
Remarks
Formerly a subspecies of Yellow-bellied Sapsucker *(Sphyrapicus varius)*.
Range
Breeds from s.-cen. B.C. and cen. Mackenzie, s. to ne. Calif., n. Nev., cen. Ariz., and s. N.Mex.

WILLIAMSON'S SAPSUCKER
Sphyrapicus thyroideus   **33:9**

Description
Size, 9½ in. (24.1 cm). Sexes very unlike, the yellow belly being the only common feature. Male: appears black in flight with white rump and shoulder patches, upperparts and crown black, face striped with white, throat shows narrow bright-red patch. Female: upperparts "zebra-striped," brownish; head brown; breast usually with black patch; sides barred; rump white.
Similarities
Other "zebra-backed" woodpecker females do not have brown head, breast patch.
Habitat
Conifer forests, burns.
Habits
Often perches quietly, unobtrusive.
Voice
Loud *kee-er, bee-er;* also *wik* and other notes.
Food
Insects, especially ants, sap.
Eggs
3–7; white; 0.9 x 0.6 in. (2.3 x 1.6 cm). Nest is tree cavity.
Range
Breeds from se. B.C. to N.Mex., excluding coastline; winters from Ariz. s. to Mexico.

HAIRY WOODPECKER
Picoides villosus   **32:4**

Description
Size, 8½–10½ in. (21.6–26.7 cm). • *White-backed,* with big bill. Black-and-white pattern above and on wings (spotted), white down center of back and below; 3 outer tail feathers usually unmarked white. Length of heavy bill twice distance from base of bill to eye. Male has red on back of head. Birds east of Rockies whiter; those of Pacific slope darker.

Similarities
Downy is almost identical, except much smaller, especially smaller bill; black spots on white outer tail feathers. Three-toed may have white back but cap is yellow.
Habitat
Woods, groves.
Habits
Tall-woods counterpart of Downy; sometimes travels in mixed groups with Downies, nuthatches, and chickadees; nearer habitations in winter; shyer, noisier, more restless than Downy.
Voice
Louder *peek* than Downy's; a loud rattle on 1 pitch; drumming usually louder than Downy's, but often indistinguishable.
Food
Borers, caterpillars, other insects.
Eggs
3–6; white; 0.9 x 0.7 in. (2.3 x 1.8 cm). Nest is hole in tree or stump.
Range
Resident throughout much of N. America from n. Canada s. to C. America; in winter, withdraws from colder mountain areas to lower elevations.

DOWNY WOODPECKER
Picoides pubescens 32:3

Description
Size, 6–7 in. (15.2–17.8 cm). White-backed, with small bill; sparrow-sized edition of Hairy, but 3 outer tail feathers have black bars; bill length equals distance from base of bill to eye. Commonest small woodpecker.
Similarities
Hairy is larger. Ladder-backed is "zebra-backed," face striped.
Habitat
Open, low woods, usually near water; groves; orchards; trees about habitations.
Habits
Easily attracted to feeders, tame; often travels in mixed flocks.
Voice
Short, sharp *pik;* a rattle of 12–15 rapid staccato notes, descending; drums in a long, unbroken roll.
Food
Borers, surface insects; some vegetation.
Eggs
4–7; white; 0.7 x 0.6 in. (1.8 x 1.5 cm). Nest is tree cavity.
Range
Resident throughout woodlands of Canada and n. U.S., from Alaska and n. Canada, s. to s. Calif. and Gulf Coast.

LADDER-BACKED WOODPECKER
Picoides scalaris 33:3

Description
Size, 6–7½ in. (15.2–19.1 cm). Common desert bird. Black-and-white "zebra-back" with • *black-and-white striped face.* Male has red-spotted crown.
Similarities
None in its range. Nuttall's west of Calif. Sierrras is blacker overall, male red restricted to hind crown and nape.

Habitat
Deserts, arid canyons, prairie groves, wood edges, foothills.
Habits
Often very vocal; may feed on tiny plants, even going to ground.
Voice
Thin high *peek;* a rattle similar to Downy's.
Food
Mainly insects, some fruit.
Eggs
4–5; white; 0.8 x 0.6 in. (2.0 x 1.5 cm). Nest is hole in agave, cactus, yucca, tree, post.
Range
Resident from s. Calif. e. to Colo., and s. to Baja Calif.

STRICKLAND'S (ARIZONA) WOODPECKER
Picoides stricklandi   33:4

Description
Size, 7–8 in. (17.8–20.3 cm). Only American woodpecker with solid • *brown back.* Upperparts brown, underparts barred and spotted, • *face striped with white;* male has red patch on nape.
Similarities
Common Flicker has barred back, white rump.
Habitat
Pine-oak forested mountains, canyons.
Habits
A loud, pecking woodpecker, often heard before seen.
Voice
Like Hairy's; a hoarse whinny; a sharp *tseek* or *spik* drums loudly.
Food
Mainly insects, some fruit.
Eggs
3–4; white; 0.9 x 0.6 in. (2.3 x 1.6 cm). Nest is tree branch (usually oak) cavity.
Range
Resident in se. Ariz. and sw. N.Mex., s. to Mexico.

WHITE-HEADED WOODPECKER
Picoides albolarvatus   33:10

Description
Size, 9 in. (22.9 cm). • *All-white head,* body and bill black, large white wing patch. Male shows red patch on nape.
Habitat
Fir and pine forests.
Habits
Feeds, often quietly, high in pines.
Voice
Repetitive sharp *chik-ik;* a rattle similar to Hairy's.
Food
Insects, spiders, pine seeds, some fruit.
Eggs
3–5; white; 1.9 x 0.7 in. (4.9 x 1.8 cm). Nest is cavity in snag.
Range
Resident from ne. Wash. and Idaho s. to s. Calif. and Nev., excluding nw. Pacific Coast.

NUTTALL'S WOODPECKER
Picoides nuttallii   33:2

Description
Size, 7–7½ in. (17.8–19.1 cm). Very similar to Ladder-backed,
usually not in Calif. west of Sierras. Male: black-and-white stripes
on face, red cap; black rear of crown connects with black upper
back; face black; nostril feathers pure white.
Similarities
Generally none in its range except Hairy and Downy, which have
white, unbarred back. Meets and occasionally hybridizes with
Ladder-backed in southern California and adjacent Mexico;
Ladder-backed has more red on head.
Habitat
Pine-oak woods, foothills, river groves, orchards.
Habits
Closely associated with oak trees.
Voice
Loud *pi-tit;* a high rattling, also drums.
Food
Mainly insects; some berries, fruit.
Eggs
4–5; white; 0.8 x 0.6 in. (2.0 x 1.6 cm).
Range
Resident in Calif. from n. Calif. to Baja Calif.; casual to s. Oreg.;
accidental to Ariz.

THREE-TOED WOODPECKER
Picoides tridactylus   32:6

Description
Size, 8–9½ in. (20.3–24.1 cm). Barred back ("ladder-backed"), or
white back and barred sides. Male has • *yellow crown patch* absent
in female. In flight, • *barred back* conspicuous and tail flashes
white.
Similarities
Black-backed has solid black back. Female Williamson's Sapsucker
has back and sides barred.
Habitat
Coniferous forests, especially spruce bogs.
Habits
Solitary, unsuspicious, sedentary; rarer, more local, less noisy;
flight less vigorous than Black-backed's.
Voice
"A loud *quip* or *queep*" (Farley); a rattle like Hairy's, but softer.
Food
Beetle larvae and other bark insects.
Eggs
4; white; 1.0 x 0.7 in. (2.5 x 1.8 cm). Nest is hole in conifer snag.
Former names
American or Northern Three-toed Woodpecker.
Range
Resident from Alaska, e. across Canada and s. to Oreg., and to
Ariz. and N.Mex.

BLACK-BACKED WOODPECKER
Picoides arcticus   32:7

Description
Size, 9–10 in. (22.9–25.4 cm). • *Solid black back* and barred sides;
male has • *yellow crown patch.*

Similarities
Three-toed is smaller; back barred or white. Male Williamson's Sapsucker has black crown, red throat.
Habitat
Boreal forests, often near water.
Habits
Solitary, unsuspicious, movements deliberate; makes periodic winter incursions south of normal range; presence in woods revealed by freshly debarked patches on dead conifers.
Voice
"A sharp shrill *chirk, chirk*" (Hardy); "*w-e-ea* . . . shrill and clear" (Knight); long screaming rattle.
Food
Mostly subsurface arboreal insects.
Eggs
4–5; white; 0.9 x 0.7 in. (2.3 x 1.8 cm). Nest is excavation in snag, tree, pole.
Former name
Arctic or Black-backed Three-toed Woodpecker.
Remarks
Look for this species in boggy areas amid dead conifers and about clearings or burns.
Range
Resident from n. Canada s. to Sierra Nevadas of cen. Calif.; also e. across Canada in boreal forests.

Perching Birds
Order Passeriformes

This order contains far more species of birds than any other order, and includes those land birds seen most commonly. The unwebbed feet have 3 toes forward and 1 behind, making them well designed for grasping a perch. The young are hatched naked, blind, and helpless and are cared for in the nest until fledged.

COTINGAS
Family Cotingidae

ROSE-THROATED BECARD
Platypsaris aglaiae   35:5
Description
Size, 6 in. (15.2 cm). Head large, bill thick, similar to flycatcher. Male: gray above, pale below, • *black cap and cheeks;* • *rose-throated.* Female: brown above, cap dark, buffy collar around nape, buff below.
Habitat
Wooded canyons, forests, riversides, large trees.
Habits
Stolid, flycatches for food.
Voice
Short, sharp *kik* or *chik.*
Food
Insects and some fruits.
Eggs
4–6; white-spotted brown; 0.9 x 0.6 in. (2.3 x 1.7 cm).
Range
Resident locally from se. Ariz. and lower Rio Grande Valley of Tex. s. to C. America.

TYRANT FLYCATCHERS
Family Tyrannidae

Flycatchers have flattened bills with a small hook at the tip and bristles about the broad base. From exposed branches they wait, quietly perching upright, then dart forth to snap up passing insects, their principal food. Their food is primarily flying insects, thus their name.

The seven species of the Genus *Empidonax* are difficult to distinguish from one another. All have dark olive-grayish backs, whitish or yellowish underparts, and two white wing bars (sometimes yellow in Western Flycatcher). Most have a conspicuous white eye-ring and pale lower mandible. Species identification is best by voice, habitat, nest and behavior. In the fall and during migration where ranges overlap, they are unfortunately usually silent and outside normal habitat.

EASTERN KINGBIRD
Tyrannus tyrannus   34:7

Description
Size, 8–9 in. (20.3–22.9 cm). Prominent white band at the tip of its fanlike tail; black above, white below; 2 very narrow white wing bars; male has a concealed crimson crown patch.
Similarities
Western and Cassin's lack white band at tail and tip and have yellow in plumage.
Habitat
Roadsides, farms, orchards, wood edges, meadows, parklands, shelter belts.
Habits
Very pugnacious; attacks crows, hawks, vultures, even alighting on their backs; flies horizontally on quivering wings.
Voice
Incisive *tzee,* alone or rapidly repeated; an excited *kipper, kipper;* *"kit-kit-kitter-kitter"* (Peterson), with descending inflection; also a dawn song.
Eggs
3–5; creamy with brown spots; 0.9 x 0.7 in. (2.3 x 1.8 cm). Nest is a bulky, twiggy, neatly lined saucer, ragged on the outside, 3–20 ft. (0.9–6.1 m) up in bush, tree stump, or structure.
Range
Breeds chiefly in E., but also throughout Rockies and w. Canada, from s. B.C., s. to Oreg.; winters in S. America; migrates through sw. states and along Pacific occasionally.

TROPICAL KINGBIRD
Tyrannus melancholicus   35:7

Description
Size, 8–9½ in. (20.3–24.1 cm). Resembles Western and Cassin's. Back olive to grayish; bright yellow of belly reaches breast, little gray on breast; head gray, dark mask through eye; • *tail dusky-brown, slightly forked.*
Similarities
Western and Cassin's have blackish unnotched tails.
Habitat
Isolated trees, scattered clumps, river groves.
Voice
Like Cassin's, but higher; a nasal *queer* or *chi-queer.*

Eggs
3–4; pinkish with brown spots; 0.9 x 0.7 in. (2.3 x 1.8 cm).
Range
Breeds locally in s. Ariz. and s. Tex., but chiefly s. of U.S.; winters
s. of U.S.; casual to Pacific Coast to s. Calif.; accidental to B.C.
and Wash.

Note: The **THICK-BILLED KINGBIRD**, *Tyrannus crassirostris* **(35:9)**,
size, 9 in. (23.9 cm), is another rare kingbird of southeastern
Arizona. It is pale-yellowish-white below and brownish above with
a heavy bill.

WESTERN KINGBIRD
Tyrannus verticalis
34:6

Description
Size, 8–9½ in. (20.3–24.1 cm). Resembles Cassin's and Tropical
kingbirds, but tail unnotched. Adult: above olive; belly yellow; head
and upper breast gray; black ear patch; • *tail black narrowly edged
with white* on each side. Immature: often lacks white tail edging, as
do worn adults also.
Similarities
Eastern is blackish above, with white tail tip. Great Crested and
Ash-throated Flycatchers have rufous tails, wing bars.
Habitat
Open country with some trees, ranches, towns, roadsides.
Habits
Flight less fluttery than Eastern's.
Voice
Noisier than Eastern, many notes similar; also "a single *kip . . .* a
quer-ich" (Stevens); a "sharp *whit* or *whit-ker-whit*" (Peterson).
Eggs
3–5; creamy, boldly spotted with brown; 0.9 x 0.7 in. (2.3 x 1.8
cm). Nest is bulky, twiggy, neatly lined saucer, but more often on
man-made structures.
Former name
Arkansas Kingbird.
Range
Breeds throughout W., from s. Canada to Great Plains and s. to
Mexico; winters in C. America.

CASSIN'S KINGBIRD
Tyrannus vociferans
34:5

Description
Size, 8–9 in. (20.3–22.9 cm). Similar to Western, but shorter,
heavier, darker; • *black tail lacks white edging* but may be faintly
tipped with whitish; • *back olive-gray; throat white,* sharply set off
by dark gray breast; body yellow.
Similarities
Immature Western lacks adult Cassin's white tail, yellow breast,
only seen in southeastern Arizona.
Habitat
Foothills, semiopen uplands, ranch groves, pine-oak mountains,
scattered trees.
Habits
Quieter, more sedentary than Western.
Voice
Harsher than Western's, less noisy; "melodious *come here, come
here*" (Bent); "also an excited *ki-ki-ki-dear, ki-dear, ki-dear,* etc."
(Peterson).

Eggs
3–5; whitish, spotted with brown; 1.0 x 0.7 in. (2.5 x 1.8 cm). Nest is bulky, twiggy cup (or of grass, wool, etc.), 8–40 ft. (2.4–12.2 m) up on limb, pole, or post.
Range
Breeds from s. Mont. and Wyo. s. to Mexico, and along s. Calif. coastline; winters in Mexico and C. America.

SCISSOR-TAILED FLYCATCHER
Muscivora forficata   34:4

Description
Size, 11–15 in. (27.9–38.1 cm). Distinctive long, scissorlike streamer tail; pink-and-gray body. Above pearly gray; wings dark; flanks and underwings pink; underparts white; scarlet patch under bend of wing and on crown, usually concealed; tail black and white above, white below.
Similarities
Western Kingbird resembles immature with short tail, but is yellowish on belly instead of pinkish, and has less white in tail and breast.
Habitat
Plains, prairies, ranches, farms, roadsides.
Habits
Often seen on wires; tail not spread when perched; in flight it wafts open and shut like pair of scissors; flight swift, graceful, low over ground; has elaborate courtship flight.
Voice
A "twittering *psee, psee, psee;* a harsh *thish-thish*" (Collins); a "harsh *keck* or *kew;* a repeated *ka-leep*" (Peterson).
Eggs
3–6; creamy, spotted with brown; 0.9 x 0.7 in. (2.3 x 1.8 cm). Nest is cup of twigs, grass 4–20 ft. (1.2–6.1 m) up on limb of deciduous tree, pole, or post.
Range
Breeds from Kans. s. through sw.-cen. U.S., including Okla., and Tex; winters in Mexico and C. America.

Note: The large **KISKADEE FLYCATCHER**, *Pitangus sulphuratus* **(35:6)**, size, 10 in. (25.4 cm), is found in the lower Rio Grande Valley of Texas. It is a big-billed flycatcher with rufous wings and tail, bright yellow underparts, and a black-and-white striped face.

GREAT CRESTED FLYCATCHER
Myiarchus crinitus   34:1

Description
Size, 8–9 in. (20.3–22.9 cm). Crested; above olive-brown; throat and breast gray, belly yellow; wings and tail cinnamon; 2 white wing bars; bill 2-toned, black above, brown below.
Similarities
Kingbirds have black or dusky tails. Wied's Crested and Ash-throated are very similar, usually not in same areas; both have all-black bill, show less belly-and-breast contrast.
Habitat
River woods, edges, orchards, farms.
Habits
Aggressive; glides from tree to tree on outspread wings; when excited, raises its slight, bushy crest; often feeds from tops of tall trees.

Voice
Raucous *wheep!* inflection rising; a whistled *whit-whit whit-whit.*

Eggs
3–8; creamy with reddish, penlike scratches; 0.9 x 0.7 in. (2.3 x 1.8 cm). Nest is tree cavity 5–60 ft. (1.5–18.3 m) up, of trash, often with discarded snakeskin.

Range
Breeds from s. Canada s. to Gulf of Mexico, chiefly in E.; winters from Mexico to S. America; migrates through w. Great Plains; accidental to Ariz.

SULPHUR-BELLIED FLYCATCHER
Myiodynastes luteiventris   **35:8**

Description
Size, 7½–8½ in. (19.1–21.6 cm). Only U.S. flycatcher • *with black streaks above and below.* Above streaked olive, • *underparts yellowish and streaked,* black stripe through eye, • *notched tail bright rufous.*

Habitat
Canyon sycamores.

Habits
A streamside flycatcher that usually escapes detection until it sallies out of the foliage.

Voice
A "high penetrating *kee-zee' ick! kee-see' ick!* by both male and female, often in duet" (Sutton).

Eggs
3–4; blotched with red; 1.0 x 0.7 in. (2.6 x 1.8 cm). Nest is cup formed of leaf stems in hole in sycamore.

Range
Breeds from se. Ariz. to C. America; winters s. of U.S.; accidental to Tex., Calif.

ASH-THROATED FLYCATCHER
Myiarchus cinerascens   **35:3**

Description
Size, 7½–8½ in. (19.1–21.6 cm). Medium-sized, 2 white wing bars, • *throat whitish, belly pale yellowish,* crown slightly crested.

Similarities
Great Crested and Wied's are brighter colored. Kingbirds are larger; some have yellow bellies, blackish tails. In Southwest, Wied's is bigger, bill larger, call different. Olivaceous is smaller, throat grayish.

Habitat
Deserts, semiarid regions, mesquite, sagebrush, pinyon-juniper slopes, open woods.

Habits
Moves about constantly over large areas; does not stay long at one perch.

Voice
A "rolling *chi-beer* or *prit-wherr*" (Peterson).

Eggs
4–5; creamy, streaked; 0.8 x 0.6 in. (2.1 x 1.6 cm). Nest is hole in tree, yucca, mesquite, post.

Range
Breeds from s. Wash. and Idaho, s. to Calif. and Mexico, e. to Colo. and Tex.

WIED'S CRESTED FLYCATCHER
Myiarchus tyrannulus   35:2

Description
Size, 8½–9½ in. (21.6–24.1 cm). Closely resembles both Ash-throated and Great Crested; has grayer throat, yellower breast, back deeper olive, black bill larger than Ash-throated; duller, less contrasting belly-and-breast border, all-dark bill, compared with Great Crested.
Habitat
Saguaros, sycamore canyons.
Voice
Very different from Great Crested's and more vigorous than Ash-throated's; "a sharp *whit* and a rolling *purreeer*" (Peterson).
Eggs
3–6; spotted, streaked; 0.9 x 0.7 in. (2.3 x 1.8 cm). Nest is hole in post or tree; woodpecker hole in saguaro.
Range
Breeds from s. Ariz. e. to s. Tex. and s. to S. America; winters s. of U.S.; casual to se. Calif., s. Nev.

OLIVACEOUS FLYCATCHER
Myiarchus tuberculifer   35:4

Description
Size, 6½–7 in. (16.5–17.8 cm). Throat grayish, belly yellow, tail rufous.
Similarities
Ash-throated is similar but larger; has white throat.
Habitat
Desert and mountain pine-oak slopes, canyons.
Habits
Often hovers in foliage to pick off insects.
Voice
Mournful, slurred *peeur*.
Eggs
3–5; streaked; 0.7 x 0.6 in. (1.8 x 1.6 cm). Nest is tree hole.
Range
Breeds chiefly in Mexico, but n. to sw. Ariz. and sw. N.Mex.; winters s. of U.S.; casual to w. Tex. and Colo.

EASTERN PHOEBE
Sayornis phoebe   34:3

Description
Size, 6¼–7¼ in. (15.9–18.4 cm). Above olive-gray; much darker on head and tail, which is notched; • *bill black;* breast grayish; underparts whitish; • *no eye-ring or wing bars,* except dull brownish bars in young. In fall, yellowish below, young with pale wing bars.
Similarities
Other small flycatchers are similar but have wing bars; also wood pewees have conspicuous wing bars, bill yellowish or whitish on lower mandible, do not pump tail.
Habitat
Farms, orchards, gardens, streamsides, canyons; usually near water, about bridges and buildings.
Habits
Tame, an early migrant; head looks blackish in flight; tail pumping diagnostic, sometimes sweeps tail sideways.

Voice

Burred *PHE-bee,* repeated, sometimes uttered in flight; a sharp *chip.*

Eggs

4–5; white, rarely dotted; 0.7 x 0.5 in. (1.8 x 1.4 cm). Nest is a thick cup of moss or mud, often under a bridge, eaves, cave, or structure.

Range

Breeds in Canada and U.S., chiefly e. of Rocky Mountains, s. to Gulf, locally in e. Colo. and e. N.Mex.; winters from Gulf Coast s. to Mexico; casual to Wyo., Ariz.

BLACK PHOEBE
Sayornis nigricans

35:13

Description

Size, 6¾–7 in. (15.9–17.8 cm). Only U.S. flycatcher with • *black breast.* All-black, except • *white belly* and outer edges of tail.

Similarities

Resemble juncos but behavior different.

Habitat

Usually near water along streams, in canyons; farmyards, towns.

Habits

Has tail-pumping and wagging habit of Eastern.

Voice

A sharp *tsip;* "song, a thin, strident *fi-bee, fi-bee,* the first 2 notes rising, the last 2 dropping" (Peterson).

Eggs

3–6; white, dotted; 0.7 x 0.5 in. (1.8 x 1.3 cm). Nest is thick cup of moss or mud, often under bridge, eaves, cave, or man-made structure.

Range

Resident from sw. U.S. and Mexico s. to S. America; casual to w. Oreg.

SAY'S PHOEBE
Sayornis saya

34:2

Description

Size, 7–8 in. (17.8–20.3 cm). Pale • *rusty belly and undertail coverts,* back pale gray-brown, coffee-brown head, • *tail black.*

Similarities

Female Vermilion Flycatcher has streaked breast. See also kingbirds.

Habitat

Open country, ranches, ravines, sagebrush plains, prairies, buttes, canyon mouths.

Habits

Very active, an early migrant; flight low, zigzaggy; wingbeats deeper, slower than Eastern; occasionally pumps tail.

Voice

Plaintive *phee-ur,* given with twitch of tail; "a swift *pit-tsee-ar*" (Hoffmann).

Eggs

4–5; white, rarely dotted; 0.7 x 0.5 in. (1.8 x 1.5 cm). Nest is cup or bracket of mud, moss, or grass on structure, ledge, or rock wall.

Range

Breeds throughout most of W. from cen. Alaska e. to n. Mackenzie, and s. to Mexico, excluding w. of Cascades and Sierras but including w. Oreg., and s.-cen. Calif.; winters from Sw. s. to Mexico.

WILLOW FLYCATCHER
Empidonax traillii   **34:12**

Description
Size, 5½–6¾ in. (14–17.1 cm). This *Empidonax* says •*fitz-BEW*.
Greenish-brown above with whiter throat than most *Empidonaces*.
Similarities
Only by its song can this species be distinguished with certainty
from the Alder.
Habitat
Bushy areas around water, swamps, brushy bogs; streamside alder
or willow thickets, near woods.
Habits
Late migrant; active, restless; sings from tops of shrubs.
Voice
Fitz-BEW, "buzzy, an even buzzier *fizz-BEW;* also a buzzy *creet*"
(R. C. Stein).
Eggs
3–4; whitish, with fine brown spots; about 0.7 x 0.5 in. (1.8 x 1.3
cm). Nest is bulky, of shreds, grasses, low in shrub, to 3 ft. (0.9 m)
up.
Range
Breeds from s. B.C. e. across Canada and s. to Calif.; winters s. of
U.S.

ALDER FLYCATCHER
Empidonax alnorum   **34:12**

Description
Size, 6 in. (15.2 cm). This *Empidonax* says *fe-BE-O*. Tends to be
grayer, less green above than Willow, with same whitish throat.
Similarities
Alder cannot be distinguished from Willow except by voice.
Habitat
Usually near water; muskegs, meadows; more open areas than
Willow on the average.
Habits
As Willow.
Voice
Fe-BE-O, the *fe* rasping, the *BE-O* whistled.
Eggs
3–4; buff with large purplish-brown spots; about 0.7 x 0.5 in. (1.8
x 1.3 cm). Nest is cottony plant material and grasses, whitish, 3–8
ft. (0.9–2.4 m) in shrub.
Range
Breeds from Alaska e. to Que., and s. to B.C.; winters s. of U.S.;
migrates through w. states.

LEAST FLYCATCHER
Empidonax minimus   **34:13**

Description
Size, 5–5¾ in. (12.7–14.6 cm). Grayest above, whitest below of the
Empidonaces; lower mandible moderately dark.
Similarities
Light lower mandible in other *Empidonaces;* noticeable at close
range.
Habitat
Orchards, streamsides, farmlands, aspen and poplar groves
(breeding), woodlands.

Habits
Jerks head and tail as it calls.
Voice
Noisy; an emphatic *che-BEC,* much repeated; call, a short *whit.*
Eggs
3–6; white or creamy; 0.6 x 0.4 in. (1.6 x 1.2 cm). Nest is neatly woven cup of plant fibers and grasses, 8–40 ft. (2.4–12.2 m) up in upright fork of tree.
Range
Breeds from s. Canada s. to cen. U.S., chiefly in E.; winters s. of U.S.; migrates through Great Plains.

HAMMOND'S FLYCATCHER
Empidonax hammondii   **35:15**
Description
Size, 5–5½ in. (12.7–14.0 cm). Hardly distinguishable, except by habitat, from Dusky. Slightly more olive, less gray; chest grayer, underparts contrastingly more yellowish.
Habitat
Upland coniferous forests; in migration other woods, thickets.
Habits
Prefers higher elevations and taller conifers than Dusky, which seeks chaparral or lowland mixed woods.
Voice
Varied; *"twur* or *tsurp* note" (Hoffmann); *"se-lip, twur, treeip"* (Peterson). The difference in voice between Hammond's and Dusky is moot.
Eggs
3–4; white; 0.6 x 0.5 in. (1.6 x 1.8 cm). Nest is woven plant fibers on limb 15–20 ft. (4.6–6.1 m) up.
Range
Breeds from cen. Alaska, s. to n. Calif. and cen. Colo.; winters from Mexico to C. America; migrates through w. lowlands.

DUSKY FLYCATCHER
Empidonax oberholseri   **35:16**
Description
Size, 5½–6 in. (14–15.2 cm). Throat whitish, underparts tinged faint yellow.
Similarities
Western is not as gray. Hammond's is distinguishable only by voice.
Habitat
Foothills, chaparral-covered slopes with some trees; brushy areas; open conifer forests of southern California mountains.
Habits
Forages quite low.
Voice
Male: "a *clip whee-Zee,* last note highest; alarm, *whit"* (Davis).
Eggs
3–5; pale cream or white; 0.7 x 0.5 in. (1.8 x 1.3 cm). Nest is of grasses and fibers, 4–7 ft. (1.2–2.1 m) up in willows, alders, etc.
Range
Breeds from s. Yukon s. to s. Calif. and N.Mex.; winters in Mexico; migrates through w. Tex.

GRAY FLYCATCHER
Empidonax wrightii   35:11

Description
Size, 5½ in. (14.0 cm). Abrupt yellow-colored lower bill, except
dark tip; back gray, with little hint of brown or olive; underparts
very little or no yellow tinge. Best identified by breeding habitat.

Similarities
Hammond's and Dusky are less gray and less yellow.

Habitat
Breeds in sagebrush, pinyon, juniper; winters in desert brush and
willows.

Voice
A two-part *chu-weet*.

Eggs
3–4; white; 0.7 x 0.5 in. (1.8 x 1.3 cm). Nest is woven grass cup in
sagebrush, juniper, or pinyon.

Range
Breeds in Great Basin region, from cen. Oreg. e. to cen. Colo., and
s. to cen. Ariz. and w. N.Mex.; winters from s. Calif. s. along
Pacific Coast and inland to Mexico; migrates through w. Tex.;
accidental to Yukon.

WESTERN FLYCATCHER
Empidonax difficilis   34:14

Description
Size, 5½–6 in. (14.0–15.2 cm). Upperparts olive-brown, wing bars
whitish or sometimes yellowish, • *underparts yellowish,* throat
especially yellow, eye-ring white.

Similarities
Other *Empidonaces* have less olive above and yellow below.

Habitat
Open deciduous woods near water; mixed or conifer forest,
canyons, groves; common.

Habits
Wings quiver in flight and at rest.

Voice
Wheezy *pee-IST,* a low *whit;* alarm, a *tsip;* song, *ps-SEET-ptsick-
sst;* "dawn song (sometimes heard all day), 3 thin notes: *pseet-trip-
seet! (seet* highest)" (Peterson).

Eggs
3–4; dull white, spotted with brown; 0.7 x 0.5 in. (1.8 x 1.3 cm).
Nest is cup of moss lined with bark, to 30 ft. (9.1 m) up on
structure, ledge, cut bank, tree trunk.

Range
Breeds from Alaska e. to sw. Alta. and s. to Baja Calif. and
N.Mex.; winters in Mexico and C. America; migrates along w.
edge of Great Plains.

BUFF-BREASTED FLYCATCHER
Empidonax fulvifrons   35:14

Description
Size, 4½–5 in. (11.4–12.7 cm). Smallest *Empidonax,* distinguished
by richly • *buffy breast;* wing bars and eye-ring white.

Habitat
Oak-pine woods, moist canyon groves.

Voice
"*Chicky-whew*" (Lusk); "*chee-lick*" (Brandt).

Eggs
3–4; creamy; 0.6 x 0.6 in. (1.6 x 1.6 cm). Nest is cup camouflaged by lichens on tree limb.
Range
Breeds from sw. Ariz. and N.Mex., s. to Mexico and C. America; winters s. of U.S.

COUES' PEWEE
Contopus pertinax   **35:12**

Description
Size, 7–7¾ in. (17.8–19.7 cm). Large; above dark gray, below uniformly lighter gray; head large, dark; slight bushy crest; • *throat gray;* flanks dusky; lower mandible conspicuously yellow. Looks like a large wood pewee, but has less conspicuous wing bars.
Similarities
Olive-sided has dark chest patches separated by white strip, throat lighter.
Habitat
Wooded desert canyons, pine-oak or pine forests.
Habits
Fast flyer, often sallies far from perch.
Voice
Note, *pip-pip* or *pil-pil;* a "thin plaintive whistle, *ho-say, re-ah,* or *ho-say, ma-re-ah?*"(Peterson), hence local nickname, "José Maria."
Eggs
3–4; creamy, spotted; 0.8 x 0.5 in. (2.1 x 1.5 cm). Nest is cup of woven vegetable matter on large tree limb.
Former name
Coues' Flycatcher.
Range
Breeds from s. Ariz. and N.Mex. s. to Mexico and C. America; winters s. of U.S.; accidental to se. Calif., se. Colo., and w. Tex.

EASTERN WOOD PEWEE
Contopus virens   **34:9**

Description
Size, 6–6¾ in. (15.2–17.1 cm). Back olive-gray, 2 white wing bars (buffy in young); below whitish, lower mandible yellow.
Similarities
Western Wood Pewee is browner, breast more brownish-gray. Eastern Pheobe has shorter wings, unbarred, tail longer, all-black, pumps. *Empidonaces* are smaller; have white eye-ring.
Habitat
River woods, shade trees.
Habits
Sits motionless, tail still; frequents middle layer of branches.
Voice
Sweet plaintive *Pee-wee,* or *PEE-a-WEE,* or *pee-AA* with rising inflection, repeated every few seconds; also a more elaborate song at dawn and twilight; sings well into August.
Eggs
2–4; creamy, with ring of brown or purple spots; 0.7 x 0.5 in. (1.8 x 1.4 cm). Nest is shallow cup of fibers covered with lichens, 6–50 ft. (1.8–15.2 m) up on horizontal branch.
Range
Breeds chiefly in E., from se. Canada s. to Gulf of Mexico; winters from C. to S. America; occasional to e. Colo., Tex. Panhandle; casual to Mont.; accidental to e. Oreg., se. Ariz.

WESTERN WOOD PEWEE
Contopus sordidulus   **34:10**

Description
Size, 6–6½ in. (15.2–16.5 cm). Sparrow-sized; above gray-brown; breast and flanks olive-gray, no eye-ring but 2 narrow white wing bars.
Similarities
Eastern Wood Pewee is not as brown above, lighter below; lower mandible light; a difficult identification.
Habitat
Open woodlands, pine-oak forest, conifers, river woods and breaks.
Habits
Similar to Eastern, but prefers somewhat more open woods.
Voice
Nasal *peeer,* suggesting Common Nighthawk, a sad *dear-me;* "*TSWEE-tee-teet, TSWEE-tee-teet, bzew, bzew,* a downward slur" (Miller); often sings at night.
Eggs
3–4; creamy, ringed with spots like Eastern's; 0.7 x 0.5 in. (1.9 x 1.4 cm). Nest is lichen-covered cup like Eastern's, grass-lined (sometimes without lichens), on horizontal limb.
Range
Breeds from s. Alaska s. to Mexico and C. America; winters s. of U.S.

OLIVE-SIDED FLYCATCHER
Contopus borealis   **34:11**

Description
Size, 7–8 in. (17.8–20.3 cm). Large bill and head, sturdy body. Above olive-gray; sides gray; white stripe down mid-breast, suggesting half-open gray vest over white shirt; white tuft often sticks out on each side of lower back, visible sometimes at rest or in flight.
Similarities
Wood pewees are smaller, with light wing bars, and lack contrast between sides and center of breast. Coues' Pewee in Ariz. or N.Mex. shows underparts less uniformly gray, throat lighter, no white "open-vest" strip.
Habitat
Conifer forests, burns (often near water); dead snags during migration; eucalyptus groves.
Habits
Makes sweeping sallies from exposed perch on dead tree or branch after passing insects.
Voice
Loud whistle, quick-*THREE BEERS* (or collegiate hip, *THREE CHEERS)*, unique among flycatchers; also "alarm note, *puip puip puip*" (Bendire).
Eggs
3; creamy, blotched with chestnut; 0.9 x 0.7 in. (2.3 x 1.8 cm). Nest is shallow twig saucer 10–50 ft. (3.0 –15.2 m) up in crotch of horizontal branch of conifer.
Range
Breeds from Alaska e. across Canada and s. to Baja Calif., Nev., Ariz., and N.Mex.; winters in S. America.

VERMILION FLYCATCHER
Pyrocephalus rubinus **35:1**

Description
Size, 5½–6½ in. (14.0–16.5 cm). Male: • *bright vermilion crown,* throat, and underparts; upperparts and tail dusky-brown to blackish; crown often erected into bushy crest. Female and immature: upperparts brownish-gray; breast white and streaked narrowly; lower belly and undertail coverts pinkish to yellow.

Similarities
Female Say's Phoebe resembles Vermilion.

Habitat
Desert stream banks, thickets, willows, mesquite, cottonwoods.

Voice
Somewhat phoebelike *p-p-pit-zeee* or *pit-a-zee.*

Eggs
2–3; creamy, blotched; 0.6 x 0.5 in. (1.7 x 1.3 cm). Nest is twig saucer in conifer.

Range
Resident from sw. states s. to S. America; casual to Colo.

NORTHERN BEARDLESS FLYCATCHER
Camptostoma imberbe **35:10**

Description
Size, 4½ in. (10.8 cm). Tiny, nondescript coloration resembling somewhat that of a kinglet or vireo. Upperparts olive-gray, underparts dingy-white; wing bars and eye-ring indistinct, bill very small.

Similarities
Bell's Vireo is larger, flanks yellower, wing bars pale to grayish, compared with buffy to brownish in some Beardless individuals. Immature Verdin above is purer gray, no wing bars.

Habitat
Stream and canyon thickets, lowland breaks, mesquite.

Habits
May feed by gleaning like vireos and warblers.

Voice
Call, "a thin *peeee-yuk;* also a series of soft *ee, ee, ee, ee, ee,* volume rising toward mid-series" (Sutton).

Eggs
2–3; creamy, speckled; 0.6 x 0.4 in. (1.6 x 1.1 cm). Nest is globe of vegetation (entrance on one side) in matted tree, palm.

Range
Resident from s. Ariz., N.Mex., and Tex. s. to Mexico and C. America.

LARKS
Family Alaudidae

Largely terrestrial birds, larks are mostly brown, streaked, have musical voices, are gregarious; the sexes are virtually alike. The hind claw is elongated and nearly straight.

HORNED LARK
Eremophila alpestris **52:7**

Description
Size, 7–8 in. (17.8–20.3 cm). Sparrow-sized; • *tiny black "horns"* (not always apparent) and black "whiskers." Male: brownish

167

above and on sides, unique face pattern, tail black, outer tail feathers and underparts white, throat light with black breast shield beneath, line over eye yellow or white. Female and immature: duller, immature lack horns.

Similarities
Water Pipit is smaller; different head pattern, pumps tail. Also longspurs when mating resemble Horned, but tail patterns differ and lark's bill is thinner.

Habitat
All types of open spaces: tundras, marshes, golf courses, fields, parkways, beaches, dunes, plains, prairies, sage flats.

Habits
Social; flocks after breeding; walks or runs instead of hopping on ground; flight light, bounding, showing from below contrast of black tail and white underparts.

Voice
Song a sustained, irregular, high-pitched tinkling and bubbling, often given as it displays in flight; flight note, *"p-seet"* (Collins). "a clear *tsee-ee* or *tsee-titi*" (Peterson).

Food
Seeds, insects.

Eggs
3–5; grayish, spotted with brown; 0.9 x 0.6 in. (2.3 x 1.5 cm).

Range
Breeds throughout most of N. America, from Arctic s. to S. America; winters from s. Canada southward.

SWALLOWS
Family Hirundinidae

These sparrow-sized, streamlined birds have short, flat bills and wide mouths; long, pointed wings; small, weak feet; and notched tails which are forked in the Barn and squared in the Cliff. They frequent the air over open country and bodies of water, spending much time in flight catching insects, their principal diet. Mixed flocks often hawk for insects low over ponds prior to bad weather. Gregarious, they migrate in large flocks by day, often along coasts, and perch on roadside wires.

VIOLET-GREEN SWALLOW
Tachycineta thalassina   **36:2**

Description
Size, 5–5½ in. (12.7–14.0 cm). • *White patches on rump,* nearly meeting over base of tail. Above violet-green, clear white below extending up over the eyes.

Similarities
Tree Swallow is steel-blue above. White-throated Swift flies with wings straight in manner of swift, black and white.

Habitat
Meadows, ranches, plains, foothills, mountains, canyons, cliffs, towns.

Habits
Wingbeats more rapid than Tree's; sometimes flies with White-throated Swifts; sings before dawn.

Voice
Fast twitter; a *"tsip-tseet tsip"* (Hoffmann); "a thin *chip* or *chi-chi;* a rapid *chit-chit-chi wheet, wheet*" (Peterson).

Eggs

4–5; white; 0.7 x 0.5 in. (1.8 x 1.3 cm). Nest is often colonial; in hole in tree, nest box, structure, other cavity.

Range

Breeds from Alaska, e. to S.Dak. and s. to Baja Calif., Tex., and Mexico; winters from Mexico s. to C. America; migrates through Great Plains.

TREE SWALLOW

Iridoprocne bicolor   **36:3**

Description

Size, 5–6½ in. (12.7–15.9 cm). Adult: above glossy greenish-blue, including area around eyes; • *white below.* Immature: above dusky-brown, sometimes with faint, broken dark band across upper breast.

Similarities

Violet-green has conspicuous white patches on rump, eye partly encircled with white. Brown Rough-winged has dingy throat; resembles young Tree in late summer. Bank has complete breastband.

Habitat

Open areas, especially near water; often breeds in dead tree in open swamps; meadows, marshes, streams, roadside wires.

Habits

Flight slightly flickering, wingbeats faster than Barn's; wings look triangular in air somewhat like Purple Martin's. Bank hugs wings close to body. An early migrant; collects in vast numbers in marshes in fall during migration, swarming over trees, wires, bushes, roads, beaches.

Voice

Chirrups, twitters, *silip, silip, silip,* much repeated; "note, *cheet* or *chi-veet . . .* song, *weet, trit, weet,* repeated with variations" (Peterson).

Eggs

3–6; white; 0.7 x 0.5 in. (1.9 x 1.3 cm). Nest is colonial or single; a feathery cup in woodpecker or other hole in tree, building, box.

Range

Breeds throughout much of N. America, from Alaska e. across Canada and s. to Calif., eastward; winters from s. U.S. s. to C. America.

ROUGH-WINGED SWALLOW

Stelgidopteryx ruficollis   **36:5**

Description

Size, 5–5¾ in. (12.7–14.6 cm). Brown-backed, dusky throat. • *Above and on throat uniform grayish-brown;* underparts all-white.

Similarities

Bank is smaller; darker, with neat dark breastband; colonial nester. Immature Tree in late summer has less dark throat.

Habitat

Open areas, creeks, ponds, waterways.

Habits

Flight more direct, less erratic than Bank's; wingbeats deeper, slower, more glides and sails; often repeatedly follows the same aerial pathway; folds wings back at end of stroke.

Voice

Twitter similar to, but lower-pitched than, Bank's; call "a *trit, trit-trit,* or *tri-ri-ri-rit*" (Saunders); a "rasping squeak . . . *quiz-z-z-zeep, quiz-z-zeep*" (Dickey); unmusical "*burp-burp*" (Cruickshank).

Eggs
4–8; white; 0.7 x 0.5 in. (1.8 x 1.3 cm). Nest is single or in small groups; hole in bank, cave, or crevice in rocks, usually near water.
Range
Breeds throughout most of N. America from cen. B.C. and cen. Alta. s. to s. U.S. and S. America; winters s. of U.S.; some winter occasionally to s. Calif.

BARN SWALLOW
Hirundo rustica   **36:6**

Description
Size, 7 in. (17.8 cm). • *Long, forked "swallow tail"* and white tail spots. Adult: above iridescent steel-blue, forehead and throat chestnut, thin blue necklace, white marks on inner tail feathers, underparts buffy to cinnamon-buff. Immature: paler, with buffy-brown throat; lacks tail streamers.
Habitat
Farms, ranches, open or semiwooded areas, often near water; fields, marshes, lakes; often around habitations.
Habits
Flight strong, swift, graceful; drives through air, wings close to and parallel with body at end of stroke; gregarious after breeding; many migrate along coast; glides little.
Voice
Various cheerful twitterings, *"sweeter-sweet, sweeter-sweet"* (Cruickshank); a *kittik, kittik,* or *kvik, kvik* repeated, soft.
Eggs
3–6; white, speckled with red-brown; 0.8 x 0.6 in. (2.0 x 1.5 cm). Nest is of mud, often inside barns, boathouses, or other structures, on rafter or against wall, lined with feathers.
Range
Breeds throughout most of N. America from n.-cen. Alaska e. across Canada, and s. to sw. U.S.; winters in S. America.

CAVE SWALLOW
Petrochelidon fulva   **36:9**

Description
Size, 5–6 in. (12.7–15.2 cm). Very similar to Cliff Swallow, but rump is darker; throat and cheeks pale or buffy; forehead dark chestnut, reversal of face colors.
Similarities
Immature Cliff Swallow in west Texas has dark throat and forehead.
Habitat
Limestone cave areas.
Habits
As Cliff Swallow.
Voice
Song, "a series of squeaks blending into a complex melodic warble, ending in double-toned notes" (Slender and Baker); a clear *weet* or *cheweet.*
Eggs
2–5; white, spotted; 0.2 x 0.5 in. (0.6 x 1.3 cm). Nest is mud or straw open cup in cavern or sinkhole.
Range
Breeds locally in caves in s. Tex. and se. N.Mex.; winters s. of U.S.

BANK SWALLOW
Riparia riparia   **36:4**

Description
Size, 4¾–5¼ in. (12.1–13.3 cm). White throat and • *brown breastband.* Above brown head; wings and tail darker; • *below white;* breastband dark and distinctive.

Similarities
Young Tree is larger, breast smudgy, no sharp band. Rough-winged has no breastband, throat dingy.

Habitat
Meadows, ponds, lakes, streams, other open areas.

Habits
Flight low, erratic, somewhat fluttering and mothlike; keeps wings close to body when sailing; rows of holes in bank mark site of nesting colony.

Voice
Notes more gritty than other swallows; "soft abrupt *ffrrutt*" (Cruickshank); song, a twitter, *"speedz-sweet, speedz-sweet"* (Dickey), much repeated.

Eggs
3–7; white; 0.7 x 0.5 in. (1.8 x 1.3 cm). Nesting is colonial; in hole lined with grass in sand or clay stream bank.

Range
Breeds throughout most of N. America from Alaska e. across Canada, and s. to s. Calif., N.Mex., and Tex.; migrates through sw. states.

CLIFF SWALLOW
Petrochelidon pyrrhonota   **36:7**

Description
Size, 5–6 in. (12.7–15.2 cm). White forehead, chestnut throat, • *buffy rump.* Adult: upperparts blue except creamy forehead, chestnut face, gray collar, underparts white, buffy rump, dark throat patch, • *tail square at tip.* Immature: browner, rump paler.

Similarities
Cave Swallow, when in southeast New Mexico or south-central Texas, has pale throat, chestnut forehead. Barn has long forked tail.

Habitat
Meadows, marshes, open to semiwooded country, cliffs, canyons, rivers, cultivated regions.

Habits
Colonial, usually quite local; has a glide in a long ellipse, ending with a steep climb.

Voice
Song, squeaky, creaking, but pleasant chirrupings, huskier than Barn's; a low *chur;* alarm note, *keer.*

Eggs
4–6; white, thickly spotted with reddish-brown; 0.7 x 0.6 in. (1.9 x 1.7 cm). Nest is gourd-shaped mud jug on outside of buildings under eaves, bridges, cliffs; colonial.

Range
Breeds throughout most of N. America from Alaska e. across Canada and s. to se. U.S. and Mexico; winters in S. America.

PURPLE MARTIN
Progne subis   **36:1**

Description
Size, 7½–8½ in. (18.4–21.6 cm). Largest swallow. Male: uniformly
• *blue-black to purple above and below* (appears black at distance).
Female and immature: throat and breast grayish, underparts
whitish, often faintly collared.

Similarities
Immature Tree Swallow is much smaller, white below. Starling is
somewhat suggestive of Martin wings when flying.

Habitat
Seashore, meadows, wide river valleys, open or logged-off
timberlands, farms, ranches, towns, saguaro deserts.

Habits
Wings triangular in flight; glides in circles with alternating quick
flaps and glides; often heard at night.

Voice
Rich, liquid, loud chirruping, a somewhat guttural *too-too* and *too-
too-too-weadle;* call, "a harsh *zhupe, zhupe,*" (Stone); alarm, a
kerp.

Eggs
3–5, white; 0.9 x 0.7 in. (2.3 x 1.8 cm). Nest is colonial; bird
boxes, hole in tree, building, woodpecker hole.

Range
Breeds from s. Canada s. to Mexico, but scarce in W.; winters in
S. America.

JAYS, MAGPIES, AND CROWS
Family Corvidae

These medium to large passerine birds, collectively called corvids,
have longish, stout, pointed bills; rounded wings; and rounded or
wedge-shaped tails. Crows and ravens are black, jays usually
colorful in blue and green, magpies black and white with long tails.
These birds are omnivorous, eating insects, berries, nuts, and seeds.

GRAY JAY
Perisoreus canadensis   **37:7**

Description
Size, 10–13 in. (25.4–33.0 cm). Adult: • *plumage gray,* darker
above, fluffy; forehead, throat, and collar white; • *black nape,* to
eye. When fluffed up, suggests a giant chickadee. Immature: dark
slate, head blackish, faint white "whisker."

Similarities
Northern Shrike has hooked bill, contrasting black wings and tail.

Habitat
Coniferous forests and clearings.

Habits
Tame, inquisitive, bold; takes food about camps, robs traps, stores
food; in flight seems to float lightly in air, glides more than most
jays, sails from top of one tree to bottom of another, then hops up
branches in a spiral around trunk, and repeats.

Voice
A "loud hawklike whistle" (Brewster), a "querulous *quee-ah, kuoo
or whah*" (Knight); many other notes, often mimics.

Eggs
3–5; greenish-gray, wreathed with brown markings at large end;

1.1 x 0.8 in. (2.8 x 2.0 cm). Nest is moss- or feather-lined bowl or twigs 4–30 ft. (1.2–9.1 m) up in conifer.

Range
Resident throughout much of Canada and n. U.S., from Alaska eastward, and in coastal rain forest, s. to n. Calif.; also in interior and Rockies to ne. Calif. and Ariz.

STELLER'S JAY
Cyanocitta stelleri   **37:3**

Description
Size, 12–13½ in. (30.5–34.3 cm). Only jay west of Rockies with a • crest. • *Foreparts black;* • *lower back, wings, and tail dark blue;* some forms have brownish tint on back.

Similarities
Blue Jay, east of Rockies, has white face, white in wings.

Habitat
Coniferous and pine-oak forests.

Habits
Aggressive, hops up tree one branch at a time in a spiral; hops.

Voice
Noisy; "a harsh *shaak, shaak;* a mellow *klook, klook*" (Dawson and Bowles); a scream like Red-tailed Hawk; various other notes; a mimic.

Eggs
3–5; greenish-blue, finely dotted with brown; 1.2 x 0.9 in. (3.0 x 2.3 cm). Nest is bowl of twigs, rootlets, pine needles 8–40 ft. (2.4–12.2 m). up in conifer.

Range
Resident throughout w. states, from Alaska, e. to Canadian prairies and s. to s. Calif. and C. America.

Note: The **BLUE JAY,** *Cyanocitta cristata* **(37:4),** size, 12 in. (30.5 cm), occurs west to foothills of Rockies in central Alberta, central Wyoming, central Colorado and Texas Panhandle. It is crested and bright blue, with black-barred and white-marked wings and tail, a white throat, and a black bib.

MEXICAN JAY
Aphelocoma ultramarina   **37:2**

Description
Size, 11½–13 in. (29.2–33.0 cm). • *Uniform gray-blue*; no crest.

Similarities
Scrub Jay is very similar, but Mexican has more uniform upper- and underparts; back is grayer; no breastband.

Habitat
Open forests, oak, pine-oak.

Voice
A "rough, querulous *wink? wink?* or *zhenk?*" (Peterson).

Eggs
4–5; green; 1.0 x 0.8 in. (2.6 x 2.1 cm). The eggs are unspotted in Arizona, spotted in western Texas. Nest is bowl of twigs in tree.

Range
Resident from s. Ariz., N.Mex., and w. Tex., s. to Mexico.

Note: Along the Mexican border of southern Texas the unmistakable **GREEN JAY,** *Cyanocorax yncas* **(37:1),** size, 11 in. (27.9 cm), is found. It is green above with a blue crown, and yellowish below with a black bib, with yellow outer tail feathers.

SCRUB JAY
Aphelocoma coerulescens 37:6

Description
Size, 11–12½ in. (27.9–31.8 cm). Back light-brown; • *head, wings, tail, blue;* underparts pale gray; variable blue band across breast.

Similarities
Steller's is crested, foreparts black. Mexican Jay only found in Mexico, border mountains.

Habitat
Foothills, oak groves, oak-chaparral, pinyon-juniper areas, brushland river breaks.

Habits
In flight, often pitches down slopes in long, shallow curves.

Voice
A rasping *"kwesh . . . kwesh;* also a harsh *check-check-check-check* and a rasping *shreek* or *shrink"* (Peterson).

Eggs
3–6; greenish or reddish, spotted; 1.0 x 0.8 in. (2.6 x 2.1 cm). Nest is of twigs, in shrub or low tree.

Other name
California Jay.

Range
Resident from Wash. e. to Wyo., and s. to Calif.,Tex., and Mexico.

YELLOW-BILLED MAGPIE
Pica nuttalli 37:11

Description
Size, 16–18 in. (40.6–45.7 cm). Yellow bill; behind each eye is a patch of bare yellow skin.

Similarities
Black-billed is larger.

Habitat
Farms, ranches, scattered oaks, streamside groves.

Habits
As Black-billed Magpie.

Voice
Resembles Black-billed's *maag?*

Eggs
5–8; olive, spotted; 1.2 x 0.9 in. (3.1 x 2.3 cm). Nesting is colonial; like Black-billed's.

Range
Resident exclusively in Calif., in Central Valley and coastal valleys between San Francisco and Santa Barbara; also s. foothills.

BLACK-BILLED MAGPIE
Pica pica 37:9

Description
Size, 17½–22 in. (44.5–55.9 cm); tail, 9½–12 in. (24.1–30.5 cm). Unmistakable; large, • *black-and-white bird* with very • *long wedge-shaped tail.* Wings short, rounded; tail glossy green; bill black; large white wing patches flash in flight.

Similarities
Yellow-billed Magpie has yellow bill.

Habitat
Foothills, roadsides, streamsides, thickets, fields, pastures, ranches.

Habits
Often found in small groups conspicuous from car approaching Rockies from east, spends much time on ground, twitches tail

constantly in its jerky walk, hops when in a hurry; flight jaylike rather than crowlike, with tail streaming behind.

Voice
Noisy; a harsh *ca-ca-ca,* higher than Common Crow; "a rapid harsh *queg queg queg queg;* also a nasal querulous *maag?* or *aag-aag*" (Peterson); other notes.

Eggs
6–9; grayish to greenish, blotched with brown; 1.3 x 0.9 in. (3.3 x 2.3 cm). Nesting is sometimes colonial; nest is huge domed mass of sticks with entry hole in 2 sides, lined with grass, often in willow thickets, tree, or bush.

Range
Resident in n. regions of N. America from Alaska s. to cen. Calif., and Ariz. and e. to Okla.

COMMON RAVEN
Corvus corax   *Fig. 12*

Description
Size, 21½–27 in. (54.6–68.6 cm). All-black, with • *wedge-shaped tail.* Larger and with longer, thicker neck than Common Crow and with longer, heavier bill; throat shaggy. In flight from below, outspread primaries, horizontal "flat" wings.

Similarities
White-necked Raven is crow-sized, white feathers at base of neck and breast rarely visible, rely on voice.

Habitat
Wild country, seashore, mountains, deserts, canyons, coastal cliffs, conifer forests, tundras.

Habits
Aggressive but wary; flight hawklike with heavy slow wingbeats, alternately flapping and soaring; sometimes high in air; spectacular aerial revolutions in courtship; hovers, drops shellfish from height to break it open; a scavenger, sometimes mobbed by crows.

Voice
Various croaks, such as *c-r-r-r-u-u-k;* a high-pitched *tok-tok-tok;* young may caw like crow.

Eggs
4–7; greenish, spotted with brown; 1.8 x 1.4 in. (4.6 x 3.6 cm). Nest is large, of sticks, bones, wool, near top of conifer or ledge of cliff.

Range
Resident throughout much of n. North America from Alaska e. across Canada and s. to C. America, but only e. in U.S. to e. foothills of Rockies.

Crow

Raven

Common Raven

Fig. 12

Common Crow, p. 176

WHITE-NECKED RAVEN
Corvus cryptoleucus

Description
Size, 19–21 in. (48.3–53.3 cm). Crow-sized, prefers lowland flats and deserts. All-black, but base of neck and breast feathers white, visible only in display or when ruffled by wind; throat shaggy, tail somewhat wedge-shaped; flies with typical flat-winged ravenlike glide.

Similarities
Common Raven is larger, no whiteness anywhere; Common Crow has shorter bill, neck more slender, and not shaggy.

Habitat
Plains, deserts, rangeland, mesquite flats, yucca plains.

Habits
Tame, inquisitive, gregarious; gathers in roosts after breeding; flight like Common Raven; soars with primaries outspread, engages in aerial maneuvers.

Voice
Guttural, lower-pitched than Common Crow's, a hoarse *quark, quark; "kraak"* (Peterson).

Eggs
4–7; greenish, scrawled and blotched with lilac and brown; 1.8 x 1.2 in. (4.6 x 3.0 cm). Nest is loose bowl of sticks, wire, in lone tree, 4–40 ft. (1.2–12.1 m) up, mesquite, telephone pole, yucca.

Range
Resident from Nebr. s. through prairies, and from Okla. Panhandle s. to Tex., N.Mex., Ariz. and Mexico.

COMMON CROW
Corvus brachyrhynchos   *Fig. 12*

Description
Size, 17–21 in. (43.2–53.3 cm). All-black, says, *"caw."* Large, stout; black glossed with purplish (in bright sun); tail gently rounded, shorter than Common Raven's; black bill and strong feet.

Similarities
White-necked Raven is often called "crow." Common Raven is larger, tail wedge-shaped.

Habitat
Fields, beaches, woods, parks, coasts, river groves.

Habits
Wary; hops and walks; frequents parkways, feeding on animals killed by cars; often seen flying overhead with steady, deep wingbeat; soars with wings in shallow V; congregates in roosts in winter.

Voice
Loud *khaaa, khaaa;* young, *car, car,* similarly pitched.

Eggs
3–6; greenish, blotched with brown; 1.7 x 1.2 in. (4.3 x 3.0 cm). Nest is bulky, well-made stick bowl 10–50 ft. (3.0–15.2 m) up in tree.

Range
Breeds throughout much of N. America from Alaska e. to sw. Mackenzie and n. Sask., s. to Baja Calif., cen. Ariz., and cen. Tex.; winters from Vancouver s. through entire U.S. except desert Sw.

PINYON JAY

Gymnorhinus cyanocephalus   **37:5**

Description
Size, 9–11¾ in. (22.9–29.8 cm). • *Dull blue;* crowlike. Adult: bill long, thin, sharp; throat white with streaks; tail short. Immature: lavender-gray.

Similarities
Scrub Jay has longer tail, patterned. Steller's Jay has crest, though depressed in flight; black and blue plumage.

Habitat
Foothills, croplands, pinyons, junipers, scrub oak, sagebrush.

Habits
Gregarious, restless, local; occurs in wandering flocks, larger in fall and winter; flight crowlike, but swifter; feeds much on ground, walking or running with head held high.

Voice
Noisy, a "mewing call, *queh-a-eh,* given in flight . . . when perched, a continual *queh, queh, queh*" (Hoffmann).

Eggs
3–4; bluish-white, speckled with brown dots; 1.2 x 0.9 in. (3.0 x 2.3 cm). Nesting is colonial; bulky twig bowl, well lined, 5–12 ft. (1.5–3.7 m) up in pinyons, junipers, or oaks.

Range
Resident from cen. Oreg. e. to w. S.Dak., and s. to n. Baja Calif., cen. N.Mex., and w. Okla.; occasionally wanders to s.-cen. Wash., Idaho, Mont., s. Calif., and w. Great Plains.

CLARK'S NUTCRACKER

Nucifraga columbiana   **37:8**

Description
Size, 12–13 in. (30.5–33.0 cm). Crowlike; • *body pale gray;* bill, wings, and tail black; • *large white wing patches;* sides of tail and undertail white, conspicuous in flight. Resembles no other high mountain bird when white patches visible.

Similarities
Gray Jay has no wing patches.

Habitat
Conifer zones of western mountains.

Habits
Short flights undulating like a woodpecker; longer flights straight, wingbeats faster than Crow's; soars, dives in air; bold, inquisitive, steals food, hops awkwardly on ground; becomes tame.

Voice
Noisy, a "guttural squawking . . . *chaar, char-r-r, chur-r-r, kra-a-a,* or *kar-r-r-r-ack*" (Bent), each syllable repeated 2–3 times; a catlike *me-ak;* other sounds.

Eggs
2–4; pale green, sparingly dotted with brown; 1.3 x 0.9 in. (3.3 x 2.3 cm). Nest is deep grass-lined bowl of twigs 7–150 ft. (2.1–95.7 m) up in conifer, well out on a branch.

Range
Resident from s. B.C. and Alta. s. to Calif. and Baja Calif.; occasional to cen. Alaska, w. Calif., s. Ariz., w. Great Plains, and w. Tex.

CHICKADEES, TITMICE, BUSHTITS, AND VERDINS
Family Paridae

These are small, plump, big-headed birds with short, straight bills, beady black eyes; rounded wings; and soft, fluffy plumage, predominantly gray, black, and white in color. They are very active, and many hang head down from a branch when feeding. They often roam in flocks, are confiding and inquisitive, and can be attracted by squeaking or by imitations of their notes. They feed on insects, insect eggs or larvae, seeds, and berries; readily come to feeding trays where they relish suet, peanut butter, sunflower and pumpkin seeds, and nut kernels. They nest in a tree hole.

BLACK-CAPPED CHICKADEE
Parus atricapillus   38:11

Description
Size, 4¾–5¾ in. (12.1–14.6 cm). • *Cap and bib solid black;* gray above with white wing edgings; white below; pale chestnut wash on flanks; • *white cheeks* noticeable at a distance.

Similarities
Mountain Chickadee has white eyebrow stripe. Chestnut-backed has rusty back. Blackpoll Warbler spring male is gray-striped, with no bob.

Habitat
Mixed forests, edges, gardens, towns.

Habits
Tame, somewhat migratory in north, may move in mixed flocks with Downy Woodpeckers, nuthatches, kinglets.

Voice
Song, a sweet whistled *Fee-bee* or *SPRING's-come,* the first note higher; a simple sibilant *sth;* a clear *chick-a-dee-dee-dee* or *dee-dee-dee.*

Eggs
4–9; white, finely spotted with reddish-brown; 0.6 x 0.5 in. (1.5 x 1.3 cm). Nest is hole in rotten snag, fur- or feather-lined.

Range
Resident throughout much of N. America from cen. Alaska and s. Mackenzie s. to n. Calif., n. Nev., s. Utah, and N.Mex.

MEXICAN CHICKADEE
Parus sclateri   39:6

Description
Size, 5 in. (12.7 cm). Similar to Black-capped, but • *black bib more extensive* across upper breast; sides • *grayish.*

Habitat
Pine-oak and conifer forests.

Voice
Low, nasal, husky *dzay-dzee;* main call usually two-noted *chick-dee;* call buzzy.

Eggs
6; white, dotted; 0.5 x 0.4 in. (1.3 x 1.1 cm). Nest is typical hole in tree.

Range
Resident from Ariz. and N.Mex. s. to Mexico.

MOUNTAIN CHICKADEE
Parus gambeli   **39:4**

Description
Size, 5–5¾ in. (12.7–14.6 cm). Similar to Black-capped, except for
• *white line over each eye;* no chestnut wash on flanks.
Similarities
No other chickadee has white line over eyes.
Habitat
In summer, mountain conifer forests; in winter, lower ranges.
Voice
"Huskier than Black-capped's, *tsick-a-zee-zee-zee*" (Peterson);
song, a high clear *fee-bee-bee.*
Eggs
7–9; white or dotted white; 0.6 x 0.4 in. (1.6 x 1.1 cm). Nest is
hole in tree.
Range
Resident from cen. B.C. s. through Rockies and Sierras to Baja
Calif. and w. Tex.

BOREAL CHICKADEE
Parus hudsonicus   **38:7**

Description
Size, 5–5½ in. (12.7–14.0 cm). • *Cap brown. Above brownish,*
below whitish; flanks a rich brown; bib black; cheeks whitish; bill
tiny.
Similarities
Black-capped has paler flanks; black cap; overall grayer color.
Habitat
Conifers.
Habits
Similar to Black-capped, but not as lively and works closer to tree
trunk.
Voice
A slower, drawling *chick-a-dee-dee-dee* than Black-capped's,
wheezier, burred; *chick-a-day-day;* song, a warble.
Eggs
5–9; white, thickly dotted with reddish-brown; 0.5 x 0.4 in. (1.3 x
1.1 cm). Nest is typical hole in snag.
Former name
Brown-capped Chickadee.
Range
Resident throughout much of wooded Canada, from Arctic tree
limit to ne. Wash., nw. Mont., and eastward.

Note: The **GRAY-HEADED CHICKADEE**, *Parus cinctus,* size, 5½ in.
(14.0 cm), is found in Alaska, the Yukon, and northwest
Mackenzie. It is similar to the Boreal Chickadee, but has a gray-
brown cap and lacks the Boreal's brown sides.

CHESTNUT-BACKED CHICKADEE
Parus rufescens   **38:8**

Description
Size, 4½–5 in. (11.4–12.7 cm). • *Back and sides chestnut,* cap and
bib black, cheeks white.
Habitat
Damp, evergreen forests and adjacent drier oaks; town shade trees.
Habits
Very common in wet conifer forests.

Voice
Like Black-capped's, but hoarser and no whistled song; *"tsick-i-see-see* or *zhee-che-che;* also a harsh *zee* or *zze-zze"* (Peterson).
Eggs
5–7; white, dotted; 0.6 x 0.4 in. (1.6 x 1.1 cm). Nest is typical hole in snag.
Range
Resident from Alaska along Pacific Coast to cen. Calif., and locally in w. Rockies of se. B.C. and w. Mont.

TUFTED TITMOUSE
Parus bicolor   38:5

Description
Size, 5–6 in. (12.7–15.2 cm). Upperparts gray; crest slender, gray or black; underparts white; flanks rusty; bill black with light spot above.
Similarities
Plain Titmouse immatures are virtually indistinguishable; crest pointed, no distinctive markings.
Habitat
Towns, groves, cedars, oak forests.
Habits
Somewhat like Black-capped Chickadee; visits feeders; after breeding season roams in mixed flocks with chickadees, woodpeckers, nuthatches; not migratory.
Voice
Similar to chickadee's; also "a whistled *peter, peter, peter,* or *hear, hear, hear"* (Peterson).
Eggs
4–7; white, spotted; 0.7 x 0.5 in. (1.8 x 1.3 cm). Nest is hole in tree or post.
Remarks
Species now includes Black-crested Titmouse.
Range
Resident chiefly in E., but w. to Nebr., Okla., and Tex.; also w. Tex. from Panhandle s. to Rio Grande.

PLAIN TITMOUSE
Parus inornatus   38:4

Description
Size, 5–5½ in. (12.7–14.0 cm). • *Small,* gray-backed, with a pointed crest; only titmouse commonly found in most of West.
Similarities
Tufted immature has short, gray crest; almost inseparable in field.
Habitat
Shade trees, river groves (locally), oak-pine forests, pinyon, junipers.
Habits
Chickadeelike; taps louder when seeking food or opening seeds.
Voice
As other chickadees; *chick-a-dee-dee;* song, "a whistled *weety weety weety* or *tee-wit tee-wit tee-wit"* (Peterson).
Eggs
6–9; white or white spotted; 0.6 x 0.5 in. (1.6 x 1.3 cm). Nest is hole in tree.
Range
Resident from s. Oreg., n. Nev., and Utah e. to Okla. and s. to Baja Calif., Ariz., s. N.Mex., and w. Tex.

BRIDLED TITMOUSE
Parus wollweberi   39:1

Description
Size, 4½–5 in. (11.4–12.7 cm). Crest black, •*face and throat
"bridled" in black and white,* upperparts gray, underparts whitish.
Habitat
Pine-oak forests, oak and sycamore canyons.
Habits
In flocks much of the time.
Voice
Like other titmice and chickadees; song of 2 syllables, repeated.
Eggs
5–7; white; 0.6 x 0.5 in. (1.6 x 1.4 cm). Nest is hole in oak.
Range
Resident from s. Ariz. and N.Mex. s. to Mexico.

BUSHTIT
Psaltriparus minimus   39:5

Description
Size, 3¾–4¼ in. (9.5–10.8 cm). A nondescript, small plain bird;
crown gray or brown, back gray, underparts pale, cheeks brownish,
bill short, tail longish. Male: with or without black mask. Female:
eyes lighter than male.
Similarities
Verdin has yellowish head.
Habitat
Mixed forests, pinyons, junipers, oak scrub, chaparral.
Habits
Highly social, active, acrobatic, often clings upside down.
Voice
"Insistent, light *tsit*'s, *lisp*'s, and *clenk*'s" (Peterson).
Eggs
5–7; white; 0.5 x 0.4 in. (1.3 x 1.1 cm). Nest is elongated woven
cup or pouch in tree or shrub.
Remarks
Black-eared Bushtit now regarded as same species.
Range
Resident from sw. B.C. s. along Pacific Coast to Baja Calif.; in
Sw., from s. Rocky Mountains e. to Okla.

VERDIN
Auriparus flaviceps   39:9

Description
Size, 4–4½ in. (10.2–11.3 cm). Distinctive; small, gray with
• *yellowish head and throat.* Adult: upperparts gray, underparts
whitish; bend of wing rusty, usually hidden. Immature: like
Bushtit, lacks yellow and rufous.
Similarities
Bushtit has longer tail than immature Verdin; usually not in
deserts. Northern Beardless Flycatcher has upperparts olive-gray,
underparts dingy-white; indistinct wing bars and eye-ring.
Habitat
Arid plains, mesquite flats, brushy deserts.
Habits
Very like chickadees and Bushtit; seems not to need water directly.

Voice
Rapid *chip-chip-chip;* an insistent *see-lip;* "song, *tsee, seesee*"
(Peterson).
Eggs
3–5; greenish, spotted; 0.5 x 0.4 in. (1.3 x 1.1 cm). Nest is rough
sphere of thorny twigs, entry on side, in low tree or shrub.
Range
Resident from se. Calif. e. to s. Tex. and s. to n. Mexico.

NUTHATCHES
Family Sittidae

Nuthatches are small, stout, tree-climbing birds with thin, sharp,
strong bills; powerful feet and legs; and short, square tails. They
clamber about tree trunks and branches, often head down, nesting
in cavities which they frequently excavate with their bills in the
soft wood of dead stumps, snags, or branches, from 4–120 ft. (1.2–
36.6 m) up. Their food consists of insects, seeds, nuts, berries, and
fruit.

WHITE-BREASTED NUTHATCH
Sitta carolinensis   **38:2**

Description
Size, 5–6 in. (12.7–15.2 cm). Face white; cap and nape black, back
blue-gray, underparts white, undertail coverts chestnut, white
markings on outer tail feathers, eyes beady-black. Female often has
grayer crown.
Similarities
Red-breasted has black eye stripe. Pygmy has brown cap. Various
chickadees have black bibs.
Habitat
Large trees in deciduous woods and suburbs, mixed forests, groves,
river breaks.
Habits
Tame; comes to feeders, likes suet; stores food in crevices;
nonbreeding birds often travel in mixed flocks with chickadees,
kinglets, Downy Woodpeckers, and creepers; flight slightly
undulating.
Voice
Nasal *yank-yank-yank;* a short, high conversational *hit-hit;* song, a
low-pitched *"to what what what what"* (Thoreau).
Eggs
4–8; white or pinkish, spotted with brown or lavender; 0.8 x 0.6
in. (2.2 x 1.6 cm). Nest is tree cavity 2–60 ft. (0.6–18.3 m) up.
Range
Resident throughout much of W. from s. Canada eastward, and s.
to Mexico.

RED-BREASTED NUTHATCH
Sitta canadensis   **38:6**

Description
Size, 4½–4¾ in. (11.4–12.1 cm). • *Broad black line through the eye.*
Male: cap black, white line over eye, upperparts dark blue-gray,
underparts reddish. Female paler.
Similarities
White-breasted has no eye stripe.

Habitat
Usually in conifer forests, except in migration. In winter, almost any place as irregular southward incursions follow depleted food supply in northern ranges.

Habits
Forages actively to ends of branches, captures insects in air; tame, frequents feeders, stores food; flight undulating; often travels in small flocks.

Voice
High-pitched nasal *ink ink ink (ank* or *enk)*, like a little tin horn; much higher than White-breasted's *yank;* song faster, higher, reedier than White-breasted's.

Eggs
4–8; white, spotted with reddish-brown; 0.6–0.8 x 0.5–0.6 in. (1.5–2.0 x 1.3–1.5 cm). Nest is hole in conifer snag.

Range
Breeds from se. Alaska, e. across boreal Canada and s. along Pacific Coast to cen. Calif., and inland to se. Ariz., Colo., w. S.Dak., and w. Nebr.; winters throughout U.S., excluding s. Calif., s. Ariz., and sw. Tex.

PYGMY NUTHATCH
Sitta pygmaea   38:3

Description
Size, 3¾–4½ in. (9.5–11.4 cm). Tiny inhabitant of pines with
• *gray-brown cap meeting black line through eye,* white spot on nape, back blue-gray, underparts whitish.

Similarities
Red-breasted lacks brown head and white underparts.

Habitat
Pines, Douglas fir, foothills.

Habits
Confiding; forages to end of branches, likes treetops; flight jerky; occurs in large flocks that keep up an endless conversation of metallic pips as they travel about, often with warblers, vireos.

Voice
"A high staccato *ti-di, ti-di, ti-di* . . ., when flying, a soft *kit, kit, kit*" (Hoffmann); an excited chattering.

Eggs
6–9; white, spotted; 0.6 x 0.4 in. (1.6 x 1.1 cm). Nest is hole in snag.

Range
Resident from s. B.C. e. to w. S.Dak. and s. to Baja Calif.

CREEPERS
Family Certhiidae

Only one of the world's species of these small, stiff-tailed, slender birds occurs in the West. These birds creep up and around tree trunks, probing for bark insects with slightly curved bills.

BROWN CREEPER
Certhia familiaris   38:1

Description
Size, 5–5¾ in. (12.7–14.6 cm). Above streaked brown, underparts white; pale band across relatively long wing; • *bill thin, curved;* tail stiff, braced for climbing.

Habitat
Woodlands, swamps, groves, shade trees.
Habits
Using tail as prop, it creeps up trunk of tree often in a spiral, or along underside of horizontal branch, feeding as it goes; then flies to base of next tree and repeats; solitary, but often travels with groups of chickadees, nuthatches, woodpeckers, kinglets; comes to feeders for suet, chopped peanuts, peanut butter.
Voice
Long, thin, high *tsee* not quickly repeated in series but longer than Golden-crowned Kinglet's; song, *seee-see-see, swee-swee;* or "thin sibilant *see-ti-wee-tu-wee* or *see-see-see-sisi-see*" (Peterson).
Food
Insects and their eggs and larvae.
Eggs
4–9; white, with reddish-brown spots; 0.6 x 0.5 in. (1.5 x 1.3 cm). Nest is under loose strip of bark low on tree trunk.
Range
Resident from Alaska e. across Canada and s. along Pacific Coast to cen. Calif., inland to s. Calif., se. Ariz., sw. N.Mex., w. Tex., and C. America.

WRENTITS
Family Chamaeidae

Regarded by some authorities as an offshoot of the Old World's 270–290 species of babblers (Family Timaliidae), the Wrentit is the only member of its family. It is a lone species found only as a resident of coastal Oregon and California west of the Sierras, the interior valleys of southern California, and southern deserts.

WRENTIT
Chamaea fasciata   39:2

Description
Size, 6–6½ in. (15.2–16.5 cm). Dark brownish, sparrow-sized, wrenlike bird with soft plumage; long, rounded tail, held slightly cocked; • *streaked brownish breast;* • *eyes white.* Secretive and often identified only by voice.
Similarities
Bushtit is smaller; white below, grayer, less secretive, travels in flocks.
Habitat
Brush, garden shrubs, parks, chaparral, coastal forest edges.
Habits
A skulker in dense cover, infrequently seen in the open; flies only short distances.
Voice
Note, a soft *prr*. Song, staccato "ringing notes on one pitch, *yip . . . yip . . . yip-yip-yip-yip-ytr-tr-tr-tr-tr-tr-r-r-r-r,* running into a trill" (Cogswell).
Food
Insects, berries.
Eggs
3–5; pale blue; 0.7 x 0.5 in. (1.8 x 1.3 cm). Nest is tightly woven cup in low bush.
Range
Resident from n. Oreg. s. along Pacific Coast to Baja Calif. and in foothills of Central Valley (Calif.).

DIPPERS
Family Cinclidae

Only one of these plump, stubby-tailed birds resembling large wrens occurs in the West. Excellent swimmers, they dive, swim under water, and walk on the bottom of streams feeding on aquatic invertebrates, small fishes, and insects.

DIPPER
Cinclus mexicanus   **37:12**

Description
Size, 7–8½ in. (17.8–21.6 cm). Short-tailed, gray, lives along mountain streams. Stocky, • *slate-gray,* blackish on wings and tail; bill horn-colored; legs yellowish.
Habitat
Mountain streams, ponds, waterfalls (lower altitudes in winter).
Habits
Each bird patrols a stretch of fast-flowing mountain stream, flying from one site to the next with buzzy, quail-like flight; bobs or dips up and down when standing still; swims, dives, submerges, walks on bottom; winks conspicuously with whitish nictitating membrane of eye.
Voice
Alarm, a sharp *jijik, jijik;* call, *zeet, zeet, zeet* (sometimes given singly); song, mockingbirdlike in repetition but higher and more wrenlike, beautifully clear and sweet, heard all year.
Eggs
3–6; white; 1.0 x 1.7 in. (2.5 x 0.8 cm). Nest is bulky sphere of moss behind waterfall or under stream bank or bridge, or on ledge.
Other name
Water Ouzel.
Range
Resident from Alaska s. along Pacific Coast to s. Calif. and C. America; also inland to e. slope of Rocky Mountains.

WRENS
Family Troglodytidae

Wrens are plump, vivacious, mostly brown birds generally smaller than sparrows that often carry the tail cocked over the back. They have thin, slightly decurved bills and short rounded wings and tails. Wrens frequent thickets and brush piles, spending much time on or near the ground. Their calls and alarm notes are harsh; some are exceptionally gifted songsters. They feed on insects and spiders.

HOUSE WREN
Troglodytes aedon   **38:15**

Description
Size, 4½–5¼ in. (11.4–13.3 cm). Stubby, plain; has longish tail, grayish underparts, and usually • *no eye stripe;* southeastern Arizona form has buffy eye stripe. Above gray-brown, flanks brownish, faint barring on wings and tail.
Similarities
Winter is smaller, with shorter tail; darker below. Bewick's is larger, with white eyebrow; longer tail, white tail corners.
Habitat
Woods, thickets, gardens, farms, towns.

Habits
Aggressive, scolds intruders; frequents bird bath and feeder; will
nest in wren house; male builds extra dummy nests.
Voice
Song, a long, gushing, bubbling melody; alarm note, a grating
chatter; a harsh scold.
Eggs
5–12; pinkish, heavily dotted with reddish-brown; 0.6 x 0.5 in. (1.5
x 1.3 cm). Nest is of twigs, etc., in virtually any cavity or bird box.
Range
Breeds from s. Canada throughout most of U.S.; winters from s.
U.S. s. to Mexico.

WINTER WREN
Troglodytes troglodytes   **38:12**

Description
Size, 4–4¼ in. (10.2–11.4 cm). Tiny, • *stub-tailed,* with a faint light
line over eye. Above and on belly dark brown, breast lighter, • *belly
barred with black.*
Similarities
House is paler, grayer, with longer tail.
Habitat
In summer, coniferous forests and underbrush; sea cliffs in Alaska;
in migration, thickets, brush piles.
Habits
Secretive, mouselike; bobs up and down, cocks tail; feeds along
stone walls, woodpiles, banks of streams; keeps near cover.
Voice
Song (one of loveliest in North America), a "rapid succession of
high tinkling warbles, trills, long sustained, often end[ing] on a
very high trill" (Peterson); call, a *tik, tik;* a hard *kip* or *kip kip.*
Eggs
4–10; white, dotted with red-brown and purple; 0.7 x 0.5 in. (1.8 x
1.3 cm). Nest is of sticks and moss in tangle near ground or in root
mass of tree; in crevice, rocks.
Range
Breeds from Alaska e. across Canada and s. through Pacific states
to cen. Calif.; also inland in n. Rockies to nw. Wyo.; winters from
Alaska s. along coast to s. Calif. and inland throughout Nev., Utah,
Colo., ne. N.Mex., and Tex., eastward.

CACTUS WREN
Campylorhynchus brunneicapillus   **40:12**

Description
Size, 7–8¾ in. (17.8–22.2 cm). Largest wren, inhabits arid regions.
Primarily distinguished from other wrens by its large size and
heavy spotting of underparts; in adults spotting clustered on upper
breast; distinctive broad, white eye stripe extending from forehead
to back of neck; white spots on outer tail.
Similarities
Sage Thrasher has gray back, unstriped, 2 white wing bars.
Habitat
Arid flats, deserts; cactus, yucca, mesquite.
Habits
Appears thrasherlike; very vocal; builds many nests.
Voice
An unbirdlike harsh loud monotone *chuh-chuh-chuh-chuh* or *chug-
chug-chug-chug.*

Eggs
4–5; white, reddish-spotted; 0.9 x 0.6 in. (2.3 x 1.6 cm). Nest is straw spheroid (like a football), entrance on one side, in cactus or thorny bush.

Range
Resident in sw. U.S. from s. Calif., s. Nev., cen. and s. Ariz., and w. Tex., s. to Mexico.

BEWICK'S WREN
Thryomanes bewickii   38:14

Description
Size, 5–5½ in. (12.7–14.0 cm). Unspotted; white eye stripe, • *white tail corners.* Brown above, whitish below, fanlike tail longer than wings. Winter colors are brighter.

Similarities
House lacks white eye stripe and tail corners. Rock is grayer, tail tips buffy, breast finely streaked.

Habitat
Woodlands, thickets, underbrush, gardens, farms, chaparral, pinyon, junipers.

Habits
Less aggressive than House Wren, more deliberate; wobbles tail from side to side, comes to feeders.

Voice
Song, loud and beautiful, variable, often confused with that of Song Sparrow, but longer, more varied, ending on a trill; alarm, a chatter; a rasping scold note.

Eggs
4–7; white or pinkish, heavily dotted with red-brown; 0.7 x 0.5 in. (1.8 x 1.3 cm). Nest is of twigs, etc., in cavities virtually anywhere; also in boxes.

Range
Resident locally from B.C. e. across Canada and s. to Mexico.

Note: The **CAROLINA WREN,** *Thyrothorus ludovicianus,* size, 6 in. (15.2 cm), is seen in central Texas. It is larger than Bewick's, with reddish-brown above and no white in tail.

LONG-BILLED MARSH WREN
Cistothorus palustris   38:10

Description
Size, 4¾–5½ in. (12.1–14.0 cm). Small, with conspicuous white eye stripe and • *white stripes on dark brown back.* Blackish unstreaked crown; whitish below, including undertail coverts.

Similarities
Bewick's has no white lines on back, longer white-cornered tail. Short-billed has buffy breastband, streaked crown, inconspicuous buffy eye stripe, brownish undertail coverts; uncommon in West.

Habitat
Freshwater cattail or salt marshes; tules, bulrushes.

Habits
Sings from stalks in marsh, often rises 6–10 ft. (1.8–3.0 m) in air and flutters down singing, climbs about reeds; males often build extra dummy nests.

Voice
Short, rich, bubbling, somewhat guttural song ending in a rattle; *cut-cut-turrrrrrrrr-ur;* a low *tsuck;* sometimes sings at night.

Eggs
3–7; pale to dark brown, often with darker dots; 0.6 x 0.5 in. (1.5 x 1.3 cm). Nest is coconut-shaped grass ball, entrance on one side, attached to marsh reeds.
Range
Breeds throughout much of s. Canada and most of U.S. from cen. B.C. e. across to cen. Sask., eastward, and s. to s. Calif., cen. N.Mex., and w. Tex.; winters along Pacific states and s. from w.-cen. Utah, sw. N.Mex.

SHORT-BILLED MARSH WREN
Cistothorus platensis   38:9

Description
Size, 4–4½ in. (10.2–11.4 cm). Tiny, with a buffy breastband, streaked crown and back, inconspicuous eye stripe. Stubby; bill and tail short; brownish above; abdomen very buffy, throat and lower breast white; undertail coverts brownish.
Similarities
Long-billed has conspicuous white eye stripe, unstreaked crown, undertail coverts white.
Habitat
Short-grass marshes, fresh or salt, wet meadows.
Habits
Secretive, often nocturnal, stays near ground, sings from weed stalks; male constructs extra dummy nests.
Voice
Short series of clinking notes, somewhat like pebbles knocked together: *chap, chap, chapper, chapper, chapper,* running down-scale and increasing in tempo.
Eggs
4–10; white; 0.6 x 0.5 in. (1.5 x 1.3 cm). Nest is round grass ball with side entrance, attached to growing stalks in wet meadow.
Range
Breeds chiefly in E., but w. to cen. Sask.; winters from Gulf Coast, southward; migrates rarely or casually to w. parts of Great Plains from Alta. to Tex.

ROCK WREN
Salpinctes obsoletus   38:17

Description
Size, 5–6¼ in. (12.7–15.9 cm). • *Finely streaked* white breast distinguishable at close range. Upperparts gray-brown; belly pale tan; white line over eye; tail rounded, with black band just before • *buffy tip,* corners conspicuously buffy.
Similarities
Bewick's is darker above, no streaks on breast. Canyon has white throat, dark belly, no eye stripe.
Habitat
Open rocky places (including deserts), mountains above timberline, plains, talus slopes, walls, rock dams.
Habits
Bobs up and down on rock, stays close to ground; flight quick, jerky.
Voice
Call, *TICK-ear,* often given with a bob; a loud purring trill on one pitch; song, a "harsh, grating *kerEE kerEE kerEE . . . chair chair chair chair chair, deedle deedle deedle, tur tur tur tur, kerEE kerEE kerEE trrrrrrr*" (Nice).

Eggs
5–8; white, lightly dotted with red-brown; 0.7 x 0.5 in. (1.8 x 1.3 cm). Nest is cup of grass or moss in rocky cranny, wall; often with rock chips on entry path.

Range
Breeds from s. B.C. e. to Sask. and s. throughout w. states to C. America; winters from Calif. southward.

CANYON WREN
Catherpes mexicanus
38:18

Description
Size, 5½–5¾ in. (14.0–14.6 cm). Conspicuous • *white throat and breast* (bib). • *Appears rusty;* no eye stripe; brown above spotted with white on back; • *belly dull chestnut;* long tail finely barred with black; bill, long, thin.

Similarities
Much paler Rock Wren has streaked breast, eye stripe; grayer above; black band near end of tail.

Habitat
Ravines, canyons, cities, cliffs, rockslides.

Habits
Creeps about rocky ledges, keeps close to ground.

Voice
Descending scale of 7–12 rich, vibrant, bell-like notes, often echoing from cliff to cliff with first notes sometimes single: *"te-you te-you te-you tew tew tew tew* or *tee tee tee tee tew tew tew tew"* (Peterson).

Eggs
4–6; white, finely dotted with red-brown; 0.7 x 0.5 in. (1.8 x 1.3 cm). Nest is cup of twigs or moss in cavity in rocks, crevice, building.

Range
Resident from s. B.C. e. to Mont. and s. to Mexico.

MOCKINGBIRDS AND THRASHERS
Family Mimidae

These slim, medium-sized birds are strong-legged, usually longer-tailed than true thrushes, have a generally more decurved bill, are excellent songsters, and are noted for their ability to mimic. They feed on wild berries, insects and fruit on the ground in thickets or in the open.

NORTHERN MOCKINGBIRD
Mimus polyglottos
40:3

Description
Size, 9–11 in. (22.9–27.9 cm). Long-tailed; gray; • *large white patch on wings* and outer tail, conspicuous in flight. Slender; gray above, whitish below; face gray; faint whitish eye stripe; outer tail feathers with much white.

Similarities
Shrikes are stouter, with hooked bills, shorter tails, black facial masks. Townsend's Solitaire is darker below, with white eye-ring, buffy wing patch.

Habitat
Gardens, farms, roadsides, towns, brush, mesquite, desert streamsides.

Habits
Active, aggressive; runs, hops, feeds on ground with tail up;
periodically raises wings to display patches; sings from fence post,
TV antenna, or roadside wire; frequents feeders (likes raisins,
nutmeats, suet, crumbs).

Voice
This species is one of the world's great songsters and an
accomplished mimic; often sings in flight and by moonlight; call, a
harsh *chak;* song varied, mellifluous, some notes often repeated 3
times.

Eggs
3–6; greenish-blue, heavily dotted with red-brown; 1.1 x 0.8 in.
(2.8 x 2.0 cm); often 2 broods. Nest is bulky, twiggy, rootlet-lined
cup on or near ground in thicket or dense tree.

Range
Resident throughout much of U.S., except northernmost areas,
from s. Oreg. e. across U.S. and s. to Mexico.

GRAY CATBIRD
Dumetella carolinensis   40:4

Description
Size, 8–9½ in. (20.3–24.1 cm). Slate-gray, with black cap;
• *undertail coverts chestnut,* tail blackish.

Similarities
In western Texas, Brown Towhee is often locally called "catbird."
Other dark all-gray songbirds are similar; see females of Brewer's
Blackbird, cowbirds, Rusty Blackbird, young Starling and
Dipper—all lack black cap, chestnut undertail coverts. Robin is
larger and heavier.

Habitat
Wet or dry thickets, hedges, underbrush.

Habits
Inquisitive, comes to feeders (likes bread, raisins, chopped peanuts,
suet), often nests near houses; usually sings from thicket; flips tail
jauntily; skulks in undergrowth.

Voice
Song, a squeaky, simple version of Northern Mockingbird's or
Brown Thrasher's, but interspersed with catlike mews (phrases
never repeated); something of a mimic; often sings at night, one of
earliest songs heard at dawn.

Eggs
3–6; greenish-blue; 1.0 x 0.7 in. (2.5 x 1.8 cm); 2 broods. Nest is
of twigs, a cup lined with rootlets 2–8 ft. (0.6–2.4 m) up in brush.

Range
Breeds from Wash. e. inland across to Great Lakes and s. inland to
e. Ariz.; winters in se. U.S. southward.

BROWN THRASHER
Toxostoma rufum   40:1

Description
Size, 10½–12 in. (26.7–30.5 cm). Long-tailed, red-brown, with a
streaked breast. Slim; • *reddish-brown above,* whitish below, with
• *brown streaks on breast;* eye yellow; bill long, slightly curved; 2
white wing bars.

Similarities
Thrushes have shorter tails, spots instead of streaks, no wing bars,
eyes brown.

Habitat
Dry thickets, brush.

Habits
Aggressive; sings from lofty perch; scratches for food on ground in thicket; walks, runs or hops; less common about habitations than Gray Catbird.

Voice
Like Gray Catbird's call or Northern Mockingbird's; repeats notes or phrases twice; few imitations: *"drop it, drop it, cover it, cover it, I'll pull it up, I'll pull it up"* (Thoreau); call, a distinctive kissing sound.

Eggs
3–6; whitish, heavily dotted with red-brown; 1.1 x 0.8 in. (2.8 x 2.0 cm); often 2 broods. Nest is bulky twig cup on or near ground in thicket.

Range
Breeds chiefly in E., but w. to Alta., and s. to Mont., e. Colo., and Tex. Panhandle; winters sparsely in e. Colo.; casual or accidental to Oreg., Utah, Calif., Ariz., N.Mex.

Note: In southern Texas the **LONG-BILLED THRASHER**, *Toxostoma longirostre* **(40:2)**, size, 10 in. (25.4 cm), occurs in situations similar to those of Brown Thrasher. It is distinguished from the latter by its less rusty color, darker crown, and gray face, with black streaks below.

BENDIRE'S THRASHER
Toxostoma bendirei

40:8

Description
Size, 9–11 in. (22.9–27.9 cm). Breast faintly spotted, bill short and straight, eye clear yellow.

Similarities
Curve-billed has longer bill, curved; eyes more orange; tail darker than back.

Habitat
Arid regions, thorny brush, cactus, desert farms. Common.

Habits
Often cocks tail.

Voice
Note, a soft *tirup;* song, steady, clear unbroken 2-syllable warble.

Eggs
3–4; light greenish, blotched; 1.0 x 0.7 in. (2.6 x 1.8 cm). Nest is cup in thorny shrub.

Range
Breeds in Sonora Desert from s. Utah and se. N.Mex. s. to Baja Calif. and Mexico; winters from s. Ariz. southward.

Note: Several quite similar slim, Robin-sized thrashers inhabit the southwest deserts. They all have gray-brown backs, long tails, and decurved bills.

CURVE-BILLED THRASHER
Toxostoma curvirostre

40:10

Description
Size, 9½–11½ in. (24.1–29.2 cm). • *Bill well-curved,* breast faintly covered with • *blurred spots,* eye reddish to pale orange, vague narrow white wing bars; tail darker than back.

Similarities
Bendire's has bill shorter, straighter; eyes paler; tail colored as back.

Habitat
Desert shrubs, suburbs, brush, slopes, cactus. Commonest of desert thrashers.
Voice
Note, a sharp *whit-wheet!;* song, a fast series of musical notes and phrases with little repetition.
Eggs
2–4; light greenish-blue, spotted; 1.1 x 0.8 in. (2.8 x 2.1 cm). Nest is cup of twigs in cholla cactus or shrub.
Range
Resident in s. Ariz., e. to w. Tex. and s. to Mexico.

LE CONTE'S THRASHER
Toxostoma lecontei   40:9

Description
Size, 10–11 in. (25.4–27.9 cm). Palest, grayest, and plainest of the desert thrashers, eyes dark, bill curved.
Similarities
Curve-billed has spots on breast, eye pale orange or reddish.
Crissal has pale eyes, rusty under tail, much darker. California is much darker, pale eyebrow, patterned face.
Habitat
Sparsely vegetated slopes and deserts, arid flats.
Habits
Shy; moves well ahead of one in its open habitat.
Voice
Note, *ti-rup;* song, like other species but infrequent, more broken, less repetitious.
Eggs
2–4; light greenish-blue, spotted; 1.1 x 0.7 in. (2.8 x 1.8 cm).
Range
Resident in Sonora Desert and from s. Calif., e. to Ariz. and s. to Baja Calif.

CRISSAL THRASHER
Toxostoma dorsale   40:11

Description
Size, 10½–12½ in. (26.7–31.8 cm). A desert thrasher, in areas where vegetation is dense. Generally dark, • *undertail patch deep chestnut,* breast uniform gray-brown, bill deeply curved, eyes yellowish.
Similarities
Curve-billed is grayer, no dark undertail patch, breast faintly spotted. California is paler under tail, darker eye.
Habitat
Mesquite or other dense streamside thickets, dense irrigated farm vegetation.
Habits
Difficult to observe in its dense habitat; rarely seen in open.
Voice
Song, rather sweet, long; note, an often repeated (2–3 times) *"pichoory* or *chideary"* (Peterson).
Eggs
2–4; solid light greenish-blue; 1.0 x 0.7 in. (2.6 x 1.9 cm). Nest is cup of twigs in desert shrub or mesquite.
Range
Resident from se. Calif. e. to w. Tex. and s. to Baja Calif. and cen. Mexico.

CALIFORNIA THRASHER
Toxostoma redivivum

40:13

Description
Size, 11–13 in. (27.9–33.0 cm). Largest of the thrashers.
Upperparts dull gray-brown, belly and • *undertail coverts pale cinnamon,* face patterned brown, pale eyebrow, tail long, • *long bill deeply curved,* eyes dark brown.

Similarities
None in its normal range. Le Conte's in desert overlap area is very pale gray, no eyebrow line. Crissal in desert overlap in darker, pale eye, undertail patch dark chestnut.

Habitat
Foothills, parks, gardens, thickets, chaparral.

Habits
Runs about on ground.

Voice
Note, "a dry *chak,* a sharp *g-leek*" (Peterson); song, a leisurely, long-sustained series of notes and phrases, with brief repetitions, some musical, some harsh.

Eggs
2–4; light blue, dotted; 1.1 x 0.8 in. (2.8 x 2.1 cm). Nest is cup of twigs in bush.

Range
Resident only in Calif. w. of Sierras and s. to Baja Calif.

SAGE THRASHER
Oreoscoptes montanus

40:5

Description
Size, 8–9 in. (20.3–22.9 cm). Robin-shaped but a bit smaller. Back gray, white breast streaked with black; tail darker, with white corners; 2 white wing bars; bill short, Robin-like; eyes yellow.

Similarities
Immature Robin has dark eye, some rusty below, shorter tail. Cactus Wren is less gray, heavy spotting concentrated on upper breast, white eye stripe. Immature Northern Mockingbird is spotted but has large white areas in wing and tail.

Habitat
Summer, sagebrush plains, mesas, brushy slopes; winter, deserts, lowland thickets.

Habits
Pumps tail when perched, feeds on ground, runs with tail high, sings from prominent perch with head raised, tail drooping.

Voice
Song, lengthy, somewhat like Brown Thrasher's, many phrases repeated but no pauses between; "a long succession of warbling phrases with very little range of pitch" (Hoffmann); also sings in flight; alarm note, a blackbirdlike *chuck, chuck.*

Eggs
4–5; deep blue, boldly spotted with brown; 1.0 x 0.7 in. (2.5 x 1.8 cm). Nest is bulky cup of bark strips, twigs, grasses in sagebrush, 1–3 ft. (0.3–0.9 m) up.

Range
Breeds from s.-cen. B.C. s. to s. Nev., Utah, and Tex. and in s. Calif.; winters from cen. Calif. s. to Mexico.

THRUSHES, BLUEBIRDS, AND ALLIES
Family Turdidae

Noted songbirds, most thrushes are slender-billed, stout-legged woodland birds that forage on the ground; many are brown-backed and have spotted breasts, as do all their young. They feed on insects, spiders, worms, grubs, wild fruits, berries and seeds.

AMERICAN ROBIN
Turdus migratorius   41:2

Description
Size, 9–11 in. (22.9–27.9 cm). Slaty back and • *brick-red breast;* common harbinger of spring. Male: bill yellow, head black, tail blackish, belly and tips of tail feathers white. Female: much paler head, tail. Juvenal: back gray, underparts rusty, buffy breast speckled with black.

Habitat
Open places, even in conifer forests; lawns, gardens, suburbs, fields, clearings, farms.

Habits
Adapts well to civilization, hops or runs on lawn, nests near houses; loose flocks spend winters in woods and swamps, sometimes gather in huge roosts; one of first species to sing at dawn; flight direct, with deep, fairly rapid wingbeats.

Voice
Song, a carol, *cheerily cheer-up, cheerily cheer-up;* notes include a querulous *cuk, cuk;* a *sssp;* a nervous *bup, bup;* a *tyeep;* others.

Eggs
3–4; Robin's-egg blue; 1.1 x 0.8 in. (2.8 x 2.0 cm); 2 broods. Nest is cup lined with grass, walled with mud, in crotch or branch of tree or in man-made nook, usually 5–20 ft. (1.5–6.1 m) up.

Range
Breeds from n. Canada s. throughout most of Canada and U.S.; winters from n. U.S. southward.

VARIED THRUSH
Zoothera naevia   41:1

Description
Size, 9–10 in. (22.9–25.4 cm). Male: cap, back, and breast like Robin, but • *black band across rusty breast;* • *orangish stripe over eye,* bill black, • *wing bars orange.* Female: faint gray breastband, duller. Immature: breastband indistinct or speckled with orange, underparts speckled with dusky.

Similarities
Immature Robin has no wing bars or eye stripe.

Habitat
Wet forests, conifers, in winter ravines, thickets, wet woods.

Habits
Robin-like.

Voice
Prolonged quavering whistle, pause, second note on different pitch.

Eggs
3–5; blue, spotted; 1.2 x 0.8 in. (3.1 x 2.1 cm). Nest is twig or moss cup in tree.

Range
Breeds from Alaska s. to n. Calif.; winters from s. B.C. southward.

HERMIT THRUSH
Catharus guttatus

41:5

Description
Size, 6½–7¾ in. (16.5–19.7 cm). Bright • *reddish-brown tail* contrasts with brown head. Russet-brown above; hint of rufous in primaries; below whitish with small, dark, wedge-shaped spots on breast.

Similarities
Other thrushes are more uniformly plumaged above. Fox Sparrow has thick bill, entire upperparts including tail reddish-brown.

Habitat
Mixed woods, forest floor; in winter, thickets, parks, woods.

Habits
Raises and slowly lowers tail on alighting; usually seen near ground, but sings from treetop; hardy, a few linger in north in winter.

Voice
Song, clear, flutelike introductory note, then various 5- to 12-note phrases, a pause, then another introductory note on a different pitch, etc.; notes are clear, bell-like; also a soft *chuck*, a catlike harsh *pay*, a scolding *tuk-tuk-tuk*.

Eggs
3–5; greenish-blue; 0.9 x 0.7 in. (2.3 x 1.8 cm). Nest is cup of moss, twigs, grasses on or near ground.

Range
Breeds from Alaska e. across Canada and s. to Calif.; winters mainly in s. U.S.

VEERY
Catharus fuscescens

41:8

Description
Size, 6½–7¾ in. (16.5–19.7 cm). Uniformly rusty back and tail.
• *Above light russet-brown or tawny,* below light buff with • *inconspicuous spots clustered on throat;* may or may not show a dull whitish eye-ring. Least spotted of the brown thrushes.

Similarities
Swainson's is same size but has broad buffy eye-ring and browner, less rufous upperparts. Hermit has reddish tail.

Habitat
Wet deciduous woodlands, swampy undergrowth, willows, ravines.

Habits
Keeps close to ground, even when singing; progresses by jumping, upright stance.

Voice
Song, a series of hollow *whree-u*'s descending the scale, or *"ta-weel'ah, ta-weel'ah, twil'ah, twil'ah"* (Ridgeway); *"vee-ur, vee-ur, veer, veer"* (Peterson); a distinctive *view* or low *phew;* a dawn chorus in breeding season.

Eggs
3–5; greenish-blue; 0.9 x 0.7 in. (2.3 x 1.8 cm). Nest is cup of grass or twigs on or near ground.

Range
Breeds from B.C. e. across s. Canada and s. from Rockies to Ariz.; winters in S. America; migrates somewhat through sw. states.

SWAINSON'S THRUSH
Catharus ustulatus   **41:9**

Description
Size, 6½–7¾ in. (16.5–19.7 cm). Olive-backed, with buffy cheeks
and eye-ring. Above uniformly dark olive-brown, breast light buff
with small dark spots, belly whitish, • *buffy eye-ring* and cheek
contrast with rest of head.

Similarities
The 3 species of *Catharus* that include Swainson's, Gray-cheeked
(see below), and Veery are quite similar. Veery is much rustier on
back and tail. Hermit has reddish tail.

Habitat
Tundra willow thickets, northern coniferous forests; aspens, river
woods, undergrowth; in migration, any woods or shrubbery.

Habits
If flushed, flies to low branch and remains motionless, then flits on
farther; sometimes pumps tail, but not habitually as Hermit does;
sings from treetop.

Voice
Song, a melodious series of ascending flutelike phrases tending to
spiral *"whao-whayo-whiyo-wheya-wheeya"* (Saunders); note, a
whit, a catlike *twee-ur twee-ur,* a short *heep.*

Eggs
3–5; greenish-blue, spotted with reddish-brown; 0.9 x 0.7 in. (2.3 x
1.8 cm). Nest is cup of grass, moss, twigs, rootlets, dead leaves,
usually in fir or spruce 4–15 ft. (1.2–4.6 m) up.

Range
Breeds from cen. Canada s. to Great Lakes, and in Nw. and
Rockies; winters s. of U.S.; migrates through sw. states.

Note: The **GRAY-CHEEKED THRUSH,** *Catharus minimus,* size, 7 in.
(17.8 cm), reaches the western Plains in migration and breeds from
northern British Columbia, Saskatchewan, and southern Alaska. It
differs from Swainson's in its grayish cheeks.

WESTERN BLUEBIRD
Sialia mexicana   **41:4**

Description
Size, 6½–7 in. (16.5–17.8 cm). Male: like Eastern, but • *head,
wings, tail blue;* breast and • *patch on back rusty-red.* Female:
paler, duller, throat gray, lacks reddish.

Similarities
Eastern has blue back, rusty breast and throat, throat of female
whitish. Mountain lacks rusty in all plumages. Male Lazuli
Bunting is smaller, with white wing bars.

Habitat
Farms, open conifer forests, scattered trees; in winter, plains,
deserts, brush, semiopen regions.

Habits
When perching, appears hunched or round-shouldered.

Voice
Note, a harsh chattering; a catlike *mew* or *pew.*

Eggs
4–6; light blue; 0.7 x 0.6 in. (2.0 x 1.6 cm).

Range
Breeds throughout much of w. U.S. from Wash. through Colo.,
Ariz., and N.Mex., s. to Mexico; winters more to w., from Wash.
s. through Calif. to Mexico.

EASTERN BLUEBIRD
Sialia sialis   **41:6**

Description
Size, 6½–7½ in. (16.5–19.1 cm). Spring male: • *bright blue above,
breast reddish,* belly white, throat rusty. Spring female: gray-brown
with bright blue on wings and tail. Juvenal: spotted above and
below. Fall adults: duller.

Similarities
Western Bluebird male has rusty patch on back, throat blue;
female has rusty breast, grayish throat.

Habitat
Orchards, wood edges, roadsides, farmlands, open country with
scattered trees.

Habits
Often seen on roadside wire, hunches when perched; catches some
insects in air, drops from perch to pounce on others.

Voice
Song, a soft mellow warble, *purity, purity;* call, *oola* or *aloola;* note,
"a musical *chur-wi* or *tru-ly*" (Peterson).

Eggs
3–7; light blue; 0.8 x 0.6 in. (2.0 x 1.5 cm); 2 broods. Nest is hole
(often abandoned woodpecker's) in stump of tree, post or bird box
3–6 ft. (0.9–1.8 m) up, entrance 1.5 in. (3.8 cm) dia.

Range
Breeds e. of Rocky Mountains from s. Canada to Gulf of Mexico,
s. to Mexico; winters from se. Ariz. southward.

MOUNTAIN BLUEBIRD
Sialia currucoides   **41:7**

Description
Size, 6½–7¾ in. (16.5–19.7 cm). Male: bright blue above,
• *turquoise breast,* belly white. Female: brownish-gray, paler below;
blue on wings, rump, and tail. Both sexes duller in winter,

Similarities
None; other bluebirds have rusty breasts. Juvenal has much less
spotting than Eastern. Western female has breast with rusty wash.
Blue Grosbeak male has brown wing bars, thick bill. Indigo
Bunting male is smaller, darker blue, breast like back.

Habitat
Open woods in mountains, scattered trees, ranches, rangeland.

Habits
Seizes insects in air, or hovers and then drops onto them on the
ground.

Voice
Usually silent; near dawn a "sweet, clear *trually trual-ly* like
Eastern, and a mellow warble" (Wheelock); note, a low *turr, chur,*
or *phew.*

Eggs
4–6; pale blue; 0.8 x 0.6 in. (2.0 x 1.5 cm); usually 2 broods. Nest
is grass-lined hole in snag, tree, cliffside or beneath eaves of
building or in bird box.

Range
Breeds from Alaska, n. B.C., and Alta. s. to mountains of s. Calif.
and w. Okla.; winters from Vancouver southward, at lower
elevations.

Note: The **COMMON WHEATEAR,** *Oenanthe oenanthe,* size, 5½–6
in. (14.0–15.2 cm), is mainly a Eurasian bird, with a white rump
and black inverted T on its tail. The male in spring has a blue-

gray back, black wings and ear patch, buff underparts, and a white stripe over the eye; the female and fall male are brown above, buff below. It inhabits the tundra barrens and rocky slopes of northern Alaska and the Yukon, south only to Mt. McKinley National Park and southwest Yukon. The **EURASIAN BLUETHROAT,** *Luscinia svecica,* size, 4¾ in. (12.0 cm), is a brown thrush with a red spot on a blue throat (males) or a white throat bordered black (females). This bird breeds in the tundra of northern Alaska.

TOWNSEND'S SOLITAIRE
Myadestes townsendi   41:3

Description
Size, 8–9½ in. (20.3–24.1 cm). Body slim, tail long, bill short. Gray with • *white eye-ring* and chin, • *outer tail feathers white, buffy wing patches* conspicuous in flight.

Similarities
The beginner is cautioned not to confuse this species with thrashers or flycatchers. Northern Mockingbird superficially is similar with light wing patches and white in tail, but Townsend's has eye-ring and darker breast.

Habitat
Open woods in mountains, canyons; in winter, ravines, canyons, brushy slopes, dryland timber.

Habits
Catches insects in air or pounces on them from perch; skulks in thickets; flight with slow-beating wings reminiscent of Say's Phoebe.

Voice
Song, a loud "prolonged, warble-like series of rapid notes, each note on a different pitch" (Saunders); sings in flight; call, a loud, metallic *tink, tink;* note, a sharp *eek.*

Eggs
3–5; whitish, spotted and scrawled with brown; 0.9 x 0.7 in. (2.3 x 1.8 cm). Nest is cup of grass or pine needles on ground (well hidden), among rocks, in stump, bank or cliff.

Range
Breeds from Alaska and n. B.C. s. to Mexico and Baja Calif., always w. of Rocky Mountains; winters from Vancouver s. to Mexico.

GNATCATCHERS, KINGLETS, AND OLD-WORLD WARBLERS
Family Sylviidae

This is a family of tiny, active woodland birds with small bills adapted to a diet of insects. Gnatcatchers are distinguishable by their comparatively long tails, kinglets by their brightly colored crowns.

BLACK-TAILED GNATCATCHER
Polioptila melanura   39:8

Description
Size, 4½ in. (11.4 cm). Very similar to Blue-gray, except breeding male has • *black cap;* less white in tail. Winter male and female: • *undertail surface mainly black,* narrow white edges.

Habitat
Desert washes, ravines, brush, sage, mesquite.
Habits
As Blue-gray Gnatcatcher.
Voice
Note, "a hard thin *chee* repeated 2–3 times, or *pee-ee-ee*" (Peterson).
Eggs
3–4; bluish-white, spotted; 0.5 x 0.4 in. (1.3 x 1.1 cm). Nest is tiny felted cup in desert shrub.
Range
Resident from s. Calif., Nev., Ariz., and N.Mex. s. to Mexico.

BLUE-GRAY GNATCATCHER
Polioptila caerulea   **38:13**
Description
Size, 4–5 in. (10.2–12.7 cm). Small, gray, with long, • *white-bordered black tail*. Slender; above blue-gray; eye-ring and underparts white; • *tail long,* mobile, contrastingly white and black, the undersurface being mainly white with a narrow black center. Breeding male: forecrown black.
Similarities
Black-tailed Gnatcatcher breeding male has black cap; winter male, female, and immature told by mainly black underside of tail.
Habitat
Thickets, swamps, open mixed woods, tall trees, oaks, chaparral, junipers, pinyon.
Habits
Active, graceful, tail always in motion and often cocked like a wren's; often hovers in front of a twig, frequently catches insects in air.
Voice
Call, a shrill, high-pitched, scolding *spee, spee, spee;* song, "rarely heard, a low, pleasant, exquisite, warbling ditty ... *zee-u, zee-u, ksee-ksee-ksee-ksee-ksee-ksu*" (Simmons).
Eggs
4–5; pale bluish-white, spotted with red-brown and slate; 0.6 x 0.5 in. (1.5 x 1.4 cm). Nest is small cup on limb, composed of plant down, bound with spider webs, shingled with lichens ... "superbly beautiful to human eyes" (Collins).
Range
Breeds from n. Calif. e. to s. Wyo. and across U.S., and s. to C. America; winters in s. U.S. and C. America.

GOLDEN-CROWNED KINGLET
Regulus satrapa   **43:10**
Description
Size, 3½–4 in. (8.9–10.2 cm). Very small, with black-bordered orange or yellow crown. Above olive-gray; underparts, 2 wing bars, and line over eye white; • *crown yellow* (female), or • *yellow with orange center* (male) (sometimes concealed), stands out when bird is excited; bill tiny; tail short, notched; plumage fluffy.
Similarities
Ruby-crowned has broken white eye-ring, no white line over eye, male crown patch scarlet.
Habitat
Conifers; other trees and suburban evergreens in winter; seashores, etc., in migration.

Habits
Tame, confiding, very active, flutters at ends of twigs, often flips out wing tips over back; catches insects in air as well as on twigs; in winter, frequently travels in mixed groups with chickadees, nuthatches and Downy Woodpeckers.

Voice
Call, a rapidly repeated high, thin *see see see.* Song, similar notes, ascending in pitch then dropping *"zee, zee, zee, zee, zee, why do you shilly-shally?"* (Stanwood).

Eggs
5–10; creamy, spotted with brown and lavender; 0.5 x 0.4 in. (1.3 x 1.0 cm). Nest is spherical, of moss, lichens, entrance at top, in conifer.

Range
Breeds throughout much of N. America from s. Alaska and cen. Canada, s. to high mountains of s. Calif., se. Ariz., n. N.Mex.; winters from s. Canada s. to C. America.

RUBY-CROWNED KINGLET
Regulus calendula   42:18, 43:12

Description
Size, 3¾–4⅓ in. (9.5–11.4 cm). Tiny, green, short-tailed, with a broken white eye-ring, giving the bird a "big-eyed" appearance. Bill tiny, tail notched; above olive-gray, below whitish; 2 wing bars, rear one with black border; male has concealed ruby patch on crown, visible when bird erects it in display.

Similarities
Any kinglet lacking a visible crown patch is a Ruby-crowned. *Empidonax* flycatchers have white eye-ring, but they sit still and flycatch from a perch; do not flit about. Hutton's Vireo has white eye-ring, heavier bill, with slight hook; has white extension to bill, thus "spectacled"; more deliberate.

Habitat
Breeds in Canadian zone bogs and conifers; in migration, found in shrubbery, orchards, second growth, swampy thickets.

Habits
Tame, active, restless; flips wings tips out over back, seems always in motion; often associated with warblers and nuthatches in migration.

Voice
Song, loud and ringing, going up and down, *"see-see-see you-you-you just-look-at-me just-look-at-me"* (Cruickshank); note, harsh, scolding, wrenlike; a husky *ji-dit.*

Eggs
5–1; creamy-white, usually finely dotted with red-brown; 0.5 x 0.4 in. (1.4 x 1.1 cm). Nest is like Golden-crown's, suspended 2–50 ft. (0.6–15.2 m) up.

Range
Breeds throughout most of N. America from Alaska e. across Canada and s. along Rockies to Ariz.; also along Pacific Coast to mountains of s. Calif.; winters from s. B.C. s. to C. America.

WAGTAILS AND PIPITS
Family Motacillidae

This family includes open country, slender larklike birds with thin bills and long hind claws. The pipits are brownish above, lighter below, some with streaked breasts and white outer feathers on their tails. The wagtails are strongly patterned, with long tails and slender legs. They sing from aloft and have a buoyant, erratic, flight. They eat insects, grubs, small mollusks, and crustaceans.

WATER PIPIT
Anthus spinoletta   **52:5**

Description
Size, 6–7 in. (15.2–17.8 cm). Unstreaked dark back and dark legs. Spring: gray-brown above, • *buffy below,* with streaks on upper breast. Fall: darker above, buffier below, breast more heavily streaked. • *Bill slender;* • *outer tail feathers white.*

Similarities
Vesper Sparrow has short, thick bill; hops, does not walk or pump tail. Horned Lark has yellow on face, with "horns." Longspurs have conical bills, do not pump tails.

Habitat
Alpine zone tundras; in migration and winter, plowed fields, dunes, flats, plains, shores.

Habits
Terrestrial, restless, nods head as it walks, usually in flocks; pumps tail more than Sprague's, dips up and down in flight. Most pipits are detected by voice when flying over.

Voice
Note, heard in migration; "a sharp *tsip-tsip, tsip-it*" (Cogswell), or "a thin *jeet* or *jee-eet*" (Peterson); song on breeding grounds is simple, pleasant often with trills; in flight, *"chwee chwee chwee chwee chwee chwee chwee"* (Peterson).

Eggs
4–7; whitish, heavily spotted with chocolate; 0.8 x 0.6 in. (2.0 x 1.5 cm). Nest is grassy cup on ground in a shelter of rock or hummock.

Range
Breeds from Arctic s. through mountains to N.Mex.; winters from B.C. along coast s. to C. America.

Note: The **EURASIAN YELLOW WAGTAIL,** *Motacilla flava,* size, 6½ in. (16.5 cm), is a long-tailed, gray and yellow, tail-pumping bird of the tundra and scrub of Alaska and n. Yukon; it winters in Asia.

SPRAGUE'S PIPIT
Anthus spragueii   **52:4**

Description
Size, 6½–7 in. (16.5–17.8 cm). • *Streaked back* and • *yellowish or pinkish legs;* head, neck, and back streaked with black and buff; white below, washed with buff, finely streaked; more white in tail than Water, bill lighter.

Similarities
Vesper Sparrow has short, thick bill; hops, does not walk or pump tail. Horned Lark has "horns," yellow face. Water has dark legs, buffier breast, grayer and less striped back, pumps tail more.

Habitat
Short-grass plains, prairies.

Habits
Flight more bouncy than Water Pipits; male spends much time in air, often occurs singly or in pairs.

Voice
Flight song delivered at average height of 300 ft. (91.4 m), a "sweet thin jingling series, *ching-a-ring-a-ring-a-ring-a,* descending

in pitch" (Salt and Wilk); call, single notes, harsher than paired
notes of Water Pipit, not as sharp.

Eggs
4–5; grayish-white, thickly blotched with purplish-brown; 0.8 x 0.6
in. (2.0 x 1.5 cm). Nest is partly domed cup of grass on ground.

Range
Breeds from Canadian prairie s. to Mont. and Daks.; winters from
s. Ariz. s. to Mexico.

WAXWINGS
Family Bombycillidae

Waxwings are slim, crested, brown birds, with sleek plumage;
black chins and foreheads; red, waxlike tips to the secondary wing
feathers; and a yellow band across the end of their squared tails.
They eat berries, especially cedar and mountain ash, and seeds and
insects.

BOHEMIAN WAXWING
Bombycilla garrulus   **41:10**

Description
Size, 7½–8¾ in. (19.1–22.2 cm). Crested, brown, with 2 white
patches on each wing. Gray above, with • *cinnamon undertail
coverts* and • *yellow area in wings,* belly grayish.

Similarities
Cedar is smaller, with yellow on belly, white undertail area, less
gray above, no yellow and white wing marks.

Habitat
Boreal zone forests, muskegs; widespread in winter.

Habits
Gregarious at all seasons; flight undulating in tight flocks; catches
insects in air, eats berries and fruit, may come to feeder for raisins.

Voice
"Alarm note, *tzee-tzee*" (Forbush) or *zreee.*

Eggs
3–6; grayish-blue, spotted with dark brown; 0.9 x 0.7 in. (2.3 x 1.8
cm). Nest is bowl of twigs, grass or moss in conifer 8–50 ft. (2.4–
15.2 m) up.

Range
Breeds from Alaska s. to n. U.S.; winters from nw. states
southward somewhat; occasional to s. Calif., Ariz., n. N.Mex., and
Tex. Panhandle.

CEDAR WAXWING
Bombycilla cedrorum   **41:11**

Description
Size, 6½–8 in. (16.5-20.3 cm). • *Crested,* brown, with little or no
white in wings. Above red brown, yellow abdomen, undertail
coverts white, yellow band at tip of tail, throat black. Adult: most
show waxy-red tips of wing secondaries. Juvenal: duller, grayer
with blurry streaking below.

Similarities
Bohemian is not as red, with undertail coverts cinnamon.

Habitat
Open woods, orchards, fruit trees.

Habits
Very social; occurs in small flocks that alight on tree in compact body; remains upright, motionless before feeding; sometimes catches insects in air; a late nester.
Voice
Thin, high hiss or lisp, *zeee,* sometimes lightly trilled.
Eggs
3–6; pale bluish-gray, spotted with blackish; 0.9 x 0.7 in. (2.3 x 1.8 cm). Nest is bulky cup of twigs, grass, moss on horizontal branch 4–40 ft. (1.2–12.2 m) up.
Range
Breeds from se. Alaska s. to cen. Calif.; winters from s. B.C. s. throughout s. U.S.

SILKY FLYCATCHERS
Family Ptilogonatidae

PHAINOPEPLA
Phainopepla nitens   **34:8**

Description
Size, 7–7¾ in. (17.8–19.7 cm). • *Crested.* Male: uniform glossy • *blue-black;* white spot on each primary forms • *white wing patches* conspicuous in flight; crest tall, pointed, slender. Female: brownish to dark gray, no conspicuous wing patches, crest slender.
Similarities
Cedar Waxwing is much browner than female Phainopepla, tail band yellow, black face mask. Northern Mockingbird has white wing patches, much white in tail, no crest.
Habitat
Greasewood, mesquite, paloverde, oak foothills, pepper trees, mistletoe.
Habits
Has flycatching habits.
Voice
Wirp.
Eggs
2–3; speckled; 0.8 x 0.6 in. (2.2 x 1.6 cm). Nest is shallow woven cup on joints in tree.
Range
Resident from cen. Calif. e. across Colo. Plateau and s. to Mexico; more northern populations migrate southward.

SHRIKES
Family Laniidae

American Shrikes are gray, black, and white predators with heavy heads, stout hooked beaks, black masks, and slim tails. They are birds of the open country, where there are scattered trees. They prey upon insects, small mammals, and birds, and impale their prey on thorns.

NORTHERN SHRIKE
Lanius excubitor   **40:7**

Description
Size, 9–10¾ in. (22.9–27.3 cm). Robin-sized, with incomplete black mask and whitish feathers just over the bill. Adult: above light-

gray; rump light; wings and tail black with some white on wings and outer tail; underparts whitish, usually showing faint bars; faint barring on breast in fall and winter; bill light below at base except in spring and summer. Immature: brownish, plainly barred below; bill brown with pale area below; this plumage worn through first winter.

Similarities
Loggerhead adult is slightly smaller; no breast barring; bill solid black; darker above, black face mask extends forward over base of upper bill. Northern Mockingbird has no black mask, is slimmer, longer-tailed, more white in wing.

Habitat
Breeds in openings in northern forests; in winter, open country, swamps.

Habits
Bold, aggressive; perches on exposed observation post or wire watching for prey; sometimes catches birds in air.

Voice
Call, "a harsh shrieking *jo-ree*" (Knight); song, prolonged, a Robin-like or Catbird-like carol with some harsh notes; also mimics; both sexes sing.

Eggs
4–7; whitish, with dark spots; 1.1 x 0.8 in. (2.8 x 2.0 cm). Nest is bulky bowl of leaves or twigs in thorny bush or on tree limb 5–20 ft. (1.5–6.1 m) up.

Range
Breeds from far north s. to s. Alaska, nw. B.C., Great Lakes; winters from cen. Canada s. to n. Calif., cen. Nev., cen. Ariz., and n. N.Mex.

LOGGERHEAD SHRIKE
Lanius ludovicianus   **40:6**

Description
Size, 8–10 in. (20.3–25.4 cm). • *Mask complete,* to and over base of bill. Adult: above gray, below whitish, bill all-black, head large, tail slim. Immature: underparts pale gray, faintly barred.

Similarities
Northern is slightly larger; black mask barely reaches bill, area over base of bill white, not black; bars on breast; young brown below and on back, not gray. Northern Mockingbird has no black mask, is slimmer, longer-tailed, white wing patches larger.

Habitat
Open country with scattered trees, low scrub, deserts.

Habits
Sits quietly on perch; drops into low, beeline flight, then rises to next lookout; flight flickering, showing small white wing patch in the black wings.

Voice
Harsh *chack-chack; tsirp-see, tsirp-see;* song, a deliberate repetitive series of half-hearted notes and phrases, somewhat Mockingbird-like, with long pauses between "*queedle, queedle,* over and over" (Peterson).

Eggs
4–8; white to greenish-gray, variously marked; 1.0 x 0.7 in. (2.5 x 1.8 cm). Nest is lined cup of rootlets, twigs, in bush or tree 5–20 ft. (1.5–6.1 m) up.

Range
Breeds from cen. Canada s. to Mexico; winters from s. U.S. southward.

STARLINGS
Family Sturnidae

This is a large and varied Old World family of gregarious, sharp-billed, short-tailed birds, many looking like blackbirds. Of the world's many species, only two (both introduced into the New World) occur in the West. They eat insects, worms, grubs.

STARLING
Sturnus vulgaris   **46:3**

Description
Size, 7½–8½ in. (19.1–21.6 cm). Very • *short-tailed;* chubby, black. Spring adult: iridescent black, white dots on back; • *bill strong, pointed, yellow;* tail squared. Fall adult: much more liberally white-spotted, some brown spots, bill black. Juvenal: grayish, blurry whitish streaking on throat and belly; bill black.

Similarities
Compare blackbirds such as Red-winged, Rusty, and Brewer's. Cowbirds have medium-long tails. Grackles have long tails.

Habitat
Farmlands with scattered trees, ranches, orchards, fields, cities, suburban lawns, even deserts.

Habits
Gregarious, active, wary; comes to feeders; walks or waddles energetically and erratically on ground; collects in great flocks in cities, suburbs, marshes; flocks in air perform intricate revolutions in concert; wings triangular or spindle-shaped in flight with pointed, fast-beating wings alternately flapping and sailing; does not rise and fall in flight like most blackbirds.

Voice
Varied, including whistles, squeaking, squealing, harsh and grating notes; mimics many other species; a "harsh *tseeeer,* a whistled *whoo-ee*" (Peterson).

Eggs
4–7; pale blue; 1.2 x 0.9 in. (3.0 x 2.3 cm). Nest is of grass, sticks, in hole in tree or other natural or man-made cavity, 10–25 ft. (3.0–7.6 m) up.

Range
Resident throughout s. Canada and most of U.S.; introduced from Europe in 1890, now from Alaska, s. to Gulf Coast and Mexico.

Note: The **CRESTED MYNA,** *Acridotheres cristatellus,* size, 10½ in. (26.6 cm), occurs only on Vancouver Island and casually in western Washington and adjacent Oregon. It is black with a short tail, white wing patch, short crest, and yellow bill and legs; it is highly gregarious.

VIREOS
Family Vireonidae

Vireos are sparrow-sized arboreal birds, olive-gray above, white or yellowish below, with thick, slightly hooked bills. They are larger but less active than warblers. The vireos divide into those with wing bars, usually having eye-rings, or "spectacles," and those without wing bars, usually having eye stripes. Their diet consists of insects.

VIREO COMPARISON CHART

Species	Wing Bars	Breast	Eye Markings
Red-eyed	no	white	black and white line
Warbling	no	white	light line
Philadelphia	no	yellowish	light line
Bell's	faint	white	white spectacles
Hutton's	yes	gray	incomplete white spectacles
Gray	one, faint	white	weak white spectacles
Black-capped	yes	white	white spectacles
Solitary	yes	white	white spectacles
Yellow-throated	yes	yellowish	yellow spectacles

BLACK-CAPPED VIREO
Vireo atricapilla   **42:15**

Description
Size, 4½–4¾ in. (11.4–12.1 cm). Above olive, rump yellowish, • *top and sides of head glossy-black* (male) or slate-gray (female), white spectacles around eye to bill and connecting across forehead, wing bars yellowish-white, underparts white with yellowish sides (male) or buffy (female).
Similarities
Female and immature Black-capped resemble the larger, white-breasted and bluish-headed Solitary.
Habitat
Brush, ravines, oak uplands.
Habits
Shy, alert, restless; often hangs head down before flitting to a lower twig.
Voice
Song, "a subdued, low, sweet, persistent musical warble . . . also interpreted as a loud, emphatic, *liquid-there-now, wait-a-bit*" (Simmons); alarm, a *chit-aa* like Ruby-crowned Kinglet's.
Eggs
3–5; white; 0.7 x 0.5 in. (1.8 x 1.3 cm). Nest is cup in shrub or small tree, 2–15 ft. (0.6–4.6 m) up.
Range
Breeds from Kans. s. to Okla., and cen. Tex.; winters in Mexico.

HUTTON'S VIREO
Vireo huttoni   **42:17**

Description
Size, 4½–4¾ in. (11.4–12.1 cm). • *Incomplete eye-ring,* broken by • *dark patch above eye;* otherwise spectacles reach over bill. Above olive-green, underparts dingy, darker tail and wings, 2 broad white bars.
Similarities
Ruby-crowned Kinglet bears close resemblance, has eye-ring incomplete but no spectacles, bill thin and tiny, call different. The *Empidonax* flycatchers have upright perching posture, flycatching habits, more complete eye-rings.
Habitat
Brush, forests, edges.
Habits
Moves by flight more than other vireos.
Voice
Hoarse, deliberate *day dee dee;* a 2-syllable *zu-weep* (rising), often repeated.

Eggs

3–4; white, dotted; 0.9 x 0.5 in. (2.2 x 1.3 cm). Nest is downy or mossy suspended cup.

Range

Resident along Pacific Coast to s. Ariz., s. to C. America.

BELL'S VIREO
Vireo bellii   42:14

Description

Size, 4½–5 in. (11.4–12.7 cm). Small, olive, nondescript vireo. Above olive, rump brighter; underparts white, but belly yellowish; faint eye-ring and spectacles; faint single or double wing bars.

Similarities

Warbling shows no sign of eye-ring or wing bars.

Habitat

Bottomland shrubbery, breaks, mesquite thickets, hedgerows.

Habits

Active, shy and retiring, fearless about nest.

Voice

Song, monotonous, *"whillowhee, whillowhee, WHEE"* (Nice), 2 versions often alternating, one rising, second falling, like question and answer; also a hoarse, scolding note.

Eggs

3–5; white, sparingly spotted with brown; 0.7 x 0.5 (1.9 x 1.4 cm). Nest is low cup in dense bush, or low tree.

Range

Breeds from cen. Calif. s. to Mexico, and in Sw. from Colo. and S.Dak., southward; winters in Mexico.

GRAY VIREO
Vireo vicinior   42:16

Description

Size, 5–5¾ in. (12.7–14.6 cm). Drab, gray-backed arid-zone vireo; • *eye-ring narrow, white;* no spectacles; no wing bars or 1 faint white bar; underparts whitish to pale gray.

Similarities

Bell's has olive above, yellowish on belly, spectacles apparent.

Habitat

Mesas, brushy mountains, junipers, scrub oak, open chaparral.

Habits

Nervous, tail frequently in motion like gnatcatcher's.

Voice

As Solitary, but more rapid.

Eggs

3–4; white, spotted; 0.7 x 0.5 in. (1.8 x 1.3 cm). Nest is in low desert shrub.

Range

Breeds from s. Calif. e. to Utah and w. Tex. and s. to Baja Calif.; winters from sw. Ariz. s. to Mexico.

SOLITARY VIREO
Vireo solitarius   42:13

Description

Size, 5–6 in. (12.7–15.2 cm). • *Bluish or grayish head* and full white spectacles. Back yellow-olive, contrasting with head, or, in Rocky Mountains, gray; 2 white wing bars; may be white or buff on breast, otherwise • *white below* with yellowish or grayish buff flanks. Western races lack bluish cast on head.

Similarities
Hutton's is grayer; Solitary Vireos resemble it, but are whiter below and have complete eye-ring. Bell's has inconspicuous wing bars.

Habitat
Mainly conifers, but also mixed woods and fringes.

Habits
Tame, sedate; an early migrant; wanders about, when singing; commonly victimized by Brown-headed Cowbird.

Voice
Song, short rising-falling whistled phrases, with longer pauses between phrases than Red-eyed and including a rising *tu-wee-tu;* a scolding note.

Eggs
3–5; white; dotted with red-brown and umber; 0.8 x 0.6 in. (2.0 x 1.5 cm). Nest is a neat basket, in tree 5–12 ft. (1.5–3.7 m) up.

Former name
Blue-headed Vireo.

Range
Breeds throughout much of wooded w. regions from cen. B.C., e. across Canada and s. to Mexico; winters from Mexico to C. America.

Note: The **YELLOW-THROATED VIREO**, *Vireo flavifrons,* size, 6 in. (15.2 cm), is casual in the Great Plains and west to Wyoming and Colorado. It is an eastern species with 2 white wing bars and a bright yellow throat and breast.

RED-EYED VIREO
Vireo olivaceus   **43:11**

Description
Size, 5½–6½ in. (14.0–16.5 cm). • *Black-and-white stripe* over the eye. Upperparts olive-green, conspicuously • *grayer on crown;* underparts white; no wing bars, red eye conspicuous only close-up.

Similarities
Warbling is grayer, head concolored with back, more uniform above, lacks black border to white eye stripe, eye dark. Philadelphia has yellowish breast, dark eye. Tennessee Warbler has similar pattern, no black eye line, dark eye, bill thin.

Habitat
Almost any woodland, shade trees, groves.

Habits
Movements deliberate, but less so than Solitary; sings through heat of day and into late summer.

Voice
Song, a continuing repetition of little phrases punctuated by pauses, remotely Robin-like, *"Hello, hello. Are you there? Can you hear me? This is Vireo. Yes, yes, Vireo, Vireo"* (Taylor). Alarm, a descending, catlike *kway,* or whining *chway.*

Eggs
3–4; white, sparingly spotted with blackish; 0.9 x 0.6 in. (2.3 x 1.5 cm). Nest is in sapling or tree fork, 4–50 ft. (1.2–15.2 m) up.

Range
Breeds chiefly in E., but common w. to e. Wash. and areas e. of Rockies, s. to Gulf Coast; winters s. of U.S.

PHILADELPHIA VIREO
Vireo philadelphicus 43:9

Description
Size, 4½–5½ in. (11.4–14.0 cm). Small, with • *yellowish breast* and
• *no wing bars.* Above olive, light white line over eye, no spectacles,
chin and belly whitish.
Similarities
Warbling has whitish or grayish breast. Tennessee has greener
back. Orange-crowned has yellow under tail. Warblers are
slimmer, have thin bills, warblerlike build and actions. See Red-
eyed.
Habitat
Edges of woods, low trees and shrubs, second growth, alders,
willow, poplars.
Habits
Tame; when feeding, rather active, sometimes flutters in front of
leaves, hangs upside down on twig, catches insects in air.
Voice
Very similar to Red-Eyed's, but a bit higher, less often repeated
and containing frequently a characteristic abrupt, rising 2-syllable
note; also a scolding note like Warbling's.
Eggs
3–5; white, sparingly spotted with umber; 0.8 x 0.5 in. (2.0 x 1.3
cm). Nest is cup in fork of shrub or tree, 9–40 ft. (2.7–12.2 m) up.
Range
Breeds chiefly in E., but w. also from s. Canada s. to n.-cen. U.S.;
winters in C. America; migrates through Mont., Colo., w. Tex.

WARBLING VIREO
Vireo gilvus 43:7

Description
Size, 4½–5½ in. (11.4–14.0 cm). Light eye line, whitish or buff
breast, and • *no wing bars;* most widespread western species lacking
eye-ring or wing bars. Above olive-gray; white spot between bill
and eye, but no spectacles; below whitish sometimes with a
yellowish wash on sides; head inconspicuously striped.
Similarities
Red-eyed and Philadelphia in California have yellowish breast.
Bell's and Gray have spectacles.
Habitat
Mixed and deciduous forests, shade trees, poplars, aspen groves;
often in upper branches of trees.
Habits
Sings often but less consistently than Red-eyed.
Voice
A pleasant continuing warble, somewhat like that of Purple Finch,
but less spirited; call, a snarling *myee;* note, a wheezy *twee.*
Eggs
4; white, often spotted with red-brown; 0.7 x 0.5 in. (1.9 x 1.4
cm). Nest is a tiny cup in tree 20–70 ft. (6.1–21.3 m) up.
Range
Breeds throughout much of N. America; from n. Canada, s. to
Mexico and C. America; winters from Mexico to C. America.

WOOD WARBLERS
Family Parulidae

These are bright-colored birds, smaller, more slender than
sparrows, with thin, straight, unhooked bills. They are more active
and restless than the sluggish, hooked-billed vireos, constantly
flitting about in search of insects and spiders. Many warblers in
fall have plumages quite different from spring, providing real
problems in identification. Many eastern warblers not normally
found in the West, or occurring only in Plains river valleys, have
turned up sporadically in the West, as far as California.

BLACK-AND-WHITE WARBLER
Mniotilta varia   **44:9**

Description
Size, 4½-5½ in. (11.4-14.0 cm). Striped crown. Male: • *striped
black and white,* throat black, 2 white wing bars. Female and
immature: lack black throat; paler, more white below.
Similarities
Black-throated Gray and Blackpoll spring males have caps solid
black.
Habitat
Woodlands, preferably deciduous, moist or dry; trunks and limbs of
trees.
Habits
Creeps along trunks and branches like a nuthatch, head up or
down; sometimes catches insects in air.
Voice
A thin, wiry, repeated *zee, zee, zee,* or *we-see, we-see,* spiraling
upward; call, a short *pip,* a louder *jink.*
Eggs
4-5; creamy, well spotted with dark brown; 0.7 x 0.5 in. (1.8 x 1.3
cm). Nest is hair-lined cup of grass and rootlets, on ground at foot
of tree, stump, or rock.
Range
Breeds from s. Canada s. to s. U.S. e. of Rockies; winters from s.
U.S. s. to S. America; migrates e. of Rockies.

TENNESSEE WARBLER
Vermivora peregrina   **43:8**

Description
Size, 4½-5 in. (11.4-12.7 cm). Spring male: plain; crown gray;
conspicuous • *white line over eye,* dark line through it; back
greenish, underparts whitish. Spring female: crown more olive,
underparts somewhat yellowish. Note light eyebrow line. Fall
adults and immature: above bright greenish, below unstreaked pale
dingy, conspicuous yellowish line over eye (grayer above, whiter
below in male), pale wing bars, undertail coverts usually white.
Similarities
Fall Orange-crowned is darker, dingier, with yellow undertail
coverts, no line over eye. Vireos are larger, bill heavier, sluggish.
Red-eyed Vireo has red eye, black eye line. Philadelphia Vireo has
thicker bill and not as green back.
Habitat
Breeds in openings in northern coniferous forest; in migration,
woodlands, edges, shade trees, brush.

Habits

Restless, prefers top branches in spring, any height in fall; often hangs upside down at tip of twig.

Voice

Song, loud, 3-part, last somewhat like Chipping Sparrow, *tenne-tenne-tenne-tenne, chip-chip, ssee-ssee-ssee-ssee-ssee;* or *"tsip-pit-tsip-PIT, tsip-PIT, tsip-PIT, tsip-pit-tsee, tsee, tsee, tsee,* more rapid and higher pitched toward the end" (Harrison); reminiscent also of Nashville's 2-part song, but louder.

Eggs

4–7; creamy, with red-brown spots; 0.6 x 0.5 in. (1.5 x 1.3 cm). Nest is cup of vegetable fibers and grass, on or close to ground.

Range

Breeds from s. Yukon e. across Canada and s. to s. Canada; winters from C. to S. America; migrates through Great Plains.

ORANGE-CROWNED WARBLER

Vermivora celata   **43:4**

Description

Size, 4½–5½ in. (11.4–14.0 cm). Nondescript, dull-colored warbler lacking wing bars. Spring adult: entirely greenish and yellow, more olive above and yellower below; yellowish eye stripe, fine black line through eye; vague streaks below; hidden patch of orange in crown; undertail coverts yellow. Fall adult and immature: duskier, often grayish, very drab; crown patch rarely visible or absent; note lack of white and any contrasty pattern.

Similarities

Fall Nashville has white eye-ring. Fall Tennessee has distinct line through eye, pale stripe over eye, no side streaks. Philadelphia Vireo is larger, bill thicker; light stripe over eye, no side streaks. Hutton's Vireo and Ruby-crowned Kinglet have light wing bars.

Habitat

Open woodlands with heavy brush, clearings, hillsides, chaparral, aspens, near water.

Habits

Forages low.

Voice

Song resembles Chipping Sparrow's, a series of 18–22 *si* notes, dropping slightly in middle, rising and fading toward end; note, a distinctive *chip.*

Eggs

4–6; white, dotted with reddish brown; 0.7 x 0.5 in. (1.8 x 1.3 cm). Nest is of grass, rootlets, on ground in shrubbery on chaparral (or other) hillside.

Range

Breeds from Alaska e. across n. Canada and s. to Baja Calif., Ariz., and N.Mex.; winters from s. U.S. s. to C. America.

NASHVILLE WARBLER

Vermivora ruficapilla   **43:2, 45:8**

Description

Size, 4–5 in. (10.2–12.7 cm). Gray head, • *white eye-ring, yellow underparts.* Spring male: back olive, often with hidden chestnut crown patch; throat yellow; no wing bars or white in tail; female duller. Fall adults duller; yellow below but with color of head and back blending. Immature: quite brownish above.

Similarities

Connecticut, Mourning, MacGillivray's all have dark throats.

Habitat
Open wood with heavy undergrowth, especially on slopes and near water.

Habits
Active; forages at all levels, relatively early spring and late fall migrant.

Voice
Song, 2-part *"see it see it see it see it, ti-ti-ti"* (Gunn); note, a *chip*.

Eggs
3–5; white, dotted with red-brown; 0.6 x 0.5 in. (1.5 x 1.3 cm). Nest is cup of grass, leaves, concealed in or under tussock on ground.

Range
Breeds from s. B.C. e. to w. Mont. and s. to s. Calif. along Cascades and Sierras inland; winters in Mexico; migrates through Pacific states.

VIRGINIA'S WARBLER
Vermivora virginiae   **42:6**

Description
Size, 4–4½ in. (10.2–11.4 cm). Adult: small, • *grayish;* • *breast, rump, undertail area yellow;* white eye-ring narrow; rufous patch on crown, visible in male at close range. Immature: gray, wash of yellow at base of tail, no rufous on crown, trace of yellow on breast.

Similarities
Lucy's has white breast, chestnut rump. Colima, in western Texas, is larger, very like Virginia's, but with orangy tone to yellow of rump, less yellow on breast.

Habitat
Brushy mountainsides, pinyons, oaks, canyons.

Habits
An active warbler, often rising to the top of brush to sing or look about.

Voice
Song resembles Yellow-rumped in quality, a series of monotonic, loose notes rising slightly toward end, *chlip-chlip-chlip-chlip-chlip-wick-wick* (Peterson).

Eggs
3–5; white, speckled; 0.6 x 0.4 in. (1.6 x 1.1 cm). Nest is vegetal cup on ground under brush or in grass.

Range
Breeds from Central Plateau and Rockies, s. to Colo. and N.Mex.; winters in Mexico; migrates along coast of s. Calif. to w. Tex.

LUCY'S WARBLER
Vermivora luciae   **42:9**

Description
Size, 4 in. (10.2 cm). Adult: tiny, with • *chestnut rump patch;* above gray, below white; crown shows small chestnut patch; eye-ring white; white undertail. Immature: similar but without chestnut patches.

Similarities
Virginia's has rump patch, undertail yellow. Colima has yellow undertail.

Habitat
Desert stream breaks: willows, cottonwoods, mesquite, paloverde, ironwood.

Habits
Closely associated with mesquite thickets.
Voice
Song, "a high, rapid *weeta weeta che che che che che*, on 2 pitches"; suggests Nashville Warbler if ending lower pitch or Yellow Warbler if ending higher (Peterson).
Eggs
4–5; white, speckled; 0.5 x 0.4 in. (1.3 x 1.1 cm). Nest is lined cup in tree, under bark or in cavity.
Range
Breeds from s. Calif., Nev., and Utah to s. Ariz. and N. Mex.; winters s. of U.S.

COLIMA WARBLER
Vermivora crissalis   **42:5**

Description
Size, 5¼ in. (13.3 cm). Very similar to Virginia's but larger, whiter below with less yellow; rump shows more orangish.
Similarities
See Virginia's.
Habitat
Piney canyons, small deciduous trees, oaks, madronas, maples.
Habits
Deliberate, vireolike movements.
Voice
Song, "a simple trill, much like that of Chipping Sparrow but rather shorter and more musical and ending in 2 lower notes" (J. Van Tyne); note, a sharp *psit*.
Eggs
4; white, spotted; 0.7 x 0.5 in. (1.8 x 1.3 cm). Nest is lined grassy cup among dead leaves on ground.
Range
Breeds from Chisos Mountains in w. Tex., s. to Mexico; winters s. of U.S.

OLIVE WARBLER
Peucedramus taeniatus   **42:4**

Description
Size, 4½–5 in. (11.4–12.7 cm). Male: distinctive; • *head and upper breast dull orange,* conspicuous • *black cheek patch,* back gray, underparts white, broad white wing bars. Female: less distinctive; crown and nape olive, breast and sides of throat yellowish, dusky ear patch but no face mask.
Similarities
Female Townsend's and Hermit resemble female Olive but lack yellow on breast and have green, not gray back.
Habitat
Mountain forests, pine or fir.
Voice
Song, a ringing "*tiddle tiddle tiddle ter* or titmouselike *peter peter peter peter*" (Peterson).
Eggs
3–4; grayish white, dotted; 0.6 x 0.4 in. (1.7 x 1.2 cm). Nest is cup high in tree.
Range
Breeds from s. Ariz. and N.Mex. southward; winters from se. Ariz. and sw. N.Mex. southward.

TROPICAL PARULA
Parula pitiayumi   42:7

Description
Size, 3¾–4 in. (9.5–10.2 cm). • *Bluish and yellow;* olive on back, throat yellow, black mask in male, white tail marks and 2 white wing bars.
Similarities
Northern has distinctive white eye-ring.
Habitat
Woods along rivers.
Voice
Song, a buzzy trill, *zzzzzzzzzz-ZIP* or *bz-bz-bz-bz-zzzzzzzzz-ZIP,* rising; call, a *chip.*
Eggs
4–5; white, brown wreathed; 0.6 x 0.4 in. (1.6 x 1.1 cm). Nest is in hanging moss cluster near tip of tree branch.
Former name
Olive-backed Warbler.
Range
Resident chiefly s. of U.S., but in n. to Rio Grande Valley (rare).

Note: The **NORTHERN PARULA,** *Parula americana,* **(42:8)** size, 4½ in. (11.4 cm), is the corresponding northern version of the Tropical Parula. It barely reaches the eastern fringe of the western region in the twilight zone of the Great Plains, and casually in West in migration. It is very like the Tropical in size but has a distinct white eye-ring in all plumages. The spring male has no face mask and shows a small rust-and-blue band across the yellow breast.

YELLOW WARBLER
Dendroica petechia   43:1

Description
Size, 4½–5¼ in. (11.4–13.3 cm). Appears all-yellow, with yellow tail patches. Male: greenish-yellow above, brighter yellow below; chestnut streaks along breast; • *wing bars yellow,* slightly duller in fall. Alaskan race in fall is duller, grayer above. Female and immature: similar but more olive above, paler below with few or no streaks.
Similarities
Female Wilson's has no yellow tail spots. Orange-crowned is very similar in some races during fall, but shows no flashing yellow tail marks. American Goldfinch has wings and tail black.
Habitat
Swamps, shrubbery, willows, or alders near water; orchards, gardens, shade trees.
Habits
Tame, forages low; an early fall migrant.
Voice
Song, *sweet, sweet, sweet, I'm so sweet,* with a goldfinchlike ending; "a cheerful, bright *tsee-tsee-tsee-tsee-titi-wee* or *weet weet weet weet tsee tsee,* given rapidly" (Peterson).
Eggs
3–5; greenish-white, spotted with brown and purple; 0.7 x 0.5 in. (1.8 x 1.3 cm). Nest is felted cup of plant fibers in fork of low bush or sapling.
Range
Breeds throughout much of N. America from n. Canada s. to s. U.S.; winters s. of U.S.

CAPE MAY WARBLER
Dendroica tigrina 44:1

Description
Size, 5–5½ in. (12.7–14.0 cm). Spring male: chestnut cheek patch, above olive with black streaks, underparts yellow with black streaks, side of neck and rump yellow, wing patch and tail corners white. Spring female, immature, and fall adults: duller, cheek patch gray (not chestnut), back unstreaked, breast nearly white with dusky streaks, • *dull yellowish patch behind ear*, 2 narrow whitish wing bars.

Similarities
Fall Yellow-rumped is browner above, back streaked, rump brighter yellow. Palm is brown above, back streaked, undertail coverts yellow, almost no wing bars; pumps tail.

Habitat
In migration seeks edges, shrubbery.

Habits
Sings from treetops, often at extreme tip of spruce or fir.

Voice
Song, a thin, high *see-see-see-see-see;* also, less commonly, a series of 6–8 *to-be*'s preceding a *see-see-see;* call, a faint *chip*. Song often confused with that of Bay-breasted.

Eggs
4–9; greenish-white, dotted with browns and lilac; 0.7 x 0.5 in. (1.8 x 1.3 cm). Nest is near top of spruce or fir, 30–60 ft. (9.1–18.3 m) up.

Range
Breeds from s. Mackenzie e. to Ont. and s. to N.Dak.; winters s. of U.S.; accidental to Calif., Colo., and Ariz.

BLACK-THROATED GRAY WARBLER
Dendroica nigrescens 44:7

Description
Size, 4½–5 in. (11.4–12.7 cm). Male: above black and gray, below white; throat, cheeks, and crown black; a seldom seen, small yellow spot before the eye. Female: similar but lacking black throat.

Similarities
Townsend's has face and underparts yellow. Black-throated Sparrow has conical bill, no stripes on sides. Chickadees have white cheeks. In California, see also Black-and-White and Blackpoll Warblers.

Habitat
Open mixed forest, junipers, pinyons, dry oak slopes.

Habits
Feeds low in brushy vegetation.

Voice
Song, variable, full of *z*'s, a buzzy chant, *zeedle zeedle zeedle zeet'che*, penultimate note higher.

Eggs
3–5; white, spotted; 0.6 x 0.5 in. (1.7 x 1.3 cm). Nest is neat cup in oak or shrub.

Range
Breeds along Pacific Coast forested areas, from s. B.C., s. to Baja Calif. and in Great Basin; winters from sw. U.S. to Mexico.

Note: Occasionally seen in the West is the distinctive **BLACK-THROATED BLUE WARBLER,** *Dendroica caerulescens,* size, 5 in. (12.7 cm), both sexes of which show a white mark on the wing. Males are blue with a black throat, females mainly brown.

YELLOW-RUMPED WARBLER
Dendroica coronata   **44:3**

Description
Size, 5–6 in. (12.7–15.2 cm). Spring male: rump, sides, crown, and
throat bright yellow; very dark, black, and gray above with
streaked back; black breast with yellow patch on sides of it; white
spot on either side of tail; belly white; throat white in eastern,
white wing patch in western, 2 white wing bars in eastern. Spring
female, immature, and fall adults: browner, females of eastern and
western forms still show respective throat colors (see male);
variable, but streaked above and on breast, 2 white wing bars; all
show yellow rump, and most have trace of yellow on sides.
Similarities
Spring Cape May has yellow rump, but also yellow underparts.
Habitat
Coniferous forests; in winter, in all types of wooded habitat, even
feeding in open on ground. One of the most common warblers.
Habits
Feeds diversely, flycatches sometimes.
Voice
Variable, a *weezy weezy see see see* or *seet-seet-seet-seet-seet,
trrrrrr;* call note, emphatic *chip.*
Eggs
3–5; purple-or brown-spotted on white; 0.7 x 0.5 in. (1.8 x 1.3
cm). Nest is cup-shaped, rather bulky, of bark shreds, twigs, in
conifer, 3–40 ft. (9.1–12.2 m) up.
Remarks
Formerly Myrtle and Audubon Warblers, now recognized as a
single species.
Range
Breeds throughout much of Canada, s. to n.-cen. B.C. and s. Alta.;
winters in sw. U.S., s. to C. America; migrates through W., but
primarily through se. U.S. to winter in sw. U.S. and C. America.

GOLDEN-CHEEKED WARBLER
Dendroica chrysoparia   **42:3**

Description
Size, 4½–5 in. (11.4–12.7 cm). Male: upperparts and • *throat black,
cheeks yellow,* black line through eye; otherwise very similar to
Black-throated Green Warbler. Female: like male, but back olive-
green, may be flecked with black; belly snowy white without
yellowish wash.
Similarities
Black-throated Green lacks black line through eye, has olive back.
Habitat
Prefers cedar ridges; junipers, oaks, stream breaks.
Habits
As Black-throated Green and Townsend's.
Voice
Song, "a hurried *tweeah, tweeah, tweesy*" (Attwater), or *"bzzzz,
layzeee, dayzeee"* (Kincaid).
Eggs
3–5; dotted; 0.6 x 0.5 in. (1.7 x 1.3 cm). Nest is lined cup in low
shrub, juniper.
Range
Breeds in s.-cen. Tex.; winters in Mexico, s. to C. America.

TOWNSEND'S WARBLER
Dendroica townsendi   **44:6**

Description
Size, 4½–5 in. (11.4–12.7 cm). Spring male: • *head conspicuously patterned in black and yellow;* • *underparts yellow,* side striped. Spring female and winter male: yellow face with • *dark cheek patch* clearly defined; throat mostly yellow, not black.

Similarities
Black-throated Green, Hermit lack Townsend's dark cheek mark, yellow of underparts.

Habitat
Conifer forests, treetops; in winter, oaks, madrones, laurels.

Habits
A treetop bird when in breeding range, but often lower on migration.

Voice
Song, like Black-throated Gray's, *dzeer, dzeer dzeer tseetsee* or *weezy, weezy, seesee,* "the first 3 or 4 notes similar in pitch, with a wheezy, buzzy quality, followed by 2 or more high-pitched sibilant notes" (Axtell).

Eggs
3–5; white, spotted; 0.6 x 0.5 in. (1.7 x 1.3 cm). Nest is cup-shaped, lined, in conifer.

Range
Breeds in forested areas along Pacific Coast, from Alaska and B.C. s. to n. Wash. and e. to Mont. and Wyo.; winters from Oreg. s. to C. America.

BLACK-THROATED GREEN WARBLER
Dendroica virens   **42:2, 44:5**

Description
Size, 4½–5¼ in. (11.4–13.3 cm). Above olive, 2 white wing bars. Spring male: underparts white, white on tail, • *black throat* and side streaks. Note • *bright yellow cheeks* against solid black throat and olive-green crown. Spring female and fall adults: similar but duller; much less black on throat. Immature: cheeks still duller, no yellow below, no black throat, back unstreaked.

Similarities
Townsend's has yellow on breast or throat in all plumages. Fall Blackburnian has gray back, darker cheek patch, yellowish breast. Hermit, in Pacific states, has yellow crown in male, back gray, less black streaking on sides. Golden-cheeked, on Edwards Plateau of central Texas, has black back, black line through eye.

Habitat
Conifers.

Habits
Forages at all levels, but usually well up.

Voice
Song, "a high lisping *zoo-zee, zoo-zoo, zee* or a more rapid *zee-zee-zee-zoo-zee*" (Peterson).

Eggs
4–5; white, spotted with various shades of brown; 0.6 x 0.5 in. (1.5 x 1.3 cm). Nest is compact cup of twigs, birchbark, bound with spider webs, on conifer limb, 15–70 ft. (4.6–21.3 m) up.

Range
Breeds from Northwest Territories e. across Canada to Ont.; winters from Tex., s. to S. America; migrates sparsely through Great Plains and w. to mountain states.

HERMIT WARBLER
Dendroica occidentalis   **42:1**

Description
Size, 4½–4¾ in. (11.4–12.1 cm). Male: only warbler with bright
• *yellow head* contrasting with • *black throat* and • *gray back;*
underparts whitish. Female: similar patterning, but black of throat
much less or lacking.
Similarities
Townsend's has black cheek, patches on crown.
Habitat
Conifers; in migration, other trees, shrubs, chaparral.
Habits
Feeds at various heights; very active.
Voice
Song, in 2 parts starting with 3 high, lisping notes and ending with
2 distinctive abrupt, terminal notes, *sweety, sweety, sweety,*
chup'chup' or *seedle* replacing *sweety.*
Eggs
3–5; white, spotted; 0.6 x 0.5 in. (1.7 x 1.3 cm). Nest is cup high
in conifer.
Range
Breeds along Pacific Coast from Wash. s. to nw. Calif.; winters in
Mexico and C. America; migrates through Calif., Ariz.

BLACKBURNIAN WARBLER
Dendroica fusca   **44:4**

Description
Size, 4½–5½ in. (11.4–14.0 cm). Orange throat and white on the
wing. Spring male: • *flaming orange throat* and breast, sides
streaked black, orange stripes on crown and sides of head, back
black with white lines, large white wing patch, tail corners white.
Spring female and fall adults: similar but duller and paler, 2 white
wing bars, some orange-yellow on throat. Immature: even duller;
similar pattern in brown and dull yellow; 2 white wing bars; head
stripes sharply yellowish.
Similarities
Black-throated Green immature lacks dark ear patch, light stripes
on back, yellow throat. American Redstart male has same colors
but very different pattern.
Habitat
Breeds in northern and mountain forests, especially conifers.
Habits
Active, forages high in trees, sings from treetops; found at all levels
in migration.
Voice
Song, a high, thin, wiry *tsee-tsee-tsee-tsee-tsee-zi-zi-zi-zi;* call, a
faint *tseeck;* also a series of *zip*'s ending on a high, rising note,
"zip zip zip zip titi tseeee" (Peterson).
Eggs
4–5; blue-green to whitish, spotted with lilac and brown; 0.7 x 0.5
in. (1.8 x 1.3 cm). Nest is cup of twigs, in conifer, near end of
limb.
Range
Breeds from se. Canada s. to ne. U.S.; winters in S. America;
migrates sparsely through Great Plains.

GRACE'S WARBLER
Dendroica graciae   **42:11**

Description
Size, 4½–5 in. (11.4–12.7 cm). • *Back gray, throat yellow,* yellowish line above eye, 2 white wing bars, sides strongly striped with black; no distinct ear patch or face mask.

Similarities
Yellow-rumped has yellow rump.

Habitat
Dry mountain pine-oak forests.

Habits
Often flycatches; very active.

Voice
"Repeated *cheedle cheedle che che che che,* [ending] in a trill" (Peterson).

Eggs
3–4; white, spotted; 0.6 x 0.4 in. (1.6 x 1.2 cm). Nest is cup high in pine.

Range
Breeds from cen. Utah and Colo. s. to sw. states and C. America; winters s. of U.S.

BLACKPOLL WARBLER
Dendroica striata   **44:8**

Description
Size, 5–5¾ in. (12.7–14.6 cm). Black with white wing bars in all plumages. Spring male: gray above, white below, black stripes on back and sides, cheek white, • *cap black.* Spring female: cap olive streaked with black, streaks less conspicuous, below white, cheeks gray. Fall adults and immature: olive-green above, white below with tinge of greenish-yellow, black streaks on back and faintly on sides, legs pale, undertail coverts white, 2 white wing bars.

Similarities
Distinguishing fall Bay-breasted is difficult. Black-and-white has crown striped black and white. Black-throated Gray has black cheeks.

Habitat
Breeds in northern and mountain forests, especially near bogs and in stunted tree areas; in migration, any trees.

Habits
Movements rather deliberate; a good fly-catcher; forages and sings chiefly at middle and upper levels in migration; a late-spring night migrant, many hit lighthouses, beacons, and TV towers.

Voice
Song, a weak, high penetrating, *zee-zee-Zee-Zee-ZEE-ZEE-Zee-Zee-zee-zee,* louder in middle, fading toward end; call, a sharp *chip.*

Eggs
3–5; whitish, boldly spotted with red-brown; 0.7 x 0.5 in. (1.8 x 1.3 cm). Nest is feather-lined cup of moss, twigs, low in spruce.

Range
Breeds throughout much of cen. Canada, from Alaska eastward; winters in S. America; regular but uncommon migrant along Pacific Coast.

CHESTNUT-SIDED WARBLER
Dendroica pensylvanica   **43:3, 45:3**

Description
Size, 4½–5¼ in. (11.4–13.3 cm). Spring male: • *yellow crown* and
• *chestnut sides;* black mark below eye, back olive with black
stripes; 2 yellow wing bars; cheeks, underparts, tail corners white.
Spring female and fall adults: similar but duller, with less chestnut
and black. Immature: bright lemon-green above, white below,
white eye-ring and face, gray cheeks, buffy wing bars.

Similarities
Spring Bay-breasted has chestnut throat and crown.

Habitat
Dry hillside thickets, second growth.

Habits
Tame; forages low where breeding, but at all levels in migration;
often victimized by Brown-headed Cowbird.

Voice
Song resembles that of Yellow but with emphatic, final double note
and many variations of theme, *pleased, pleased, pleased t'*
MEETcha.

Eggs
Usually 4; white, spotted chiefly at larger end with brown and
lavender; 0.6 x 0.5 in. (1.7 x 1.3 cm). Nest loose, of bark strips and
plant fibers, in low bush.

Range
Breeds from s. Canada e. across Canada and s. into n.-cen. states;
winters in C. America; rare but regular visitor to Calif.

BAY-BREASTED WARBLER
Dendroica castanea   **45:2**

Description
Size, 5–6 in. (12.7–15.2 cm). Spring male: • *chestnut (or bay) on*
back of head, crown, throat, sides; face mask black; gray above with
black stripes; wing bars, underparts, outer tail tips white. Spring
female: similar but paler; buff instead of chestnut, face grayish.
Fall adults and immature: above olive, below buffy, cheek patch
dusky, stripes on back, legs usually black, often trace of chestnut on
flanks, undertail coverts buffy.

Similarities
Fall Blackpoll has trace of streaks below, pale legs, white undertail
coverts. Chestnut-sided adult has brown sides, not head.

Habitat
Coniferous forests, breeds in bogs and forest openings.

Habits
In breeding season, forages at all levels; movements somewhat
deliberate.

Voice
Song, a thin, high *"seetzy-seetzy-seetzy-seetzy-see"* (Gunn);
reminiscent of Blackburnian, Cape May, and Golden-crowned
Kinglet; call, a fine *tsip.*

Eggs
4–7; often bluish-green, spotted with brown; 0.7 x 0.5 in. (1.8 x
1.4 cm). Nest of twigs and moss on branch of conifer.

Range
Breeds in se. Canada and ne. U.S.; winters in C. and S. America;
some migrate through Great Plains.

PALM WARBLER
Dendroica palmarum

44:2

Description
Size, 4½–5½ in. (11.4–14.0 cm). • *Chestnut cap;* pumps its tail; sexes similar. Spring adult: olive-brown above; rump greenish-yellow; yellow line over eye; throat, upper breast, and undertail coverts yellow; belly gray-white; tail corners white; cheek patch olive, light chestnut streaks on breast. Fall adult and immature: above browner, concealing chestnut cap; line over eye, throat whitish; breast dull white, brownish streaks below. In western race, belly always white; chestnut brighter and belly always yellow in eastern race.

Similarities
Cape May young has white undertail coverts.

Habitat
Brushy edges of forest muskegs; in migration, lawns, fields, swamps, undergrowth.

Habits
Tail pumping habit very distinctive, spends much time on ground or in low bushes.

Voice
Song, a "short rapid series of thin, lisping notes, similar to that of a Chipping Sparrow, *thi, thi, thi, thi, thi*" (Griscom); call, a *chip.*

Eggs
4–5; creamy white spotted with brown and lilac; 0.7 x 0.5 in. (1.8 x 1.3 cm). Nest is cup of grass, of moss, in moss at foot of bush or tree in or near dry muskeg.

Range
Breeds from cen. Canada to n. U.S.; winters from s. U.S. s. to C. America; migrates casually through Great Plains.

OVENBIRD
Seiurus aurocapillus

45:4

Description
Size, 5½–6½ in. (14.0–16.5 cm). A sparrow-sized ground warbler. Olive above, with a black-bordered • *rufous-buff crown;* eye-ring and underparts white; breast and sides heavily streaked with black; legs pinkish; no wing bars.

Similarities
Waterthrushes lack rusty crown.

Habitat
Deciduous or mixed woodlands; in migration, thickets, garden shrubbery, parks.

Habits
Suggests a small thrush, feeds and walks daintily on forest floor, sings from branch.

Voice
Song, a loud ringing crescendo or *tea-cher- tea-cher, TEA-CHER;* alarm note, a sharp, hard *chik.*

Eggs
3–6; creamy white, finely spotted with red-brown and lilac; 0.8 x 0.6 in. (2.0 x 1.5 cm). Nest is leafy-roofed cup with side entrance, bulky, of grass and leaves, concealed, on forest floor.

Range
Breeds from cen. Canada s. to n. Gulf states; winters from Gulf of Mexico s. to S. America; migrates sparsely through Great Plains.

NORTHERN WATERTHRUSH
Seiureus noveboracensis   **45:6**

Description
Size, 5½–6½ in. (14.0–16.5 cm). Sparrow–sized, with a
conspicuous yellowish or white eyebrow stripe and heavily striped
underparts. Above dark olive-brown; below greenish-yellow to
whitish; heavy, • *dark streaks on sides and breast;* legs pinkish; no
wing bars.

Habitat
Breeds in northern bogs and swamps and areas of quiet water;
streamsides, lakeshores; in migration, any shrubbery.

Habits
Feeds along water's edge, often teeters like Spotted Sandpiper as it
walks or runs, often along edge of pond or watercourse; often sings
at middle level, birds chase each other in fast zigzag flights; early
migrant.

Voice
Song, a loud, carrying *"sweet sweet sweet swee-wee-wee-chew-
chew-chew-chew"* (Gunn), last notes lower, diagnostic; call, a
sharp distinctive *clink.*

Eggs
4–5; pinkish-white, spotted and scrawled with red-brown and
lavender; 0.8 x 0.6 in. (2.0 x 1.5 cm). Nest is cup of moss in moss
at base of stump or tree, near water.

Range
Breeds from nw. Alaska s. to Mont. and e. across n. U.S.; winters
from Mexico to S. America; migrates e. of Rockies.

CONNECTICUT WARBLER
Oporornis agilis   **45:11**

Description
Size, 5¼–6 in. (13.3–15.2 cm). Olive above, yellow below; long,
• *yellow undertail coverts;* • *complete white eye-ring.* Male: • *hood
gray.* Female: duller. Immature: even duller; crown, throat, and
upper breast brownish; eye-ring slightly buffy; in fall, always at
least a suggestion of a hood.

Similarities
Nashville has yellow throat. Mourning male is blacker on breast in
all plumages, lacking eye-ring; Fall Mourning and MacGillivray's
have incomplete eye-ring; spring male has black throat.

Habitat
Mixed woods near water, muskeg; in migration, any underbrush.

Habits
If flushed, flits to low branch and sits motionless.

Voice
Song, a loud ringing *"chuckety chuckety chuckety chuck"* (Allen);
call, a metallic *pink.*

Eggs
3–5; creamy white, with a few dark spots around larger end; 0.8 x
0.6 in. (2.0 x 1.5 cm). Nest is frail grass cup, in moss, on ground
in swamp.

Range
Breeds from s.-cen. Canada s. to Mich., Wis., and n. Minn.;
winters in S. America; accidental to Utah; migrates casually w. to
Calif., Colo., Tex. Panhandle.

MOURNING WARBLER
Oporornis philadelphia
45:10

Description
Size, 5–5¾ in. (12.7–14.6 cm). Spring male: • *gray hood; black apron*; above olive, below yellow; legs pinkish; throat blacker than male MacGillivray's. Spring female and immature: similar to female and immature Connecticut, but no eye-ring in spring and eye-ring in fall broken.

Similarities
MacGillivray's is inseparable in fall; Rockies, west. Spring male Connecticut has gray, not black, apron.

Habitat
Slash, brush, forest clearings.

Habits
A late migrant.

Voice
Song, *"chirry chirry, chorry chorry"* (Peterson); call, a distinctive *chip* somewhat like Common Yellowthroat's.

Eggs
See MacGillivray's.

Remarks
Mourning is the eastern counterpart of, and perhaps the same species as, MacGillivray's.

Range
Breeds from Alta., e. across cen. Canada and s. to n. U.S.; winters in S. America.

MacGILLIVRAY'S WARBLER
Oporornis tolmiei
45:12

Description
Size, 4¾–5½ in. (12.1–14.0 cm). Spring male: very similar to, and probably conspecific with, Mourning, but has broken, • *white ring around eye*, more black between bill and eye, and less black on breast; otherwise, like Mourning. Immature: eye-ring complete. Spring female and fall adults: indistinguishable from Mourning.

Similarities
See Mourning. Nashville has bright yellow throat. Connecticut has complete eye-ring.

Habitat
Brushy undergrowth, moist thickets.

Voice
Song loud, resembles Mourning's; "a rolling *chiddle-chiddle-chiddle, turtle-turtle"* (Peterson), voice dropping; call, like Mourning.

Eggs
3–5; white, brown-spotted; 0.7 x 0.5 in. (1.8 x 1.3 cm). Nest is cup of grass, in briars, weeds, or low brush.

Range
Breeds from Alaska e. to Rockies and in forests s. along Pacific Coast to Baja Calif.; winters in Mexico and S. America; migrates to w. edge of Great Plains.

COMMON YELLOWTHROAT
Geothlypis trichas
43:6, 45:9

Description
Size, 4½–5¾ in. (11.4–14.6 cm). • *Yellow throat*, furtive; upperparts olive, belly white, undertail coverts yellow, no white wing bars or tail corners. Male: distinctive black mask. Female:

223

browner, with whitish eye-ring and no black mask. Immature: very nondescript, throat often buffy and sides brownish; male may show trace of black mask.

Similarities
Nashville has gray crown, underparts fully yellow. Connecticut and Mourning have dark, not yellow, throat.

Habitat
Swamps, fresh- and saltwater marshes, damp undergrowth, thickets, streamsides.

Habits
Active, inquisitive, wrenlike; skulks near ground and in underbrush.

Voice
Song, a distinctive, rapid *witch-i-ty, witch-i-ty, witch-i-ty, WITCH,* also a flight song; alarm note, a chatter; call, a distinctive *check* or *chep.*

Eggs
3–5; creamy white, spotted with red-brown and lilac; 0.7 x 0.5 in. (1.8 x 1.3 cm). Nest is large, loose, grass cup under bush in marsh.

Range
Breeds throughout much of Canada and U.S. from cen. Canada s. to Mexico; mostly s. of U.S.

YELLOW-BREASTED CHAT
Icteria virens   45:5

Description
Size, 6½–7½ in. (16.5–19.1 cm). Largest and most unwarblerlike warbler, with rather long tail, unusual song, and behavior. • *Yellow throat;* conspicuous • *white spectacles;* no wing bars; olive above; bright yellow breast; mustache, belly, and undertail coverts white. Sexes similar.

Similarities
Yellow-throated Vireo has white wing bars.

Habitat
Dry hillsides with second growth, thickets, briars, willows, damp canyons.

Habits
Rises in air from thicket and sings with wings flopping and feet and tail dangling; secretive, but attracted by hand-kissing.

Voice
Song, a highly variable medley of calls, clucks, mews, whistles, and gurgles, often with long pauses, remotely suggesting a Northern Mockingbird; often sings at night; note, buglelike, distinctive *whoit* or *kook.*

Eggs
3–5; white, spotted with brown and lilac; 0.9 x 0.7 in. (2.3 x 1.8 cm). Nest is large cup of grass, leaves, well hidden in thicket, 1–5 ft. (0.3–1.5 m) up.

Range
Breeds locally throughout most of U.S. from s. Canada s. to n. Mexico; winters from Mexico to C. America.

RED-FACED WARBLER
Cardellina rubrifrons   42:10

Description
Size, 5–5¼ in. (12.7–13.3 cm). Bright • *red face,* back gray, belly white, black patch on head, nape white, breast bright red.

Habitat
High mountains, open forests.

Habits
Flycatches at times, often in treetops.
Voice
Song like Yellow Warbler, clear, sweet.
Eggs
3–4; white, spotted; 0.6 x 0.5 in. (1.6 x 1.3 cm). Nest is ground cup under tree, bush, grassy clump.
Range
Breeds from s.-cen. Ariz. and sw. N.Mex. s. to Mexico; winters in Mexico and C. America.

WILSON'S WARBLER
Wilsonia pusilla   43:5
Description
Size, 4¼–5 in. (10.8–12.7 cm). Male: bright yellow below; • *black "skullcap";* olive-green above; no streaks, wing bars, or white in tail. Female and immature: yellow line over beady black eye; olive about ears; trace of skullcap in female, none in immature.
Similarities
Female Yellow Warbler has tail corners yellow.
Habitat
Woodland stream brush, alders, willows, undergrowth, damp tangles.
Habits
Twitches tail, flips wings.
Voice
Song, a quick dry chatter, last half faster, lower, *chee chee chee chee-chee, chi-chi-chi-chi-ch-ch;* call, a *chip.*
Eggs
3–6; white, with brown spots, wreathed around large end; 0.6 x 0.5 in. (1.5 x 1.3 cm). Nest is loose grassy cup on ground amid brush.
Range
Breeds from Alaska e. across Canada and s. to s. Calif. and cen. Nev. and n. N.Mex.; winters from Mexico to C. America.

CANADA WARBLER
Wilsonia canadensis   45:7
Description
Size, 5–5¾ in. (12.7–14.6 cm). • *Black necklace of spots on a yellow throat;* no wing bars; slate-blue above, bright yellow below; undertail coverts white; no white on wings or tail. Male: black forehead, • *yellow spectacles.* Female and immature: duller, with faint necklace and spectacles.
Habitat
Woodland undergrowth.
Habits
Active, catches flies.
Voice
Song, a short jumble, introduced by a *chip,* same as call note: *"chip, chupety swee-ditchety"* (Gunn).
Eggs
3–5; white, speckled with red-brown, mainly around large end; 0.7 x 0.5 in. (1.8 x 1.3 cm). Nest is cup of grass, leaves, rootlets, hidden in woods on ground.
Range
Breeds from s. Canada to n. U.S., chiefly in E. always e. of Rockies, winters in S. America.

AMERICAN REDSTART
Setophaga ruticilla   45:1

Description
Size, 4½–5¾ in. (11.3–14.6 cm). In any plumage, an interrupted broad orange or yellow band across mid-tail. Adult male: black, with bright • *salmon-orange patches on shoulder, wings, and tail;* belly and undertail coverts white. Adult female: salmon-orange replaced by yellow, head gray, upperparts olive, underparts white. First-year male: plumage intermediate between female and adult male, yellow areas of female replaced with orange.

Habitat
Deciduous woodlands, shrubbery.

Habits
Restless, frequently spreads tail, showing dark, inverted T pattern, droops wings, flits about butterflylike; good flycatcher.

Voice
Various, sometimes alternated, including a thin series, *zee-zee-zee-zee-Zeezee-oo* (the last descending), and a double-noted *teetza-teetza-teetza-teetza-teetza;* a sharp *chik;* a clear *tseet.*

Eggs
3–5; whitish, spotted with lilac and brown, wreathed at larger end; 0.6 x 0.5 in. (1.7 x 1.3 cm). Nest is cup of bark shreds, leaf stalks, in fork of bush or saplings, 3–30 ft. (0.9–9.1 m) up.

Range
Breeds from se. Alaska, cen. Man., cen. Que. s. to Utah and Colo.; winters from Mexico to S. America; migrates through Great Plains.

PAINTED REDSTART
Myioborus pictus   42:12

Description
Size, 5–5½ in. (12.7–14.0 cm). Sexes similar. Head, bill, throat, back solid black; • *bright red lower breast;* large • *white wing patches;* sides of tail white.

Habitat
Mountain pine-oak woods, watered oak canyons.

Habits
Often half spreads wings, fans tail like American Redstart.

Voice
Song, repetitive *weeta weeta wee* or *weeta weeta chilp chilp chilp;* note, a *clee-ip.*

Eggs
3–4; white, spotted; 0.6 x 0.5 in. (1.7 x 1.3 cm). Nest is cup of grass on ground or steep bank.

Range
Breeds from s. Ariz., N.Mex., and w. Tex. s. to C. America; winters chiefly s. of U.S.; some strays recorded in s. Calif. in winter.

WEAVER FINCHES
Family Ploceidae

This is a varied family of widespread Old World sparrowlike birds, including the introduced House Sparrow, that feed on grain, fruit, seeds, garbage, and insects.

HOUSE SPARROW

Passer domesticus

Fig. 13

Description

Size, 5¾–6¼ in. (14.6–15.9 cm). Bill stout; tail short, slightly notched. Male: • *throat black,* mixed with gray in winter and in immature; cheek and wing bars white, nape chestnut, red-brown and gray above, grayish-white below. Female: lacks black throat; streaked, buffy, gray and brown above; pale brownish-gray below; buffy line over eye.

Similarities

Purple Finch is redder, with lighter streaks on sides and belly.

Habitat

Cities, towns, suburbs, farms, ranches. Common.

Habits

Social, tame, gregarious, hardy, prolific; hops; often discolored by soot, grime.

Voice

Calls, a *chissik, chissik;* a chirp; alarm note, *tell, tell;* various chattering and twittering notes.

Eggs

5–6; gray-white, speckled with brown and gray; 0.9 x 0.6 in. (2.3 x 1.5 cm). Nest is conglomeration of straw, debris, in ivy, tree, building cavity, or cranny.

Range

Resident throughout N. America from Alaska s. to S. America; introduced from England.

Fig. 13

House Sparrow

MEADOWLARKS, BLACKBIRDS, AND ORIOLES

Family Icteridae

This family comprises a varied group of birds with more or less conical, pointed bills and rounded, not notched, tails. Sexes often are dissimilar. Most species are gregarious. They feed on seeds and insects.

BOBOLINK

Dolichonyx oryzivorus

52:1

Description

Size, 6–8 in. (15.2–20.3 cm). Bill rather short, tips of tail feathers look worn. Spring male: white above, solid black below, • *head and underparts black;* nape buffy; large • *white areas on wings,* back, and rump. Female and fall male: very different, sparrowlike; rich buff, darker above; crown black with buff stripe down middle; buff and brown stripes on back.

Similarities
Female Dickcissel has rusty shoulder patch, slightly notched tail. Female Red-winged Blackbird is duskier, with heavy breast streaks. Male Lark Bunting has white only on wings.

Habitat
Breeds in grassy fields, meadows, and along river valleys; in migration, marshes, grain fields.

Habits
Sings as it flies low or hovers over meadow; perches on grass stalk, shrub, or fence post; gathers in large flocks in migration.

Voice
Song, a rich, bubbling medley given in both hovering flight and quivering descent, beginning with low, reedy notes, *"bob-o-link, bob-o-link, spink, spank, spink"* (William Cullen Bryant); call or flight note, a metallic *pink*.

Eggs
4–7; variable, gray to red-brown, spotted and blotched; 0.8 x 0.6 in. (2.0 x 1.5 cm). Nest is grassy cup, hidden in tall meadow grass.

Range
Breeds from inland B.C. e. across Canada and s. to ne. Calif., Utah, Ariz.; winters s. of U.S.; casual to N.Mex., Tex. Panhandle; accidental to Calif.

EASTERN MEADOWLARK
Sturnella magna

Description
Size, 8½–11 in. (21.6–27.9 cm). Note the big, black crescent on yellow breast. Plump; bill rather long; tail short, wide, rounded; wide, black stripes on crown; white cheeks; brown upperparts; streaked sides; • *white outer tail feathers* conspicuous in flight.

Similarities
Western Meadowlark is nearly identical; closeup, yellow of throat extends onto face farther, voice differs. Dickcissel is much smaller, with white throat, slimmer, short bill. Other birds with white outer tail lack black-and-yellow pattern below.

Habitat
Hayfields, pastures, meadows, plains, prairies.

Habits
Flicks tail as it walks on ground; sings from tree or fence post, early spring to late fall; flight quail-like, frequently sailing with wings outstretched and pointed slightly down; silhouette in flight like a Starling's; flocks in winter.

Voice
Song, a sweet, whistled *"ah-tick-seel-yah"* (Thoreau), or rendered as *spring-o!-the YE-ar* and as *tee-you, tee-yair;* call, an emphatic *dzhert;* alarm note, a rapid guttural chatter (often as it flies away).

Eggs
3–7; white, spotted with brown and lavender; 1.0 x 0.8 in. (2.5 x 2.0 cm). Nest is arched saucer of grasses and weeds, under tufts of grass in field.

Range
Resident from se. Canada e. across U.S., and from Ariz. s. to Mexico and S. America; northern populations move s. in winter.

WESTERN MEADOWLARK
Sturnella neglecta   **52:2**

Description
Size, 8½–11 in. (21.6–27.9 cm). Nearly identical to Eastern
Meadowlark, but trifle paler, yellow of chin carried on to cheeks.
Similarities
Eastern cannot be safely distinguished from Western in field except
by voice.
Habitat, Habits, Food
Like Eastern; where both occur the Western prefers higher, drier
grassy areas, and the Eastern, wetter, lower sites.
Voice
Song very unlike Eastern's; a liquid, loud, warbled *"tung-tung-
tung-ah, tillah'-tillah', tung"* (Ridgeway); call, a metallic *tuk* or
tchuck, sharper, harsher, than Eastern's; alarm note, a chatter.
Eggs
Same as Eastern. Nest is partly domed grassy saucer, in grass.
Range
Breeds throughout w. U.S., from s. Canada s. to Mexico; winters
in C. America.

YELLOW-HEADED BLACKBIRD
Xanthocephalus xanthocephalus   **46:1**

Description
Size, 8–11 in. (20.3–27.9 cm). Male: • *yellow head and neck;* black
between eye and bill; body all-black except white spot on wing,
visible in flight. Female and immature male: dark brown; yellowish
cheek and line over eye; throat and breast yellow; female smaller
than male, lacks white wing spot.
Similarities
Grackles have longer tails.
Habitat
Fields, marshes, tules, farmyards; forages in open country, fields.
Habits
Gregarious; flight slow, deliberate, undulating; flocks long, loose,
not wide like Red-winged, often flocks with Red-winged, cowbirds.
Voice
Song, like rusty hinge, *"oka WEE wee"* (Bent), *oka* guttural,
WEE wee loud whistles; alarm a vehement *"klookoloy, klookoloy
klook ooooo"* (Dawson); call, a low *kek* or *kruk*.
Eggs
3–5; white to greenish, heavily blotched with gray and red-brown;
1.0 x 0.7 in. (2.5 x 1.8 cm). Nest is cup woven of grass and sedge
attached to cattails, reeds, 1–3 ft. (0.3–0.9 m) over water in marsh.
Range
Breeds throughout much of w. U.S. from cen. B.C., Alta., and n.-
cen. Sask. s. to s. Calif., Ariz., n. N.Mex.; winters from se. Oreg.,
cen. Ariz., s. N.Mex., and w. Tex. s. to Mexico.

RED-WINGED BLACKBIRD
Agelaius phoeniceus   **46:2**

Description
Size, 7–9 in. (17.8–22.9 cm). Adult male: black, with • *red-and-
yellow shoulder patch.* The yellow is always visible, but the red
may sometimes be concealed (populations of eastern Calif. valley
area have red epaulets lacking yellow border). Immature male:
sooty-brown, mottled; reddish shoulder patches. Female and young:

229

brownish, heavily streaked below; bill sharp, pointed; light stripe over eye; like a large, dark sparrow but with blackbird appearance.

Similarities
Tricolored, in California, has darker red epaulets, visible mostly in flight, with conspicuous white margin.

Habitat
Breeds in swamps, marshes, muskegs, hayfields; forages in farmlands, shorelines, fields.

Habits
Noisy, gregarious; sings from reeds, tree or fence post, spreading wings and tail; comes to feeders; walks, runs, or hops; flight undulating; often occurs in mixed flocks with cowbirds, Starlings, and grackles.

Voice
Song, a "pleasing *conk-er-EEE* or *oolong TEA*" (Collins); call, a loud *chak* or *check,* also a high *tee-urr.*

Eggs
3–5; pale bluish-green, spotted or scrawled with brown, purple, and black; 1.0 x 0.7 in. (2.5 x 1.8 cm). Nest is bulky grass cup attached to grass or marsh reeds or bush.

Range
Resident from n. Canada s. to C. America; northern populations migrate southward in winter to s. B.C., s. Canada, southward.

TRICOLORED BLACKBIRD
Agelaius tricolor

Description
Size, 7½–9 in. (19.1–22.9 cm). Black with dark red shoulder patch, mainly visible in flight, conspicuously • *margined with white.* Female: streaked brown and white, appearing very dark.

Similarities
Red-Winged male has scarlet shoulder bordered with buff, not white. Female Red-winged is difficult to distinguish, but is lighter and with streaked belly.

Habitat
Tule marshes, cattails; forages in wet fields.

Habits
Territorial; highly gregarious, nests in dense colonies numbering into thousands.

Voice
Song, like Red-winged's, but more nasal, *"on-ke-kaaaangh"*; note, a nasal *"kemp"* (Cogswell).

Eggs
3–4; greenish, scrawled; 0.9 x 0.6 in. (2.4 x 1.7 cm). Nesting is colonial; woven cup attached to marsh vegetation.

Range
Resident from s. Oreg. s. through Calif., w. of Sierras, along Pacific Coast to Baja Calif.; populations move about depending upon availability of food.

NORTHERN ORIOLE
Icterus galbula 47:1

Description
Size, 7–8 in. (17.8–20.3 cm). Adult male: eastern version has all-black head and throat; • *bright orange back*; wings black with white band and orange shoulder patch; tail patterned black and orange in a Y, the Y base at top so corners are orange; rump and underparts orange. Western male has patterned head; orange with black eye,

back, line, crown, and narrow throat patch; wings black with white patch; tail reversed from eastern, forming T with band (top) of T at tip; rump and underparts as eastern, but averaging yellower. Female: black only in wings, which have 2 white bars; otherwise olive or grayish above and in tail; back streaked in eastern form; below whitish, tinged orange-yellow in eastern and grayish and yellowish in western. First-year males: intermediate between females and males, having black on throat.

Similarities
Female Scarlet Tanager has no wing bars. Female Western Tanager is greener. Other orioles have no males with similar pattern and orange color; females and immatures of Hooded, Scott's, and Orchard are yellower, greener-tinted, with no orange tone.

Habitat
Deciduous woods; river and stream trees, shade trees, ranches, shelterbelts.

Habits
Sings from treetop or upper branch; migrates by day, flying high.

Voice
Song whistled, variable, 4 to 8 notes per song, often in paired notes, may sing on wing; call a whistled *peter,* titmouselike; also a chattering alarm call.

Eggs
3–6; grayish-white, streaked with brown and black; 0.9 x 0.6 in. (2.3 x 1.5 cm). Nest is of fibers, elongated, hanging pouch near end of branch.

Remarks
Former Baltimore and Bullock's Orioles are now regarded as 1 species.

Range
Breeds from s. Canada s. through U.S. to Mexico; winters in C. America.

HOODED ORIOLE
Icterus cucullatus   **47:4**

Description
Size, 7–7¾ in. (17.8–19.7 cm). Adult male: black and orange, white wing bars, throat black, • *crown or "hood" orange.* Female: back olive-gray, underparts yellowish, head and tail yellowish-gray, 2 white wing bars. Immature male: like female, but throat black.

Similarities
Female and immature Northern have whitish belly; shorter, less curved bill. Immature male Scott's has ill-defined black throat not reaching to front of eyes.

Habitat
Shade trees, open woodlands, palms, thickets.

Habits
Often feeds low in bushes, although sings and also feeds up high in trees as well.

Voice
Song, *"chut chut chut whew whew,* opening notes throaty" (Peterson), ending in piping whistles; note, a sharp *eek.*

Eggs
3–5; whitish, spotted; 0.8 x 0.6 in. (2.2 x 1.6 cm). Nest is woven pouch under palm frond, in Spanish moss, in yucca.

Range
Breeds from cen. Calif. e. to N.Mex., and s. Tex., and s. to C. America; winters s. of U.S.

ORCHARD ORIOLE
Icterus spurius   47:3

Description
Size, 6–7 in. (15.2–17.8 cm). Slender. Adult male: head, neck,
back, tail black; • *chestnut breast*, belly, and rump; 1 fine white
wing bar. Female: olive-green above, underparts and rump green-
tinged yellow, 2 white wing bars. Immature male: like female but
with black face and chin. Later intermediate plumages show
scattered chestnut; may breed in this stage.
Similarities
Female and young male Hooded are larger, bills longer; yellow has
orangish tone. Female and immature male Scott's have streaks on
back. Males of other orioles are orange, or yellow, and black.
Female tanagers have heavy bill, lack wing bars, except Western.
Habitat
Orchards, farmlands, scattered trees, towns.
Habits
Active; pumps tail, may hang head downward.
Voice
Song, a lively, pleasant warble, quite variable, "a fast-moving
outburst interspersed with piping whistles, guttural notes"
(Peterson); call, a *chak* like Red-winged's, a longer rattle.
Eggs
4–6; bluish-white, scrawled with browns and purples; 0.8 x 0.6 in.
(2.0 x 1.5 cm). Nest is woven pouch of grass 4 in. (10.2 cm) deep,
hung from small branch, usually of fruit tree.
Range
Breeds chiefly in e. U.S., with some w. to Daks., Nebr., and Colo.;
winters from Mexico to S. America; casual to Calif., s. Ariz.

SCOTT'S ORIOLE
Icterus parisorum   47:2

Description
Size, 7¼–8¼ in. (18.4–21.0 cm). Adult male: lemon-yellow; back,
head, wings, • *tail yellow* with inverted black T up center and tip;
• *head solid black*. Female: underparts more greenish-yellow than
most species, back streaked. Immature male: throat black;
distinguishable from young male Hooded and Orchard Orioles by
grayer, unstreaked back, more black on face, dingier underparts.
Similarities
Female Orchard is smaller, bill shorter. Female Hooded has
underparts yellower. Female Northern has belly whitish.
Habitat
Desert mountains, scrub and dry woods; Joshua trees, yuccas, oak
slopes, pinyons.
Habits
Closely associated with yuccas and agaves, feeding about their
blossoms.
Voice
Song, like Western Meadowlark's; whistles.
Eggs
2–4; whitish, streaked and blotched; 0.9 x 0.6 in. (2.4 x 1.7 cm).
Nest is grass pouch in small oak, pinyon, yucca, Joshua tree.
Range
Breeds from se. Calif., s. Nev., Utah, Ariz., and s. N.Mex. south to
Mexico; winters s. of U.S.

Note: The **BLACK-HEADED ORIOLE**, *Icterus graduacauda*, of south
Texas, is yellow with all-black head, wings, and tail.

RUSTY BLACKBIRD
Euphagus carolinus   **46:5**

Description
Size, 8½–9¾ in. (21.6–24.8 cm). Eye yellow, tail of medium
length. Spring male: black head with dull green sheen. Female:
gray. Fall and winter adult: barred with rusty head and body.
Immature: similar but even rustier.

Similarities
Brewer's male is more iridescent, with greener sheen; female has
dark eyes. Grackles are larger, tails longer, more iridescent.
Cowbirds are smaller, bill shorter, eyes dark.

Habitat
Swamps, marshes, fields; muskeg in summer, river groves.

Habits
Gregarious, sometimes in mixed flocks; walks and runs, nodding
head.

Voice
Song, an even rhythmic *"totalee-eek-totalee-eek;* or a rapid
kawicklee kawicklee" (Saunders); call, a *kik* and a rattling, a *chuk*
or loud *chack.*

Eggs
3–6; light bluish-green, blotched with chocolate and gray; 1.0 x 0.8
in. (2.5 x 2.0 cm). Nest is bulky cup of leaves, grass, in alder or
willow, 1–2 ft. (0.3–0.6 m) above water in swamp.

Range
Breeds throughout much of Canada from Alaska to cen. B.C., cen.
Alta. and cen. Sask.; winters in E.; migrates through cen. B.C.;
casual to Calif., Idaho, Utah, Ariz., N.Mex.

BREWER'S BLACKBIRD
Euphagus cyanocephalus   **46:6**

Description
Size, 8–10 in. (20.3–25.4 cm). Medium-tail. Male: yellow eyes;
black, with • *iridescent purplish head;* body glossed with greenish
in strong light, appears black at distance; in winter, may show
some rusty barring. Female: gray, paler over eye and on throat,
eyes dark. Immature: males may show slight grayish edgings.

Similarities
Spring male Rusty, east of Rockies, shows little iridescence; female
darker gray, eye yellow. Brewer's prefers fields and farms; Rusty,
river groves and swamps. Female cowbirds have short bill.

Habitat
Fields, farmyards, ranches, parks, cities, open country, lakeshores.

Habits
When not breeding, wanders in flocks over countryside, gathers in
roosts at night.

Voice
Song, a hoarse, whistled *squee* or *que-ee* or *ksh-eee* like rusty
hinge; a scolding *check* or *tshup,* a *kit-tit-tit-tit* and, when elevating
tail, a *chug-chug-chug, tucker,* or *tit-tit-tit* (Laidlaw Williams).

Eggs
4–6; grayish, heavily spotted with brown; 1.0 x 0.7 in. (2.5 x 1.9
cm). Nests in small colonies; bulky grass-lined cup of twigs, bark
and mud, in bush or tree, 1–30 ft. (0.3–9.1 m) up.

Range
Breeds from sw. B.C. and Man. e. to e. U.S. and s. to Baja Calif.
and w. Tex.; winters from Wash., Idaho, and Mont. southward.

GREAT-TAILED GRACKLE
Quiscalus mexicanus   46:9

Description
Size, male, 16–17 in. (40.6–43.2 cm); female, 12–13 in. (30.5–33.0
cm). Largest blackbird, with a very • *long, wide V-shaped*
("keeled") *tail;* eyes yellow. Male: iridescent hues of blue and
purple. Female: much smaller than male, dark brown, paler below.
Immature: paler, browner.
Similarities
Common Grackle is much smaller; female Great-tailed much
lighter brown on breast.
Habitat
Towns, groves, river breaks, thickets.
Habits
Gregarious, especially in winter; walks with tail high, alert
carriage, trace of waddle; tail often blown about by wind.
Voice
Noisy, male song a 4-part crackling, hissing, ending in 1 or more
piercing *cha-we* notes; warning note, a clack; whistled calls also
(Selander and Giller).
Eggs
3–5; pale-blue, dotted and scrawled with purple; 1.3 x 0.9 in. (3.3
x 2.3 cm). Nest is bulky grass-lined cup of sticks, grass and mud,
in bush or tree, 1–40 ft. (0.3–12.2 m) up or in swamp reeds;
colonial.
Range
Resident from Calif., Ariz., and s. Tex. s. to S. America; in winter
some northern populations move southward.

COMMON GRACKLE
Quiscalus quiscula   46:8

Description
Size, 11–13½ in. (27.9–34.3 cm). Male: medium-length, keeled or
• *V-shaped tail;* eye yellow; all-black with iridescent hues of purple,
blue, green, bronze, especially on head and back. Female: smaller,
tail shorter; duller, iridescent only on forepart of body. Immature:
dull brown, eye brown.
Similarities
Rusty and Brewer's have shorter tails. Boat-tailed, along Mexican
border, is much larger, tail longer, female brown.
Habitat
Lawns, fields, shade trees, farmlands, streamsides.
Habits
Flight more even than other blackbirds, not rising and falling;
walks.
Voice
A medley of harsh squeaking and guttural noises (some like rusty
hinge), not altogether unmusical; call, *chuck* or *chack*, louder,
lower than Red-winged's.
Eggs
4–6; greenish-white to reddish-brown, variously scrawled or
blotched with brown; 1.2 x 0.9 in. (3.0 x 2.3 cm). Nest is bulky
cup of twigs, grass and mud on branch or in fork of conifer, or tree
cavity 5–40 ft. (1.5–12.2 m) up; or in bush, reeds; colonial.
Range
Breeds from s. Mackenzie, Alta., and Mont. e. across Canada and
U.S., and s. to Great Plains; winters from Kans. s. to Tex.; casual
w. of Rocky Mountains; accidental to Calif.

BROWN-HEADED COWBIRD
Molothrus ater **46:4**

Description
Size, 6–8 in. (15.2–20.3 cm). Male: rather small, bill sparrowlike, black with iridescent reflections of purple and green on upper back, • *head and neck coffee-brown*, eyes dark. Female and immature: • *all gray;* rather streaked below in juvenal plumage.

Similarities
Female Rusty and Brewer's are larger, bills longer. Young Starling has shorter tail, longer bill. Catbird is darker than female cowbird, crissum chestnut.

Habitat
Fields, farms, open woods, edges, barnyards, river groves.

Habits
Gregarious, often flocks with Starlings, Red-wingeds, grackles; associates with cattle; spreads wings and tail as it squeaks; flight like Red-winged's.

Voice
Male, a characteristic *glee*, or song a squeaky *glu-glu-gleeee*, like swinging an unoiled gate; female, a rattling chatter; flight call, "*weee-titi* (high whistle, 2 lower notes)" (Peterson); call, a *chuck*.

Eggs
4–5; white, speckled with brown; 0.9 x 0.7 in. (2.3 x 1.8 cm). Parasitic, no nest, lays eggs in other birds' nests.

Range
Breeds throughout much of s. Canada and U.S. from cen. Canada, s. to Mexico; winters from cen. Calif. and sw. states southward.

BRONZED COWBIRD
Molothrus aeneus **46:7**

Description
Size, 6½–8¾ in. (16.5–22.2 cm). Conspicuous • *ruff on nape* during breeding season, eyes red. Male: like Brown-headed, but latter is smaller, with contrasting brown head, dark eyes and no neck ruff. Female: smaller, duller, ruff less conspicuous, but very like male.

Habitat
Semiopen country, brushland, farms.

Habits
As Brown-headed.

Voice
"High-pitched mechanical creakings" (Kincaid).

Eggs
1–4; bluish-green, pale; 0.9 x 0.7 in. (2.3 x 1.8 cm). Parasitic; lays eggs in nests of orioles and other birds, which raise the Cowbird young.

Range
Resident from sw. U.S., s. to Mexico and C. America; casual to se. Calif.

TANAGERS
Family Thraupidae

Tanagers are arboreal birds of tropical origin. Males are brilliantly hued, especially with bright red, and the females are greenish above and yellow below, somewhat like large warblers or vireos, but having heavier, blunter bills that are usually notched or "toothed." Tanagers are deliberate, even sluggish in action and feed on insects, fruits, and berries.

WESTERN TANAGER
Piranga ludoviciana   **47:6**

Description
Size, 6¼–7½ in. (15.9–19.1 cm). Both sexes have wing bars. Adult
male: yellow rump with • *red head* and black back; black also on
wings and tail; 2 wing bars, pale yellowish; face mostly yellow in
fall, replacing earlier red. Female and immature plumages like
those of Scarlet Tanager, dull-olive above, yellow below, but pale
yellow wing bars.

Similarities
Female orioles are similar to female Western, but sides of face, tail
lighter; bill more sharply pointed. Females of other tanagers have
no wing bars.

Habitat
Open conifer or mixed woods, edges; widespread in migration.

Habits
Often catches insects on wing; tame around people; visits feeders.

Voice
Call, a dry *pit-tik* or *pit-i-tik*; song, short phrases, like a hoarse
Robin or Black-headed Grosbeak but less sustained, rougher.

Eggs
3–5; pale greenish-blue, spotted with brown; 1.0 x 0.7 in. (2.5 x
1.8 cm). Nest is shallow saucer of grass, bark shreds, on low
branch of pine, fir, oak.

Range
Breeds throughout much of w. U.S. from Alaska and s. Mackenzie
s. to Baja Calif., excluding desert areas; winters from Mexico to C.
America; migrates throughout range and e. to Great Plains.

HEPATIC TANAGER
Piranga flava   **47:5**

Description
Size, 7–7¾ in. (17.8–19.7 cm). Male: dull orange-red all over,
• *dark ear patch, bill blackish*, no crest. Female: above dusky, below
yellowish.

Similarities
Female Hepatic is distinguishable from female orioles by lack of
wing bars and heavier bill, and from female Summer Tanager by
blackish bill, gray ear patch, more orange-yellow throat. Summer
Tanager male is rosier-red all over, bill yellowish. Hepatic prefers
upland woods; Summer Tanager, lowland streamside trees.

Habitat
Open pine-oak forests of mountains.

Habits
Active; moves from tree to tree.

Voice
Note, a *chuck*; song, like Black-headed Grosbeak's.

Eggs
Like Western's.

Range
Breeds from nw. Ariz. and N.Mex., s. to Mexico and S. America;
winters from Mexico, southward.

Note: The eastern **SCARLET TANAGER**, *Piranga olivacea*, size, 7
in. (17.8 cm), is similar to the Hepatic. The summer male is all-
red with black wings and tail. It occurs along the eastern fringe of
the Great Plains.

SUMMER TANAGER
Piranga rubra   **47:7**

Description
Size, 7–7¾ in. (17.8–19.7 cm). Adult male: entirely • *rose-red* all
over with • *no black*, bill yellowish. Female: above olive-green
tinged with yellow; below yellow with orange tinge, overall more
orange-yellow, less yellow-green than female Scarlet. Immature
male: like female, or with varying patches of rose-red and green.

Similarities
Orioles have wing bars. Female Western has wing bars. Cardinal
has crest, face black. Hepatic is darker, bill blackish.

Habitat
Deciduous woods, shade trees, riverine woods, willows,
cottonwoods, pine and mixed forests.

Habits
May catch insects in air.

Voice
Song, a sweet, long carol suggesting both a Robin and Rose-
breasted Grosbeak; call, a *chicky-chucky-tuck*, a *"pi-tuck or pik-i-
tuck-i-tuck"* (Peterson); phrases less nasal and resonant than
Western's.

Eggs
3–4; greenish-blue, spotted with purplish-brown; 0.9 x 0.7 in. (2.3
x 1.8 cm). Nest is thin, shallow cup of bark shreds and grass, near
end of limb 5–30 ft. (1.5–9.1 m) up.

Range
Breeds in sw. U.S. from se. Calif. and Ariz. s. to Mexico and
across s. U.S.; winters in C. America

GROSBEAKS, FINCHES, SPARROWS, LONGSPURS, AND BUNTINGS
Family Fringillidae

This is a very large family characterized by strong, short, conical
bills adapted for eating seeds (other dietary items include buds,
fruits, berries, insects). Grosbeak-type bills are large, somewhat
rounded in outline, and thick at the base; finches, sparrows, and
buntings have canarylike bills; the crossbills have bills that are
crossed at the tip. The sparrows, juncos, and longspurs are usually
grouped together. Most are, in one plumage, streaked brown above,
pale and often streaked below. They forage and nest near or on the
ground, but often sing from an elevated perch.

CARDINAL
Cardinalis cardinalis   **48:1**

Description
Size, 7½–9 in. (19.1–22.9 cm). Bill red-orange; crested. Male: • *all
red;* • *pointed crest;* face and throat black. Female: reddish hue
confined to crest, wings, and tail; back brownish-gray; face
blackish; head and underparts buff-brown.

Similarities
Male Summer Tanager has no crest. Female Pyrrhuloxia has gray
back, stubby yellow bill.

Habitat
Gardens, thickets, towns, woodland edges.

Habits
Tame, frequents feeders, likes sunflower seeds; sings from high perch.

Voice
Song, a loud, clear whistle, *wheat wheat wheat, what-cheer what-cheer what-cheer;* call, a sharp *tik* or *chip.*

Eggs
3–4; whitish, blotched with brown and lilac; 1.1 x 0.8 in. (2.8 x 2.0 cm). Nest is ragged cup of twigs, bark shreds in low bush.

Range
Resident from se. Calif. e. to s. Ariz., s. N.Mex., and w. Tex., and s. to Baja Calif. and Mexico; introduced into s. Calif.; widespread in E.

PYRRHULOXIA
Cardinalis sinuatus   48:2

Description
Size, 7½–8¾ in. (19.1–21.0 cm). Male: slender; • *back gray;* • *breast rose-red* extending down front; crest red and pointed; heavy • *bill yellow,* almost parrotlike. Female: back gray, breast buffy-brown, bill yellow, some red in crest and on wings.

Similarities
Female Cardinal is browner, bill reddish.

Habitat
Desert scrub, brush, mesquite, paloverde, ironwood.

Habits
Much as Cardinal but less often on ground.

Voice
Song, clear, monotone series *quink quink quink quink quink;* "also a slurred, whistled *what-cheer, what-cheer,* etc., thinner and shorter than Cardinal's" (Peterson).

Eggs
3–4; white, spotted; 0.9 x 0.7 in. (2.3 x 1.8 cm). Nest is neat cup in desert thorny scrub.

Range
Resident chiefly of Mexico, but n. to s. Ariz., N.Mex., and s. Tex.

ROSE-BREASTED GROSBEAK
Pheucticus ludovicianus   47:14

Description
Size, 7–8½ in. (17.8–21.6 cm). Spring male: heavy whitish bill, head and upperparts black, white on wings and lower back, large • *rose-red triangle on breast.* Fall male: streaked dark brown, rose and white of underparts spotted with dusky. Female: like a large, brown sparrow; streaked, white line over eye, 2 white wing bars, heavy buff bill.

Similarities
Female Purple Finch is much smaller, with smaller bill. Female Black-headed Grosbeak has browner breast, fine streaks only on sides, shows yellow below.

Habitat
Open woods, edges, shade trees, thickets, aspens.

Habits
Sluggish; male sings from high perch (also on nest), feeds from near ground to treetops.

Voice
Song, a rolling sugary warble, like Black-headed Grosbeak's; call, a sharp *ink,* distinctive.

Eggs
See Black-headed.
Range
Breeds from cen. Canada s. to cen. U.S. and e. across U.S.; winters from Mexico to S. America; some hybrids occur where it meets the Black-headed.

BLACK-HEADED GROSBEAK
Pheucticus melanocephalus   **47:12**

Description
Size, 6½–7¾ in. (16.5–19.7 cm). Male: black head; orange-brown underparts, nape, and rump; • *white wing bars*, conspicuous in flight; bill large, pale; breast rusty; yellowish belly; yellow wing linings (both sexes); tail pattern in flight shows black and white. Female: streaked brown above, buffy-brown below, sides lightly streaked.
Similarities
Female Rose-breasted, in East, is whiter, with more streaked breast, shows no yellowish below.
Habitat
Deciduous and mixed woods, especially pine-oak, and edges; groves, parks, gardens; pinyons, chaparral, river valley woods.
Habits
Comes to feeders and picnic tables.
Voice
Song, a loud, long warble with rising and falling passages, somewhat like Rose-breasted or Robin, but more varied; call, a sharp *ik* or *tik*.
Eggs
3–4; bluish-white, spotted with brown; 0.9 x 0.7 in. (2.3 x 1.8 cm). Nest is frail saucer of twigs, plant stems, in bush or tree, 5–20 ft. (1.5–6.1 m) up.
Range
Breeds from s. B.C. and Sask. s. to Mexico; winters s. of U.S.; hybridizes with Rose-breasted in Plains.

BLUE GROSBEAK
Guiraca caerulea   **47:8**

Description
Size, 6–7½ in. (15.2–19.1 cm). Bill heavy. Male: purplish-blue (looks black in poor light) with • *2 rufous wing bars*. Female and immature: brown above, often with some bluish; buffy-brown below; bill tan-colored; wings brown; • *2 buffy wing bars*. Spring immature male is like female, but with a mixture of brown and blue.
Similarities
Indigo Bunting is smaller; bill small, no wing bars, male all-blue. Female Lazuli Bunting is smaller; whitish wing bars.
Habitat
Scattered shrubs in dry fields, thickets near water, farms, willows, weeds.
Habits
Sluggish, shy, sings from bush top.
Voice
Song, a finchlike warble, like Indigo Bunting's or Orchard Oriole's; short phrases rising and falling, but slower and more guttural; call, a loud *chuck* or *chink*.

Eggs
3–5; light blue; 0.9 x 0.6 in. (2.3 x 1.7 cm). Nest is cup of grass, rootlets, snakeskin in shrub or on low branch.
Range
Breeds from Calif., s. Colo., and S.Dak. s. throughout U.S. to Mexico; winters from Mexico to C. America.

INDIGO BUNTING
Passerina cyanea   47:10

Description
Size, 5½–5¾ in. (14.0–14.6 cm). Bill small. Spring male: • *rich blue*; looks black in poor light; in fall, strong mixture brown on back and head, and whitish below (molting birds mixed blue and brown). Female and immature: sparrowlike; warm brown above, paler below, faintly streaked; often some blue in tail and wings in adult female.
Similarities
Blue Grosbeak is larger, bill heavier, with buffy brown wing bars.
Habitat
Dense brush on open hillsides, clearings, edges, roadsides.
Habits
Sings from exposed perch on wire or near top of tree, persistently, through heat of day.
Voice
Loud song, high, strident, paired notes *"sweea sweea sit sit seet seet sayo"* (Saunders), well-measured phrases at different pitches; call note, a sharp *chit* or *tsip*.
Eggs
3–4; pale blue; 0.8 x 0.6 in. (2.0 x 1.5 cm). Nest is cup of grass, weeds, in bush.
Range
Breeds chiefly in E. but is spreading w. to Alta., Utah and s. to Tex.; winters from Mexico to C. America. Has bred Calif.

VARIED BUNTING
Passerina versicolor   47:13

Description
Size, 4½–5½ in. (11.4–14.0 cm). Male: small, dark, with a blue crown and • *bright red patch on nape;* plum-purple, colored like an Easter egg. Female: all-over gray-brown, underparts lighter, no distinctive marks of any kind.
Similarities
Male Painted has rump, breast bright red. Female Indigo is more rusty, brown, faint wing bars and breast streaks.
Habitat
Brush, chaparral, thickets, cactus.
Habits
Frequents dense tangles but male sings in open.
Voice
Song, like Painted Bunting's, but more distinctly phrased and less warbled, thin and bright; note, like Indigo song but notes less distinctly paired.
Eggs
3–4; bluish; 0.7 x 0.5 in. (2.0 x 1.4 cm). Nest is grassy cup in bush.
Range
Breeds from s. Ariz., sw. N.Mex., w. Tex., southward; winters s. of U.S.; accidental to se. Calif.

LAZULI BUNTING
Passerina amoena   **47:11**

Description
Size, 5–5½ in (12.7–14.0 cm). Adult male: • *azure-blue head and rump,* orange-brown breastband, white belly and wing bars. Immature male: often has blue on head only. Female: head and back unstreaked gray-brown, nondescript; some blue on wings, rump, and tail, light wing bars, underparts pale with buffy wash on breast; no streaking.

Similarities
Female Blue Grosbeak is larger, paler; much bigger bill, brownish wing bars. Western Bluebird is larger; lacks wing bars, bill more slender. Female Indigo Bunting has no wing bars.

Habitat
Dry brushy areas, sagebrush, burns, streamsides.

Habits
Sings from top of trees.

Voice
Song, like Indigo Bunting's.

Eggs
3–4; pale blue-green; 0.7 x 0.5 in. (1.8 x 1.4 cm). Nest is loose cup of grass, leaves, in low bush near water.

Remarks
Observers in plains should be aware that Indigo and Lazuli hybridize; mixed traits of the two may be seen in hybrids on breeding grounds or migration in eastern Plains states.

Range
Breeds from s. B.C. to N.Dak., s. to Baja Calif. and Okla.; winters in Mexico; sporadically hybridizes with Indigo in plains.

PAINTED BUNTING
Passerina ciris   **47:9**

Description
Size, 5–5½ in. (12.7–14.0 cm). Male: • *red, purple, and green;* one of the most gaudily colored birds in North America. No wing bars, head and nape violet, back yellow-green, rump and underparts red. Female: distinctive as our only small, all-green bird; bright green above, paler lemon-green below, no wing bars or streaks. Fall immature: like female. Spring immature male: like female, but some blue on head.

Similarities
Female Lesser Goldfinch is also greenish, wings blackish, with wing bars.

Habitat
Dry thickets, woods edges, roadsides, gardens.

Habits
Shy; sings from elevated perch.

Voice
Song suggests that of Indigo, but weaker, call note, a sharp *chip.*

Eggs
3–5; white, spotted with red-brown; 0.7 x 0.5 in. (1.9 x 1.4 cm). Nest is woven cup of grass, leaves, in bush or tree in river flood plain.

Range
Breeds from N.Mex. and Okla. e. across s. U.S. and s. to Mexico; winters from Gulf states to C. America.

DICKCISSEL
Spiza americana   **52:6**

Description
Size, 6–7 in. (15.2–17.8 cm). Like House Sparrow and
meadowlark. Spring male: black crescent on yellow breast, white
throat, no white outer tail feathers. Streaked brown above, head
gray, bill bluish, throat white, yellow line over eye, chestnut on
shoulder, belly whitish. Fall male: often lacks bib. Female and
immature: paler, no black V, breast with some streaks plus yellow
wash, chestnut shoulder, suggestion of male head pattern.

Similarities
Female House Sparrow lacks rusty shoulder patch. Fall male,
female Bobolink are larger; buffier, with head stripes; lack chestnut
on shoulder.

Habitat
Dry open places; fields, especially alfalfa; prairies, roadsides.

Habits
Breeds in loose colonies; sings from wire, bush, or stalk persistently
into late summer; gathers in flocks in migration.

Voice
Song, a staccato *dick, dick dick-cissel,* much repeated.

Eggs
3–5; greenish-blue; 0.8 x 0.6 in. (2.0 x 1.5 cm). Nest is of grass, on
ground by grassy tussock or in low bush.

Range
Breeds in n.-cen. U.S. s. to Gulf Coast; winters from Mexico to S.
America. Reaches e. edge of our region.

EVENING GROSBEAK
Coccothraustes vespertina   **48:3**

Description
Size, 7–8½ in. (17.8–21.6 cm). Chunky body with short, notched
tail; very large, yellowish-white bill; big white wing patches
conspicuous in flight. Male: gold and dull yellow, head dusky-
brown with yellow eye stripe, wings and tail black. Female: much
of brown and gold replaced by gray; dingier white on wings, white
tail; yellow, black, and white patterning enough to be recognizable.

Similarities
Female Pine Grosbeak has longer bill, stubby, black; no white in
tail.

Habitat
Conifer forests; in winter, very diverse trees.

Habits
Gregarious, wanders widely in winter; tame, frequents feeders,
likes sunflower and hemp seeds; attracted by salt.

Voice
Song, a sweet warble; also a metallic cry, "a ringing *cleer* or *clee-
ip*" (Peterson); call, like a loud House Sparrow's; a chatter.

Eggs
3–4; blue-green, marked with gray and brown; 0.9 x 0.6 in. (2.3 x
1.7 cm). Nest is saucer-shaped; of twigs, grass; in top of conifer
15–20 ft. (4.6–6.1 m) up.

Range
Breeds from s. Canada through mountain forests to Mexico, and e.
to Great Lakes; winters along Pacific Coast and inland e. to Great
Plains.

PURPLE FINCH
Carpodacus purpureus   **48:6**

Description
Size, 5½–6¼ in. (14.0–15.9 cm). Stout bill, notched tail. Male: dull
• *raspberry-colored all over,* brightest on head and rump, no streaks
on side or belly. Female and immature: sparrowlike in heavy,
brown streaking all over; gray-brown above, white below; pale line
over eye; dark stripe below dark cheek.

Similarities
Pine Grosbeak is much larger, wings dark, with 2 white bars.
Male House is more orange-red, dark streaks on sides and belly;
female House is paler, more solidly grayish, streaks lighter, no face
pattern, stubby bill. Cassin's has square bright red crown
contrasting with brown neck and back. Common Redpoll is gray-
brown with streaked, black chin.

Habitat
Mixed, but especially conifer woods, edges, shade trees.

Habits
Sings from treetops; wanders in fall, frequents feeders, likes seeds
of sunflower, hemp, millet.

Voice
Song, a mellifluous warble, like Warbling Vireo, but louder; also
like House Finch but lower, shorter, less disjointed; call, a dull,
metallic *tick, tick;* in flight, a sharp *pink.*

Eggs
4–5; blue-green, with dark spots chiefly around larger end; 0.8 x
0.6 in (2.0 x 1.5 cm). Nest is frail, neat cup of bark shreds, rootlets
in conifer.

Range
Resident in forests throughout w., from B.C., s. to Mexico, and
farther e. in northern regions; northern populations migrate
southward in winter.

CASSIN'S FINCH
Carpodacus cassinii   **48:10**

Description
Size, 6–6½ in. (15.2–16.5 cm). Male: like Purple Finch, but red
crown contrasts sharply with brown of neck, tail squared, breast
paler. Female: finely streaked version of Purple Finch, whiter
below, much paler head.

Similarities
House Finch, male has belly-streakings; female much grayer, less
distinct stripes above and below, face less patterned. Common
Redpoll is smaller; more pinkish than red, chin black.

Habitat
Open pine woods and edges in mountains.

Habits
Feeds from ground to treetops.

Voice
Song, a warble reminiscent of both Purple and House Finches,
lively; flight note, a *gee-d'yip.*

Eggs
4–5; blue-green, lightly spotted with purplish; 0.8 x 0.6 in. (2.0 x
1.5 cm). Nest is thin saucer of grass, twigs on branch of pine.

Range
Resident, chiefly from s. Canada s. to Baja Calif. and e. to e. edge
of Rockies; some breed n. to cen. B.C., others winter s. to Mexico.

HOUSE FINCH
Carpodacus mexicanus   48:8

Description
Size, 5–5¾ in. (12.7–14.6 cm). Nearly size of House Sparrow.
Male: brownish; bright orange-red forehead, stripe over eye, rump,
and breast (some birds show almost orange); • *narrow dark streaks
on sides and underparts* (diagnostic). Female and immature:
nondescript gray-brown above, underparts streaked with dusky,
• *head pattern lacking strong stripes,* bill stubby.

Similarities
Purple and Cassin's males are redder, less orange and lack
streaking on white abdomen; females more contrastingly streaked,
discrete head markings.

Habitat
Cities, towns, open country, ranches, deserts, canyons, coastal
scrub.

Habits
Gregarious, comes to feeders.

Voice
Song, a continuous, variable warble, higher, longer, than Purple
Finch's, often ending in a harsh nasal *wheer* or *che-urr;* notes
various, some like House Sparrow's chirping.

Eggs
4–5, pale blue, some black spots; 0.7 x 0.5 in. (1.8 x 1.4 cm). Nest
is compact cup of grass, paper, rags almost anywhere, usually near
a house; in vines, cactus.

Range
Resident throughout W. from s. Canada s. to Mexico and e. to
Nebr.; introduced in e. U.S.

PINE GROSBEAK
Pinicola enucleator   48:4

Description
Size, 8–10 in. (20.3–25.4 cm). Big and chunky (nearly Robin-
sized) with 2 white wing bars; tail long, notched; bill, wings, tail
black. Adult male: rose-red and gray. Female: olive-yellow and
gray, replacing male's red, head and rump yellowish, 2 white wing
bars. Immature male: like female, but often with some red or
orange brown on head and rump.

Similarities
White-winged Crossbill is smaller; has crossed bill. Purple and
Cassin's Finches lack wing bars. Evening is stockier, tail short.

Habitat
Conifer woods, edges, shrubbery; in winter, mixed woods, orchards.

Habits
Tame; flocks in winter come south irregularly, sometimes not at
all.

Voice
Song, loud warbles and whistles reminiscent of Robin or Purple
Finch but interspersed with a *twang;* a whistled *yew, yew, yew* or
tee-tee-tew somewhat like Greater Yellowlegs's; flight call, *pee-ah;*
alarm note, a musical *chee-vli.*

Eggs
3–5; greenish-blue spotted with purple; 0.9 x 0.6 in. (2.3 x 1.5
cm). Nest is loose cup of twigs, rootlets, grass in conifer.

Range
Breeds from n. Canada s. to mountains of Ariz. and N.Mex.;
winters to Great Basin, Great Plains, eastward.

GRAY-CROWNED ROSY FINCH
Leucosticte tephrocotis   **52:14**

Description
Size, 5¾–6¾ in. (14.6–17.1 cm). Sparrow-sized, • *chestnut-brown, pinkish rump;* tail notched; bill yellowish; forehead and sometimes chin black; • *crown gray;* foreparts deep chestnut-brown; wings, belly, rump pinkish and brown. Female is duller, gray head patch reduced or lacking.

Similarities
Black and Brown Rosy Finches are similar; many authorities consider these conspecific with Gray-crowned.

Habitat
Mountains above timberline, rocks, cirques, snowfields, tundra islands (Alaska); in winter, open areas on mountains, foothills, plains east of Sierras and Cascades.

Habits
Active, catches insects in air; often feeds near snowbanks; gathers in flocks in fall, wanders widely in winter.

Voice
Song, a high-pitched series of *chips;* flight note, a *"chee-chee-chi-chi-chi"* (Hoffmann).

Eggs
3–5; white; 0.9 x 0.6 in. (2.3 x 1.5 cm). Nest is grass cup in rock crevice above timberline.

Range
Resident in mountains from Alaska through Cascades to s. Calif.; in winter, moves to lower elevations.

BLACK ROSY FINCH
Leucosticte atrata   **52:15**

Description
Size, 6 in. (15.2 cm). Like Gray-crowned, may be same species, but body blackish instead of chestnut.

Habitat, Habits, Voice, Eggs
Similar to Gray-crowned.

Range
Breeds in Rocky Mountains of sw. Mont., Idaho, Wyo., ne. Nev., and n. Utah; winters s. to n. Ariz. and N.Mex.

BROWN ROSY FINCH
Leucosticte australis   **52:13**

Description
Size, 5¾–6¼ in. (14.6–15.4 cm). Like Gray-crowned (may be same species), but body lighter brown, no gray head patch; crown dusky.

Habitat, Habits, Voice, Eggs
Like other rosy finches.

Range
Breeds in s. Rockies from se. Wyo. to N.Mex.; winters in lower elevations.

COMMON REDPOLL
Carduelis flammea   **48:9**

Description
Size, 5–5½ in. (12.7–14.0 cm). Tail notched. Male: • *red cap on forecrown;* streaked with gray-brown above, paler on rump; 2 white wing bars; whitish below; • *chin black;* dark streaks on

flanks; breast pink. Female and immature: similar, but no pink on breast; browner in spring plumage.

Similarities
Purple, Cassin's, House finches are larger; redder, rumps red, no black chin. Pine Siskin is darker, no red, more heavily streaked.

Habitat
Tundra bushes in summer, bushy fields in winter.

Habits
Gregarious, active; clings to weed stems; sings from perch or in air; flocks twitter as they feed; frequents feeders (likes rolled oats and hemp seeds).

Voice
Call, a dry *ch-ch*, repeated; song, a trill followed by a rattle; call in flight, a *chip chip chip.*

Eggs
3–6; pale greenish blue, brown and lavender spotted; 0.7 x 0.5 in. (1.8 x 1.4 cm). Nest of grass and down low in bush or tree.

Range
Breeds from Arctic s. to n. Canada; winters from n. Canada s. to cen. U.S.

Note: The **HOARY REDPOLL**, *Carduelis hornemanni*, size, 5 in. (12.7 cm), is a high-Arctic redpoll, irregularly wintering to British Columbia, Montana, South Dakota. It is difficult to distinguish, being paler, with • *unstreaked white rump.*

PINE SISKIN
Carduelis pinus   **48:11**

Description
Size, 4½–5¼ in. (11.4–13.3 cm). • *Heavily streaked,* with a flash of • *yellow in wings and tail,* deeply notched tail, wings with 2 white wing bars and yellow notch at base of primaries; base of outer tail yellow.

Similarities
Goldfinches are never streaked; much yellower American Goldfinch is most alike in size and actions. Redpoll is paler, pinkish, streaks lighter and none on breast. Female Purple, Cassin's, House finches are larger, lack yellow, bills heavier.

Habitat
Breeds in conifers; at other times found almost anywhere—mixed woods, alders, weedy areas near woods, treetops.

Habits
Active, gregarious, often in "winter finch" flocks with goldfinches, crossbills, redpolls; flight goldfinchlike; frequently recognized by flight call; frequents feeders (likes millet).

Voice
Song, like American Goldfinch, but lower, longer, rougher; call, a soft *tit-i-tit; it-it-it;* a rasping, rising *shre-e-e-e-e;* a louder *klee-ip* or *chlee-ip;* notes from large flock in concert make a humming sound.

Eggs
3–6; pale greenish-blue, spotted with brown; 0.7 x 0.5 in. (1.8 x 1.3 cm). Nest is neat saucer of grass, twigs in conifer 8–30 ft. (2.4–9.1 m) up.

Range
Breeds from s. Alaska and Canada s. through w. U.S. to Mexico; winters from Alaska s. to Mexico.

AMERICAN GOLDFINCH
Carduelis tristis   48:13

Description
Size, 4½–5½ in. (11.4–14.0 cm). Summer male: • *yellow* with
• *black wings and tail,* forehead black, tail notched, no streaks,
white markings on wings and tail, white rump; bill pink. Female
and immature: yellow replaced by brownish olive yellow,
conspicuous wing bars on blackish wings, no black on forehead, bill
conical. Winter male: like female, but with yellow shoulder, grayer.

Similarities
Common Redpoll and Pine Siskin are streaked. Warblers have thin
bill. Yellow Warbler is yellow all over. Female Lesser Goldfinch
has greener back than female American, no white on rump.

Habitat
Open country, weedy fields, edges, gardens, orchards, river groves.

Habits
Lively, gregarious; characteristic "roller-coaster" undulating flight,
utters distinctive flight note at top of rise; nests late, often well into
August; in winter, flocks with redpolls, Siskins, crossbills.

Voice
Song, long, pleasing, somewhat canarylike; also a medley of trills
and other notes, often including a *swee,* or *dee-ar;* flight note, *per-
CHIK-o-ree; "ti-dee-di-di"* (Peterson).

Eggs
3–7; bluish-white; 0.6 x 0.5 in. (1.7 x 1.4 cm). Nest is neat, felted
cup of grass, bark shreds, thistledown in fork of bush or sapling.

Range
Resident from s. B.C. along Pacific Coast and mountains s. to Baja
Calif.; northern populations move s. in winter.

LESSER GOLDFINCH
Carduelis psaltria   48:14

Description
Size, 3¾–4½ in. (9.5–11.4 cm). Adult male: all-black above
(eastern form) or • *greenish above* with black cap (most of West);
• *yellow below;* bold white wing and tail marks conspicuous in
flight; tail notched. Female: like female American Goldfinch, but
smaller, greener, rump dark. Immature male: like adult male but
only crown black.

Similarities
Female American Goldfinch is not as green on back.

Habitat
Open country, brush, woods, streams, gardens.

Habits
Gregarious, a late nester.

Voice
Not as loud as American's; call, a sweet plaintive *tee-yee* (second
note rising) or *tee-yer* (second note dropping); notes often paired;
alarm, a "shivering note like the jarring of a cracked piece of
glass" (Hoffmann).

Eggs
4–5; bluish-white; 0.5 x 0.4 in. (1.4 x 1.3 cm). Nest is small cup,
like American's, in bush or low tree.

Range
Breeds from Oreg., n. Nev., Utah, and Colo. s. to Mexico and C.
America; winters along Pacific Coast and inland, southward;
migrates mainly in Rockies, otherwise resident.

LAWRENCE'S GOLDFINCH

Carduelis lawrencei   **48:12**

Description
Size, 4–4½ in. (10.2–11.4 cm). Male: black-faced gray bird with
• *yellow on wings,* rump, and breast; head gray with black face;
back grayish; breast and rump yellowish; wing bars broad, yellow.
Female: similar but more olive and lacks black face.

Habitat
Open woodlands of oak or oak-pine; chaparral, edges.

Habits
Gregarious; even nests may be in groups.

Voice
Song, like American Goldfinch's; call, doubly accented, distinctive
tink-oo.

Eggs
4–5; white; 0.6 x 0.4 in. (1.7 x 1.3 cm). Nest is small, neat cup in
low bush or tree.

Range
Breeds from cen. and s. Calif. s. to Baja Calif.; winters from cen.
Calif., cen. Ariz., and w. Tex. s. to Mexico.

WHITE-WINGED CROSSBILL

Loxia leucoptera   **48:7**

Description
Size, 6–6¾ in. (15.2–17.1 cm). Crossed mandibles, 2 broad • *white
wing bars.* Adult male: • *rose-red,* brightest on rump; wings, tail
black. Female: olive-gray, yellowish rump; more streaky than Red
Crossbill, wing bars evident in flight. Immature male: intermediate
between female and adult male in color.

Similarities
Pine Grosbeak is much larger, bill uncrossed. Red Crossbill has
wings all-dark.

Habitat
Conifers, especially spruce.

Habits
Travels in small flocks; occurs irregularly winter or summer.

Voice
Song, loud, long, varied, sometimes given in air, a succession of
loud trills on varied pitches; calls, a clear *sheep,* a dry *chif-chif,* a
liquid *peet.*

Eggs
2–4; pale bluish-green, spotted with brown and lavender; 0.8 x 0.6
in. (2.0 x 1.5 cm). Nest is cup of twigs, birchbark, moss, feather-
lined in fork of conifer.

Range
Resident from cen. Canada s. to s. U.S.; irregular visitor to Wash.,
Colo.; casual to Oreg., N.Mex.

RED CROSSBILL

Loxia curvirostra   **48:5**

Description
Size, 5¼–6½ in. (13.3–16.5 cm). Crossed mandibles; bill at a
distance looks slender; tail short, notched. Adult male: • *brick-red,*
brightest on rump; wings and tail dusky. Female: olive-gray; dull;
crown, rump, and breast olive-yellowish; wings and tail dusky.
Immature male: more orangish, intermediate between male and
female.

Similarities
White-winged has white wing bars. Hepatic Tanager in mountains of southwestern states is larger, bill not crossed.

Habitat
Conifers, especially spruce.

Habits
Tame, actions parrotlike, extracts seeds from pine cones with its crossed mandibles; sings from treetop; flight undulating, travels in small flocks; nests from January to July.

Voice
Song, *"too-tee too-tee, too-tee, tee, tee"* (Hoffmann), or warbled passages and *chips;* call, a *pip,* or *pip pip;* "a hard *jip-jip* or *jip-jip-jip*" (Peterson).

Eggs
3–5; pale greenish, spotted with brown and lavender; 0.8 x 0.6 in. (2.0 x 1.5 cm). Nest is saucer of evergreen twigs and moss, feather-lined, usually in conifer 5–20 ft. (1.5–6.1 m) up.

Range
Mainly resident from cen. Canada s. to Baja Calif. and e. to Great Plains.

GREEN-TAILED TOWHEE
Pipilo chlorurus   **49:7**

Description
Size, 6¼–7 in. (15.9–17.8 cm). Back plain olive-gray; wings and tail olive-green; tail long, rounded; • *cap red-brown;* • *throat clear white;* breast gray; wing linings yellow.

Similarities
Brown Towhee is browner, throat buffy.

Habitat
Dry brushy mountain slopes, open pines, sagebrush.

Habits
Forages low, often runs instead of flying away.

Voice
Song, variable, opening with a sweet *wee chirr,* then some high notes, ending in a weak trilling *weet-chirr-cheeeeee—chirrr;* a plaintive *mew,* also a *chink* (like Brown Towhee).

Eggs
3–4; bluish-white, spotted with red-brown; 0.8 x 0.6 in. (2.0 x 1.5 cm). Nest is large, of grass, bark shreds in underbrush.

Range
Breeds from cen. Oreg. s. to s. Calif. and se. N.Mex.; winters s. from s. Ariz. and w. Tex.; migrates w. to Calif. coast and e. to Great Plains.

Note: The **OLIVE SPARROW,** *Arremonops rufivirgatus,* size, 6 in. (15.2 cm), is a smaller, olive-colored finch similar to the Green-tailed Towhee. It lacks the rufous cap and has brown stripes on crown. It is seen in south Texas.

RUFOUS-SIDED TOWHEE
Pipilo erythrophthalmus   **49:8**

Description
Size, 7–8½ in. (17.8–21.6 cm). Tail long, rounded; eyes red. Male: hood, upperparts, breast black; • *rufous flanks;* belly white; rows of white spots on back and wings; • *white tips of outer tail feathers* conspicuous in flight. Female: black replaced by brown. Juvenal (summer): like a large, slender sparrow with white tail corners, 2 buffy wing bars, streaked below.

Habitat
Dry woods, especially second growth, edges, thickets, chaparral, city shrubbery.

Habits
Forages on ground, scratching noisily among dead leaves; sings from bush or low branch; opens and shuts tail.

Voice
Song, a buzzy, drawn-out *drink your tee-e-e-e, see towhee,* or sometimes just *teeeeee* or *chweeeee;* call, a *che-WINK!* a *chwee,* or *shrenk.*

Eggs
3–6; white, finely dotted with red-brown; 1.0 x 0.8 in. (2.5 x 2.0 cm). Nest is loose cup of leaves, bark shreds well hidden on ground in woods.

Range
Breeds from s. Canada to s. U.S. and C. America; winters from sw. states s. to Mexico, southward.

BROWN TOWHEE
Pipilo fuscus   49:9

Description
Size, 7¼–10 in. (18.4–25.4 cm). Nondescript brown; tail long; gray-brown above; grayer in Rocky Mountain form; whitish below, • *undertail coverts rusty;* throat buffy, streaked; large dark breast spot.

Similarities
Thrashers are larger, longer-tailed; bills slim, curved. Abert's is browner, with black face, underparts buffy-brown. Green-tailed is greenish, throat conspicuously white.

Habitat
Canyons, brushy and rocky areas, pinyons, thickets, open woods.

Habits
Common, much like Rufous-sided, but less conspicuous.

Voice
Song, a spiritless *"tsip tsip tsip sip, churr, churr, churr"* (Hoffmann), a "rapid *chink-chink-ink-ink-ink-ink-ink,* on one pitch" (Peterson), often ending in a trill, sometimes a repeated *chilp;* call, a metallic *chink,* like Green-tailed, often repeated.

Eggs
3–4; bluish-white; lightly blotched with blackish, often in 2 broods; 0.9 x 0.7 in. (2.3 x 1.8 cm). Nest is large, deep cup of grass, rootlets in bush or low tree.

Range
Resident along Pacific Coast from Oreg. to Baja Calif., and sw. from Ariz. to Tex., s. to Mexico.

ABERT'S TOWHEE
Pipilo aberti   49:10

Description
Size, 8–9 in. (20.3–22.9 cm). Very similar to Brown Towhee, but browner; a uniformly brown desert species with • *black patch at base of bill.*

Similarities
Brown Towhee lacks black face.

Habitat
Dense vegetation of desert washes and rivers; mesquite, paloverde, ironwood.

Habits
Much like Brown Towhee.

Voice
"Song like Brown, but terminal trill more guttural" (Robbins, et al.); note, a "single sharp *peel*" (Marshall).

Eggs
3–4; light bluish-green, scrawled; 0.9 x 0.7 in. (2.4 x 1.8 cm). Nest is of grass in low tree or desert bush.

Range
Resident in Ariz. and parts of Utah, N.Mex., and Calif., s. to Baja Calif. and Mexico.

LARK BUNTING
Calamospiza melanocorys   52:3

Description
Size, 6 in. (15.2 cm). Spring male: unique, all-black, large • *white wing patch* conspicuous in flight. Female: streaked brown above, white below; brown on cheeks; white wing patch smaller than in male; streaks on breast. Fall male: similar to female, but chin, wings, tail may be blackish.

Similarities
Male Bobolink has white patches on back and base of wings, not on wings, also has yellow nape. Female Purple Finch is suggestive of streaked brown Lark Bunting, but lacks wing patch, has notched tail. Various sparrows resemble female Lark Bunting, but latter is chunkier, with wing patch.

Habitat
Prairies, plains; in winter, open desert regions, desert scrub, brush.

Habits
Shy, sings from fence, or, from air; gregarious when not nesting; flocks wheel in unison.

Voice
Song, a series of warbled trills, Cardinal-like, or Chat-like (unmusical) *chug*'s; flight call, a sweet *whoo-ee*.

Eggs
4–6; pale blue; 0.8 x 0.6 in. (2.0 x 1.7 cm). Nest is loose cup of grass, plant down in tussock of grass on ground.

Range
Breeds in Canadian prairies and s. to w.-cen. U.S.; winters from s. Calif. and cen. Ariz., and cen. Tex. s. to Mexico.

SAVANNAH SPARROW
Passerculus sandwichensis   50:5

Description
Size, 4½–5¾ in. (11.4–14.6 cm). Streaked open country bird with a • *short, notched tail* and (usually) pale • *yellow stripe over eye;* light stripe through crown; breast streaked with black and with dark central spot; underparts whiter than most sparrows; legs usually pinkish. Tone varies greatly in different populations, from dark brown with heavy ventral streaks to very pale brown with paler ventral streaks.

Similarities
Song has larger breast spot, wider and browner streaks, longer, rounded tail, never any yellow over eye. On Great Plains, Baird's has streaks below very confined to breast; crown stripe broader, ochre-colored. Vesper has white outer tail feathers.

Habitat
Fields, fresh or salt meadows, beaches, dunes, prairies, open country.

Habits
Hops, rarely walks; runs through cover with head low; if flushed,

makes short zigzag, undulating flight revealing notched tail, then drops back into meadow; sings from low perch or wire.

Voice
Song, a weak buzzy *tsip tsip tsip seee saaay* or *tsip tsip tsip saaay seee,* last note lower; call, a light *tsip* or *thlip.*

Eggs
4–6; pale greenish-white, spotted with red-brown and purple brown; 0.8 x 0.6 in. (2.0 x 1.5 cm). Nest is a grassy hollow on ground.

Range
Resident from n. Alaska and Canada s. to Mexico, excluding Sw.; northern populations move s. in winter.

SHARP-TAILED SPARROW
Ammospiza caudacuta   50:8

Description
Size, 5–6 in. (12.7–15.2 cm). • *Gray ear patch bordered with buff;* short, spiky-tipped tail. Adult: face deep ochre-yellow, completely surrounding ear patch; cap entirely dark (no obvious central stripe); nape gray; back dark with light streaks; more or less streaked on buffy breast. Young: much buffier, with dark streaking above and below. Subspecies (interior and coastal races) are variable in brightness of facial lines, amount of breast streaking, and overall dark or light coloration.

Similarities
Young Bobolink is vaguely similar, but much larger. Le Conte's has white center crown stripe.

Habitat
Marshes, muskegs, reedy margins.

Habits
Scurries mouselike through grass; if flushed, soon drops again into marsh tail down.

Voice
Song, a "wheezy *tsup tsup shreeeeeeeee* with a *sh* sound in the gradually fading trill" (Saunders).

Eggs
4–5; grayish-white, finely dotted with brown; 0.7 x 0.6 in. (1.9 x 1.5 cm). Nest is loosely woven grass cup in marsh.

Range
Breeds locally from cen. Canada e. across U.S.; winters in E., passes through Plains states.

LE CONTE'S SPARROW
Ammospiza leconteii   50:9

Description
Size, 4½–5½ in. (11.4–14.0 cm). Tail short, spiky-tipped; central crown stripe white; hind neck and nape with • *purple-chestnut,* in young buffy, unstreaked; back very strongly striped; eye line • *buffy-ochre;* throat and breast yellow-buff; sides finely streaked.

Similarities
Sharp-tailed is very similar, but crown unstriped. Grasshopper has no orange face stripe, sides unstreaked.

Habitat
Prairie tall grass, marshes, fields with matted cover, wet meadows.

Habits
Secretive, prefers running to flight; if flushed, flies jerkily for a short distance, then drops back into cover; sings from top of weed or bush.

Voice
Song, 2 thin, high grasshopperlike hisses, first note almost inaudible, ending with a *chip.*

Food
Seeds, insects, berries.

Eggs
4–5; whitish, heavily spotted with red-brown; 0.7 x 0.5 in. (1.8 x 1.4 cm). Nest is grassy cup in grass.

Range
Breeds from s.-cen. Canada to n.-cen. U.S.; winters chiefly in e. U.S.

BAIRD'S SPARROW
Ammodramus bairdii   **50:7**

Description
Size, 5–5½ in. (12.7–14.0 cm). Resembles commoner Grasshopper Sparrow in shape but not as flat-headed or bob-tailed. Head and neck yellow-brown closely streaked with black, prominent ochre crown stripe, breast buffy with necklace of sharp black streaks, tail notched.

Similarities
Savannah has lighter, narrower crown stripe; breast streakings more extensive.

Habitat
Dry, long-grass prairies.

Habits
Sings from weed stalk.

Voice
Song, somewhat like Savannah's, but more musical, beginning with 2 or 3 high *zips,* several *chips,* then a trill.

Eggs
3–5; whitish, blotched with red-brown and black; 0.8 x 0.6 in. (2.0 x 1.5 cm). Nest is like Grasshopper's.

Range
Breeds from Alta., Sask., and Man. s. to S.Dak.; winters from s. prairies s. to Mexico; migrates through Great Plains.

GRASSHOPPER SPARROW
Ammodramus savannarum   **50:11**

Description
Size, 4½–5¼ in. (11.4–13.3 cm). Adult: small, bob-tailed, large flat head, buff breast; crown with pale median stripe on large head, back heavily streaked, tail bristly tipped; yellow spot before eye and on bend of wing, throat and • *breast unstreaked buff.* Late summer young: whitish breasts with streaks. Immature: breast buffier than adult, loses streaks.

Similarities
Savannah has notched tail, streaked breast.

Habitat
Grasslands, meadows, prairies, dry fields, numerous in Plains.

Habits
Runs persistently; flight feeble, fluttering, close to ground.

Voice
Song, an insectlike buzzing *tit-zeeeeeeee,* often passing unnoticed; call, a *tlik.*

Eggs
3–5; white, sparingly spotted with brown; 0.7 x 0.5 in. (1.8 x 1.4 cm). Nest is cup of grass in grass.

Range
Breeds from s. Canada s. to deserts of Sw.; winters s. from cen. Calif., s. Ariz., and s. N.Mex., to C. America.

VESPER SPARROW
Pooecetes gramineus   **50:2**

Description
Size, 5–6½ in. (12.7–16.5 cm). • *White outer tail feathers,*
conspicuous in flight; streaked grayish-brown above, whitish below,
with dark streaks ending sharply on lower breast; no breast spot;
• *bend of wing chestnut,* not very conspicuous; • *pale eye-ring* and
rather distinct ear patch.

Similarities
Song is browner above, dark breast spot, lacks white outer tail
feathers. Meadowlarks are larger, chunkier. Juncos are mostly
slate-gray. Lark Sparrow has tail corners white.

Habitat
Upland fields, pastures, roadsides, sagebrush.

Habits
Hops, does not walk; sings more toward evening and at dawn, has
courtship flight song; seen in small flocks in migration.

Voice
Song begins with 2 low, clear minor notes, then 2 higher notes, *"ah
ah ay ay tetetetetetatatatata toto tu"* (Saunders); call, a short *tsi.*

Eggs
4–6; grayish-white, dotted with red-brown; 0.8 x 0.6 in. (2.0 x 1.5
cm). Nest is grass cup in grass.

Range
Breeds throughout much of N. America, from n. Canada s. to s.-
cen. U.S.; winters from breeding range s. to Mexico.

LARK SPARROW
Chondestes grammacus   **49:3**

Description
Size, 5½–6¾ in. (14.0–17.1 cm). Adult: fan-shaped • *tail black with
white sides and corners;* • *chestnut crown stripes* and ear patch, eye
line and mustache white; black line on either side of chin; dark
spot in center of unmarked white breast. Juvenal: fine dark streaks
replace dark breast spot, head pattern duller but distinguishable.

Similarities
Vesper has white outer tail feathers.

Habitat
Open country with some trees; pastures, open woods; roadsides,
ranches.

Habits
When sitting, raises crown at intervals; sings from elevated perch,
or while hovering in air.

Voice
Song, broken, "somewhat like that of Indigo Bunting but louder,
clearer, much finer" (Forbush); "clear notes and trills with pauses
in between, best characterized by buzzing and churring passages"
(Peterson).

Eggs
3–5; white, spotted and scrawled with black; 0.8 x 0.6 in. (2.0 x
1.5 cm). Nest is grassy cup on ground or in low bush.

Range
Breeds from s. Canada s. to Mexico; winters from cen. Calif., s.
Ariz., w. Tex. s. to C. America.

RUFOUS-WINGED SPARROW
Aimophila carpalis

51:6

Description
Size, 5–5½ in. (12.7–14.0 cm). Grayish, characterized by • *red-brown shoulder* (difficult to observe); back grayer than Chipping Sparrow, crown with light median stripe, tail not notched, rather pronounced "whiskers."

Similarities
Chipping has rufous cap, no whiskers. Rufous-crowned is much browner, cap solid red-brown, back stripes brown not black. Brewer's has no median crown stripe, notched tail. Botteri's is much browner, breast buffier, no median crown stripe, whiskers fainter.

Habitat
Desert thornbrush and tall grass (tubosa).

Habits
A dry grassland inhabitant; local.

Voice
Tseep note and chipping of phrased song.

Eggs
4; pale-blue; 0.7 x 0.5 in. (1.8 x 1.4 cm). Nest is woven cup of grass in bush.

Range
Resident from se. Ariz. s. through desert to Mexico.

BOTTERI'S SPARROW
Aimophila botterii

50:6

Description
Size, 5¼–6¼ in. (13.3–15.9 cm). Nondescript; breast buffy-gray. • *tail brown.* No crown stripe.

Similarities
Cassin's is almost identical; back streaks more broken, less brown and tail gray—the only other sparrow breeding in the same very local range.

Habitat
Coarse tall desert grass.

Habits
Stays on or near ground, running along it; requires tall grass.

Voice
Song diagnostic, very unlike Cassin's; a steady "tinkling and 'pitting,' sometimes running into a dry rattle" (Peterson), always given while bird is perched.

Eggs
3–5; white; 0.7 x 0.5 in. (1.9 x 1.4 cm). Nest is grassy cup on ground.

Range
Breeds chiefly in Mexico, but n. to se. Ariz. and s. Tex.; winters s. of U.S.

RUFOUS-CROWNED SPARROW
Aimophila ruficeps

51:9

Description
Size, 5–6 in. (12.7–15.2 cm). Adult: reddish-brown and gray streaks on back, appears dark; crown solid rufous; eye line gray; black whiskerlike streaks bordering sides of throat; breast unstreaked grayish-white, tail rounded. Juvenal: crown brown, streaked; thin, dark streaks on breast.

255

Similarities
Chipping has white line over eye, notched tail, black back stripes.
Swamp is more rufous and with black stripes above, grayer below.
Habitat
Dry, grassy hillsides with low scrub, open pine-oak woods.
Habits
Sometimes occurs in small, loose colonies; skulks mouselike on
ground, sings from low perch, looks hunched when sitting.
Voice
Song, several *mew*'s, then a gurgling warble, first part rising, last
part falling, somewhat like House Wren's; call, a musical *deer,
deer,* or a nasal *chur, chur, chur.*
Eggs
3–5; bluish-white; 0.8 x 0.6 in. (2.0 x 1.5 cm). Nest is grassy cup
on ground.
Range
Resident from Calif., s. Ariz., and s. N.Mex. s. to Mexico;
northern populations migrate southward in winter.

CASSIN'S SPARROW
Aimophila cassinii   50:10

Description
Size, 5¼–5¾ in. (13.3–14.6 cm). Adult: plain gray above,
indistinctly marked with brown and black; plain whitish or pale
buffy below, unmarked or with trace of streaking on lower sides;
breast unmarked; • *tail gray.* Immature: breast streaked.
Similarities
Botteri's is very similar. Grasshopper is browner, crown light, back
more distinctly marked, breast buffier. Brewer's is slimmer, tail
notched, 2 wing bars, more clearly striped above with black and
buff.
Habitat
Dry short-grass, arid regions with low brush.
Habits
When breeding, sings through heat of day; utters flight song;
Botteri's does not sing in flight; fluttering up several feet with head
up and wings outspread.
Voice
Song, diagnostic, "skylarking," beginning with "one or two short
opening notes, a high sweet trill, and 2 lower notes, *ti ti
tseeeeeeeeee tay tay*" (Peterson).
Eggs
3–5; white, 0.7 x 0.5 in. (1.9 x 1.4 cm). Nest is deep grass cup on
ground.
Range
Breeds from Ariz. and N.Mex. s. to Mexico; winters s. from U.S.–
Mexican border.

SAGE SPARROW
Amphispiza belli   49:4

Description
Size, 5–6 in. (12.7–15.2 cm). Predominantly gray, with • *single
distinct breast spot;* • *"whiskers" on side of throat dark,* cheek dark,
eye-ring and line over eye white, face pattern of Black-throated
without the black throat.
Similarities
In all plumages Lark Sparrow is browner with strong white and
brown pattern atop head.

Habitat
Sagebrush plains, dry brushy foothills, open chaparral; deserts in winter.

Habits
Skulks, moves from bush to bush.

Voice
Song, "a simple set pattern, *tsit-tsoo-tseee-tsay*, 3rd note highest" (Peterson); "4 to 7 notes forming a jerky but somewhat melodic phrase; higher notes with a squeaky, sibilant tone, lower notes tinkling" (Axtell).

Eggs
3–4; white, speckled; 0.7 x 0.5 in. (1.6 x 1.4 cm). Nest is loose grassy cup in shrub.

Range
Breeds from Wash. s. to Baja Calif. and through Great Basin; winters from s. Calif., Ariz., N.Mex. s. to Mexico.

BLACK-THROATED SPARROW
Amphispiza bilineata   **49:1**

Description
Size, 4¾–5¼ in. (12.1–13.3 cm). Adult: • *white face stripes* and • *black throat;* upperparts unmarked gray-brown, tail black with white edges, sides of head gray-brown, underparts white. Immature: throat white or grayish, but with same face pattern, breast streaked.

Similarities
Black-throated Gray Warbler has warblerlike bill, wing bars. Sage is similar to young of Black-throated, with central breast spot.

Habitat
Sagebrush and creosote (greasewood) bush deserts; cholla cactus flats.

Habits
Forages low, sings from ground or bush top.

Voice
Song, a tinkling, somewhat burred *tra-REE-rah, REE-rah-ree,* "a sweet *cheet cheet cheeeeeeee*" (Peterson) with 2 short, clear opening notes and a fine trill on a different pitch.

Eggs
3–4; bluish-white; 0.7 x 0.5 in. (1.8 x 1.4 cm). Nest is loose grassy cup in bush or cactus.

Range
Breeds from nw. Nev. and s. Wyo. s. through desert to Mexico; winters from Sw., s. to Mexico.

DARK-EYED JUNCO
Junco hyemalis   **49:5**

Description
Size, 5½–6½ in. (14.0–16.5 cm). Solid gray, brown-gray, or blackish head, pinkish-white bill. Females, and especially immatures, tend to be duller and browner; juvenals are streaked below. Dark tail with white outer feathers, white belly. Variously gray above and on breast and sides (eastern "slate-colored" form); gray with 2 white wing bars ("white-winged" Black Hills form); pinkish or brown sides with red-brown back and, in males, gray (northern Rocky Mountain "pink-sided" form) to black (western "Oregon" form) hood.

Similarities
Yellow-eyed closely resembles southern Rocky Mountain form of "gray-headed" Dark-eyed, with dark upper bill, eyes yellowish, breast and throat paler tending to whitish on throat, and vague black mask often evident.

Habitat
Coniferous and mixed deciduous woods and forests (in north and in mountains), dry or wet, edges, second growths, brush above timberline; in migration and winter all woodlands, gardens, fields, bird feeding stations.

Habits
Abundant; rather tame; often sing in winter and early spring while still in winter flocks.

Voice
A Chipping Sparrow-like trill, but more musical, less mechanical.

Eggs
3–6; gray-white, spotted with lilac and brown. 0.8 x 0.6 in. (2.0 x 1.5 cm). Nest is grass cup, lined, on ground in woods.

Former names
White-winged Junco, Oregon Junco, Slate-colored Junco, Gray-headed Junco.

Remarks
The several forms of this complex have long been considered separate species, although it is well known that these forms hybridize wherever their breeding ranges meet. Modern taxonomists, specialists in bird relationships, feel the evidence justifies but two western species, the widespread Dark-eyed Junco, including the various hybridizing forms, and the Yellow-eyed Junco of the southwest.

Range
Breeds from Alaska e. across Canada and U.S. and s. to Baja Calif.; winters from Nw. s. to s. U.S.

YELLOW-EYED JUNCO
Junco phaeonotus   **49:6**

Description
Size, 6 in. (15.2 cm). • *Yellow eyes;* otherwise similar to southern Rocky Mountain form of Dark-eyed Junco. Throat and breast pale, showing some whitish; back rufous; sides gray; lacks hooded effect, shows faint eye mask, dark lores; upper bill is dark, lower is pale.

Similarities
Gray-headed form of Dark-eyed lacks yellow eye.

Habitat
In mountain pine-oak or conifer forests.

Habits
Walks rather than hopping as does Dark-eyed.

Voice
Song complex, 3-part, as *"chip chip chip, wheedle wheedle, che che che che che"* (Peterson).

Eggs
3; unmarked bluish white; 0.7 x 0.6 in. (1.8 x 1.6 cm).

Range
Resident from se. Ariz. and sw. N.Mex. s. to Mexico and C. America.

TREE SPARROW
Spizella arborea

51:7

Description

Size, 5½–6½ in. (14.0–16.5 cm). Solid • *red cap* and • *black spot on plain breast;* bill dark above, yellow below; tail slightly notched; brown line through eye, whiskers below eye; streaked brownish above; gray on sides of face; 2 white wing bars; breast whitish; a little white at sides of tail.

Similarities

Field has pink bill, buffy face, no breast spot. Lark Sparrow has no wing bars, rounded white-edged tail. Chipping has black line through eye, white line above.

Habitat

Arctic scrub, low trees, weedy fields, gardens, roadsides.

Habits

Gregarious in winter, visits feeders, often sings in concert.

Voice

Song, a sweet, variable *"eee eee tay titititee tay"* (Saunders), opening on 1 or 2 high notes; call, a *tee-lo,* a *tseet.*

Eggs

4–5; light greenish-blue, dotted with light brown; 0.7 x 0.5 in. (1.9 x 1.4 cm). Nest is grass cup lined with feathers on ground or in bush.

Range

Breeds in n. N. America from Alaska e. across Canada and s. to Calif., nw. Tex.; winters from n. U.S. s. to s. Nev., n. Ariz., cen. N.Mex., and w. Tex.

CHIPPING SPARROW
Spizella passerina

50:14

Description

Size, 5–5¾ in. (12.7–14.6 cm). Notched tail; trim and slim. Spring adult: streaked brown above, chestnut cap; breast grayish-white, unmarked; black line through eye, white line over eye; 2 thin white wing bars. Winter adult: cap and eyebrow stripe are duller, crown is streaked, bill paler. Immature: cap brown with pale central stripe and black streaks, buffy eye line, nape streaked brown, brown ear patch not bordered with black, breast buffier, gray rump.

Similarities

Clay-colored has gray nape, whitish breast, brown rump. Rufous-crowned is heavier, tail rounded, brown stripes, with whiskers. Brewer's is pale, gray-brown back not contrasting with rump, crown finely streaked.

Habitat

Likes open ground under or near trees; open pinewoods, edges, lawns, gardens, roadsides, orchards, towns.

Habits

Tame, feeds on ground, sings from wire or high branch, often victimized by Brown-headed Cowbird.

Voice

Song, a simple, unmusical, dry trill on 1 pitch; call, a short *tsip* or *chip.*

Eggs

3–5; greenish-blue, wreath of blackish dots about larger end; 0.7 x 0.5 in. (1.8 x 1.4 cm). Nest is hair-lined cup in tree or bush.

Range

Breeds throughout much of N. America from n. B.C., cen. Yukon, s. Mackenzie, s. to s. U.S.; winters from cen. Calif., s. Nev., cen. Ariz., s. N.Mex., and w. Tex., southward.

CLAY-COLORED SPARROW
Spizella pallida   50:12

Description
Size, 5–5½ in. (12.7–14.0 cm). Head mainly white with black malar streak; • *white central crown stripe* separating brown areas, black-bordered • *brown ear patch,* white stripe over eye, nape unmarked, pale gray; back gray-brown striped with black; 2 fine, white wing bars; entire underparts white. Immature: much browner.

Similarities
Immature Chipping has brown cheek patch, more bordered with black, and buffier rump. Brewer's has crown finely streaked, ear patch indistinct. Lark Sparrow is larger, has somewhat similar head pattern, but with dark spot on breast.

Habitat
Grasslands, fields, brush often near water.

Habits
Feeds on ground; in breeding season sings persistently from bush.

Voice
Song, unbirdlike, 3 to 4 slow insectlike buzzes, *bzzzz, bzzzz, bzzzz,* etc.; call, a *chip.*

Eggs
3–5; light greenish-blue, brown spots about larger end; 0.6 x 0.5 in. (1.7 x 1.4 cm). Nest is like Chipping Sparrow's.

Range
Breeds inland from B.C. e. to s. Ont. and s. through Mont., Wyo., Colo., and prairie states; winters from Tex. s. to Mexico; migrates through Great Plains.

BREWER'S SPARROW
Spizella breweri   50:13

Description
Size, 5–5½ in. (12.7–14.0 cm). Plain breast and • *finely streaked crown;* very similar to Clay-colored Sparrow, but crown of Brewer's evenly, finely streaked, ear mark less distinct; tail notched, whitish below with a buffy tinge on sides and breast.

Similarities
See Clay-colored. Immature Chipping is darker, whiter eye stripe, darker crown with pale central stripe.

Habitat
Breeds among sagebrush; winters on desert; also bushes above tree line in northern Rockies.

Habits
Likes sagebrush and other small bushes.

Voice
Song, a series of metallic insectlike trills on different pitches, "sounds like a Chipping Sparrow trying to sing like a Canary" (Peterson); call, a *tsip.*

Eggs
3–5; eggs like Chipping, but with spots yellowish-brown; 0.7 x 0.5 in. (1.8 x 1.4 cm). Nest is cup on ground in sagebrush or low conifer.

Range
Breeds from B.C. e. across Canada to Sask., and s. to s. Calif., Ariz., and N.Mex., excluding Pacific nw. coastline; winters from s. Calif. e. to se. Ariz., s. N.Mex., and w. Tex., and s. to Mexico.

FIELD SPARROW
Spizella pusilla   51:5

Description
Size, 5¼–6 in. (13.3–15.2 cm). • *Pink bill;* tail slightly notched;
reddish ear patch; rusty patch on side in front of wing; no spot on
breast; rather rufous above, unmarked below; narrow eye-ring
bordered with dull gray; 2 white wing bars.
Similarities
Tree has dark spot on breast, broader wing bars, dark upper bill.
Chipping is grayer, white line over eye, dark bill. Rufous-crowned
has rounded tail, brown stripes, whiskers.
Habitat
Dry brushy hillsides, overgrown fields, pastures.
Habits
Sings from low perch, sometimes by moonlight; gregarious in
winter.
Voice
Song, a sweet, musical *twee-twee-twee-te-te-te-te-te-te,* first notes
slow, trill becoming faster and sometimes rising or dropping in
pitch; call, a *tsip* or *tsee,* a bit querulous.
Eggs
4–5; grayish-white, dotted with red-brown and lilac; 0.7 x 0.5 in.
(1.8 x 1.4 cm). Nest is hair-lined cup in low bush, grass.
Range
Breeds from nw. Mont., n. N.Dak., e. across U.S. and s. to cen.
Tex.; winters from cen. U.S. s. to Gulf of Mexico.

BLACK-CHINNED SPARROW
Spizella atrogularis   49:2

Description
Size, 5–5½ in. (12.7–14.0 cm). Distinctive, juncolike, with streaked
brown back, • *gray head and underparts,* white wing bars. Male: bill
pinkish; base of bill surrounded conspicuously by black patch on chin
and to eyes. Female: lacks face patch; identifiable by brown
back, unmarked gray head and breast.
Similarities
Juncos lack streaked brown back.
Habitat
Sagebrush, chaparral, brushy mountain slopes.
Habits
Secretive; flies long distances, difficult to approach.
Voice
Song, single-pitched series of notes sometimes descending slightly;
"starts with several high, thin, clear notes and ends in a rough trill,
sweet, sweet, sweet, weet-trrrrrr" (Peterson).
Eggs
3–4; bluish-white or spotted; 0.6 x 0.5 in. (1.7 x 1.4 cm). Nest is
neat grass cup in low bush.
Range
Breeds from s. Calif. e. across Sw., and s. to Mexico; winters from
Mexico, southward.

HARRIS' SPARROW
Zonotrichia querula   51:1

Description
Size, 7–7¾ in. (17.8-19.7 cm). Spring adult: • *black cap,* face,
throat, and upper breast; bill pinkish; brownish above, whitish
belly; sides streaked; cheeks gray; 2 white wing bars; black bib

encircling bill to eyes and crown. Fall adult: black crown partly veiled with gray. Immature: crown spotted brown; buffy face, flanks, and undertail coverts; throat white; necklace of brown streaks; streaked whiskers.

Similarities
Spring male Lapland Longspur has similar head pattern, but smaller eye, chestnut nape. House in fall plumage has remote resemblance. White-crowned has streaked crown.

Habitat
Breeds in dwarf timber near tundra; in winter, thickets, brushy edges, open woodlands.

Habits
Habits similar to White-crowned and White-throated.

Voice
Song, quavering like White-throated's, 1–5 clear whistles, usually same pitch, then a pause, followed by several notes on another pitch; alarm note, "a loud *weenk* or *wink*" (Sutton); "winter songs interspersed with chuckling sounds" (Peterson).

Eggs
3–5; greenish-white, spotted and blotched with brown; 0.8 x 0.6 in. (2.4 x 1.7 cm). Nest is grassy cup on ground.

Range
Breeds from n. Mackenzie and n. Man. s. through to Great Plains; winters in s.-cen. prairies.

WHITE-CROWNED SPARROW
Zonotrichia leucophrys   51:3

Description
Size, 5½–7 in. (14.0–17.8 cm). Adult: crown with • *5 white stripes enclosing 4 black ones,* center of crown white, also stripe over eye and under black stripe through eye; pearly-gray breast; face and nape grayish; brown above with 2 white wing bars; • *bill variable, yellowish to pinkish,* may be dusky-tipped. Southern Rocky Mountain form has white stripe over eye ending at eye; other races have that stripe continuing to bill, with black eye stripe as well reaching bill. Immature: similar but crown striped with red-brown and buff; underparts washed with brown.

Similarities
White-throated is much darker billed, adults have throat conspicuously white, set off by black lines at sides, yellow spot before eye; immature much dingier on breast. See Golden-crowned.

Habitat
Boreal scrub, shrubby mountain slopes, dwarf willows; in winter or migration, edges, thickets, roadsides, gardens, towns, open scrub.

Habits
Can puff up crown; forages on ground, sings from elevated perch, sometimes at night.

Voice
Song varies, with many local "dialects"; one version a clear, whistled *aaaa ee aay;* another, a husky, descending *see say so;* song usually "followed by a husky trill or series of trills and *chillip*'s" (Peterson); call, a loud *chink* or *pink.*

Eggs
3–5; greenish-white, heavily spotted with red-brown; 0.8 x 0.6 in. (2.2 x 1.7 cm). Nest is grassy cup, well-lined, on ground or in bush.

Range
Breeds throughout much of Canada from Alaska s. to Calif. and N.Mex.; winters in sw. states and Mexico; some coastal populations are year-round residents.

GOLDEN-CROWNED SPARROW
Zonotrichia atricapilla

51:2

Description
Size, 6–7 in. (15.2–17.8 cm). Adult: closely resembles White-crowned, but in place of head stripes has a black-bordered, dull, golden-yellow median crown stripe, as if side black and white stripes of White-crowned were eliminated and central white stripe colored yellow; crown pattern takes several years to develop; bill dark above, pale below. Immature: like some winter adults, but larger, tail longer; has whisker marks and white wing stripes.

Similarities
Immature White-crowned has buffy median crown stripe, not yellow; eyebrow line broad, buffy; bill pink or yellowish.

Habitat
Spruce forest, boreal scrub; in winter, edges, dense shrubbery, like White-crowned.

Habits
Similar to White-crowned.

Voice
Song, plaintive, melodic, down-scaling, beginning with 3 high, whistled notes and sometimes ending with a final faint trill.

Eggs
4–5; white, speckled; 0.8 x 0.6 in. (2.2 x 1.6 cm). Nest is grassy cup, lined with rootlets, under bush.

Range
Breeds from Alaska and B.C. inland to n. Rockies; winters along Pacific Coast from s. B.C. s. to Baja Calif.

WHITE-THROATED SPARROW
Zonotrichia albicollis

51:4

Description
Size, 6–7 in. (15.2–17.8 cm). • *Black-and-white striped crown* with yellow spot before eye and a distinct • *white throat* set off by black malar line. Adult: back brownish streaked with black, breast gray, white throat patch bordered by black whiskers, cheeks and underparts gray, 2 white wing bars, bill grayish. Immature: duller, with some blurry streaking below; crown stripes brown and buffy, throat patch duller but distinct. There is considerable variation with sharply marked, brighter, and duller, less white-faced phases.

Similarities
White-crowned may have grayish to whitish throat but not sharply bordered in black, pink or yellow bill. Adult Swamp has rusty crown.

Habitat
Brushy pastures, thickets, slash piles.

Habits
Active, tame, common; visits feeders, forages noisily on ground; sings from ground or low perch, sometimes at night; gathers in small flocks in fall.

Voice
Song, a set of clear whistles, *old Sam PEAbody, PEAbody, PEAbody;* notes, a distinctive, longish *ssst,* or a hard *chink;* also a slurred *tseet,* a *chip.*

Eggs
4–5; grayish-white, dotted with red-brown; 0.8 x 0.6 in. (2.0 x 1.5 cm). Nest is like Golden-crowned's.

Range
Breeds from cen. Canada e. across Canada and n. U.S.; winters chiefly in E.; migrates sparsely through Great Plains.

FOX SPARROW
Passerella iliaca   **50:1**

Description
Size, 6¼–7¼ in. (15.9–18.4 cm). Large, stocky; streaked • *red-brown* and gray above; rusty brightest on wings, rump, and tail; white below, heavily streaked with red-brown, forming a more or less conspicuous spot on the breast; dark races almost unstreaked deep brown above with blackish spot-streaks on white breast; rustier eastern and southern races show more markings above, rusty streaks on gray, and brown to rufous streaks below. All are distinguished by a slightly notched tail and pale lower bill.

Similarities
Hermit Thrush has dark, thinner, longer bill, spotted below, tail not notched.

Habitat
Open woods, edges, thickets, mountain chaparral, underbrush.

Habits
Scratches noisily on ground with both feet at once, sings from elevated perch, visits feeders; gregarious except when breeding.

Voice
Song, "loud, beautiful, whistled melody *hear hear I sing-sweet sweeter most-sweetly*" (Cruickshank); alarm note, a smack; call, *sssp*.

Eggs
4–5; pale green, heavily spotted with rusty-brown; 0.9 x 0.7 in. (2.3 x 1.8 cm). Nest is grass cup, feather lined, in bush or on ground.

Range
Breeds from far north in Alaska s. to s. Calif. along Pacific Coast and Rockies; winters from s. B.C., southward and inland.

SWAMP SPARROW
Melospiza georgiana   **51:8**

Description
Size, 5–5¾ in. (12.7–14.6 cm). • *Rusty crown;* gray-bordered • *white throat;* tail rounded. Spring adult: dark rufous above; very faint wing bars; rusty crown; cheeks, nape, and underparts gray. Fall adult: often buffy on sides and underparts, crown streaked with black, light center stripe. Immature: crown brown, striped, no red; breast buffy, outlining white throat, dimly streaked.

Similarities
Song has heavily streaked breast, tail longer, more rounded, lighter above. Chipping is more slender, tail forked, black and white eye stripes. Field and Tree have prominent white wing bars. Rufous-crowned is less rusty, habitat differs.

Habitat
Muskegs, swamps, bushy fresh marshes; in migration, also weedy fields.

Habits
Sings from reed or bush in marsh.

Voice
Song, a 1-pitch trill, slower, louder, sweeter than Chipping Sparrow's, sometimes 2 pitches simultaneously; a hard *chink* or *chip*, like White-throated.

Eggs
4–5; variable, often bluish-white, heavily spotted with brown; 0.8 x 0.6 in. (2.0 x 1.5 cm). Nest is grassy cup on marsh hummock.

Range
Breeds from e.-cen. Canada s. to e.-cen. U.S.; winters from Great Lakes s. to Gulf of Mexico; casual to all w. states.

LINCOLN'S SPARROW
Melospiza lincolnii
50:3

Description
Size, 5–6 in. (12.7–15.2 cm). Appearance like a slender, elegant Song Sparrow. Body slender; tail squared; finely streaked grayish-brown above, white below; face gray; • *fine dark streaks on buffy breastband,* streaks seldom merging into dark spot; crown brown, with light gray central stripe; eye-ring white, narrow.

Similarities
Song is heavier, rustier brown above, wide, dark mustaches, center spot on breast, broader streaks below; juvenal Song is finely streaked below but has no eye-ring or buffy wash on breast. Swamp is browner above, darker below; gray, not buff, on breast; immature has indistinct breast streaks. Baird's has notched tail, short, buffy crown stripe. Savannah has notched tail, yellow before eye, no buff on breast.

Habitat
Breeds beside bogs and water in muskeg region; willows, alder thickets, wet areas; in migration, edges, stone walls, bushy fences.

Habits
Secretive, skulking, does not sing in migration, hard to see, but responds to hand-kissing.

Voice
Song, sweet and gurgling, "suggests the bubbling, guttural notes of the House Wren, combined with the sweet rippling music of the Purple Finch" (Dwight); starts low, rises abruptly, drops; call, a low *tsip.*

Eggs
4–5; white, heavily blotched with brown; 0.8 x 0.6 in. (2.0 x 1.5 cm). Nest is grassy cup in bog or muskeg.

Range
Breeds from n. Canada e. across Canada and U.S. and s. along mountains to s. Calif. and N.Mex.; winters from sw. states to C. America; migrates through lowlands.

SONG SPARROW
Melospiza melodia
50:4

Description
Size, 5–7 in. (12.7–17.8 cm). One of the commonest sparrows. Brown streaked below with streaks converging into spot on breast. Adult: tail long, rounded, drooped so back is humped when landing; streaked brownish above; light line across top of brown crown and over each eye; white below, occasionally lacking central black spot. Juvenal: streaked buffy band across breast, often lacks central spot; hard to separate from young Lincoln's, latter has fine back streaks. Numerous subspecies show much variation in size and color, from pale in desert areas to darker and larger in the humid northwest and Alaskan islands—extremes look like totally different species.

Similarities
Savannah has notched tail, yellow over eye, pinker legs. Fox is larger, streaks usually broader below. Lincoln's is trimmer, side of face grayer, fine breast streakings in sharply defined area.

Habitat
Tundra, marshes, thickets, shrubbery, roadsides, gardens, beaches (Alaska).

Habits
Tame, visits feeder; likes to bathe; pumps tail up and down in flight; forages on ground, sings from elevated perch.

Voice

Song, "variable series of notes, some musical, some buzzy" (Peterson); *"Maids! Maids, Maids! hang up your teakettle-ettle-ettle"* (Thoreau); call, a *tsak*, or low, nasal *tchep*.

Eggs

3–7; variable, often whitish, spotted with red-brown; 0.8 x 0.6 in. (2.2 x 1.5 cm). Nest is grassy cup on ground or in low bush.

Range

Breeds throughout much of Canada and U.S. from Alaska, e. across Canada, and s. to Baja Calif. and Mexico; winters from Alaska along coastline s. to Baja Calif. and inland from Idaho and Mont. s. to Mexico.

McCOWN'S LONGSPUR
Calcarius mccownii   **52:12**

Description

Size, 5¾–6 in. (14.6–15.2 cm). Black inverted T on rather short white tail, diagnostic in all plumages. Spring male: black crown, whisker, small breast patch; white eye line, throat, belly; gray collar, flanks, lower breast; streaked brown-gray above, chestnut shoulder patch. Fall male: tawnier above, black largely replaced by gray. Female and immature: streaked brownish above, buffy below, like young Chestnut-collared, but tail pattern different.

Similarities

Spring male Chestnut-collared has chestnut collar, much larger black patch below. Male Lapland has 2 white wing bars. Horned Lark has thin bill, black sideburns.

Habitat

Short-grass plains, prairies.

Habits

Flight undulating; after breeding season, gregarious; often found with Horned Larks and Chestnut-collared Longspurs.

Voice

Song, in display flight, a pleasant, clear, sweet warble, with twitters, uttered hovering in air with wings seemingly straight up; call note, a *chirrup-chirrup;* also a dry rattle.

Eggs

3–4; pale greenish; dotted with brown; 0.8 x 0.6 in. (2.0 x 1.5 cm). Nest is grassy saucer on ground.

Range

Breeds from s. Alta. and Man. s. through n. Colo. and Nebr.; winters in prairies from s. Colo., s. to Mexico.

LAPLAND LONGSPUR
Calcarius lapponicus   **52:9**

Description

Size, 6–7 in. (15.2–17.8 cm). Legs dark; white outer tail feathers in all plumages. Spring male: crown, face, throat, spots on sides, tail black; face outlined by white stripes from eyes down sides; collar chestnut; upperparts brownish, sparsely streaked on sides with black; nape reddish; underparts and outer tail feathers white. Fall male and female: black replaced by dark smudge on sides of lower throat and ear; legs black. In flight, wings look dark.

Similarities

Other longspurs have more white on tail, different tail patterns. Pipits and Horned Larks have thin bills, longer tails. Snow Bunting is much lighter; wings light, not dark.

Habitat
Tundra, beaches, plains, fields, prairies.

Habits
Gregarious, except when breeding; hard to detect when motionless in field; flies like Snow Bunting and often seen with it.

Voice
Song, short and gushing, *"tee-tooree, tee-tooree, teereeoo"* (Snyder), uttered on wing; in winter, a hoarse *churrr;* also a musical *teew* or a rattle followed by a whistled *dicky-dick-do.*

Eggs
6; variable, often greenish-gray, heavily blotched with red-brown; 0.8 x 0.6 in. (2.2 x 1.5 cm). Nest is grassy depression, feather-lined, on tundra.

Range
Breeds from Arctic s. to n. Canada; winters from s. B.C., Mont., S.Dak. s. to s. Calif., Mexico.

Note: Another far northern longspur is **SMITH'S LONGSPUR**, *Calcarius pictus* **(52:11)**, size, 6 in. (15.2 cm). It breeds in coastal northern Alaska and Mackenzie, and winters in the West to Alberta and Saskatchewan. This species is buffy throughout with white shoulder patches in winter plumage; the outer tail feathers are broadly white, more so than the Lapland.

CHESTNUT-COLLARED LONGSPUR
Calcarius ornatus   **52:10**

Description
Size, 5½–6½ in. (14.0–16.5 cm). Distinctive black triangle on white tail in all plumages. Spring male: • *cap and underparts (except throat) solid black,* collar (nape) chestnut, 2 black face stripes at rear, streaked brown above, throat and cheeks whitish. Fall male: black and chestnut replaced by brown. Female and immature: streaked brown above, buffy below, faint streaks on sides and breast; nondescript, best identified by tail pattern of dark triangle on white.

Similarities
Other longspurs such as Lapland are similar. Also Vesper Sparrow, pipits have straight white sides on tails. McCown's has no streaks on sides and breast, tail black forms inverted T.

Habitat
Plains, prairies; in winter, dry grasslands.

Habits
Gregarious, often found with other longspurs; flight undulating, showing tail pattern.

Voice
Song, short, high, weak, twittery, uttered on wing, somewhat like Western Meadowlark's; flight call, a twitter; note, "a finchlike *ji-jiv*" (Peterson).

Eggs
3–5; pale greenish, speckled with brown; 0.8 x 0.6 in. (2.0 x. 1.5 cm). Nest is grassy hollow in grass.

Range
Breeds from Canadian prairies s. to w.-cen. U.S., including Colo. and Nebr.; winters from n. Ariz., and Kans. s. to Sw. and Mexico.

267

SNOW BUNTING
Plectrophenax nivalis   **52:8**

Description
Size, 6–7¼ in. (15.2–18.4 cm). Whitest of all land birds. Summer
male: • *white*; only black bend of wing, wing tips, and middle tail
feathers. Summer female: similar, but black replaced by dusky.
Winter male: white; cap, ear patch, band on breast, spots on rump
rusty; bend of wing, wing tips, middle tail feathers, spots on back
black. Winter female and immature: the rustiest, least white of all.
In flight (in winter), bird appears all-white below, including tail,
dark above; wings always flash some white.

Similarities
In flight, Water Pipit and Horned Lark show black tails and dark
wings.

Habitat
Summer, tundras, mountain slopes; otherwise diverse, open
country, prairies, shores, fields, salt marshes and flats.

Habits
Walks, runs, hops; gregarious except when breeding; compact
flocks wheel in unison, and settle on field like snowflakes; often
seen with Lapland Longspurs; in north, frequents camps, villages;
called "House Sparrow of the Arctic."

Voice
Song, *"turee-turee-turee-turiwee"* (Snyder); notes include a high,
whistled *teer* or *tew;* also a rough, purring *brrt.*

Eggs
4–7; variable, whitish spotted with brown; 0.8 x 0.6 in. (2.0 x 1.5
cm). Nest is feather-lined tundra depression.

Range
Breeds in Arctic; winters from s. Canada s. to n. U.S.

Mammals

Consulting Editors
Donald Patten
Curator of Mammalogy
Los Angeles County Museum of Natural History
and
Sydney Anderson
Curator of Mammalogy
American Museum of Natural History

Illustrations
Plates 53–56, 60, 62–64, 66, 67 Nina L. Williams
Plates 57, 58, 68 Jennifer Emry-Perrott
Plates 59, 61, 65, 69 John Hamberger
Text Illustrations by Nina L. Williams and Jennifer Emry-Perrott

Mammals
Class Mammalia

The last to develop and in some ways the highest biologic class of vertebrates are the warm-blooded mammals that have a body covering of hair or fur, breathe by means of lungs, and produce milk for their young. All mammals, except the platypus and echidna, which lay eggs, bring forth their young alive and provide them with a period of preliminary parental care.

Most mammals have four feet and a tail. The seal, whale, walrus, and manatee have forefeet that, through evolution, have been transformed into flippers. The original hind limbs of the whales have totally disappeared except for two small bones deeply buried in the body near the base of the tail. In a few species, such as bears and human beings, the tail is greatly reduced or missing entirely. Geologically early mammals had five digits on each foot, but environmental adaptations caused a reduction in many species, such as the horse where only one digit, that bearing the hoof, remains.

Evolution

Throughout the entire 110 million years of the Mesozoic era, which began 180 million years ago, the cold-blooded reptiles dominated all the land and sea fauna. Reptiles reached their evolutionary climax with the gigantic dinosaurs, but it is not from these that the earliest mammals derived. Rather, mammals sprang from a primitive reptilian stock that first appeared during the Triassic period, about 153 million years ago. The pelycosaurs and therapsid reptiles had many characteristics approaching those of the most primitive mammals, such as heterodont mammalian teeth; the same skull bones that are absent from the mammalian skull, much reduced, or lacking; and a general posture not unlike that of the earliest mammals. Because of the world dominance of the larger, more ferocious reptiles, the Mesozoic mammals remained numerically few and little larger than a modern chipmunk.

With a shifting of the earth's distribution of land masses, a colder climate and perhaps other changes wiped out the dinosaurs completely. The entire earth was left to the relatively unhindered evolution of the mammals, undoubtedly helped by the flowering plants, fruits, and grasses that began to clothe the land areas with the onset of the Cenozoic era.

Evolutionary development has given rise to three quite different groups of mammals. It is presumed that the subclass Prototheria arose in the Jurassic period, 120 million years ago. Modern representatives are the primitive egg-laying monotremes, the platypus and echidna, that are confined to Australia and New Guinea.

The Jurassic pantotheres gave rise to two new groups in the succeeding Cretaceous period, sixty-five million years ago. The less advanced were the Metatheria, the marsupials or pouched mammals that bring forth living young but in an embryonic condition. These are kept and nourished in an external pouch for a protracted period. The more advanced were the Eutheria, the placental mammals, that retain the young for a longer period within the body of the mother and that lack any outer pouch.

With the onset of the Cenozoic era, the placentals soon proved to be superior to the marsupials, which were ultimately driven to the ends of the earth and have survived almost exclusively in Australia and South America. Placental mammals are named from the placenta through which the young are nourished inside the body of the mother. This provides a safe first home in which more complex mammalian organisms can develop for a longer time. This characteristic has enabled these mammals to concentrate on producing a few offspring, even one young in some cases, instead of the dozens or hundreds of young or eggs that reptiles produce in order to assure the perpetuation of their kind.

Adaptation
Over a period of seventy million years the mammals have invaded and adapted themselves to every type of environment: the sea, land surfaces as well as underground, treetops, and even the air. In their environmental adaptation the mammals have successfully invaded all altitudes, latitudes, and climates as well.

In addition to achieving this wide geographic distribution, mammals have developed great variation in size. The largest mammal is the one-hundred-foot, one-hundred-ton Blue Whale, which, in length and weight, outranks by 50 percent the largest Mesozoic dinosaur. The smallest mammal in America is the Pygmy Shrew, which when adult, weighs barely 1/12 of an ounce, less than the weight of a dime.

Field Study
The field study of mammals differs greatly from that of birds or seashells. Aside from a few common creatures, such as rabbits, squirrels, and skunks, most mammals are not often seen. Most are nocturnal and nearly all are secretive. The larger mammals are quite wary, while sea mammals are seen only by accident or in very restricted areas where they congregate along the coast. Mammal study is usually part of a general field trip on which some mammals and the tracks, scats, or signs of others may be observed. Taking a "stand," that is, sitting quietly at a likely place in the woods, will often afford a view of some of the shyer creatures.

It is important to look for signs of mammals, even where there is little likelihood of seeing the animals. Becoming expert at interpreting tracks in mud, dust, or snow; at analyzing scats; at recognizing claw marks, gnaw marks, nests, houses, mounds, ridges, food piles, and other telltale indicators is sometimes the only way to identify species. Bones, skulls, and teeth can also aid in the identification of their former owners.

Some species vary widely in color over different parts of their range, and they vary as well in coloration because of age, sex, and season. The most common or typical color forms are illustrated, while the text describes the range of variation.

Habitat
The best places to see mammals on land include woods, wood edges, stone walls, the unkempt corners of suburbs and countryside (small mammals); the edges of lakes and streams (moose, beaver, otter, muskrat); deep woods and swamps (bear, bobcat); caves and old mines (bats); old houses and outbuildings (mice, squirrels).

Mammals in general may be seen throughout the year. Many, however, appear to be most active in spring and fall. The hibernators, of course, are hidden away in winter, although if their roosts are known, they may be readily seen.

Teeth and Food

Of all the distinguishing characteristics the teeth seem to be the most definitive. The primitive placental mammalian tooth count was forty-four, each jaw consisting of three incisors, one canine, four premolars, and three molars on either side. This dental formula is written 3143/3143. The first set of numbers refers to the upper jaw, the second set to the teeth on each side of the lower jaw. Incisors are for cutting, canines for tearing, premolars and molars for grinding or shearing. Omnivorous man has the formula 2123/2123. The rodents, which gnaw but do not tear, have lost their canines; armadillos, which neither gnaw nor tear, have lost both incisors and canines; many mice have no premolars; and some whales have no teeth at all, but substitute baleen or whalebone.

The food of an animal determines its habits; its teeth reflect its food. Thus, when useful, the dental formula is given with each genus within the text. It can be a useful tool in the identification of the skulls that the naturalist sometimes encounters in the field, cave, or pellet of a raptor.

Taxonomy and Nomenclature

The names and organization of the Class Mammalia in this chapter follow the most recent checklist of J. Knox Jones, Jr., Dilfold C. Carter, and Hugh H. Genoways of Texas Tech University, entitled *Revised Checklist of North American Mammals North of Mexico, 1979.*

Range and Scope

There are about 4000 mammal species in the world, representing 122 families. Of these, there are about 410 species in 40 families in North America north of Mexico. This chapter describes nondomesticated species that occur in the region of North America west of the 100th meridian, from the northern tip of Alaska to the Mexican/U.S. border.

Ranges apply to the species as a whole. Subspecies, of which there are very many in some genera, are not described, except where they are regionally widespread and characteristically rather distinct.

USEFUL REFERENCES

Armstrong, D. M. 1972. *Distribution of Mammals in Colorado.* Monograph of the Museum of Natural History, University of Kansas.

Barker, W. 1956. *Familiar Animals of America.* New York: Harper.

Bowles, J. B. 1975. *Distribution and biogeography of mammals of Iowa.* Special Publications of the Museum, Texas Tech University. (No. 9, 1–184 pp.)

Burt, W. H., and Grossenheider, R. P. 1976. *A Field Guide to the Mammals.* 3rd ed. Boston: Houghton Mifflin.

Cahalane, V. H. 1947. *Mammals of North America.* New York: Macmillan.

Findley, J. A.; Harris, A. H.; Wilson, D. F.; and Jones, C. 1975. *Mammals of New Mexico.* Albuquerque: University of New Mexico Press. xxii + 360 pp.

Hall, E. R. 1946. *Mammals of Nevada.* Berkeley: University of California Press. xii + 710 pp.

Hall, E. R., and Kelson, K. R. 1959. *The Mammals of North America*. 2 vols. New York: Ronald Press.

Hamilton, W. J., Jr. 1939. *American Mammals*. New York: McGraw-Hill.

Ingles, L. G. 1954. *Mammals of California and Its Coastal Waters*. 2nd ed. Stanford: Stanford Univ. Press.

Palmer, E. L. 1949. *Fieldbook of Natural History*. New York: McGraw-Hill.

Palmer, R. S. 1952. *The Mammal Guide*. New York: Doubleday.

Turner, R. W. 1974. *Mammals of the Black Hills of South Dakota and Wyoming*. University of Kansas, Mus. Nat. Hist., Misc. Publ. 60:1–178.

Marsupials
Order Marsupialia

Marsupials are characterized by simple brain form, the presence of epipubic bones on the pelvis of both sexes, and a unique angular process of the jaw, which turns inward. In most species the females have a pouch, or *marsupium,* located on the lower abdomen in which the young finish their embryonic development. A single species extends northward from the New World tropics into the United States.

NEW WORLD OPOSSUMS
Family Didelphidae

VIRGINIA OPOSSUM   55:6
Didelphis virginiana   *Fig. 14*

Description
Size: head and body, 15–20 in. (38.1–50.8 cm); tail, 9–13 in. (22.9–33 cm); weight, 6–12 lb. (0.27–0.54 kg); 50 teeth, 5134/4134; mammae, usually 13. Pelage of underfur and white-tipped over-hairs coarse. Upperparts usually gray, underparts paler; face white; ears, feet, and inner part of tail black. Snout long, legs short, 5 toes on each foot; 1st toe on hind foot opposable, lacks claw; tail long, scaly, basal $\frac{1}{10}$ furred, prehensile.

Habitat
Woodlands in farming areas, forest edges.

Habits
Nocturnal; climbs well, feigns death if frightened; makes nest of vegetation in cavity.

Reproduction
2 litters per year; nest located in a burrow, hollow log, hollow tree; gestation 12½ days; 8–18 (occasionally more) young that weigh $\frac{3}{10}$–$\frac{7}{100}$ oz. (1–2 g), born at an early stage of development; young travel immediately to pouch, where they attach firmly to a teat and remain for 55–70 days; pouch young usually number 6–9.

Range
Widespread, well established, from B.C. to Baja Calif., w. of Cascade-Sierra mountains (introduced in 1920s); also in e. Oreg. Blue Mountains; se. Ariz.; w.-cen. Colo.

Other name
Common Opossum.

Virginia Opossum

Fig. 14

Nine-banded Armadillo, p.295

Insectivores
Order Insectivora

The insectivores are generally small mammals with many small, sharp teeth, long pointed snouts, and flat, claw-bearing plantigrade feet. They live on or under the ground. Of all placental mammals living today, these are regarded as the most primitive or most like the ancestral mammalian stock.

SHREWS
Family Soricidae

The generally mouselike shrews have long, pointed noses, small eyes often partly hidden in fur, and inconspicuous ears provided with pinnae (an exterior process). The feet are normally developed with five toes. (Most mice have only four toes on the front feet.) A shrew lacks the strong digging forelimbs of the mole, and it is a swift runner above ground. Because most shrews are small (three and one-half to six inches long, or 8.9–15.2 cm) they are secretive, spending their lives beneath leaf litter, grass, fallen logs, and in the runways of other animals. They are usually grayish or browning, generally lighter beneath; both sexes look alike. Females have six mammae, usually two abdominal and four inguinal. Except for the Least Shrew, all have thirty-two brown-tipped teeth, formula 3133/1113.

Shrews do not hibernate. They are aggressive, irascible, and nervous. Because of their small size, shrews have very high metabolic rates and are voracious feeders, eating at a minimum their own weight in insects, insect larvae and pupae, worms, and snails (sometimes other invertebrates and even mice) every twenty-four hours. They are therefore of considerable ecological importance and are valued as a control of insects. Shrews breed early in the year and many species have more than one litter; young are brought forth in a little round nest of shredded vegetation concealed in leaves, rocks, logs, or a burrow. Their gestation averages eighteen to twenty-two days, and litters range from four to ten. The young are often independent within three weeks. Maximum age rarely exceeds two years. Many utter tiny, high-pitched squeaks. Young shrews have a pencil of hairs at the tip of the tail. These wear off with age and the tail tip of an old shrew is nearly naked. It is difficult to distinguish individual species of living shrews in the wild, as they scurry by so fast.

LONG-TAILED SHREWS
Genus *Sorex*

MASKED SHREW
Sorex cinereus   **53:14**

Description
Size: head and body, 2–2½ in. (5.1–6.4 cm); tail, 1¼–2 in. (3.2–5.1 cm); weight, ¹⁄₁₀–¹⁄₅ oz. (2.8–5.6 g). Skull with relatively narrow rostrum; braincase rather high. Teeth narrow, 5 bicuspids on each side of upper jaw, third and fourth slightly smaller than first and second. Grayish-brown above, paler below; tail dark above, buffy below. Pelage paler below in winter.

Similarities
Vagrant, Arctic, Trowbridge's, and Dusky Shrews are larger.
Merriam's is pale grayish with whitish underparts. Pygmy Shrew
is slightly smaller, with single-cusped teeth in upper jaw, having 3
instead of 5 unicuspids on each side.

Habitat
Moist open areas near water from salt marsh to alpine meadow;
may be common in northern and mountainous portions of range.

Range
Alaska, Aleutians; Canada; Wash.; e. Oreg.; Idaho; Mont.; Wyo.;
cen. Utah; Rocky Mts. of Colo. and n.-cen. N.Mex.

Other name
Cinerous or Common Shrew.

VAGRANT SHREW
Sorex vagrans   *Fig. 15*

Description
Size: head and body, 2⅓–2⅘ in. (5.8–7.1 cm); tail, 1½–1⅘ in.
(3.8–4.6 cm). Tail is slightly more than ⅓ to ½ total length. Third
unicuspid smaller than fourth; unicuspids, except fifth, with
pigmented ridge extending from near apex. Color varied; pattern
tricolored through bicolored to almost unicolored. Summer: reddish
to grayish. Winter: black to pale gray.

Habitat
Varied, primarily forests and wet meadows and marshes.

Range
Alaska, s. of Brooks Range; Aleutians; nw. Canada and offshore
islands; B.C.; w. ¾ Alta.; w.-cen. Sask.; nw. ¾ Wash.; w. Oreg.;
nw. Calif. and Sierra Mts.; w. ½ Mont.; cen. Idaho; Wyo.; ne. and
cen. Utah; w. ½ Colo.; e. Ariz.; w. N.Mex.

Note: A subspecies, *Sorex vagrans parvidens,* occurs in the San
Gabriel Mountains of Southern California. It is possible that *S.
obscurans, S. ornatus,* and *S. pacificus* hybridize with, and are
therefore subspecies of, *S. vagrans.*

Fig. 15

Vagrant Shrew

ORNATE SHREW
Sorex ornatus   **53:12**

Description
Size: head and body, 2⅓–2½ in. (5.8–6.4 cm); tail, 1½–1⅘ in.
(3.8–4.6 cm). Skull has braincase flattened on top, cranium
relatively narrow. Unicuspids relatively narrow; first and second
larger than third and fourth, third smaller than fourth. Grayish-
brown above, pale underparts; tail indistinctly bicolored.

Similarities
Trowbridge's is larger with dark underparts. Desert is pale ash-
gray.

Habitat
Coastal and inland marshes, streamsides, damp earth on hill slopes,
sometimes dry slopes beneath chaparral.

Range

Calif. coastal and inland valley marshes to cen. Sierras, s. from Tehama Co. to Baja Calif., with overlap into w.-cen. Nev.; also e., cen., s. Sierras into sw. Nev. boundary region.

Note: The subspecies *Sorex ornatus californicus* hybridizes with the Vagrant Shrew in California, hence, Ornate Shrew may possibly be considered a further subspecies of the Vagrant Shrew.

WATER SHREW
Sorex palustris   *Fig. 16*

Description
Size: head and body, 3⅓–3⅖ in. (8.3–8.7 cm); tail, 2½–3 in. (6.4–7.6 cm); weight, ⅓–½ oz. (9.5–14.2 g). Hind feet large, fringed along sides with stiff hairs. Third unicuspid smaller than fourth. Back black with "frosted," gray-tipped hairs; underparts white, gray, or brownish; tail markedly bicolored.

Similarities
Pacific Water Shrew is larger and brownish. Large body size and stiff hairs along sides of hind feet distinguish Water Shrew from all other shrews except Pacific Water Shrew.

Habitat
Borders of or in ponds and streams in meadows, marshes, and wooded areas.

Habits
Swims, dives, can run on surface of water.

Range
In s.-cen. Mackenzie, se. across Canada; se. Alaska; B.C. and Vancouver Is.; Cascade-Sierra Mts. of Wash.; Oreg. to s.-cen. Calif.; cen. to e. Oreg.; Idaho; w. Mont.; nw. to se. Wyo.; ne. Nev.; cen.-s. Utah; cen.-sw. Colo.; n.-cen. N.Mex.; Rockies generally; e.-cen. Ariz.

Other name
Northern Water Shrew.

Pygmy Shrew, p. 281

Merriam's Shrew

Fig. 16

Water Shrew

PACIFIC WATER SHREW
Sorex bendirii

Description
Size: head and body, 3½–3⅘ in. (8.9–9.7 cm); tail, 2½–3⅕ in. (6.4–8.1 cm). Hind feet with weak, stiff bristlelike hairs along sides adapted for swimming. Dark brown or black above, sometimes frosted underparts slightly paler; tail unicolored.

Similarities
Water Shrew is smaller and blackish.
Habitat
Near streams or beaches, under logs, in swamps, marshes, wet
wooded area, and humid coastline.
Remarks
Largest shrew of Genus *Sorex.*
Range
In sw. B.C.; Wash.; Oreg., both w. of Cascades; nw. Calif., along
Pacific Coast.

DWARF SHREW
Sorex nanus

Description
Size: head and body, 2½ in. (6.4 cm); tail, 1¾ in. (4.4 cm). Skull
small, slightly narrow. Pale grayish-brown above, underparts
grayish; tail indistinctly bicolored.
Similarities
Merriam's is larger and pale gray with distinctly bicolored tail.
Masked, Vagrant, Dusky are larger.
Habitat
Arid regions; quite rare.
Range
Scattered localities in s. Utah and n. Ariz., ne. from Kingman,
Ariz., on border; also se. Wyo. to e.-cen. Colo., e. of Rockies.

ARCTIC SHREW
Sorex arcticus   **53:13**

Description
Size: head and body, 2¾–3 in. (7–7.6 cm); tail, 1¼–1⅔ in. (3.2–4.1
cm); weight, ⅙–⅓ oz. (7–11 gm). Third unicuspid larger than
fourth; ridges extending from apices medially toward cingula but
incomplete, weakly pigmented. Tricolored in most pelages; dorsal
region darkest, grayish to brownish; flanks brownish or tan;
underparts pale.
Similarities
Masked is smaller, with underparts paler than upper. Dusky is not
tricolored and light brown. Pygmy is smaller and light brown.
Habitat
Tundras in northern range, swamps and bogs in southern.
Remarks
Each color area contrasting, this is most brilliantly colored shrew.
Range
In w., n., cen. Alaska; n. Yukon; Mackenzie; extreme ne. B.C.; e.
across Canada; s. into se. Sask.; n. N.Dak.

MERRIAM'S SHREW
Sorex merriami   *Fig. 16*

Description
Size: head and body, 2¼–2½ in. (5.7–6.4 cm); tail, 1½ in. (3.8 cm).
Third unicuspid larger than fourth; unicuspid row crowded;
unicuspids higher than long, lack heavily pigmented internal ridge.
Pale grayish-drab above, underparts and feet nearly white; tail
bicolored.
Similarities
Masked is slightly larger, brownish. Dusky is larger, brownish.

Vagrant is larger, feet dark. Dwarf is smaller, with indistinctly bicolored tail.

Habitat
Sagebrush and bunchgrass in open, arid regions; quite rare.

Range
Great Basin drainage; Wyo.; w. Colo.; ne. ¼ Ariz.; nw. ¼ N.Mex.

TROWBRIDGE'S SHREW
Sorex trowbridgii   *Fig. 17*

Description
Size: head and body, 2½–2⅝ in. (6.4–7.1 cm); tail, 2–2½ in. (5.1–6.4 cm). Third unicuspid smaller than fourth; internal ridge of unicuspids weakly pigmented, not ending in internal cusplet. Dark gray or blackish above and below; sometimes hued in brown; tail sharply bicolored, nearly white below.

Similarities
Pacific is larger, tail unicolored; Vagrant is light brown, with shorter tail. Ornate is smaller; Dusky is light brown, underparts whitish.

Habitat
Dry coniferous forests, occasionally in moist forests in absence of other shrews; in southern range occurs in chaparral, moist timbered canyons, thick vegetation near water.

Range
In sw. B.C.; w. of Cascades in Wash. and Oreg.; nw. and n. Calif., coastally s. to Los Angeles Co.; down Sierras to s.-cen. Calif.

Fig. 17

Trowbridge's Shrew

PACIFIC SHREW
Sorex pacificus

Description
Size: head and body, 3⅓ in (8.3 cm); tail, 2–2¾ in (5.1–7 cm). Cinnamon-brown, including tail, feet, underparts.

Similarities
Trowbridge's and Dusky are smaller; tails bicolored. Water Shrew is larger; blackish, with stiff hairs on sides of hind feet.

Habitat
Damp coastal redwood and spruce forests.

Remarks
Considered by some experts to be a subspecies of *S. vagrans*.

Range
In sw. Oreg.; nw. Calif.

DUSKY SHREW
Sorex obscurus   **53:11**

Description
Size: head and body, 2½–3 in. (6.4–7.6 cm); tail, 1⅝–2½ in. (4.1–6.4 cm). Dull-brown above, underparts whitish; tail bicolored.

Similarities
Very difficult to differentiate from other shrews occurring in same areas. Vagrant is reddish-brown or blackish. Pacific is larger.

Pygmy and Dwarf are smaller. Arctic is tricolored. Masked is smaller, grayish. Merriam's is smaller and pale gray. Trowbridge's has dark underparts.

Habitat

Wet meadows and swamps and near streams in forests.

Remarks

Sometimes considered a subspecies of *S. vagrans.*

Range

In w., cen., se. Alaska; s. ½ Yukon; s. Mackenzie; B.C.; Alta.; n. and w. Wash., Idaho; w. Mont.; w. Oreg.; n. Sierras of Calif.; also Rockies in w. Wyo., w. Colo.; ne. to sw. Utah; cen. Ariz.

OTHER SHREWS

Genera *Microsorex* and *Notiosorex*

PYGMY SHREW

Microsorex hoyi   *Fig. 16*

Description

Size: head and body, 2–2½ in. (5.1–6.4 cm); tail; 1–1⅖ in. (2.5–3.6 cm); weight, ¹⁄₁₂ oz. (2.4 g); 32 teeth, 3133/1113. Skull resembles that of *Sorex* but narrower, more flattened. Nose long, pointed; eyes tiny, black. First and second unicuspids with distinct internal ridge terminating in pronounced internal cusp; apices of unicuspids curved posteriorly; third unicuspid disklike, fourth normal, fifth minute. Brownish above, paler below; tail indistinctly bicolored.

Similarities

Masked has longer tail, 5 upper unicuspids on each side of jaw. Dusky and Vagrant are larger. Arctic is larger, more brightly colored.

Habitat

Boreal; wet and dry woods adjacent grass clearings; rare.

Remarks

The smallest living mammal.

Range

In cen. Alaska; Yukon drainage; s. Yukon; s. Mackenzie; B.C.; Alta.; w.-cen. Sask.; Wash.; Idaho; w. Mont.; w. Oreg., including Cascade–Sierras to s. Calif.; w. Wyo.; w. Colo.; ne. Utah, with sw. extension into sw. corner; cen. N.Mex., n. to s.

DESERT SHREW

Notiosorex crawfordi

Description

Size: head and body, ±2 in. (5.1 cm); tail, ±1 in. (2.5 cm); weight, ⅙–¼ oz. (5–8 gm); 28 teeth, 3113/2013. Ears extend visibly beyond fur. Three unicuspid teeth in each upper tooth row. Grayish above, similar or paler below.

Similarities

Merriam's is slightly larger, darker, and with longer tail. Other regional shrews inhabit more moist and mountainous sections.

Habitat

Low desert, among sagebrush, creosote bush, other desert shrubs, and chaparral slopes.

Range

Desert areas of s. Calif.; extreme s. Nev.; Ariz., s. of Grand Canyon; N.Mex.; se. Colo.; Okla.; w. ⅔ Tex.

MOLES
Family Talpidae

Moles are small, burrowing animals with dark, soft, velvety fur to which dirt will not readily cling and which can lie forward almost as well as backward. Their length from nose to tip of tail ranges between four and nine inches; they weigh from one and one-half to four ounces (43–113 g), usually lack functional eyes and external ears, and their outwardly turned forefeet make them powerful diggers. They utter a high-pitched squeak. Moles occupy all humid habitats with suitable soils from sea level to above timberline and are active at all hours and seasons. They do not hibernate, and spend their life almost entirely underground.

The presence of moles can be determined by the low ridges, which they push up along the earth's surface, and mounds of earth above a green lawn. The mole's diet is largely earthworms and other invertebrates, with some roots. Their nest is a six-inch (15.2 cm) agglomeration of dry vegetation in a sheltered or underground earth mound. The female brings forth her two to six naked young from April to June.

SHREW-MOLE
Neurotrichus gibbsii   **53:15**

Description
Size: head and body, 3–3½ in. (7.6–8.9 cm); tail, 1–1½ in. (2.5–3.8 cm). Smallest American mole. Skull scarcely constricted interorbitally; braincase broad, nose naked, nostrils open to sides, eyes small but apparent. Tail stiffly haired, about ½ body length. Front feet longer than broad, 6 tubercles on sole of hind foot. Upper molars with bicuspidate internal edge; first and second subequal, third smaller. Gray or blackish mouse-gray.

Similarities
Water Shrew's front feet not conspicuously broad, nose not naked. Trowbridge's Shrew is smaller and front feet thin.

Habitat
Subterranean runways in humid areas of redwood, fir, and pine forests.

Range
Nw. Coast, from se. B.C. to San Francisco Bay, Calif.

COAST MOLE
Scapanus orarius

Description
Size: head and body, 5–5½ in. (12.7–14 cm); tail, 1⅓ in. (2.2 cm). Tail slightly haired; front feet broader than long; nose naked, nostrils open above. Teeth are evenly spaced, uncrowded, unicuspid. Blackish-brown to black.

Similarities
Townsend's is darker and larger. Broad-footed is difficult to distinguish in field or hand; teeth more crowded, less evenly spaced.

Habitat
Well-drained soil, in fairly deep burrows; enters deciduous woods.

Range
Nw. Coast, from sw. B.C., w. Wash., with overlap into cen. Wash., coastal Oreg., and nw. Calif.; also ne. ⅓ Oreg. and extreme se. Wash.; extreme w.-cen. Idaho.

TOWNSEND'S MOLE
Scapanus townsendii   **53:17**

Description
Size: head and body, 6–7 in. (15.2–17.8 cm); tail, ±2 in. (5.1 cm), slightly haired. Front feet broader than long; nose naked, nostrils open upward. Blackish-brown to nearly black.

Similarities
Coast is smaller.

Habitat
Damp, humid, easily worked soils in forests, fields, and gardens.

Range
Nw. Coast, from extreme sw. B.C. to nw. tip of Calif., w. of Cascades.

BROAD-FOOTED MOLE
Scapanus latimanus   **53:16**

Description
Size: head and body, 5–6 in. (12.7–15.2 cm); tail, 1½ in. (3.8 cm). Tail nearly completely haired; front feet broader than long; nose with short hairs almost to end of snout, nostrils open upward. Teeth often are unevenly spaced and crowded. Blackish-brown to black.

Similarities
Coast is difficult to differentiate. Townsend's is larger, blacker.

Habitat
Valleys and mountain meadows with moist soils containing an abundance of invertebrate life.

Range
S. Oreg.; n. Calif., coastal to Baja Calif., Sierras to Los Angeles Co.; absent from San Joaquin Valley.

Other name
California Mole.

Bats
Order Chiroptera

Bats are unique placental mammals in that they are modified for flight. They fly by virtue of greatly lengthened finger bones supporting thin wing membranes made of a relatively naked, double layer of skin and extending to and usually between the hind legs. The middle finger is the longest and the thumb is free, bearing a claw. The forearm length is the distance from the elbow to the wrist. The sexes look alike, but the young are frequently darker. The membranous structures are peculiar in that the interfemoral membrane joins the legs and tail. The calcar is a cartilage that extends from the foot along the outer edge of this membrane and acts as a brace. If some interfemoral membrane extends beyond the calcar, the latter is said to be keeled. Inside the ear is a leaflike formation termed the tragus.

Bats utter high-pitched squeaks, which are inaudible to human beings. They hear the echoes of these sounds as they bounce back off insects and nearby obstacles; thus, bats can find their prey in the dark and avoid hitting obstructions. Most bats in the United States feed on insects, caught either in flight or on the ground. Leaf-nosed bats eat nectar and fruit. Bats are twilight-flying (crepuscular) and nocturnal. They roost by day, hanging upside down by their feet in caves, mine shafts, rocky crevices. Some species are solitary, others are colonial. In winter bats either hibernate, often in colonies in caves, or like birds, migrate south. Some species both migrate and hibernate.

Females often assemble in "maternity roosts" away from the males before giving birth to the young, usually one or two, which are born from May to July. Young babies often cling to their mother when she flies; later they hang by themselves at their roost until they are able to fly, which is usually in five to six weeks.

LEAF-NOSED BATS
Family Phyllostomatidae

This group of bats, primarily tropical and subtropical, is characterized by a leaflike, triangular flap of thick skin (called the noseleaf) that projects upward from the tip of the nose. The tail is variable or absent.

CALIFORNIA LEAF-NOSED BAT
Macrotus californicus   *Fig. 18*

Description
Size: forearm, 2 in. (5.1 cm); 34 teeth, 2123/2133. Skull, limbs, general form slender; ears large, subovate; nose has distinct leaflike skin-flap. Upperparts vary from buffy-gray to dark brown; underparts pale drab to buffy-brown, usually with silvery wash. Tail extends to edge of interfemoral membrane.

Similarities
Mexican Long-nosed has long rostrum, no tail. Long-tongued has long, slender rostrum; ears small, dark brown.

Habitat
Desert caves and mine tunnels.

Remarks
The only big-eared bat in the United States with a triangular noseleaf.

Range
Deserts of se. Calif.; extreme s. Nev.; w. and s. Ariz.

Other name
Leafnose Bat.

Fig. 18

Long-tongued   California   Mexican
Bat   Leaf-nosed Bat   Long-nosed Bat

LONG-TONGUED BAT
Choeronycteris mexicana   *Fig. 18*

Description
Size: forearm, 1¾ in. (4.4 cm); 30 teeth, 2123/0133. Nose long, slender, triangular noseleaf on top of snout projects upward; ears small, inconspicuous; tail small, extends less than halfway to edge of interfemoral membrane. Dark brown to sooty gray.

Similarities
California Leaf-nosed has large ears, tail extends to edge of interfemoral membrame. Mexican Long-nosed has no tail.

Habitat
Desert caves, mine tunnels.
Range
Extreme s. Calif.; s. Ariz.; s. into Mexico.
Other name
Hognose Bat.

MEXICAN LONG-NOSED BAT
Leptonycteris nivalis

Fig. 18

Description
Size: forearm, 2⅛ in. (5.2 cm); 30 teeth, 2122/2132 (molars elongated). Nose very long, slender noseleaf on tip; ears medium-large, extend well above head; interfemoral membrane narrow; tail absent. Upperparts medium brown in posterior, paler over shoulders; underparts paler than posterior (about like shoulder region).
Similarities
Long-tongued and California Leaf-nosed have tails.
Habitat
Dry caves, tunnels.
Range
Big Bend Region of Tex.
Other name
Long-nosed Bat.

Note: The closely related and slightly smaller species *Leptonycteris sanborni* reaches into se. Ariz. and sw. N.Mex.

COMMON OR PLAIN-NOSED BATS
Family Vespertilionidae

The tail, in this largest and most widely distributed family of bats, does not protrude beyond the interfemoral membrane. Except in the Red and Hoary Bats, there are two mammae. In the Genus *Myotis,* the tragus (a thin, fleshy projection arising from the inner base of the ear) is pointed, and the interfemoral membrane is not furred, but may be scantily haired. In the family, teeth vary from twenty-eight to thirty-eight with the dentition formula 1113/2123 to 2133/3133, and except in the Long-eared Bats, the ears will not extend beyond the nose if they are laid forward. There is usually one offspring, born in late spring or early summer, and it weans in about six weeks.

All members of this family are relatively small, most are colored brown, and all have simple snouts without a noseleaf. The membranes are complete and the tail reaches to the edge of the interfemoral membrane, but not beyond. Many species are very difficult to differentiate in the hand or in the field. Most bats found north of Mexico belong to this family; hence, they are called "common bats." They have a nearly worldwide distribution, and are insect eaters. Their nocturnal feeding complements the diurnal feeding of birds, and at times, the plain-nosed bats are as numerous as birds.

MOUSE-EARED BATS
Genus *Myotis*

LITTLE BROWN MYOTIS
Myotis lucifugus   **54:5**

Description
Size: forearm, 1½ in. (3.8 cm). Small. Skull has a braincase that
rises gradually from rostrum. Ears moderate-sized; when laid
forward, reach almost to nostril. Pelage long, silky; individual hairs
shiny or glossy, almost metallic at tips. Upperparts cinnamon-buff
to dark brown; underparts buffy to pale gray, often with lighter
wash; buffy shoulder spot occasionally shows.

Similarities
Yuma, California and Small-footed are smaller. Cave is larger.
Keen's and Long-eared have large ears which reach beyond nose
when laid forward. Fringed has conspicuous fringe of hairs along
edge of interfemoral membrane. Long-legged is larger; fur dull. Big
Brown Bat is larger.

Habitat
Associated with but not restricted to timbered areas, caves,
buildings, hollow trees by day, near water or forest at dusk; very
common.

Habits
Roosts singly or in clusters in caves, rock crevices, holes in trees,
behind shutters. Begins evening flight early, sometimes in late
afternoon.

Remarks
Commonest and most widely distributed *Myotis*.

Range
In w.-cen., s., se. Alaska; s. ½ Yukon and e. across all Canada and
s.; all U.S., n. of s. Calif.; extreme s. Nev. and Ariz.; also in
extreme n. N.Mex.

Other name
Little Brown Bat.

CAVE MYOTIS
Myotis velifer

Description
Size: forearm, 1⅝–1⅘ in. (4.1–4.6 cm). Medium-sized. Skull
large, robust, well-developed sagittal crest; rostrum broad, when
viewed from above barely less than that of braincase. Wing
membrane arises from base of toes; calcar well-developed, not
keeled. Pelage moderate length, dull on back. Dull sepia to drab
above, underparts paler.

Similarities
Little Brown, Yuma, Long-Legged, California, Small-footed are
smaller. Long-eared is smaller, ears larger. Fringed has hair fringe
along edge of tail membrane. Big Brown Bat is larger.

Habitat
Dry desert caves; common.

Range
Sw. desert states; se. ¼ Utah; s. ½ Colo.; extreme se. Calif.; Ariz.,
s. of Grand Canyon; N.Mex.; w. ½ Tex.; also w. ½ Okla.

YUMA MYOTIS
Myotis yumanensis

Description
Size: forearm, 1⅓–1½ in. (3.8 cm). Skull has braincase rising abruptly from rostrum level; sagittal crest usually absent. Interfemoral membrane haired nearly to knees; foot relatively large, robust; tail barely reaches beyond membrane. Calcar with lobe at end. Upperparts tawny, buffy, or brown; darker subspecies often with buffy wash. Underparts paler, buffy to yellowish-white; membranes pale brownish; fur dull.

Similarities
Cave, Long-legged and Big Brown Bat are larger. Little Brown is larger; hair glossy. Keen's and Long-eared have large ears, can extend beyond nose when laid forward. Fringed has hair fringe along edge of tail membrane. California and Small-footed are smaller.

Habitat
Rather open areas from below sea level to 11,000 ft. (3352.8 m).

Habits
Usually found in colonies, or roosting singly, by day in caves or little-used buildings; they fly at dusk near water or forest edges.

Remarks
One of the most common western *Myotis*.

Range
In sw. B.C. and Vancouver Is. through Pacific states to Baja Calif.; Idaho; sw. ½ Mont.; nw. ½ Wyo.; w. Nev.; e.-cen.-s. Utah; Ariz.; N.Mex.; w. ⅓ Tex.

LONG-EARED MYOTIS
Myotis evotis

Description
Size: forearm, 1⅖–1⅗ in. (3.6–4.1 cm). Skull very similar to Keen's, upper profile curving gradually from long rostrum to low summit of braincase; sagittal crest often present, never large; braincase viewed from above oval and bulging posteriorly. Ears blackish and large, when laid forward extend about ⅕ in. (5.2 mm) beyond nose; tragus large. Pelage long, glossy. Upperparts light to medium brown; ears conspicuously darker, blackish. Interfemoral membrane with fringe of inconspicuous hairs at edge.

Similarities
Fringed has smaller ears, distinct fringe of hairs on edge of interfemoral membrane. Little Brown, Yuma, and Long-legged have smaller ears. Cave is larger, ears smaller. Keen's has slightly smaller ears, dark brown. California and Small-footed are smaller. Big Brown Bat is larger.

Habitat
Thinly forested to semidesert areas; uncommon; not a cave bat.

Habits
Prefers to roost singly or in small clusters in secluded niches of buildings and probably in trees.

Range
In s. B.C.; Vancouver Is.; s. Alta.; sw. Sask.; s. through w. states and e. as far as Black Hills of S.Dak.; w. ⅔ Colo.; N.Mex.; except se. corner.

Note: A population of these bats from se. Ariz. and sw. N.Mex. is considered by some to be a distinct species, *M. auriculus*.

KEEN'S MYOTIS

Myotis keenii

Description

Size: forearm, 1⅖–1⅗ in. (3.6–4.1 cm). Small. Skull relatively lightly built, slender, sagittal crest sometimes present; ears long, when laid forward, extend about 1/16 in. (1.6 mm) beyond end of nose. Tragus long, narrow, and pointed. Length of upper row of teeth slightly exceeds greatest palatal breadth including molars. Pelage long, silky, dull. Dark brown. Ear length distinguishes this from all other *Myotis* within its range.

Similarities

Little Brown, Yuma, Cave, Long-legged and Small-footed have smaller ears. Long-eared has larger ears, extending ⅕ in. or more beyond nose when laid forward. Fringed has hair fringe on edge of tail membrane. California is smaller. Big Brown Bat is larger.

Habitat

Humid forests.

Habits

Roosts singly or in small colonies in obscure places; usually flies late at night.

Remarks

Some experts believe *M. keenii* to be the same species as *M. evotis.*

Range

Nw. Coast, from n. B.C. to s. Wash., w. of Cascades, including major offshore islands.

FRINGED MYOTIS

Myotis thysanodes

Description

Size: forearm, 1⅗–1⅘ in. (4.1–4.6 cm). Skull like that of Long-eared but larger, more robust, broader; sagittal crest well-developed; ear large, projects beyond muzzle when laid forward. Length of upper tooth row exceeded by greatest breadth of palate including molars. Wing membrane runs to base of toes. Interfemoral membrane conspicuously fringed with stiff hairs along free edge of tail. Upperparts yellowish-brown to darker olivaceous tones; underparts same or barely lighter.

Similarities

Little Brown, Yuma, Keen's, Long-eared, Long-legged, and California are smaller and lack conspicuous fringe on tail membrane. Cave has no fringe. Big Brown Bat is larger with no fringe.

Habitat

Abandoned buildings; probably a cave dweller.

Range

In s.-cen. B.C., a n. extension from Wash.; extreme e. and s. Wash.; Oreg.; s. ½ Idaho; s. ¾ Wyo., including Black Hills of S.Dak.; all states w. of Great Plains.

LONG-LEGGED MYOTIS

Myotis volans

Description

Size: forearm, 1½–1⅗ in. (3.8–4.1 cm). Skull small, delicate; rostrum short, braincase abruptly elevated from rostral level, globose in profile; sagittal crest low, poorly defined. Ears conspicuously short, rounded, barely reach rostrum when laid forward; foot small. Calcar distinctly keeled. Pelage long, soft; underwing and membranes lightly furred down to elbows and knees. Upperparts ochreous-tawny to dark smoke-brown; underparts smoke-brown to dull yellowish-white washed with buff; tips of hairs above slightly burnished.

Similarities
Little Brown, Yuma, California, Small-footed are smaller. Keen's, Long-eared have larger ears. Fringed is larger, hair fringe on tail membrane edge. Cave and Big Brown Bat are larger.

Habitat
Does not inhabit caves; prefers open forest; fairly common.

Habits
Not social in roosting.

Range
In sw. ¾ B.C.; s. ½ Alta., except extreme e. border; s. to Mexico; e. to Black Hills of S.Dak.

Other name
Hairy-winged Myotis.

CALIFORNIA MYOTIS
Myotis californicus

Description
Size: forearm, 1⅕–1⅖ in. (3.0–3.6 cm). Skull delicate, slender; rostrum relatively short, tapering; braincase rising abruptly from rostral level (high profile), flat-topped; sagittal crest obsolete or absent. Ear extends beyond muzzle when laid forward. Foot relatively small. Calcar keeled. Color variable, upperparts brown to distinctly yellowish; underparts usually paler. Bases of hairs much darker than tips. Ears and membranes dark, contrasting with body fur.

Similarities
Yuma is usually larger with larger foot. Small-footed has sharp black mask across face. Little Brown, Cave, Keen's, Long-eared, Fringed, Long-legged Myotis and Big Brown Bat are larger.

Habitat
Hibernates in caves, old mines; dwells principally in open semiarid to arid regions.

Habits
Flight highly erratic; abrupt changes of direction both vertically and laterally.

Range
From w.-cen. B.C., w. of Cascades; in all w. states, e. to Idaho; e. ¾ Colo. and N.Mex.; w. Tex. into Mexico.

SMALL-FOOTED MYOTIS
Myotis leibii

54:3

Description
Size: forearm, 1⅕–1½ in. (3–3.8 cm). Skull small, delicate; braincase sloping gradually up from rostral level (low profile); sagittal crest low when present. Ear barely exceeds muzzle when laid forward. Foot small; calcar long, slender, keeled. Pelage long, silky, frequently glossy-tipped. Ears black and face with black mask; upperparts light buff to golden brown; underparts buffy to nearly white; wings and interfemoral membrane dark brown, almost black.

Similarities
Yuma is larger with no black mask. Little Brown, Cave, Keen's, Long-eared, Fringed, Long-legged Myotis and Big Brown Bat are larger. California is often difficult to differentiate: brown mask, ears dark brown, braincase of skull rising more sharply from rostrum.

Habitat
Caves, old mines, abandoned buildings; not colonial, usually hangs singly; relatively common but rare in some parts of range.
Range
In s.-cen. B.C. to extreme sw. Sask.; s. through all w. states e. of Cascade-Sierra Mts. to Mexico.
Other name
Sometimes called *Myotis subulatus*.

OTHER PLAIN-NOSED BATS
Genera *Lasionycteris, Pipistrellus, Eptesicus, Lasiurus, Euderma, Plecotus,* and *Antrozous*

SILVER-HAIRED BAT
Lasionycteris noctivagans   **54:6**

Description
Size: forearm, 1⅔ in. (4.2 cm); 36 teeth, 2123/3133. Skull flattened; rostrum broad, upper surface concave on each side; sagittal crest obsolete. Ears short, nearly as broad as long, rounded and naked. Interfemoral membrane furred on basal half above. Color is unique among bats: upperparts dark brownish-black, strongly washed with silver-tipped hairs down middle of back; underparts slightly lighter, silvery wash less pronounced.
Similarities
Hoary is larger, throat buffy. Red is brick- or rusty-red.
Habitat
Forested regions; relatively abundant.
Habits
Begins flying in late afternoon or early evening, flight slow and erratic; hunts mainly along streams and, where water is scarce, along edges of timber; gregarious, congregating in great numbers, males may be solitary by season.
Range
In se. Alaska; sw. Yukon; B.C., except ne. corner; e. across Canada and s. through w. states, except Great Central Valley and s. ½ of Calif., sw. Ariz., and sw. Tex.

WESTERN PIPISTRELLE
Pipistrellus hesperus   **54:8**

Description
Size: forearm, 1–1⅛ in. (2.5–3 cm); 34 teeth, 2123/3123. Skull nearly straight in dorsal profile. Inner upper incisors unicuspidate; other upper incisor with accessory cusp. Tragus blunt, tip bent forward. Calcar keeled. Color smoke-gray to buff-brown.
Similarities
California Myotis is larger; buffy to brown, tragus pointed. Yuma Myotis is larger; tragus pointed. Small-footed Myotis has black mask, pointed tragus. Other bats are larger.
Habitat
Low, arid regions; days in crevices in cliffs or buildings.
Habits
Takes flight early in evening, sometimes abroad in late afternoon; easily recognized in flight by contrast between grayish back and blackish membranes, ears, feet, and nose and by erratic flight pattern.
Remarks
The smallest of western bats.

Range

In se. Wash.; e. Oreg.; sw. corner Idaho; n., w., s. Nev.; Utah, except nw. and ne. corners; Calif., except nw. coast and n. and ne. parts; Ariz.; N.Mex., except extreme e. part; also Okla. panhandle; sw. Colo.; w. Tex.

BIG BROWN BAT
Eptesicus fuscus   54:2

Description

Size: forearm, 1⅘–2 in. (4.6–5.1 cm); wingspread, 12 in. (30.5 cm); 32 teeth, 2113/3123. Females larger than males. Skull similar to Western Pipistrelle; rostrum flattish, usually rounded off above. Ears small, nose broad, tragus broad and blunt. Upper incisors well-developed, inner larger than outer, usually has distinct secondary cusp; outer incisor separated from canine by space equal to greatest diameter of incisor; lower incisors subequal, trifid, closely crowded. Interfemoral membrane thick, with sprinkling of hairs above on basal quarter. Calcar keeled. Upperparts brown, usually dark, sometimes reddish-brown; underparts paler, sometimes cinnamon or even buffy; ears, nose, feet, and membranes blackish. Fur long and glossy.

Similarities

All *Myotis* are smaller.

Habitat

Buildings, crevices, caves.

Habits

Wholly insectivorous; tolerates humans well, often roosting in occupied buildings; hibernates singly or in clusters.

Remarks

One of the commonest, most widely distributed bats.

Range

B.C.; Alta.; sw. ⅔ Sask.; e. across Canada, s. of Hudson Bay, and s. throughout U.S.

RED BAT
Lasiurus borealis   54:10

Description

Size: forearm, 1½–1⅔ in. (3.8–4.1 cm); 32 teeth, 1123/3123; mammae, 4. Ears low, broad, rounded, naked inside, densely furred outside on basal ⅔; tragus triangular. Tail longer than forearm. Interfemoral membrane fully furred. Upperparts brick- to rusty-red washed with white, males usually brighter than females; underparts slightly paler; anterior part of shoulder has buffy-white patch.

Similarities

Hoary is larger. Silver-haired is blackish-brown. Southern Yellow has interfemoral membrane not densely furred to edge.

Habitat

Forested areas; common.

Habits

Solitary; in summer roosts mostly in trees or shrubs, often near or even on the ground; begins flying early in evening, usually hunts along watercourses or about trees.

Range

In sw. B.C.; Wash. and Oreg., w. of Cascades; Calif., except extreme ne. corner; sw. Nev.; s. and e. Utah; Ariz.; extreme sw. N.Mex.

HOARY BAT
Lasiurus cinereus

54:4

Description
Size: forearm, 2+ in. (5.1 cm); wingspread, 14 in. (35.6 cm) 32 teeth, 1123/3123; mammae, 4. Skull robust; rostrum broad, short; ears short, rounded and edged with black. Tail membrane heavily furred on top to edges. Upperparts varied, usually yellowish- to mahogany-brown, strongly frosted with silver, giving hoary appearance; underparts whitish on belly, pale brown on chest, yellowish on throat; white patches of fur at wrist and elbow. Fur long and soft.
Similarities
Silver-haired and Red are smaller.
Habitat
Forested areas; solitary, hangs in trees or shrubs by day.
Habits
Migratory, usually moving to warmer climate in winter.
Remarks
Size and color distinguish this species.
Range
In sw. Mackenzie; sw. ¾ B.C.; Alta.; s. Sask., e. across Canada and s. throughout U.S.

SOUTHERN YELLOW BAT
Lasiurus ega

54:9

Description
Size: forearm, 1⅘–2⅛ in. (4.6–5.6 cm); 30 teeth, 1113/3123; mammae, 4. Ear short, tapering, and furred up ½ outside surface. Tail membrane heavily furred only on basal ⅓. Color highly variable, ranging from yellow-brownish washed with black to buffy-white.
Similarities
Hoary and Red have completely furred tail membranes.
Habitat
Upland desert mountains.
Range
Mountains of se. Calif.; s. ⅓ Ariz.

SPOTTED BAT
Euderma maculatum

54:1

Description
Size: forearm, 2 in. (5.1 cm); 34 teeth, 2123/3123. Skull with low, rounded, large braincase; rostrum markedly reduced; ears extraordinarily large. Upperparts black or dark sepia; 2 "saddle-marks" and spot at base of tail white; underparts white.
Habitat
Arid and semiarid regions; rare.
Range
In se. Oreg.; s. Idaho; sw. Mont., through desert areas of e. Calif.; Nev.; Utah; w. Wyo.; Ariz.; far w. Colo.; w. N.Mex.

TOWNSEND'S BIG-EARED BAT
Plecotus townsendii

54:7

Description
Size: forearm, 1½–1¾ in. (3.8–4.4 cm); 36 teeth, 2123/3133. Males smaller than females. Skull slender, highly arched; rostrum

greatly reduced; ears greatly enlarged (over 1 in.), joined basally across forehead; muzzle bears 2 conspicuous glandular masses in front of eyes. Interfemoral membrane naked. Upperparts pinkish-buff to blackish; underparts buffy to brownish.

Similarities
Pallid has no distinct lumps on nose; ears separate.

Habitat
Primarily a cave dweller, often found in attics and barns.

Habits
Usually roosts in small groups in semilight areas, easily disturbed; takes flight readily, emerges in late dusk, flies at considerable elevation, descending near ground only after dark.

Range
In s. B.C., w. of Rockies; s. ½ Vancouver Is.; all w. states except e. Mont., se. Wyo., and ne. Colo.

Other name
Lump-nosed Bat.

PALLID BAT
Antrozous pallidus

54:11

Description
Size: forearm, 2-2⅖ in. (5.1-6.1 cm); 28 teeth, 1113/2123. A large pale bat with big ears; females larger than males. Skull with high, smooth braincase; rostrum large, more than ½ braincase length; muzzle simple, with low horseshoe-shaped ridge. Ears large, over 1 inch tall, separate, extend well beyond muzzle when laid forward. Tragus with wavy edge. Upperparts creamy, yellowish, even light brown; underparts paler, almost white. Color palest in desert, darkest on northwest coast.

Habitat
Both sexes roost together by day in crevices, houses, barns; common.

Habits
Migratory; flight relatively slow, 10–11 wingbeats per sec., often near ground. In desert regions often hawks back and forth in an arroyo below level of surrounding desert; may alight and feed on ground-dwelling insects; can be caught in a collector's mousetrap.

Range
In s.-cen. B.C.; e. Wash.; e. Oreg.; also sw. Oreg., e. of Coast Range; Calif.; nw., w., s. Nev.; s. ½ Utah; sw. ½ Colo., and s. to Mexico; absent from n. Sierras in Calif.

FREE-TAILED BATS
Family Molossidae

Bats in this family have tails that extend beyond the interfemoral membrane, usually for half or more of their length; also hair that is short, dense, dark brown, and exudes a musty odor. Ears project forward and are often joined near their bases. The tragus is very small. All are primarily cave bats, but may inhabit buildings. They are colonial. The family is widely distributed through the warmer parts of both the Old and New Worlds.

BIG FREE-TAILED BAT
Tadarida macrotis

Description
Size: forearm, 2⅓-2½ in. (5.8-6.4 cm); 30 teeth, 1123/2123. Ears connected at base and large, extending beyond end of rostrum.

Skull large, robust; rostrum relatively long. Fur glossy. Upperparts reddish-brown to dark brown; underparts paler.

Similarities
Western Mastiff is larger.

Habitat
Desert caverns; rare.

Range
From s. edge of B.C., s. through s. Wash.; e. Oreg., except ne. corner; sw. corner of Idaho; e. Nev.; s. ¼ Calif.; e. across s. Utah, s. Colo., Kans., Okla. panhandle, and s. Tex. and s. through Ariz. and N.Mex. into Mexico.

BRAZILIAN FREE-TAILED BAT
Tadarida brasiliensis   *Fig. 19*

Description
Size: forearm, 1⅔–1⅘ in. (4.1–4.6 cm); 32 teeth, 1123/3123. Upper lip characterized by deep vertical grooves. Ears separate, but meet at midline. Pelage short, velvety. Upperparts dark brown, bases of hairs whitish; underparts slightly paler. Wings long and narrow.

Similarities
Big Free-tailed has ears connected at base, larger. Western Mastiff is larger.

Habitat
Caverns, buildings.

Habits
Congregates in huge colonies; migrates south in winter.

Remarks
The common western free-tailed bat famed for its large cave colonies at Carlsbad, New Mexico.

Range
All sw. states, s. from sw. Oreg., n.-cen. Nev., n. Colo., and s. S.Dak.

Fig. 19

Brazilian Free-tailed Bat

WESTERN MASTIFF BAT
Eumops perotis   *Fig. 20*

Description
Size: forearm, 2⅞–3⅛ in. (7.1–7.9 cm); 30 teeth. Largest of all U.S. bats. Ears large, projecting forward and united above rostrum. Tail with distal ½ free, extends well beyond membrane. Upperparts sooty-brown, slightly paler below.

Similarities
All other free-tailed bats are smaller.
Habitat
Roosts in small colonies in cracks and small holes, seeming to prefer man-made structures; uncommon.
Habits
Leaves roost only after full darkness; rarely seen on wing.
Range
In s.-cen. and s. Calif.; s. Ariz.; sw. N.Mex.; sw. Tex. and into Mexico.

Fig. 20

Western Mastiff Bat

Edentates
Order Edentata

Members of this chiefly tropical group, which includes sloths, anteaters, and armadillos, have incomplete teeth in one or another sense. The dentition is deciduous only in armadillos. Teeth are single-rooted, lack a covering enamel layer, and are absent from the anteriormost parts of the jaws.

ARMADILLOS
Family Dasypodidae

NINE-BANDED ARMADILLO
Dasypus novemcinctus

55:8
Fig. 14

Description
Size: head and body, 15–17 in. (38.1–43.2 cm); tail, 14–16 in. (35.6–40.6 cm); weight, to 17 lb. (7.7 kg) (usually ⅔ as much); 28–36 teeth, 00-7/00-7, degenerate, premolars indistinguishable from molars. Feet strong, 4 toes on forefoot, 5 toes on hind foot; all digits clawed. Body, tail, top of head covered with "horny" (actually keratin) material consisting of 9 flexible bands in the center; underparts and ears naked. Short, scattered hairs grow between the plates. Upperparts tan or yellowish.
Habitat
Low brushy areas, open woodlands and rock outcrops.
Habits
Most active at night; a burrower; sometimes seen along roads and often killed by automobiles.
Voice
Grunt.
Food
Insects, other small invertebrates, some vegetable matter.

Remarks
Only "armored" mammal; unique.
Range
Extreme se. N.Mex.; Tex., except panhandle.

Hares, Rabbits, and Pikas
Order Lagomorpha

This group is geologically old. Fossilized remains have been found in Eocene strata. The order is characterized by the presence of four upper incisors, instead of two, as in rodents. A longitudinal groove occurs on the anterior face of each of the first upper incisors. All lagomorphs are herbivorous, eating mainly leaves and nonwoody stems.

PIKAS
Family Ochotonidae

Characteristics that differentiate this family from the hares and rabbits (Leporidae) are hind legs scarcely longer than forelegs; short ears, about as wide as high; and dentition 2032/1023.

PIKA
Ochotona princeps   *Fig. 21*

Description
Size: head and body, 6⅛–8½ in. (15.7–21.6 cm); weight, 3⅗–4½ oz. (105–130 g). Skull flattened; interorbital region wide. Ears round, shorter than head, and with white edges. Tail not visible. Upperparts grayish to cinnamon-buff; underparts washed with buff.
Habitat
Mountain heights and rocks, particularly talus slopes and lava beds; presence recognizable by small piles of fresh hay in rock slides.
Habits
Active during day, inactive at night.
Voice
Series of peculiar short squeaks, "*chickck-chickck*" (Hall & Kelson).
Range
In se. B.C.; e. of Cascade-Sierras in Wash.; Oreg.; and Calif.; e. through w. ½ Mont.; w. ⅔ Wyo., including high mts. of Nev.; Utah; Colo.; cen. N.Mex.
Other name
Cony.

Fig. 21   Pika

HARES AND RABBITS
Family Leporidae

This family is characterized by members with hind legs longer than forelegs, ears longer than wide, soft fur, and short cottony tails that are usually white underneath. The larger species are called hares or jackrabbits, the smaller ones rabbits or cottontails. Their fur is generally brownish or grayish above (sometimes white in the north or in winter) and paler or white below. They have twenty-eight teeth, formula 2033/1023, with one pair of upper incisors directly behind the other.

These vegetarian mammals occupy all habitats from desert to moist forest, from sea level to above timberline, and from Mexico to the limit of Arctic land. They do not hibernate. Hares generally prefer open country, while rabbits like shrubby cover. By breeding early and often they withstand the heavy tolls levied by fox and owl, gun and auto. Many species have periodic fluctuations in population numbers.

COTTONTAILS
Genus *Sylvilagus*

Cottontail, Brush, and Pygmy rabbits are smaller and have shorter ears and hind legs than do hares; they usually seek safety by hiding and rarely feed far from their cover. A fur-lined nest is especially constructed for the four or five altricial babies that are born, after a gestation period of about twenty-eight days, naked with their eyes closed and helpless. They are weaned in less than three weeks. Females often breed after nine months of age and produce several litters a year.

PYGMY RABBIT
Sylvilagus idahoensis

Description
Size: head and body, 8½–11 in. (21.6–27.9 cm); ear, 2¼–2½ in. (5.7–6.4 cm); weight, ½–1 lb. (228–455 g). Smallest of rabbits. Tail short, dusky above and gray below. Slate-gray with pinkish tinge.
Similarities
Cottontails are larger, with conspicuous white tail.
Habitat
Sagebrush regions at lower elevations; lives in burrows; requires relatively moist soils and clumps of rabbit brush (*Chrysothamnus*) and sagebrush (*Artemesia*).
Remarks
Difficult to see in dense cover.
Range
In se. Wash.; e. Oreg.; sw., s., se. Idaho; extreme ne. Calif. and Lake Tahoe area of Sierras; n. Nev.; w. Utah.

Note: Some authors use the name *Brachylagus idahoensis*.

BRUSH RABBIT
Sylvilagus bachmani

56:14

Description
Size: head and body, 11–13 in. (27.9–33 cm); ear, 2–2⅜ in. (5.1–6.6 cm); weight, 2½–3½ lb. (1.1–1.6 kg). Ears and tail relatively small for rabbit. Body uniformly dark brown or brownish-gray; tail whitish underneath; hair at midventer part gray at base.

Similarities
Desert Cottontail is larger, ears and legs longer; grayish. Black-tailed Jackrabbit is larger, prefers open areas.
Habitat
Heavy brush, chaparral.
Habits
Feeds near cover in early morning and evening.
Range
Coastal, s. from nw. Oreg. to Baja Calif.; also in Sierras, e. of Great Central Valley in Calif.

DESERT COTTONTAIL
Sylvilagus audubonii   **55:1, 56:15**

Description
Size: head and body, 12–15 in. (30.4–38.1 cm); ear, 3–4 in. (7.6–10.1 cm); weight, 1⅖–2¾ lb. (0.62–1.2 kg). Long hind legs, long ears, sparseness of hair on ears, shortness of hair on feet. Color is pale grayish with yellowish cast.
Similarities
Brush Rabbit is smaller, dark brown, unicolored ears. Nuttall's has shorter ears; Eastern Cottontail is larger, ears shorter. Snowshoe Hare is dark brown or white, and seen in high mountains. Jackrabbits are larger, and found in open areas. Pygmy Rabbit is smaller, and seen in heavy brush.
Habitat
Arid lowlands and valleys with some brushy cover; common.
Range
Desert W., s. from s. Mont. and sw. N. Dak.; w. of 100th meridian, to Mexico; also s. and cen. Calif. coast through Central Valley; s. Nev.; s. Utah; Ariz.; N. Mex.

EASTERN COTTONTAIL
Sylvilagus floridanus   **55:1, 56:12**

Description
Size: head and body, 14–17 in. (35.6–43.2 cm); ear, 2½–3 in. (6.4–7.6 cm); weight, 2–4 lb. (0.9–1.8 kg). Grayish-brown with underside of tail cotton-white; feet whitish; nape rusty; rump fur more brownish than grayish.
Similarities
Desert is smaller, ears longer; not found in timber. Snowshoe Hare is larger; dark brown in summer, white in winter. All native cottontail rabbits have rump fur that is more gray than brown.
Habitat
Brush, brier patches; likes taller cover than does Desert Cottontail.
Habits
Often feeds in late afternoon or early morning; active at night.
Range
In Wyo.; ne. Colo.; cen. and s. N. Mex.; extreme se. Ariz.; w. Tex.; introduced into cen. and e. Wash. and ne. Oreg.

NUTTALL'S COTTONTAIL
Sylvilagus nuttallii   55:1

Description
Size: head and body, 12–14 in. (30.4–35.6 cm); ear, 2⅛–2⅜ in.
(5.6–6.6 cm); weight, 1½–3 lb. (0.7–1.4 kg). Hind feet densely
haired; ears with dense and long hairs over inner surfaces. Grayish
with yellowish cast dorsally, venter and tail white.

Similarities
Desert is found in valleys, low deserts; ears longer. Snowshoe Hare
is brown. Pygmy Rabbit is smaller; found in low deserts.
Jackrabbits are larger, ears longer.

Habitat
Mountains; seldom below area of pines; in north range, occurs in
brush and rocks of sagebrush areas principally, but also in
timbered areas and thick cover along streams and hillsides.

Range
Mountains of all w. states, e. of Cascade-Sierra crest.

Other name
Mountain Cottontail.

HARES AND JACKRABBITS
Genus *Lepus*

Hares and jackrabbits are larger and have longer ears and hind
legs than do cottontails; they prefer to seek safety in flight, rather
than by hiding. After a gestation period of thirty to forty-three
days, the three to eight precocial young of *Lepus* are born fully
furred and with their eyes open. There are usually two or more
litters a year.

SNOWSHOE HARE
Lepus americanus   55:3, 56:16

Description
Size: head and body, 13–18 in. (33–45.7 cm); ear, 3½–4 in. (8.9–
10.2 cm); weight, 2–4 lb. (0.9–1.8 kg). Feet large, long-furred,
permits rapid travel over snow; ears small for a hare, barely longer
than head. Summer: brown above, white below; tail dark above;
feet brownish, not whitish as in cottontails. Winter: white all over
with bases of hairs dark; ear tips dark; tail all white.

Similarities
Arctic's tail always white, fur (tip to skin) all-white in winter.
White-tailed Jackrabbit is larger, ears longer; tail always white.
Cottontails are brownish throughout year, feet whitish, nape patch
rusty. Black-tailed Jackrabbit has black stripe down rump and on
top of tail.

Habitat
Swamps, forests, brush of North and mountains; laurel and
rhododendron thickets.

Habits
Spends day in cover; feeds at night in open; uses own trails; easily
snared.

Range
Alaska, s. of Arctic Slope; nw. Canada, s. through Wash.; Oreg.;
Idaho; Mont.; n. Calif. Cascades; n. and w. Wyo.; ne. Utah; w.
Colo.; n.-cen. N. Mex.

Other name
Varying Hare.

ARCTIC HARE
Lepus arcticus   *Fig. 22*

Description
Size: head and body, 17–24 in. (43.2–61 cm); ear, 3¾–4½ in. (9.5–11.4 cm); weight. 6–12 lb. (2.7–5.4 kg). Upperparts gray in summer in south range, white throughout year in north range. In winter, white in all subspecies except tips of ears black, fur white to base; tail always all-white.

Similarities
Snowshoe is smaller; tail brown in summer, fur dark at base in winter.

Habitat
Barren grounds, tundras, rocky slopes north from tree limit.

Habits
Frequently hops on hind feet without touching forefeet to ground.

Range
Arctic Slope of Canada and offshore islands of Arctic Ocean.

Fig. 22

winter

summer

Arctic Hare

ANTELOPE JACKRABBIT
Lepus alleni   **55:2**
                                                           Fig. 23

Description
Size: head and body, 16–21 in. (40.6–53.3 cm); ears, approximately ¼ body length, very broad; weight, 6–13 lb. (2.7–5.9 kg). Above dark buff sharply outlined by whitish or iron-gray sides and rump; belly white.

Similarities
All cottontails and the Brush Rabbit have smaller body and ears, brownish or grayish on sides. Black-tailed is brownish on sides and hips.

Habitat
Deserts, dry valley slopes, and mesas far from water; conspicuously common.

Fig. 23

Antelope Jackrabbit

Habits
When frightened, can draw the loose skin of either side over its back, thereby shifting almost completely to one side or the other the dark dorsal area, leaving a flashing white side always toward the intruder.

Remarks
Has a striking color pattern and huge ears.

Range
In sw. Ariz.

Note: A population of related Mexican jackrabbits that barely reaches extreme southwestern New Mexico is of a distinct species, *Lepus callotis*.

WHITE-TAILED JACKRABBIT
Lepus townsendii

55:2
Fig. 24

Description
Size: head and body, 18–22 in. (45.7–55.9 cm); ear, 5–6 in. (12.7–15.2 cm); weight, 5–8 lb. (2.3–3.6 kg). Large; ears noticeably longer than head. Summer: upperparts grayish-brown. Winter: white to pale gray. Tail all-white below and above or with dusky or buffy middorsal stripe that does not extend onto back. Ear tips black.

Similarities
Snowshoe Hare is smaller, with shorter ears; dark brown in summer; prefers forests. Black-tailed Jackrabbit has black on top of tail. Cottontails are smaller; do not turn white in winter.

Habitat
Open country and exposed mountain slopes; sagebrush and grassy areas at lower elevations.

Remarks
This species is sometimes called the Snowshoe Rabbit, presumably because it has at least 2 annual molts. It should not be confused with Snowshoe Hare.

Range
In s. Alta.; s. Sask.; sw. Man., s. through mountain and desert states as far as s. Utah; s. Colo.; sw. Kans.; extreme n.-cen. N. Mex.; also e. Wash.; e. Oreg.; Idaho; e. of Cascade-Sierra crest to cen. Calif.; extreme w. and n. Nev., plus ne. ¼.

Fig. 24

summer

winter

White-tailed Jackrabbit

BLACK-TAILED JACKRABBIT
Lepus californicus   55:2, 56:13

Description
Size: head and body, 17–21 in. (43.2–53.3 cm); ear, 6–7 in. (15.2–17.8 cm); weight, 3–7 lb. Large. Upperparts gray to blackish; tail with black middorsal stripe extending onto back and brownish below; ears large and tipped with black on outside (recognition clue), flanks white.

Similarities
Antelope Jackrabbit has no black on ears and white sides; White-tailed Jackrabbit is whitish in winter, usually no black on top of tail. Snowshoe Hare is white in winter. Cottontails and Brush Rabbit are much smaller; ears not black-tipped.

Habitat
Widespread in grasslands and open areas; common.

Habits
Active in early morning and early evening, when feeding.

Range
In se. Wash.; Oreg., e. of Willamette Valley; s. Idaho; Calif., except extreme nw. coast; Nev.; Utah, except extreme ne. corner; w. and e. Colo. to w. Nebr.; s. through Ariz., N. Mex., Tex.

Rodents
Order Rodentia

Rodents are small- to medium-sized gnawing mammals distinguished by having only four incisors, two above and two below, that continue to grow throughout life. In place of canines, rodents have a conspicuous space between the incisors and the grinding cheek teeth. Most rodents have four toes on each front foot, five on each rear foot. An extremely successful order of mammals, they are widely and abundantly distributed from the tropics to the tundras, from the sea beach to alpine meadows. Rodents have adapted to almost all ecological situations, except true flight and marine waters. Various species are terrestrial, scansorial, arboreal, volant, saltatorial, fossorial, and semiaquatic. In genera, species, and individuals they probably outnumber any other mammalian order three to one.

MOUNTAIN BEAVER
Family Aplodontidae

The oldest known group of living rodents (records from the Eocene), the aplodontids are thought by some to include the ancestors of all later rodents which evolved after the Paleocene epoch. A single species exists today along the West Coast of North America. Each foot has five toes, but thumbs are small and clawless. Females have three pairs of mammae and produce two to four young. Gestation is twenty-eight to thirty days. Food consists of ferns, tree leaves, and various plants.

MOUNTAIN BEAVER
Aplodontia rufa   *Fig. 25*

Description
Size: head and body, 12–17 in. (30.4–43.2 cm); tail, 1–1⅕ in. (2.5–3 cm); weight, 2–3 lb. (0.9–1.4 kg). Body compact; legs short,

stout; ears and eyes small and rounded; tail small, not readily visible. Fur pinkish-cinnamon to blackish-brown (graying with age), relatively uniform; small white spot below ear.

Habitat

Forests and densely vegetated thickets, mostly at moist lower elevations.

Habits

Makes burrows extensively into moist earth; seldom seen more than a few yards from cover.

Range

Pacific Nw. coastal forests from s.-cen. B.C. to nw. Calif.; also down Cascade-Sierras to cen. Calif.

Other name

Seweller.

Fig. 25

Mountain Beaver

SQUIRRELS
Family Sciuridae

This family includes the marmots; woodchucks; prairie dogs; ground, tree, and flying squirrels; and chipmunks. All have hairy, sometimes bushy tails. The tooth formula is 1023/1013, except for the chipmunks and Red Squirrel, for which the formula is 1013/1013. Incisors are yellow, except for the Woodchuck and Marmot where they are white. All species have four toes on the front foot, five on the rear. All are active by day, except the Flying Squirrel, which emerges only at night. All nest in the ground in burrows or under logs or rocks except the tree-nesting flying squirrels and tree squirrels. Most members of the family are capable of vocalization, such as a high-pitched whistle, barking, or chattering.

CHIPMUNKS
Genus *Eutamias*

LEAST CHIPMUNK
Eutamias minimus

56:10, 60:8

Description

Size: head and body, 3½–4½ in. (8.9–11.4 cm); tail 3–4½ in. (7.6–11.4 cm); weight, 1–1⅘ oz. (30–52 g). Small. Skull has high, narrow braincase. Washed-out yellowish with pale fulvous dark stripes to rich grayish-fulvous with black stipes; stripes continue to base of tail. Tail long and round, grayish to lemon-yellow underneath; ear almost unicolored on back side.

Similarities

Lodgepole and Colorado have ears blackish in front, whitish behind. Townsend's is larger, stripes indistinct. Cliff has indistinct side stripes.

Habitat

Brushy, semiopen desert areas; high elevation coniferous forests; and northern forests.

Habits
When running, carries tail straight in air.
Range
In s. Yukon; s. Mackenzie; ne. B.C.; Alta., Sask.; e.-cen. Wash., e. Oreg.; s. Idaho; e. Mont.; Wyo.; ne. Calif., down Sierras to cen. Calif.; n. and e. Nev.; ne., e., cen. Utah; w. ⅔ Colo.; n.-cen. N.Mex.; Ariz., Kaibab Plateau, White Mts.

Note: A closely related species, the **YELLOW PINE CHIPMUNK,** *Eutamias amoenus* (**57:10**), ranges mostly to the west of *E. minimus* in southern British Columbia, southwest Alberta, Washington, Idaho, west Montana, Oregon, northwest Wyoming, northeast California, northern Nevada, and northern Utah. No single feature separates the two species at all places, but at any one place where they occur together there are real differences. This and other cases noted in other species accounts below are not readily resolved by the field observer. Detailed study with the animals in hand is usually needed.

TOWNSEND'S CHIPMUNK

Eutamias townsendii **57:13, 60:8**

Description
Size: head and body, 5⅓–6 in. (13.5–15.2 cm); tail, 3⅗–4⅗ in. (9.1–11.7 cm); weight, 2½–4¼ oz. (70–123 g). Skull massive, broad, flattened; braincase smallish. Tail relatively slender, sparsely haired, reddish-brown or tawny underneath. Color variable from tawny or olivaceous with obscure stripes and tawny underparts to grayish-ochreous with conspicuous stripes and white underparts. Light and dark markings usually weakly contrasting, tending to blend at borders; in general has dull yellowish or grayish light stripes along sides and back, and the dark stripes are blackish. Backs of ears are bicolored, blackish anteriorly, grayish posteriorly.
Similarities
Least, Colorado, Lodgepole are smaller, stripes distinct.
Habitat
Dense forests.
Range
Nw. Coast, w. of Cascade crest, s. B.C. to n. Calif., and down Sierras to cen. Calif.; also extreme ne. Calif. with overlap into extreme s.-cen. Oreg.

Note: This species has been recently divided into four species, the others being the **YELLOW-CHEEKED CHIPMUNK,** *Eutamias ochrogenys*; **ALLEN'S CHIPMUNK,** *E. senex*; and the **SISKIYOU CHIPMUNK,** *E. siskiyou*. These formerly were subspecies of *E. townsendii*. They occupy the southern part of the overall range given above, and cannot be readily distinguished in the field. Also in northwest California another species occurs, the **SONOMA CHIPMUNK,** *E. sonomae*, which differs from the above in being paler, and in having longer legs and ears and a broader, longer, and bushier tail.

MERRIAM'S CHIPMUNK

Eutamias merriami **60:8**

Description
Size: head and body, 5–6 in. (12.7–15.2 cm); tail, 4–5½ in. (10.1–14 cm); weight, 1⅘–3$\frac{1}{10}$ oz. (53–88 g). Feet and ears long, slender; ears sparsely furred on convex surfaces in summer. Grayish-brown,

with indistinct stripes. Back side of ear unicolored grayish or buffy; all stripes on head brownish; edge of tail hairs white to light buff.

Similarities
Colorado has distinct white stripes.

Habitat
Mixed oak and pine forests and chaparral slopes of foothills and brushy areas.

Range
Along sw. Calif. coast, s. from Bay Area; also in Sierras, e. of San Joaquin Valley through s. Calif. mts. to Baja Calif.

COLORADO CHIPMUNK
Eutamias quadrivittatus   **56:11, 60:8**

Description
Size: head and body, 5 in. (12.7 cm); tail, 3⅛–4½ in. (8.1–11.4 cm); weight, 1⅗–2⅓ oz. (50–65 g). Head, rump, and sides gray with fulvous wash on sides; ears black in front, white behind; tail fulvous below, tipped with black, bordered with white or pale fulvous.

Similarities
Least is smaller, dorsal stripes continue to base of tail. Townsend's, Merriam's, and Cliff have indistinct side stripes.

Habitat
Ponderosa pine belt and higher on ridges and rocky slopes; common.

Range
In se. Utah; extreme n. Ariz.; w. ½ Colo.; n. N.Mex.

Other name
Say's Chipmunk.

Note: The **UINTA CHIPMUNK,** *Eutamias umbrinus (Fig. 25a),* a similar species, occurs mainly to the west and north of *E. quadrivittatus* in California (White Mountains), Nevada, north Arizona, Utah, northwest Wyoming, north Colorado, and nearby areas. The **PALMER'S CHIPMUNK,** *E. palmeri,* another similar species, occurs only near Charleston Peak in south Nevada. The **RED-TAILED CHIPMUNK,** *E. ruficaudus (Fig. 25a),* of northeast Washington, north Idaho, southeast British Columbia, southwest Alberta, and west Montana, has deep tawny upperparts and sides and the tail of ochreous tawny beneath. The **GRAY-COLLARED CHIPMUNK,** *E. cinereicollis,* of southeast Arizona, and southwest New Mexico, resembles *E. quadrivittatus* but is grayer on the shoulders. The **GRAY-FOOTED CHIPMUNK,** *E. canipes,* of south New Mexico and the Guadaloupe Mountains of Texas, resembles *E. cinereicollis.*

Fig. 25a

Red-tailed Chipmunk   Uinta Chipmunk

CLIFF CHIPMUNK
Eutamias dorsalis   **57:12, 60:8**

Description
Size: head and body, 5–6 in. (12.7–15.2 cm); tail, 3⅘–4½ in. (9.7–
11.4 cm); weight, 1⁹⁄₁₀–2³⁄₁₀ oz. (55–67 g). Upperparts smoke-gray;
dorsal stripe moderately distinct, dark, in some stages obsolete;
other stripes indistinct.

Similarities
Colorado and Least have distinct dark and light stripes.

Habitat
Lower edge of ponderosa pine forest in juniper-pinyon pine belt.

Range
In w. Nev.; extreme se. Oreg.; cen.-sw. Utah, with extension ne.
into extreme sw. Wyo. and ne. Colo.; Ariz., in broad diagonal band
from nw. to se.; sw. N.Mex.

LODGEPOLE CHIPMUNK
Eutamias speciosus   **60:8**

Description
Size: head and body, 4½–5¼ in. (11.4–13.3 cm); tail, 2⅓–4½ in.
(5.8–11.4 cm); weight, 2¹⁄₁₀ oz. (60 g). Head and shoulders grayish
to brownish, top of head brown; dark stripes more blackish than
reddish, lateral dark stripes indistinct, light stripes conspicuously
white, contrasting sharply with dark stripes; ears white behind,
black in front; edges of tail fur buffy.

Similarities
Townsend's and Merriam's are larger, with indistinct lateral
stripes. Least is smaller.

Habitat
In chaparral and dense stands of lodgepole pines in damper, more
sheltered basins of the Sierra Nevada.

Range
Sierra Nevada Mts. from n. of Lake Tahoe to s. Calif.; Mt. Pinos,
San Gabriel, and San Bernardino ranges.

Note: Two other, similar species of chipmunks that occur in the
same general area are the **LONG-EARED CHIPMUNK**, *Eutamias
quadrimaculatus*, (*Fig. 25b*), and the **PANAMINT CHIPMUNK**, *E.
panamintinus* (*Fig. 25b*). *E. quadrimaculatus* is slightly larger than
E. speciosus and resembles *E. townsendii* somewhat, though it has
longer ears. *E. panamintinus* occurs along the California-Nevada
border to the southeast of the range of *E. speciosus*.

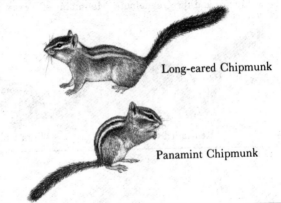

Long-eared Chipmunk

Fig. 25b

Panamint Chipmunk

MARMOTS
Genus *Marmota*

WOODCHUCK
Marmota monax

56:6, 60:2

Description
Size: head and body, 16–20 in. (40.6–50.8 cm); tail, 4–7 in. (10.1–17.8 cm); weight, 5–10 lb. (2.3–4.5 kg). Large; heavy-bodied; ears small, legs and tail short. Above dark brown to yellowish-brown; grizzled; longer guard hairs white-tipped; paler, sometimes rusty, below. No white between eyes, sides of neck same color as back; feet dark brown to black.
Similarities
Hoary Marmot has black and white on head and shoulders.
Habitat
Dry woods and adjacent open spaces; brushy ravines, rocky slopes, fields.
Habits
Most active in early morning or late afternoon; hibernates in winter, not too deeply. Can climb, swim; often hunted. Main den entrance frequently has fresh dirt about it, especially in spring.
Voice
Shrill whistle.
Range
In e.-cen. Alaska, s. through s. Yukon; s. Mackenzie; nw., cen., se. B.C.; n. Alta.; n. ¾ Sask.; extreme n. Idaho panhandle and nw. Mont.
Other name
Ground Hog.

YELLOW-BELLIED MARMOT
Marmota flaviventris

57:3

Description
Size: head and body, 14–19 in. (35.6–48.2 cm); tail, 4½–9 in. (11.4–22.9 cm); weight, 5–10 lb. (2.3–4.5 kg). Heavy-bodied. Resembles Woodchuck but grizzled yellowish-brown above with white-tipped hairs; face black, usually with white between eyes behind dark band; sides of neck buffy; feet buff to dark brown (never black); underparts dull yellow.
Similarities
Hoary has conspicuous white and black on head and shoulders.
Habitat, Voice
Same as Woodchuck.
Range
High elevations of Rocky Mts. and Cascade-Sierras, among rocks and talus slopes, from s.-cen. B.C., e. of Cascade-Sierra crest, to s.-cen. Calif.; n. Nev.; Idaho; w. Mont.; nw., s.-cen. Wyo.; mts. and plateaus of Utah; w. ⅔ Colo.; n.-cen. N.Mex.

HOARY MARMOT
Marmota caligata

57:1

Description
Size: head and body, 18–21 in. (45.7–53.3 cm); tail, 7–10 in. (17.8–25.4 cm); weight, to 30 lb. (2.5–4.5 kg). Mixed black and white on head and shoulders, body brownish with yellowish tinge; feet black or blackish-brown, venter whitish, tail often reddish-

brown; frequently with white on forefeet. Distinct narrow black bar behind white nose.

Similarities
Woodchuck, Yellow-bellied Marmot have no black and white on head and shoulders.

Habitat
Rock slides and talus slopes near meadows in high mountains.

Habits
Hibernates in winter.

Range
In nw., sw., cen., se. Alaska; Yukon; B.C. and Vancouver Is.; sw. Alta.; Wash., Cascades; extreme ne. Idaho panhandle; w. Mont., Bitterroot Mts.

ANTELOPE SQUIRRELS
Genus *Ammospermophilus*

WHITE-TAILED ANTELOPE SQUIRREL
Ammospermophilus leucurus   **57:9**

Description
Size: head and body, 5½–6½ in. (14–16.5 cm); tail, 2–3 in. (5.1–7.6 cm). Upperparts brownish or cinnamon; 2 white stripes on back extend from sides to hips, no dark stripes; tail broadly white or whitish below, bordered with fuscous-black. Tail carried curved over back, exposing white fur below. In winter pelage more grayish.

Habitat
Scrub and juniper vegetation of arid deserts and foothills.

Range
In sw. Idaho; se. Oreg.; extreme ne., se. ¼ Calif.; nw., w., s. ½ Nev.: w., s. ½, e. Utah; n. Ariz.; far w. Colo.; arm into nw. N.Mex.

Note: Three closely related species are **NELSON'S ANTELOPE SQUIRREL**, *Ammospermophilus nelsoni*, of south-central California; **HARRIS' ANTELOPE SQUIRREL**, *A. harrisii*, of southern Arizona and southwest New Mexico; and the **TEXAS ANTELOPE SQUIRREL**, *A. interpres*, of southern New Mexico and west Texas. *A. harrisii* lacks the median whitish area of the other species.

GROUND SQUIRRELS
Genus *Spermophilus*

RICHARDSON'S GROUND SQUIRREL
Spermophilus richardsonii   *Fig. 26*

Description
Size: head and body, 7¾–9½ in. (19.7–24.1 cm); tail, 2–4½ in. (5.1–11.4 cm). Upperparts drab or smoke-gray more or less shaded with fuscous and dappled with cinnamon-buff; underside of tail clay-color, cinnamon-buff, sayal brown, or ochreous-buff; belly pale buff or whitish. Tail bordered with white or yellowish.

Similarities
Spotted has distinct spots. Thirteen-lined has stripes on body. Belding's usually has brownish median area of back with tail rufous below.

Habitat
Plains, in grasses and sagebrush.
Range
In se. B.C.; s. Alta.; sw. Sask.; Mont., except nw. and se. corners; e.-cen. Idaho; w. ⅔ Wyo.; also an island embracing se. Oreg.; sw. Idaho; and ne. Nev.
Other name
Picket Pin.

Note: A similar species, the **UINTA GROUND SQUIRREL,** *Spermophilus armatus (Fig. 26a)*, occurs also in eastern Idaho, southwestern Montana, western Wyoming, and north-central Utah. The underside of the tail, however, is grayish rather than buffy.

Fig. 26

Richardson's Ground Squirrel   Northern Pocket Gopher, p. 318

Franklin's Ground Squirrel, p. 311

Mexican Ground Squirrel, p. 311

Fig. 26a

Uinta Ground Squirrel

TOWNSEND'S GROUND SQUIRREL
Spermophilus townsendii   **57:4**
Description
Size: head and body, 5½–7 in. (14–17.8 cm); tail, 1½–2⅓ in. (3.8–5.8 cm). Skull has stout rostrum, its sides nearly parallel; supraorbital borders slightly elevated. Upperparts smoke-gray shaded with pinkish-buff, with pale spots; underparts whitish: tail short and underside reddish. Animal blends well with surroundings.
Similarities
White-tailed Antelope Squirrel has stripes on sides, undertail white. Belding's Ground Squirrel is larger, undertail reddish.
Habitat
Sagebrush and grasses of arid valleys.
Range
Great Basin drainage of se. Wash.: e. Oreg.; s. Idaho; Nev.; w. Utah.

GROUND SQUIRRELS

BELDING'S GROUND SQUIRREL
Spermophilus beldingi   57:5

Description
Size: head and body, 8–9 in. (20.3-22.9 cm); tail, 2⅛–3 in. (5.6–7.6 cm). Upperparts smoke-gray mixed with reddish-brown; broad brownish or grayish band down back. Forehead pinkish-cinnamon; underparts grayish washed with pinkish-cinnamon, most pronounced on pectoral region, forelegs, forefeet, and hind feet. Tail slightly darker than back on upper side with black tip and white to buff border, distinctly reddish beneath.
Similarities
Townsend's has smaller tail, not fulvous beneath. Richardson's tail is pale buff or clay-colored beneath.
Habitat
Fields and meadows of arid upland valleys and mountain slopes.
Range
In e. Oreg.; sw. Idaho; ne. Calif.; extreme nw. corner and n.-cen. Nev.

COLUMBIAN GROUND SQUIRREL
Spermophilus columbianus   57:8

Description
Size: head and body, 9⅘–11⅘ in. (25–30 cm); tail, 3¹/₁₀–4⅓ in. (8–11.6 cm). Hind foot longer than 1½ in. (4.3 cm). Skull has short and broad rostrum. Nose and face tawny or hazel; neck gray; upperparts buff or brownish with buffy spots.
Similarities
Townsend's, Richardson's, and Belding's Ground Squirrels are unspotted.
Habitat
Grasslands of high valleys.
Range
Se. B.C.; sw. Alta.; e. Wash.; ne. Oreg.; n. Idaho; and w. Mont.

Note: The similar but larger **ARCTIC GROUND SQUIRREL,** *Spermophilus parryii,* inhabits northern British Columbia, Alaska, and northern Canada east to Hudson Bay. Similar species that occur with *S. columbianus* are the **IDAHO GROUND SQUIRREL,** *S. brunneus,* in southwest Idaho, and the **WASHINGTON GROUND SQUIRREL,** *S. washingtoni,* in southwest Washington and northeast Oregon. The hind foot is less than 1½ in. (4.3 cm) in these two species.

THIRTEEN-LINED GROUND SQUIRREL
Spermophilus tridecemlineatus   56:8

Description
Size: head and body, 4½–6½ in. (11.4–16.5 cm); tail, 2½–5¼ in. (6.4–13.3 cm). Skull long, narrow, lightly built. Upperparts marked with series of 13 whitish, longitudinal stripes sometimes divided into rows of nearly square white spots, sometimes continuous; belly whitish. Base color varies from light to dark brown.
Similarities
Chipmunks have stripes on sides of face. Spotted Ground Squirrel has spots but not arranged in rows.
Habitat
Well-drained prairies and areas of short grass; solitary.

310

Habits
Hibernates 6 months each year.
Range
Great Plains states, s. from e.-cen. Alta.; s. ⅓ Sask.; e. of Rockies, s. to cen. N.Mex.; Tex., with long extension through ne. N.Mex. into extreme e.-cen. Ariz.

Note: A similar species, the **MEXICAN GROUND SQUIRREL,** *Spermophilus mexicanus* (*Fig. 26a*), ranges mostly to the south of *S. tridecemlineatus* in southeast New Mexico and southwest Texas as well as in Mexico. There are usually nine rows of nearly square white spots in *S. mexicanus.*

SPOTTED GROUND SQUIRREL
Spermophilus spilosoma   56:9

Description
Size: head and body, 5–6 in. (12.7–15.2 cm); tail, 2¼–3½ in. (5.7–8.9 cm); long, slender, not bushy. Grayish-brown or reddish-brown with squarish white or buff spots on back, sometimes indistinct; belly whitish.
Similarities
Thirteen-lined has distinct stripes on body. Richardson's has no distinct spots.
Habitat
Semiarid prairies and open forests; prefers sandy soil.
Range
Extreme se. Wyo.; w. Nebr., and Kans.; Okla. panhandle; e. ½ and sw. Colo.; se. Utah; n. and se. Ariz.; N.Mex., except extreme w. -cen. part; Tex. panhandle and w. part.

FRANKLIN'S GROUND SQUIRREL
Spermophilus franklinii   *Fig. 26a*

Description
Size: head and body, 9–10 in. (22.9–25.4 cm); tail, to 6 in. (15.2 cm). Head grayish, back tawny-olive or clay, tail blackish and with buff and creamy-white longer hairs, underparts buffy or whitish. Hind foot is 2⅛–2⅓ in. (5.3–5.7 cm) long.
Similarities
Thirteen-lined and Spotted have spots, Richardson's has relatively shorter tail, all 3 are smaller.
Habitat
Dense, high grass or weedy vegetation.
Habits
Eats some animal material as well as plant material; hibernates several months; 5–10 young per litter.
Range
E.-cen. Alta.; s. Sask.; s. Man.; s. through e. Great Plains to Kans., Mo., Ill.; e. to Wis. and w. Ind.

GOLDEN-MANTLED GROUND SQUIRREL
Spermophilus lateralis   57:11

Description
Size: head and body, 6–8 in. (15.2–20.3 cm); tail, 2½–4¾ in. (6.4–12.1 cm). Chipmunklike; head coppery, without stripes; white stripe on each side of back bordered with black from shoulder to hip; remaining back gray, buff, cinnamon, or fawn; tail short and haired, but not bushy.

Similarities
All chipmunks have stripes on sides of face. Red and Spruce Tree
Squirrels, and Douglas' Squirrel are larger, with no contrast
between color of body and head, no lateral white stripes.
Habitat
High mountains, in pine, fir, and spruce forests and chaparral.
Habits
Lives in ground burrows; hibernates in winter.
Range
In se. B.C. and bordering sw. Alta.; extreme se. Wash.; Idaho; w.
Mont.; Oreg., e. of Cascades, except for extension to extreme sw.
coast; n. Calif.; n. Nev.; Utah, extreme nw. area and nw. corner
with arm extending to sw. corner; w. and s. Wyo.; w. ⅔ Colo.; ne.
N.Mex.; Ariz., n.-cen. to e.-cen. in pine belt.

Note: The **CASCADE GOLDEN-MANTLED GROUND SQUIRREL**,
Spermophilus saturatus, is found in the Cascades of British
Columbia and Washington.

ROCK SQUIRREL
Spermophilus variegatus   56:5
Description
Size: head and body, 10–11 in. (25.4–27.9 cm); tail, 7–10 in.
(17.8–25.4 cm). The largest ground squirrel in its range with
bushy tail nearly as long as its body. Upperparts variegated black
and white, slightly mottled, often with buff; head and forepart of
back black (in many subspecies); tail mixed black or brown and
buffy-white, giving slightly mottled effect.
Similarities
California has light gray shoulders and sides of head and dark
band of fur spreading over middle of back. Other ground squirrels
are smaller, tails shorter. Prairie dogs have short tails and are
found in open prairies only.
Habitat
Rocky areas in pinyon pines and junipers of arid regions.
Habits
Often seen foraging in open or perched on top of a boulder;
hibernates in winter, nests beneath boulders.
Range
Sw. deserts from s. Nev., n. Utah, Wyo., except nw. and ne.
corners; w. Okla. panhandle; extreme nw. Tex. panhandle; s. into
Mexico.

CALIFORNIA GROUND SQUIRREL
Spermophilus beecheyi   57:6
Description
Size: head and body, 9–11 in. (22.9–27.9 cm); tail, 5–9 in. (12.7–
22.9 cm) Bushy tail. Head cinnamon or brown; upperparts brown
flecked with whitish or buffy; back with conspicuous dark band
running from head and spreading over middle of back; sides of
neck and shoulders whitish, extending backward in 2 divergent
stripes separated by dorsal triangular area of dark; belly buff.
Similarities
Rock has dark sides of head and shoulders, no dark dorsal band.
Other ground squirrels are smaller; tails shorter, less bushy.
Western Gray Squirrel has white belly, no buff, tail very bushy.
Habitat
Grasslands and oak-savannahs of valleys and foothills, preferring
semiopen country with low vegetation.

Habits
Colonial; burrows often conspicuous on open hillsides; hibernates in winter, aestivates in summer.
Range
W. Coast, s. from s.-cen. Wash.; w. Oreg.; Calif., generally w. of Sierra crest s. of Lake Tahoe; into Baja Calif.; also extreme w.-cen. Nev. (e. of Lake Tahoe area).

ROUND-TAILED GROUND SQUIRREL
Spermophilus tereticaudus **57:7**

Description
Size: head and body, 6⅕–6⅓ in. (15.4–16 cm); tail, 2⅖–4⅕ in. (6–10.7 cm). Upperparts pinkish-cinnamon, underparts white; tail long, slender, and not broadly haired. Hind foot to 1¼–1⅓ in. (3.2–4 cm).
Habitat
Near mesquite or creosote bushes in arid areas.
Habits
Shy and secretive.
Range
Se. Calif., s. Nev., and sw. Ariz.; Mexico.

Note: The **MOHAVE GROUND SQUIRREL**, *Spermophilus mohavensis*, occurs only in the Mohave Desert of southern California. It has a white undersurface on the tail and is larger than the Round-tailed.

PRAIRIE DOGS
Genus *Cynomys*

BLACK-TAILED PRAIRIE DOG
Cynomys ludovicianus **56:7**

Description
Size: head and body, 11–13 in. (27.9–33 cm); tail, 3–4 in. (7.6–10.2 cm); weight, 2–3 lb. (0.9–1.4 kg). Ears small; tail comparatively long, averaging more than ⅕ total length. Skull massive, occipital region ovoid from posterior aspect. Upperparts in summer dark pinkish-cinnamon finely laced with black and buff to give yellowish appearance; tail above like back, terminal ⅓ black or blackish-brown; underparts of body whitish or buffy-white.
Similarities
Rock Squirrel and California Ground Squirrel are smaller, tail longer.
Habitat
Short-grass prairies.
Habits
Builds mounds at burrow entrance, 25–75 ft. (7.6–22.9 cm) apart, each mound 1–2 ft. (0.3–0.6 m) high; frequently sits on mound; does not hibernate.
Range
In e. Mont.; sw. N.Dak.; w. ⅔ S.Dak.; n. and e. Wyo.; e. Colo.; s. ⅔ N.Mex.; extreme sw. Ariz.; w. Tex. and panhandle.

WHITE-TAILED PRAIRIE DOG
Cynomys leucurus   **57:2**

Description
Size: head and body, 11⅘–12⅕ in. (30–31 cm); tail, 1½–2⅓ in.
(4–6 cm). Upperparts pinkish-buff mixed with black, end of tail
white. Hind foot 2⅓–2½ in. (6–6.5 cm). Black-tailed has black-
tipped tail. Ground squirrels are smaller and more slender. Utah
Prairie Dog and Gunnison's Prairie Dog have complimentary
ranges as noted below.
Habitat
At higher elevations than Black-tailed. Short grass in high valleys.
Habits
Colonial but tends to form smaller colonies than Black-tailed.
Range
Mostly to w. of range of Black-tailed, in w. Wyo., nw. Colo., and
ne. Utah.

Note: In southwest Utah there is a related species, the **UTAH
PRAIRIE DOG**, *Cynomys parvidens*, and in southwest Colorado,
southeast Utah, northwest New Mexico, and northeast Arizona is
another species, the **GUNNISON'S PRAIRIE DOG**, *C. gunnisoni*,
(Fig. 26b). Both have white-tipped tails.

Fig. 26b  Gunnison's Prairie Dog

TREE SQUIRRELS
Genera *Sciurus, Tamiasciurus,* and *Glaucomys*

Tree squirrels are primarily arboreal, but are also seen on the
ground. They do not hibernate. Their food, some of which they
store, includes nuts, berries, fruits, seeds, buds, twigs, bark, eggs,
fungi, and insects. Gestation takes forty to forty-five days. The one
to seven young are born naked and blind; there are one to two
litters a year.

ABERT'S SQUIRREL
Sciurus aberti   **56:4**

Description
Size: head and body, 11–12 in. (27.9–30.4 cm); tail, 8–9 in. (20.3–
22.9 cm). Most colorful tree squirrel and only species of this
coloration. Back dark grizzled iron-gray; sides gray to black;
median dorsal stripe indistinct, varying from rufous to chocolate-
brown; tail either all-white or white beneath and broadly bordered
with white; belly white or black; prominent ear tufts black or
blackish.
Habitat
Yellow pine forests of mountains and desert plateaus.
Range
In n. Ariz.; nw. and n.-cen. N.Mex., to cen.-sw. Colo.
Other name
Tassel-eared Squirrel.

Note: The squirrel found north of the Grand Canyon, *Sciurus
kaibabensis*, is considered by some specialists to be a distinct
species. It has an all-white tail.

WESTERN GRAY SQUIRREL
Sciurus griseus   **56:2, 60:11**

Description
Size: head and body, 12 in. (30.4 cm); tail, 10–12 in. (25.4–30.4 cm), very bushy. Upperparts vary from dark gray to light gray with yellowish wash; underparts, from white to gray with tawny suffusion; tail gray but often with blackish or tawny suffusion; feet dark. Easily recognized in its range by white belly and dusky feet.

Similarities
California Ground Squirrel has less bushy tail, shoulders whitish. Douglas' has yellowish or rusty belly.

Habitat
Forested areas of mountains and lowlands among oaks and pines.

Habits
Arboreal, but forages on ground; migratory in years of food scarcity, so numbers may vary locally.

Range
In cen. Wash.; cen. to w. Oreg., except coast; most of Calif. except San Joaquin Valley, ne. corner, and desert areas.

Note: A reddish Mexican species of tree squirrel, the **NAYARIT SQUIRREL**, *Sciurus nayaritensis (Fig. 26c)*, is known from extreme southwestern New Mexico. The **GRAY SQUIRREL**, *S. carolinensis (Fig. 26c)*, has been introduced in some western city parks.

Fig. 26c

Nayarit Squirrel   Gray Squirrel

RED SQUIRREL
Tamiasciurus hudsonicus   **56:1**

Description
Size: head and body, 7–8 in. (17.8–20.3 cm); tail, 4–6 in. (10.2–15.2 cm), bushy. Only predominantly red squirrel. Upperparts rusty-reddish, brownish or grayish, usually purest on sides; paler in winter and with ear tufts; in summer, black line along side; underparts white to grayish-white or faintly tinged with yellow. Ears with long hair on tips in winter.

Habitat
Coniferous forests, less common in hardwoods.

Habits
Leaves piles of cone cuttings on rocks, logs.

Voice
Noisy, ratchetlike scolding chatter, usually heard first, accompanied by flicking of tail and twitching of body.

Range
Alaska, s. of Brooks Range; Yukon; Mackenzie; B.C.; Alta., except se. ¼; s. through Rockies to se. N.Mex.

DOUGLAS' SQUIRREL
Tamiasciurus douglasii   *Fig. 27*

Description
Size: head and body, 7 in. (17.8 cm); tail, 5–6 in. (12.7–15.2 cm).
Color varies individually, geographically, and seasonally.
Upperparts dark olivaceous-brown to brownish-gray, usually with
broad median band of dark rusty to chestnut; distinct black line on
sides in summer, becomes indistinct or absent in winter; underparts
strong buffy-gray through ochreous tones to reddish-orange with
strong black wash; tail below grizzled-rusty bordered with black
and edged with buffy- or white-tipped hairs. Long hairs on tips of
ears in winter.
Similarities
Western Gray Squirrel is larger; gray with white belly.
Habitat
Coniferous forests.
Range
Nw. Coast from sw. B.C.; Oreg., e. to include Blue Mts.; n. Calif.
and in Cascade-Sierras to below midpoint of state.
Other name
Chickaree.

Fig. 27   Douglas' Squirrel

NORTHERN FLYING SQUIRREL
Glaucomys sabrinus   **56:3**

Description
Size: head and body, 5½–6 in. (14–15.2 cm); tail, 4⅓–5½ in.
(10.9–14 cm). Upperparts vary from cinnamon to pecan-brown;
tail above cinnamon to fuscous, even blackish; underparts white or
creamy-white at tips, lead-colored at base; sides of head and face
gray, often washed with buff, cinnamon, or fuscous.
Habitat
Conifers and mixed hardwood forests.
Habits
Nocturnal, gregarious; can glide up to 125 feet (38.1 m), lands
with an audible thump.
Voice
Whistled, birdlike *tseet;* sparrowlike twitterings.
Remarks
Flying squirrels have a fold of loose, furred skin along their sides
from wrist to ankle; when extended, this enables them to glide (not
fly) from tree trunk to tree trunk.
Range
Alaska, s. of Brooks Range; Yukon; Mackenzie; B.C.; Alta., except
se. ¼; s. through Rockies to se. N.Mex.

POCKET GOPHERS
Family Geomyidae

Chunky, toothy, and big-headed, these brownish rat-sized burrowers are noted for their underslung jaws and external, fur-lined cheek pouches that they use for carrying food to storage. They have twenty teeth, 1013/1013. The long-clawed, sturdy forefeet and the front teeth are adapted for fast and protracted digging. Eyes and ears are small; the mouth closes with the upper incisors outside. The tail is sensitive to touch, nearly naked, and shorter than the body.

Seldom seen aboveground, pocket gophers are easily detected by their fan-shaped mounds of earth, which they push out as they excavate their tunnels. They are unsociable mammals that fight on meeting one another except when mating. They cannot swim and rarely drink. Species are often difficult to differentiate, especially where ranges interdigitate—rarely does more than one species occupy a given area. While most pocket gophers are some shade of brown, color may vary from almost white to nearly black. Gestation is presumed to be about twenty eight days; the litter size is usually one to five with one to three litters a year.

SOUTHERN POCKET GOPHER
Thomomys umbrinus

Description
Size: head and body, 4⅝–5 in. (11.7–12.7 cm); tail, 2–2⅜ in. (5.1–6.1 cm); weight, 2⅝–4½ oz. (75–130 g); mammae, 6. Males larger than females. Color varies from black to almost white, but usually yellowish-brown to deep chestnut.
Similarities
Botta's is larger, prefers lowlands. Other gophers with which Southern might be confused usually occur at much lower elevations.
Habitat
Mountains only.
Range
Huachuca Mts. in Sw.
Other name
Pygmy Pocket Gopher.

TOWNSEND'S POCKET GOPHER
Thomomys townsendii

Description
Size: head and body, 7–7½ in. (17.8– 19 cm); tail, 2–3⅘ in. (5.1–9.7 cm); weight, 8½–10⅛ oz. (240–290 g); mammae, 8. Skull has sphenoidal fissure; ears short, rounded. Two color phases; upperparts either grayish- or slaty-black, underparts lighter but not sharply bicolor. Mouth area, feet, and tail may be whitish. Black patch behind ear same size as or smaller than ear.
Similarities
Resembles somewhat the Southern in appearance, but geographic distributions are different. Northern is smaller; found in high mountains. Botta's is smaller; brownish.
Habitat
River valleys and old lake beds with deep soils.
Remarks
Largest pocket gopher in its range.
Range
In se. Oreg.; s. Idaho, ne. Calif.; n. Nev.

NORTHERN POCKET GOPHER
Thomomys talpoides   *Fig. 26*

Description
Size: head and body, 5–6½ in. (12.7–16.5 cm); tail, 1¾–3 in. (4.4–7.6 cm); weight, 2⅗–4⅗ oz. (75–130 g); mammae, 10. Males larger than females. Skull robust, sphenoidal fissure absent. Color variable, but usually rich dark brown sometimes highly washed with blackish; less often pale grayish to lead-color; underparts paler, usually washed with buff; black patches behind rounded ears about 3 times size of ear.

Similarities
Botta's is found in lower elevations, usually not grayish, and with 8 mammae. Townsend's is larger; found in river valleys. Plains is larger; in lowlands; with distinct groove in anterior face of each upper incisor.

Habitat
High mountains and mountain meadows, in north range also in lowlands, where soil is thin or overgrazed.

Range
In se. B.C.; se. ¼ Alta.; s. ½ Sask.; sw. Man.; e. ⅔ of Wash. and Oreg.; Idaho; Mont.; N.Dak.; Wyo., S.Dak.; extreme ne. Calif.; n. Nev.; n. and e. Utah; w. ⅔ Colo.; extreme n.-cen. Ariz.; nw. N.Mex.

Note: The smaller ranges of two related species lie to west of *Thomomys talpoides.* The **MOUNTAIN POCKET GOPHER,** *T. monticola,* of north-central and east California and the **WESTERN POCKET GOPHER**, *T. mazama,* lies north of *T. monticola* in northern California in a belt through Oregon in the Cascade Mountains and in northwest Oregon. Another and smaller species, the **IDAHO POCKET GOPHER,** *T. idahoensis,* has a small range in southeastern Idaho, southwest Wyoming, and extreme northeast Utah. Another and larger species with a small range is the **CAMAS POCKET GOPHER,** *T. bulbivorus,* of the Willamette Valley area in northwest Oregon.

BOTTA'S POCKET GOPHER
Thomomys bottae

Description
Size: head and body, 4⅘–7 in. (12.2–17.8 cm); tail, 2–3¾ in. (5.1–9.5 cm); weight, 2½–8⅓ oz. (71–235 g), mammae, 8. Color usually some shade of brown. Must be identified according to habitat. Dark patch behind ear about same size as ear.

Similarities
Northern is smaller; with 10 mammae, grayish; found in high mountains. Townsend's is larger; gray. Southern is smaller; in mountains. Plains is larger; with 2 distinct grooves on each incisor.

Habitat
Varied, but usually lowland valleys.

Remarks
May interbreed with the Southern Pocket Gopher. Both vary extremely in coloration and size—small on some desert mountains, large in valleys—and almost white in south-central California deserts, while nearly black along parts of coast.

Range
Extreme sw. Oreg.; Calif.; s. ½ Nev.; extreme w., s., se.-cen. Utah; Ariz.; Wyo.; N.Mex., except ne. corner; w. Tex.

PLAINS POCKET GOPHER
Geomys bursarius

Description
Size: head and body, 5½–9 in. (14–22.9 cm); tail, 2–4½ in. (5.1–11.4 cm); mammae, 6. Larger in northern range, smaller in southern. Teeth with 2 distinct grooves on front of each upper incisor. Color ranges from yellowish-tawny in West, browns toward East. Occasional albinos and melanistic (black) individuals found.

Similarities
Northern and Botta's Pocket Gophers have one distinct groove on front near inside of each incisor.

Habitat
Deep soil of the Great Plains.

Range
In se. Wyo.; s. S.Dak.; e. Colo.; Nebr.; Kans.; Okla.; Tex.; e.-cen. N.Mex.

Note: Two closely related species are the **DESERT POCKET GOPHER,** *Geomys arenarius,* of south-central New Mexico and near El Paso, Texas; and the **TEXAS POCKET GOPHER,** *G. personatus,* of south Texas. Their ranges do not overlap that of *G. bursarius.*

POCKET MICE, KANGAROO MICE, AND KANGAROO RATS
Family Heteromyidae

This family of mostly tiny, small-eared rodents is found only west of the Mississippi. They are long-tailed, nocturnal, burrowing mammals with hind limbs that are much larger than forelimbs, and external fur-lined cheek pouches that they use to carry food to storage. Their tails are nearly as long or longer than the head and body. They have twenty teeth, 1013/1013. They make underground tunnel systems with sleeping, nesting, and food storage chambers. The entrance is an inconspicuous surface hole that is, in some species, plugged during the day. Their principal foods are seeds and greens. They are inactive in cold weather. Populations are subject to periodic fluctuations. The one to eight, usually four, young are born in spring or summer; there are one or more litters a year.

POCKET MICE
Genus *Perognathus*

Pocket Mice are the smallest family members, with moderately long and untufted tails. They vary in color from pale yellowish to dark gray, with paler underparts. Their faces are unmarked, but a buffy lateral stripe separates the darker back from the white belly. These pocket mice are poor jumpers; they inhabit areas of sandy soil, rocks, and gravel. Their voice is a thin, high squeak. The soles of the hind feet are naked.

PLAINS POCKET MOUSE
Perognathus flavescens **58:1**

Description
Size: head and body, 2¼–3¾ in. (5.7–9.5 cm); tail, 2– 2⅗ in. (5.1–6.6 cm); weight, ⅓ oz. (10 g). Pale yellowish; belly white. No distinct yellow spots behind ears.
Similarities
Silky has clear yellow patches behind ears. Hispid is larger.
Habitat
Prairies in sandy soil.
Range
In w. Nebr., w. Kans., w. Okla., n. Tex., e. N.Mex.

Note: A similar species, the **OLIVE-BACKED POCKET MOUSE,** *Perognathus fasciatus* (*Fig. 28*) lies chiefly to the northwest of *P. flavescens* in extreme southeast Alberta, southern Saskatchewan, southwest Manitoba, eastern Montana, western North Dakota, eastern two-thirds of Wyoming, western South Dakota, extreme northeast Utah and northwestern Colorado, along a belt of Colorado east of the mountains to western Nebraska. Another similar species, the **APACHE POCKET MOUSE,** *P. apache* (*Fig. 28*) lies to the west of *P. flavescens* in southeast Utah, southwest Colorado, northeast Arizona, and western New Mexico.

Fig. 28

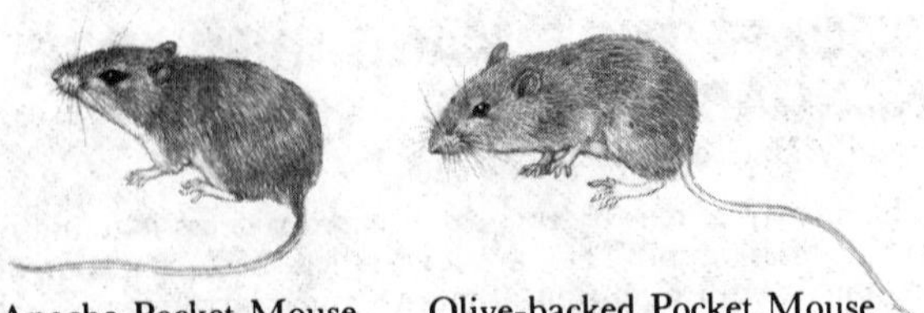

Apache Pocket Mouse — Olive-backed Pocket Mouse

SILKY POCKET MOUSE
Perognathus flavus **58:3**

Description
Size: head and body, 2–2½ in. (5.1–6.4 cm); tail, 1¾–2¼ in. (4.4–5.7 cm); weight. ⅕–⅖ oz. (7–10 g). Skull light; rather short, broad. Upperparts pale yellow, finely sprinkled or lined with black; underparts white, sometimes with faint tawny wash; clear yellow patch behind ears. Tail without crest, slightly shorter than head and body.
Similarities
Plains has no ear patch. Tail longer than 2½ in. (6.4 cm) in all other similar-appearing species.
Habitat
Prairie sandy soils; uncommon.
Range
Extreme se. Wyo.; w. Nebr.; e. ½ Colo.; w. Kans.; w. Okla.; w. Tex. and panhandle; Ariz., ne, and se. corners; N.Mex.

LITTLE POCKET MOUSE
Perognathus longimembris **58:2**

Description
Size: head and body, 2½–2⅘ in. (6.4–7.1 cm); tail, 2–3½ in. (5.1–8.9 cm); weight, ⅕–⅖ oz. (6–10 g). Fur soft, no bristles or spines. Color buffy to grayish-buff; underparts white or pale tawny to

buffy; tail usually bicolored. Two small but distinct patches at base of ear.

Similarities

Great Basin is larger, dark olive-gray; all other similar pocket mice larger and with crested tails. Dark Kangaroo Mouse is brownish, tail swollen in middle.

Habitat

Gravelly desert bench terraces in area of thinly scattered scrub; very numerous.

Remarks

Easily caught in snap traps baited with seeds or rolled oats.

Range

In se. Oreg.; Calif., extreme ne., Central Valley, se.-cen., s. mountains; Nev.; extreme w. Utah; w. Ariz.

Note: Two similar species are the **ARIZONA POCKET MOUSE,** *Perognathus amplus,* in Arizona, chiefly to the east of *P. longimembris* and with nonoverlapping range, and the **SAN JOAQUIN POCKET MOUSE,** *P. inornatus,* with smaller and partly overlapping range in the San Joaquin Valley of central California.

LONG-TAILED POCKET MOUSE

Perognathus formosus

58:6

Description

Size: head and body, 3⅕–3⅘ in. (8.1–9.7 cm); tail, 3⅘–4⅘ in. (9.7–12.2 cm); weight, 1⅓–1⅗ oz. (36–45 g). Long tail conspicuously crested with long hairs on terminal third. Fur soft; no spines or bristles. Skull distinctly crested distally. Upperparts white, sometimes faintly buffy; tail distinctly bicolored. No distinct spot at base of ear, long hairs originating at front edge of ear and reaching nearly across length of ear.

Similarities

Little Pocket Mouse is yellowish, tail not crested. Desert is yellowish. Great Basin has uncrested tail. Spiny has rump with long, spinelike hairs.

Habitat

Rocky, low-desert slopes with gravelly soil.

Range

In w., e., s. ⅓ Nev.; se. Calif.; w. Utah; extreme ne.-cen. Ariz.

CALIFORNIA POCKET MOUSE

Perognathus californicus

58:5

Description

Size: head and body, 3⅕–3⅖ in. (8.1–8.6 cm); tail, 4–5⅘ in. (10.2–14.7 cm); weight, ⅖–⁷⁄₁₀ oz. (12–20 g). Ears much elongated; black to buffy hairs at anterior base of ear nearly as long as ear. Skull with markedly vaulted braincase. Pelage markedly hispid; strong white and spinelike hairs on rump and flanks; tail crested, longer than head and body. Upperparts brownish-gray flecked with fulvous; underparts and feet yellowish-white; distinct fulvous stripe along sides of body; tail bicolored.

Similarities

Other pocket mice in range have no spines on rump.

Habitat

Coastal; common.

Range

In s. Calif., coastal range from San Francisco Bay to Mexico and San Joaquin Valley region w. of Sierra Nevada.

GREAT BASIN POCKET MOUSE
58:4

Perognathus parvus

Description

Size: head and body, 2½–3 in. (5.1–7.6 cm); tail, 3¼–4 in. (8.3–10.1 cm); weight, ⅗–1¹⁄₁₀ oz. (20–30 g). Skull large, slightly rounded in dorsal profile. Tail long, moderately penicilate, bicolored; fur soft, no bristles or spines. Upperparts pinkish to ochreous buff, thinly overlaid with blackish; underparts white to buff. Some dark hairs inside ears.

Similarities

Long-tailed and Desert have tail distinctly bushy near tip. Little Pocket Mouse is smaller, buffy. Kangaroo mice have white bellies, tails fat at middle.

Habitat

Ponderosa and yellow pines, pinyon-juniper belt, chaparral and sagebrush.

Range

Great Basin drainage of s.-cen. B.C.; e. Wash.; Oreg., except w. ⅓ and ne. corner; s. Idaho.; ne., e.-cen. Calif.; Nev., except s. tip; w. ½ Utah; extreme sw. Wyo.; extreme nw. Ariz.

Note: Two similar species with small ranges in south-central California are the **WHITE-EARED POCKET MOUSE**, *Perognathus alticola*, and the **YELLOW-EARED POCKET MOUSE**, *P. xanthonotus*.

SPINY POCKET MOUSE

Perognathus spinatus

Description

Size: head and body, 3–3⅜ in. (7.6–9.1 cm); tail, 3–4½ in. (7.6–11.4 cm); weight, ⅖–⁷⁄₁₀ oz. (12–20 g). Tail long, with long crest. Skull rather slender, flattened. Pelage harsh, conspicuous white and brown spines on rump, sometimes extending to shoulders. Upperparts brownish to pale buffy-yellow; underparts white or buffy-white; lateral line usually obsolete or very pale; tail brownish above, white below.

Similarities

All other pocket mice in range lack spinelike hairs on rump.

Habitat

Hot, low deserts; common.

Range

Extreme se. Calif.

Note: Another species in extreme southwest California is the **SAN DIEGO POCKET MOUSE**, *P. fallax*, which has a well-marked lateral line and less hispid pelage than does *P. spinatus*.

HISPID POCKET MOUSE
53:8

Perognathus hispidus

Description

Size: head and body, 4½–5 in. (11.4–12.7 cm); tail, 3½–4½ in. (8.9–11.4 cm). Skull large, rostrum robust. Pelage harsh. Distinguished by large size and uncrested tail, shorter than head and body. Upperparts ochreous mixed with blackish hairs, sides usually only slightly paler than back; underparts white; tail tricolored, blackish above, white below, lateral line conspicuous.

Similarities

Other pocket mice are smaller or have crested tail.

Habitat

Prairies; burrow mounds conspicuous on nearly bare ground.

Range
In s.-cen. N.Dak.; w. ⅔ S.Dak.; extreme se. Mont.; far e. Wyo.; e. ½ Colo.; s. to Mexico and e. through Tex.; also far se. Ariz.; sw., e. N.Mex.

Note: Another species of similar size but with longer and hairier (crested) tail and grayer color is **BAILEY'S POCKET MOUSE,** *Perognathus baileyi,* of extreme southern California and southern Arizona.

DESERT POCKET MOUSE

Perognathus penicillatus

58:8

Description
Size: head and body, 3–3⅘ in. (7.6–9.7 cm); tail, 3½–4⅘ in. (8.9–12.2 cm); weight, ½–1$\frac{1}{10}$ oz. (14–32 g). Skull moderate size; rostrum robust, high. Crested tail is longer than head and body. No spines on rump, some inconspicuous bristles. Small, indistinct light spot at base of ear. Upperparts yellowish-brown to yellowish-gray; underparts white to buff; lateral line obscure or absent; tail indistinctly bicolored, upper side and tuft dusky, white below to tuft.
Similarities
Long-tailed is slate-gray. Rock Pocket Mouse prefers rocks. Spiny has spinelike hairs on rump. Other pocket mice have uncrested tails.
Habitat
Brushy or shrubby deserts, usually on sand, less often on fine silts and gravel; uncommon.
Range
In se. Calif.; extreme s. Nev.; w. and s. Ariz.; far s. N.Mex.

ROCK POCKET MOUSE

Perognathus intermedius

58:7

Description
Size: head and body, 3–3⅘ in. (7.6–9.7 cm); tail, 3⅕–4 in. (8.1–10.2 cm); weight, ⅖–$\frac{7}{10}$ oz. (11-20 g). Tail long and crested; moderate spines on rump. Skull with well-arched braincase; rostrum slender, depressed. Upperparts usually gray but highly variable from pale buffy-gray to nearly black on some lava outcroppings, sides paler than back; underparts varying from buffy-white to much darker; tail generally much darker distally than proximally, lighter below than above.
Similarities
Desert Pocket Mouse prefers sand. All others within range have no crest on tail.
Habitat
Lava and rocky slopes with gravelly soil and sparse vegetation.
Range
Ariz., except nw. and ne. corners; w. ½ N. Mex.; extreme w. Tex.

Note: A similar species with slightly heavier rump spines and larger size is **NELSON'S POCKET MOUSE,** *Perognathus nelsoni,* which occurs in west Texas, east of the range of *P. intermedius.*

KANGAROO MICE
Genus *Microdipodops*

Kangaroo mice are characterized by tails that appear swollen along the middle, smaller at the base and tip, and never tufted. They are small, silky-haired mammals with large heads for their size. The soles of the hind feet are densely haired.

DARK KANGAROO MOUSE
Microdipodops megacephalus   58:9

Description
Size: head and body, 2⅘–3 in. (7.1–7.6 cm); tail, 2⅔–4 in. (6.7–10.2 cm); weight, ⅖–⅗ oz. (10–17 g). Tail is short-haired, swollen in middle, lacks terminal tuft. Upperparts brownish, blackish, or grayish; underparts lead-colored and hairs white-tipped; tail tip blackish.
Similarities
Great Basin Pocket Mouse has underparts washed with fulvous; tail not swollen in middle. Little Pocket Mouse is yellowish.
Habitat
Sandy desert soils with sagebrush.
Range
In se. Oreg.; nw. and ne.-cen. Nev.; far w. Utah.

Note: The **PALE KANGAROO MOUSE**, *Microdipodops pallidus*, is a similar species with a smaller range to the south of that of *M. megacephalus*.

KANGAROO RATS
Genus *Dipodomys*

Kangaroo rats all have long tails tufted at the tip, very long hind legs, and distinct facial markings. They prefer arid or semiarid country and easily worked soil. The belly is always white and the soles of the hind feet are moderately haired.

PANAMINT KANGAROO RAT
Dipodomys panamintinus   60:3

Description
Size: head and body, 5 in. (12.7 cm); tail, 6⅖–7⅗ in. (16.3–19.3 cm); weight 2³⁄₁₀–3³⁄₁₀ oz. (64–94 g). Head with light cheek patches and large white spots just behind ears. Hind feet bear 5 toes. Dark ventral tail stripe running to tip. Lower incisors awl-shaped.
Similarities
Chisel-toothed has lower incisors chisel-shaped, prefers sagebrush and greasewood. Ord's is smaller, tail shorter. Merriam's is smaller, with 4 toes on hind feet. Desert is larger, paler, with no black markings, and with 4 toes on hind feet.
Habitat
Areas of scattered pinyon pines, yuccas and sagebrush, sandy or gravelly soils.
Range
In e.-cen. Calif. and w.-cen. Nev.; essentially limited to boundary line overlapping into Death Valley.

Note: Four other species of kangaroo rats with relatively small ranges in southern California are **STEPHEN'S KANGAROO RAT**, *Dipodomys stephensi* (Fig. 28a), just south of *D. panamintinus*;

the **NARROW-FACED KANGAROO RAT**, *D. venustus (Fig. 28a)*, along the coast south of San Francisco; the **BIG-EARED KANGAROO RAT**, *D. elephantinus (Fig. 28a)*, in Bear Valley, San Benito County; and the **AGILE KANGAROO RAT**, *D. agilis*, **(58:11)** along the coast north to the Los Angeles area and inland to overlap the range of *D. panamintinus*.

Narrow-faced
Kangaroo Rat

Stephen's
Kangaroo Rat

Fig. 28a

Big-eared Kangaroo Rat

ORD'S KANGAROO RAT
Dipodomys ordii

53:5, 60:3

Description
Size: head and body, 4–4½ in. (10.2–11.4 cm); tail, 5–6 in. (12.7–15.2 cm); weight, 1⅖–2⅓ oz. (39–65 g). Tail crested. Whisker patch either lacking or small and black. Lining of cheek pouches white. Eyes large; hind feet bear either 4 or 5 toes; fur silky. Lower incisors awl-shaped (rounded), not flat across front. Above pale to bright orange-brown; white below; dorsal and ventral dark tail stripes broader than lateral white ones; ventral stripe tapers to a point near tip of tail.

Similarities
Panamint has longer tail, larger feet. Merriam's has only 4 toes, lateral light tail stripes broader than dark ones. Chisel-toothed has flat lower incisors, chisellike. Banner-tailed and Desert have white-tipped tails.

Habitat
Hard desert soils; most widely distributed of the kangaroo rats.

Habits
Nocturnal; runs by leaping like a kangaroo; seldom seen above-ground in very cold or very hot weather.

Range
In se. Alta.; sw. Sask.; s. to Mexico and e. to 98th meridian; also s.-cen. Wash.; e. Oreg., except far ne. corner; s. Idaho; n. ¾ Nev.; Utah; Ariz., except sw. ¼; e. ¾ Wyo.; Colo.; n. Mex.; w. Tex.

CHISEL-TOOTHED KANGAROO RAT
Dipodomys microps

60:3

Description
Size: head and body, 4–5 in. (10.2–12.7 cm); tail, 5⅗–7½ in. (14.2–19 cm); weight, 1⁷⁄₁₀–3⅛ oz. (48–90 g). Lining of cheek pouches sometimes black. Hind feet bear 5 toes (fifth toe small, on side of hind foot); fur silky. Lower incisors chisel-shaped (flat

325

across front). Upperparts buff to brownish; white stripe from flank to base of tail on each side; base of tail white. Lateral white tail stripe narrower than dorsal and ventral dark stripes; ventral dark stripe runs to tip of tail.

Similarities
Panamint has larger hind feet, prefers pinyon-yucca belt. Merriam's is smaller with 4 toes. Ord's has rounded lower incisors. Desert is colored lighter, tail white-tipped.

Habitat
Sagebrush or greasewood; gravelly or sandy soil and rocky slopes. Rarely among pinyon pines or on low open flats.

Range
In se. Oreg.; extreme ne., e.-cen. Calif.; Nev., except extreme w.-cen. and ne. corner; Utah, nw. and extreme sw. corner; extreme nw. Ariz.

Other name
Great Basin Kangaroo Rat.

HEERMANN'S KANGAROO RAT
Dipodomys heermanni   60:3

Description
Size: head and body, 4–5 in. (10.2–12.7 cm); tail, 6½–8½ in. (16.5–21.6 cm); weight, 1⅘–3¾₁₀ oz. (50–94 g). Skull broad. Differentiation of this species from others in some parts of its range requires study of the skull. Ears moderately long and frequently blackish; hind feet have either 4 or 5 toes, normally 4. Color yellowish-brown; lateral stripe white. Tail light gray, with white tip, slight or no crest.

Similarities
Giant is larger, with 5 toes.

Habitat
Low valleys in grasslands, open gravelly slopes with chaparral and pine and live oak areas.

Range
In sw.-cen. Oreg.; Calif., n.-cen. to coast above San Francisco; Sierra and Coast Range foothills to s.-cen. part.

DESERT KANGAROO RAT
Dipodomys deserti   58:10, 60:3

Description
Size: head and body, 5–6½ in. (12.7–16.5 cm); tail, 7–8½ in. (17.8–21.6 cm); weight, 2⅑₁₀–4⅑₁₀ oz. (82–140 g). Easily recognized by large size, pale colors, white tail tip. Hind feet bear 4 toes. Distal one-third of tail crested with long dusky hair; ventral dark tail stripe sometimes absent or indistinct. Upperparts pale yellowish, rest of body white; no dark markings except a dusky band in front of the white tail tip.

Habitat
Wind-drifted sand in alkali sinks and creosote-sagebrush vegetation; in burrows to 20 in. (50.8 cm) deep.

Habits
Well-beaten trails lead away from burrows.

Range
Nev., extreme sw. and s. tip; se. Calif.; far sw. Ariz.

GIANT KANGAROO RAT
Dipodomys ingens

58:13, 60:3

Description
Size: head and body, 5⅝–6 in. (14.2–15.2 cm); tail, 7–8 in. (17.8–20.3 cm); weight, 3⁷⁄₁₀ oz. (105 g). Largest kangaroo rat in its area. Tail and ears short relative to head and body. Skull broad, ears short. Hind feet bear 5 toes. Color yellowish-brown. All tail stripes prominent, tip of tail dark.
Similarities
Heermann's has smaller head and body with 4 toes.
Habitat
Grasslands of valleys and semiarid lowlands, sandy loam soils.
Range
Along sw. borders of San Joaquin Valley; also Carrizo Plain and Cuyama Valley; all in s. Calif., n. of Los Angeles Co.

BANNER-TAILED KANGAROO RAT
Dipodomys spectabilis

58:12, 60:3

Description
Size: head and body, 5–6 in. (12.7–15.2 cm); tail, 7–9 in. (17.8–22.9 cm); weight, 3½–5⅙ oz. (100–150 g). Large body size. Hind feet bear 4 toes. Color most spectacular of any kangaroo rat; head and body dark brownish, lateral stripe white; tail with prominent white tip and bears narrow white side-stripes which end about ⅔ distance to tip, then a black band just before white tip.
Similarities
Ord's and Merriam's are smaller; no white tail tip.
Habitat
Semiarid grasslands and deserts, firm but not too sandy soils.
Range
Extreme ne. Ariz.; also se. corner, s., cen., and nw. N.Mex.; w. Tex.

MERRIAM'S KANGAROO RAT
Dipodomys merriami

58:14, 60:3

Description
Size: head and body, 4 in. (10.2 cm); tail, 5–6⅓ in. (12.7–16 cm); weight, 1⅙–1⁷⁄₁₀ oz. (33–47 g). Skull narrow. Hind feet bear 4 toes. Color variable, from pale yellowish to dark brownish above. Lateral white tail stripes wider than dorsal and ventral dark stripes. Tip of tail crested and tail tuft brown or black. Dark whisker patches not connected across nose.
Similarities
Ord's has 5 toes; ventral tail stripe broad at base, tapering to point near tail tip. Banner-tailed and Desert are larger, with white-tipped tails. Others in range have 5 toes.
Habitat
Alkali sinks and creosote sagebrush scrub vegetation, desert soils.
Range
Nev., except n. border and ne. ¼; se. Calif.; s. ½ Ariz.; s. ⅓ N.Mex.; e. Tex.

Note: Two other species with four toes on hind feet and with relatively small ranges that do not overlap that of *Dipodomys merriami* are the **TEXAS KANGAROO RAT**, *D. elator,* at the eastern edge of the area of this manual in southwest Oklahoma and north-central Texas, and the **FRESNO KANGAROO RAT**, *D. nitratoides,* in central California.

BEAVER
Family Castoridae

The beaver is the largest rodent in most of North America and the only land mammal with a broad flat tail. The body is thickset and compact, the legs short, ears small, hind feet large, toes webbed.

BEAVER
Castor canadensis

55:5
Fig. 29

Description
Size: head and body, 25–30 in. (63.5–76.2 cm); tail, 9–10 in. (22.9–25.4 cm); weight, 30–60 lb. (13.6–27.2 kg); 20 teeth, 1013/1013. Incisors large and yellowish anteriorly, grinding teeth high-crowned. Skull massive; fur lustrous. Color rich brown; tail gray, flat, and scaly, shaped like a paddle and nearly hairless.

Similarities
Muskrat is much smaller; tail slender, flattened from side to side. River Otter has tail covered with fur.

Habitat
Streams or lakes.

Habits
Builds watertight dam of sticks and mud across a stream, a large cone-shaped house in the pond; trees gnawed or cut a foot from the ground are telltale signs.

Voice
Various sounds made only within the lodge; outside, slaps water with its tail as sign of warning.

Food
Bark of aspen, alder, birch, maple, willow, and other vegetation.

Range
Throughout N.A., s. of Arctic tundras, except in coastal and desert Calif.; desert Nev.; far w. Utah; and sw. corner Ariz.

Fig. 29

Porcupine, p. 350

Beaver

NEW WORLD RATS AND MICE
Family Cricetidae

This family includes an enormous assemblage of rodents differing widely in habitats, habits, and structures. They have sixteen teeth, 1003/1003. The hind foot has five toes. Mice and rats have large eyes and ears, long tails, and four toes on the front foot. Voles and lemmings have small eyes and ears, short tails, and either four or five toes on the front foot. All groups have one, sometimes two or more, litters a year, of three to four young, usually in a nest of vegetation on the ground; none hibernate. They are mostly vegetarians. Females possess three pairs of mammae. The family is often combined with the Muridae.

HARVEST MICE
Genus *Reithrodontomys*

These are brownish mice with a longitudinal groove on each incisor, conspicuous ears, no external cheek pouches, and medium-sized, thinly haired tails. The young are darker than the adults. They are nocturnal and good climbers. Their voice is a high ventriloquistic bugling. The diet consists of plant cuttings, seeds, and insects.

PLAINS HARVEST MOUSE
Reithrodontomys montanus

53:10

Description
Size: head and body, 2–3 in. (5.1–7.6 cm); tail, 2–2⅝ in. (5.1–6.6 cm). Teeth have distinct groove down front of each upper incisor. Color pale gray above with a faint tawny cast, middle of back often darker; feet, underparts of body, and tail white.

Similarities
Western is difficult to differentiate; tail usually slightly longer, and with less distinct dorsal tail stripe.

Habitat
Brier patches, roadside ditches, bogs, pastures; also areas of short grass and prickly pear cactus.

Range
Great Plains as far w. as cen. Colo.; cen. and s. N.Mex.; se. Ariz.; w. Tex.

FULVOUS HARVEST MOUSE
Reithrodontomys fulvescens

59:5

Description
Size: head and body, 2⅘–3⅕ in. (7.1–8.1 cm); tail, 3⅓–4 in. (8.5–10.2 cm). Largest and most colorful of the harvest mice. Upperparts grayish-brown, sides bright orange-buff, underparts whitish, tail darker above.

Similarities
Plains has shorter tail. Western lacks bright fulvous sides; tail shorter.

Habitat
Grass and brushy lowlands along streams.

Range
In se. Ariz.; extreme sw. and s. Tex.

WESTERN HARVEST MOUSE

Reithrodontomys megalotis *Fig. 30*

Description
Size: head and body, 2⅘–3 in. (7.1–7.6 cm); tail, 2⅓–3⅕ in. (5.8–8.1 cm). Skull broad. Front of upper incisors distinctly grooved. Upperparts varied geographically from various shades of buff mixed with dark brown or blackish, often with darker stripe down back; underparts of tail and body white to gray; tail indistinctly bicolored.
Similarities
Plains has thin-striped and somewhat shorter tail.
Habitat
Grassy areas, arid regions.
Range
Extreme s.-cen. B.C.; e. Wash., except extreme ne. corner; e. and s. Oreg. to coast; s. Idaho; s. to Mexico; also se. Alta.; e. Mont.; s. through w. Tex. to Mexico. Absent from Rockies of Idaho panhandle to s.-cen. Colo.

Note: The **SALT-MARSH HARVEST MOUSE**, *Reithrodontomys raviventris*, is similar to *R. megalotis* but generally darker and confined to salt marshes around San Francisco Bay.

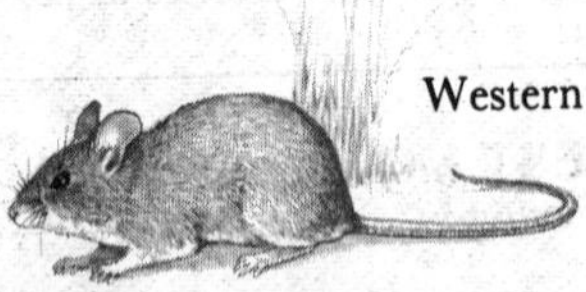

Western Harvest Mouse

Fig. 30

WHITE-FOOTED MICE

Genus *Peromyscus*

These common mice have no grooves on their upper incisors. Those mice that live in woodlands are dark; those that live in the open are pale. Most are terrestrial; all are nocturnal. Food consists of seeds, berries, fruit, insects, carcasses.

CALIFORNIA MOUSE

Peromyscus californicus **59:3**

Description
Size: head and body, 3⅘–4⅗ in. (9.7–11.7 cm); tail, 5–5⅘ in. (12.7–14.7 cm). Largest of its genus. Ears and feet large. Skull large, braincase well inflated. Premaxillary and nasal bones extending posteriorly to same level. Upperparts russet mixed with dark brown; underparts pale or nearly white including feet; tail blackish, may be indistinctly bicolored.
Similarities
Canyon, Cactus, Deer, Brush, and Pinyon are all smaller, have tails less than 5 in. (12.7 cm) long.
Habitat
Oak- or chaparral-covered hillsides and ravines.
Habits
Nest elaborately built of sticks and twigs containing a grass-lined chamber; often employs man-made shelter for nest site.
Range
Along s. Calif. coast and in s. Sierras.

CACTUS MOUSE
Peromyscus eremicus

59:4

Description
Size: head and body, 3⅕–3⅗ in. (8.1–9.1 cm); tail, 3⅘–4⅖ in. (9.7–11.2 cm). Skull average; high braincase, somewhat inflated; premaxillary bones extend conspicuously beyond posterior ends of nasal bones; ears large, thin, no white edge. Pelage soft, silky. Tail usually longer than head and body and thinly haired. Underparts pale gray faintly tinged with fulvous; underparts and feet whitish; sole of foot naked to heel; tail faintly bicolored, broad dorsal brown stripe, longer hairs at tip.

Similarities
California is larger, tail longer. Canyon has tuft at tail tip. White-footed has shorter tail. Deer has well-haired tail, distinctly bicolored. Brush has tail with long hairs near tip. Pinyon has large ears; prefers foothills.

Habitat
Low, hot deserts, often in cactus stands.

Range
In s. Calif.; s. tip Nev.; Ariz.; nw. corner and s. ⅖; s. N.Mex.; w. Tex.

Note: A related species, **MERRIAM'S MOUSE**, *Peromyscus merriami*, ranges into south Arizona. These species can not be distinguished readily in the field.

DEER MOUSE
Peromyscus maniculatus

53:3, 60:4

Description
Size: head and body, 2⅘–4 in. (7.1–10.2 cm); tail, 2–5 in. (5.1–12.7 cm); weight, ⅖–1⅕ oz. (10–35 g). Most widely distributed and variable of genus. Tail slightly pencillike, clothed with short hairs. Skull smooth, delicate; braincase somewhat arched, well inflated, premaxillary bones not extending posteriorly to ends of nasals. Upperparts usually pale grayish-buff to deep reddish-brown overlaid with some dusky; underparts white; feet white; tail usually less than 90% of head and body, sharply bicolored, dark above, white below; ears sometimes with fine light edge. When present, tufts in front of ears often whitish.

Similarities
Cactus's tail is not sharply bicolored; scantily haired. California is larger. Canyon's tail is longer than head and body; fur long, lax. White-footed in south lacks sharply bicolored tail. Brush has tail as long as, or longer than, head and body. Pinyon has large ears, to 1 in. (2.5 cm) high.

Habitat
Widely varied; in grasslands, prairies, mixed vegetation, and woods.

Remarks
This species constitutes a long series of intergrading populations.

Range
In cen. to e.-cen. Alaska; e. across Canada and s. to Mexico, including se. Alaskan islands but not Vancouver Is.; w. Tex.; absent from Tex., s. of panhandle.

Note: A related species, the **SITKA MOUSE**, *Peromyscus sitkensis*, occurs on some of the islands in extreme south Alaska. Another species, *P. melanotis*, ranges from Mexico into south Arizona where it occupies some mountain ranges. These species cannot be readily distinguished from *P. maniculatus* in the field.

CANYON MOUSE

Peromyscus crinitus

59:7

Description
Size: head and body, 3–3⅗ in. (7.6–9.1 cm); tail, 3½–4⅓ in. (8.9–10.9 cm). Tail is well haired, tufted at end. Ear is about as long as hind foot, no whitish edge. Pelage long, lax. Posterior limits of premaxillary and nose bones meet at same level. Upperparts mixed ochreous and brown or black, hairs basally lead-colored; underparts much paler, sometimes white; feet white; tail weakly bicolored, usually longer than head and body.

Similarities
Cactus has untufted tail. Brush is brown. Deer has shorter tail. California is larger, has longer tail. Pinyon has large ears, nearly 1 in. (2.5 cm) high; head and body larger.

Habitat
Rocks on arid slopes, canyons, and old lava; distribution is therefore discontinuous.

Habits
Makes nests in crevices and clefts (more commonly in burrows under rocks). Lives at altitudes from sea level to over 10,000 ft. (3048 m).

Remarks
In some individuals tail is not longer than head and body.

Range
In se. ¼ Oreg.; sw. Idaho; far ne., e.-cen., se. ⅓ Calif.; Nev.; Utah, except n.-cen. to s.-cen. strip; w. and n. Ariz.; extreme sw. Wyo., n. of e. border of Utah; w. ¼ Colo.; far nw. N. Mex.

BRUSH MOUSE

Peromyscus boylii

Description
Size: head and body, 3⅗–4⅕ in. (9.1–10.7 cm); tail, 3⅗–4⅖ in. (9.1–11.2 cm). Tail somewhat pencillike, well haired, equal to or longer than head and body. Skull medium; rostrum depressed; premaxillary bones extend posteriorly to end of nasal bones. Upperparts vary from dark, rich tawny or brownish to grayish-buff or cinnamon, purest on sides; underparts white or creamy; tail bicolored, brownish above, white below; feet white, rear portion of sole haired.

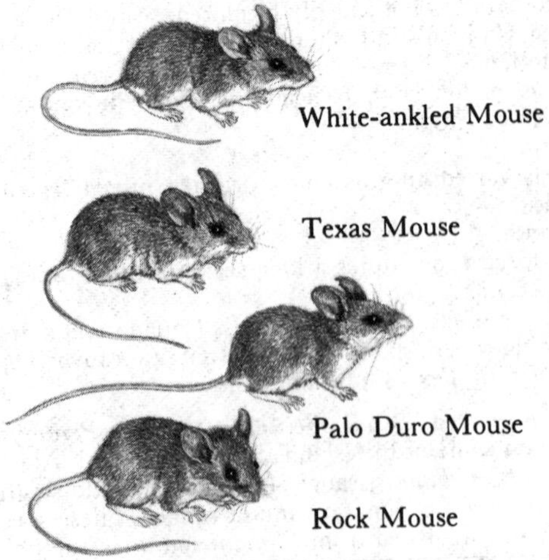
White-ankled Mouse
Texas Mouse
Palo Duro Mouse
Rock Mouse

Fig. 30a

Similarities
Cactus has scantily haired tail; California is larger; tail much longer. Canyon is pale gray or buffy. Deer and White-footed have tails shorter than head and body. Pinyon has 1-in. (2.5 cm) high ears; in California some may be difficult to differentiate.
Habitat
Arid and semiarid chaparral areas, rocky situations.
Habits
Usually nests under a rock, in a crevice, or in a pile of sticks or brush.
Range
Calif., except Nw. Coast, n. Sierras, se. corner; s. tip Nev.; e. ⅔ Utah; Ariz., except sw. corner; w. and s. Colo.; N.Mex.; extreme w. Okla. panhandle; w. ½ Tex.

Note: Two related species are the **TEXAS MOUSE,** *Peromyscus attwateri* (*Fig. 30a*), that ranges west through central Oklahoma and to parts of west Texas, and the **WHITE-ANKLED MOUSE,** *P. pectoralis* (*Fig. 30a*), that ranges from Mexico into west Texas.

PINYON MOUSE
Peromyscus truei

59:6

Description
Size: head and body, 3⅗–4 in. (9.1–10.2 cm); tail, 3⅖–4⅘ in. (8.6–12.2 cm). Ears large, usually longer than hind foot. Skull medium; braincase vaulted; premaxillary bones as long as or extending slightly posterior to nasals. Pelage long, silky. Upperparts grayish-brown; underparts whitish; lateral line usually distinct; feet white; tail slightly shorter or longer than head and body, distinctly bicolored, brownish or dusky above, whitish below, dorsal tail stripe covering one-third the circumference.
Similarities
Cactus has indistinctly hairy tail; found in low desert. Deer and White-footed have smaller ears. California is larger, tail over 5 inches. Canyon is pale gray or buff; smaller. Brush has smaller ears; dorsal tail stripe covers one-half the circumference.
Habitat
Rocky situations among pygmy conifers, especially in arid or semiarid regions; occasionally at higher elevations.
Range
Oreg., s. Cascades and sw. border; Calif., except nw. coast, Central Valley, extreme se.; s. ¾ Nev.; Utah, except n.-cen. to cen. strip; w. ⅓, se. ¼ Colo.; ne. ¾ Ariz.; nw. ¾ N.Mex.

Note: Two other species of *Peromyscus* with relatively large ears are **PALO DURO MOUSE,** *P. comanche* (*Fig. 30a*), of the Palo Duro Canyon region of north Texas, and the **ROCK MOUSE,** *P. difficilis* (*Fig. 30a*), of southeast Utah, west Colorado, extreme south Wyoming, and North Mexico.

WHITE-FOOTED MOUSE
Peromyscus leucopus

53:1

Description
Size: head and body, 3½–4⅕ in. (8.9–10.7 cm); tail, 2–4 in. (5.1–10.2 cm). Upperparts variable from pale to rich reddish-brown; feet, underparts white; tail generally shorter than head and body and bicolored, usually not sharply; ears usually dusky, narrowly edged with whitish.

Similarities
Cactus has scantily haired tail, distinct tail tuft of longer hairs. Deer always has sharply bicolored tail. Brush has longer tail than head and body. Pinyon has nearly 1-in. (2.5 cm) high ears.
Habitat
Woods, brush.
Range
Extreme se. Alta.; s. Sask.; e. Mont.; far ne. Wyo.; w. Nebr.; Okla.; se. Colo.; cen.-se. Ariz.; se. ¾ N.Mex.; Tex.

NORTHERN GRASSHOPPER MOUSE
Onychomys leucogaster

53:7, 60:5

Description
Size: head and body, 4–5 in. (10.2–12.7 cm); tail, 1⅗–2½ in. (4.1–6.4 cm); weight, ⅘–1⅖ oz. (24–40 g). Stocky, heavy-bodied. Upperparts brownish to pinkish-cinnamon or buff, most intense along dorsal areas; underparts pure white, sharply demarcated from dorsum; tail white-tipped, bicolored, less than 50% of head and body. Ear edges white; feet white.
Similarities
Southern Grasshopper is smaller; tail is 50–60% of body. Mice of Genus *Peromyscus* have more slender bodies, longer tail.
Habitat
Sagebrush scrub.
Food
Chiefly insects.
Range
S. Alta.; s. Sask.; sw. Man.; s. to s. Tex.; also se. Wash.; e. Oreg.; sw. and s. Idaho; extreme ne. Calif.; n. ⅔ Nev.; n. and e. Ariz.; N.Mex.; absent from Rocky Mountains of Idaho; w. Mont.; nw. Wyo.; cen. Colo.; n. Utah.

SOUTHERN GRASSHOPPER MOUSE
Onychomys torridus

60:5
Fig. 31

Description
Size: head and body, 3½–4 in. (8.9–10.2 cm); tail, 1⅗–2 in. (4.1–5.1 cm); weight, 7/10–⅘ oz. (20–25 g). Upperparts grayish or pinkish-cinnamon; underparts white; tail white-tipped, 50–60% of head and body.
Similarities
Northern Grasshopper is larger; tail less than 50% of head and body; usually at higher elevations where ranges overlap.
Habitat
Low, hot valleys of grassland and desert scrub.
Voice
Shrill whistle.
Food
Chiefly insects.
Range
In s. ⅓ Calif., Great Central Valley; nw., s. ⅓ Nev.; Ariz., except ne. ¼; s. N.Mex.; w. Tex.

Fig. 31

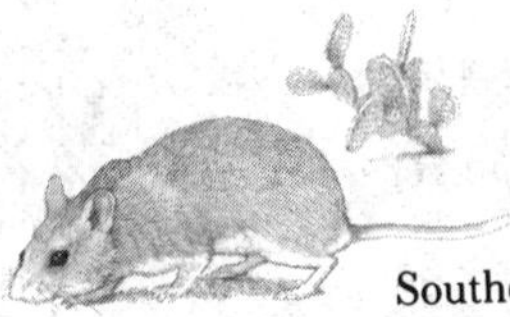

Southern Grasshopper Mouse

OTHER MICE
Genera *Onychomys, Baiomys,* and *Sigmodon*

NORTHERN PYGMY MOUSE
Baiomys taylori

59:1

Description
Size: head and body, 2–2½ in. (5.1–6.4 cm); tail, 1⅖–1⅘ in. (3.6–4.6 cm). The smallest mouse, resembling a young House Mouse. Tail covered with short hairs. Upperparts pale drab to nearly black; underparts grayish to creamy-buff; tail paler below than above.

Similarities
House Mouse is larger; tail naked. Harvest mice have grooved front teeth.

Habitat
Grassy areas.

Habits
Makes small runways in grooves.

Food
Mostly seeds.

Range
Extreme se. Ariz. and adjoining N.Mex.

HISPID COTTON RAT
Sigmodon hispidus

Description
Size: head and body, 5–8 in. (12.7–20.3 cm); tail, 3–5 in. (7.6–12.7 cm); weight, 2⅘–8½ oz. (80–240 g). Tail scaly, sparsely haired, coarsely annulated, shorter than head and body. Ears large and rounded, but nearly covered by long hairs growing anterior to base of ear. Skull relatively long, narrow. Grinding surfaces of molars flat and with S- or Σ-shaped ridges. Upperparts coarsely grizzled, blackish or dark brownish hairs mixed with buffy or grayish hairs; underparts usually pale to dark grayish, sometimes faintly washed with buff. Feet gray.

Similarities
Grasshopper, Harvest, and Deer Mice are much smaller, with large membranous conspicuous ears. Woodrats are grayish, with large membranous ears. Black and Norway Rats have large membranous ears not covered with long hairs.

Habitat
Low altitudes in grasslands and weeds, sedges and cattail marshes along rivers.

Range
In s. N.Mex.; Okla. panhandle; Tex.

Note: Two other species of cotton rats are the *Sigmodon hispidus*-like **ARIZONA COTTON RAT,** *S. arizonae,* of extreme southeast California, southern Arizona, and extreme southeastern Arizona and southwest New Mexico; and the **YELLOW-NOSED COTTON RAT,** *S. ochrognathus,* of extreme southeast Arizona, extreme southwest New Mexico, and the Big Bend area of west Texas.

WOODRATS
Genus *Neotoma*

These rats have conspicuous ears and eyes, soft fur, white feet, hairy tails, and flat molars. In comparison, Old World rats have smaller ears, dusky feet, scaly tails, and cusped molars. When alarmed, woodrats thump with their hind feet. Scats are frequently deposited in large piles. Their food is seeds, fruits, leaves, berries, cactus pulp, grass, and insects. These nocturnal rats collect unusual objects, such as cans, silver, belt buckles, and so forth at their nest sites and sometimes replace objects they take with other items. Other names include pack rat and trade rat.

SOUTHERN PLAINS WOODRAT
Neotoma micropus   59:12

Description
Size: head and body, 7½–8½ in. (19–21.6 cm); tail, 5½–6½ in. (14–16.5 cm). Skull robust, sculptured. Upperparts steel-gray; belly gray; feet white; white on throat and breast; tail bicolored, blackish above, gray below.

Similarities
White-throated has back mixed with fulvous. Desert is smaller; throat hairs dark at bases, back with mixed fulvous. Mexican has throat hairs slate-gray at bases.

Habitat
Plains.

Habits
Builds stick houses.

Range
In s. ¾ N.Mex., w. ½ Tex.

Note: The **EASTERN WOODRAT**, *Neotoma floridana*, ranges westward north of the range of *N. micropus* as far as the foothills in central Colorado and in western Nebraska. It differs from *N. micropus* in its less steel-gray color.

WHITE-THROATED WOODRAT
Neotoma albigula   59:9

Description
Size: head and body, 7½–8½ in. (19–21.6 cm); tail, 5½–7⅓ in. (14–18.6 cm). Skull with relatively broad rostrum. Upperparts grayish washed with fulvous to ochreous mixed with dusky; underparts white or grayish, hairs lead-colored basally, except on throat; feet white; tail bicolored, brownish above, whitish below.

Similarities
Desert, Mexican, Stephens' have throat hairs dark at bases. Dusky-footed is larger; hind feet dusky above; tail blackish dorsally. Bushy-tailed has squirrellike tail.

Habitat
Arid to semiarid valleys and plains; deserts. Found mostly around rocks and under mesquite trees in deserts.

Range
Extreme se. Calif.; se. Utah; Ariz., except nw. corner; N.Mex., except ne.-cen.; sw. and se. Colo.; w. Tex.

DESERT WOODRAT
Neotoma lepida

59:11

Description
Size: head and body, 5⅘–7 in. (14.7–17.8 cm); tail, 4⅓–6⅖ in. (10.9–16.3 cm). Skull robust, sculptured. Color pale to dark gray, washed variously with fulvous; underparts grayish or faintly buffy, all hairs slate-gray basally. Tail dark gray or dusky above, pale gray below; about three-fourths length of head and body.

Similarities
White-throated has throat hairs white to bases. Mexican is hard to differentiate without examining skull; tail white below. Dusky-footed is larger; hind feet dusky above. Bushy-tailed has squirrellike tail. Stephens' has dusky patch on top of hind foot below ankle, tail slightly bushy.

Habitat
Low, hot, cactus-covered, arid deserts.

Habits
Houses often on level ground, made of sticks and debris at base of cactus or shrub; entrance usually almost paved with spines; occasionally burrows into clay. Animals climb among sharp cactus spines.

Range
In se. Oreg.; sw. Idaho; Nev.; s. ½ Calif.; Utah, w. ½, s., with ne. extension to ne. boundary; extreme nw. Colo.; w. ½ Ariz.

STEPHENS' WOODRAT
Neotoma stephensi

Description
Size: head and body, 6–6¾ in. (15.2–17.1 cm); tail, 6 in. (15.2 cm). Skull similar to that of Desert Woodrat, but smaller, less angular; braincase more smoothly rounded; frontal region broader, more flattened. Tail notably hairier. Upperparts yellowish to grayish-buff, dusky on back; underparts white or creamy; feet usually white with dark patch on top of hind foot below angle. Tail pale gray to grayish-brown above, paler below; slightly bushy at end.

Similarities
White-throated, Desert, Mexican lack bushy tails, top of hind foot white to ankle. Dusky-footed has top of foot with dark hairs; larger. Bushy-tailed tail is black above and bushy.

Habitat
Typically arid and semiarid situations.

Range
Ariz., ne. quadrant plus w. extension from cen.; far w.-cen. to nw. N.Mex.

MEXICAN WOODRAT
Neotoma mexicana

Description
Size: head and body, 6½–7¾ in. (16.5–19.7 cm); tail, 6–6½ in. (15.2–16.5 cm). Upperparts grayish, grayish-buff, dull brown, russet, or bright rufous, according to subspecies; normally gray with a fulvous wash; underparts grayish-white to yellowish, hairs slate-gray basally. Tail distinctly bicolored; white below, black above.

Similarities
White-throated has throat hairs white to bases; prefers valleys, plains. Bushy-tailed has squirrellike tail. Desert is difficult to

distinguish without skull, tail less markedly bicolored. Stephens' has hind foot with dark below ankle, slightly bushy tail.

Habitat
Rocky areas and cliffs of plateaus and high deserts.

Range
In se. Utah; far sw. and n., n.-cen., se. Colo.; e. ⅗ Ariz.; w. ¾ N.Mex.; w. Tex.; absent from Rockies.

DUSKY-FOOTED WOODRAT

Neotoma fuscipes

59:10

Description
Size: head and body, 7⅗–9 in. (19.3–22.9 cm); tail, 6¾–8⅔ in. (17.1–22 cm). Large arboreal woodrat. Skull large, long, relatively narrow. Upperparts grayish-brown; underparts grayish to whitish; tail nearly as long as head and body, slightly paler below than above, short hairs over scales; hind feet sprinkled on top with dusky hairs.

Similarities
White-throated is usually smaller; hind feet white above. Desert is smaller; hind feet white. Bushy-tailed has squirrellike tail, hind feet white.

Habitat
Dense chaparral, mixed vegetation, live-oak forests, and riparian vegetation.

Habits
Often builds large, conspicuous cone-shaped stick houses either on the ground or in lower branches of oak trees; some houses 6 ft. (1.8 m) high and 12 ft. (3.65 m) in diameter.

Range
In w. Oreg., except coast; w. ½ Calif., except Central Valley, from Sierra crest to coast.

BUSHY-TAILED WOODRAT

Neotoma cinerea

59:13

Description
Size: head and body, 7–9⅔ in. (17.8–24.5 cm); tail, 5⅕–7⅖ in. (13.2–18.8 cm). Unique for its very bushy, squirrellike tail. Skull with relatively short braincase. Upperparts vary from pale gray lightly washed with buff to dark brownish-black; underparts vary from white to pinkish or buff; hind feet white; tail dusky above, whitish below.

Similarities
White-throated, Desert, Mexican, Dusky-footed have short-haired tails, tapering toward tip.

Habitat
Rimrocks and pine belt of higher elevations.

Habits
Builds stick house in crevice in ledge rock or in abandoned cabin or mine tunnel; nest globular mass or cup-shaped, of shredded bark, dry grass, moss, leaves.

Range
In se. Yukon; B.C., except ne. corner; s. through w. states to n. and e.-cen. Calif.; s. Nev.; n. Ariz.; w. ⅔ Colo.; and n. N.Mex.

Other name
Mountain Packrat.

VOLES
Genera *Clethrionomys* and *Phenacomys*

With the exception of the Muskrat and tree mice, microtine rodents are generally small mice with tails that are never as long as the head and body length. Long hairs growing in front of the ears tend to cover the anterior ear surfaces, and the grinding surfaces of the cheek teeth have a characteristic pattern of triangles and loops. Each foot bears five toes. Red-backed mice (Genus *Clethrionomys*) are thickset, volelike mammals about five inches long with short tails usually less than one inch long. They inhabit forest floors, bogs, and meadows surfaced with a thick mat of ground vegetation. Several species of the Genus *Phenacomys* are arboreal, building their nests high in fir, spruce, and hemlock trees, upon whose needles they feed. Other species live and nest on heather and grasses growing on the floor of coniferous forests.

SOUTHERN RED-BACKED VOLE
Clethrionomys gapperi   **53:9, 61:6**

Description
Size: head and body, 3⅔–4⅔ in. (9.3–11.8 cm); tail, 1⅛–2 in. (3–5.1 cm); weight, ½–1⅖ oz. (15–40 g). Tail almost one-half length of head and body; slender, short-haired except at tip. No grooves on upper incisors. Two color phases occur in North, red and gray; usually identifiable by reddish back and gray sides, although in gray phase the reddish may be absent.

Similarities
Sometimes difficult to differentiate from other area voles without examining skulls. Northern Bog and Collared Lemmings are larger in body size, shorter tails; Western Red-backed, Heather, and other voles have no color contrast between back and sides.

Habitat
Coniferous, deciduous, and mixed forests where it is cool and moderately damp; dwells on ground. Also in grassy meadows, chaparral, and rocky areas in southern part of range.

Range
B.C., except far n.; e. across Canada and s. coastally to nw. Calif., including Wash., ne. Oreg. and coast; Idaho, except sw. corner; w. ½ and ne. corner Mont.; N.Dak.; w. ¾ Wyo. and Blackhills; ne. Utah; w. Colo., except w. border; extreme e.-cen. Ariz.; nw. ¼ N.Mex., except extreme nw. corner.

Other name
Boreal Red-backed Mouse.

Note: To the north of the range of *Clethrionomys gapperi* in Alaska and northern Canada, east to Hudson Bay lies the range of a related species, the **NORTHERN RED-BACKED VOLE**, *C. rutilus.*

WESTERN RED-BACKED VOLE
Clethrionomys occidentalis

Description
Size: head and body, 5⅛–6½ in. (13.2–16.5 cm); tail, 1⅓–2 in. (3.8–5.1 cm). Tail slender, short-haired except at tip. Upperparts somber; dorsal stripe ill-defined, obscured by intermixed black hairs, varying from light hazel to deep chestnut; sides light to dark buffy-gray; feet whitish to dusky; tail bicolored, indistinctly to sharply.

Similarities
Southern Red-backed has color of sides contrasting with that of back; tail unicolored. Heather has short bicolored tail; prefers high mountains. Mountain Vole is grayish, prefers high mountain meadows. Townsend's, Long-tailed Voles have larger body size.

Habitat
Under logs on moist floor of dense forests.

Range
Nw. Coast from s. B.C., not including Vancouver Is., to n. Calif., w. of Cascade Mts.

Other name
California Red-backed Mouse.

RED TREE VOLE
Phenacomys longicaudus   **61:1**

Description
Size: head and body, 4–4⅓ in. (10.2–10.9 cm); tail, 2⅖–3⅓ in. (6.1–8.5 cm). A tree-dwelling mouse; well haired. Bright reddish-brown, tail blackish.

Habitat
Humid forests.

Habits
Builds nest among branches.

Range
Coastal cen.-s. Oreg.; n. Calif.

Other name
Tree Phenacomys.

Note: Two other species of relatively long-tailed *Phenacomys* occur in the same region with *P. longicaudus*. These are the gray-colored **WHITE-FOOTED VOLE,** *P. albipes* (*Fig. 31a*), and the cinnamon-brown **DUSKY TREE VOLE,** *P. silvicola;* (*Fig. 31a*).

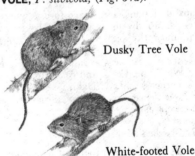

Dusky Tree Vole

Fig 31a

White-footed Vole

HEATHER VOLE
Phenacomys intermedius   **61:2**

Description
Size: head and body, 3½–4¾ in. (8.9–12.1 cm); tail, 1–1⅔ in. (2.5–4.2 cm). Tail less than ½ length of head and body. Upperparts gray to brownish, face in some subspecies yellowish; underparts silver-white; tail sharply bicolored.

Similarities
Mountain Vole has longer tail; difficult to differentiate. Long-tailed Vole has much longer tail. Southern Red-backed and Western Red-backed Voles have reddish backs, tail either longer or unicolored, nose not yellowish.

Habitat
Open grassy parks in forests, rocky slopes of high mountains and tundra.
Range
In s. Yukon; s. Mackenzie; B.C.; Alta., except se. corner; n. ¾ Sask.; Wash.; Oreg., except coast and se. ¼; Idaho, except sw. corner; w. Mont.; w. Wyo.; n.-cen. and ne.-cen. Calif.; Rockies of Wyo. and Colo.; n.-cen. N.Mex.
Other name
Mountain Phenacomys.

MEADOW MICE AND OTHER VOLES

Genera *Microtus* and *Lagurus*

Meadow mice (Genus *Microtus*) are generally found wherever there is good grass cover, although some species live on the forest floor among litter or in rocky areas where grasses are not abundant. They have long, grayish-brown fur; short ears and tails; and beady eyes. Their upper incisors are ungrooved, and their rather dull-colored tails are more than an inch long. They make two-inch-wide runways, leaving cut grass stems. They are active both day and night, and can swim and dive. In winter they make round holes to the surface through the snow. Their voice is a high-pitched squeak. They eat grass, roots, bark, and seeds. The Sagebrush Vole (*Lagurus curtatus*) is associated with sagebrush-scrub vegetation in arid, semidesert country.

MEADOW VOLE

Microtus pennsylvanicus **53:4, 60:7, 61:4**

Description
Size: head and body, 3½–5 in. (8.9–12.7 cm); tail, 1⅖–2⅗ in. (3.6–6.6 cm). Most widely distributed vole. Ears nearly hidden in fur. Color varies from gray faintly washed with brown (in West) to dark brown in East; fur grizzled, underparts vary from silvery to buff to dark gray; tail bicolored.
Similarities
Mountain is hard to distinguish, prefers high mountain meadows. Tundra is larger; yellowish. Long-tailed has longer tail. Yellow-cheeked has yellow nose. Water has larger head and body. Prairie has shorter tail. Red-backed have reddish back contrasting with pale grayish sides, tail blackish. Heather Vole's tail is usually shorter; pale gray.
Habitat
Varied; grasslands, low moist areas.
Habits
Populations fluctuate greatly.
Range
Alaska, except tundra regions; all n. Canada, s. to ne. Wash.; e. Idaho; n.-cen. Utah; Mont.; Wyo.; Colo., except nw. and se. corners; nw. N.Mex.

MOUNTAIN VOLE

Microtus montanus

Description
Size: head and body, 4–5½ in. (10.2–14 cm); tail, 1⅕–2⅗ in. (3–6.6 cm); weight, 1$\frac{1}{10}$–3 oz. (30–85 g). Above grayish-brown to

blackish; belly white; feet usually dusky; tail bicolored to nearly
unicolored.

Similarities
Water is larger. Long-tailed's tail is longer; hard to differentiate.
Meadow is difficult to differentiate; usually not in mountains.
California prefers low flats and valleys; difficult to differentiate.
Townsend's has black tail. Heather Vole is hard to differentiate;
nearer mountain tops. Red-backeds have reddish back contrasting
with grayish sides.

Habitat
Around springs and meadows of intermontane valleys.

Range
Great Basin drainage; s.-cen. B.C.; e. Wash., except ne. corner;
Oreg., except coast and Willamette Valley; sw. Mont.; w. ⅔ Wyo.;
ne. ¼ Calif.; n. Nev.; Utah; w. Colo.; extreme nw. and se. Ariz.;
nw. N.Mex.

CALIFORNIA VOLE
Microtus californicus   *Fig. 32*

Description
Size: head and body, 4¾–5⅔ in. (12.1–14.4 cm); tail, 1⅗–2⅘ in.
(4.1–7.1 cm). Grayish-brown, blackish in West toward Coast,
reddish in deserts; tail bicolored, feet pale. Ears project noticeably
above fur.

Similarities
Mountain has dusky feet; found in high mountain meadows. Long-
tailed has longer tail, is above foothills. Townsend's has dusky feet,
tail blackish. Western Red-backed Vole has reddish back, buffy
sides. Heather Vole has shorter tail; in high mountains. Red Tree
Mouse is reddish, tail black.

Habitat
Dry, wet and coastal grassy meadows, between seashore and high
mountains.

Range
Nw. Coast from s.-e. Oreg. to Baja Calif., w. of Cascades in n.
Calif. but throughout Sierras in cen. and s. Calif.

California Vole

Fig. 32

Long-tailed Vole

Townsend's Vole

TOWNSEND'S VOLE
Microtus townsendii   *Fig. 32*

Description
Size: head and body, 4¾–6⅖ in. (12.1–16.3 cm); tail, 2–3 in. (5.1–
7.6 cm). Above blackish-brown; belly gray; tail blackish, slightly
bicolored; feet dusky. Ears extend noticeably beyond fur.

Habitat
Moist marshes, fields, and meadows from sea level to mountains,
usually near water.

Range

Nw. Coast from s. B.C. to nw. Calif., including Vancouver Is. and
other islands off sw. B.C.

Note: A much smaller species, the **CREEPING VOLE,**
Microtus oregoni, has a range about the same as that of
M. townsendii.

TUNDRA VOLE
Microtus oeconomus   **61:5**

Description

Size: head and body, 5–6¾ in. (12.7–17.1 cm); tail, 1⅝ in. (4.1
cm). Distinguishable by fairly uniform color above and size. Color
dull brown washed with buff or fulvous; underparts grayish; tail
bicolored.

Similarities

Yellow-cheeked is larger; nose yellow. Meadow is smaller, where
ranges meet. Long-tailed has longer tail. Red-backeds have
reddish backs. Lemmings have much shorter or brightly colored
tails.

Habitat

Moist tundras.

Range

In Alaska and the Yukon, w. and n. of Lake Athabasca, Canada.

Note: Two other species of *Microtus* have ranges similar to that of
M. oeconomus. These are the narrow-skulled **SINGING VOLE,** *M.
gregalis,* and the closely related **INSULAR VOLE,** *M. abbreviatus,*
which is known only from two islands in the Bering Sea west of
Alaska.

LONG-TAILED VOLE
Microtus longicaudus   *Fig. 32*

Description

Size: head and body, 4½–5⅓ in. (11.4–13.2 cm); tail, 2–3½ in.
(5.1–8.9 cm); weight, 1⅗–2 oz. (37–58 g). Skull relatively smooth,
not heavily ridged. Dark gray washed with brown or blackish; feet
grayish-white; tail indistinctly bicolored.

Similarities

Prairie and Tundra have tails under 2 inches, Mountain has
whitish belly. Meadow's tail is usually shorter. California prefers
foothills and valleys; tail shorter. Townsend's is larger; tail black.
Water is larger overall. Heather Vole is reddish with black tail or
shorter tail. Red-backeds have reddish back contrasting with
grayish sides.

Habitat

Variable, moist and wet meadows and streambanks among willows
and pines to drier sagebrush-scrub vegetation.

Range

Extreme se. Alaska; s. Yukon; sw. Mackenzie; B.C.; w. and s.
Alta.; Wash. and Oreg., except Palouse region; Idaho; w. Mont.;
Wyo., except se. corner; n. Calif. and Sierras; n. ¾ Nev.; Utah,
except extreme sw. corner; w. ½ Colo.; nw. and e. Ariz.; nw. ¾
N.Mex.

Note: A closely related species, the **CORONATION ISLAND VOLE,**
M. coronarius, is known only from Coronation, Warren, and
Forrester islands in southeast Alaska.

YELLOW-CHEEKED VOLE

Microtus xanthognathus **61:8**

Description
Size: head and body, 6⅖–7 in. (16.3–17.8 cm); tail, 1⅘–2 in. (4.6–5.1 cm). A tundra vole easily identified in its range by its yellow cheeks, large size, and long tail. Dull brown upperparts, gray belly; rusty-yellow or chestnut nose and ear patch; tail indistinctly bicolored.
Habitat
Spruce and birch forests; tundras.
Range
In ne. Alaska; nw. Canada as far s. as cen. Alta. and cen. Sask.; absent from s. Yukon and B.C., present only in extreme ne. corner.

WATER VOLE

Microtus richardsoni

Description
Size: head and body, 5⅗–6½ in. (14.2–16.5 cm); tail, 2⅖–3⅗ in. (6.1–9.1 cm). Largest vole in its range. Above dull grayish-brown; underparts pale gray; tail bicolored.
Similarities
Long-tailed, Meadow, and Mountain have shorter heads and bodies. Western Red-backed Vole has reddish back, buffy sides.
Habitat
Wet mountain meadows, marshes, and streambanks, semiaquatic.
Range
Cascades of s. B.C., Wash., Oreg.; Rockies of se. B.C., sw. Alta., far e. Wash., ne. Oreg. (Blue Mts.), Idaho (n. ⅔ and far e.), nw. Wyo., n.-cen. Utah.

PRAIRIE VOLE

Microtus ochrogaster **61:7**

Description
Size: head and body, 3½–5 in. (8.9–12.7 cm); tail, 1⅕–1⅗ in. (3–4.1 cm); weight, $\frac{7}{10}$–1⅖ oz. (20–40 g); mammae, 6. Common vole of the prairies. Color grayish to blackish-brown, mixed with fulvous-tipped hairs; belly whitish or fulvous; base of tail yellowish-rusty.
Similarities
Meadow has longer tail; difficult to differentiate. Long-tailed has longer tail. Mountain and Heather prefer mountains. Western Red-backed has gray sides, red back; prefers mountains. Sagebrush is ash-gray; upper incisors grooved.
Habitat
Prairies of Great Plains.
Range
To w. of 100th meridian as far as se. Alta., s. Sask., Mont. foothills of Rockies, ne. Wyo., extreme ne. Colo., w. Nebr.

SAGEBRUSH VOLE

Lagurus curtatus **61:3**

Description
Size: head and body, 3⅘–4½ in. (9.7–11.4 cm); tail, ⅗–1⅛ in. (1.5–2.8 cm). Palest vole, and commonest in sagebrush regions. Face slender. Upper incisors grooved and differentially colored.

Pale gray above; underparts and feet white; tail usually less than 1 in. (2.5 cm).

Similarities
Prairie Vole is dark gray; tail longer. All other voles have longer tails and/or are not found in sagebrush.

Habitat
Short grass, sagebrush, arid regions.

Range
In s. Alta.; extreme sw. Sask.; se. Wash.; e. Oreg.; extreme ne.-cen., Calif.; Idaho, except n. panhandle; Mont., except nw. ¼; w. ⅔ Wyo.; n. ⅘ Nev.; w. and ne. Utah.

WATER RATS
Genus *Ondatra*

MUSKRAT
Ondatra zibethicus

55:4
Fig. 33

Description
Size: head and body, 10–14 in. (25.4–35.6 cm); tail, 8–11 in. (20.3–27.9 cm); weight, 2–4 lb. (0.9–1.8 kg); mammae, 8–10. The only mammal adapted for aquatic life with a vertically flattened tail. Pelage dense, thick, rather coarse guard hairs. Hind feet partly webbed and larger than forefeet. Upperparts varying from dark brownish and blackish to brightly reddish; underparts silvery; tail scaly, black, and nearly naked.

Habitat
Fresh and saltwater marshes; lakes, ponds, watercourses.

Habits
Highly aquatic, active at any hour; often seen swimming with head barely appearing above water and wedge-shaped. Builds a cone-shaped, beaverlike lodge 5 ft. (1.5 m) in diameter at base to 3 ft. (0.9 m) above water, of mud and sticks in marsh or tundra lake with an underwater entrance; burrows in mud banks, has feeding platform on a mat of rushes.

Voice
Moans, squeals, chatters from within lodge.

Food
Stems of cattails, grasses; mussels.

Range
Throughout N.A., except Arctic Alaska, n. of Brooks Range; Cascade Mts. and s.-cen. Oreg.; sw.-cen. Nev.; and extreme s. Ariz.; also occurs in far ne. Calif. tule marsh area.

Muskrat

Fig. 33

Nutria, p. 350

LEMMINGS
Genera *Lemmus* and *Synaptomys*

These small, volelike mammals have very short tails, less than one inch in length. Their long, soft, grizzled grayish-brown fur almost hides their small ears. They are colonial and active both day and night, and eat plants of various sorts.

BROWN LEMMING
Lemmus sibiricus   **61:9**

Description
Size: head and body, 4½–5½ in. (11.4–14 cm); tail, ⅘–1+ in. (2–2.5+ cm). Upper incisors not grooved. Pelage thick, long. Head grayish, body reddish, rump brown; underparts creamy to medium brown. No dorsal stripe.

Similarities
Collared lemmings have dark dorsal median stripe. Northern Bog Lemming is grayish-brown, incisors grooved.

Habitat
Tundra and adjacent forests.

Range
Alaska, except Aleutians and s. and se. parts; Yukon, except sw.; n. ⅔ Mackenzie; n.-cen. B.C.

NORTHERN BOG LEMMING
Synaptomys borealis   **53:6**

Description
Size: head and body, 4–4¾ in. (10.2–12.1 cm); tail, ⅘–1 in. (2–2.5 cm). Long, thick claws on front toes in winter. Upper incisors grooved. Upperparts pale to dark brown; underparts lighter.

Similarities
Voles of the Genera *Microtus* and *Clethrionomys* have ungrooved upper incisors. Heather Vole has longer tail; is an alpine species. Sagebrush Vole has whitish belly; found in sagebrush regions.

Habitat
Locally distributed in open or wooded, moist or dry areas.

Range
In Alaska, except sw. and n. of Brooks Range; s. ¾ Yukon; s. ½ Mackenzie; B.C.; extreme parts of n. Wash.; n. Idaho panhandle; and nw. Mont.

COLLARED LEMMINGS
Genus *Dicrostonyx*

This is a circumpolar group of lemmings that turn white in winter. They have one to eight young (usually three to four), born in each of two litters annually (June and July). The voice has been reported as a squeal or chuckling note.

COLLARED LEMMING
Dicrostonyx torquatus   **61:10**

Description
Size: head and body, 4–5½ in. (10.2–14 cm); tail, ⅖–⅘ in. (1–2 cm). Ears and tail very short; third and fourth foreclaws enlarged. Upper incisors not grooved. Summer: above brownish-black with

some buff, dark stripe down back, tawny collar across throat, creamy-buff below.
Similarities
Brown Lemming lacks dark streak down back; is brown in winter.
Habitat
Dry gravelly tundras or sandy areas.
Food
Leaves and other plants, especially cotton grass.
Range
Tundras of Alaska coast; n. Canada; and Arctic Ocean islands e. to Hudson Bay.
Other name
Greenland Collared Lemming.

OLD WORLD RATS AND MICE
Family Muridae

Old World rats and mice are grayish-brown to black above, usually grayish below. They have even-colored, nearly naked, long scaly tails; sixteen teeth, 1003/1003; and molars with three rows of tubercles. They frequent buildings, dumps, ships or fields; are active at all hours; swim; and do not hibernate. Their varied diet includes grain, groceries, garbage, and meat. They build nests of anything soft in almost any kind of hole. Gestation takes from nineteen to twenty-three days; there are five to nine young, and several litters a year.

BLACK RAT
Rattus rattus
Fig. 34

Description
Size: head and body, 7 in. (17.8 cm); tail, 9 in. (22.9 cm); weight, 2⅖–10½ oz. (70–300 g). Tail noticeably longer than head and body. Three phases: brown above and white below; brown above and gray below; or black above and gray below.
Similarities
Norway Rat has shorter tail than head and body. Woodrats are bicolored with haired tail.
Habitat
Mostly around buildings; rare in North, common in South.
Voice
A squeal or squeak.
Remarks
Fleas that infest this rat may be carriers of the plague or "black death."
Range
Along W. Coast from sw. B.C. to Mexico; far sw. Nev.; extreme sw. Ariz.

Fig. 34

Norway Rat, p. 348   Black Rat

NORWAY RAT
Rattus norvegicus

60:10
Fig. 34

Description
Size: head and body, 7–10 in. (17.8–25.4 cm); tail, 5–8 in. (12.7–20.3 cm); weight, 6$\frac{7}{10}$–10$\frac{1}{5}$ oz. (190–290 g). Pelage coarse; ears short; tail naked, scaly. Color grayish-brown; belly grayish, not white.

Similarities
Woodrats usually have white underparts and feet; tail haired. Black Rat has tail longer than head and body.

Habitat
Cities and farmyards.

Habits
Very destructive.

Voice
A squeal.

Range
In w., s., se. coast Alaska; B.C., except ne. $\frac{1}{2}$; s. Alta.; far s. Sask.; all U.S.

HOUSE MOUSE
Mus musculus

59:2

Description
Size: head and body, 3$\frac{1}{6}$–3$\frac{2}{5}$ in. (8.1–8.6 cm); tail, 2$\frac{4}{5}$–3$\frac{4}{5}$ in. (7.1–9.7 cm); weight, $\frac{2}{5}$–$\frac{4}{5}$ oz. (12–24 g). Pelage short, ears nearly naked. Dull gray to grayish-brown above, underparts grayish or buffy. Tail scantily haired, scales show clearly, unicolored.

Similarities
White-footed and Deer Mice have white bellies. Voles have short-haired tails. Harvest mice have incisors grooved. Jumping mice have white bellies.

Habitat
In or near buildings.

Voice
A squeak.

Range
In w., s., se. coast Alaska; B.C., except ne. $\frac{1}{2}$; s. Alta.; far s. Sask.; all U.S.

JUMPING MICE
Family Zapodidae

Members of this family are small, delicate mice with very long tails and hind legs that allow them to leap more than six feet. They differ from pocket mice and kangaroo rats by the absence of external cheek pouches. Their tails are scantily haired and are never tufted at the tip. They are yellow- or orange-brown above with darker, somewhat bristly hairs in a band along the back. The underparts are whitish. The tail is dark above and white below. Each upper incisor has a groove down the front. They prefer damp meadows and forests, and hibernate from late autumn until late spring. They are primarily nocturnal, although not uncommonly seen by day. Insects, seeds, berries, and other fleshy fruits make up their diet. Dentition is 1013/1003. Usually two litters of four to six young are produced each summer. The round nest is made of grasses and leaves and lined with finer materials, placed under logs and other debris, in tufts of grass or clumps of shrubs.

MEADOW JUMPING MOUSE

Zapus hudsonius

53:2

Description

Size: head and body, 3–3½ in. (7.6–8.9 cm); tail, 4–5¾ in. (10.2–14.6 cm); weight, ½–1³⁄₁₀ oz. (15–37 g). Skull narrow; tail scantily haired; upper premolars small; hind feet large. Sides grayish-yellow, dorsal band blackish.

Similarities

Western is larger; prefers mountains.

Habitat

In grass; often seen jumping like a frog.

Range

In cen., s., e. Alaska; n., ne., e. B.C.; Alta., except extreme sw. corner; Sask.; Mont., e. of Rockies; ne. and e. Wyo.; ne. Colo.; Nebr., except sw. corner; e. ½ Kans.

WESTERN JUMPING MOUSE

Zapus princeps

59:8
Fig. 35

Description

Size: head and body, 3–4 in. (7.6–10.2 cm); tail, 5–6⅕ in. (12.7–15.7 cm); weight, ⁷⁄₁₀–1 oz. (19–27 g); hind feet large. Sides yellowish; back darker, not sharply contrasting with sides; belly white or buff.

Similarities

Meadow is smaller; not in mountains.

Habitat

Mountains, except in northwest.

Range

In s.-cen. Yukon; B.C., except ne. ¼; s. ½ Alta.; sw. ¼ Sask.; w. ¾ Wyo.; w. ⅗ Colo.; nw. coast, n., ne., e.-cen. Calif; n. ½ Nev.; ne. ⅔ Utah; extreme e.-cen. Ariz.; n.-cen. and w.-cen. N.Mex.

Fig. 35

Western Jumping Mouse

PACIFIC JUMPING MOUSE

Zapus trinotatus

Description

Size: head and body, 3½–4⅕ in. (8.9–10.7 cm); tail, 5–6½ in. (12.7–16.5 cm); weight, ⅘–⁹⁄₁₀ oz. (23–26 g). Skull broad, deep; upper premolars large. Color various shades of ochreous and tawny, sides paler than back; lateral line distinct and bright; belly white; tail brown above, lighter below.

Habitat

Moist grassy areas or wooded areas with an understory of weeds and ferns.

Range

In s. B.C.; w. Wash.; coastal and Cascades; Oreg.; nw. Calif. coast.

PORCUPINE

Family Erethizontidae

The spiny-quilled porcupines are large, blackish rodents about the size of a small dog. Most of the body, especially the rump and tail, is thickly set with long, sharp, needle-tipped spines.

PORCUPINE

55:7
Erethizon dorsatum
Fig. 29

Description
Size: head and body, 18–22 in. (45.7–55.9 cm); tail, 7–9 in. (17.8–22.9 cm); weight, 10–30 lb. (4.5–13.6 kg); 20 teeth, 1013/1013.
Skull compact, broad, heavily constructed, heavily ridged. Underfur soft, covered with longer coarse guard hairs among which grow the quills. Underfur blackish; guard hairs and quills often light-tipped.
Habitat
Forests, preferably with conifers or poplars.
Habits
Usually nocturnal, but can sometimes be seen by day shuffling over the forest floor or hunched into a large ball in a tree; fearless, never attacks; slow, nests in a rocky den, burrow, hollow log.
Voice
Snort and bark; groaning and crying sounds; a high-pitched squeal.
Food
Twigs, leaves, and buds in summer; inner bark of conifers and hardwoods in winter.
Reproduction
1 young; well-developed, can climb trees and eat leaves when 2 days old.
Range
Alaska, except n., w., sw. coasts; all Canada; all w. states, except coastally from Vancouver Is. to s. Calif.; also absent from se. Calif. and extreme sw. Ariz.

NUTRIAS

Family Capromyidae

NUTRIA

Fig. 33
Myocastor coypus

Description
Size: head and body, 22–25 in. (55.9–63.5 cm); tail, 12–17 in. (30.5–43.2 cm); weight, 15–18 lb. (6.8–8.2 kg). A large, compact, robust Muskrat-like mammal strongly specialized for aquatic life. Skull heavy, well ridged. Brownish underfur fine, covered by longer coat of coarse overhair.
Similarities
Beaver and Muskrat have naked, flattened tails.
Habitat
Aquatic locales, marshes, bogs.
Habits
Swims well, makes shallow burrows in banks with an enlarged nesting chamber at rear.
Food
Aquatic vegetation.
Reproduction
2–8 young.

Range
Widely introduced into Oreg. and Calif. Native in South America.
Other name
Coypu.

Cetaceans
Orders Odontoceti and Mysticeti

Whales and their allies, collectively known as cetaceans, are the most completely aquatic mammals. They are distinguished by nostrils (blowholes) set high on the head; forelimbs modified into flippers; a horizontal tail (the flukes); and an absence of hind limbs. The habitat of most cetaceans is the open sea and few large whales are seen unless washed ashore. Smaller whales are sometimes visitors to bays, estuaries, and near-shore lagoons. Baleen whales are filter feeders, feeding on animal plankton (krill), which becomes abundant in the nutrient-rich colder waters. In the Northern Hemisphere, baleen whales feed to a greater extent on small, schooling fishes. Most toothed whales feed primarily on fishes or squids and, to a much lesser extent, on other invertebrates.

Some cetaceans indulge in extensive courtship before mating. Toothed whales are equipped with teeth, have a single blowhole, an asymmetrical skull, and nasal bones that are not part of the roof of the narial passage. In contrast, baleen whales lack teeth, being equipped instead with dense sheets of baleen that grow downward from the upper plates and form an effective sieve by which microorganisms are strained from the sea. The narial openings, or blowholes, are paired.

BEAKED WHALES
Family Ziphiidae

The beaked whales have tapering heads and elongated beaks, and a dorsal fin placed behind the midpoint of the back. Each side of the lower jaw bears one tooth, sometimes two. They are characteristically dark above and paler on the sides and underparts. Preferring the cold waters of open oceans, they are seldom observed near shore. Their food probably consists of squid and other cephalopods, but some fish are also eaten. Beaked whales of the Genus *Mesoplodon* are particularly poorly known, and information is based on a small number of stranded specimens. Consequently, the exact numbers of species and their characteristics have not been delimited.

NORTH PACIFIC BOTTLE-NOSED WHALE
Berardius bairdii
62:2
Description
Size: length, 35–42 ft. (10.7–12.8 m). Largest of the beaked whales. Skull symmetrical, 4 functional teeth in lower jaw; rostrum or beak slender. Dorsal fin small, set far back on body. Color black, with a whitish area on the lower belly.
Remarks
Rare, little known.
Range
Pacific Coast from Alaska to Calif.

NORTH PACIFIC BEAKED WHALE
Mesoplodon stejnegeri

Description
Size: length, 17 ft. (5.2 m). Snout elongated into a beak; dorsal fin small. Color blackish, gray on belly and head. Two throat grooves; 1 large tooth near tip on each side of lower jaw.
Remarks
Some experts feel that some stranded specimens of this type actually constitute a second species, *Mesoplodon carlhubbsi*.
Range
Pacific Coast, Alaska to Oreg.

GOOSE-BEAKED WHALE
Ziphius cavirostris   **62:7**

Description
Size: length, 18–28 ft. (5.5–8.5 m); teeth, 2. Body thick; distinct keel from dorsal fin to tail; no notch between flukes. Grooves on throat converge toward front. Color variable; back gray to black, occasional white on head and back; sides brownish or spotted; belly usually whitish.
Habits
Travels in gams of 30 or more, but often solitary.
Range
Pacific Ocean.

NORTH ATLANTIC BOTTLE-NOSED WHALE
Hyperoodon ampullatus   **62:3**

Description
Size: length, 20–30 ft. (6.1–9.1 m). Dorsal fin well back on body. Forehead of male rises abruptly from the short beak; old males recognizable by white dorsal fin and whitish patch on forehead. Color black to light brown or yellowish; whitish about head; belly whitish.
Habits
Highly sociable; migratory.
Range
Arctic Ocean.

GIANT AND PYGMY SPERM WHALES
Families Physeteridae and Kogiidae

SPERM WHALE
Physeter macrocephalus   **62:5**

Description
Size: length, of males, 40–60 ft. (12.2–18.3 m), of females, to 30 ft. (9.1 m); weight, to over 110,000 lb. Head large, square-snouted; lower jaw relatively small, narrow, and with 24 teeth; teeth in upper jaw sometimes form but do not erupt. Spout a single hole on left, directed forward. Head contains huge spermaceti organ, or oval sac containing unique waxy substance, above bones in front that serves in part to counteract pressures in deep diving and apparently in sound production. Dorsal fin absent; back slightly humped. Above bluish-gray, paler below.
Habits
Migratory, polygamous.

Food

Principally squid (including giant squid) and deep-sea sharks, rays, skates and bony fishes.

Range

Pacific, Arctic oceans, southward; most numerous between latitudes 50° N and 50° S.

PYGMY SPERM WHALE
Kogia breviceps

62:4

Description

Size: length, 9–13 ft. (2.7–4 m); weight, to 881 lb. (400 kg). The only very small whale with a protruding snout. Head blunt, rounded; lower jaw narrow, 12–16 teeth. Dorsal fin small, curved, toward rear. Color blackish above, light below.

Habits

Probably solitary, rarely seen, occasionally stranded.

Food

Squid, octopi.

Range

Pacific Coast, in warm waters; seldom observed.

Note: The **DWARF SPERM WHALE,** *Kogia simus,* a similar species, has 8 to 11 teeth.

WHITE WHALE AND NARWHAL
Family Monodontidae

These are medium-sized whales without a dorsal fin. They possess a single blowhole. They occur only in Arctic and north temperate waters.

WHITE WHALE
Delphinapterus leucas

62:1

Description

Size: length, to 18 ft. (5.5 m); weight, to 4000 lb. (1814.4 kg). Both jaws have 8–10 teeth on each side. The only white whale. Young are dark gray, then mottled, then yellowish; adults white.

Habits

Migratory; travels in gams, often ascends rivers.

Food

Squid, fish, crustaceans.

Range

Arctic and subarctic seas.

Other name

Beluga Whale.

NARWHAL
Monodon monoceros

Description

Size: length, to 18 ft. (5.5 m); weight, to 3000 lb. (1360.8 kg). Male recognizable by its forward projecting, twisted tusk up to 9 ft. (2.7 m) long; female has no visible tusk. Snout blunt, eyes small; ridge along midback. Above mottled gray, below white. The gray young may be confused with young White Whale.

Habits

Can dive to 1200 ft. (365.8 m); males may use tusk in fighting.

Voice
Females may roar in calling young.
Food
Shrimp, cuttlefish, fish.
Range
High Arctic seas.

PORPOISES AND DOLPHINS
Family Delphinidae

Members of this family are small- to medium-sized whales with teeth in both jaws. They have a single blowhole far back from the snout, and the dorsal fin is near the middle of the back. They are generally dark above and lighter below. They feed on fish or squids.

ROUGH-TOOTHED PORPOISE
Steno bredanensis

Description
Size: length, to 8 ft. (2.4 m); 20–27 teeth, with fine vertical wrinkles, in each side of upper and lower jaws. Beak not distinctly set off from forehead; rostrum long, narrow, compressed. Purplish-black dorsally, sides with yellowish-white spots; beak and ventral surface white, tinged with rose and purple.
Range
Known in continental U.S. only from Stinson Beach, Marin Co. in Calif.; common around Hawaii.

PACIFIC BOTTLE-NOSED DOLPHIN
Tursiops truncatus   **63:1**

Description
Size: length, 10–12 ft. (3–3.7 m); 20–26 teeth on each side of upper and lower jaws. Color grayish-black above, white below except for dark area from vent to fluke; white on upper lip.
Range
W. Coast, Calif. to Baja Calif.
Other name
Gill's Bottle-nosed Dolphin.

NORTHERN RIGHT-WHALE DOLPHIN
Lissodelphis borealis   **62:6**
                                           Fig. 39

Description
Size: length, 5–8 ft. (1.5–2.4 m); 43–45 teeth on each side of upper and lower jaws. Large and streamlined body; no dorsal fin. Color black; narrow white belly stripe from breast to tail.
Range
W. Coast.

COMMON DOLPHIN
Delphinus delphis   **63:6**

Description
Size: length, 6½–8½ ft. (2–2.6 m); weight, to 180 lb. (81.6 kg); 40–50 small teeth on each side of upper and lower jaws. The average-sized dolphin that plays about ships. Beak is about 6 in. (15.2 cm).

Back and flippers black, flanks yellowish, belly white. White eye-ring around each eye connected across groove which separates beak from forehead by 2 white lines.

Habits
Travels in schools, follows ships, makes graceful leaps out of water.

Range
Pacific Coast.

Other name
Saddle-backed Dolphin.

Note: The **PACIFIC DOLPHIN**, *Delphinus bairdi*, is similar to the Common Dolphin, but its flanks are banded with golden. It ranges from British Columbia, south along the coast to Baja California, and has recently been considered to be a localized subpopulation of the Common Dolphin.

PACIFIC WHITE-SIDED DOLPHIN
Lagenorhynchus obliquidens

63:3

Description
Size: length, 7–9 ft. (2.1–2.7 m); 29–31 teeth on each side of upper and lower jaws. Nose blunt; rostrum short. Greenish-black above, pale stripe along sides, belly white.

Range
Pacific Coast, from Alaska s. to Calif.

KILLER WHALE
Orcinus orca

63:5

Description
Size: length, 15–30 ft. (4.6–9.1 m); 10–15 teeth on each side of upper and lower jaws. Dorsal fin large, exposed by cutting water (indication of whale's presence); nose blunt. Flippers very broad. Color jet-black, with white extending up on sides posteriorly; clear white spot behind each eye; belly white.

Habits
Occurs in schools of up to 40 or more.

Food
Seals and fish.

Range
Pacific Coast; very common in Aleutian waters.

GRAMPUS
Grampus griseus

63:7

Description
Size: length, 9–13 ft. (2.7–4 m); 2–7 teeth per jaw side. Nose rather blunt; flippers slender. Body dark gray or blackish marked with numerous irregular streaks; head tinged with yellow; belly grayish-white; flippers mottled grayish.

Range
Pacific Ocean.

Other name
Risso's Dolphin.

FALSE KILLER WHALE
Pseudorca crassidens   *Fig. 36*

Description
Size: length, 13–18 ft. (4–5.9 m); 8–10 teeth per side of jaw. A small, slender whale. Snout blunt, rounded; head flattened. Dorsal fin relatively small, recurved, just in front of midback. Color black.

Range
Pacific Ocean.

False Killer Whale

Fig. 36

COMMON PILOT WHALE
Globicephala melaena   **63:8**

Description
Size: length, 14–28 ft. (4.3–8.5 m); 7–12 teeth per side of jaw. Skull largest of North America species in proportion to body size; forehead high, bulges forward. Dorsal fin rather large, recurved, well forward of midback; flippers about ⅕ body length. Uniformly black, sometimes white blaze on back behind dorsal fin.

Habits
Travels in large schools; occasionally becomes stranded on shore.

Remarks
Southern California and Mexican Pilot Whales may belong to separate species, *Globicephala scammoni*.

Range
Pacific Ocean.

HARBOR PORPOISE
Phocoena phocoena   **63:4**

Description
Size: length, 4–6 ft. (1.2–1.8 m); weight, to 50 lb. (22.7 kg); 23–27 teeth in each tooth row. Skull small; rostrum short, broad. Teeth small, compressed, spadelike. Dorsal fin triangular in profile. Slate-gray to black above; flanks grayish fading to white below.

Habits
Frequents shorelines, in harbors; common.

Range
Pacific Coast, from Alaska to Mexico.

Other name
Common Porpoise.

DALL'S PORPOISE
Phocoenoides dalli   **63:2**

Description
Size: length, 5–6 ft. (1.5–1.8 m); 23–27 teeth per side of jaw. Strikingly marked. Head with relatively short, flat beak. Color black, except for large white area across the vent region, extending slightly over halfway up each side.

Habits
Frequently enters open channels in summer; usually occurs in groups of 2–12.
Range
Pacific Coast, from Alaska to Santa Barbara Is., Calif. (rarely).

GRAY WHALE
Family Eschrichtiidae

GRAY WHALE
Eschrichtius robustus
Fig. 37
Description
Size: length, 35–45 ft. (10.7–13.7 m); baleen, 1 ft. (0.3 m) whitish.
A large, blotched whale often seen fairly close inshore, especially in migration. Rather slender; with 2–5 longitudinal folds on throat. Series of dorsal bumps, but no dorsal fin on back. Color blotched grayish-black.
Habits
Spouts are quick, not more than 10 ft. (3 m) high; migrates between Arctic Ocean where it feeds in summer to Baja California where it calves and breeds in winter.
Range
Pacific Coast.

Fig. 37

Gray Whale

FINBACK WHALES
Family Balaenopteridae

Members of this family have short, broad plates of whalebone or baleen hanging in rows from each side of the upper jaw, and they possess no teeth as adults. A dorsal fin and a double row of blowholes are present. The throat has many furrows. The largest species, the Blue Whale, may exceed one hundred feet in length. The skulls have a flat, broad rostrum. They are commonly called fin whales or rorquals.

BLUE WHALE
Balaenoptera musculus

Description
Size: length, to 106 ft. (32.3 m); weight, to 140 tons (127 t); baleen, to 40 in. (101.6 cm), bluish-black; 80–100 throat furrows. Largest of the whales, identifiable by its great size and U-shaped snout. Dorsal fin small, set far back. Color slaty- to bluish-gray above, yellowish or whitish below.
Habits
Spouts to 20 ft. (6.1 m) vertically; migrates to warm water to breed, travels singly or in gams; at nonbreeding times frequents waters near pack ice.

Food
Mainly small, shrimplike invertebrates (krill).
Remarks
The largest animal ever known to exist, 2 to 3 times as large as the greatest Mesozoic dinosaur, *Brontosaurus*.
Range
Pacific Ocean.

FIN WHALE
Balaenoptera physalus *Fig. 38*

Description
Size: length, 60–80 ft. (18.3–24.4 m); weight, to 70 tons (63.5 t); baleen, to 3 ft. (0.9 m); 70–80 throat furrows. The only whale with whitish-, yellowish-, or purplish-streaked whalebone. Head flat; dorsal fin small, with slightly concave rear edge, just in front of flukes. Color gray above, white below.
Habits
Spout 15–20 ft. (4.6–6.1 m) high, inclined forward, narrow at start then elliptical, made with loud whistling sound. Fastest of all whales; comes near ships.
Food
Plankton, small crustaceans.
Range
Pacific Coast.

Fig. 38 Fin Whale

Minke Whale

MINKE WHALE
Balaenoptera acutorostrata *Fig. 38*

Description
Size: length, 20–33 ft. (6.1–10.1 m); weight, to about 3 tons (2.7 t); baleen, to 8 in. (20.3 cm), white; 50–70 throat furrows. The only cetacean with a broad white band across the flipper. Snout triangular from above; dorsal fin far back, tip curved. Color varying shades of dark above, white below.
Habits
Spout faint, vertical; migratory, travels singly or in small groups; frequents coastal waters.
Food
Crustaceans and small fish.
Range
Pacific Ocean.
Other name
Little Piked Whale.

HUMP-BACKED WHALE
Megaptera novaeangliae   *Fig. 39*

Description
Size: length, 40–50 ft. (12.2–15.2 m); weight, 25–45 tons (22.7–40.8 t); baleen plates black with black or olive-black bristles, short, coarse. No other finback has a humped back and scalloped flippers and flukes. Body relatively short, thick; pectoral fins long, about ⅓ body length, dorsal fin small; throat furrowed. Flippers have fleshy knobs along front edge; fluke is irregular, scalloped, in outline on its posterior border. Color black above, white below.

Habits
Spout is a 20-ft. (6.1-m) expanding column; travels singly or in small schools, can leap clear of (broach) the water.

Food
Small crustaceans.

Voice
Complex songs have been recorded.

Range
Oceans of the world.

Northern Right-Whale Dolphin, p. 354

Fig. 39

Hump-backed Whale

RIGHT WHALES
Family Balaenidae

These are the only whales with long, narrow, and flexible plates of whalebone, or baleen, hanging in rows from each side of the upper jaw, smooth throats, double blowholes, and no dorsal fin. When sounding (diving deeply), the flukes are thrown clear of the water.

BLACK RIGHT WHALE
Balaena glacialis

Description
Size: length, 60–70 ft. (18.3–21.3 m); weight, to 60 tons (54.4 t); baleen, to 9 ft. (2.7 m), black. Head large, about ⅕ body length; baleen numerous, dorsal fin absent. Skull has convex rostrum in lateral view. Color blackish, often pale below.

Habits
Spout 10–15 ft. (3–4.6 m) high, in 2 columns which diverge to form a V; slow-moving; travels singly or in small gams; uncommon.

Food
Plankton and other small invertebrates.

Range
Pacific Coast, from Alaska, Aleutian Is., s. to Baja Calif.

Other name
Pacific Right Whale.

BOWHEAD WHALE
Balaena mysticetus

Description
Size: length, 50–65 ft. (15.2–19.8 m); weight, to over 50 tons (45.4 t); baleen, to 14 ft. (4.3 m), black. The only whale with a head more than ⅓ its body. Body very stout; lower jaw bowed gently upward, not arched. Color dark grayish-brown above; belly may be spotted with white; lower jaw and throat white with string of black spots.

Habits
Spout V-shaped; does not migrate; not gregarious. When basking, part of back may project above the water.

Range
Circumpolar, polar, and subpolar seas.

Other name
Greenland Right Whale.

Carnivores
Order Carnivora

Members of this order are distinguished by their large canine teeth and strong jaws, legs, and claws. There are three incisor teeth in each of the lower jaws. They are primarily flesh-eaters, although many also eat berries, nuts, and fruits. All have five toes on the front foot and four or five on the hind. The smallest terrestrial carnivore, the Least Weasel, weighs less than two ounces; the largest, the Grizzly Bear, weighs up to 1500 or more pounds (680.4 kg). The females are smaller than the males, sometimes by as much as one-third. Carnivores occur in all latitudes and at all elevations; they occur in low, hot deserts, humid rain forests, and up to timberline. Females breed once or twice a year, bringing forth a litter of blind and usually furred young in a secluded den or burrow. These usually stay with the mother into the summer or fall or, in the case of some species, into the next year. Some muskolids have a prolonged gestation period (up to one year), but most of the growth of the embryo takes place in the final few weeks.

COYOTES, WOLVES, AND FOXES
Family Canidae

All members of this family are doglike, with bushy tails. They have four toes on each hind foot. The dentition is 3142/3143. All have a scent gland on the top of the tail near the base, its position revealed by black-tipped hairs without underfur. Their claws are nonretractable. They walk on their toes and are adapted to running.

COYOTE
Canis latrans   65:1, 66:5, 66:8

Description
Size: head and body, 32–54 in. (81.3–137.2 cm); tail, 11–16 in. (27.9–40.6 cm); height at shoulder, to 26 in. (66 cm); weight, 20–50 lb. (9–22.7 kg). Distinguished by long, bushy tail that droops when running; appearance is very doglike. Nose and ears pointed, nose pad narrow, less than 1 in. (2.5 cm); legs long. Color gray or reddish-gray; legs, feet, ears rusty; throat and belly whitish; tail black-tipped.

Similarities
Foxes are smaller and hold tail horizontally when running. Gray
Wolf is larger and holds tail high when running; has wider nose
pad.
Habitat
Brush country, ranches, farmlands; prefers open spaces.
Habits
Mostly nocturnal but often active by day.
Voice
Wild, doglike howl, often ending in a series of *yap-yap-yap*'s; in
evening utters a series of high-pitched *yap*'s, frequently in chorus.
Food
Small mammals and ground-nesting birds; in summer, also fruits
and berries.
Range
From cen. Alaska to Mexico.; e. from the Coast (except Vancouver
Is.) to Great Lakes.
Other name
Prairie Wolf.

GRAY WOLF
Canis lupus

66:3, 66:9
Fig. 40

Description
Size: head and body, 43–48 in. (109.2–121.9 cm); tail, 12–19 in.
(30.5–48.2 cm); height at shoulder, 26–28 in. (66–71.1 cm);
weight, 70–170 lb. (31.8–77.1 kg). Largest of canines. Pelage long;
ears more rounded, nose pad wider, more doglike appearance than
Coyote. Color varies from dark gray and almost black to nearly
white in Arctic; outside Alaska, gray.
Similarities
Coyote is smaller and carries tail low when running; nose pad
narrower.
Habitat
Wilderness only; tundras, plains, forests.
Habits
Carries tail high when running; travels singly in pairs, or in packs;
clever, hunts cooperatively.
Voice
Loud howl, various barks, answer one another.
Food
Young, sick, and aged deer; mountain sheep, bison, and musk oxen;
small mammals; also berries and fruits.
Range
Alaska; Canada, except Great Plains area; formerly an inhabitant
of most of U.S.; occurs now only in s.-cen. Cascades of Oreg., se.
Utah, sw. Colo., and extreme sw. Ariz.
Other name
Timber Wolf.

Fig. 40

Gray Wolf

ARCTIC FOX
Alopex lagopus

66:4
Fig. 41

Description
Size: head and body, to 31+ in. (78.7+ cm); tail, 11+ in. (27.9+ cm); height at shoulder, to 12 in. (30 cm); weight, 7–15 lb. (3.2–6.8 kg). The only fox with short rounded ears and, in the winter white phase, all-white color. Feet heavily furred. Blue phase in summer is brownish or gray above, with yellowish-white sides and belly; in winter, slate-blue, sometimes with brown on head and feet, no white tip to tail. White phase in summer is like blue phase; in winter is all-white.

Similarities
Red Fox is reddish-yellow with white tip on tail.

Habitat
Tundra and Arctic coast; introduced successfully on various Aleutian Islands.

Habits
Unsuspicious; burrows in snow for temporary den; solitary or travels in pairs, but never in packs.

Voice
High-pitched yapping bark.

Food
In winter, dead seals, walruses, whales; in summer, rodents, birds, eggs, berries.

Remarks
The Blue Fox is a color phase occurring chiefly in western Alaska and Aleutians, where large numbers have been commercially ranched.

Range
In w., n. Alaska, Aleutians; n. Canada and islands of Arctic Ocean.

Fig. 41
Arctic Fox

summer   winter

RED FOX
Vulpes vulpes

65:2; 66:4, 66:7

Description
Size: head and body, 22–42 in. (55.9–106.7 cm); tail, 14–16 in. (35.6–40.6 cm); height at shoulder, 14–16 in. (35–40 cm); weight, 8–15 lb. (3.6–6.8 kg). The only fox with a white-tipped tail. Red phase is reddish above, white below; feet black. Black phase is all-black, white tip on very bushy tail. Cross phase is intermediate between Red Fox and Silver Fox; reddish-brown, heavy black markings on shoulder.

Habitat
Dry uplands with open areas; suburbs; less often lowlands. Also damp but not frozen tundras, as of Aleutian Islands.

Habits
Has no winter den, sleeps in open; home range about 2-mi. (3.2-km) diameter.

Voice
Male, "a short yelp, ending in a *yurr,* as if gargling"; female, "a yapping scream" (Palmer).

Food
Rabbits, rodents, snakes, berries, and fruit.

Remarks
Silver Fox is a black Red Fox and may occur in the same litter with red pups; the black phase dominates the red phase in the northern range.

Range
Alaska; Canada, except w. coast and offshore islands; Cascades of Wash.; w. ½, e. ¼ Oreg.; Idaho, except far s.; Mont.; Wyo.; cen. Nev.; Utah, except extreme nw. and sw. corners; w. ⅔ Colo.; extreme ne. Ariz.; nw. ⅔ N.Mex.

SWIFT FOX
Vulpes velox

65:3; 66:4

Description
Size: head and body, 15–31 in. (38.1–78.7 cm); tail, 9–12 in. (22.9–30.5 cm); height at shoulder, 10–12 in. (25–30 cm); weight, 4–6 lb. (1.8–2.7 kg). Smallest fox and only one with a black-tipped tail; no black stripe down top of bushy tail. Body slender, ears large. Color gray to pale yellowish-brown; blackish spot on either side of snout.

Habitat
Deserts, plains, dry foothills.

Habits
Nocturnal, shy; rarely seen.

Voice
Weak bark.

Food
Mice and other small mammals, insects, some fruit.

Range
In se. Alta.; sw. and s. Sask.; extreme sw. Man.; Mont., w. of Rockies; Wyo., except nw. ¼ and extreme sw. corner; n. and e. Colo.; n. and w. Tex.; also se. Oreg.; sw. Idaho; cen. and s. Calif.; Nev., except extreme ne. and w. corners and nw. ⅛; w. Utah; w., s., ne. Ariz.; sw. ½ N.Mex.

Other name
Kit Fox.

Note: The **KIT FOX**, *Vulpes macrotis,* is a closely related western desert representative, differing from the Swift Fox in having a narrower skull, larger ears, and being slightly smaller.

GRAY FOX
Urocyon cinereoargenteus

65:5, 66:4

Description
Size: head and body, 21–44 in. (53.3–111.8 cm); tail, 11–16 in. (27.9–40.6 cm); height at shoulder, 14 in. (35 cm); weight, 7–13 lb. (3.2–5.9 kg). The only fox with a black streak down the middle of its tail. Upperparts salt-and-pepper gray; sides, legs, feet rusty; underparts gray; tail long, bushy.

Similarities
Red Fox has white-tipped tail. Swift Fox has black on tail only at tip. Coyote is larger with black on tail only at tip.

Habitat
Brush, wooded lowlands, chaparral, swamps; common.

Habits
Largely nocturnal, solitary; can climb trees.

Voice
Harsh; rarely heard.

Food
Rabbits, rodents, reptiles, berries, and fruits.
Range
In w. Oreg.; Calif., except extreme ne. corner; s. ½ Nev.; s. ½ Utah; Colo., except e. ⅓; Ariz.; N.Mex.; Tex., except ne. corner of panhandle.

BEARS
Family Ursidae

Bears are the largest living carnivores. They are heavily built, with very short tails and nonretractile claws. They have five toes on each foot, and their small ears are almost concealed in their long fur. The tooth formula is 3142/3143. They walk on the entire foot, like humans, and often stand erect on their hind legs. The gestation period is seven to nine months.

BLACK BEAR
Ursus americanus   **66:1, 67:3**

Description
Size: length, 5–6 ft. (1.5–1.8 m); height at shoulder, 2–3 ft. (0.6–0.9 m); weight, 200–400 lb. (90.7–181.4 kg). The smallest, most common bear and, in the black phase, the only black bear. Western form is cinnamon or black to nearly white or blue (called "Blue" or "Glacier" Bears). Face is always brown; patch of white usually adorns the breast.
Similarities
Grizzly is larger with hump on shoulders.
Habitat
Forests, swamps, mountains.
Habits
Solitary, quarrelsome; hibernates in winter, or dormant; marks home range by clawing boundary trees.
Voice
Various whines, grunts, huffs.
Range
From s. of tundra throughout Alaska and Canada; Wash., except se. corner; w. Oreg.; nw. Calif. and Cascade-Sierras; w. Mont.; Idaho, except extreme sw.; w. ⅔ Wyo.; Utah, except w. deserts; w. ½ Colo.; n. and e. Ariz.; w. ½ N.Mex.; far w. Tex.
Other names
Brown Bear, Cinnamon Bear.

GRIZZLY BEAR
Ursus horribilis   *Fig. 42*

Description
Size: length, 6–7 ft. (1.8–2.1 m); height at shoulder, 3–3⅓ ft. (0.9–1.1 m); weight, 325–850 lb. (147.4–385.6 kg). Only bear with hump on its shoulders; largest bear outside Alaska. Claws on front feet to 4 in. (10.2 cm) long, twice as long as on hind feet, curved; face profile "dished" in. Color deep brown, darker along spine, limbs, and ears; grizzled or light-tipped hairs on upper parts. Color may vary to yellowish, grayish, or blackish.
Similarities
Black is smaller; claws on front foot not noticeably large; no hump on shoulders.

Habitat
In forested areas of oak and beech trees.
Habits
Wanders widely within home range, swims well; cubs, but not adults, can climb trees.
Voice
Various whines, grunts, coughs, huffs, and roar.
Range
Alaska, s. of Brooks Range; Yukon; Mackenzie; B.C.; w. Alta.; w. Mont.; w. Wyo.; w. Colo.; n.-cen. N.Mex.
Other name
Silver Tip.

Polar Bear

Fig. 42

Grizzly Bear

POLAR BEAR
Ursus maritimus

Fig. 42

Description
Size: length, 6½–8 ft. (2–2.4 m); height at shoulder, 3–4 ft. (0.9–1.2 m); weight, 600–1100 lb. (272.2–499 kg). Only white bear. Head small, neck long; claws not strongly curved but with well-developed cutting edges. Pelage dense; underfur waterproof. Color white, but often with a yellowish tinge; young whiter than adults; eye, nose pad, foot black.
Habitat
Spends most of time on ice floes; when on shore remains near water.
Habits
Roams widely, swims excellently, a powerful fighter; female becomes dormant in winter.
Voice
A roar.
Food
Seals, young walruses, stranded whales.
Range
Along n. coast of Alaska and Canada; also Arctic Ocean floes and islands.

EARED SEALS
Family Otariidae

Primarily marine carnivorous mammals that come ashore to breed, these seals have small external ears and hind flippers (legs) that can be turned forward to assist in movement over land. The males are up to four and one half times as large as the females. Their bodies are slender and elongate, and they have thirty-four to thirty-eight teeth. The females have four mammae.

NORTHERN FUR SEAL
Callorhinus ursinus   68:1

Description
Size: length of male, to 6 ft. (1.8 m), of female, to 4½ ft. (1.3 m); weight of male to 700 lb. (317.5 kg), of female to 135 lb. (61.2 kg); 38 teeth, 3142/3141. Pelage soft, lustrous. Skull facially broad, convex, slightly depressed. Muzzle short in male and forehead with distinctive crest. Males blackish above, gray on shoulders and front of neck, belly reddish, face brownish; females gray above, reddish below.

Similarities
Northern Sea Lion is larger, not reddish below. Harbor Seal is spotted.

Habitat
Open sea 6–8 months each year; rocky islands and shores, especially when breeding.

Habits
Gregarious; fast swimmer (to 17 mph.), dives to 30 fathoms (54.9 m).

Voice
Roar and bellow.

Food
Mostly small fish.

Reproduction
1 young; breeds shortly after pups are born; gestation about 11 months.

Remarks
Economically important; the Pribilof Islands herd is carefully managed and harvested for pelts, oil, and meat.

Range
Pacific Coast from Pribilof Islands to s. Calif.

Other name
Alaska Fur Seal.

NORTHERN SEA LION
Eumetopias jubatus   68:4

Description
Size: length of male, to 13 ft. (4 m), of female, to 9 ft. (2.7 m); weight of male, to 2000 lb. (907.2 kg), of female, to 600 lb. (272.2 kg); 34 teeth, 3141/2141. Pelage harsh, no underfur; males with mane on neck. Skull large; forehead of male without crest (low profile). Color buff or yellowish-tan, rather dark, but lighter right after molt; naked parts of skin black.

Similarities
California Sea Lion is smaller; darker, with high forehead. Alaska Fur Seal is much smaller; reddish below, face brown. Harbor Seal is spotted. Elephant Seal has no external ears.

Habitat
Open ocean.
Habits
Quiet unless molested.
Voice
As in other sea lions but a bit deeper bellow.
Food
Squid, pollack, sand lances, flounders, sculpin, cod, herring, small sharks, perch, some salmon, halibut, and sablefish.
Reproduction
Usually 1 pup, on a beach; 3 months nursing; breed June and July.
Range
Pacific Coast from Bering Sea and Aleutians (common) to off Santa Rosa Is., Calif.; often seen off coast at San Francisco.
Other name
Steller's Sea Lion.

CALIFORNIA SEA LION
Zalophus californianus

68:3

Description
Size: length of male, to 8 ft. (2.4 m), of female, to 6 ft. (1.8 m); weight of male, to 1000 lb. (453.6 kg), of female, 150–200 lb. (68.0–90.7 kg); 36 teeth, 3142/2141. Skull slender, elongated, high forehead; muzzle broad, heavy; males with conspicuous crest on top of head (conspicuous in profile). Ears small, pointed; eyes large. Color blackish when wet; varying from light buff to deep sepia when dry.
Similarities
Northern Sea Lion is larger, paler, low forehead, seldom barks. Elephant Seal is much larger, no external ears, usually quiet. Harbor Seal is spotted.
Habitat
Open to onshore seas.
Habits
Gregarious, occasionally comes ashore on sandy beaches; can swim to 10 mph. Estimated North American population over 100,000.
Voice
Bark, bellow.
Food
Squid, octopi, variety of fishes.
Reproduction
1 young.
Remarks
Barking or honking almost continual during breeding season.
Range
Pacific Coast, s. from B.C. to Baja Calif. and adjacent mainland Mexico; principal breeding grounds on offshore islands of s. Calif., Baja Calif., and Gulf of Calif.

HAIR SEALS
Family Phocidae

These earless seals have hind flippers that extend behind the tail and cannot be turned forward, so that, on land, they must wriggle their way forward. They have short, thick necks and claws on all digits; the sexes are nearly alike. They inhabit coastal and offshore marine waters and ice floes, feeding on fish, squid, crustaceans, and mollusks. The dentition is usually 3141/2141. The fur is coarse and without underfur in adults. Incisors are simple with pointed crowns; molars and premolars are not distinguishable.

HARBOR SEAL
Phoca vitulina   68:5

Description
Size: length, to 5 ft. (1.5 m); weight, to 255 lb. (115.7 kg).; 34–36 teeth; mammae, 2. Cheek teeth large and often set obliquely in jaw. The only spotted seal. Color highly variable, usually yellowish-gray above varied with irregular dark brown or black spots; sometimes brown with gray spots, or uniform silvery-gray or brownish-black; ventral area paler, lacking spots.

Similarities
Sea lions and fur seals have no spots, external ears, can rotate hind flippers forward. Northern Elephant Seal is larger, no spots.

Habitat
Coastal waters, frequenting harbors, bays, mouths of rivers, even inland lakes.

Habits
Forms into loosely organized colonies but does not form harems; travels seasonally. North American population estimated between 40,000 and 100,000.

Voice
Unimpressive grunts and barks.

Range
Arctic and Pacific oceans as far s. as Baja Calif.; also freshwater lakes in Alaska.

BEARDED SEAL
Erignathus barbatus   68:6

Description
Size: length of male, to 12 ft. (3.7 m), of female, to 8 ft. (2.4 m); weight of male, to 1000 lb. (453.6 kg), of female, to 500 lb. (226.8 kg); mammae, 4. Distinguished by thick tufts of bristles on each side of its muzzle. Forehead high, muzzle broad. Color uniformly dark grayish to yellowish, slightly darker on back.

Habitat
Shallow water, 2½–4⅙ feet (.7610–1.2710 m) at edge of ice pack; mouths of small bays and rivers.

Habits
Solitary, except in breeding season; sluggish, only slightly migratory; scrapes mollusks from bottom with claws.

Range
Arctic Ocean, s. to Bering Sea.

Other name
Square Flipper.

NORTHERN ELEPHANT SEAL
Mirounga angustirostris   **68:7**

Description
Size: length of male, to 20 ft. (6.1 m), of female, to 11 ft. (3.4 m); weight of male, to 8000 lb. (3628.7 kg), of female, to 2000 lb. (907.2 kg); 30 teeth, 2141/1141; mammae, 2–4. Largest of the seals and the only with a large, overhanging, proboscislike snout on the male. Snout inflatable, hind feet bilobed, claws rudimentary when present. Color brown to grayish, lighter on belly; nearly naked.

Similarities
Sea lions and fur seals are much smaller with external ears; can rotate hind flippers forward. Harbor Seal is much smaller, usually spotted.

Habitat
Warmer waters, sandy beaches.

Habits
Gregarious, solitary when at sea, lie close together on sandy beaches; nocturnal; bulls maintain a harem of several females. Population estimated at 8000 to 10,000.

Voice
Loud bellow.

Range
Pacific Ocean, s. from B.C. to islands off s. Calif. and Baja Calif.; once nearly extirpated, except for small population on Guadalupe Is.

WALRUS
Family Odobenidae

Walruses have hind flippers that can be turned forward to assist in land locomotion. Both sexes have large ivory tusks projecting downward from the upper jaw. The body is thick and heavy; the head relatively short; and the muzzle is blunt, broad, with coarse bristles. They have no external ears. There are eighteen to twenty-four teeth, and on the females, four mammae. Males are much larger than females. The hide is nearly naked.

WALRUS
Odobenus rosmarus   **68:2**

Description
Size: length of male, to 12 ft. (3.7 m), of female to 9 ft. (2.7 m); weight of male, to 3000 lb. (1360.8 kg), of female, to 1800 lb. (816.5 kg). Skull thick, heavy, swollen anteriorly. Teeth 1130/0130. The only marine mammal with 2 large white ivory tusks. Hair short, becoming sparse with age; skin thick, wrinkled; feet bear 5 toes, each with a nail. Color black when wet, bay when dry.

Habitat
Ice floes and Arctic islands; offshore waters of Bering Sea coasts and Aleutian Islands; estimated population 45,000–90,000.

Habits
Usually found in groups; uses tusks as "digging sticks"; bottom feeder in 50 fathoms (91.4 m) or less.

Voice
Bellow; elephantlike trumpeting.

Food
Clams; occasionally seals.

Remarks
Economically important to Eskimos for meat, hides, and ivory carvings.
Range
Arctic Ocean and ne. Bering Sea; occasionally s. to Umnak and Unalaska islands of Aleutians.

RACCOONS AND ALLIES
Family Procyonidae

Members of this family are small- to medium-sized carnivores about the size of a dog, usually with long tails. The molars are 2/2 or 2/3, low-crowned, broad, and multituberculate; carnassial teeth not well developed. The tooth formula is 3142/3142. They walk on the entire foot, which has five toes and semiretractile or nonretractile claws. The tail has distinct yellowish-white rings or very indistinct rings.

RACCOON
Procyon lotor   **60:6, 67:7**

Description
Size: head and body, 18–33 in. (45.7–83.8 cm); tail, 8–12 in. (20.3–30.5 cm); weight, 6–35 lb. (2.7–15.4 kg). The only mammal with black face mask and bushy, ringed tail. Body stout, fur long, snout and ears pointed. Color grizzled gray, brown, and black; black mask on face; tail with 4–6 black rings. Nonretractile claws.
Habitat
Woods, swamps; lives in trees, feeds along water's edge.
Habits
Nocturnal; washes food; a good fighter, climber, and swimmer.
Voice
Varied, includes barks, growls, a throaty cry, a whine, and an owllike quaver; often a shrill night cry.
Food
Fish, crayfish, birds, eggs, corn, vegetables, fruit.
Range
In s. B.C., plus Vancouver Is.; Wash.; Oreg.; Idaho; Calif., except se.-cen.; n., w.-cen. Nev.; s., e. Utah; Ariz.; N.Mex.; s. Alta.; s. Sask.; ne., e. Mont.; e., s. Wyo.; Colo.; s. to Mexico.

RINGTAIL CAT
Bassariscus astutus   *Fig. 43*

Description
Size: head and body, 14–17 in. (35.6–43.2 cm); tail, 19 in. (48.2 cm); weight, 30-45 oz. (870-1300 g). Feet thickly furred between pads, claws semiretractile. Ears and eyes large. Upperparts light buff to pinkish-buff, overcast with black or dark brownish overhairs; underparts white, may be washed with pale buff; eye ringed black or dark brown; tail long and ringed with whitish and blackish-brownish rings.

Fig. 43

Ringtail Cat

Similarities
Raccoon has shorter tail and black mask. Coati has indistinctly ringed tail.
Habitat
Arid regions, in rough country; broken hillsides in chaparral.
Habits
Good climbers; nocturnal.
Food
Omnivorous; eats small mammals, birds, eggs, berries, other fruit.
Range
In sw. Oreg.; Calif., except ne. corner and Central Valley; s. Nev.; s. ½ Utah; w. Colo.; Ariz.; sw. ¾ N.Mex.; w. Tex.

COATI
Nasua narica   *Fig. 44*

Description
Size: head and body, 20–25 in. (50.8–63.5 cm); tail, 20–25 in. (50.8–63.5 cm); weight, 15–25 lb. (6.8–11.3 kg). Long-snouted, tough nose pad, aids in rooting out grubs and tubers. Nonretractile claws. Upperparts pale brown to reddish, often overlaid with yellow; neck and shoulders yellowish; eyes masked with pale umber to brown; muzzle, chin, throat whitish; white spots above and below each eye; ears white-tipped, nose whitish; tail indistinctly ringed with whitish and blackish-brown rings.
Similarities
Raccoon has shorter tail, distinctly ringed; black face mask.
Habitat
Arid country and open forest.
Habits
Active day or night; forages in groups; tail often carried aloft.
Voice
Loud grunts when alarmed.
Food
Chiefly invertebrates and small vertebrates.
Range
In sw. ⅓ Ariz.; extreme sw. N.Mex.; sw. Tex.

Fig. 44

Coati

WEASELS, SKUNKS, AND ALLIES
Family Mustelidae

These mammals are often predominantly brown, but vary considerably in color and size. They usually have long, slender bodies; short legs; short, rounded ears; and anal scent glands, which are paired and large. Each foot has five toes. Fur of most species is of fine quality and often very valuable.

MARTEN
Martes americana   **64:1, 65:7**

Description
Size: head and body of males, 16–17 in. (40.6–43.2 cm), of females, 14–15 in. (35.6–38.1 cm); tail of males, 8–9 in. (20.3–22.9 cm), of females, 7–8 in. (17.8–20.3 cm); weight of males, 2–4 lb. (0.9–1.8 kg); 38 teeth, 3141/3142. Distinguished by a pale buff patch on throat and chest. Color yellowish- to dark brown above; head, belly paler; legs, tail, ears dark.

Similarities
Mink has white patch on chin. Fisher is larger; dark brown with head and back grizzled.

Habitat
Coniferous forests.

Habits
In winter, terrestrial; in summer, arboreal; does not hibernate.

Voice
High-pitched screams and squeals.

Food
Variable, including squirrels, mice, birds, eggs, and fish.

Range
Alaska, except coastal tundras; Yukon; B.C.; n. and w. Alta.; w ½ Wash.; n. ⅔ Idaho; w. Mont.; nw., s.-cen. Wyo.; w. and ne. Oreg.; nw., n. Calif. and Cascade Sierras; ne. Utah; w. Colo.; extreme n.-cen. N.Mex.

Other name
Sable.

FISHER
Martes pennanti   **64:6, 65:9**

Description
Size: head and body, 20–25 in. (50.8–63.5 cm); tail, 13–15 in. (33–38.1 cm); weight, 4½–10 lb. (2–4.5 kg); 38 teeth, 3141/3142. The only furbearer that is solid blackish-brown. Body long, slim; tail bushy, tapering. Color dark brown; head and shoulders grizzled; foreparts grayish; rump, legs, tail blackish; tail tip black.

Similarities
Marten is smaller, with buffy patch on throat and breast. Wolverine is larger, with yellowish stripes on sides and rump.

Habitat
Large forests.

Habits
Mainly nocturnal; more arboreal than ground-dwelling; solitary.

Voice
Scream and hiss.

Food
Squirrels, mice, raccoons, rabbits, some vegetable matter, carrion.

Range
B.C., except far nw. corner and extreme se.; w. Wash.; w. Oreg.; n. Calif.; n. ⅓ and w. Alta.; n. ½ Sask.; e. Idaho panhandle; w. Mont.

WEASELS
Genus *Mustela*

Members of this genus have small heads, long necks, slender bodies, and dentition formula of 3131/3132. They are nocturnal and ferocious for their size. They are carnivorous, feeding chiefly on small animals, both warm- and cold-blooded.

ERMINE
Mustela erminea   **65:6**

Description
Size: head and body of male, 6–9 in. (15.2–22.9 cm), of female, 5–7½ in. (12.7–19 cm); tail of male, 2¼–4 in. (5.7–10.2 cm), of female, 2–3 in. (5.1–7.6 cm); weight of male, 2½–3⅔ oz. (70.9–95 g); of female, 1½–2½ oz. (42.5–70.9 g). The only weasellike mammal with a black-tipped tail. Males considerably larger than females. In summer, brown above; feet and underparts white; white line down hind leg. In winter, white. Spring and fall molts show transition between brown and white upperparts. End of tail always black.

Similarities
Mink is larger and of uniform color. Long-tailed Weasel of both sexes is larger; tail longer; no white line on hind leg.

Habitat
Field borders, open woodlands, brushy and rocky places; nowhere common.

Habits
A good swimmer and climber; sometimes hunts in pairs.

Voice
Varied squeals, barks, hisses, and chatters.

Range
Alaska; Canada, except se. Alta. and sw. Sask.; Wash.; Ore.; n. Calif.; Idaho; w. Mont.; w. Wyo.; n. Nev.; n. Utah; nw. Colo.; extreme n.-cen. N.Mex.

Other name
Short-tailed Weasel.

LEAST WEASEL
Mustela nivalis   **64:9, 65:4**

Description
Size: head and body of male, 6–6½ in. (15.2–16.5 cm), of female, 5½–6 in. (14–15.2 cm); tail of male, 1⅛–1½ in. (3–3.8 cm); of female, 1–1⅛ in. (2.5–3 cm); weight of male, 1⅖–1⁷⁄₁₀ oz. (41–50 g), of female, 1⅗–1⅘ oz. (45–50 g). The smallest living carnivore and the only weasel without a black-tipped tail. In summer, brown above; white below. In winter, white; sometimes a few black hairs but no black tip to tail.

Similarities
Ermine, Long-tailed Weasel have black-tipped tails.

Habitat
Open woods, lawns, grassy areas.

Habits
Can swim.

Voice
Weak bark or shriek.

Range
Alaska, except extreme s. and se.; Yukon; Mackenzie; ne. corner B.C.; Alta.; n. Sask.; extreme nw.-cen. Mont.

LONG-TAILED WEASEL
Mustela frenata   64:8, 67:4

Description
Size: head and body of male, 9–10½ in. (22.9–26.7 cm), of female,
8–9 in. (20.3–22.9 cm); tail of male, 4–6 in. (10.2–15.2 cm), of
female, 3–5 in. (7.6–12.7 cm); weight of male, 6–8¾ oz. (170.1–
248.1 g), of female, 3–3½ oz. (85.1–99.2 g). Most common weasel
and with widest distribution. In summer, brown above; yellowish-
white below; tail black-tipped. In winter, in northern range, white
with black-tipped tail; in southern range, like summer. In some
parts of its range it wears a white bridle across the face, or white
spot between eyes and in front of ears; head is usually darker
brown than the body; long, slender body; long neck, head slightly
larger than neck.
Similarities
Ermine of both sexes are smaller; white line down inside of hind
leg. Mink is more uniform dark brown.
Habitat
Farmlands, prairies, woodlands; in fact, all possible land habitats.
Habits
Nocturnal, but often seen by day; usually solitary, climbs well.
Voice
Hisses, screams, purrs.
Range
In se. and s. B.C.; s. Alta.; s. Sask.; sw. Man.; all U.S., except
extreme se. Calif., s. Utah; also in w. and n. Ariz.; cen.-w. N.Mex.

BLACK-FOOTED FERRET
Mustela nigripes   65:8

Description
Size: head and body, 15–18 in. (38.1–45.7 cm); tail, 5–6 in. (12.7–
15.2 cm). The only weasel with a black mask and black feet.
Above yellowish-brown to buff; forehead and feet black; tail black-
tipped.
Similarities
Swift Fox has bushy tail, feet not black.
Habitat
Prairies, coinciding with range of prairie dogs on which it feeds.
Habits
More active at night than by day; wary.
Voice
Hiss and chatter.
Remarks
In the Old West it was abundant and widespread; today it is one
of the rarest North American mammals.
Range
In se. Alta.; s. Sask.; Mont. e. of the Rockies; w. N.Dak.; w. ⅚
S.Dak.; e. Wyo.; e. Colo.; extreme e. and se. Utah; ne. Ariz.;
N.Mex., except extreme sw. corner.

MINK
Mustela vison   64:7, 67:5

Description
Size: head and body of male, 13–17 in. (33–43.2 cm), of female,
12–14 in. (30.5–35.6 cm); tail of male, 7–9 in. (17.8–22.9 cm), of
female, 5–8 in. (12.7–20.3 cm); weight, 1¼–2¼ lb. (0.6–1 kg). The
only uniformly dark brown weasellike mammal with a white patch

on its chin. Color usually rich dark brown; no seasonal change; sometimes scattered small white spots on belly; white chin patch. Tail slightly bushy.

Similarities
Weasels have white or yellowish underparts. Marten has yellow patch on throat and breast. River Otter is larger.

Habitat
Near streams, marshes; in winter, woods.

Habits
Nocturnal, solitary, wary; spends much time in water; emits strong odor when cornered; stores food in den.

Voice
Various hisses, screams, barks, purrs.

Range
Alaska, except Arctic slope; Canada; U.S. as far s. as s. border of Oreg.; n. ½ Calif.; extreme w. and ne. Nev.; nw. ⅔ Utah; Colo.; nw. N.Mex.

WOLVERINE
Gulo gulo   **64:2, 65:10**

Description
Size: head and body, 29–32 in. (73.7–81.3 cm); tail, 7–9 in. (17.8–22.9 cm); weight, 20–40 lb. (9.1–18.1 kg); 38 teeth, 3141/3142. Largest and fiercest of the mustelids, this bearlike furbearer has a broad light stripe on each side and a strong skunklike odor. Pelage long; ears small, back arched, tail bushy; feet large for body size. Color varies from yellowish-brown to almost black, paler on head; 2 broad yellow stripes from shoulders join on the rump.

Similarities
Fisher is smaller, tail longer; lacks yellowish stripes.

Habitat
Wilderness, chiefly brushlands, forests, mountains; tundra of North.

Habits
Active at all hours and seasons; mainly terrestrial, but can climb.

Voice
Snarl and growl.

Food
Any animal it can kill; also carrion.

Range
Alaska, Canada to Great Plains; B.C., except extreme sw. and se.; Vancouver Is.; Wash., Oreg.; cen. Calif. Sierras; possibly e. Colo.

Other names
Glutton, Skunk Bear, Carcajou, Loup Garou.

BADGER
Taxidea taxus   **60:1, 67:8**

Description
Size: head and body, 18–22 in. (45.7–55.9 cm); tail, 7–12 in. (17.8–30.5 cm); weight, 13–25 lb. (5.9–11.3 kg); 34 teeth, 3131/3132. Distinguished by white stripe from nose back, over top of its head. Body heavy, flattened, stout; short-legged, front claws very long. Pelage short, coarse, guard hairs long. Upperparts grayish-yellow to reddish with grizzled appearance, underparts paler; head black and white, feet black, tail yellowish.

Habitat
Dry, open country.

Habits
Nocturnal, but also abroad by day; a powerful digger, lives in an underground den, a good fighter.
Voice
Grunts, growls; usually silent.
Food
Ground squirrels, other small mammals, birds, eggs, reptiles.
Range
In se. B.C.; e. Wash.; Oreg., except coast; se. ¼ Alta.; s. Sask.; sw. Man.; all other w. states except Idaho panhandle.

WESTERN SPOTTED SKUNK
Spilogale gracilis   **64:3, 67:2**

Description
Size: head and body, 9–13½ in. (22.9–34.3 cm); tail, 4½–9 in. (11.4–22.9 cm); weight, 1–2 lb. (0.5–0.9 kg); 34 teeth, 3131/3132. Fur black with 4–6 white stripes broken into spots on upperparts; white spots on head, 1 on forehead and 1 under each ear; tip of tail white. Geographic races differ in proportions of white to black.
Habitat
Brushy areas, edges of woods, wastelands.
Habits
Nocturnal, terrestrial; a good swimmer, climber, burrower; playful.
Food
Rodents, birds, eggs, insects, fruits.
Voice
Comparatively silent but may hiss, snarl, grunt.
Range
All w. states except ne. Wash.; extreme n. Idaho panhandle; Mont.; n. ⅔ Wyo.; ne. S.Dak.; N.Dak.

STRIPED SKUNK
Mephitis mephitis   **64:4, 67:1**

Description
Size: head and body, 13–18 in. (33–45.7 cm); tail, 7–10 in. (17.8–25.4 cm); weight, 6–10 lb. (2.7–4.5 kg); 34 teeth, 3131/3132. The only small black mammal with a large white V along the top of its back; about the size of a house cat, this skunk is probably best known by its characteristic odor. Color black; narrow white stripe up middle of forehead, broad white area on nape forms V at shoulders dividing into 2 white lines that continue to base of tail; tip of bushy tail may or may not be white.
Similarities
Spotted is smaller; white stripes broken into spots. Hooded has longer tail; white V on back rare. Hog-nosed has undivided white back stripe.
Habitat
Practically all land habitats.
Habits
Hunts at night; protects itself by using scent glands if molested.
Voice
Low *churr*'s and growls.
Food
Rats, mice, chipmunks, insects, fruits, berries.
Remarks
Presence usually first determined by odor.
Range
All U.S. and w. Canada except nw. ⅔ B.C.

HOODED SKUNK
Mephitis macroura   *Fig. 45*

Description
Size: head and body, 12–16 in. (30.5–40.6 cm); tail, 14–15 in. (35.6–38.1 cm). Hair on neck forms a ruff. Color in 2 patterns with intermediate variants: entire back white, including tail; or back nearly all-black, with 2 white tail stripes. Belly black.

Similarities
Striped has shorter tail; white V on back. Hog-nosed has long, bare snout; tail shorter; entire back and tail white.

Habitat
Arid areas; deserts.

Food
Insects and other invertebrates, some plant food.

Range
In se. Ariz.; extreme sw. N.Mex.

Fig. 45

Hooded Skunk

Hog-nosed Skunk

HOG-NOSED SKUNK
Conepatus mesoleucus   *Fig. 45*

Description
Size: head and body, 14–19 in. (35.6–48.2 cm); tail, 7–12 in. (17.8–30.5 cm); weight, 5–10 lb. (2.3–4.5 kg); 32 teeth, 3121/3132. Snout long, naked for about 1 in. (2.5 cm) on top, piglike. Pelage short, coarse. Color 2-toned; entire back and tail white; underparts and lower sides black.

Similarities
Striped has white blaze on forehead. Hooded has longer tail; any white on back is mixed with black.

Habitat
Arid places.

Habits
Roots for food.

Food
Insects and other invertebrates.

Range
In se. ¼ Ariz., with extension to nw. boundary; se. Colo.; N.Mex., except nw. ½ and far e.-cen.; extreme w. Okla. panhandle; w., s. ½ Tex. and extreme nw. panhandle.

RIVER OTTER
Lutra canadensis   **64:5, 67:6**

Description
Size: head and body, 26–30 in. (66–76.2 cm); tail, 12–19 in. (30.5–48.2 cm); weight, 10–30 lb. (4.5–13.6 kg); 37 teeth, 3141/3142.

The only weasellike mammal with webbed feet. Skull flattened dorsally, ears small, snout broad. Tail round, thick at base, tapering toward tip. Upperparts brown to grayish; chin and throat grayish-white; underparts with silvery sheen.

Similarities
Beaver has flat, scaly tail. Mink is smaller; feet not webbed. Sea Otter prefers Pacific Ocean along West Coast; head grayer, tail shorter.

Habitat
Watercourses, rivers, lakes, marshes; not oceanic.

Habits
Makes earth slides into the water on which it slides in sport; active at all times of day and year.

Voice
Varied, including grunts, chatters, chuckles.

Food
Crayfish, frogs, and fish.

Range
Alaska, except Arctic slope; all Canada and U.S., except s. Calif.; s. ⅔ Nev.; se. ½ Ariz.; s. ¾ N.Mex.

SEA OTTER
Enhydra lutris

Fig. 46

Description
Size: head and body, 30–36 in. (76.2–91.4 cm); tail, 11–13 in. (27.9–33 cm); weight, 30–85 lb. (13.6–38.6 kg); 32 teeth, 3131/2132. Skull large, high, inflated, blunt; neck short, thick; ears short, pointed; legs and tail short; hind feet webbed, flattened into broad flippers. Color varies from black to almost red, usually uniform dark brown, glossy, with white-tipped hairs; head and neck grayish or yellowish, paler on throat and chest.

Similarities
Seals and sea lions are larger, tails shorter, fur shorter; flippers well-developed. River Otter is smaller; head dark brown; tail longer.

Habitat
Rocky shores, among kelp beds.

Habits
Spends much time sleeping on back on sea's surface or feeding; hauls out onto shore during storms; gregarious.

Voice
Soft growl or hiss when warning; loud squeal at other times.

Food
Sea urchins and mollusks.

Remarks
Almost exterminated by fur trade, Sea Otter is protected and making a gradual comeback.

Range
Aleutian Islands to Calif.; most often seen off Amchitka, w. Aleutians, and Pt. Lobos, Calif.

Fig. 46

Sea Otter

CATS
Family Felidae

Members of this family, which include the domestic house cat, have short faces, small rounded ears, five toes on each front foot and four behind, and rectractile claws. The dentition in most cases is 3131/3121. (In the Bobcat and Lynx it is 3121/3121.) Carnassial teeth are well developed. Cat nature and appearance, except for color and size, is homogeneous for all forms. Gestation is fifty-six to one hundred days, with one to six young, usually two or three. Most litter once a year.

MOUNTAIN LION
Felis concolor

66:11
Fig. 47

Description
Size: head and body, 42–54 in. (106.7–137.2 cm); tail, 30–36 in. (76.2–91.4 cm); height at shoulder, 26–31 in. (66–78.7 cm); weight, 80–260 lb. (36.3–117.9 kg); mammae 8, 6 functional. The largest North American cat and the only that has uniformly colored adults and long tail. Skull broad, short, round. Pelage short, whiskers prominent. Color tawny to grayish; tip of tail, backs of ears, sides of nose brown; underparts whitish.
Habitat
Rough mountains, rimrocks, forests, swamps.
Habits
Active at all hours, ranges widely for food, climbs trees when chased.
Voice
Prolonged scream.
Food
Mainly deer, also smaller animals.
Range
W. of Rockies and s. of Yukon, with extension into sw. Sask.; absent from nw. and ne. B.C. and Calif. Central Valley.
Other names
Cougar, Puma, Panther.

Fig. 47

Mountain Lion

LYNX
Felis lynx

Fig. 48

Description
Size: head and body, 32–40 in. (81.3–101.6 cm); tail, 4 in. (10.2 cm); height at shoulder, 20–29½ in. (50–75 cm); weight, 15–30 lb. (6.8–13.6 kg); mammae, 4. Unique for its combination of tufted ears and bobbed tail with a completely black tip. Skull large; legs long, feet large, pads well furred; prominent ear tufts. Color grayish-buff, lightly spotted, guard hairs whitish; eyelids white, ears buffy-brown at base with central white spot and black tip.

Similarities
Bobcat is smaller; tail tip black only on top.
Habitat
Forests, swamps.
Habits
Primarily nocturnal, solitary; climbs and swims well; large feet enable it to travel on snow; dens in hollow log or other sheltered place; ranges widely; a valued fur bearer.
Voice
Catlike, seldom heard.
Food
Small mammals, Snowshoe Hare, and small rodents.
Range
Alaska, Canada; e. Wash.; ne. Oreg.; Idaho panhandle; extreme nw. Mont.; also w. Wyo.; cen.-ne. Utah; extreme nw. Colo.
Other name
Canada Lynx.

Fig. 48

Bobcat

Lynx

BOBCAT
Felis rufus

66:10
Fig. 48

Description
Size: head and body, 25–30 in. (63.5–76.2 cm); tail, 5 in (12.7 cm); height at shoulder, 20–23½ in. (50–60 cm); weight, 15–35 lb. (6.8–15.9 kg); mammae, 4. Short-tailed, with tail tip black only on top. Skull robust; ear tufts inconspicuous; feet large. Above pale to reddish-brown with black streaks and spots; below whitish with dark spots; tail white at extreme tip, after black topping.
Similarities
Lynx is larger; conspicuously tufted ears, footpads well furred, tail black tipped. Other cats have longer tails.
Habitat
Forests, swamps, deserts, mountains; prefers scrub, thickets, broken country.
Habits
Nocturnal; swims, climbs well.
Voice
Catlike.
Food
Hares, rabbits, other small mammals; ground birds.
Range
Entire W., s. of s.-cen. B.C. and including sw. and se. Alta.; s. Sask.; extreme s. Man.

Odd-Toed Ungulates
Order Perissodactyla

Members of this order have the main axis of the foot directly through the third digit, which is always longer than any of the other toes on both forefeet and hind feet.

HORSE AND BURRO
Family Equidae

HORSE
Equus caballus

Description
Size: height at shoulder, to 5½ ft. (1.7 m); weight, to 1100 lb. (499 kg); teeth, 3133/3133. Color variable, but usually gray or brown.
Habitat
Open plains, foothills, arid regions.
Habits
In the wilds, runs in herds usually led by a stallion.
Remarks
Became wild when escaped from early Spanish explorers. Formerly scores of thousands roamed the western states; today, only a few sparse bands inhabit the remoter regions of the Southwest; easily tamed and thereafter indistinguishable from domestic horses.
Range
Remote desert areas of Calif.; Nev.; Utah; Wyo.; Colo.; Ariz.; N.Mex.; w. Tex.
Other name
Wild Horse.

WILD BURRO
Equus asinus

Description
Size: height at shoulder, 30 in. (76.2 cm). Ears long; tail sparsely haired; eyes deep-set; mane wiry, uneven, coarse. Legs medium-long, feet small. Color variable from white to black, usually fairly uniform and with white on nose, belly, and flanks.
Similarities
Horse is larger, body more robust.
Habitat
Arid regions of the Southwest, thinly timbered slopes, mountain valleys.
Voice
Typical donkey's bray; in the wilds usually a dawn chorus, hence the nickname "Rocky Mountain Canary."
Remarks
The wild burro, like the wild horse, is an escapee from domestic service following early Spanish explorations into the Southwest. Protected in some states; easily tamed and make excellent pack animals.
Range
Desert areas of the Sw.

Even-Toed Ungulates
Order Artiodactyla

Mammals in this order have their weight equally distributed over either two or, rarely, four toes on each foot. The main axis falls between the third and fourth digits; each toe ends in a nail-like hoof. They are medium-sized to large animals, and the young are able to walk and run within minutes of birth.

OLD WORLD SWINE
Family Suidae

Members of this family were introduced from Old World farms and went wild. A few feral European forms have been released to provide big game hunting.

WILD BOAR
Sus scrofa   *Fig. 49*

Description
Size: length, 3½–5 ft. (1.1–1.5 m); height at shoulder, to 3 ft. (0.9 m); weight, to 400 lb. (181.4 kg); 44 teeth if all present, usually 3143/3143; mammae, normally 12. The only wild pig in America. Tusks to 1 ft. (0.3 m), upcurved. Legs long, 4 toes on each foot; hair coarse. Color variable, depending on ancestry, but generally pale gray to blackish.
Similarities
Peccary has upper tusks pointing downward, 3 toes on each hind foot.
Habitat
Mountains, forests.
Habits
Good swimmer and fierce fighter; will breed with domestic swine.
Voice
Various piglike grunts.
Food
Roots, tubers.
Reproduction
1–2 litters per year, mature in less than 1 year; gestation 115–140 days; average 5–6 young.
Range
In Oreg.; Calif., Monterey, Butte and San Luis Obispo counties, Santa Cruz Is.; Tex.

Fig. 49

Wild Boar

PECCARIES
Family Tayassuidae

Members of this family, with their mobile elongated snouts, are truly the wild pigs of the New World. They are small, piglike mammals with a dentition of 2133/3133.

COLLARED PECCARY
Dicotyles tajacu   *Fig. 50*

Description
Size: length, 34–36 in. (86.4–91.4 cm); height at shoulder, 20–24 in. (50.8–61 cm); weight, 40–50 lb. (18.1–22.7 kg); 38 teeth; mammae, 2. Skull has narrow rostrum, sides not flattened. Hair coarse, hind feet bear only 3 toes; large, well-developed musk gland on middorsal line of rump. Color mixed black and gray, lighter over front of shoulder; young are reddish with black stripe down back.

Similarities
Wild Boar is larger; upper tusks curve upward; 4 toes on each hind foot.

Habitat
Brushy, semidesert of chaparral, mesquite, cacti (especially saguaro), oaks; along cliffs, near waterholes.

Habits
Most active mornings and late afternoons; gregarious, usually in bands of 2–25.

Food
Omnivorous.

Remarks
Much hunted for hides and flesh.

Range
In s. Ariz.; extreme s. N.Mex.; w. to s. Tex.

Other name
Javelina.

Fig. 50

Collared Peccary

DEER
Family Cervidae

These are the only hoofed mammals that have antlers which are shed every year. Females are appreciably smaller than males and lack antlers, except for the Caribou. Deer hear well and have an excellent sense of smell. The lower canines are incisorlike. Cervids are browsers and lack upper incisors. They have a complex stomach and chew the cud.

ELK
Cervus elaphus   69:2

Description
Size: length, to 9½ ft. (2.9 m); height at shoulder, 4–5 ft. (1.2–1.5 m); weight of male, 700–1100 lb. (349.3–499 kg), of female, 500–650 lb. (226.8–294.8 kg); 34 teeth, 0133/3133; mammae, 4. The male is the second largest deer, with large narrow antlers. Antlers long—record spread 74 in. (188 cm), at least 5 tines, unpalmated, branching. Hair short, mane slight. In summer, light brown, head and limbs darker, rump buffy. In winter, grayish-brown, head and limbs dark, rump buffy, mane longer and darker. Calf primarily brown with light spots till early fall, rump buffy.

Similarities
Moose has palmated antlers, large overhanging snout, brown rump. Mule Deer is smaller, with black on tail. Caribou has whitish neck.

Habitat
Semiopen woodlands, mountain meadows in summer, foothills, plains, valley.

Habits
Summers in mountains, winters in valleys; often destructive of ranch forage; alert, curious, most active mornings and evenings, usually seen in groups of 25 or more, both sexes together in winter, old bulls separate in summer.

Voice
Far-carrying bugle; a loud bark of alarm.

Food
Grass, leaves, twigs.

Range
Alaska, introduced to Afognak Is. (Gulf of Alaska) only; Vancouver Is. and far se. B.C., also ne. mountains; sw. Alta.; s. Sask., except extreme s.; s. Man.; w. Wash.; w. Oreg.; nw. Calif., and e.-cen. Sierras; Idaho panhandle; w. Wyo., especially abundant in Yellowstone National Park and Jackson Hole, and Bighorn Mts.; ne. to sw. Utah; n.-s., cen. Colo.; extreme s.-cen. N.Mex.

Other name
Wapiti.

MULE DEER
Odocoileus hemionus   60:9, 69:5

Description
Size: length, to 6½ ft. (2 m); height at shoulder, 3–3½ ft. (0.9–1.1 m); weight of male, 125–400 lb. (61.2–181.4 kg), of female, 100–150 lb. (45.4–68 kg); 32 teeth, 0033/3133; mammae, 4. The only deer with black on its tail. Antler record spread, in Rocky Mountains, to 47½ in. (120.7 cm); tines branch equally, dichotomous branching, not prongs from common base. Ears large, tail ropelike, scent glands on legs large. Color reddish in summer, brownish-gray in winter; belly, throat patch, rump patch white; tail either black tipped or black on top.

Similarities
White-tailed has broad tail, white below; antlers with main beam and prongs rising from it. Elk is larger with no black on tail. Caribou has larger antlers, neck whitish, no black on tail; hoofs click when walking. Moose is larger, with overhanging snout. Pronghorn has white sides, no black on tail.

Habitat
Forests, brushy areas, rocky uplands, desert shrubs, chaparral.

Habits
Most active mornings, evenings, and on moonlit nights; occurs singly or in small groups, more gregarious in winter; has a jumping gait, carries tail down at all times; summers in mountains, winters in valleys.

Voice
Bucks, guttural grunt, especially during rut; both sexes snort when alarmed; fawns and does, a bleatlike *baaa* seldom heard.

Food
Grass, forbs, moss, leaves, twigs.

Range
Extreme s. and se. Alaska; s. Yukon; s. Mackenzie; e. to Hudson Bay and s. to Mexico throughout all w. states.

Note: The smaller **PACIFIC COAST BLACK-TAILED DEER** is considered to be a subspecies of the Mule Deer. The former is all-black above with a bushy tail.

WHITE-TAILED DEER
Odocoileus virginianus

60:9, 69:6

Description
Size: length, to 6 ft. (1.8 m); height at shoulder, to 3¾ ft. (1.1 m); weight of male, 75–400 lb. (34–181.4 kg), of female, 50–250 lb. (22.7–113.4 kg); 32 teeth, 0033/3133. The only deer with the tail white below and the same color above as in back; in woods, raised tail shows as large, white flag. Antlers have erect, unbranched tines rising from the main beam, record spread 33½ in. (85.1 cm). Adults tawny above in summer, blue-gray in winter, white below; fawn with white spots on reddish coat for 3½ months.

Similarities
Mule Deer has black tip on tail, prongs of antlers not from main beam. Elk is larger; rump patch yellowish. Caribou has rump patch and neck whitish, antler spread greater. Moose is larger; antlers palmated, overhanging snout, no white. Pronghorn has large white rump; horns, not antlers.

Habitat
Low mixed woodlands, forest edges, second growth.

Habits
Secretive, alert; in flight raises tail to show white flag; in winter in heavy snow congregates in groups of 25 or more and keeps packed-down "yards" at feeding grounds.

Voice
Whistling snort when startled; fawns utter low bleat; old bucks in rut give guttural grunts.

Food
Grass, leaves, twigs.

Range
In se. ¼ B.C.; e. across Canada and s. into Wash., except nw. corner; Oreg., except extreme sw. and se. corners; extreme ne. Calif.; nw. Nev.; Idaho, except far sw.; Mont.; Wyo.; extreme ne. Utah; extreme nw., e. ½ Colo.; s. ½ Ariz.; N.Mex., except nw. corner.

Other name
Virginia Deer.

MOOSE

Alces alces

Description
Size: length, to 10 ft. (3 m), height at shoulder, to 7½ ft. (2.3 m); weight of male, to 1400 lb. (635 kg), of female, 600–800 lb. (272.1–362.9 kg); 32 teeth, 0033/3133; mammae, 4. Largest deer, distinguished by its broadly palmated antlers in the male and overhanging snout in both sexes. Body heavy, legs long, tail short, muzzle broad and overhanging, ears large, neck short; dewlap or "bell" of hair and skin hanging from throat. Antlers (male only) massive, broadly palmate and flat, with small prongs projecting from the palms; record spread 77⅝ in. (197.1 cm). Coat blackish or brownish, legs lighter; calf dull reddish-brown.

Habitat
Forests near shallow lakes, wilderness, marshes, swamps.

Habits
Feeds in shallow water at dawn and dusk; good swimmer, found singly or by 2's or 3's, bull with cow and/or calf, ranges within 10 mi. (16.1 km) of birth site; hunters call bull on moose horn by imitating cow's voice.

Voice
Bull, a rising *moo*; cow, softer, more like a domestic cow's *moo*.

Food
Water plants, leaves, twigs.

Range
Alaska, except extreme coastal tundra; all Canada in w. range except se. Alta. and sw. Man.; Rocky Mountains of Idaho, w. Mont., ne. Utah, w. Wyo., nw. Colo.

CARIBOU

Rangifer tarandus

Description
Size: length, to 8 ft. (2.4 m); height at shoulder, 3½–4 ft. (1.1–1.2 m); weight of male, 250–600 lb. (113.4–272.2 kg), of female, 150–300 lb. (68–158.8 kg); 32–34 teeth, 0133/4033. Distinctive for antlers that project forward toward nose. Neck maned below; belly shaggy; feet large; hoofs rounded, click characteristically when walking. The males and most females sport antlers that are semipalmated, with 1 prominent brow-tine down over the nose; beams flattened, record spread 60 in. (152.4 cm). Body dark chocolate-brown, whitish on neck and rump; white above each hoof; antlers dark mahogany-brown, velvet dark brown.

Similarities
Mule Deer has black-tipped tail. White-tailed Deer has tawny neck and rump. Elk has chestnut-brown neck. Moose is larger; antlers broadly palmated, overhanging snout; hoofs pointed.

Habitat
Tundras, muskegs, coniferous forests; mountains to above timberline.

Habits
Migratory, travels in great herds; gait loping and running or bounding.

Voice
Various snorts and grunts; when in rut, bucks roar.

Food
Lichens, reindeer moss, grasses, browse of willows and birches.

Range
In se. Alaska; s. Yukon; sw. ½ Mackenzie; ne. and extreme B.C.; n. ½ and sw. Alta.; n. ⅔ Man.; extreme n. Idaho panhandle, with slight overlap into adjoining Wash. and Mont.

PRONGHORNS
Family Antilocapridae

Only one species of this family exists, and it is strictly a North American mammal. Both sexes have true horns with a bone core covered with horny sheath composed of agglutinated hair. The sheaths are shed annually. Pronghorns lack upper incisors. They are cud chewers and have a complex stomach.

PRONGHORN
Antilocapra americana

69:3

Description
Size: length, to 4½ ft. (1.4 m); height at shoulder, to 3½ ft. (1.1 m); weight, 75–140 lb. (34–63.5 kg); 32 teeth, 0033/3133; mammae, 4. The only deerlike animal with a white rump and 2 white throat bands. Horns 2-pronged; bucks have horns longer than ears and prongs directed forward; horns of doe seldom as long; each prong projects forward, slightly curved. Hoofs pointed, 2 on each foot; front hooves longer than back. Above tan, grayish in winter; below white; rump patch large, white; 2 broad white bands across throat. Buck has black face and patch on side of neck; doe has black mask and patch almost absent; kid (to 3 months) is gray.
Similarities
Mountain Sheep has massive coiled horns, no white bands across throat; prefers mountains.
Habitat
Plains, prairies, sagebrush flats, deserts.
Habits
Active at all times; travels in bands; erects white hairs of rump patch when disturbed; clocked by airplane at 84 mph, average speed about 40 mph.
Food
Browse plants, sagebrush, weeds, grasses.
Range
Extreme s.-cen. Wash.; se. Oreg.; s. Idaho; ne. and se. Calif.; Nev.; Utah; Ariz., except extreme s.; extreme se. Alta.; sw. Sask.; e. ⅔ Mont.; sw. N.Dak.; w. ½ S.Dak.; Wyo.; Colo.; w. Nebr.; Okla. panhandle; w. ½ Tex.
Other name
Antelope.

WILD CATTLE
Family Bovidae

Members of this family have unbranched hollow horns over bony cores and are not shed every year. Domestic cattle, sheep, and goats belong to this family. Upper incisors are absent. All have a complex stomach and are cud chewers.

BISON
Bison bison

60:12
Fig. 51

Description
Size: length, to 11½ ft. (3.5 m); height at shoulder, 5–6 ft. (1.5–1.8 m); weight, 800–2000 lb. (362.9–907.2 kg); 32 teeth, 0033/3133; mammae, 4. Only wild ox with huge head, high shoulder hump, long shaggy hair on shoulders and front legs, and short, usually scabby, hair on its sides. Tail tufted, ropelike; head massive. Cows are smaller, more evenly colored, less bearded, have a smaller

hump and more slender and curved horns. Bull is dark brown, with lighter head, shoulders, legs, tail; calf is reddish-yellow, with lighter legs and belly.

Habitat
Grasslands and open woodlands; in central Alaska introduced successfully to spruce and birch forests.

Habits
Highly gregarious, formerly migratory; eyesight poor, hearing and sense of smell keen; likes to wallow in mud or dust.

Voice
A bellow.

Remarks
Bison has been successfully crossbred with domestic cattle to produce a "cattalo."

Range
In e.-cen. Alaska; ne. and s.-cen. Alta.; Wyo., in Yellowstone National Park and in cen. on reserve at Thermopolis; Mont., reservations n. of Missoula and on s.-cen. border; S.Dak. Black Hills; se.-cen. Utah; nw. Colo.; n.-cen. and se.-cen. Ariz.

Other name
Buffalo.

Note: These inhabit restricted reserves, since the Bison survives only as a ward of man. Where introduced into the subarctic, they are free-roamers and are gradually increasing in numbers. An occasional albino is seen.

Bison

Fig. 51

MUSKOX
Ovibos moschatus

Fig. 52

Description
Size, length, to 6 ft. (1.8 m); height at shoulder, to 5½ ft. (1.7 m); weight, to 900 lb. (408.2 kg); 32 teeth, 0033/3133; mammae, 4. Distinctive for long fur that hangs down almost to its feet. Horns broad, flat, plastered close to skull, tips curved upward; shoulders with slight hump; legs and neck short; tail very short. Bull is deep brown or blackish; nose and patch behind shoulders pale. Cow frequently has paler face and more slender horns.

Fig. 52

Muskox

Habitat
Arctic and subarctic tundras and foothills.
Habits
Gregarious; habit of group forming circle with calves inside, if attacked.
Voice
A bellow.
Food
Sedges, grasses, leaves, twigs.
Range
Extreme ne. Alaska coast; n. Yukon, n. to e. Mackenzie.

MOUNTAIN SHEEP

Ovis canadensis

Fig. 53

Description
Size: height at shoulder, 2½–3½ ft. (0.8–1.1 m]; weight of male, 125–275 lb. (56.7–124.7 kg), of female, 75–150 lb. (34–68 kg); 32 teeth, 0033/4033; mammae, 2. Horns massive, coiled (males only) with backward spiral that slopes outward and then forward to complete an arc; record spread 33 in. (83.8 cm). Color brown to grayish-brown; rump creamy-white.
Similarities
Mountain Goat is white, horns black. Pronghorn has branched horns, white bands across throat.
Habitat
Rugged mountains and slopes with sparse timber.
Habits
Gregarious; sexes usually separate in summer.
Food
Grasses and forages.
Range
Rocky Mts. of se. B.C.; sw. Alta.; Idaho; w. ⅞ Mont.; extreme sw. N.Dak. and w. S.Dak.; Wyo.; Calif. Sierras and se. desert mountains; Utah; w. ⅔ Colo.; Ariz.; w. ¾ N.Mex.; w. Tex.
Other name
Bighorn Sheep.

Fig. 53

black phase ♂

Mountain Sheep

♂

Dall's Sheep

♂

♀

DALL'S SHEEP
Ovis dalli   *Fig. 53*

Description
Size: height at shoulder, 3–3⅓ ft. (0.9–1 m); weight, 125–300 lb.
(56.7–136.1 kg); 32 teeth, 0033/4033; mammae, 2. Body stocky;
horns of adult males massive, smaller in females, narrow at base,
to 15½-in. (39.4-cm) circumference; record spread 35 in. (88.9 cm).
Nose narrow, pointed; ears small, pointed; tail, including hair,
shorter than ear; no beard on chin. Color white or whitish to
nearly black in southern range; neck white or grizzled; horns
yellowish.
Similarities
Mountain Goat has small horns, black, slightly back-curved.
Habitat
High and rugged mountains, cliffs and slopes.
Habits
Small herds.
Food
Grasses.
Range
Alaska, except Arctic coast and w.-sw.; Yukon, except Arctic slope;
sw.-cen. Mackenzie; n.-cen. B.C.
Other name
White Sheep.

MOUNTAIN GOAT
Oreamnos americanus   *Fig. 54*

Description
Size: head and body, to 5 ft. (1.5 m); height at shoulder, 3⅓ ft. (1.1
m); weight, to 276 lb. (125.2 kg); 32 teeth, 0033/4033. Saberlike
horns to 1 ft. (0.3 m), pointing backward. Hair long, relatively
coarse; beard under chin characteristic. Color white, hoofs and
horns black.
Similarities
Some 5 subspecies occur in total range, all unmistakably goatlike.
Mountain and White Sheep are yellowish, massive, with spiral-
shaped horns.
Habitat
High mountains in summer, lower elevations in winter.
Habits
Not gregarious.
Range
Mountains of Alaska; Yukon; Mackenzie; B.C.; w. Alta.; Mont.;
Idaho; in continental U.S. found mainly in sanctuaries of national
parks.

Fig. 54   Mountain Goat

Reptiles

Consulting Editor
Robert L. Bezy
Associate Curator of Herpetology
Los Angeles County Museum of Natural History

Illustrations
Plates 70–71 John C. Yrizarry and Biruta Akerbergs
Plates 75, 77, 78 Biruta Akerbergs
Plates 72–74, 76, 79–80 Jennifer Emry-Perrott
Text Illustrations John C. Yrizarry and
Jennifer Emry-Perrott

Reptiles

Class Reptilia

A reptile is a cold-blooded or ectothermic vertebrate; that is, its temperature depends on its surrounding environment because it can produce little or no heat of its own. Reptiles reproduce by means of amniotic eggs that are laid on land (oviparity), or are retained and hatched within the mother's body (ovoviviparity). Unlike amphibians, reptiles are not dependent on a return to water in order to breed. Fertilization is internal, and there is no free-living larval stage. The young as well as the adult have a dry skin bearing scales or horny plates. Reptiles breathe by means of lungs, and in aquatic turtles breathing is supplemented by passing water in and out of the pharynx and vascular sacs connected with the cloaca.

The reptiles evolved from the amphibians of the Devonian period of the Paleozoic era some 410 million years ago. The extinct progenitors of the reptiles were the cotylosaurs, or "stem reptiles," whose fossils were first discovered in the coal swamp sediments of the Mississippian period, formed 355 million years ago. The dinosaurs dominated the entire Mesozoic. This great era, spanning a hundred million years, has come to be known as the "Age of Reptiles" and culminated in the great dinosaurs, or "hot-blooded" giant reptiles. Most of the dinosaurs became extinct with the onset of the Cenozoic period, about 70 million years ago, their only living survivors being the birds. The Tuatara, crocodilians, turtles, lizards, and snakes are the remaining members of the reptile stock.

Size

The sizes given for identification are the range in adult size of a particular species.

Range and Scope

Because nearly all their body heat comes directly from their environment, reptiles cannot survive perpetually frozen polar climates and are most abundant in tropical regions. In the United States, turtles and snakes are most abundant in the Southeast, while lizards abound in the Southwest. In North America west of the 100th meridian, there are 155 species of reptiles: 15 turtles, 64 lizards, and 76 snakes. All of these species, plus some subspecies, are included in this chapter. The area of coverage is the land area west of the 100th meridian, from the northern tip of Alaska to the Mexican-American border.

Nomenclature

The common and scientific names of the species included in this chapter are in accord with those of either Stebbins (1966) or Conant (1975).

USEFUL REFERENCES

Carr, A. 1952. *Handbook of Turtles.* Ithaca, N.Y.: Comstock.

Cochran, D. M., and Goin, C. J. 1970. *A New Field Book of Reptiles and Amphibians.* New York: Putnam.

Conant, R. 1975. *A Field Guide to Reptiles and Amphibians of Eastern and Central North America.* 2d ed. Boston: Houghton Mifflin.

REPTILES

Ditmars, R. L. 1936. *Reptiles of North America*. New York: Doubleday.

Ernst, C. H., and Barbour, R. W. 1973. *Turtles of the United States*. Lexington: University Press of Kentucky.

Leviton, A. E. 1972. *Reptiles and Amphibians of North America*. New York: Doubleday.

Oliver, J. A. 1955. *Natural History of North American Amphibians and Reptiles*. Princeton, N.J.: Van Nostrand.

Pope, C. H. 1946. *Turtles of the United States and Canada*. New York: Knopf.

Schmidt, K. P., and Davis, D. D. 1944. *Field Book of Snakes*. New York: G. P. Putnam's Sons.

Shaw, C. E., and Campbell, S. 1974. *Snakes of the American West*. New York: Knopf.

Smith, H. M. 1946. *Handbook of Lizards*. Ithaca, N.Y.: Comstock.

Stebbins, R. C. 1954. *Amphibians and Reptiles of Western North America*. New York: McGraw-Hill.

_____. 1966. *A Field Guide to Western Amphibians and Reptiles*. Boston: Houghton Mifflin.

Wright, A. H., and Wright, A. A. 1957. *Handbook of Snakes*. 2 vols. Ithaca, N.Y.: Comstock.

Zweifel, R. G.; Zug, G. R.; McCoy, C. J.; Rossman, D. A.; and Anderson, J. D. (eds.), 1963–Present. *Catalogue of American Amphibians and Reptiles*. Soc. for Study of Amphibians and Reptiles, New York.

GLOSSARY

Anterior Toward the front (of the body).

Carapace Upper shell of a turtle or tortoise, including bony plates and horny shields.

Caudal scale Straplike scale extending across ventral surface of tail.

Cloaca Area through which internal wastes discharge.

Dewlap Skin fold hanging from neck region.

Diurnal Daytime.

Dorsolateral Pertaining to the upper sides of the animal.

Exfoliation Scaling off in flakes.

Frontals Bony membranes that form the forehead.

Interspace Area of merging of two dorsal color patterns in lizards and snakes.

Lateral Of or pertaining to the side of the body.

Maxillary bone Bone on each side of head, forming side border of upper jaw and bearing most of the upper teeth.

Occipitals Bony membranes that form the posterior part of the skull.

Parietal bones A pair of membrane bones in the roof of the skull between the frontals and occipitals.

Plastron Underpart of the shell of a turtle or tortoise.

Posterior Toward the rear (of the body).

Postocular Behind the eye.

Postorbital Behind the eye.

Reticulate Having the form or appearance of a net.

Riparian Relating to the bank of a river, lake, or pond.

Rugose Rough, wrinkled.

Shield In turtles, any one of the horny plates that cover the shell.

Temporal horns In horned lizards, horns toward the sides of the crown.

Tubercle Any of various small knoblike prominences.

Vent Opening on the surface of the cloaca.

Venter Belly.

Vertebral stripe Stripe down the midline of the back.

Vertical pupil Eye in which pupil is elliptical; long axis is vertical.

Turtles
Order Testudines

The turtles are distinguished for their ancient ancestry, biological conservatism, and characteristic shell, or armor, which the order has worn since the Triassic period. Into this shell the turtle can withdraw head and limbs for protection. Indeed, this extraordinary creature is especially bizarre in having its ribs outside its hip and shoulder girdles—the only vertebrate in which this occurs—and its enveloping armor made of an upper, rounded carapace connected on each side by a bridge to a flatter ventral plastron. The shell itself is composed of sutured bony plates overlaid by a regularly patterned series of horny shields. Although growth rings are present in the shields of some species, they cannot reliably be used as a method of determining the age of any given animal.

Features of Turtles
Simply stated, a turtle is a reptile with a shell. All species have four limbs, which are usually somewhat flattened, with five toes on both front and hind feet. In sea turtles the limbs are modified into flippers on which the claws are either reduced in numbers or are absent. Turtles are air breathers and have lungs, although in aquatic forms respiration is aided by the mouth and cloaca, each of which, like a gill, is able to absorb some oxygen from the water. This permits the animal to stay under water a considerable time. Turtles lack teeth, but the jaws are covered with a horny sheath. The sense of smell is moderately well developed, and turtles see amazingly well. Although ears are usually evident, turtles are thought to respond more to mechanical ground vibrations than to those that are airborne.

Fig. 55
Parts of a Typical Turtle

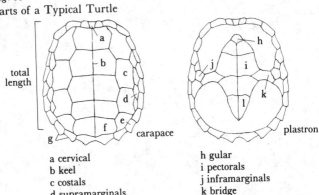

a cervical
b keel
c costals
d supramarginals
e marginals
f vertebrals
g toothed rear margin

h gular
i pectorals
j inframarginals
k bridge
l anal shield

Habitat
Turtles like the edges of bodies of water, and they enjoy basking on offshore logs or rocks. Look for land turtles in open woods, at the edges of fields, in abandoned pastures, gardens, natural parks, and semiwild areas. Throughout the Southwestern states, desert species are frequently seen on and near highways. Sea turtles are usually seen only in regional marine tanks or when accidentally caught by fishermen.

Turtles are most likely to be seen in spring, when they are moving about looking for mates or nesting sites after their winter hibernation. They are also likely to be seen after rains. In early fall, they are again on the move and more readily seen than in midsummer.

Measurements

The sizes given for the species described below are the minimum and maximum lengths of the *carapace*.

SNAPPING TURTLES, MUSK AND MUD TURTLES, AND ALLIES
Family Chelydridae

In this family are two subfamilies: Chelydrinae, which includes the snapping turtles, and Kinosterninae, which comprises the musk and mud turtles. All have twenty-three or twenty-five marginal shields, and the plastron may or may not be reduced (cruciform). The snapping turtles have a very small plastron, a large head, powerful jaws with a hooked beak, paired barbels, and a tail with erect bony scales. The musk and mud turtles have ten or eleven plastral plates, jaws with smooth edges, and barbels on the chin. Their limbs are not paddlelike; they have four or five claws on each foot and no scales on the tail.

SNAPPING TURTLE
Chelydra serpentina   **70:1**

Description

Length, 8–18 in. (20.3–45.7 cm). The largest freshwater turtle in the West. Carapace low, with 3 moderately high longitudinal ridges; rear margin usually saw-toothed. Plastron cross-shaped, much reduced (consists of 4 pairs of shields with 1 unpaired anterior shield). Tail very long, usually more than ½ length of carapace. Carapace shields varied in light and dark striations radiating from growth centers, pattern may be obscure; generally dull brown to black above, dull whitish to cream below, unpatterned in large individuals; head marked with small spots, blotches; bars less marked on underside. Head covered with skin instead of plates, eyes visible from above, jaws strong and hooked anteriorly; neck, legs warty; feet flattened, with toes fully webbed. Male averages smaller, has anal opening nearer tip of tail. Young rougher above than adult. Juveniles show greater color contrast than adults, with ventral surface generally darker.

Habitat

Any aquatic situation, preferably with mud; often crawls on land; common even near civilization.

Habits

Lies in wait for prey at bottom; snaps with quick lunge; sometimes hibernates in muskrat house; not a particularly good swimmer but closely restricted to water.

Food

Omnivorous, commonly feeding beneath water.

Remarks

Vicious to handle on land, less so in water.

Range

Extr. s. Sask. and s. Man., s. to Gulf of Mexico and e. to Atlantic Ocean; in West includes e. Mont., e. Wyo., e. Colo., e. N. Mex.; absent from far w. Tex.

YELLOW MUD TURTLE
Kinosternon flavescens   70:2

Description
Length, 4–5¾ in. (10.2–14.6 cm). Carapace uniformly olive,
yellowish-olive, yellowish-green, or yellowish-brown; posterior
margin and sides of each shield narrowly margined with blackish;
plastron yellowish to buff; limbs unpatterned; neck, head, throat
showing considerable bright yellow. Carapace smooth, unmarked,
unkeeled, low, broad, 9th and 10th edge plates elevated, counting
from front of shell. Male has horn-tipped tail and 2 patches of
horny scales on inner surface of hind leg. Juveniles have somewhat
symmetrically dark-spotted plastron, and carapace shields are not
margined with black.

Similarities
Sonoran Mud Turtle has head and neck intricately mottled; 9th
edge plate not elevated above 8th; supraorbital ridges absent.

Habitat
In almost any body of water, preferably with mud bottom, to
elevation of 5000 ft. (1524 m).

Habits
Basks; may leave water and seek food on land, particularly in
rainy season; migrates overland if pools dry up; has strong odor.

Range
Far s.-sw. Nebr., extr. w. Kans., far e. Colo., extr. w. Okla., se.
Ariz., se. N. Mex., w. Tex.

SONORAN MUD TURTLE
Kinosternon sonoriense   70:4

Description
Length, 4–6½ in. (10.2–16.5 cm). Head and neck intricately
mottled with light and dark flecking; carapace olivaceous, may be
marked with scattered spots and lines radiating from growth
centers; limbs gray or gray-brown; plastron yellowish variously
suffused with brown, or unmarked. Carapace about ½ again as
long as wide, shields relatively smooth, growth lines usually faintly
discernible; sides round off toward plastron, no flaring margin.
Plastron with 5 pairs of shields and 1 unpaired gular; pectorals
triangular, anterior and posterior plastral lobes hinged, separated
by abdominal plates. Toes webbed; nails slender, pointed. Male
larger than female, tail longer and with stronger terminal nail,
upper jaw with more pronounced hook.

Similarities
Yellow Mud Turtle has head and neck less mottled, supraorbital
ridges present; 9th edge plate more elevated than 8th and of near-
triangular shape. See also sex differences under Yellow Mud
Turtle.

Habitat
Highly aquatic, in ponds, streams, waterholes, in arroyos, rivers.

Habits
Frequents woodlands.

Range
Extr. se. Calif. and sw. Ariz., along lower Colo. R.; se.-cen. Ariz.,
principally the Gila R. drainage to 5000 ft. (1524 m) elevation;
extr. sw. N. Mex. and w. border of Tex.

FRESHWATER AND BOX TURTLES, LAND TORTOISES, AND THEIR ALLIES
Family Testudinidae

Divided into two subfamilies, this group of turtles bears from twenty-three to twenty-five marginal shields, and the plastron is never cruciform. In the subfamily Emydinae there are twenty-five marginal shields, and the toes show varying degrees of webbing. The temporal region of the skull is only weakly roofed with bone, or not at all. These turtles are largely aquatic or semiaquatic, although the box turtles are primarily terrestrial. In the subfamily Testudininae the members are robust chelonians with a deep shell showing conspicuous growth rings and from twenty-three to thirty-five marginal shields. The toes are short and lack webbing, the tail is short, and the anterior surface of the forelimbs and head supports conspicuous scales. The roof of the skull is quite incomplete. These turtles are tortoises, land-dwelling forms that rarely enter water.

WESTERN POND TURTLE
Clemmys marmorata   70:10

Description
Length, 3–7 in. (7.6–17.8 cm). Carapace olive, dark brown, to blackish (each shield marked with network of spots, lines, or dashes of brown or black, tending to radiate from growth centers), or may be obscurely marked and uniform in color; head, dorsal surface of limbs, and tail variously marked with dark brown or blackish spots, lines, or flecks on lighter undercolor. Carapace low, unkeeled, posterior border smooth; surface smooth in old adults. Plastron has 6 pairs of shields and broad bridge; forelimbs prominently scaled, hindlimbs less so, toes webbed, nails slender and prominent. Jaw crushing surface smooth or undulating. Male has a lower carapace, more marbled, usually concave plastron, vent at or beyond carapace margin when tail is extended, tail more slender and tapered. Shields ridged concentrically (in immatures with concentric and radiating ridges); irregular longitudinal dusky bands on body may show.
Habitat
Quiet ponds, small lakes, slow streams; also found in brackish or salt water.
Habits
Thoroughly aquatic, but may bask on logs or shore rocks; hibernates in bottom muds (except in south of range).
Range
Extr. sw. B.C., Puget Sound area; coastal belt from mouth of Columbia R., Oreg., into n. Baja Calif.; introduced into Truckee, Carson, and Walker rivers in Nev.

WESTERN BOX TURTLE
Terrapene ornata   *Fig. 56*

Description
Length, 4–5¾ in. (10.2–14.6 cm). Plastron of 6 pairs of shields (exclusive of bridge), front section hinged so it can "box" all extremities within shell. Carapace flat-topped, unkeeled, growth rings pronounced, costal and vertebral shields large; color dark brown to black, sometimes reddish-brown, lightening on marginals. Conspicuous yellowish lines, bars, spots radiate from growth center

of each shield; plastron yellowish, with or without darker markings; head mottled with brown or dusky and yellowish, neck streaked with whitish or yellow. Limbs stout, toes not webbed; nails long, sharp (unless worn). Female vent nearer tail base; male plastron flat to slightly concave and 1st hind toe sharply inturned. Juvenile shell almost round; flatter, darker than adult.

Habitat
Short-grass plains, prairies; sandy, semiarid regions; woodlands, swamps; to 6000 ft. (1829 m) elevation.

Habits
Mild disposition; burrows, emerges in morning, at dusk, during rains.

Remarks
This species has an interesting courtship pattern. The male chases the female (if a turtle may be said to "chase"); upon reaching her, he raises himself on his hind legs and hurls the front of his plastron at the rear of her carapace, emitting from each nostril a stream of fluid, which he sprays on her back. After half an hour of such wooing, the female yields.

Range
As far w. as far se. Ariz., se. N. Mex.; s. Nebr., Kans., e. Colo., s. into Mexico.

Fig. 56

Western Box Turtle

RIVER COOTER
Pseudemys concinna   Fig. 57

Description
Length, 9–16 in. (22.9–40.6 cm). Carapace brown or olive with alternating dark and light whorls on each scute; marginals with eyelike spots below. Plastron has narrow black lines along sutures. Carapace flat with longitudinal furrows and saw-toothed rear margin. Upper jaw notched in front, flanked by cusp on each side. Male has long toenails on front legs; shell flatter than in female.

Similarities
Pond Slider has carapace with yellow streaks and bars, rather than whorls.

Habitat
Rivers, tanks, and ditches.

Habits
Basks, chiefly in rivers.

Range
Fla., e. through Tex. to se. N. Mex.

Fig. 57   River Cooter

SLIDER
Pseudemys scripta   70:8

Description
Length, 5–8 in. (12.7–20.3 cm). Color olive, brown, to black;
patterned with yellowish streaks and bars; second and third costal
shields with streaks parallel to long axis of shields; underside of
both shells, yellow with dark eyespots, arranged more or less
symmetrically; head and limbs with yellow stripes, a yellow spot or
broad red strips behind eye. Carapace has longitudinal wrinkles,
rear margin toothed; beak notched, edges of both jaws smooth.
Plastron not hinged. Male has long nails on front feet; darker than
female.

Similarities
Painted Turtle has carapace smooth, not wrinkled, margin not
saw-toothed; River Cooter has light carapace with dark whorls on
rings.

Habitat
Lakes, canals, ponds.

Habits
Basks alone or in groups; seldom seen on land.

Range
100th meridian w. to extr. e. and s. N. Mex.

PAINTED TURTLE
Chrysemys picta   70:6

Description
Length, 3½–9⅞ in. (8.9–25.1 cm). Carapace dark, sooty brown,
deep olive-gray to almost black, front edge of shields bordered with
yellow, network often with light yellow or red line down center;
plastron yellow or buff, may be marked with dark blotches; head
and limbs have yellow lines; red mark behind eyes; Carapace
smooth, low, gently arched, edges rounded, rear margin smooth.
Plastron composed of 6 pairs of shields (exclusive of bridge), which
may or may not show growth lines; toes webbed, claws sharp and
slender. Upper jaw notched, crushing surface ridged. Male shell
lower, forenails longer than female.

Similarities
Western Pond Turtle has network of dark lines radiating from
center of shields; Pond Slider has carapace with longitudinal
wrinkles and saw-toothed rear margin.

Habitat
Quiet waters, slow streams, marshes, ditches; brackish tidal water;
to 6000 ft. (1829 m) elevation in slow streams, marshes, ditches;
brackish tidal water.

Habits
Timid; often basks in groups or floats with head sticking up
through water amid vegetation; sometimes migrates short distances
across land.

Remarks
Easily tamed.

Range
Far s. Canada, coast to coast; e. Wash. and Columbia R. to mouth,
including Oreg. side; n. Idaho, Mont., Wyo. (except sw.); e. Colo.;
N.Mex. (except extreme w. and far se.), extr. nw. Tex. panhandle,
Okla. panhandle, thence e. across U.S.

DESERT TORTOISE
Gopherus agassizi   **70:3**

Description
Length 6–14½ in. (15.2–36.8 cm). Carapace growth rings pronounced but not reliable for age determination. Plastron with 6 pairs of shields, also with well-defined growth lines, broad bridge, no hinge. Above dull brown or horn, individual shields usually centrally light brown or yellowish; plastron yellowish, unpatterned. Head scaly, jaw margins roughly serrate, eardrum moderately distinct; limbs elephantine, nails blunt, no webbing. Male larger, gular shields and tail longer, plastron concave.

Habitat
In desert, with available water; frequents washes, sandy and gravelly flats, canyon bottoms, rocky hillsides, etc., to 3500 ft. (1067 m).

Habits
Active by day, especially in morning and afternoon; in hot season active at night; a burrower into banks or beneath dry shrubs, creates "dens" reaching 30 ft. (9.1 m) in length; colonial, up to 17 individuals in a den.

Range
Se. Calif., s. tip Nev., extr. sw. Utah, w. Ariz. (including far s.-cen. border).

SEA TURTLES
Family Cheloniidae

These marine turtles range up to four feet (1.2 m) in shell length and have limbs modified into flippers. The carapace has become lightened, and the short, heavy neck cannot be completely drawn back into the shell. The female returns to land to lay eggs in holes that she excavates in the sand above high-water mark. They are widespread in warm seas.

LOGGERHEAD
Caretta caretta

Description
Length 28–45 in. (71.1–114.3 cm), and weight, 300–500 lb. (136.1–226.8 kg). Carapace slender, 5 or more costal shields, not overlapping (except sometimes slightly in young). Head very broad, scaly, 2 pairs of prefrontals. Carapace reddish-brown, often yellowish at shield margins; head shields yellowish- to olive-brown; below cream, clouded. Limbs paddle-shaped, forelimbs with 2 claws. Male has tapering shell, longer tail. Young have 3 keels above, 2 below.

Habitat
Uncommon; coastal bays; brackish streams; high seas.

Habits
Will attack if molested; an oceanic wanderer but adaptable to different environments; often floats on surface a good deal; water speed to 1 mph. (1.6 km/hr.).

Range
Off s. Calif. coast and islands; also upper end of Gulf of Calif.

Note: The **PACIFIC RIDLEY**, *Lepidochelys olivacea* (**70:5**), has 6–8 costal shields on each side and two pairs of prefontals. It has been recorded once off the coast of Humboldt County, California.

Plates

Note on the Bird Plates

On each of the bird plates following (Plates 1 through 52) each species is designated by a different number. These guidelines will be of help in using the numbering system:

1. When a simple number (**1, 2, 3,** etc.) appears with an illustration, the bird depicted is either an adult male (designated ♂) or an adult female (designated ♀); in many instances, there are no immediately visible distinctions between males and females.

2. When the number is followed by a letter (**1a, 1b,** etc.) the bird depicted is a variant form. The major variants are immature or juvenal plumages, geographical races or subspecies, or color phases.

PLATE 1
LOONS AND GREBES

1 Common Loon, **1a** summer, **1b** winter, p. 10. **2** Arctic Loon, **2a** summer, **2b** winter, p. 11. **3** Red-throated Loon, **3a** summer, **3b** winter, p. 11. **4** Red-necked Grebe, **4a** summer, **4b** winter, p. 12. **5** Western Grebe, p. 14. **6** Horned Grebe, **6a** summer, **6b** winter, p. 12. **7** Eared Grebe, **7a** winter, **7b** summer, p. 13. **8** Pied-billed Grebe, **8a** winter, **8b** summer, **8c** immature, p. 14. **9** Least Grebe, **9a** summer, p. 13.

PLATE 2
PELAGIC BIRDS

1 Parasitic Jaeger, **1a** light phase immature, **1b** light phase,
1c dark phase, p. 102. **2** Pomarine Jaeger, **2a** light phase, p. 103.
3 Long-tailed Jaeger, **3a** light phase, p. 103. **4** Sooty Shearwater, p. 16.
5 New Zealand Shearwater, p. 16. **6** Leach's Storm-Petrel, p. 17.
7 Northern Fulmar, **7a** a dark phase, **7b** light phase, p. 15. **8** Pink-footed
Shearwater, p. 16. **9** Black Storm-Petrel, p. 18. **10** Ashy Storm-Petrel, p. 18.
11 Fork-tailed Storm-Petrel, p. 17. **12** Manx Shearwater, p. 17.

PLATE 3 PELICANS, CORMORANTS, ALBATROSS, AND ALLIES

1 Magnificent Frigatebird, **1a** immature, p. 22. **2** Anhinga, p. 22.
3 Blue-footed Booby, p. 20. **4** Brown Pelican, **4a** immature, p. 19.
5 Brandt's Cormorant, **5a** immature, **5b** breeding, p. 21. **6** Black-footed
Albatross, p. 15. **7** Double-crested Cormorant, **7a** immature,
7b breeding, p. 21. **8** Pelagic Cormorant, **8a** immature, **8b** breeding, p. 21.

PLATE 4
HERONS

1 Cattle Egret, **1a** immature, **1b** breeding, p. 26. **2** Great Egret, p. 24.
3 Louisiana Heron, p. 24. **4** Reddish Egret, **4a** dark phase,
4b white phase, p. 25. **5** Snowy Egret, **5a** immature, **5b** breeding, p. 25.
6 Little Blue Heron, **6a** immature, p. 24. **7** Yellow-crowned Night Heron,
7a immature, p. 26. **8** Black-crowned Night Heron, **8a** immature, p. 26.
9 Green Heron, **9a** immature, p. 23. **10** Least Bittern, p. 26.
11 American Bittern, p. 27.

PLATE 5
HERONS, IBISES, STORKS, CRANES

1 Great Blue Heron, **1a** breeding, **1b** immature, p. 23.
2 Great Egret, **2a** breeding, p. 24.**3** Sandhill Crane, p. 75.
4 Whooping Crane, p. 75. **5** Wood Stork, p. 28.**6** White-faced Ibis,
6a breeding, **6b** immature, p. 28. **7** Roseate Spoonbill,
7a breeding, **7b** immature, p. 29.

PLATE 6
GEESE AND SWANS

1 Snow Goose, **1a** "White form," **1b** "White Form" immature,
1c "Blue form," p. 32. **2** Ross' Goose, p. 33. **3** White-fronted Goose,
3a immature, p. 32. **4** Brant, p. 31. **5** Whistling Swan, **5a** immature, p. 30.
6 Canada Goose, **6a** large race, **6b** small race, p. 31.

**PLATE 7
POND DUCKS**

1 Wood Duck, p. 39. **2** Fulvous Whistling-Duck, p. 33.
3 Eurasian Wigeon, p. 37. **4** American Wigeon, p. 38. **5** Common Teal, p. 36.
6 Northern Shoveler, p. 38. **7** Blue-winged Teal, p. 36. **8** Gadwall, p. 35.
9 Cinnamon Teal, p. 37. **10** Mallard, p. 34. **11** Northern Pintail, p. 35.

PLATE 8
POND DUCKS IN FLIGHT—FEMALES

1 Black Duck, p. 35. **2** Mallard, p. 34. **3** Gadwall, p. 35. **4** Northern Pintail, p. 35. **5** Common Teal, p. 36. **6** Blue-winged Teal, p. 36.
7 Northern Shoveler, p. 38. **8** American Wigeon, p. 38.
9 Wood Duck, p. 39.

PLATE 9
BAY AND SEA DUCKS IN FLIGHT—FEMALES

1 Redhead, p. 40. **2** Canvasback, p. 39. **3** Greater Scaup, p. 41.
4 Lesser Scaup, p. 41. **5** Common Goldeneye, p. 42. **6** Bufflehead, p. 43.
7 Hooded Merganser, p. 47. **8** Red-breasted Merganser, p. 47.
9 Oldsquaw, **9a** winter, p. 43. **10** King Eider, p. 44.
11 White-winged Scoter, p. 46. **12** Surf Scoter, p. 45.

PLATE 10
BAY AND SEA DUCKS

1 Barrow's Goldeneye, p. 42. **2** Bufflehead, p. 43. **3** Common Goldeneye, p. 42. **4** Harlequin Duck, p. 44. **5** Oldsquaw, **5a** winter, **5b** summer, p. 43. **6** Black Scoter, p. 45. **7** Surf Scoter, **7a** subadult, p. 45. **8** White-winged Scoter, **8a** subadult, p. 46. **9** King Eider, **9a** immature, p. 44.

PLATE 11
POCHARDS, MERGANSERS, RUDDY DUCK

1 Redhead, p. 40. **2** Canvasback, p. 39. **3** Ring-necked Duck, p. 40.
4 Lesser Scaup, p. 41. **5** Greater Scaup, p. 41. **6** Hooded Merganser, p. 47.
7 Red-breasted Merganser, p. 47. **8** Common Merganser, p. 48.
9 Ruddy Duck, **9a** summer, **9b** winter, p. 46.

PLATE 12
VULTURES, EAGLES, CARACARA, OSPREY

1 Turkey Vulture, p. 49. **2** Black Vulture, p. 49.
3 Bald Eagle, **3a** immature, p. 59. **4** Crested Caracara, p. 61.
5 Osprey, p. 60. **6** Golden Eagle, **6a** immature, p. 59.

PLATE 13
ACCIPITERS AND BUTEOS

1 Northern Goshawk, **1a** immature, p. 52. **2** Cooper's Hawk,
2a immature, p. 52. **3** Sharp-shinned Hawk, **3a** immature, p. 53.
4 Gray Hawk, **4a** immature, p. 57. **5** Red-shouldered Hawk,
5a immature, p. 54. **6** Ferruginous Hawk, **6a** light phase, p. 57.
7 Red-tailed Hawk, **7a** immature, p. 53. **8** Swainson's Hawk, **8a** immature,
8b light phase, p. 55. **9** Rough-legged Hawk, **9a** light phase, p. 56.

PLATE 14
HAWKS IN FLIGHT

1 Red-shouldered Hawk, p. 54. **2** Swainson's Hawk, **2a** dark phase, **2b** light phase, p. 55. **3** Gray Hawk, p. 57. **4** Red-tailed Hawk, **4a** "Harlan's" race, dark phase, **4b** typical western race, **4c** eastern race, p. 53. **5** Ferruginous Hawk, **5a** dark phase, **5b** light phase, p. 57. **6** Rough-legged Hawk, **6a** light phase, **6b** dark phase, p. 56.

PLATE 15
HAWKS IN FLIGHT

1 Crested Caracara, p. 61. **2** Osprey, p. 60. **3** Common Black Hawk, p. 58.
4 White-tailed Hawk, p. 55. **5** Harris' Hawk, p. 58.
6 Zone-tailed Hawk, p. 56. **7** Northern Harrier, p. 60.

PLATE 16
FALCONS, KITES, GROUSE

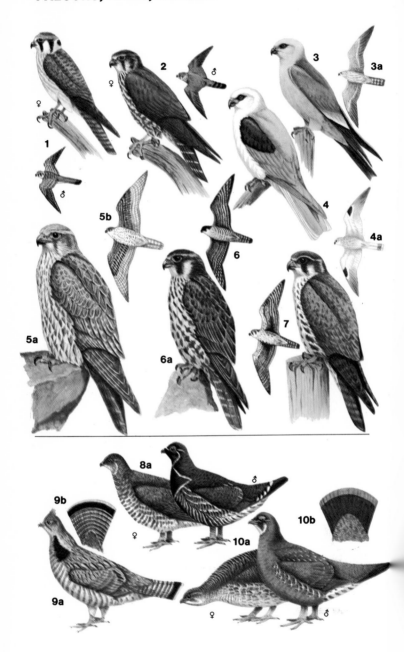

1 American Kestrel, p. 64. 2 Merlin, p. 63. 3 Mississippi Kite,
3a immature, p. 51. 4 White-tailed Kite, 4a immature, p. 51.
5 Gyrfalcon, 5a gray phase immature, 5b gray phase, p. 61.
6 Peregrine Falcon, 6a immature, p. 62. 7 Prairie Falcon, p. 62.
8 Spruce Grouse, 8a no. Rocky Mt. - Cascades race, p. 66.
9 Ruffed Grouse, 9a gray phase, 9b tail of rufous phase, p. 66.
10 Blue Grouse, 10a no. Rocky Mt. forms, 10b tail of other forms, p. 65.

PLATE 17
OPEN COUNTRY GROUSE AND PHEASANT

1 Rock Ptarmigan, **1a** winter, **1b** summer, p. 67. **2** Willow Ptarmigan, **2a** summer, **2b** winter, p. 68. **3** Greater Prairie Chicken, **3a** displaying, p. 68. **4** White-tailed Ptarmigan, **4a** summer, p. 67. **5** Sharp-tailed Grouse, **5a** displaying, p. 69. **6** Sage Grouse, **6a** displaying, p. 69. **7** Ring-necked Pheasant, p. 73.

**PLATE 18
RAILS**

1 Purple Gallinule, **1a** immature, p. 79. **2** American Coot, **2a** immature,
2b chick, p. 79. **3** Common Gallinule, **3a** immature, **3b** chick, p. 78.
4 Clapper Rail, **4a** chick, p. 76. **5** Virginia Rail, **5a** immature, p. 76.
6 Sora, **6a** immature, p. 77. **7** Black Rail, p. 78. **8** Yellow Rail, p. 77.

PLATE 19
SHOREBIRDS—FULL BIRDS IN SPRING PLUMAGES

1 Piping Plover, **1a** fall, p. 81. **2** Semipalmated Plover, **2a** fall, p. 80.
3 Killdeer, p. 82. **4** Snowy Plover, p. 81. **5** Mountain Plover,
5a fall, p. 82. **6** Black Turnstone, p. 85. **7** Black-bellied Plover, p. 83.
8 American Golden Plover, p. 83. **9** Ruddy Turnstone, p. 84.
10 Upland Sandpiper, p. 87. **11** Buff-breasted Sandpiper, p. 96.
12 Marbled Godwit, p. 98. **13** Willet, p. 90. **14** Whimbrel, p. 86.
15 Long-billed Curlew, p. 85. **16** Black-necked Stilt, p. 102.
17 Black Oystercatcher, p. 80. **18** American Avocet, **18a** fall, p. 101.

PLATE 20
SHOREBIRDS—MOST BIRDS IN SPRING PLUMAGES

1 Western Sandpiper, p. 95. **2** Least Sandpiper, p. 93.
3 Semipalmated Sandpiper, p. 94. **4** White-rumped Sandpiper, p. 92.
5 Baird's Sandpiper, p. 93. **6** Sanderling, p. 95. **7** Dunlin, p. 94.
8 Pectoral Sandpiper, p. 91. **9** Surfbird, **9a** fall, p. 98.
10 Rock Sandpiper, p. 91. **11** Red Knot, p. 92. **12** Hudsonian Godwit, p. 99.
13 Common Snipe, p. 85. **14** Short-billed Dowitcher, p. 97.
15 Wilson's Phalarope, p. 100. **16** Wandering Tattler, **16a** fall, p. 90.
17 Red Phalarope, p. 99. **18** Northern Phalarope, p. 100.

PLATE 21
SHOREBIRDS–MOST BIRDS IN FALL PLUMAGES

1 Semipalmated Sandpiper, p. 94. **2** Least Sandpiper, p. 93.
3 Western Sandpiper, p. 95. **4** Sanderling, p. 95. **5** Ruddy Turnstone, p. 84.
6 White-rumped Sandpiper, p. 92. **7** Rock Sandpiper, p. 91.
8 Black Turnstone, p. 85. **9** Dunlin, p. 94. **10** Red Knot, p. 92.
11 American Golden Plover, p. 83. **12** Black-bellied Plover, p. 83.
13 Stilt Sandpiper, **13a** spring, p. 96. **14** Short-billed Dowitcher, p. 97.
15 Hudsonian Godwit, p. 99. **16** Solitary Sandpiper, p. 88.
17 Willet, p. 90. **18** Spotted Sandpiper, **18a** spring, p. 87.
19 Wilson's Phalarope, p. 100. **20** Lesser Yellowlegs, p. 89. **21** Greater
Yellowlegs, p. 88. **22** Northern Phalarope, p. 100. **23** Red Phalarope, p. 99.

PLATE 22
FALL SHOREBIRDS IN FLIGHT

1 White-rumped Sandpiper, p. 92. **2** Western Sandpiper, p. 95.
3 Pectoral Sandpiper, p. 91. **4** Rock Sandpiper, p. 91. **5** Dunlin, p. 94.
6 Spotted Sandpiper, p. 87. **7** Northern Phalarope, p. 100.
8 Sanderling, p. 95. **9** Solitary Sandpiper, p. 88. **10** Red Phalarope, p. 99.
11 Red Knot, p. 92. **12** Short-billed Dowitcher, p. 97. **13** Wilson's
Phalarope, p. 100. **14** Stilt Sandpiper, p. 96. **15** Lesser Yellowlegs, p. 89.

PLATE 23
SHOREBIRDS IN FLIGHT

1 Semipalmated Plover, **1a** spring, p. 80. **2** Snowy Plover, **2a** spring, p. 81.
3 Killdeer, p. 82. **4** Buff-breasted Sandpiper, p. 96.
5 Upland Sandpiper, p. 87. **6** American Golden Plover, **6a** fall, p. 83.
7 Mountain Plover, **7a** fall, p. 82. **8** Common Snipe, p. 85. **9** Black-bellied
Plover, **9a** fall, p. 83. **10** Wandering Tattler, **10a** fall, p. 90.
11 Black Turnstone, **11a** fall, p. 85. **12** Surfbird, **12a** fall, p. 98.

PLATE 24
LARGER GULLS

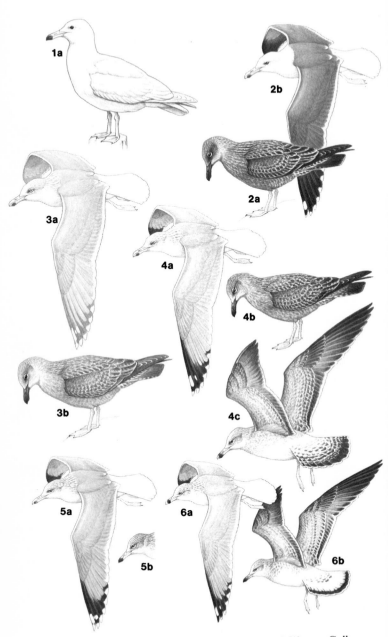

1 Glaucous Gull, **1a** second winter immature, p. 105. **2** Western Gull,
2a first winter immature, **2b** winter northern race, p. 106.
3 Glaucous-winged Gull, **3a** winter, **3b** first winter immature, p. 105.
4 Herring Gull, **4a** winter, **4b** first winter immature,
4c second winter immature, p. 106. **5** California Gull, **5a** winter,
5b first winter immature, p. 107. **6** Ring-billed Gull, **6a** winter,
6b first winter immature, p. 108.

PLATE 25
SMALLER GULLS

1 Heermann's Gull, **1a** summer, **1b** first winter immature, p. 110.
2 Franklin's Gull, **2a** winter, **2b** summer, **2c** first winter immature p. 108.
3 Bonaparte's Gull, **3a** summer, **3b** winter, **3c** first winter immature, p. 109.
4 Sabine's Gull, **4a** first winter immature, **4b** summer, p. 111.
5 Mew Gull, **5a** winter, **5b** first winter immature, p. 109.
6 Black-legged Kittiwake, **6a** winter, **6b** first winter immature, p. 110.

PLATE 26
TERNS AND GULLS

1 Least Tern, **1a** fall, **1b** spring, p. 114. **2** Black Tern, **2a** fall, p. 116.
3 Common Tern, **3a** fall, **3b** spring, p. 113. **4** Forster's Tern, **4a** fall,
4b spring, p. 112. **5** Gull-billed Tern, **5a** fall, **5b** spring, p. 112.
6 Arctic Tern, **6a** spring, p. 113. **7** California Gull, **7a** spring, p. 107.
8 Mew Gull, **8a** spring, p. 109. **9** Elegant Tern, **9a** fall, p. 114.
10 Black-legged Kittiwake, **10a** spring, p. 110. **11** Ring-billed Gull,
11a spring, p. 108. **12** Herring Gull, **12a** spring, p. 106.
13 Royal Tern, **13a** fall, p. 115. **14** Caspian Tern, **14a** fall, p. 115.

PLATE 27
TERNS—ADULTS IN SPRING PLUMAGES

1 Royal Tern, p. 115. **2** Caspian Tern, p. 115. **3** Black Tern,
3a immature, p. 116. **4** Elegant Tern, p. 114. **5** Gull-billed Tern,
5a immature, p. 112. **6** Forster's Tern, **6a** immature, p. 112.
7 Arctic Tern, p. 113. **8** Common Tern, **8a** immature, p. 113.
9 Least Tern, **9a** immature, p. 114.

PLATE 28
ALCIDS

1 Common Murre, **1a** summer, **1b** winter, p. 117. **2** Pigeon Guillemot,
2a summer, **2b** immature, **2c** winter, p. 117. **3** Tufted Puffin, **3a** summer,
3b winter, p. 121. **4** Marbled Murrelet, **4a** winter, **4b** summer, p. 118.
5 Ancient Murrelet, **5a** winter, **5b** summer, p. 119.
6 Xantus' Murrelet, p. 118. **7** Cassin's Auklet, p. 119.
8 Rhinoceros Auklet, **8a** immature, **8b** summer, p. 120.

PLATE 29
DOVES AND QUAILS

1 Spotted Dove, p. 125. **2** Mourning Dove, p. 124. **3** White-winged
Dove, p. 123. **4** White-fronted Dove, p. 124. **5** Inca Dove, p. 125.
6 Mountain Quail, p. 70. **7** Common Ground Dove, p. 124.
8 Gray Partridge, p. 73. **9** California Quail, p. 71. **10** Bobwhite, p. 70.
11 Gambel's Quail, p. 71. **12** Scaled Quail, p. 72. **13** Montezuma Quail, p. 71.

PLATE 30
OWLS

1 Burrowing Owl, p. 132. **2** Elf Owl, p. 131. **3** Pygmy Owl, p. 130.
4 Screech Owl, **4a** rufous phase, **4b** gray phase, p. 128.
5 Saw-whet Owl, p. 135. **6** Hawk Owl, p. 129. **7** Short-eared Owl, p. 134.
8 Long-eared Owl, p. 134. **9** Spotted Owl, p. 132. **10** Barn Owl, p. 127.
11 Barred Owl, p. 133. **12** Snowy Owl, p. 130.
13 Great Horned Owl, p. 129. **14** Great Gray Owl, p. 133.

PLATE 31
PIGEONS, SWIFTS, NIGHTJARS

1 Rock Pigeon, p. 122. **2** Band-tailed Pigeon, p. 122. **3** Vaux's Swift, p. 139.
4 Black Swift, p. 138. **5** White-throated Swift, p. 139. **6** Lesser
Nighthawk, p. 137. **7** Pauraque, p. 136. **8** Common Nighthawk, p. 137.
9 Poor-will, **9a** tail, p. 136. **10** Whip-poor-will, **10a** tail, p. 136.

PLATE 32
WOODPECKERS

1 Pileated Woodpecker, p. 148. **2** Common Flicker,
2a Eastern "Yellow-shafted" form, **2b** Western "Red-shafted" form, p. 147.
3 Downy Woodpecker, p. 152. **4** Hairy Woodpecker, p. 151. **5** Red-bellied
Woodpecker, **5a** immature, p. 149. **6** Three-toed Woodpecker,
6a "White-backed" form,   **6b** "Ladder-backed" form, p. 154.
7 Black-backed Woodpecker, p. 154. **8** Red-naped Sapsucker,
8a immature, p. 150. **9** Lewis' Woodpecker, p. 150.
10 Red-headed Woodpecker, **10a** immature, p. 149.

PLATE 33
WOODPECKERS, CUCKOOS, TROGON

1 Common Flicker, **1a** southwestern "Gilded" form, p. 147.
2 Nuttall's Woodpecker, p. 154. **3** Ladder-backed Woodpecker, p. 152.
4 Strickland's (Arizona) Woodpecker, p. 153. **5** Golden-fronted Woodpecker,
5a immature, p. 149. **6** Gila Woodpecker, **6a** immature, p. 148.
7 Acorn Woodpecker, p. 149. **8** Red-breasted Sapsucker, p. 150.
9 Williamson's Sapsucker, p. 151. **10** White-headed Woodpecker, p. 153.
11 Elegant Trogon, p. 145. **12** Roadrunner, p. 126.
13 Groove-billed Ani, p. 127.

PLATE 34
FLYCATCHERS AND PHAINOPEPLA

1 Great Crested Flycatcher, p. 158. 2 Say's Phoebe, p. 161.
3 Eastern Phoebe, p. 160. 4 Scissor-tailed Flycatcher, p. 158.
5 Cassin's Kingbird, p. 157. 6 Western Kingbird, p. 157.
7 Eastern Kingbird, p. 156. 8 Phainopepla, p. 203. 9 Eastern Wood
Pewee, p. 165. 10 Western Wood Pewee, p. 166.
11 Olive-sided Flycatcher, p. 166. 12 Willow/Alder Flycatcher, p. 162.
13 Least Flycatcher, p. 162. 14 Western Flycatcher, p. 164.

PLATE 35
FLYCATCHERS AND BECARD

1 Vermilion Flycatcher, p. 167. 2 Wied's Crested Flycatcher, p. 160.
3 Ash-throated Flycatcher, p. 159. 4 Olivaceous Flycatcher, p. 160.
5 Rose-throated Becard, p. 155. 6 Kiskadee Flycatcher, p. 158.
7 Tropical Kingbird, p. 156. 8 Sulphur-bellied Flycatcher, p. 159.
9 Thick-billed Kingbird, p. 157. 10 Northern Beardless Flycatcher, p. 167.
11 Gray Flycatcher, p. 164. 12 Coues' Pewee, p. 165.
13 Black Phoebe, p. 161. 14 Buff-breasted Flycatcher, p. 164.
15 Hammond's Flycatcher, p. 163. 16 Dusky Flycatcher, p. 163.

PLATE 36
SWALLOWS AND CUCKOOS

1 Purple Martin, p. 172. **2** Violet-green Swallow, **2a** immature, p. 168.
3 Tree Swallow, **3a** immature, p. 169. **4** Bank Swallow, p. 171.
5 Rough-winged Swallow, p. 169. **6** Barn Swallow, p. 170.
7 Cliff Swallow, p. 171. **8** Yellow-billed Cuckoo, p. 126.
9 Cave Swallow, p. 170. **10** Black-billed Cuckoo, p. 126.

PLATE 37
CORVIDS, KINGFISHERS, DIPPER

1 Green Jay, p. 173. **2** Mexican Jay, p. 173. **3** Steller's Jay, p. 173.
4 Blue Jay, p. 173. **5** Pinyon Jay, p. 177. **6** Scrub Jay, p. 174.
7 Gray Jay, **7a** juvenal, p. 172. **8** Clark's Nutcracker, p. 177.
9 Black-billed Magpie, p. 174. **10** Belted Kingfisher, p. 146. **11** Yellow-billed Magpie, p. 174. **12** Dipper, p. 185. **13** Green Kingfisher, p. 146.

PLATE 38
NUTHATCHES, TITMICE, WRENS AND OTHERS

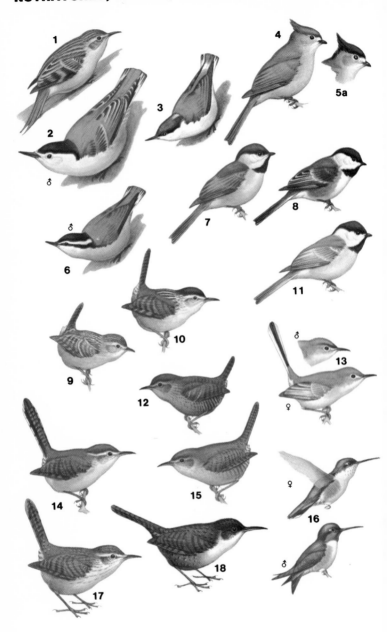

1 Brown Creeper, p. 183. **2** White-breasted Nuthatch, p. 182.
3 Pygmy Nuthatch, p. 183. **4** Plain Titmouse, p. 180. **5** Tufted Titmouse,
5a "Black-crested" form, p. 180. **6** Red-breasted Nuthatch, p. 182.
7 Boreal Chickadee, p. 179. **8** Chestnut-backed Chickadee, p. 179.
9 Short-billed Marsh Wren, p. 188. **10** Long-billed Marsh Wren, p. 187.
11 Black-capped Chickadee, p. 178. **12** Winter Wren, p. 186.
13 Blue-gray Gnatcatcher, p. 199. **14** Bewick's Wren, p. 187.
15 House Wren, p. 185. **16** Ruby-throated Hummingbird, p. 140.
17 Rock Wren, p. 188. **18** Canyon Wren, p. 189.

PLATE 39
TITMICE, HUMMINGBIRDS AND OTHERS

1 Bridled Titmouse, p. 181. **2** Wrentit, p. 184.
3 Blue-throated Hummingbird, p. 145. **4** Mountain Chickadee, p. 179.
5 Bushtit, **5a** "Black-cheeked" form, p. 181. **6** Mexican Chickadee, p. 178.
7 Rivoli's Hummingbird, p. 144. **8** Black-tailed Gnatcatcher, **8a** spring, p. 198.
9 Verdin, **9a** immature, p. 181. **10** Broad-billed Hummingbird, p. 144.
11 Costa's Hummingbird, p. 142. **12** Black-chinned Hummingbird, p. 141.
13 Lucifer Hummingbird, p. 140. **14** Calliope Hummingbird, p. 143.
15 Allen's Hummingbird, p. 143. **16** Broad-tailed Hummingbird, p. 142.
17 Anna's Hummingbird, p. 141. **18** Rufous Hummingbird, p. 142.

PLATE 40
MIMIDS, SHRIKES, CACTUS WREN

1 Brown Thrasher, p. 190. **2** Long-billed Thrasher, p. 191.
3 Northern Mockingbird, **3a** immature, p. 189. **4** Gray Catbird, p. 190.
5 Sage Thrasher, p. 193. **6** Loggerhead Shrike, p. 204. **7** Northern Shrike,
7a winter, **7b** immature, p. 203. **8** Bendire's Thrasher, p. 191.
9 LeConte's Thrasher, p. 192. **10** Curve-billed Thrasher, p. 191. **11** Crissal
Thrasher, p. 192. **12** Cactus Wren, p. 186. **13** California Thrasher, p. 193.

PLATE 41
THRUSHES AND WAXWINGS

1 Varied Thrush, p. 194. **2** American Robin, **2a** juvenal, p. 194.
3 Townsend's Solitaire, **3a** juvenal, p. 198.
4 Western Bluebird, **4a** juvenal, p. 196. **5** Hermit Thrush, p. 195.
6 Eastern Bluebird, **6a** juvenal, p. 197. **7** Mountain Bluebird,
7a juvenal, p. 197. **8** Veery, p. 195. **9** Swainson's Thrush, p. 196.
10 Bohemian Waxwing, p. 202. **11** Cedar Waxwing, **11a** juvenal, p. 202.

PLATE 42
WARBLERS AND VIREOS

1 Hermit Warbler, **1a** immature, p. 218. **2** Black-throated Green Warbler,
2a spring, **2b** immature, p. 217. **3** Golden-cheeked Warbler, **3a** immature, p. 2
4 Olive Warbler, **4a** immature, p. 213. **5** Colima Warbler, p. 213.
6 Virginia's Warbler, **6a** immature, p. 212. **7** Tropical Parula, p. 214.
8 Northern Parula, p. 214. **9** Lucy's Warbler, **9a** immature, p. 212.
10 Red-faced Warbler, **10a** immature, p. 224. **11** Grace's Warbler, p. 219.
12 Painted Redstart, p. 226. **13** Solitary Vireo, **13a** Rocky Mt. race,
13b other races, p. 207. **14** Bell's Vireo, p. 207. **15** Black-capped Vireo, p. 206.
16 Gray Vireo, p. 207. **17** Hutton's Vireo, **17a** Pacific Coast race,
17b Rocky Mt. race, p. 206. **18** Ruby-crowned Kinglet, p. 200.

PLATE 43
WARBLERS, VIREOS, KINGLETS

1 Yellow Warbler, **1a** spring, **1b** immature, p. 214. **2** Nashville Warbler,
2a fall, **2b** immature, p. 211. **3** Chestnut-sided Warbler, **3a** immature, p. 220.
4 Orange-crowned Warbler, **4a** spring, **4b** fall, **4c** immature, p. 211.
5 Wilson's Warbler, p. 225. **6** Common Yellowthroat, **6a** immature, p. 223.
7 Warbling Vireo, p. 209. **8** Tennessee Warbler, **8a** fall, **8b** immature, p. 210.
9 Philadelphia Vireo, p. 209. **10** Golden-crowned Kinglet, p. 199.
11 Red-eyed Vireo, p. 208. **12** Ruby-crowned Kinglet, p. 200.

PLATE 44
WARBLERS

1 Cape May Warbler, **1a** immature, **1b** spring, p. 215. **2** Palm Warbler,
2a fall, p. 221. **3** Yellow-rumped Warbler, **3a** spring eastern "Myrtle" form,
3b "Myrtle" immature, **3c** spring western "Audubon's" form,
3d "Audubon's" immature, p. 216. **4** Blackburnian Warbler, **4a** spring, p. 218.
5 Black-throated Green Warbler, **5a** spring, **5b** immature, p 217.
6 Townsend's Warbler, **6a** immature, **6b** spring, p. 217.
7 Black-throated Gray Warbler, p. 215. **8** Blackpoll Warbler,
8a spring, p. 219. **9** Black-and-White Warbler, p. 210.

PLATE 45
WARBLERS—SPRING PLUMAGES

1 American Redstart, p. 226. 2 Bay-breasted Warbler, p. 220.
3 Chestnut-sided Warbler, p. 220. 4 Ovenbird, p. 221.
5 Yellow-breasted Chat, p. 224. 6 Northern Waterthrush, p. 222.
7 Canada Warbler, p. 225. 8 Nashville Warbler, p. 211.
9 Common Yellowthroat, p. 223. 10 Mourning Warbler, p. 223.
11 Connecticut Warbler, p. 222. 12 MacGillivray's Warbler, p. 223.

PLATE 46
BLACKBIRDS AND STARLING

1 Yellow-headed Blackbird, p. 229. **2** Red-winged Blackbird,
2a immature, p. 229. **3** Starling, **3a** fall, **3b** spring, **3c** immature, p. 205.
4 Brown-headed Cowbird, **4a** juvenal, p. 235. **5** Rusty Blackbird,
5a spring, **5b** fall immature, p. 233. **6** Brewer's Blackbird, **6a** spring,
6b fall immature, p. 233. **7** Bronzed Cowbird, p. 235.
8 Common Grackle, p. 234. **9** Great-tailed Grackle, p. 234.

PLATE 47 ORIOLES, TANAGERS, FINCHES—SPRING PLUMAGES

PLATE 48
FINCHES

1 Cardinal, p. 237. 2 Pyrrhuloxia, p. 238. 3 Evening Grosbeak, p. 242.
4 Pine Grosbeak, p. 244. 5 Red Crossbill, p. 248. 6 Purple Finch, p. 243.
7 White-winged Crossbill, p. 248. 8 House Finch, p. 244.
9 Common Redpoll, p. 245. 10 Cassin's Finch, p. 243. 11 Pine Siskin, p. 246.
12 Lawrence's Goldfinch, p. 248. 13 American Goldfinch,
13a summer, p. 247. 14 Lesser Goldfinch,
14a eastern "Black-backed" form, 14b western "Green-backed" form, p. 247.

PLATE 49
SPARROWS, JUNCOS, TOWHEES

1 Black-throated Sparrow, p. 257. **2** Black-chinned Sparrow, p. 261.
3 Lark Sparrow, p. 254. **4** Sage Sparrow, **4a** coastal Calif. "Bell's"
race, p. 256. **5** Dark-eyed Junco, **5a** Black Hills "White-winged" form,
5b eastern "Slate-colored" form, **5c** central Rocky Mt. "Gray-headed" form,
5d so. Rocky Mt. "Gray-headed" form, **5e** no. Rocky Mt. "Pink-sided"
form, **5f** western "Oregon" form, p. 257. **6** Yellow-eyed Junco, p. 258.
7 Green-tailed Towhee, p. 249. **8** Rufous-sided Towhee, p. 249.
9 Brown Towhee, **9a** Rocky Mt. form, **9b** Pacific Coast form, p. 250.
10 Abert's Towhee, p. 250.

PLATE 50
SPARROWS

1 Fox Sparrow, **1a** "Dusky-brown" form, **1b** "Slaty" form, p. 264.
2 Vesper Sparrow, p. 254. **3** Lincoln's Sparrow, p. 265.
4 Song Sparrow, p. 265. **5** Savannah Sparrow, p. 251. **5a** "Belding's" race, p. 251.
6 Botteri's Sparrow, p. 255. **7** Baird's Sparrow, p. 253.
8 Sharp-tailed Sparrow, p. 252. **9** LeConte's Sparrow, p. 252.
10 Cassin's Sparrow, p. 256. **11** Grasshopper Sparrow, p. 253.
12 Clay-colored Sparrow, **12a** immature, p. 260. **13** Brewer's Sparrow, p. 260.
14 Chipping Sparrow, **14a** spring, **14b** immature, p. 259.

PLATE 51
SPARROWS

1 Harris' Sparrow, **1a** spring, **1b** immature, p. 261. **2** Golden-crowned
Sparrow, **2a** immature, p. 263. **3** White-crowned Sparrow, **3a** immature, p. 262.
4 White-throated Sparrow, **4a** immature, p. 263. **5** Field Sparrow, p. 261.
6 Rufous-winged Sparrow, p. 255. **7** Tree Sparrow, p. 259.
8 Swamp Sparrow, **8a** spring, p. 264. **9** Rufous-crowned Sparrow, p. 255.

PLATE 52 OPEN COUNTRY SONGBIRDS—MALES IN SPRING PLUMAGES

1 **Bobolink**, p. 227. 2 **Western Meadowlark**, p. 229. 3 **Lark Bunting**, p. 251.
4 **Sprague's Pipit**, p. 201. 5 **Water Pipit**, 5a fall, p. 201. 6 **Dickcissel**, p. 242.
7 **Horned Lark**, 7a "Yellow-faced" form, 7b "White faced" form,
7c juvenal, p. 167. 8 **Snow Bunting**, 8a winter, 8b immature, p. 268.
9 **Lapland Longspur**, 9a fall, p. 266. 10 **Chestnut-collared Longspur**,
10a fall, p. 267. 11 **Smith's Longspur**, 11a fall, p. 267.
12 **McCown's Longspur**, 12a fall, p. 266. 13 **Brown Rosy Finch**, p. 245.
14 **Gray-crowned Rosy Finch**, 14a immature, p. 245.
15 **Black Rosy Finch**, p. 245.

PLATE 53
MICE, SHREWS, MOLES, ALLIES

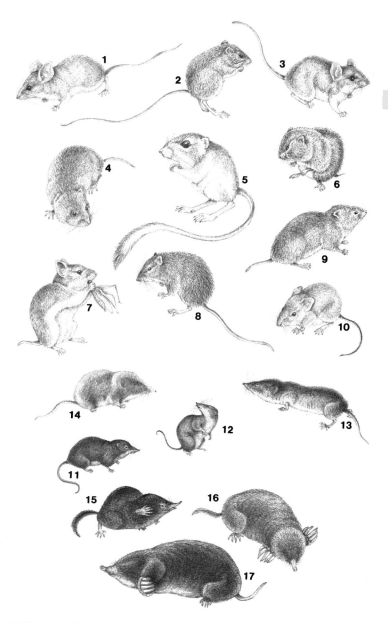

1 White-footed Mouse, p. 333. **2** Meadow Jumping Mouse, p. 349.
3 Deer Mouse, p. 331. **4** Meadow Vole, p. 341. **5** Ord's Kangaroo Rat, p. 325.
6 Northern Bog Lemming, p. 346. **7** Northern Grasshopper Mouse, p. 334.
8 Hispid Pocket Mouse, p. 322. **9** Southern Red-backed Vole, p. 339.
10 Plains Harvest Mouse, p. 329. **11** Dusky Shrew, p. 280.
12 Ornate Shrew, p. 277. **13** Arctic Shrew, p. 279. **14** Masked Shrew, p. 276.
15 Shrew-mole, p. 282. **16** Broad-footed Mole, p. 283.
17 Townsend's Mole, p. 283.

**PLATE 54
BATS**

1 Spotted Bat, p. 292. **2** Big Brown Bat, p. 291. **3** Small-footed Myotis, p. 289. **4** Hoary Bat, p. 292. **5** Little Brown Myotis, p. 286. **6** Silver-haired Bat, p. 290. **7** Townsend's Big-eared Bat, p. 292. **8** Western Pipistrelle, p. 290. **9** Southern Yellow Bat, p. 292. **10** Red Bat, p. 291. **11** Pallid Bat, p. 292.

PLATE 55
TRACKS—SMALLER MAMMALS

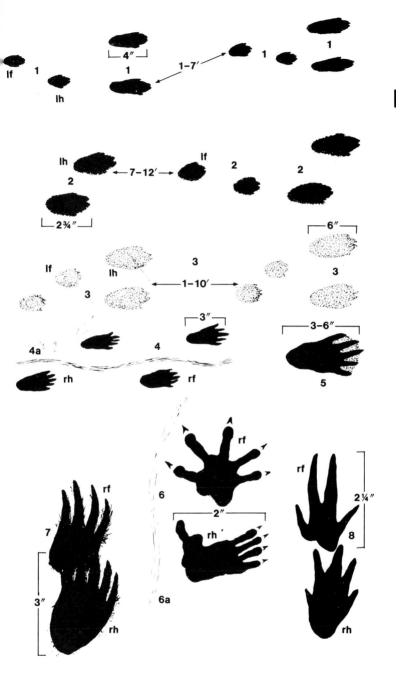

1 Cottontail, p. 298. 2 Jackrabbit, p. 300.
3 Snowshoe Hare, p. 299. 4 Muskrat, **4a** tail mark, p. 345.
5 Beaver, hind foot covers 4 in. (10.2 cm) front to next track, p. 328.
6 Virginia Opossum, **6a** tail mark, p. 275. 7 Porcupine, p. 350.
8 Armadillo, p. 295.

PLATE 56
RABBITS AND SQUIRRELS

1 Red Squirrel, p. 315. **2** Western Gray Squirrel, p. 315. **3** Northern Flying Squirrel, p. 316. **4** Abert's Squirrel, p. 314. **5** Rock Squirrel, p. 312. **6** Woodchuck, p. 307. **7** Black-tailed Prairie Dog, p. 313. **8** Thirteen-lined Ground Squirrel, p. 310. **9** Spotted Ground Squirrel, p. 311. **10** Least Chipmunk, p. 303. **11** Colorado Chipmunk, p. 305. **12** Eastern Cottontail, p. 298. **13** Black-tailed Jackrabbit, p. 302. **14** Brush Rabbit, p. 297. **15** Desert Cottontail, p. 298. **16** Snowshoe Hare, **16a** summer, **16b** winter, p. 299.

PLATE 57
SQUIRRELS AND CHIPMUNKS

1 Hoary Marmot, p. 307. **2** White-tailed Prairie Dog, p. 314.
3 Yellow-bellied Marmot, p. 307. **4** Townsend's Ground Squirrel, p. 309.
5 Belding's Ground Squirrel, p. 310. **6** California Ground Squirrel, p. 312.
7 Round-tailed Ground Squirrel, p. 313. **8** Columbian Ground Squirrel, p. 310.
9 White-tailed Antelope Squirrel, p. 308. **10** Yellow Pine Chipmunk, p. 304.
11 Golden-mantled Ground Squirrel, p. 311. **12** Cliff Chipmunk, p. 306.
13 Townsend's Chipmunk, p. 304.

PLATE 58
POCKET MICE AND KANGAROO RATS

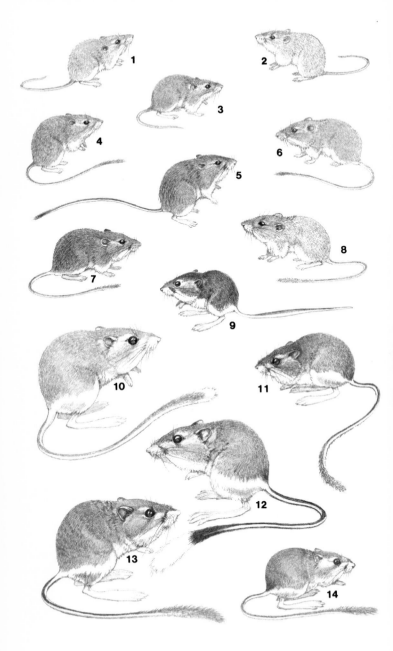

1 Plains Pocket Mouse, p. 320. 2 Little Pocket Mouse, p. 320.
3 Silky Pocket Mouse, p. 320. 4 Great Basin Pocket Mouse, p. 322.
5 California Pocket Mouse, p. 321. 6 Long-tailed Pocket Mouse, p. 321.
7 Rock Pocket Mouse, p. 323. 8 Desert Pocket Mouse, p. 323.
9 Dark Kangaroo Mouse, p. 324. 10 Desert Kangaroo Rat, p. 326.
11 Agile Kangaroo Rat, p. 325. 12 Banner-tailed Kangaroo Rat, p. 327.
13 Giant Kangaroo Rat, p. 327. 14 Merriam's Kangaroo Rat, p. 327.

PLATE 59
MICE AND RATS

1 Pygmy Mouse, p. 335. **2** House Mouse, p. 348. **3** California Mouse, p. 330.
4 Cactus Mouse, p. 331. **5** Fulvous Harvest Mouse, p. 329.
6 Pinyon Mouse, p. 333. **7** Canyon Mouse, p. 332.
8 Western Jumping Mouse, p. 349. **9** White-throated Woodrat, p. 336.
10 Dusky-footed Woodrat, p. 338. **11** Desert Woodrat, p. 337.
12 Southern Plains Woodrat, p. 336. **13** Bushy-tailed Woodrat, p. 338.

PLATE 60
TRACKS—CARNIVORES

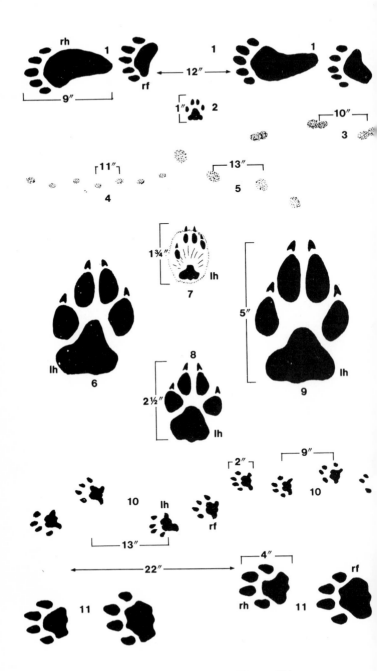

1 Black Bear, p. 287. 2 House Cat. 3 Wolf, walking, p. 284.
4 Fox, trotting, p. 285. 5 Coyote, trotting, p. 283. 6 Dog. 7 Red Fox, p. 286.
8 Coyote, p. 283. 9 Gray Wolf, p. 284. 10 Bobcat, p. 299.
11 Mountain Lion, p. 298.

PLATE 61
VOLES AND LEMMINGS

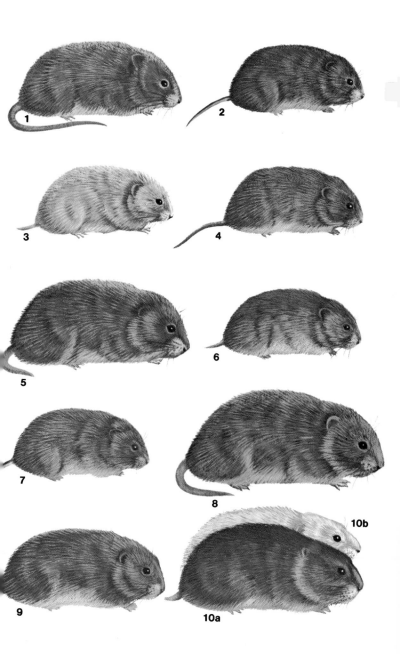

1 Red Tree Vole p. 340. **2** Heather Vole, p. 340. **3** Sagebrush Vole, p. 344.
4 Meadow Vole, p. 341. **5** Tundra Vole, p. 343. **6** Southern Red-backed
Vole, p. 339. **7** Prairie Vole, p. 344. **8** Yellow-cheeked Vole, p. 344. **9** Brown
Lemming, p. 346. **10** Collared Lemming, **10a** summer, **10b** winter, p. 346.

PLATE 62
TOOTHED WHALES

1 White Whale, p. 353. **2** North Pacific Bottle-nosed Whale, p. 351.
3 North Atlantic Bottle-nosed Whale, p. 352. **4** Pygmy Sperm Whale, p. 353.
5 Sperm Whale, p. 352. **6** Northern Right-Whale Dolphin, p. 354.
7 Goose-beaked Whale, p. 352.

PLATE 63
DOLPHINS AND OTHER WHALES

1 Pacific Bottle-nosed Dolphin, p. 354. **2** Dall's Porpoise, p. 356.
3 Pacific White-sided Dolphin, p. 355. **4** Harbor Porpoise, p. 356.
5 Killer Whale, p. 355. **6** Common Dolphin, p. 354. **7** Grampus, p. 355.
8 Common Pilot Whale, p. 356.

PLATE 64
TRACKS—WEASELS AND ALLIES

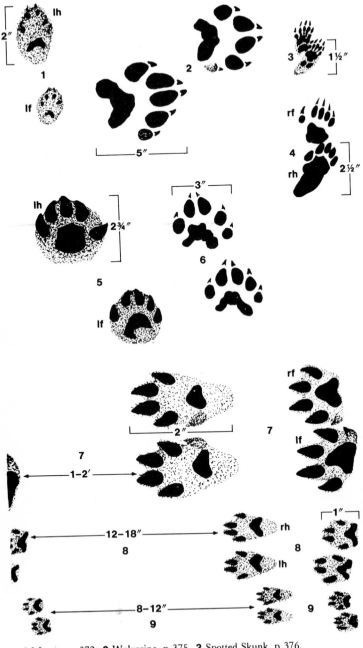

1 Marten, p. 372. 2 Wolverine, p. 375. 3 Spotted Skunk, p. 376.
4 Striped Skunk, p. 376. 5 River Otter, p. 377. 6 Fisher, p. 372.
7 Mink, p. 374. 8 Long-tailed Weasel, p. 374. 9 Least Weasel, p. 373.

**PLATE 65
WEASELS, FOXES, ALLIES**

1 Coyote, p. 360. **2** Red Fox, **2a** "cross" phase, p. 362. **3** Swift Fox, p. 363.
4 Least Weasel, **4a** summer, **4b** winter, p. 373 **5** Gray Fox, p. 363.
6 Ermine, **6a** summer, **6b** winter, p. 373. **7** Marten, p. 372.
8 Black-footed Ferret, p. 374. **9** Fisher, p. 372. **10** Wolverine, p. 375.

PLATE 66
TRACKS—CARNIVORES

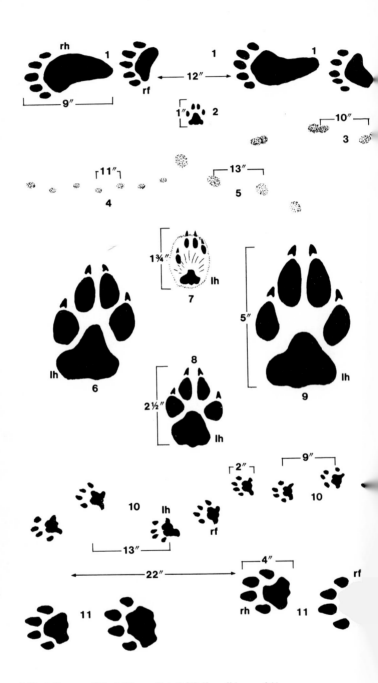

1 Black Bear, p. 364. 2 House Cat. 3 Wolf, walking, p. 361.
4 Fox, trotting p. 362. 5 Coyote, trotting, p. 360. 6 Dog. 7 Red Fox, p. 362.
8 Coyote, p. 360. 9 Gray Wolf, p. 361. 10 Bobcat, p. 380.
11 Mountain Lion, p. 379.

**PLATE 67
CARNIVORES**

1 Striped Skunk, p. 376. **2** Spotted Skunk, p. 376.
3 Black Bear, **3a** black phase, **3b** cinnamon phase, p. 364.
4 Long-tailed Weasel, **4a** winter, **4b** intermediate, **4c** summer, p. 374.
5 Mink, p. 374. **6** River Otter, p. 377. **7** Raccoon, p. 370. **8** Badger, p. 375.

**PLATE 68
SEALS**

1 Northern Fur Seal, p. 366. **2** Walrus, p. 369. **3** California Sea Lion, p. 367.
4 Northern Sea Lion, p. 366. **5** Harbor Seal, p. 368. **6** Bearded Seal, p. 368.
7 Northern Elephant Seal, p. 369.

PLATE 69
DEER AND PRONGHORN

1 Moose, p. 386. **2** Elk, p. 384. **3** Pronghorn, p. 387. **4** Caribou, p. 386.
5 Mule Deer, **5a** Northwest, **5b** Rocky Mountain, p. 384.
6 White-tailed Deer, **6a** summer, **6b** winter, p. 385.

PLATE 70
TURTLES

1 Snapping Turtle, p. 397. **2** Yellow Mud Turtle, p. 398. **3** Desert Tortoise, p. 402. **4** Sonoran Mud Turtle, p. 398. **5** Pacific Ridley, p. 402. **6** Painted Turtle, p. 401. **7** Leatherback, p. 404. **8** Slider, p. 401. **9** Spiny Softshell, p. 403. **10** Western Pond Turtle, p. 399.

PLATE 71
LIZARDS AND SKINKS

1a

2

3

1

4

5

6

7

8

9

10

11

12

13

1 Great Plains Skink, **1a** immature, p. 423. **2** Collared Lizard, p. 411.
3 Mountain Skink, p. 424. **4** Western Skink, p. 425. **5** Many-lined
Skink, p. 424. **6** Gilbert's Skink, p. 425. **7** Four-lined Skink, p. 426.
8 Western Fence Lizard, p. 415. **9** Striped Plateau Lizard, p. 416.
10 Canyon Lizard, p. 416. **11** Eastern Fence Lizard, p. 415.
12 Lesser Earless Lizard, p. 408. **13** Greater Earless Lizard, p. 409.

PLATE 72 HORNED AND FRINGE-TOED LIZARDS, DESERT IGUANA

1 Texas Horned Lizard, p. 418. **2** Desert Horned Lizard, p. 420.
3 Flat-tailed Horned Lizard, p. 419. **4** Round-tailed Horned Lizard, p. 420.
5 Short-horned Lizard, p. 419. **6** Coast Horned Lizard, p. 418.
7 Regal Horned Lizard, p. 420. **8** Mojave Fringe-toed Lizard, p. 411.
9 Colorado Desert Fringe-toed Lizard, p. 410. **10** Desert Iguana, p. 407.

PLATE 73
SPINY LIZARDS AND TREE LIZARDS

1 Yarrow's Spiny Lizard, p. 413. **2** Crevice Spiny Lizard, p. 413.
3 Granite Spiny Lizard, p. 414. **4** Desert Spiny Lizard, p. 414.
5 Clark's Spiny Lizard, p. 414. **6** Bunch Grass Lizard, p. 412.
7 Tree Lizard, p. 417. **8** Sagebrush Lizard, p. 416.
9 Long-tailed Brush Lizard, p. 417. **10** Small-scaled Lizard, p. 417.

PLATE 74
GECKOS, WHIPTAILS, NIGHT LIZARDS

1 Texas Banded Gecko, p. 407. 2 Banded Gecko, p. 407.
3 Leaf-toed Gecko, p. 406. 4 Desert Night Lizard, p. 422.
5 Granite Night Lizard, p. 423. 6 Island Night Lizard, p. 422.
7 New Mexican Whiptail, p. 429. 8 Plateau Whiptail, p. 428.
9 Little Striped Whiptail, p. 426. 10 Texas Spotted Whiptail, p. 427.
11 Chihuaha Whiptail, p. 427. 12 Checkered Whiptail, p. 429.

PLATE 75
WHIPTAILS, ALLIGATOR LIZARDS, GILA MONSTER

1 Giant Spotted Whiptail, p. 429. **2** Western Whiptail, p. 428.
3 Six-lined Racerunner, p. 428. **4** Desert-grassland Whiptail, p. 428.
5 Orange-throated Whiptail, p. 426. **6** Arizona Alligator Lizard, p. 431.
7 Northern Alligator Lizard, p. 430. **8** Southern Alligator Lizard, p. 430.
9 Gila Monster, p. 432.

PLATE 76
SNAKES

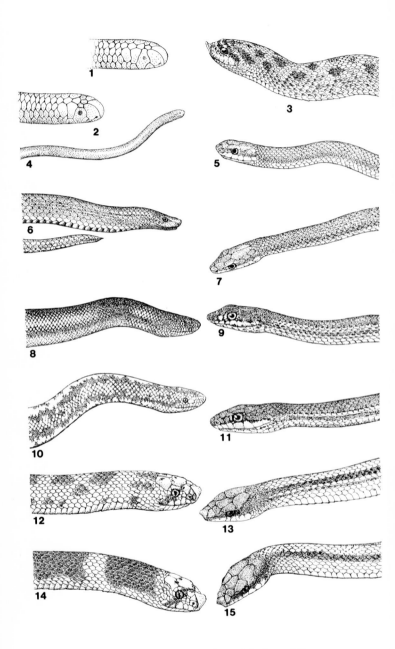

PLATE 77

RINGNECK, RACERS, WHIPSNAKES, KINGSNAKES

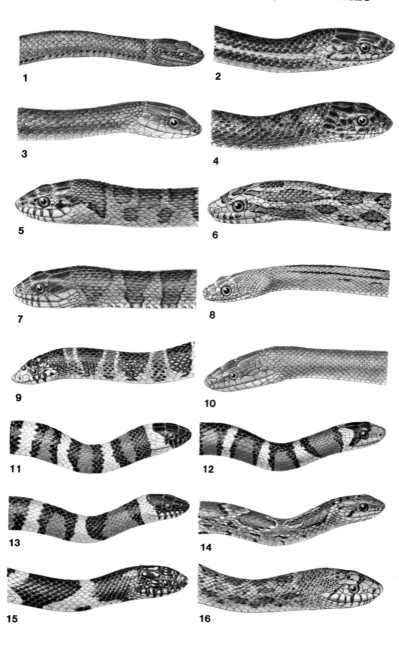

1 Ringneck Snake, p. 436. 2 Striped Racer, p. 439. 3 Racer, p. 438.
4 Coachwhip, p. 439. 5 Plain-bellied Water Snake, p. 446.
6 Corn Snake, p. 441. 7 Northern Water Snake, p. 446. 8 Trans-pecos
Rat Snake, p. 442. 9 Long-nosed Snake, p. 445. 10 Green Rat Snake, p. 442.
11 California Mountain Kingsnake, p. 444. 12 Sonora Mountain
Kingsnake, p. 444. 13 Milk Snake, p. 444. 14 Gray-banded
Kingsnake, p. 445. 15 Common Kingsnake, p. 443. 16 Gopher Snake, p. 443.

PLATE 78
GARTER, BLACK-HEADED, CORAL SNAKES

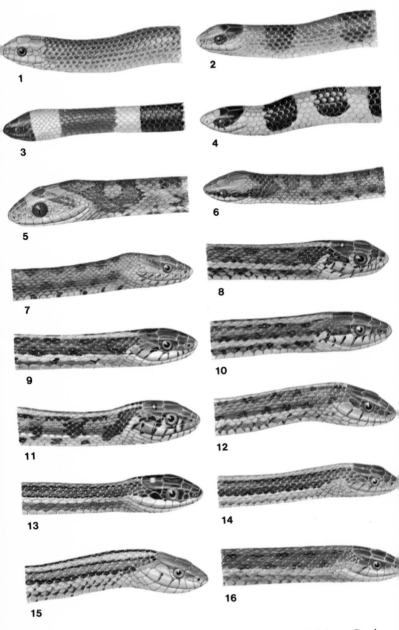

1 Ground Snake, p. 451. **2** Western Ground Snake, p. 451. **3** Arizona Coral Snake, p. 456. **4** Banded Sand Snake, p. 452. **5** Sonora Lyre Snake, p. 455. **6** Night Snake, p. 456. **7** Narrow-headed Garter Snake, p. 447. **8** Mexican Garter Snake, p. 449. **9** Black-necked Garter Snake, p. 448. **10** Plains Garter Snake, p. 450. **11** Checkered Garter Snake, p. 449. **12** Western Terrestrial Garter Snake, p. 447. **13** Western Ribbon Snake, p. 450. **14** Western Aquatic Garter Snake, p. 448. **15** Common Garter Snake, p. 450. **16** Northwestern Garter Snake, p. 448.

**PLATE 79
SNAKES**

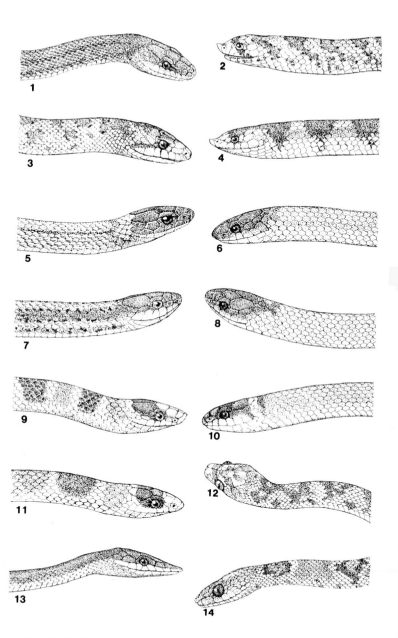

1 Baird's Rat Snake, p. 442. **2** Western Hook-nosed Snake, p. 453.
3 Glossy Snake, p. 442. **4** Desert Hook-nosed Snake, p. 453.
5 Red-bellied Snake, p. 446. **6** Western Black-headed Snake, p. 454.
7 Lined Snake, p. 453. **8** Plains Black-headed Snake, p. 454.
9 Western Shovel-nosed Snake, p. 452. **10** Huachuca Black-headed
Snake, p. 454. **11** Sonora Shovel-nosed Snake, p. 452. **12** California Lyre
Snake, p. 454. **13** Vine Snake, p. 455. **14** Texas Lyre Snake, p. 455.

PLATE 80
POISONOUS SNAKES

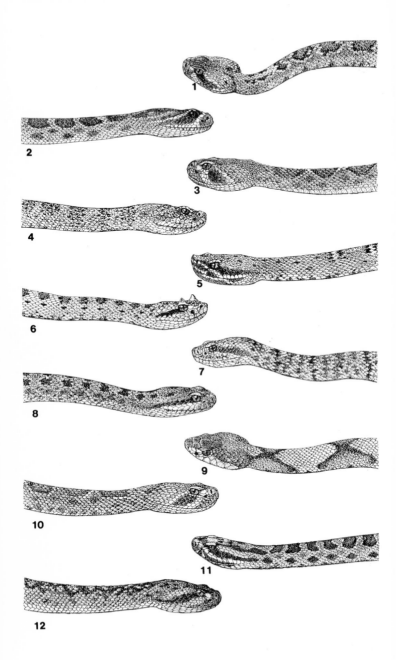

1 Mojave Rattlesnake, p. 461. **2** Western Rattlesnake, p. 461.
3 Red Diamond Rattlesnake, p. 458. **4** Speckled Rattlesnake, p. 459.
5 Ridge-nosed Rattlesnake, p. 462. **6** Sidewinder, p. 458.
7 Rock Rattlesnake, p. 459. **8** Twin-spotted Rattlesnake, p. 460.
9 Copperhead, p. 457. **10** Western Diamondback Rattlesnake, p. 458.
11 Massasauga, p. 457. **12** Black-tailed Rattlesnake, p. 460.

PLATE 81
SALAMANDERS AND NEWTS

1 Olympic Salamander, p. 472. **2** Long-toed Salamander, p. 471.
3 Ensatina, **3a** "Monterey" form, **3b** Ensatina, "Yellow-blotched" form, p. 475.
4 California Newt, terrestrial stage, p. 473. **5** Rough-skinned Newt, p. 472.
6 Red-bellied Newt, terrestrial stage, p. 473. **7a** Tiger Salamander,
"California" form, **7b** Tiger Salamander, "Barred" form, p. 470.
8 Northwestern Salamander, p. 470. **9** Pacific Giant Salamander, p. 471.

PLATE 82
LUNGLESS SALAMANDERS

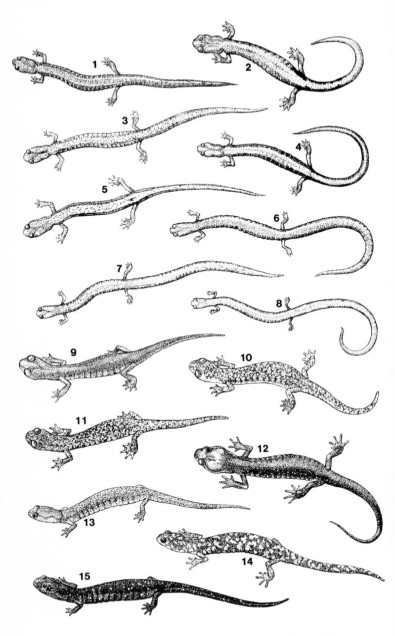

1 Del Norte Salamander, p. 474. **2** Van Dyke's Salamander, p. 475.
3 Jemez Mountains Salamander, p. 474. **4** Western Red-backed
Salamander, p. 475. **5** Dunn's Salamander, p. 474. **6** Pacific Slender
Salamander, p. 476. **7** Oregon Slender Salamander, p. 477.
8 California Slender Salamander, p. 476. **9** Limestone Salamander, p. 479.
10 Shasta Salamander, p. 479. **11** Mount Lyell Salamander, p. 479.
12 Arboreal Salamander, p. 478. **13** Sacramento Mountain Salamander, p. 478.
14 Clouded Salamander, p. 477. **15** Black Salamander, p. 477.

PLATE 83
TRUE TOADS AND SPADEFOOT TOADS

1 Plains Spadefoot, p. 482. **2** Couch's Spadefoot, p. 483.
3 Western Spadefoot, p. 483. **4** Gulf Coast Toad, p. 488. **5** Southwestern
Toad, p. 487. **6** Green Toad, p. 486. **7** Great Plains Toad, p. 485.
8 Red-spotted Toad, p. 487. **9** Western Toad, p. 485. **10** Texas Toad, p. 486.
11 Woodhouse's Toad, p. 488. **12** Colorado River Toad, p. 484.

PLATE 84
FROGS

1 Chorus Frog, p. 489. **2** Northern Cricket Frog, p. 490.
3 Canyon Treefrog, p. 491. **4** Arizona Treefrog, p. 490.
5 Burrowing Treefrog, p. 492. **6** Pacific Treefrog, p. 491.
7 Barking Frog, p. 484. **8** Cliff Frog, p. 484.

PLATE 85
FROGS

1 Tailed Frog, p. 481. **2** Great Plains Narrow-mouthed Toad, p. 496.
3 Foothill Yellow-legged Frog, p. 492. **4** Northern Leopard Frog, p. 494.
5 Mountain Yellow-legged Frog, p. 492. **6** Red-legged Frog, p. 493.
7 Spotted Frog, p. 495. **8** Wood Frog, p. 495. **9** Bullfrog, p. 494.

PLATE 86
SHARKS

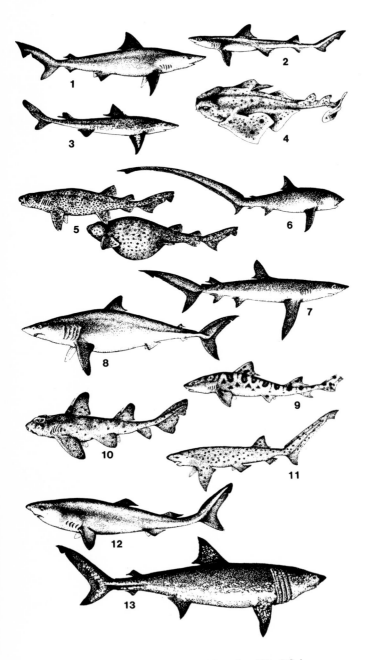

1 Soupfin Shark, p. 513. **2** Gray Smoothhound, p. 512. **3** Spiny Dogfish, p. 514. **4** Pacific Angel Shark, p. 514. **5** Swell Shark, distended with air, p. 511. **6** Common Thresher Shark, p. 510. **7** Blue Shark, p. 513. **8** Mako Shark, p. 510. **9** Leopard Shark, p. 512. **10** Horn Shark, p. 509. **11** Sevengill Shark, p. 509. **12** Pacific Sleeper Shark, p. 514. **13** Basking Shark, p. 511.

PLATE 87
SKATES AND RAYS

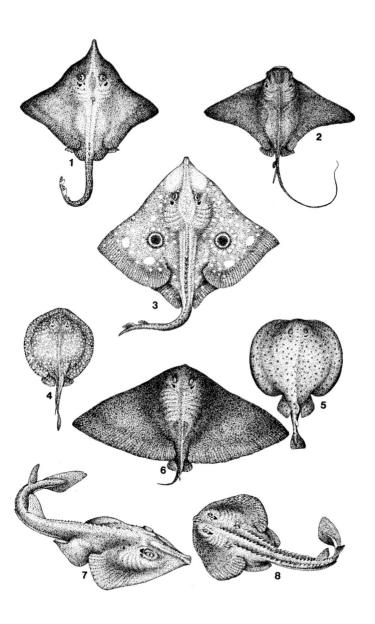

1 Longnose Skate, p. 516. 2 Bat Ray, p. 517. 3 Big Skate, p. 516.
4 Round Stingray, p. 517. 5 Pacific Electric Ray, p. 518.
6 California Butterfly Ray, p. 517. 7 Shovelnose Guitarfish, p. 515.
8 Thornback, p. 515.

PLATE 88
BONY FISHES

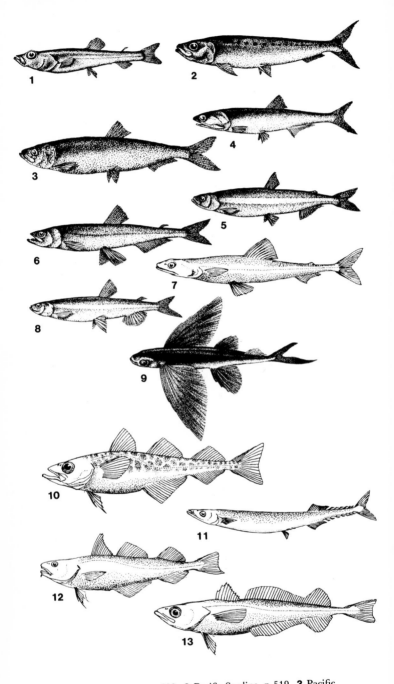

1 Pacific Smoothtongue, p. 522. **2** Pacific Sardine, p. 519. **3** Pacific Herring, p. 519. **4** Northern Anchovy, p. 520. **5** Surf Smelt, p. 521. **6** Eulachon, p. 521. **7** California Lizardfish, p. 522. **8** Night Smelt, p. 521. **9** California Flyingfish, p. 524. **10** Walleye Pollock, p. 526. **11** Pacific Saury, p. 523. **12** Pacific Tomcod, p. 525. **13** Pacific Hake, p. 526.

PLATE 89
SILVERSIDES, SEA BASSES AND OTHERS

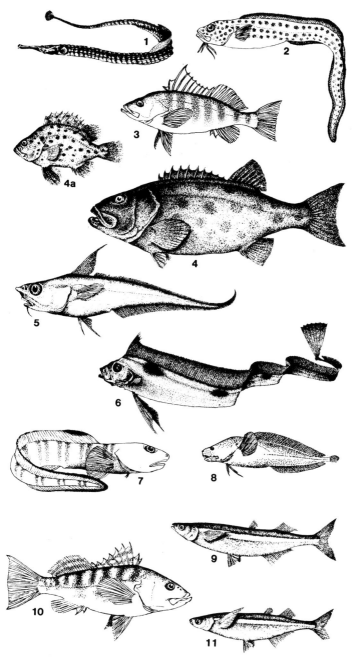

1 Kelp Pipefish, p. 528. **2** Spotted Cusk-eel, p. 526.
3 Barred Sand Bass, p. 531. **4** Giant Sea Bass, **4a** immature, p. 531.
5 Pacific Grenadier, p. 527. **6** King-of-the-Salmon, p. 529.
7 Blackbelly Eelpout, p. 527. **8** Red Brotula, p. 527. **9** Jacksmelt, p. 524.
10 Kelp Bass, p. 530. **11** Topsmelt, p. 524.

**PLATE 90
MARINE FISH**

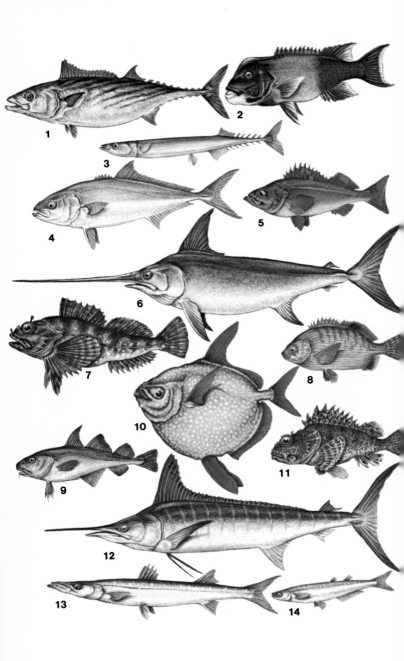

1 Pacific Bonito, p. 533. **2** California Sheephead, p. 542.
3 Pacific Saury, p. 523. **4** California Yellowtail, p. 532.
5 Chilipepper, p. 544. **6** Swordfish, p. 534. **7** Cabezon, p. 550.
8 Rainbow Perch, p. 539. **9** Pacific Tomcod, p. 525. **10** Opah, p. 528.
11 California Scorpionfish, p. 544. **12** Striped Marlin, p. 534.
13 California Barracuda, p. 530. **14** California Grunion, p. 525.

PLATE 91
CROAKERS AND OTHERS

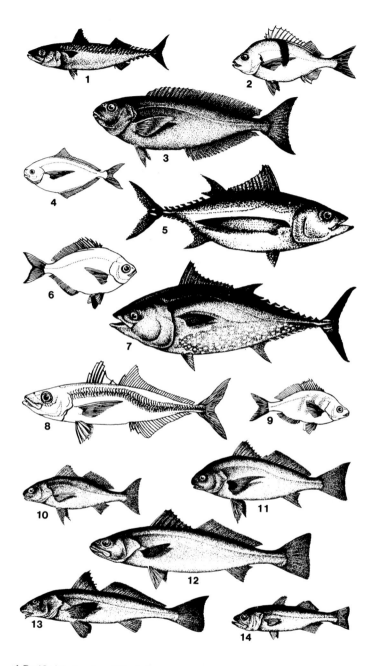

1 Pacific Mackerel, p.533. **2** Sargo, p.535. **3** Ocean Whitefish, p.537.
4 Pacific Butterfish, p.532. **5** Albacore, p.533. **6** Walleye
Surfperch, p.539. **7** Bluefin Tuna, p.534. **8** Jack Mackerel, p.532.
9 Shiner Perch, p.538. **10** White Croaker, p.536.
11 Spotfin Croaker, p.536. **12** White Seabass, p.537.
13 California Corbina, p.535. **14** Queenfish, p.536.

PLATE 92
SCORPIONFISHES AND OTHERS

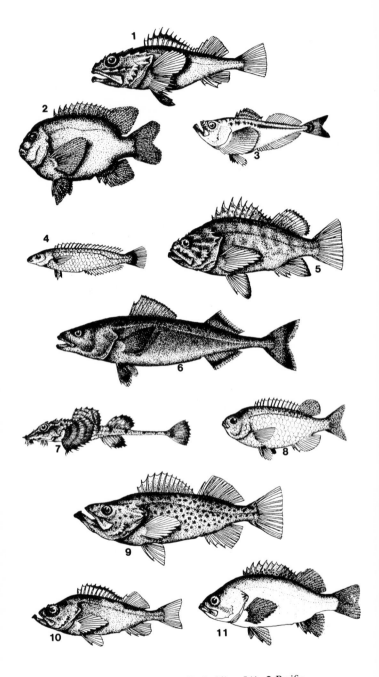

1 Shortspine Thornyhead, p. 548. **2** Garibaldi, p. 541. **3** Pacific Sandfish, p. 541. **4** Señorita, p. 541. **5** Vermilion Rockfish, p. 546. **6** Sablefish, p. 548. **7** Sturgeon Poacher, p. 543. **8** Blacksmith, p. 540. **9** Bocaccio, juvenal, p. 547. **10** Pacific Ocean Perch, p. 544. **11** Black Rockfish, p. 545.

PLATE 93
PERCHES AND ROCKFISH

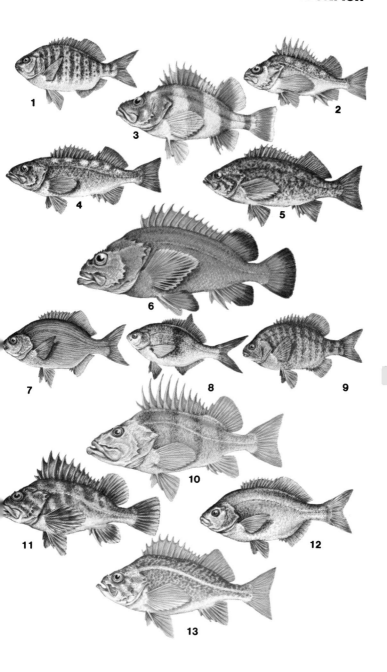

PLATE 94
GREENLINGS, SCULPINS, CLINIDS AND OTHERS

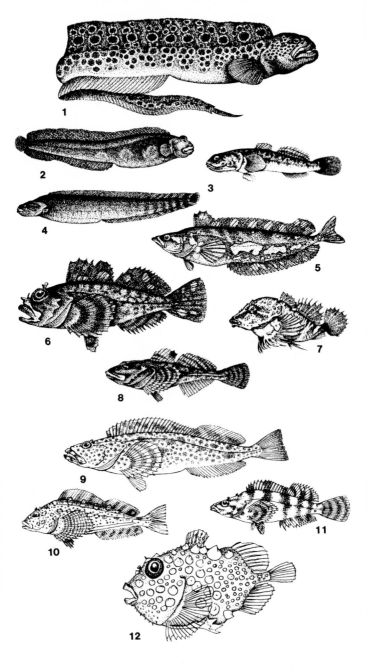

1 Wolf-eel, p. 553. **2** Monkeyface-eel, p. 553. **3** Longjaw Mudsucker, p. 551.
4 Rock Prickleback, p. 554. **5** Giant Kelpfish, p. 552. **6** Cabezon, p. 550.
7 Grunt Sculpin, p. 550. **8** Staghorn Sculpin, p. 550. **9** Lingcod, p. 549.
10 Kelp Greenling, p. 549. **11** Painted Greenling, p. 549.
12 Pacific Spiny Lumpsucker, p. 552.

PLATE 95
FLATFISHES

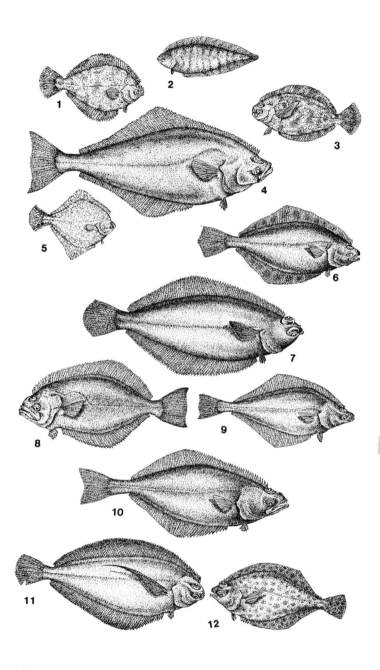

1 Hornyhead Turbot, p. 558. **2** California Tonguefish, p. 558.
3 Fantail Sole, p. 555. **4** Pacific Halibut, p. 556. **5** Diamond
Turbot, p. 557. **6** Petrale Sole, p. 556. **7** Dover Sole, p. 557.
8 California Halibut, p. 555. **9** English Sole, p. 557. **10** Arrowtooth
Flounder, p. 556. **11** Rex Sole, p. 555. **12** Pacific Sanddab, p. 554.

PLATE 96
STURGEON, HERRING, TROUT

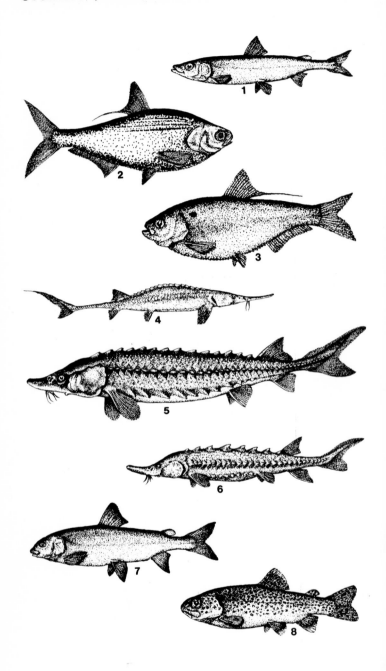

1 Bonneville Cisco, p. 568. **2** Gizzard Shad, p. 564. **3** Threadfin Shad, p. 564. **4** Shovelnose Sturgeon, p. 562. **5** White Sturgeon, p. 562. **6** Green Sturgeon, p. 562. **7** Mountain Whitefish, p. 568. **8** Arizona Trout, p. 569.

**PLATE 97
SALMON, TROUT, CHAR**

1 Rainbow Trout, p. 569. **2** Brook Trout, p. 571. **3** Brown Trout, p. 570.
4 Golden Trout, p. 569. **5** Cutthroat Trout, p. 570. **6** Chinook Salmon, p. 567.
7 Sockeye Salmon, p. 567. **8** Dolly Varden, p. 571. **9** Chum Salmon, p. 566.
10 Pink Salmon, p. 566. **11** Coho Salmon, p. 567. **12** Lake Trout, p. 572.

PLATE 98
MINNOWS

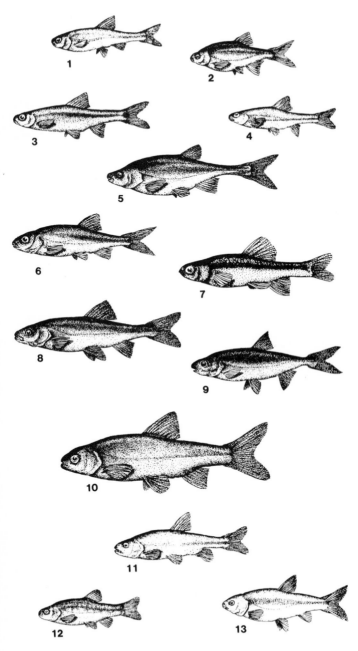

1 Sand Shiner, p. 584. 2 Red Shiner, p. 584. 3 Spottail Shiner, p. 584.
4 Bigmouth Shiner, p. 583. 5 Roundtail Chub, p. 580. 6 California
Roach, p. 578. 7 Longfin Dace, p. 576. 8 Sacramento Blackfish, p. 584.
9 Chiselmouth, p. 575. 10 Utah Chub, p. 579. 11 Leatherside Chub, p. 580.
12 Arroyo Chub, p. 580. 13 Tui Chub, p. 580.

PLATE 99
OTHER MINNOWS

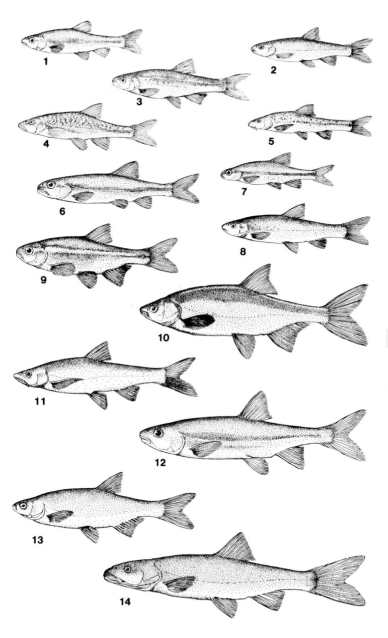

1 Brassy Minnow, p. 579. **2** Western Silvery Minnow, p. 578. **3** Virgin Spinedace, p. 582. **4** Speckled Dace, p. 585. **5** Speckled Chub, p. 580. **6** Lake Chub, p, 576. **7** Roundnose Minnow, p. 577. **8** Plains Minnow, p. 579. **9** Redside Shiner, p. 586. **10** Hitch, p. 581. **11** Flathead Chub, p. 581. **12** Peamouth, p. 582. **13** Golden Shiner, p. 583. **14** Northern Squawfish, p. 585.

PLATE 100
SMELTS, SUCKERS, CATFISH

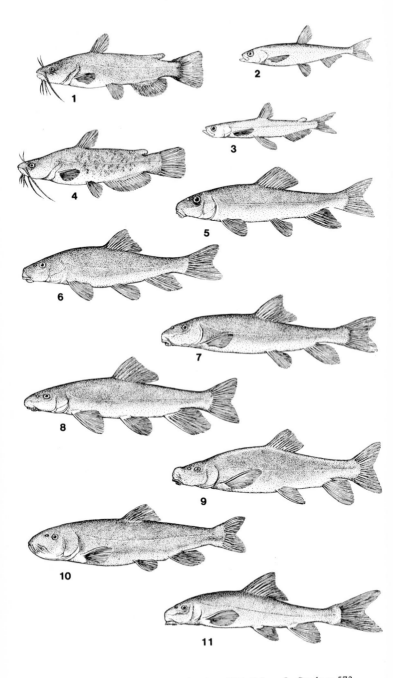

1 Black Bullhead, p. 592. **2** Pond Smelt, p. 573. **3** Longfin Smelt, p. 573.
4 Brown Bullhead, p. 593. **5** White Sucker, p. 588. **6** Sonora Sucker, p. 588.
7 Utah Sucker, p. 587 **8** Tahoe Sucker, p. 589. **9** Lost River Sucker, p. 589.
10 Cui-ui, p. 590. **11** Flannelmouth Sucker, p. 588.

PLATE 101
FRESHWATER FISHES

1 Northern Pike, p. 574. **2** Largemouth Bass, p. 602. **3** Razorback
Sucker, p. 591. **4** Bigmouth Buffalo, p. 591. **5** White Crappie, p. 602.
6 Bluegill, p. 601. **7** Walleye, p. 603. **8** Mexican Tetra, p. 575.
9 Yellow Bullhead, p. 592. **10** Longnose Sucker, p. 587. **11** Sacramento
Perch, p. 600. **12** Desert Pupfish, p. 596. **13** Channel Catfish, p. 593.

PLATE 102
KILLIFISH, SUNFISH, PERCH, SCULPINS

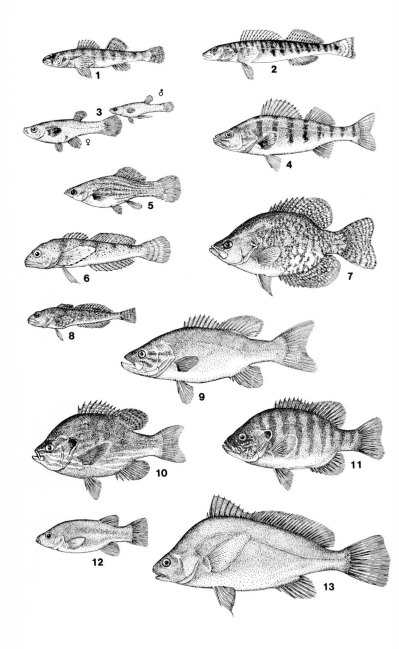

1 Iowa Darter, p. 603. **2** Logperch, p. 604. **3** Mosquitofish, p. 597.
4 Yellow Perch, p. 603. **5** Sailfin Molly, p. 598. **6** Piute Sculpin, p. 608.
7 Black Crappie, p. 602. **8** Riffle Sculpin, p. 606. **9** Smallmouth
Bass, p. 601. **10** Pumpkinseed, p. 601. **11** Green Sunfish, p. 600.
12 White River Killifish, p. 595. **13** Freshwater Drum, p. 604.

**PLATE 103
CHITONS**

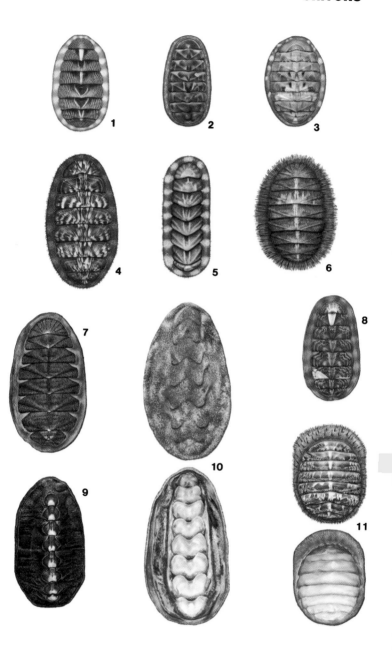

1 Lined Chiton × ½, p. 616. **2** Gould's Baby Chiton × 1½, p. 617.
3 Hartweg's Chiton × 1, p. 617. **4** Hairy Chiton × ½, p. 618.
5 California Nuttall Chiton × 1, p. 617. **6** Mossy Chiton × ½, p. 618.
7 Hinds' Chiton × ½, interior and exterior, p. 618.
8 Woody Chiton × ½, p. 618. **9** Black Katy Chiton × ½, p. 619.
10 Giant Chiton × ⅛, p. 619. **11** Veiled Chiton × ½, p. 619.

PLATE 104 OTHER CHITONS, TUSK SHELLS, SQUID AND OCTOPUS

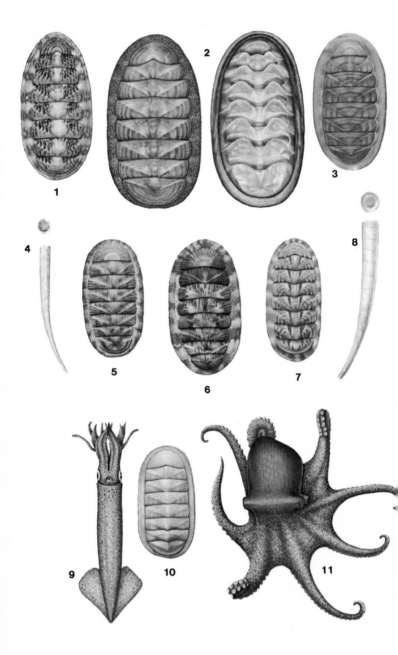

1 Heath's Chiton × ½, p. 620. 2 Conspicuous Chiton × ½, p. 620.
3 Regular Chiton × 1, p. 620. 4 Six-sided Tusk × 1, p. 658.
5 Trellised Chiton × 1, p. 621. 6 Merten's Chiton × 1, p. 621.
7 Red Chiton × 1½, p. 616. 8 Indian Money Tusk × 1, p. 658.
9 Common Pacific Squid × ¼, p. 699. 10 White Chiton × 1½, p. 620.
11 Common Pacific Octopus × ¹⁄₁₀, p. 699.

PLATE 105
ABALONES (ALL × ½)

1 Black Abalone, exterior and interior, p. 622. **2** Flat Abalone, p. 624.
3 Pinto Abalone, p. 624. **4** Green Abalone, p. 623.
5 Threaded Abalone, p. 624. **6** White Abalone, p. 624.
7 Pink Abalone, p. 623. **8** Red Abalone, p. 623.

PLATE 106 KEYHOLE LIMPETS, TURBANS, TOP AND CHINK SHELLS

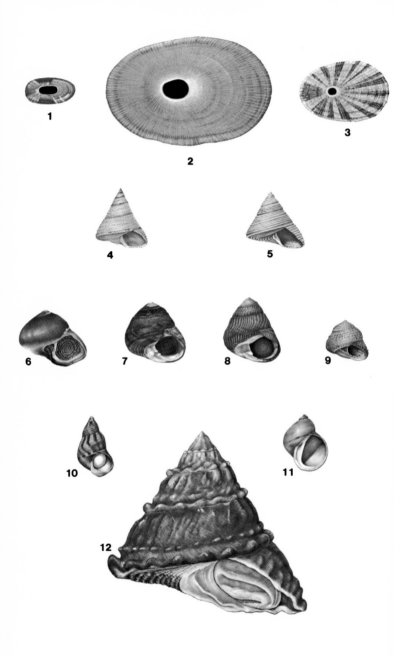

1 Two-spotted Keyhole Limpet × 1, p. 626. **2** Giant Keyhole Limpet × ½, p. 625. **3** Rough Keyhole Limpet × ½, p. 625. **4** Ringed Top Shell × ½, p. 629. **5** Channeled Top Shell × ½, p. 629. **6** Norris Top Shell × ½, p. 630. **7** Black Top Shell × ½, p. 630. **8** Speckled Top Shell × ½, p. 630. **9** Banded Top Shell × ½, p. 631. **10** Banded Pheasant Shell × 1½, p. 631. **11** Carinate Chink Shell × 1½, p. 633. **12** Wavy Turban × ½, p. 631.

PLATE 107
LIMPETS

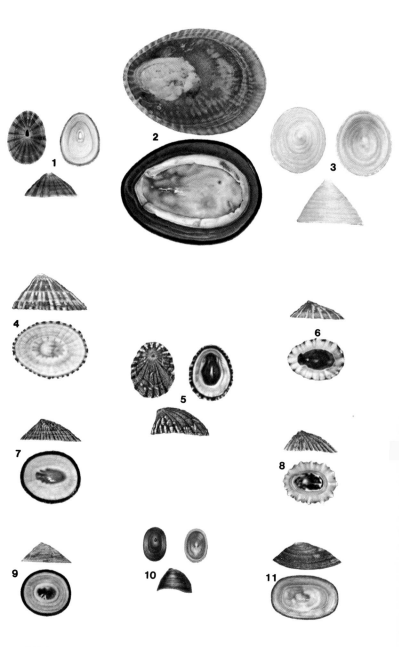

PLATE 108 PERIWINKLES, SCREW SHELLS, AND ALLIED FAMILIES

1 Checkered Periwinkle × 1½, p. 632. **2** Flat Periwinkle × 1, p. 632.
3 Scaled Worm Shell × ½, p. 634. **4** Cooper's Turret × ½, p. 633.
5 Wroblewski's Wentletrap × ½, p. 634. **6** California Horn Shell × ½, p. 634.
7 Striate Cup-and-Saucer Shell × ½, p. 635. **8** Half Slipper Shell × ½, p. 635.
9 Hooked Slipper Shell × ½, p. 636. **10** Onyx Slipper Shell × ½, p. 635.
11 Apple Seed × 6, p. 636. **12** Chestnut Cowrie × ½, p. 637.
13 California Coffee Bean × 1, p. 636. **14** Baby's Ear Shell × ½, p. 637.
15 Recluz's Moon Shell × ½, p. 637.

PLATE 109
FROG AND ROCK SHELLS

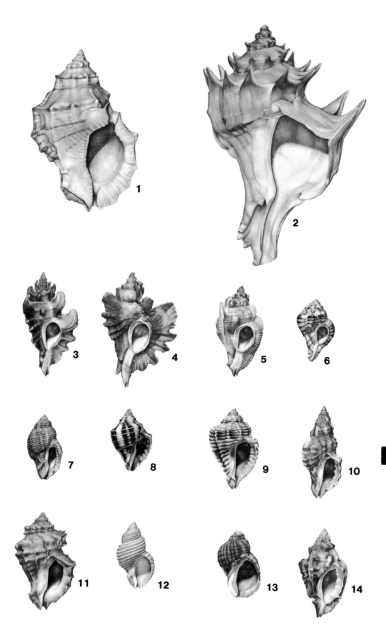

1 California Frog Shell × ½, p. 638. **2** Belcher's Chorus
Shell × ½, p. 638. **3** Three-winged Rock Shell × ½, p. 639.
4 Frill-winged Rock Shell × ½, p. 639. **5** Festive Rock Shell × ½, p. 639.
6 Gem Rock Shell × ½, p. 641. **7** Angular Unicorn × ½, p. 641.
8 Checkered Unicorn × 1, p. 641. **9** Circled Rock Shell × 1, p. 640.
10 Poulson's Rock Shell × ½, p. 640. **11** Frilled Dogwinkle × ½, p. 642.
12 Channeled Dogwinkle × ½, p. 642. **13** Emarginate Dogwinkle × ½, p. 642.
14 Nuttall's Hornmouth × ½, p. 640.

PLATE 110
DOVE SHELLS, WHELKS, DOG WHELKS

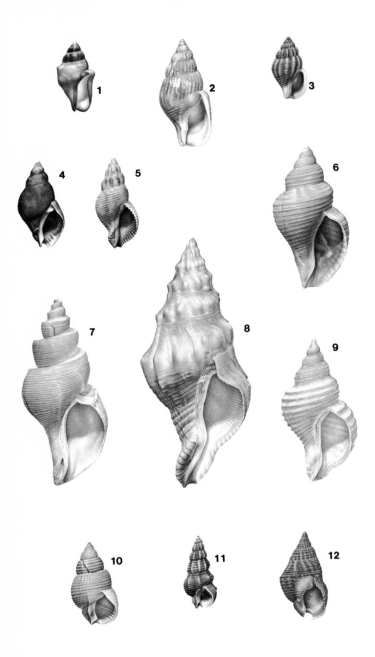

1 Keeled Dove Shell × 2, p. 643. **2** Columbian Amphissa × 1, p. 643.
3 Joseph's Coat Amphissa × 1½, p. 643. **4** Livid Macron × 1, p. 645.
5 Dire Whelk × ½, p. 645. **6** Phoenicean Whelk × ½, p. 644.
7 Tabled Whelk × ½, p. 644. **8** Kellet's Whelk × ½, p. 645.
9 Ridged Whelk × ½, p. 644. **10** Fat Dog Whelk × 1, p. 646.
11 Lean Dog Whelk × 1, p. 646. **12** Channeled Dog Whelk × ½, p. 646.

**PLATE 111 OLIVES
AND ALLIES; BUBBLES, PYRAMS, PTEROPODS**

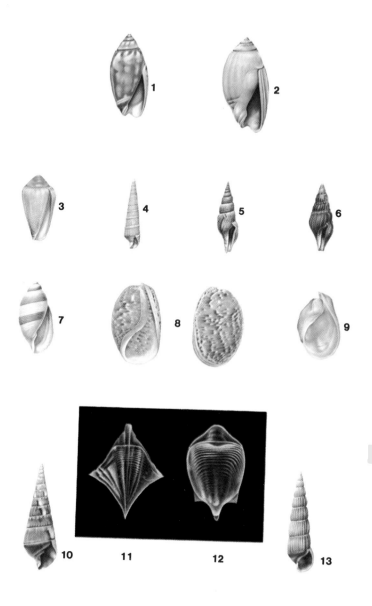

PLATE 112
SEA SLUGS, THE NUDIBRANCHS

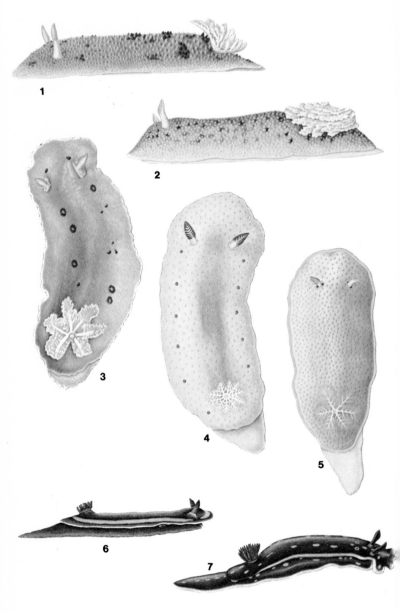

1 Monterey Doris × 1, p. 653. **2** Noble Pacific Doris × ½, p. 654.
3 San Diego Doris × 1, p. 654. **4** Yellow-spotted Doris × 5, p. 654.
5 Yellow-rimmed Doris × 1½, p. 654. **6** Porter's Blue Doris × 2½, p. 655.
7 California Blue Doris × 1, p. 655.

PLATE 113
OTHER SEA SLUGS

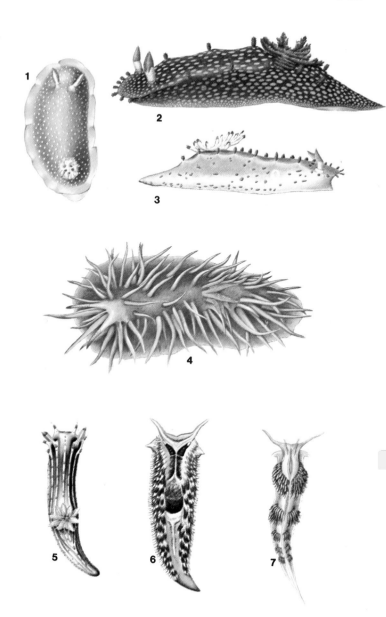

1 Common Yellow Doris × 2, p. 655. **2** Maculated Doris × 1, p. 656.
3 Carpenter's Doris × 1, p. 656. **4** Hopkins' Doris × 2½, p. 657.
5 Orange-spiked Doris × 2, p. 656. **6** Papillose Eolis × ½, p. 657.
7 Long-horned Hermissenda × 1, p. 657.

PLATE 114 NUT SHELLS, ARKS, BITTERSWEET, FILE AND JINGLE SHELLS

1 Taphria Nut × 1½, p. 660. 2 Smooth Nut Shell × 2½, p. 660.
3 Almond Yoldia × 1½, p. 660. 4 Comb Yoldia × 1, p. 661.
5 Cooper's Yoldia × 1, p. 661. 6 Bittersweet × ½, p. 662.
7 Baily's Miniature Ark × 3, p. 661. 8 Many-ribbed Ark × ½, p. 662.
9 Hemphill's File × 1, p. 669. 10 Jingle Shell × ½, p. 669.
11 Pearly Monia × ½, p. 669.

PLATE 115
MUSSELS

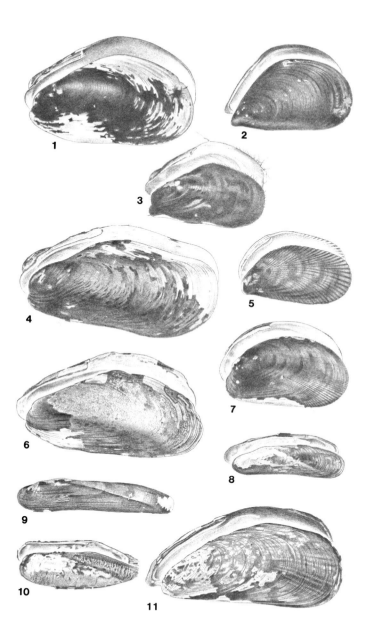

1 Horse Mussel × ½, p. 663. **2** Blue Mussel × ½, p. 665. **3** Carpenter's Horse Mussel × 1, p. 663. **4** Fat Horse Mussel × ½, p. 663. **5** Platform Mussel × 1, p. 664. **6** Straight Horse Mussel × ½, p. 664. **7** Little Black Mussel × ¾, p. 665. **8** California Pea-pod Shell × 1, p. 665. **9** Pea-pod Shell × 1, p. 665. **10** Rock Borer Mussel × 1, p. 666. **11** California Mussel × ½, p. 664.

PLATE 116
SCALLOPS AND OYSTERS

1 Hinds' Scallop × ½, p. 667. **2** Pink Scallop × ½, p. 666. **3** Iceland Scallop × ½, p. 667. **4** Speckled Scallop × ½, p. 667. **5** Kelp-weed Scallop × ½, p. 668. **6** Giant Rock Scallop × ⅛, p. 668. **7** California Oyster × ¾, p. 670. **8** Japanese Oyster × ¾, p. 670.

**PLATE 117 ASTARTES,
CARDITAS, THYASIRAS, LUCINES, JEWEL BOXES**

PLATE 118
COCKLES, QUAHOGS, ROCK DWELLERS

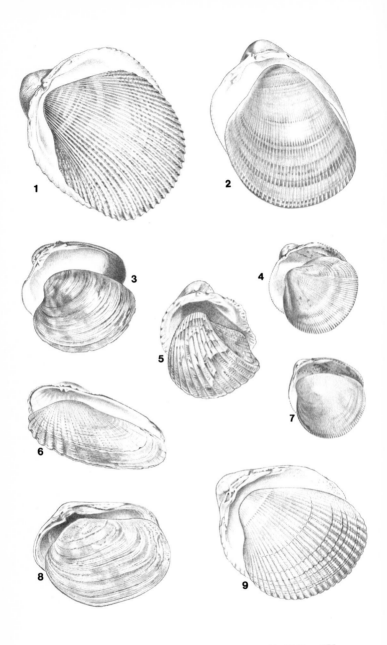

1 Giant Pacific Cockle × ⅓, p. 674. **2** Giant Egg Cockle × ⅓, p. 675.
3 Northern Quahog × ⅓, p. 676. **4** Hundred-lined Cockle × ¼, p. 674.
5 Strawberry Cockle × 1, p. 674. **6** False Angel Wing × ¾, p. 680.
7 Little Egg Cockle × 1, p. 675. **8** Heart Rock Dweller × 1, p. 679.
9 Nuttall's Cockle × ⅓, p. 675.

PLATE 119
VENUS CLAMS

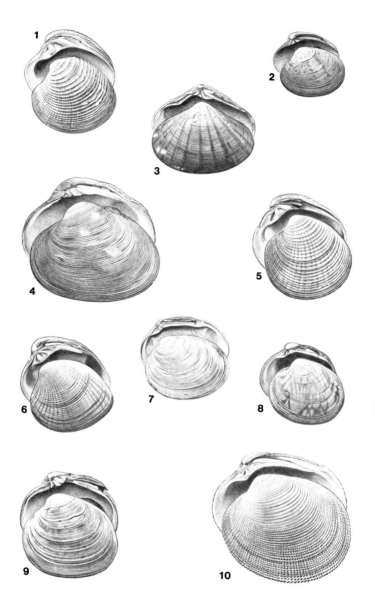

1 California Venus × ½, p. 676. **2** Philippine Littleneck × ½, p. 678.
3 Pismo Clam × ½, p. 678. **4** Washington Clam × ⅜, p. 679. **5** Frilled
California Venus × ½, p. 676. **6** Smooth Pacific Venus × ½, p. 676.
7 Rough-sided Littleneck × ⅜, p. 678. **8** Pacific Littleneck × ½, p. 677.
9 Smooth Washington Clam × ³⁄₁₀, p. 679.
10 Thin-shelled Littleneck × ⅜, p. 677.

PLATE 120
TELLINS AND SEMELES

1 Carpenter's Tellin × 3, p. 682. **2** Modest Tellin × ½, p. 682.
3 Ida's Tellin × 2½, p. 682. **4** Bent-nosed Macoma × ½, p. 683.
5 Indented Macoma × ½, p. 683. **6** White Sand Macoma × ½, p. 683.
7 Grooved Macoma × ½, p. 684. **8** Bark Semele × ½, p. 685.
9 Rock-dwelling Semele × 1, p. 685. **10** California Cumingia × 1, p. 685.

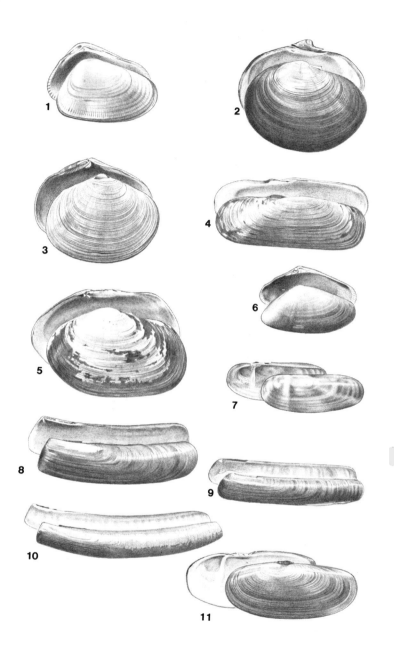

PLATE 121
BEAN CLAMS, GARI SHELLS, RAZOR CLAMS

1 Little Bean Clam × 1½, p. 684. **2** Purple Clam × ½, p. 686.
3 False Donax × 1¼, p. 686. **4** Jackknife Clam × ¾, p. 686.
5 Sunset Shell × ½, p. 687. **6** California Bean Clam × 1, p. 684.
7 Transparent Razor Clam × 1, p. 680. **8** Blunt Razor Clam × ¾, p. 681.
9 Rosy Razor Clam × ¾, p. 681. **10** Myra's Razor Clam × ¾, p. 681.
11 Pacific Razor Clam × ½, p. 680

PLATE 122
SURF, GAPER, SOFT-SHELLED AND BASKET CLAMS

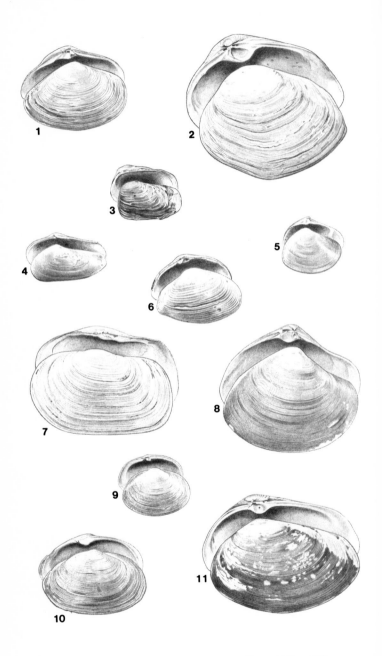

PLATE 123
PIDDOCKS, SHIPWORMS, PAPER SHELLS

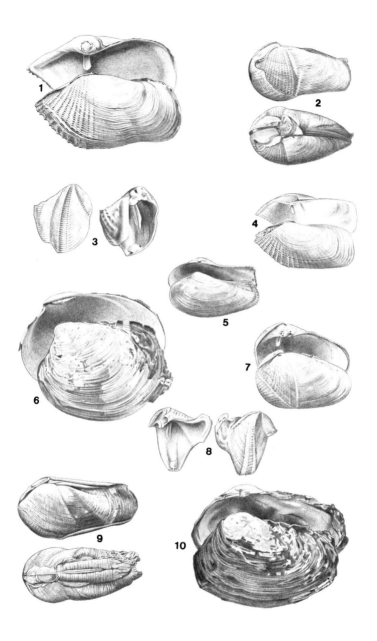

1 Pilsbry's Piddock × ½, p. 693. **2** Common Piddock × ¾, p. 692.
3 Naval Shipworm × 1½, p. 694. **4** Pacific Piddock × ¾, p. 692.
5 California Lyonsia × 1, p. 695. **6** Sea Bottle Shell × 1, p. 695.
7 Oval Piddock × 1, p. 693. **8** Feathery Shipworm × 1½, p. 694.
9 California Piddock × ⅖, p. 693. **10** Rock Entodesma × ½, p. 695.

PLATE 124 PANDORAS, THRACIAS, SPOON AND DIPPER SHELLS

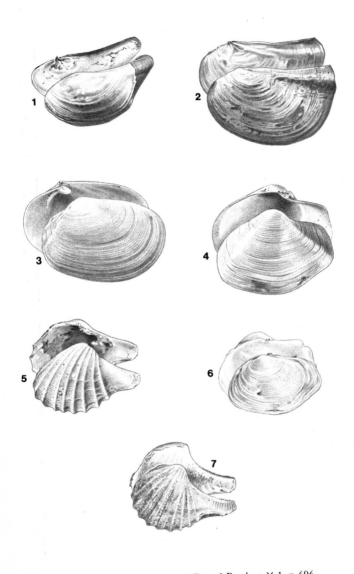

1 Western Pandora × 1¼, p. 696. **2** Dotted Pandora × 1, p. 696.
3 Western Spoon Clam × ¾, p. 697. **4** Pacific Thracia × 1, p. 697.
5 Oldroyd's Dipper × 4, p. 698. **6** Short Thracia × 1, p. 697.
7 California Dipper × 4½, p. 698.

**PLATE 125
SEASHORE LIFE**

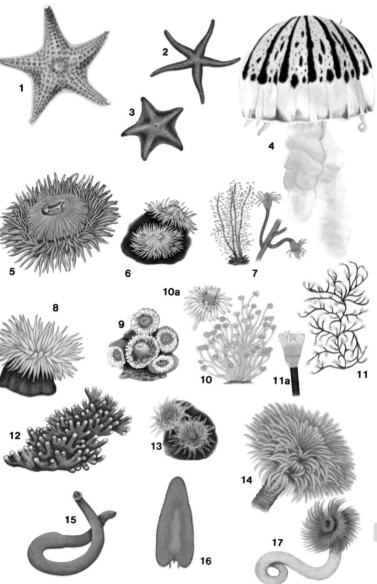

1 *Dermasterias imbricata*, Leather Star, p. 726. **2** *Henricia leviuscula* (Starfish), p. 726. **3** *Patiria miniata*, Bat Star, p. 727. **4** *Pelagia noctiluca*, Purple-striped Jellyfish, p. 714. **5** *Anthopleura xanthogrammica*, Great Green Anemone, p. 716. **6** *Anthopleura elegantissima*, Aggregating Anemone, p. 717. **7** *Garveia annulata* (Hydroid), p. 711. **8** *Tealia crassicornis* (Anemone), p. 716. **9** *Corynactis californica* (Anemone), p. 717. **10** *Tubularia crocea* (Hydroid), **10a** pod detail, p. 710. **11** *Eudendrium californicum* (Hydroid), **11a** pod detail, p. 711. **12** *Stylantheca porphyra*, Hydrocoral, p. 713. **13** *Balanophyllia elegans* (Stony Coral), p. 717. **14** *Eudistylia polymorpha* (Marine Worm), p. 736. **15** *Amphiporus bimaculatus* (Ribbon Worm), p. 720. **16** *Polychoerus carmelensis* (Flatworm), p. 719. **17** *Serpula vermicularis* (Marine Worm), p. 736.

**PLATE 126
SEASHORE LIFE**

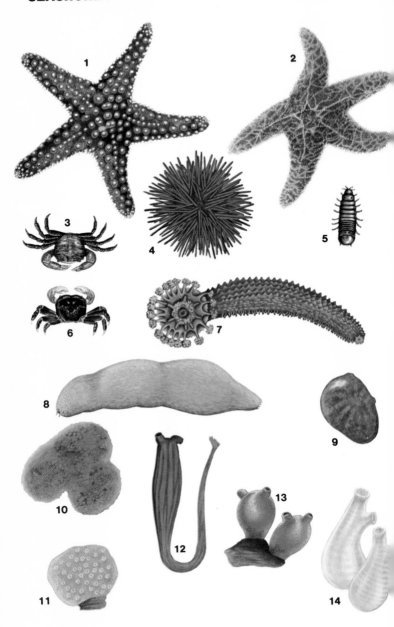

1 *Pisaster giganteus* (Starfish), p. 728. **2** *Pisaster ochraceus,* Purple Starfish, p. 728. **3** *Pachygrapsus crassipes,* Rock Crab, p. 759. **4** *Strongylocentrotus franciscanus,* Giant Red Urchin, p. 731. **5** *Limnoria tripunctata* (Gribble), p. 745. **6** *Hemigrapsus nudus,* Purple Shore Crab, p. 758. **7** *Stichopus californicus,* Sea Slug, p. 734. **8** *Urechis caupo,* Fat Innkeeper, p. 739. **9** *Ascidia ceratodes* (Sea Squirt), p. 760. **10** *Aplidium californicum,* Sea Pork, p. 759. **11** *Polyclinum planum* (Sea Squirt), p. 760. **12** *Styela montereyensis* (Sea Squirt), p. 761. **13** *Cnemidocarpa finmarkiensis* (Sea Squirt), p. 760. **14** *Ciona intestinalis,* Sea Vase, p. 760.

PLATE 127 HYDROIDS, JELLYFISH, SEA ANEMONES, COMB JELLIES

1 *Aurelia aurita*, Moon Jelly, p.714. **2** *Velella velella*,
By-the-wind-sailor, p.713. **3** *Beroe forskali* (Comb Jelly), p.718.
4 *Polyorchis penicillatus* (Hydroid), p.712. **5** *Chrysaora melanaster*
(Jellyfish), p.714. **6** *Syncoryne mirabilis* (Hydroid), p.710.
7 *Pleurobrachia bachei*, Sea Gooseberries, p.718. **8** *Metridium senile*,
Brown Sea Anemone, p.717. **9** *Epiactis prolifera* (Anemone), p.716.
10 *Obelia longissima* (Cup Hydroid), p.711. **11** *Abietinaria*
(Hydroid), p.712. **12** *Haliclystus auricula* (Jellyfish), p.714.
13 *Hydractinia*   (Hydroid), p.711.

PLATE 128
SEA WORMS

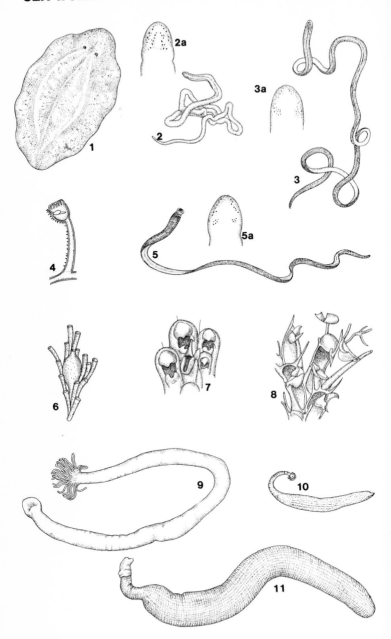

1 *Kaburakia excelsa* (Polyclad Worm), p.719. **2** *Amphiporus imparispinosus* (Ribbon Worm), **2a** head detail, p.720. **3** *Emplectonema gracile* (Ribbon Worm), **3a** head detail, p.720. **4** *Pedicellina cernua* (Entoprocta), p.721. **5** *Paranemertes peregrina* (Ribbon Worm), **5a** head detail, p.720. **6** *Crisia occidentalis* (Entoprocta), p.722. **7** *Phidolopora pacifica*, Lace Coral, p.723. **8** *Bugula californica* (Moss Animal), p.722. **9** *Phoronopsis viridis* (Tube-dwelling Worm), p.724. **10** *Phascolosoma agassizii* (Sipunculid Worm), p.739. **11** *Sipunculus nudus* (Sipunculid Worm), p.739.

PLATE 129
ANNELIDS

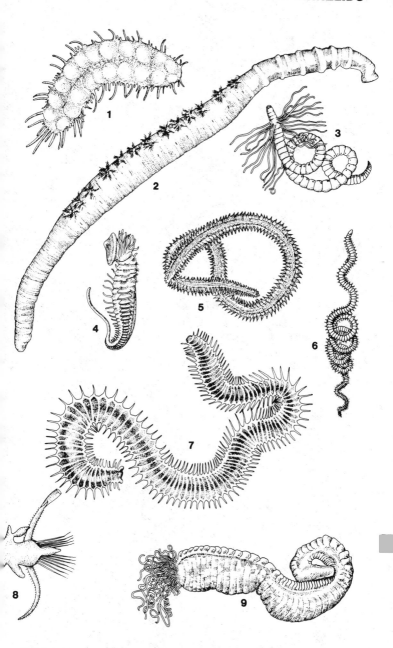

1 *Halosydna brevisetosa* (Scale Worm), p. 736. **2** *Abarenicola pacifica,*
Lugworm, p. 735. **3** *Cirratulus cirratus* (Marine Worm), p. 737.
4 *Phragmatopoma californica* (Marine Worm), p. 738. **5** *Nephtys californiensis*
(Marine Worm), p. 737. **6** *Arabella iricolor* (Marine Worm), p. 737.
7 *Nereis vexillosa,* Clam Worm, p. 735. **8** *Hemipodus borealis*
(Marine Worm), p. 736. **9** *Thelepus crispus* (Marine Worm), p. 738.

GREEN TURTLE
Chelonia mydas

Description
Length, 30–48 in. (76.2–121.9 cm), and weight to 650 lb. (294.8 kg). Only sea turtle on West Coast with 4, rather than 5, costal shields on each side of upper shell. Carapace low relief, slightly convex dorsally, somewhat heart-shaped, smooth, unkeeled; plates do not overlap. Head has only 1 pair of prefontals, single mandibular scale. Carapace brown to greenish-olive, mottled or blotched with dark brown; head plates brownish to olive, edged with yellow. Limbs paddlelike, each foot has 1 or 2 claws. Male longer, carapace narrower; tail greatly elongate, prehensile, nail-tipped; front flipper claw enlarged, curved. Juveniles have slight overlap of carapace scuts; paddles relatively larger than in adults; whitish to yellowish edges on shell and flippers.
Similarities
Loggerhead Turtle has 2 pairs of prefontals, 5 costal shields.
Habitat
Shoal waters with submarine vegetation; migrates over long distances in open ocean.
Habits
Sometimes basks on surface; swimming speed to 1.4 mph. (2.3 km/hr.); occasionally comes ashore on remote rocks and beaches.
Range
Extr. s. Calif. coast (occasional).

SOFTSHELL TURTLES
Family Trionychidae

Members of this peculiar family have a low and leathery carapace instead of a higher, horny shell, and the plastron is covered with smooth skin. The nostrils are at the end of a long tubelike extension of the snout, and the feet are three-clawed, webbed. The males are smaller than the females. They occur in North America, Asia, and Africa.

SPINY SOFTSHELL
Trionyx spiniferus   **70:9**

Description
Length, 3½–18 in. (8.9–45.7 cm). Carapace low, oval to circular in outline, often wider behind than in front, tubercles on anterior border (reduced in southwest of range), covered with leathery skin, often soft, flexible, without sutures. Skin of head, limbs, tail relatively smooth; ridge projects from nasal septum. Feet broad, flattened, toes fully webbed, nails well developed. Above greenish-olive, olive-brown, brown, to grayish; carapace spotted with blackish, obscured with age; shell margined with pale yellow or whitish, fading with increase in size; head and limbs variously spotted with blackish; eyestripe whitish to pale yellow, outlined with black; underparts unmarked whitish to yellowish. Male, tail stouter, longer; tubercles on carapace more abundant and prominent.
Habitat
Large, slow rivers; small streams, ponds.
Habits
Highly aquatic, good swimmer; can submerge for long periods, extends neck to surface periodically to breathe; ambushes prey in bottom mud or sand; primarily carnivorous.

Remarks
Hisses violently when annoyed, may strike and bite.

Range
Gila R. and Colo. R. drainages of extreme se. Calif.; s. tip Nev.;
nw. Ariz., Gila R. across s. and far se.; N. Mex., far sw. and s.,
se., e.; e. Mont., (except Canadian border and far ne.), far sw. N.
Dak., ne. Wyo., S. Dak., e. Colo., Nebr., Kans., Okla., Tex., and
across cen. U.S. In the West it occurs primarily in 3 major
drainages: upper tributaries of Mississippi-Missouri rivers, the Rio
Grande, and the Colorado R.

Note: The **SMOOTH SOFTSHELL,** *Trionyx muticus* (*Fig. 58*), has a
more pointed snout, and the nostrils can be seen from below. The
marks on the dorsal surface of its limbs are less contrasting. It
occurs from Alabama west through northern Texas to the Conchos
River, above the Conchos Dam in northeastern New Mexico.

Fig. 58

Smooth Softshell

LEATHER-BACKED TURTLES
Family Dermochelyidae

There is only one species in this family, a giant marine form that
is widespread in the oceans, ranging considerable distances from
shore.

LEATHERBACK
Dermochelys coriacea   **70:7**

Description
Length, 48–72 in. (121.9–182.9 cm), and weight, to 1600 lb.
(725.6 kg). Largest living turtle and the only sea turtle with
leathery shell. Scaleless, with 7 nodular longitudinal ridges on
carapace, 5 ridges on plastron. Shell consists of polygonal bones
imbedded in skin, and neither ribs nor vertebrae are connected to
shell. Carapace dark brown, slaty, or black; uniform or blotched
with whitish or pale yellow; shell ridges and margins of flippers
light. Limbs are perfect flippers and lack claws; forelimbs long,
flattened. Male shell tapers posteriorly, carapace depressed in
profile (female carapace flat or convex).

Habitat
Pelagic.

Habits
Strong, fights with both jaws and flippers; seldom observed.

Voice
Utters loud sounds when agitated.

Range
Widely distributed in tropical and subtropical seas, ranges along
West Coast as far n. as B.C. and Vancouver Is. (Nootka Sound).

Lizards and Snakes
Order Squamata

Although lizards and snakes differ quite markedly from one another, they are grouped into a single order containing the suborders Sauria for the lizards, Serpentes for snakes, and Amphisbaenia (amphisbaenians). Common characteristics of snakes include scales, a pair of eversible copulatory organs, and a jaw suspended from the skull by a movable element. Although most lizards have limbs and most snakes do not, each group includes some that reverse this characteristic to some degree; some lizards have only vestigial limbs or lack them entirely, and some snakes show vestigial limbs. Almost every snake characteristic can be matched in some species of lizard, such as lack of limbs and eyelids, and reduced ear structure (but lizards do not have expandable jaws). Snakes probably evolved out of a group of antecedent lizards sometime in the Cretaceous period.

Lizards
Suborder Sauria (Lacertilia)

There are about 3000 lizard species in the world, representing twenty families; of these, sixty-five species in eight families occur in North America west of the 100th meridian.

Features of Lizards
A lizard typically is a scaly reptile with ears, eyelids, a nonexpandable jaw, and four limbs, each with five clawed toes. The single legless species in the West has movable eyelids. Lizards also have several rows of scales on the underside. The tails are often fragile and readily break off when a lizard is caught; a new one will soon be regenerated. The discarded tail may wriggle for several minutes, serving to distract an enemy while its owner escapes. Many lizards have excellent eyesight and a good sense of smell; their teeth are generally all alike, often used more to grab and hold their prey than for chewing. Most lizards are terrestrial and diurnal, feeding largely on insects, spiders, worms, and other lizards.

Fig. 59

Parts of a Typical Lizard

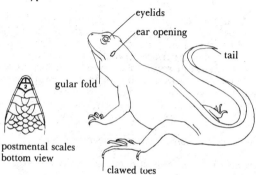

Habitat
Lizards are often quite conspicuous, especially such forms as the collared, horned, and fence lizards. Look for lizards basking or

405

running along stone walls and fences, on stumps and logs, among brush piles and dead leaves, around old piles of sawdust, in deserted buildings, clearings, in canyons, on limestone hills and rocky outcrops, and on flat, sandy or rocky soil with sparse vegetation. Some occur on tree trunks, others in moist situations near water. Many lizards can be approached to within a few feet, and from there binoculars can bring them into excellent scale-counting view.

Habits
Most lizards hibernate in the colder months, emerging in the spring. Lizards can stand a good deal of heat and prefer sunny days, particularly mornings. Cool or cloudy days keep them under cover, and many retreat to their hiding places in late afternoon even while the sun is still up. Nevertheless, it is a popular misconception that lizards thrive on higher temperatures than do other animals. A basking lizard soon changes its position in order to control its temperature, and some species achieve temperature control to some degree by changing color, assuming a light phase when warm and a darker phase when chilly. Experiments show that the normal activity range of many lizards is at about the same temperature as the blood warmth of man.

Measurements
Length given in the descriptions that follow are for the *snout-vent* length.

GECKOS
Family Gekkonidae

Members of this family have soft skins, large eyes, usually with a vertically elliptical pupil and often without movable lids, a broad fleshy tongue, teeth attached to the sides of the jaws, and toes broadened at the tips. Most geckos are nocturnal, although some are active by day. Many are good climbers and, by means of their adhesive toepads and claws, can run along on the underside of objects as well as on vertical surfaces. About seventy-five genera are recognized worldwide, most in the subtropical to tropical latitudes.

LEAF-TOED GECKO
Phyllodactylus xanti   **74:3**

Description
Length, 2–2½ in. (5.1–6.4 cm). Eyes large, pupil vertical, no movable lids; toe tip has 2 large plates separated by claw. No other western gecko has this characteristic. Body and tail flattened (tail especially at base); dorsal surfaces granular, interspersed with scattered keeled scales in irregular rows; ventral surfaces covered with larger flat scales. Above pale brown, gray, or nearly cream, marked irregularly with bars, crossbands, or spots of dark brown; below whitish to yellowish, often variously brown-flecked. Male has slightly broader tail base.

Habitat
On boulders in desert canyons, usually near springs.

Habits
Hides in crevices by day, feeds at night; tail easily discarded.

Range
Imperial Valley of s. Calif., s. of Palm Springs to tip of Baja Calif.; also offshore islands.

BANDED GECKO
Coleonyx variegatus   74:2

Description
Length, 2½–3 in. (6.4–7.6 cm). Distinctive, one of two geckos with
both vertical pupil and movable eyelids. Skin soft, delicate; all
scales granular, tubercles on back; slender toes; preanal pores,
usually 5 or more, form uninterrupted series. Body and tail
rounded, tail constricted at base. Above flesh, cream, or yellowish
with variously broken brown crossbands on body and tail; below
whitish, mostly unmarked. Male has cloacal bones underlying
protuberances on each side of tail base.

Similarities
Texas Banded Gecko usually has 4 or less preanal pores, separated
by unpitted scales medially.

Habitat
Under boards, fallen road signs, yucca stems, dried animal
carcasses, stones, rock flakes; often on highway at night.

Habits
Nocturnal; may run with tail curled over back or to one side; when
molested, may stand stiff-legged, head and tail elevated, back
swayed, and emit squeaking sounds.

Range
S. Calif., s. Nev., sw. Ariz. (except far se. corner), extr. sw. corner
of Utah, very extr. sw. N.Mex.

TEXAS BANDED GECKO
Coleonyx brevis   74:1

Description
Length, 1¾–2¼ in. (4.4–5.7 cm). Very similar to Banded Gecko,
but has a total of 3–6 (usually 4 or less) preanal pores, interrupted
medially by 1 or more unpitted scales. Dark bars on body of adult
are wider than interspaces, often replaced by spots.

Similarities
Banded Gecko has 4–12 preanal pores (usually 5 or more)
uninterrupted medially.

Habitat
Rocky outcrops, boulders, canyons.

Range
N. Mex. (an inverted V range from n.-cen. to s.-se.), e. to s. Tex.

IGUANAS
Family Iguanidae

This is the dominant family of New World lizards. Its members
exhibit great variety in form, habits, and habitat, but all have the
teeth lying in grooves on the inner surface of the jaws. They also
have a characteristic habit of bobbing up and down.

DESERT IGUANA
Dipsosaurus dorsalis   72:10

Description
Length, 4–5½ in. (10.2–14.0 cm). Body and tail rounded, tail long,
almost twice body length, with keeled scales; head relatively small,
short, rounded, rather high; dorsal surfaces weakly keeled, midline
of back from head to well out on tail supports row of slightly
enlarged, keeled scales; sides bear small, granular scales; ventral
scales smooth, overlapping. Above grayish-brown variously barred

laterally with brown to rusty (centrally slate) and blotched and spotted with light gray or white. Brown markings may unite to form lateral longitudinal lines; tail spots may form encircling bands of brown on whitish. Below, whitish, with grayish streaks on throat. Male head broader, more angular; male usually with large postanal scales.

Habitat
Primarily desert with creosote bushes; frequents low, sandy plains, retreats to rodent burrows; occasionally found in rocky areas and up to 3200-ft. (972.8-m) elevations.

Habits
Primarily diurnal, a late riser; timid, alert; when running, may raise forelegs and use only hindlegs; chiefly herbivorous.

Range
Se. Calif., far s. Nev., sw. Ariz., very extr. sw. corner of Utah.

CHUCKWALLA
Sauromalus obesus   *Fig. 60*

Description
Length, 5½–8 in. (14.0–20.3 cm). Robust, differing from all other western forms in lacking rostral scale. Body somewhat flattened, with loose folds of skin on neck and sides, except when inflated; tail thick at base, tapering to blunt tip, rather flattish, not easily lost; scales on back and sides small, rounded; ventral scales similar. Color varies with age, sex, locality. Male generally with light gray or red body; black head, neck, shoulders, chest, limbs, uniform or speckled with gray or whitish; tail grayish or straw. Female grayish or brownish-gray; head and limbs dark; variously retained juvenile crossband pattern. Male head broader, neck folds more conspicuous. Juvenile distinctly crossbanded, particularly conspicuous on tail with black bands on olive-gray or yellow.

Habitat
Closely restricted to rocky hillsides in desert areas.

Habits
When molested, can inflate body to wedge itself in a crevice; seeks shelter in crevices among massive rocks.

Range
Se. Calif., s. tip Nev., extreme sw. Utah, w. Ariz. including cen. bulge into Gila Co. and Monument Valley into Utah.

Fig. 60

Chuckwalla

LESSER EARLESS LIZARD
Holbrookia maculata   **71:12**

Description
Length, 2–2½ in. (5.1–6.4 cm). Body short, plump, somewhat flattened; tail about body length; scales tiny; fleshy fold on chest (gular fold); no ear opening. Above brownish, gray, to white, typically with brown to sooty spots prominently arranged in 4 longitudinal rows; spots edged with blackish posteriorly and margins, in turn, edged with whitish; sides commonly show

yellowish; below white to cream, may show yellow or pink
suffusion; throat shows orange or yellow spot; pair (sometimes 3)
black bars low on each side between limbs; underside of tail
unmarked or with small dark bars. Male has enlarged postanal
scales; tail longer, basically broader; femoral pores larger; bars on
lower sides more distinct.

Similarities
Side-blotched Lizard has ear opening present, dark spot in armpit,
no overlapping upper labials; Zebra-tailed Lizard has limbs, tail
longer, ear opening present externally; Greater Earless Lizard has
black bars on underside of tail.

Habitat
Sandy or gravelly flats and dry rocky areas with sparse vegetation.

Habits
Active, particularly on hot days; makes short dashes; inquisitive,
not wary, usually found in pairs.

Range
Extr. s.-sw. S. Dak., extr. se. Wyo., w. Nebr., Colo., e. and far sw.
corner; w. Kans.; far se. Utah, e. Ariz., with extension along
Grand Canyon; N. Mex. (except extr. n.-cen.), w. Okla., w. Tex.

GREATER EARLESS LIZARD
Holbrookia texana
71:13

Description
Length, 2–3 in. (5.1–7.6 cm). Body flat; tail flat, with black
crossbars on underside; blue patch with 2 black bars on each side
of belly. Above slate-gray with numerous small flecks. Below
white. Long legs, no ear opening, gular fold present, diagonal
troughs between upper labials. Male tail base broader, tail longer,
femoral pores larger, large postanal scales.

Similarities
Zebra-tailed Lizard has ear opening; Lesser Earless Lizard has no
black bars beneath tail.

Habitat
Among hillside rocks, along stream courses, in desert or semiarid
areas.

Habits
When disturbed, seeks shelter under rocks; active by day, curls and
wags tail.

Remarks
Can be caught by tossing a hat for lizard to crawl under.

Range
Se. corner Ariz. and nw. extension into cen. Yavapai Co.; far s. N.
Mex. with e.-cen. arm extending into Torrance Co.; w. Tex.
(except panhandle).

ZEBRA-TAILED LIZARD
Callisaurus draconoides
Fig. 61

Description
Length, 2½–3½ in. (6.4–8.9 cm). External ear opening present;
body and tail flattened; limbs and tail longer than in Greater
Earless Lizard. Above pale gray to brownish-gray dotted with
whitish to yellowish; 2 longitudinal rows of dark spots on back,
dark bands on tail; sides often lemon-yellow; below white with
dusky throat patch, center pink or orange. Gular fold(s) present;
dorsal scales granular, unkeeled; ventral scales larger, flat, smooth;
lacks projecting border scales. Male has distinct ventrolateral blue
or blue-green patch marked with wedge-shaped black bars (vague

or absent in female); broader tail base, less definite dark dorsal spotting, darker throat, large postanal scales.

Similarities
Earless Lizards lack external ear openings.

Habitat
Sandy and gravelly areas in deserts; favors sand floors of washes, margins of dunes.

Habits
One of the fastest reptiles, clocked to 18 mph (29.0 km), may run to 50 yards (45.7 m), veering from side to side without using forelimbs and with tail curled forward well off ground; wags tail nervously when alerted.

Range
Sw. Nev. (except w.-cen. boundary area); se. Calif., extr. sw. corner of Utah, w. and s. Ariz.

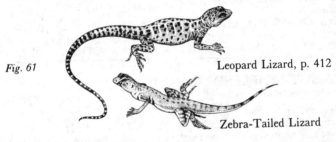

Fig. 61   Leopard Lizard, p. 412

Zebra-Tailed Lizard

COACHELLA VALLEY FRINGE-TOED LIZARD
Uma inornata

Description
Length, 2¾–4½ in. (7.0–11.4 cm). Very flat; toes fringed with prominent pointed scales; countersunk lower jaw. Skin velvety, dorsal scalation granular, internasal scales usually in 3 rows. Above whitish with many closely set black marks tending to outline circular areas, may show black centers; below white, underside of tail with black bars, dark streaks on throat not forming crescents; black spot on side of belly absent or reduced to small dot(s). Male tail base broader, large postanal scales.

Similarities
Mojave Fringe-toed Lizard has black crescents on throat; Colorado Desert Fringe-toed Lizard has large black spot on each side of belly.

Habitat
Restricted to fine, loose, wind-blown sand of dunes, flats, edges of washes in low desert.

Habits
When frightened, burrows rapidly into sand, or hides in rodent burrow or beneath a bush.

Range
Coachella Valley to San Gorgonio Pass, cen. Riverside Co., Calif.

COLORADO DESERT FRINGE-TOED LIZARD
Uma notata   **72:9**

Description
Length, 2¾–4½ in. (7.0–11.4 cm). Ground color commonly pale yellow to cream, orange on lower sides, conspicuous ventrolateral black mark on each side between limbs; streaks on throat not

forming crescents; 25–40 fringe scales on external side of 4th toe and along sole to base of 5th toe. Body and tail flattened; scalation smooth, granular, dorsal texture velvety; ventral scales smooth, overlapping; head oval from above, wedge-shaped in profile; eyelids with projecting, often pointed, scales.

Similarities

Mojave Fringe-toed Lizard has well-defined black gular crescents; Coachella Valley Fringe-toed Lizard lacks conspicuous spots on side of belly.

Habitat, Habits

See Coachella Valley Fringe-toed Lizard.

Range

Far se. Calif., Colorado Desert; extr. sw. Ariz.

MOJAVE FRINGE-TOED LIZARD

Uma scoparia

72:8

Description

Length, 2¾–4½ in. (7.0–11.4 cm). Color characterized by eyelike dorsal marks scattered instead of tending toward rows; well-defined black gular crescents; conspicuous black spot on sides of belly.

Similarities

Coachella Valley and Colorado Desert Fringe-toed Lizards have streaks rather than crescents on throat.

Habitat

Wind-blown sand.

Habits

Surprisingly fast; clocked to 23 mph (37.0 km/hr.).

Range

Mojave Desert area of s. Calif.; one small locality in Ariz.

COLLARED LIZARD

Crotaphytus collaris

71:2

Description

Length, 3–4½ in. (7.6–11.4 cm). Two black collars, thick body, thin neck, and large head. Tail long, commonly twice snout-vent length, often laterally somewhat flattened. Males often green or yellow with yellowish or red dots, throat patch bluish (orange east of Rockies); usually 2 transverse black collars broken in front and separated by light area in shoulder region; 4–5 dull yellow to rusty transverse stripes may show on dorsal body surface; blue spots on back, dark bands on tail and on back, dark bands on tail and on big, froglike hindlegs. Females often gray, with red spots when breeding. Both sexes lighter in breeding season. Young duller, crossbands more conspicuous. Male has broader tail base, larger femoral pores, usually loose dewlap skin on throat, broader head.

Similarities

Leopard Lizard has narrower head, rounder tail, no black collar marks; Banded Rock Lizard has one black collar.

Habitat

Arid and semiarid regions, limestone-topped hills and bluffs, prairie rocks and canyons, unshaded hillsides.

Habits

Wary, agile, good jumper and runner; pugnacious, aggressive, often eat other lizards.

Range

Se. Oreg.; sw.-s. Idaho; Calif., extr. ne., far e., and se.; Utah (except ne. and cen. corridor); Colo., far w.-sw., and extr. se.-cen.; Ariz., N. Mex., Okla., nw. Tex.

LEOPARD LIZARD
Crotaphytus wislizeni

Fig. 61

Description
Length, 3½–5 in. (8.9–12.7 cm). Similar to Collared Lizard but with leopard spots and no collar. Above light to dark gray with large dusky spots and whitish crossbars, may meet or alternate at midline; below whitish to yellow, with dusky longitudinal streaks or spots on throat, sometimes dusky suffusion on chest and sides of belly; underside of tail often banded with dark gray. Can change color from light phase (conspicuous dark "leopard" spots and obscure light bars) to dark phase (spots obscure, light bars prominent). Body robust, head relatively large, tail long, distinct neck; dorsal scales granular; ventral scales larger, overlapping, sometimes weakly keeled; 1 or more transverse gular folds. Male has broader tail base, larger femoral pores, large postanal scales.

Similarities
Collared Lizard has 2 collars, laterally flattened tail.

Habitat
Arid and semiarid plains of sparse vegetation; likes sandy and coarse gravelly soil and hardpan; occasionally among sand dunes and in rocky areas.

Habits
Diurnal, wary, fast; may bite if captured, or holds mouth open in threatening manner; often eats other lizards.

Range
Oreg., cen. on Columbia R., se.; s. Idaho; in Calif., extr. ne., San Joaquin Valley, se. deserts; Nev., Utah, w. Colo., Ariz., N. Mex. (except extr. sw. and ne.), far w. Tex.

Note: The **BLUNT-NOSED LEOPARD LIZARD**, *Crotaphytus silus*, is sometimes recognized as a separate species. It occurs in the San Joaquin Valley in California.

SPINY LIZARDS
Genus *Sceloporus*

This is one of the largest groups of lizards, and includes the "swifts" and "blue bellies." All species have keeled, pointed, overlapping scales on their backs. They further differ from *Uta* and *Urosaurus* in having an incomplete gular fold. Males usually have a blue patch on the throat and on both sides of the belly. Most are climbers, living on trees, fences, rocks, etc.; a few are ground dwellers. They range from Canada to Central America.

BUNCH GRASS LIZARD
Sceloporus scalaris

73:6

Description
Length, 1¾–2½ in. (4.4–6.4 cm). The only spiny lizard in the West with scale rows on either side of body parallel to dorsal rows, rather than extending diagonally upward toward them. Above usually light brown with 4 longitudinal whitish stripes and blackish-brown crescent-shaped marks interspersed with smaller crescents, lower sides occasionally orange to reddish; black spot usually in front of base of forelimb. Male usually nearly uniform in dorsal coloration, with orange lateral stripes and ventrolateral blue patches on belly; tail base broader, postanal scales large, larger femoral pores.

Similarities
Striped Plateau Lizard without blue belly marking in male; lateral scales in diagonal rows.
Habitat
Between 4000 ft. (1219.2 m) elevation (Mexico) to 9500 ft. (2895.6 m) (se. Ariz.); prefers bunch grass areas in open coniferous forests.
Habits
Secretive, seeks shelter in grass clumps.
Range
Spotty localities in extr. se. Ariz. (Huachuca, Dragoon, Santa Rita, Chiricahua Mts.) and N. Mex. (Animas Mts.).

YARROW'S SPINY LIZARD
Sceloporus jarrovi

73:1

Description
Length, 2⅓–3⅓ in. (5.9–8.5 cm). In light phase, has striking pattern of whitish scales above, which may show flesh or bluish cast, conspicuously edged with black; collar broad, black, white-edged posteriorly. Body and base of tail flattish; 5–7 ear scales, tips pointed, largest in center; over 40 dorsal scales. Male has tail base broader, blue ventrolateral markings and throat patch.
Similarities
Crevice Spiny Lizard has collar band with white border front and back; dorsal scales larger, less than 40 from behind head scales to above vent.
Habitat
In oak and conifer belts; likes rocky places and cliff faces.
Habits
Agile climber.
Range
Spotty localities in se. Ariz. mts., mostly above 5000 ft. (1520.0 m); extr. sw. corner of N. Mex.

CREVICE SPINY LIZARD
Sceloporus poinsetti

73:2

Description
Length, 3¾–4⅝ in. (9.5–11.7 cm). Body and tail flattened. Broad collar (2½–5 scales wide in middle), black, margined with light; tail with dark bands toward tip; below whitish to cream. Dorsal scales large, keeled, pointed, projecting; less than 40 from behind head scales to above vent. Male has enlarged postanal scales, broader tail base, ventrolateral blue belly patches, and blue throat patch. Patterning of juveniles usually more distinct, especially tail rings.
Similarities
Yarrow's Spiny Lizard has more than 40 scales between head scales and area above vent; black collar lacking white border in front.
Habitat
To 8400 ft. (2553.6 m) elevation (Mexico); in granite, limestone, and other rock outcrops in arid and semiarid regions.
Habits
A wary crevice dweller; can be easily extracted from shallow retreats.
Range
Cen.-s. N. Mex. (except extr. w. boundary and e.), e. to cen. Tex.

DESERT SPINY LIZARD
Sceloporus magister   73:4

Description
Length, 3½–5½ in. (8.9–14.0 cm). Black shoulder mark wedge-shaped, edged with whitish or pale yellow. Above pale yellow to yellowish-brown with obscure to distinct crossbands or dusky spots; sides often rusty; below whitish to pale yellow. Dorsal scales 29–37 (average 32) between interparietal plate and line connecting rear bases of thighs; ear scales 5–7, pointed, largest at center; no wrist bands. Male has enlarged postanal scales, tail base broader, throat patch blue, blue ventrolateral area edged medially with black.

Similarities
Granite Spiny Lizard has darker body with less conspicuous black wedge, less keeled scales; Clark's Spiny Lizard has 3 ear scales, crossbands on forearms, body gray to green above.

Habitat
Semiarid regions among rocks, yuccas, creosote brush, cacti, etc.

Habits
Active, agile, wary, difficult to capture; good climber on rocks, tree trunks, sides of buildings; hides beneath yucca spines or in a rodent burrow, crevices, woodrat nests, niches in old buildings.

Range
Calif. (inner Coast Range n. to Panoche Pass; also e. and se.); w.-cen. and s. Nev.; far sw., s., se. Utah; Ariz., sw. N.Mex., far w. Tex.

CLARK'S SPINY LIZARD
Sceloporus clarki   73:5

Description
Length, 3–5 in. (7.6–12.7 cm). Similar to Desert Spiny Lizard, but only 3 ear scales with or without rounded tips, upper scale usually longest. Above gray, green, or bluish-green; forearms, wrists usually show conspicuous dark brown to black transverse marks. Male has enlarged postanal scales.

Similarities
Desert Spiny Lizard has no crossbars on forearms, 5–7 pointed ear scales, is yellow or brown above; Granite Spiny Lizard is smaller and darker, has poorly defined black neck wedge, less-keeled scales.

Habitat
Arid and semiarid regions, among rocks or in wooded areas; frequents both trees and boulders.

Habits
Diurnal, often found in trees, climbs to escape enemies; also seeks shelter in rock crevices, or in rodent burrow.

Range
Se. Ariz., as far w. as Ajo Mts., far sw. corner of N. Mex.

GRANITE SPINY LIZARD
Sceloporus orcutti   73:3

Description
Length, 3¼–4 in. (8.3–10.2 cm). Similar to Desert Spiny Lizard, but color generally darker; dorsal scales have shorter points and keels weak or absent; ear scales shorter. When a dark mark is present on each side in front of forelimb (usually absent), it extends well up onto shoulders, as in Desert Spiny Lizard. Male has enlarged postanal scales and extensive ventral blue suffusion and, often, a broad purple middorsal stripe; dorsal scales blue,

ventral surfaces blue-tinted. Juveniles have rusty head, body, and tail with light crossbanding, dark shoulder mark often distinct.

Similarities
Clark's and Desert Spiny Lizards have dorsal scales more pointed and keeled; lighter above, shoulder patch more distinct.

Habitat
Boulders in arid and semiarid regions; especially chaparral-covered hillsides.

Habits
Likes to bask in sun; when hot, seeks relief beneath rocks or in crevices.

Range
Extr. s.-cen. Calif. s. from n. side San Gorgonio Pass on both sides of mts. to tip of Baja Calif.

WESTERN FENCE LIZARD
Sceloporus occidentalis
                                                                    71:8

Description
Length 2½–3½ in. (6.4–8.9 cm). One of the best-known western lizards. Above black (old males), brown, gray, usually with 2 rows down back of blackish scallops or blotches; below whitish, often dark-spotted; rear surface of limbs yellow to orange. Body and tail rounded, tail long; dorsal scales projecting, pointed, keeled, overlapping, 35–51 between interparietal plate and line connecting posterior base of thighs; ventral scales smaller, smooth, overlapping; ear opening distinct, toothed in front. Male has enlarged postanal scales and prominent blue patch on each side of belly, throat usually blue.

Similarities
Sagebrush Lizard has rust on side of neck and body, black bar on shoulder, smaller scales on back, white or cream marks on blue throat patch; Eastern Fence Lizard has 2 small lateral throat patches marked with black in front; Side-blotched Lizard has black spot in armpit, complete gular fold.

Habitat
Wooded rocky areas; frequents talus and rocky outcrops of hillsides, canyons, along streams; also around old buildings, woodpiles, fences, woodrat nests, gopher burrows; to 9000 ft. (2743.2 m) elevations.

Habits
Diurnal, good climber, easily snared.

Other names
Swift Lizard, Blue-bellied Lizard.

Range
S.-cen. and extr. se. Wash.; Oreg. (except coastally n. from Rogue R.); sw. Idaho; se. Calif.; Nev. (except extreme s. tip); far w. Utah (except nw. corner).

EASTERN FENCE LIZARD
Sceloporus undulatus
                                                                    71:11

Description
Length, 2⅛–3¼ in.(5.4–8.3 cm). Similar to Western Fence Lizard. Above dark gray; blue throat markings often form wedge on each side. Male has blue under sides and neck, female more boldly marked on upperparts. Male has enlarged postanal scales.

Similarities
Western Fence Lizard usually has single blue throat patch.

415

Habitat
Widely various, from plains to high mountains, areas involving rocks, sand, hardpan, loose soil, dry piny and deciduous woods; fences, clearings, old houses, brush heaps, etc.

Habits
Fast runner, artful dodger about tree trunks, good climber; most active on sunny days.

Range
Se. Utah, ne. Ariz., Colo. (except n.–s. corridor); N.Mex.; far sw.-s. S.Dak., w. Nebr., w. Kans., Okla., Tex., on e. to Atlantic Ocean.

Note: The **STRIPED PLATEAU LIZARD,** *Sceloporus virgatus* **(71:9),** has a plain white belly without blue patches. The sides have a broad, dark stripe separating the unbroken white stripe above and below. It occurs above 4000 ft. (1219.2 m) elevation in southeastern Arizona (Chiricahua Mountains) and southwestern New Mexico (Peloncillo, Guadalupe, and Animas mountains).

SAGEBRUSH LIZARD
Sceloporus graciosus   73:8

Description
Length, 2–2½ in. (5.1–6.4 cm). Similar to Western Fence Lizard, but smaller; scales on back of thigh smooth, 42–68 dorsal scales. Throat powder-blue with white speckling, backs of thighs gray; black bar on shoulder; ventrolateral patches light blue. Characteristic rusty in armpits, and often rust, light orange, or yellowish on sides. Male has enlarged postanal scales.

Similarities
Western Fence Lizard has scales on back of thigh smooth, no rusty armpits, no black bar on shoulder.

Habitat
Sagebrush flats, piñon-juniper slopes, open coniferous forests; attracted to rocks, brush, stumps, surface litter; reaches 10,000 ft. (3048.0 m). elevations (higher than Western Fence Lizard), does not descend to lowlands.

Habits
Primarily a ground dweller, but may climb boulders or trees; hides in crevices, rodent burrows, or in bushes.

Range
Se.-cen. Wash., cen. and e. Oreg., s. Idaho, extr. s.-cen. Mont., cen. and w. Wyo., Calif. mts.; Nev. (except extr. s. tip); Utah, w. Colo.; Ariz., n. of Grand Canyon; far nw. corner of N. Mex.

CANYON LIZARD
Sceloporus merriami   71:10

Description
Length, 1½–2¼ in. (3.8–5.7 cm). Only spiny lizard in the West with tiny, granular scales on sides of body. Scales on back small but keeled. Gular fold incomplete. Black bar on shoulder just in front of foreleg. Above gray with vague dark and light markings. Male has 2 large blue belly patches broadly outlined with black.

Habitat
Boulders and rock walls of canyons.

Range
Big Bend area of Tex., w. to Presidio Co.

LONG-TAILED, SMALL-SCALED, AND TREE LIZARDS

Genus *Urosaurus*

LONG-TAILED BRUSH LIZARD

Urosaurus graciosus   **73:9**

Description
Length, 1⅞–2¼ in. (4.8–5.7 cm). Body slim, tail long, often twice as long as snout-vent length; broad middorsal band of enlarged, keeled, overlapping scales. Above dark to light ash-gray with dark crossbars on back; whitish line between upper jaw and groin; below white with some gray flecks, throat patch yellow or reddish. Gular fold present; dorsolateral fold usually with enlarged scales. Male has powder-blue to blue-green ventrolateral belly patches medially white-flecked.

Similarities
Tree Lizard has shorter tail, dorsal scale strip broken by central row of small scales; Small-scaled Lizard has shorter tail, blue throat.

Habitat
Desert areas with loose sand.

Habits
Diurnal; spends most of time in bushes, especially among creosote bush branches; capable of marked color changes.

Range
Se. Calif., far s. tip of Nev., far w. Ariz. with e. bulge through Maricopa Co.

SMALL-SCALED LIZARD

Urosaurus microscutatus   **73:10**

Description
Length, 1½–3 in. (3.8–7.6 cm). Similar to Tree Lizard, but has abruptly enlarged median dorsal scales; no central strip of small scales along dorsal band; tail less than twice snout-vent length. Male has ventrolateral blue patches on belly, which may be united; throat pale blue-green or blue with central patch of yellow or orange, female has yellow or orange throat, lacks ventrolateral markings.

Similarities
Tree Lizard has strip of abruptly enlarged dorsal scales broken by midline of small scales.

Habitat
A rock dweller.

Habits
Diurnal, good climber.

Range
Very extr. s.-cen. Calif. on each side of mts., s. from Borrego Palm Canyon.

TREE LIZARD

Urosaurus ornatus   **73:7**

Description
Length, 1⅞–2¼ in. (4.8–5.7 cm). Coloration as for Small-scaled Lizard. A moderately slender lizard with complete gular fold and dorsal band of enlarged scales broken by central longitudinal row of small scales, conspicuous enlarged scales in dorsolateral folds.

Male throat blue, pale blue-green, yellow, or greenish; blue ventrolateral patches on belly.

Similarities
Small-scaled Lizard has no middorsal strip of small scales; Long-tailed Brush Lizard has long tail, middorsal band of enlarged scales unbroken; Side-blotched Lizard has dark spot in armpit, no enlarged dorsal scales.

Habitat
On rocks and in trees such as mesquite, oak, pine, and juniper. From near sea level to 9000 ft. (2743.2 m) elevations.

Habits
Diurnally active; good climber, usually found high on rocks, ledges, cliffs, in bushes or trees, on sides of buildings.

Range
Extr. e. Calif., extr. se. Nev., se. Utah, extr. sw. Wyo., far w. Colo., Ariz., sw. N. Mex., e. to cen. Tex.

HORNED LIZARDS
Genus *Phrynosoma*

TEXAS HORNED LIZARD
Phrynosoma cornutum   **72:1**

Description
Length, 2½–5½ in. (6.4–13.0 cm). Sides of face have dark stripes radiating from eye. Above gray-brown, reddish-brown, tan, to gray; vertebral light streak and back blotched with dark brown to sooty, edged with light; below pale with black spots. Body squat, spiny, tail short; head spines slender, prominent; gular scales small, smooth, close-set, 1 enlarged row on each side of throat; eardrums distinct; ventral scales weakly keeled; 2 rows of pointed fringe scales on each side of body spines. Male has enlarged postanal scales and swollen tail base.

Similarities
Regal Horned Lizard has 4 prominent occipital horns; Round-tailed Horned Lizard has no fringe of scales on sides of body.

Habitat
Open flat areas with scrubby vegetation.

Habits
Most active during heat of day; when irritated, may eject thin stream of blood from eyes as far as 7 ft. (2.1 m).

Range
Extr. se. Ariz., se. N. Mex.; far se. Colo., Kans., Okla., Tex.

COAST HORNED LIZARD
Phrynosoma coronatum   **72:6**

Description
Length, 2½–4 in. (6.4–10.2 cm). Body has 2 rows of projecting fringe spines at sides. Head has 2 elongate occipital horns, bases usually not in contact; 2–3 longitudinal rows of enlarged, overlapping, pointed gular scales on each side of throat; nostrils open on line between snout and supraocular ridge; eardrum not covered with scales. Above gray, brown, reddish, or yellow, generally matching habitat soil; neck has large blackish mark on each side; back has undulating dark brown to black marks posteriorly edged with light; below yellow to cream, dusky-spotted. Male has swollen tail base and large postanal scales.

Similarities
Desert Horned Lizard has blunter snout, shorter horns, 1 row of lateral fringe scales on body.
Habitat
Sea level to 6000 ft. (1828.8 m) altitude in valleys, foothills, semiarid mountains; often in sand, wind-blown deposits, flood plains, washes, as well as in grassland, brushland, coniferous forests, broadleaf woodland.
Habits
Burrows into loose soil by wriggling head from side to side and kicking alternately with hindlegs.
Range
Calif. w. of Sierras and s. mts. to just n. of San Francisco Bay, but including Sacramento Valley.

SHORT-HORNED LIZARD
Phrynosoma douglassi   72:5

Description
Length, 2½–4½ in. (6.4–11.4 cm). Readily recognized by short head spines, small chin shields, and single row of lateral abdominal spines. Gular scales small, may be slightly enlarged in a row on each side of throat; eardrums exposed, may sometimes be concealed, or nearly so, by scales. Body squat, spiny, short-tailed. Above slaty, brownish, buff, reddish, or yellowish; back marked with either rows of dark splotches or transverse bars posteriorly margined with light; large dark brown blotch behind head on each side of neck; below white or yellowish, variously suffused with grayish or dark spotting. Male has enlarged postanal scales.
Habitat
Semiarid short-grass plains, hardpan, sandy and rocky terrains. Most widely distributed lizard in North America, and reaches 10,400 ft. (3169.9 m) elevations.
Range
Extr. s.-cen. B.C., se. Wash., Oreg. Cascades, extr. ne.-cen. Calif. Cascades, s. Idaho, far ne. Nev., Utah, e. Ariz.; very extr. sw. Alta., Mont. (except nw. fifth, extr. w. and far sw., and far ne.-cen.), sw. N. Dak., w. S. Dak., Wyo. (except nw. fifth and extr. s-cen.); far nw. corner and ne. Colo. (also broken localities in extr. se.-cen.); nw. Kans., N. Mex. (except extr. nw. and extr. e.); far w. Tex.

FLAT-TAILED HORNED LIZARD
Phrynosoma m'calli   72:3

Description
Length, 2¾–3½ in. (6.9–8.3 cm). Above gray with distinctive dark, narrow, vertebral stripe, which does not occur in any other western species of horned lizard; 1 longitudinal row of widely spaced dusky spots on each side of midline; below white, unmarked. Body has 2 rows of lateral abdominal fringe scales, gular scales granular, except for 1 row on each side. Head spines longer, more slender than in other species; tail flatter.
Habitat
Restricted to sandy areas.
Habits
Likes wind-blown sand, buries self by diving headfirst into sand or by sidewise movements of body; disappears at temperatures above 106°F (41.1°C).
Range
Far se. Calif., extr. sw. Ariz.

ROUND-TAILED HORNED LIZARD
Phrynosoma modestum   72:4

Description
Length, 1½–2¾ in. (3.8–6.9 cm). Body lacks lateral abdominal
fringe scales; tail slender, rounded, broadening abruptly near base.
Head shows no ear openings, horns short, gular scales small.
Above ash-white, gray to light brown; usually matches soil of
habitat; dark blotch on each side of neck and above groin, another
on each side of tail base; tail barred. Below unmarked whitish
except for spotting on throat and grayish tint between limbs. Sexes
differ as in Coast Horned Lizard.
Habitat
In arid regions with scattered scrubby growth.
Range
N. New Mex., n. Tex. to se. Ariz., and small population in Okla.

DESERT HORNED LIZARD
Phrynosoma platyrhinos   72:2

Description
Length, 2¾–3¾ in. (7.0–9.5 cm). Similar to Coast Horned Lizard,
but with blunter snout, shorter horn, 1 row of lateral fringe scales
on body; gular scales small, granular, with one row of enlarged
scales on each side of throat; eardrums may or may not be covered
with scales. Coloration as for Coast Horned Lizard; 2 rows of
pointed fringe scales on each side of body. Male tail base broader,
large postanal scales, femoral pores larger.
Habitat
Arid regions; sandy, gravelly flats; wind-blown sand, along washes
in areas of sparse, low-growing desert shrubs.
Habits
Sunbathes on rocks or open sandy patches between bushes; color
camouflage effective.
Range
Great Basin deserts of far se. Oreg., extr. sw. Idaho, Nev., w. Utah,
very extr. ne. Calif., Mojave and Colorado deserts of Se.; w. Ariz.

REGAL HORNED LIZARD
Phrynosoma solare   72:7

Description
Length, 3–4½ in. (7.6–11.4 cm). Head has 4 close-set, prominent,
large occipital horns. Body has 1 row of lateral abdominal fringe
scales, chest scales weakly keeled; row of enlarged pointed scales on
each side in gular region; scales on underside of tail prominently
keeled. Tail short, about ½ snout-vent length. Above light gray,
sometimes tinted with buff or reddish; neck blotch either side
behind head sooty to dark brown; blotches may join and extend
along sides nearly to groin. Light vertebral stripe commonly
present. Underparts whitish, sometimes buff-tinged, spotted. Sexes
differ as in Desert Horned Lizard.
Habitat
In arid regions of both plains and mountains, often among
Mesquite and Saguaro cacti.
Habits
When disturbed may eject blood either forward or backward from
eyes.
Range
S.-cen. Ariz. from Yuma Co. e. to near N. Mex. boundary and s.
into Mexico.

BANDED ROCK AND SIDE-BLOTCHED LIZARDS
Genera *Petrosaurus* and *Uta*

BANDED ROCK LIZARD
Petrosaurus mearnsi   *Fig. 62*

Description
Length, 3–4 in. (7.6–10.2 cm). Single black collar, often bordered behind with white; banded tail. Below whitish with bluish tint on belly and limbs; throat bluish-gray spotted with whitish, light gray, or pink. Body flattened dorsoventrally; dorsal scales granular, except for keeled pointed scales on tail and fore- and uppersurfaces of limbs; ventral scales smooth, somewhat overlapping; complete gular fold. Male has more intense gular pattern and bluish underparts.

Similarities
Collared Lizard is not flattened, has 2 black collars, unbanded, nonspiny tail.

Habitat
Restricted to arid and semiarid regions, favors canyons with massive boulders.

Habits
Active by day, climbs easily and fast; inquisitive, peers over or around boulders.

Range
Mts. of extr. s.-cen. Calif.

Fig. 62

Banded Rock Lizard

SIDE-BLOTCHED LIZARD
Uta stansburiana   *Fig. 63*

Description
Length, 1½–2⅓ in. (3.8–5.9 cm). A small ground-dwelling lizard with black spot on side behind armpit, small unpointed dorsal scales and complete gular fold. Above brown with chevron blotches or blue and yellow speckles. Male speckled with pale blue flecks.

Similarities
Tree Lizard has enlarged scales in middle of back, no armpit spot; Western Fence Lizard has large pointed scales, incomplete gular fold.

Habitat
Arid and semiarid regions of sand, gravel; rocky places, hardpan (occasionally), washes, arroyos, rocky flats and hillsides, on beaches behind strand.

Fig. 63

Side-Blotched Lizard

421

Habits
Primarily ground dwelling, seldom climbs except on boulders to
bask; very abundant in some parts of range.
Range
S.-cen. Wash., e. Oreg., s. Idaho, sw. Wyo., far w. Colo.; Calif.
(except Central Valley and nw. quadrant); Nev., Utah, Ariz., extr.
sw. N. Mex., w. Tex. (except n. panhandle).

NIGHT LIZARDS
Family Xantusiidae

Members of this family lack eyelids, the eyes being covered with an
immovable transparent covering, as in snakes. Species found in
United States have elliptical pupils, small granular dorsal scales,
and large rectangular belly scales; there is a gular fold. The
eardrums are exposed and quite large. In spite of their name, they
are more secretive than nocturnal. They bear live young.

ISLAND NIGHT LIZARD
Xantusia riversiana   **74:6**

Description
Length, 2¾–3¾ in. (7.0–9.5 cm). Body large, stout; 16 longitudinal
rows of belly scales at middle of body; two rows of scales above
eyes. Tail a bit shorter than snout-vent length to a bit longer.
Above gray to brownish, spotted or reticulated with sooty,
occasionally forming dorsolateral stripes; occasionally a brownish
vertebral stripe margined with black; head unmarked or mottled
cinnamon spotted with dusky; below whitish with various spottings
on sides of head, body, and beneath tail. Male has somewhat larger
femoral pores.
Habitat
Under surface object in areas where loose rocks intermingle with
cacti and sparse brush.
Habits
Secretive.
Range
Confined to only 3 islands off coast of s. Calif.: San Clemente,
Santa Barbara, and San Nicolas.

DESERT NIGHT LIZARD
Xantusia vigilis   **74:4**

Description
Length, 1½–1¾ in. (3.8–4.4 cm). Above gray, yellowish-brownish,
or olive, variously spotted according to individual or color phase;
below whitish, dull greenish-yellow, or pale gray, usually slightly
spotted laterally in gular area, on body sides, and ventrally along
tail. Small body and tail rounded, limbs relatively short, tail to 1⅓
times snout-vent length; 12 longitudinal rows of ventral plates at
midbody. Male thighs have more angular contour (female's
rounded); male shorter, tail stouter.
Similarities
Granite Night Lizard has 14 longitudinal rows of belly plates;
body flatter.
Habitat
Arid and semiarid situations associated with plants of Genus *Yucca*
(Joshua Tree, Spanish Dagger, Spanish Bayonet), under surface
litter, from below sea level to 9300 ft. (2834.6 m) elevations.

Habits
Active at all seasons both day and night under cover; good climber; tail readily lost; easily changes color from light olive (most common in evening) to dark brown (early morning to late afternoon); rarely seen unless covering litter removed.
Range
S. Calif, e. of Sierras and San Gabriel Mts. with narrow mt. corridor s. to Baja Calif. (absent from Imperial Valley and Ariz. border region); s. Nev., very extr. sw. Utah, far w. Ariz.

GRANITE NIGHT LIZARD
Xantusia henshawi
74:5

Description
Length, 2–2¾ in. (5.1–7.0 cm). Body flattened; skin soft, pliable. Above yellowish with large dark brown to sooty spots, tail narrowly barred. In usual daytime dark phase, dark spots so extensive as to reduce yellowish to network of narrow lines. Below white, may show slight purplish cast, and can be darkened to light gray. Gular fold present; back covered with granular scales, belly with 14 longitudinal rows of large rectangular plates. Male has tail base more swollen, elongate; whitish oval area against bluish-white background in region of femoral pores.
Similarities
Desert Night Lizard has 12 longitudinal rows of belly scales.
Habitat
On rocky slopes in arid and semiarid regions, under exfoliation on boulders in canyons.
Habits
When held, wraps tail about the fingers.
Range
Mts. of extr. s.-cen. Calif.

SKINKS
Family Scincidae

The skinks are ground lizards with shiny, flat cycloid scales and short legs. The back and belly scales are approximately the same size. They feed on insects and spiders, hibernate in winter, and lay eggs, which the female often broods. Most skinks are secretive, some are burrowers, and they constitute one of the most widely distributed of all lizard families.

GREAT PLAINS SKINK
Eumeces obsoletus
71:1

Description
Length, 3–5⅝ in. (7.6–14.3 cm). The only western spotted skink with lateral scales of trunk in diagonal rows extending upward and backward. Above ash-gray to very light beige, usually extensively marked with small black spots or a network; sides cream, occasionally flecked with pinkish; below unmarked cream or pale yellow. Tail 1½ times snout-vent length; frequently broken or regenerated; limbs of adults overlap when adpressed to sides. Juveniles markedly different from adults; very small, uniformly black with orange spots on top and whitish spots on sides of face, tail blue.

Habitat
In rocky areas beneath rock slabs or in crevices, under surface litter, in both wooded and grassland areas of plains to 6800 ft. (2072.6 m) altitude in mountains.

Habits
Secretive; abroad by day but seldom observable.

Remarks
May bite when captured.

Range
Se. and pine belt of n.-cen. Ariz.; N. Mex. (except far nw.); s. Nebr.; e. Colo.; Kans.; extr. w. Okla., cen. and w. Tex.

MOUNTAIN SKINK
Eumeces callicephalus   71:3

Description
Length, 2–2½ in. (5.1–6.4 cm). Above yellowish-brown to tan becoming grayish on tail, may grade into dark purplish-blue at tip in some individuals; often has Y-shaped mark on back of head; narrow dorsolateral whitish stripe, may be broken, on each side 4 scales below midline, below which is a sooty brown lateral band. Chest, abdomen whitish to gray or bluish-gray; throat, and sometimes chest, cream; underside of tail light bluish or light gray with yellowish cast; side of face pale orange in some individuals. Body stoutish, rounded; tail 1½–2 times snout-vent length, limbs of adults do not overlap when adpressed to sides, postnasal scale may or may not be present. Young have more distinct light striping, especially Y-mark on head.

Habitat
Rocky situations.

Range
Extr. s. Ariz. mts., Pajarito, Santa Rita, Chiricahua, Huachuca.

MANY-LINED SKINK
Eumeces multivirgatus   71:5

Description
Length, 2¼–3 in. (5.7–7.6 cm). Body slender, tail more than 1½ times snout-vent length, limbs of adults do not overlap when adpressed to sides. Above clearly marked with numerous dorsal and lateral stripes, dorsolateral stripe on 3rd row below middle of back; prominent, relatively broad vertebral stripe, lighter than adjacent color, bordered on each side by dark stripe, underparts generally gray, throat cream. Juveniles have less striping than adults.

Similarities
Mountain Skink has light dorsolateral stripe on 4th row below middle of back.

Habitat
Under surface litter or running about among shrubs in short-grass prairies; also in pine and spruce forests, in urban vacant lots.

Range
N.-cen. and e. pine belt of Ariz.; N.Mex. except extr. ne. and sw., and far e.; extr. sw.-cen. S. Dak.; cen. and w. Nebr.; extr. se.-e. Wyo.; cen. and e. Colo.; extr. nw. Kans.; w. Tex.; N.Mex. except extreme ne. and sw. N.Mex.

WESTERN SKINK
Eumeces skiltonianus   71:4

Description
Length, 2½–3¼ in. (6.4–8.3 cm). Above, broad dorsal stripe brown edged with black. On each side, there is a conspicuous white dorsolateral stripe originating on nose, extending over the eye, and passing along side of body on 2nd and 3rd scale row, below mid-back. Below this is a broad dark brown stripe, bordered below by a second light stripe that originates on upper jaw and extends to tail. Tail is dull blue or gray. Tail about 1½ times snout-vent length; limbs of adults may or may not overlap when adpressed to sides; 24–28 scales encircling middle of body. Young more vividly striped and with bright blue tail.

Similarities
Gilbert's Skink is unstriped or, if striped, scales of light stripes are edged with brown or gray.

Habitat
Under surface litter, inside rotten logs; woodland, forests, grasslands with herbaceous cover.

Habits
Secretive, most active in late afternoon; moves rapidly by snakelike body undulations; tail readily lost and wiggles violently when severed.

Range
Extr. s.-cen. B.C.; e. Wash.; Idaho (except extr. e.); far w. Mont.; Oreg. (except far nw. corner); n. Calif. and coastal belt to Baja Calif.; n. Nev., with island on Charleston Mt. in sw.; w. and cen. Utah.

GILBERT'S SKINK
Eumeces gilberti   71:6

Description
Length, 2½–4 in. (6.4–10.2 cm). Very similar to Western Skink. Adults are plain olive or with some dark spotting that may form network. In certain areas adults are striped with lateral stripes variegated, scales edged with gray or brown. Tail of adults is usually red. Young are often very similar to Western Skink, but with pink instead of blue tails, in certain parts of range.

Similarities
In Western Skink broad lateral stripes are uniformly dark brown, rather than absent or variegated; blue or blue-gray tail persists in adults and all young are blue-tailed. Where the two species live together differentiation is difficult, but young of Gilbert's Skink have pink tails with dark lateral stripe stopping at base of tail; young Western Skinks have blue tails, and dark lateral stripe extends well onto tail.

Habitat
In situations comparable to those of Western Skink, to 8000 ft. (2438.4 m) elevation.

Range
Cen. Sierras and coast range s. of San Francisco Bay area in Calif., mts. of s. Calif., plus scattered Death Valley region mts.; Clark Co., Sheep Mts., Nye Co., Nev.; Yavapai Co., Ariz., vicinity of Wickenberg, Yarnell, Castle Hot Springs.

FOUR-LINED SKINK
Eumeces tetragrammus   71:7

Description
Length, 2–2⅝ in. (5.1–6.7 cm). Only skink with 4 light lateral
stripes ending on shoulder. Back brown, gray, or olive. Male has
orange wash on sides of throat. Young have bright blue tail, stripes
as in adult.
Habitat
In brush, trash piles, clumps of cacti, and packrat nests in scrub,
grasslands, and woodlands.
Range
Tex., s. and w. to Big Bend.

WHIPTAILS
Family Teiidae

This is a varied, largely South American family whose members
are generally slender with very long tails. Western species usually
have small dorsal scales and quadrangular, nonoverlapping ventrals
arranged in longitudinal and transverse rows. A complete gular
fold is present. The tongue is forked and is frequently protruded
by the lizards, whose movements are jerky. The family includes
unisexual (all-female) species that reproduce parthenogenetically.

ORANGE-THROATED WHIPTAIL
Cnemidophorus hyperythrus   75:5

Description
Length 2–2½ in. (5.1–6.4 cm). Distinguished from other whiptails
by having a single frontoparietal. Tail 2–3 times snout-vent length;
14–18 femoral pores, 2–4 scales between ends of femoral pore
rows. Top of head olive-gray, back has 2 beige stripes down center
that may unite to form single stripe except in front; top of tail
dusky gray, bluish in juveniles; dorsolateral stripes yellowish.
Ground color between stripes dark brown to blackish. Underparts
light gray to yellowish-white, often orange-hued, throat area may
be deep orange.
Similarities
Western Whiptail has vague, or no, stripes, divided frontoparietal.
Habitat
Chaparral with rocks in patches of loose sand or soil.
Habits
Diurnal, wary, rather secretive; hides in rodent burrows or bushes;
movements nervous, jerky.
Range
Extr. sw. Calif. counties of San Bernardino (w. of mts.), Riverside,
and San Diego.

LITTLE STRIPED WHIPTAIL
Cnemidophorus inornatus   74:9

Description
Length, 2–2¾ in. (5.1–7.0 cm). Above ground color dark brown,
without spots; 7 dorsal stripes, median stripe sometimes obscure, of
light yellow on neck grading to orangish posteriorly; tail slightly
darkish, powder-blue, dusky toward base; face and lower sides of
neck bluish; below powder-blue, lighter on throat, deeper on tail.
Dorsal granules, 52–72; scales on posterior surface of forearm and

anterior to gular fold only slightly enlarged; circumorbital semicircles normal. Tail 2–2¼ times snout-vent length. Juveniles have less ventral blue.

Similarities
Desert-Grassland Whiptail is unisexual, has greenish, rather than blue, tail; Plateau Whiptail is unisexual, less blue, and has 65–85 dorsal granules.

Habitat
Usually on sandy soil of bottom lands and grasslands, also in juniper grasslands.

Range
Se. and ne. Ariz., extr. s. and cen. N. Mex., w. Tex.

CHIHUAHUA WHIPTAIL
Cnemidophorus exsanguis   74:11

Description
Length, 3–3¾ in. (7.6–9.5 cm). Unisexual species. Above, 6 or 7 light stripes on brown field with light spots; below whitish to cream. Tail green-brown. Scales enlarged on posterior surface of forearm; scales in front of gular fold abruptly enlarged; circumorbital semicircles normal; 65–86 dorsal granules. Juveniles and immatures striped and, in western part of range, have much less evident spotting than adults.

Similarities
Texas Spotted Whiptail has dark chest in male, paravertebral stripes farther apart; in Giant Spotted Whiptail there are no stripes in large adults, 90 or more dorsal granules; Checkered Whiptail has checkered pattern on back, smaller scales on back of forearm.

Habitat
Canyons, sandy bottoms of washes, on rocky slopes; also among broad-leafed trees, piñons, and junipers.

Range
Cen. and se. Ariz., cen. and sw. N. Mex.

Note: The **RUSTY-RUMPED WHIPTAIL,** *Cnemidophorus septemvittatus,* has stripes that fade into a rusty patch on the rump. It has both males and females and occurs in the Chisos Mountains area of western Texas.

TEXAS SPOTTED WHIPTAIL
Cnemidophorus gularis   74:10

Description
Length, 2½–3½ in. (6.4–8.9 cm). Above, 7 or 8 light stripes on dark field with brown spots. Vertebral stripe broader and less distinct than others. Tail brown or red-brown. Scales on anterior margin of gular fold and posterior surface of forearms enlarged; 78–96 dorsal granules, 10–21 between paravertebral stripes; circumorbital semicircles normal. Male has chest and belly purple and black.

Similarities
Chihuahua Whiptail has only faded markings on throat, 3–7 granules between paravertebral stripes; Checkered Whiptail has no dark markings on chest, smaller scales on posterior surface of forearm.

Habitat
Acacia-mesquite scrub.

Range
S. Okla.; cen., w., and s. Tex.; extr. se. N. Mex.

DESERT-GRASSLAND WHIPTAIL
Cnemidophorus uniparens   75:4

Description
Length, 2–2¾ in. (5.1–7.0 cm). Unisexual species. Above 6 or 7
light stripes on dark field without spots. Tail light green. Below,
white. Has 60–75 dorsal granules and scales on back of forearm
slightly enlarged; scales anterior to gular fold, abruptly enlarged
circumorbital semicircles, normal.
Similarities
Little Striped Whiptail has tail and belly bluish, 52–72 dorsal
granules; males and females present.
Habitat
Primarily Mesquite grasslands.
Range
Se. Ariz., sw. N. Mex.

Note: The **PLATEAU WHIPTAIL,** *Cnemidophorus velox* **(74:8)**, has a
light blue tail and 65–85 dorsal granules. It is also unisexual, but
occurs at higher elevations in piñon-juniper and Yellow Pine
woodlands of central Arizona and New Mexico, to southern Utah
and western Colorado.

SIX-LINED RACERUNNER
Cnemidophorus sexlineatus   75:3

Description
Length, 2½–3⅜ in. (6.4–8.6 cm). Above, 7 light stripes on an
unspotted dark background; green wash anteriorly. Below white.
Scales in front of gular fold abruptly enlarged; scales on posterior
surface of forearm only slightly enlarged; 68–110 dorsal granules;
circumorbital semicircles normal.
Similarities
Little Striped Whiptail and Desert-grassland Whiptail lack green
foreparts.
Habitat
In great variety of situations; prefers drier, more open localities
with substratum of sandy loam or other loose soil: plains and hills,
but not mountains.
Habits
Very fast, stops and starts suddenly, tail whips about when
running; most active during warm mornings; puts tongue out
frequently; tends to be colonial, but digs own burrows.
Range
W. of 100th meridian only to extr. se. Wyo., e. Colo., very extr.
ne. N. Mex., n. Tex. panhandle with spotty records s. to Mexico.

WESTERN WHIPTAIL
Cnemidophorus tigris   75:2

Description
Length, 3–3¾ in. (7.6–9.5 cm). Dorsal surface patterned with
many close-set black or sooty brown transverse bars that are
commonly separated by 4 longitudinal stripes and which may be
faded toward rear. Below marked with slaty to black spots, most
abundant on upper abdomen and chest. Scales anterior to gular
fold only very slightly enlarged, grading into smaller granules of
fold; scales on posterior surface of forearm not enlarged;
circumorbitals normal; frontoparietal divided. Juvenile tail often
light powder-blue.

Similarities
Checkered Whiptail has scales anterior to gular fold abruptly and strongly enlarged, less black on throat and chest.
Habitat
In wide variety of habitats; e.g., dry sandy or gravelly washes, rocky areas, in both loose and firm soil, often in sagebrush or creosote bush desert, grassland, on brushy slopes.
Habits
Diurnal; very swift runner, stops and starts suddenly, hides in bushes or rodent burrow, may run on hindlegs only, tail tip lashes about when running.
Range
Se. Oreg., sw. Idaho, Calif. (except ne. San Francisco Bay area and Cascade-Sierras); Nev., Utah, Ariz., sw. N. Mex., w. Tex.

GIANT SPOTTED WHIPTAIL
Cnemidophorus burti   75:1

Description
Length, 3⅓–5½ in. (8.5–14.0 cm). Largest whiptail in the West. Above, 6 or 7 stripes on dark background with spots. In large adults stripes are faint or absent. Red on head and neck. Below white. Scales enlarged on posterior surface of forearm and in front of gular fold; 90–120 dorsal granules; circumorbital scales normal.
Similarities
Chihuahua Whiptail is smaller, stripes are always present; has less than 95 dorsal granules.
Habitat
In canyons.
Range
Se. Ariz., w. to Ajo Mts.; sw. N. Mex.

CHECKERED WHIPTAIL
Cnemidophorus tesselatus   74:12

Description
Length, 3¼–4⅛ in. (8.3–10.7 cm). Unisexual species. Scales just in front of gular fold abruptly and conspicuously enlarged, those on posterior surface of forearm small; circumorbitals normal. Above conspicuous checkerboard of yellow-brown with black bars and spots. Below white with few faded spots scattered on throat, chest, and belly.
Similarities
Western Whiptail has scales anterior to gular fold only very slightly enlarged.
Habitat
In canyons, grassy hill terrain, on base slopes of mesas and in sparsely vegetated areas of few trees; also rocky situations.
Range
Arm from cen. to se. corner of Colo.; extr. w. Okla. panhandle, N. Mex., w. Tex., s to Mexico.

NEW MEXICAN WHIPTAIL
Cnemidophorus neomexicanus   74:7

Description
Length, 2½–3 in. (6.4–7.6 cm). Unisexual species. No other whiptail has wavy middorsal stripe and circumorbitals that extend beyond point above center of eye, usually to point above front of eye. Scales on posterior surface of forearm granular, unenlarged;

scales in front of gular fold moderately enlarged; dorsal granules 71–85. Seven distinct yellow longitudinal body stripes; dark ground color with light spots; tail blue. Juveniles have black ground color, yellow stripes, yellow spots in dark fields distinct, pattern more contrasting than in adults; bright blue tail. This species has no males.

Habitat
Desert playas. Coexists with Western, Checkered, and Little Striped Whiptails.

Range
Rio Grande Valley from n. of Santa Fe, N.Mex., to extr. w. Tex.

ALLIGATOR LIZARDS AND RELATIVES
Family Anguidae

Members of this family have elongate bodies, relatively short limbs, long tails, and a fold along each side of the body. Under the fold are granular scales that separate the large square scales of the back from those of the belly. Although most species lay eggs, some give birth to live young. They are widely distributed in North America and Europe.

NORTHERN ALLIGATOR LIZARD
Gerrhonotus coeruleus   75:7

Description
Length, 3½–5½ in. (8.9–14.0 cm). Above olive, greenish, or bluish ground color; below whitish to cream, with longitudinal sooty lines between scale rows. Back has 14 or 16 lengthwise rows of scales.

Similarities
Southern Alligator Lizard has 14 dorsal scale rows, belly stripes down middle of scale rows.

Habitat
In more humid and cooler habitats within or near coniferous forests; found under surface objects in shrubby or grassy situations.

Habits
Diurnal; gives birth to live young.

Range
Far s. tip Vancouver Is., far s. B.C., irregularly w. and n.-cen. Wash. to e. of Cascades; n. Idaho; extr. w. Mont.; w. Oreg., (except upper Willamette Valley); coastal s. Calif. to Monterey area, n.-cen. part on w. slope of Cascades, down Sierras to Kern Co.; small island in very extr. ne. corner of Calif.; absent from Sacramento Valley.

SOUTHERN ALLIGATOR LIZARD
Gerrhonotus multicarinatus   75:8

Description
Length, 4–6½ in. (10.2–16.5 cm). Tail slightly over twice snout-vent length. Scales on back usually in 14 rows, interoccipital plate usually single. Above light brown, olive-gray, to dull yellowish; body and tail usually marked with irregular transverse dark bands, 9–13 on body; below pale gray to dull pale yellow, often with sooty longitudinal lines down center of scale rows. Male head somewhat broader proportionately than female's.

Similarities
Northern Alligator Lizard has dark longitudinal stripes between scale rows on belly, dorsal scale rows usually 14 or 16.

430

Habitat
In well-wooded areas, open grassland, and regions of widely
scattered shrubs; oak and chaparral coastal belt of valleys and
foothills.
Habits
Active by day, dusk, and at night; good climber, uses tail
prehensilely, movements generally slow and deliberate; will enter
water, swims well by lateral body undulations; aggressive, plays
possum, exudes strongly odorous excrement when captured.
Range
Extr. s.-cen. Wash.; Oreg., cen. Columbia R. stretch and corridor
between Coast and Cascade ranges s.; Calif., w. of Cascade-Sierra
crest to Baja Calif.; absent from San Joaquin Valley.

Note: The **PANAMINT ALLIGATOR LIZARD,** *Gerrhonotus
panamintinus,* has 7–8 dark crossbands. It occurs in the Panamint,
Nelson, and Inyo mountains of California.

ARIZONA ALLIGATOR LIZARD
Gerrhonotus kingi   **75:6**
Description
Length, 3–5 in. (7.6–12.7 cm). Above gray or tan with 8–11 wavy
brown crossbars; a row of large white spots on upper jaw; below,
where marks are present, they are more irregular in distribution,
often as scattered flecks. Scales of back usually in 14 longitudinal
rows. Juveniles markedly different, with strongly contrasting
crossbands.
Habitat
To over 7000 ft. (2133.6 m) elevations, in canyons, scattered loose
rocks, sparse oaks and grass, grassland, brush.
Range
Se. Ariz., far sw. N. Mex.

TEXAS ALLIGATOR LIZARD
Gerrhonotus liocephalus   *Fig. 64*
Description
Length, 5–8 in. (12.7–20.3 cm). Distinguished from other western
lizards by its very elongate tail and body, which reaches 20 in.
(50.8 cm) in total length; small weak limbs, and a median,
unpaired scale behind nostril. Above tan with 7–8 irregular dark
crossbands.
Habitat
Dry areas with piñon and juniper forests.
Habits
Climbs in bushes.
Range
S. and w. Tex.

Fig. 64

Texas Alligator Lizard

California Legless Lizard, p. 432

CALIFORNIA LEGLESS LIZARDS
Family Anniellidae

CALIFORNIA LEGLESS LIZARD
Anniella pulchra   *Fig. 64*

Description
Length, 4½–6½ in. (11.4–16.5 cm). The only legless lizard in the
West. Body slender, cylindrical, snakelike; 8–9 in. (20–22.5 cm)
total length and ¼ in. (0.6 cm) in diameter; limbless, vestigial
girdles present; no ears, eyes small but with movable lids (unlike
snakes). Scales over all of body except head overlapping, smooth,
cycloid. Teeth largish, few, recurved, bases swollen. Above silvery
to brownish to black; light individuals exhibit thin, dark, median
line from head to tail, plus several fainter longitudinal lines. Below
yellow. Male somewhat slimmer than female.
Habitat
In loose sandy soils to 6400 ft. (1950.7 cm) elevation.
Habits
Good burrowers; active at dusk and night; commonly lie buried at
shallow depths; when lost tail is regenerated, tip resembles head.
Range
Calif. coast s. of San Joaquin River into Baja Calif.; also s. end of
Sierras on e. side of San Joaquin Valley.

VENOMOUS LIZARDS
Family Helodermatidae

GILA MONSTER
Heloderma suspectum   **75:9**

Description
Length, 12–16 in. (30.5–40.6 cm). Body large, heavy, tail ½–⅓
snout-vent length. Scales beadlike; venom glands at sides of lower
jaw not connected with teeth. Eyelids well developed, pupil round;
ear opening large, border untoothed; gular fold present. Tail short,
blunt-tipped, stout, base often constricted; limbs relatively short.
Above with conspicuous spots, bars, reticulations, dots of buff,
orange, or pinkish, and blackish-brown; tail marked usually with
rather broad rings of light against dark ground; below yellowish-
white with irregular brownish markings; sides of face, underside of
head, feet generally blackish-brown to black.
Habitat
On desert flats, wide canyons and washes in rocky areas; in
vegetated regions of cacti, ocotillo, mesquite clumps, and on
creosote bush desert.
Habits
Most often active at dusk and night; gait slow, awkward, but good
climber, powerful digger; likes to lie submerged in water when
available.
Remarks
When biting, a Gila Monster chews slowly with bulldog tenacity,
and grooves in teeth facilitate flow of venom into wound by
capillary action; venom is primarily neurotoxic, affecting
respiratory and circulatory centers, but only occasionally fatal.
Range
Extr. se. Nev., very extr. sw. Utah, se. Calif., w. Ariz. (where
protected by state law); extr. sw. N.Mex.

Snakes

Suborder Serpentes

There are approximately 3000 species of snakes worldwide, arranged in 10 families. In the West there are 76 species.

Features of Snakes

A snake, or serpent, is usually a limbless reptile with expandable jaws, slender, hooked teeth, no ear openings or movable eyelids, and usually a single row of large belly scales. The boas have flaps at each side of the vent, a condition that is in contrast to the limblessness of most others. The flicking of its forked tongue picks up dust particles that give the snake sensations of both taste and smell, even though the sense of smell itself is well developed, as is the sense of sight. A snake hears by means of vibrations that its body picks up through the ground, not through ears. Snakes cast off their skins, usually several times a year, by crawling out of them headfirst, leaving the old skin behind.

Fig. 65

Parts of a Typical Snake

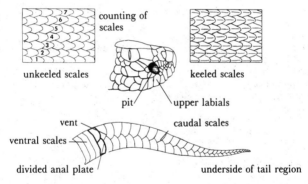

Snakes have no voice, though some are able to make a hissing sound, while others can buzz with rattles on the ends of their tails. They are exclusively carnivorous, eating frogs, toads, salamanders, rabbits, rats, mice, shrews, birds, and worms. (There are no vegetarian serpents.) Snakes digest slowly and, after feeding, they will often be sluggish for several days. Their meals may exceed 50 percent of their own body weight.

Identification

Individual snakes of the same species often vary widely in color in different parts of their range or even in the same locality. Thus, pattern—lines, blotches, or spots—shape of body (long, short, slender, stout), and shape of head are important for identification, often more so than color. Indeed, colors may vary from light in young to dark in old individuals, and from buffish to blackish or from greenish to reddish-brown.

Measurements

Sizes given for the species described below are for *total length,* including tail.

BLIND SNAKES
Family Leptotyphlopidae

The pinkish, semitransparent members of this family of small, harmless snakes, under 15 inches (38.1 cm) in length, look a good deal like enlarged earthworms. Their tiny degenerate eyes are covered by scales and their scales are the same size all around the body, not enlarged on the belly as in other snakes. They are burrowing and egg-laying.

WESTERN BLIND SNAKE
Leptotyphlops humilis   76:1

Description
Length, 9–16 in. (22.9–40.6 cm). Body very slender, 3/16 in. (0.5 cm) or less in diameter, continuous with head and neck, without constriction; eyes vestigial as black spots beneath ocular scales. Scales smooth, shiny, uniform in size; no enlarged belly scales, but 1 or 2 enlarged scales in front of anus. Tail tipped with small spine. Above dark or light brown to pinkish on 5–7 dorsal scale rows; below cream to light pink, uncommonly light gray.
Similarities
California Legless Lizard has movable eyelids, dark vertebral stripe.
Habitat
Under rocks and bushes; frequents stony and sandy deserts, brush- and boulder-strewn slopes.
Habits
Subterranean, can burrow rapidly into loose soil or sand; may emerge at night, often collected from highways.
Range
Extr. s. Calif., s. Nev. tip; extr. sw. Utah; far w., s.-cen., and far s. Ariz.; extr. s. N. Mex., w. Tex.

TEXAS BLIND SNAKE
Leptotyphlops dulcis   76:2

Description
Length, 6–13 in. (15.2–33.0 cm). Very similar to Western Blind Snake, but distinguished by having 3 scales between oculars, and 2, instead of 1, labials between ocular and lower nasal.
Habitat
See habitat for Western Blind Snake.
Habits
A burrower.
Reproduction
Snakes 8 or more inches (20.3 cm) long may contain eggs.
Range
Extr. se. Ariz., extr. s. and cen. N. Mex.; w. Okla., w. Tex.

BOAS
Family Boidae

Members of this family in the West are small, stout, and harmless. They have well-developed eyes that have a vertically elliptical pupil, smooth, shiny scales (but no enlarged chin scales), and vestiges of a pelvis and hindlimbs. The belly scales are not as wide as those of most snakes. They bear their young alive.

ROSY BOA
Lichanura trivirgata   76:10

Description
Length, 24–42 in. (61.0–106.7 cm). Head triangular, relatively small, no enlarged scales between eyes, neck moderately distinct. Body heavy; spurs may be present. Above slaty, bluish, or brownish-gray; 3 broad, longitudinal, reddish-brown stripes, may be irregular in outline or ill defined in coastal forms; below yellowish-white, spotted with brown or gray. Spurs larger in male, sometimes lacking in female.

Habitat
On desert and coastal slopes with rocks and chaparral; in rocky or sandy canyons; in arid sections under 4500 ft. (1371.6 m) elevations, restricted to stream borders.

Habits
Nocturnal, largely crepuscular, occasionally seen by day; docile, slow, fearless; good climber.

Range
S. Calif. (except Imperial Valley and extr. se.), sw. Ariz.

RUBBER BOA
Charina bottae   76:8

Description
Length, 14–29 in. (35.6–73.7 cm). Tail short, blunt, almost as blunt as head, hence popular name "two-headed snake." Scales on back of head enlarged, varied in number, size, shape; no distinct neck; spurs small, usually distinct in male, Above pale tan to dark brown, often with tinge of yellowish, greenish, or bluish; tip of tail slightly darker than body; below orange-yellow, yellow, to cream; chin and throat may be clouded with gray or brown.

Habitat
In moist situations near or in coniferous forest; under surface objects, inside rotten logs.

Habits
Largely crepuscular, also nocturnal, but occasionally out by day; ready burrower, good climber and swimmer.

Range
Extr. s. B.C., Wash. (except far w. and cen.-s. island); Idaho; far w. Mont.; extr. nw. Wyo., plus n. Bighorn Mts.; Oreg. (except very extr. nw. and se.); n. Calif. coast to s. of Monterey Bay, Sierras; scattered sites in s. mts. e. of Los Angeles; n. Nev., far n.-cen. Utah.

COMMON HARMLESS SNAKES
Family Colubridae

The members of this nonpoisonous family of snakes have conspicuous eyes and broad belly scales, and the head plates are large and regular in shape and arrangement. Most have a spine on the end of the tail, and none have pits between the nostrils and the eyes, nor permanently erect front upper fangs. With the exception of the worm snakes, boas, the coral snakes, and the rattlesnakes, all snakes found in the West belong to this family. These snakes differ from the worm snakes in having functional eyes, enlarged belly scales, and upper-jaw teeth. They differ from the boas in having broader ventral scales, no sign of vestigial hindlimbs, and loosely jointed skull bones; from the coral snakes in lacking fixed hollow fangs in the front of the upper jaw; and from rattlesnakes in lacking the rattles and facial pit. Included are both live-bearing and egg-laying species.

RINGNECK SNAKE
Diadophis punctatus 77:1

Description
Length, 12–30 in. (30.5–76.2 cm). The only western snake with yellow neck ring and bright orange to red belly. Scales smooth, 13–15 rows at midbody, anal plate divided; caudals 54–76 in males, 48–66 in females. Above uniform olive, blue-gray, to blackish-slate, slightly darker on head; neck ring distinctive red, orange, or yellowish; below orange, yellow-orange, or coral red, brighter on tail, with belly usually spotted. Adult males have tubercles on scales above vent.

Habitat
In wooded areas along canyon breaks, usually under surface objects but occasionally in the open; often in gardens and occasionally in salt marshes.

Habits
Coils and waves tail when caught, giving rise to common name "thimble snake."

Range
Wash., island in extreme se.; w. Oreg. (except extr. nw. and Columbia R. at Portland); Calif. w. of Cascade-Sierra crest (except Great Central Valley), s. to Baja Calif.; Idaho, island along Snake R. in Boise Co.; Utah, se. Ariz., N. Mex., Tex.

SHARP-TAILED SNAKE
Contia tenuis 76:6

Description
Length, 8–18 in. (20.3–45.7 cm). Above brown, reddish- or yellow-brown, grayish, rarely weakly spotted; upper tail surface sometimes reddish; below white to cream with distinctive alternating black crossbars. Body moderately stout; tail short, conical, ending in a pointed scale. Scales smooth, loreal present, preocular single, anal plate divided.

Habitat
Usually under surface objects and often near water.

Habits
Secretive, rarely in open, active only during rains.

Range
Wash., extr. se. Vancouver Is., s. of Tacoma; Oreg., island in w.-cen. in Jackson and Benton cos., also far sw. in Siskiyou Mts.; Calif., w. of Sierra crest and s. to midpoint of state (but absent from Great Central Valley).

WESTERN HOGNOSE SNAKE
Heterodon nasicus 76:3

Description
Length, 16–32 in. (40.6–81.2 cm). Body heavy, head broad and barely distinct from neck, tail very short. Much enlarged and upturned rostral scale; scales keeled, prefrontals separated by small scales, anal plate divided. Above light olive, olive-gray to yellowish, with row of darker blotches down vertebrae (may be edged with light); below, throat and sides cream to white, central area of black with yellow markings may extend full body length.

Similarities
Western Hook-nosed Snake has dorsal scales smooth, rostral concave rather than keeled above; Eastern Hognose Snake, snout less upturned, underparts grayish or yellowish with black markings.

Habitat
In dry, sandy areas, prairies.
Habits
Diurnal, a fast burrower; feigns death when first captured; when threatened, puffs up, hisses, strikes—hence popular name of "blow snake" or "puff adder"—but is harmless.
Range
Extr. se. Alta., extr. s. Sask., extr. sw. Man. Primarily a Great Plains inhabitant with w. boundary along foothills of e. Rockies through Mont., Wyo., Colo., N. Mex., and far se. corner of Ariz.

EASTERN HOGNOSE SNAKE
Heterodon platyrhinos

Description
Length, 24–42 in. (61.0–106.7 cm). Similar to Western Hognose Snake, both forms being the only snakes with a keeled, upturned rostral; averages larger, has less-upturned rostral, prefrontals in contact. Color variable; if blotched, patterns may be gray, black, brown, even reddish, on a lighter ground; dark line across top of head, dark line from eye to corner of mouth; below whitish or yellowish with black markings, lightest under tail. Occasional dark-phase individuals are nearly uniformly black above except for white lips, under snout, chin, and throat; underparts gray.
Habitat
Dry, sandy areas, prairies, open woods, uplands, hillsides, fields, along sandy river banks.
Habits
Similar to those of Western Hognose Snake.
Range
Great Plains e. of Rockies, in extr. se. Mont., extr. sw. N. Dak., very extr. ne. Wyo., S. Dak. (except far sw. and far ne. parts); possibly in very extr. se. Colo.; Okla. (except very extr. w. panhandle), Tex. e. of line from nw. panhandle to s. tip.

SPOTTED LEAF-NOSED SNAKE
Phyllorhynchus decurtatus

76:12

Description
Length, 12–20 in. (30.5–50.8 cm). A much-enlarged rostral scale gives rise to name. Above light tan, gray-brown, pinkish; vertebral row of more than 17 brown blotches; commonly somewhat rectangular, with long axis crosswise; sides with 1–3 rows of smaller blotches, may have dark stripe across head; below unmarked white to cream. Head barely wider than neck; eyes large, slightly protuberant. Scales smooth, anal plate single. Male tail much longer, broader; males tend to fewer ventrals and more caudals than females.
Similarities
Saddled Leaf-nosed Snake has dorsum with fewer than 17 dark blotches.
Habitat
In rocky, sandy deserts with sparse brush.
Habits
Active only at night; good burrower; when annoyed, hisses violently, may flatten neck into dewlap.
Range
Se. Calif., s. Nev., w. Ariz., as far e. as Patagonia Mts. along Mexican border.

SADDLED LEAF-NOSED SNAKE
Phyllorhynchus browni   76:14

Description
Length, 12–30 in. (30.5–76.2 cm). Very similar to Spotted Leaf-nosed Snake, but fewer than 17 dorsal blotches, exclusive of those on tail. Above creamy white to pinkish; saddlelike blotches on body and tail usually elongate-oval, gray-brown to reddish-brown to chocolate; dark head stripe more pronounced than in Spotted Leaf-nosed Snake; below white.

Habitat
In desert area with mesquite, some saltbush, often under stones.

Habits
Nocturnal; when annoyed, hisses, flattens throat into dewlap.

Remarks
Best collected by night from black-top roads.

Range
S.-cen. Ariz.

RACER
Coluber constrictor   77:3

Description
Length, 22–78 in. (55.9–198.1 cm). Body long, slender, tail very long. Scales satin-smooth, lustrous; 15–17 rows; 2–3 anterior temporals, usually 2 preoculars, loreal present, rostral scale unmodified, anal plate divided. Head elongate, moderately distinct from neck, eyes relatively large. Above unpatterned brown, olivaceous, bluish, greenish-blue to dark gray or black; below unmarked whitish, gray, light blue, pale yellow, orange-yellow, to yellow-green. Juveniles marked with brown or blackish saddles or spots.

Similarities
Smooth Green Snake is smaller, has single anterior temporal; young Gopher Snake has keeled scales, 4 prefrontals; Night Snake has vertical pupils; see also whipsnakes, which have 11–15 dorsal scale rows.

Habitat
In open country, forest openings, fields, grass-bordered streamsides, meadows, thin brush, stone walls, rocky outcrops, trash piles, roadsides, etc.

Habits
Agile, will fight when cornered; when foraging, holds head and neck well off ground; good climber.

Range
Far sw. B.C.; e. Wash., Idaho (except extr. n. panhandle); Mont. (except extr. nw.); Oreg. (except coastal belt); Calif. (except San Joaquin Valley and se.); nw. Nev. and Lake Mead area; Utah (except far w., but including nw. and sw. corners); Ariz., extr. nw. and far ne.; Wyo. (except Yellowstone Park region); Colo. (except broad n.–s. center belt); N. Mex. (except far sw. corner and se.); Tex. (except w.-cen. two-fifths), thence, e.

SMOOTH GREEN SNAKE
Opheodrys vernalis   76:5

Description
Length, 11–26 in. (27.9–66.0 cm). Only green snake with smooth scales. Above uniformly green, below unmarked white to cream. Body slender, head distinct, tail long; smooth dorsal scales in 15 longitudinal rows, anal plate divided.

Similarities
Green Rat Snake has 25–37 scale rows, dorsal scales weakly keeled; Racer is larger, has small, wedged preoculars, 2 instead of 1, anterior temporals; Rough Green Snake has keeled dorsal scales.
Habitat
Foothills to mountains, in meadows, marshes, grassy openings, or other moist situations.
Habits
Secretive, gentle.
Range
Discontinuous in w. part of range; extr. s. Man., s. to Tex., w. to Utah. Isolated populations in mountains of this area.

Note: The **ROUGH GREEN SNAKE,** *Opheodrys aestivus* **(76:7)**, resembles the Smooth Green Snake, but has keeled dorsal scales and occurs in the West in central Texas.

STRIPED RACER
Masticophis lateralis   **77:2**

Description
Length, 30–60 in. (72.2–152.4 cm). Above uniformly blackish-brown to black, lighter on tail, each side with 1 light longitudinal line, about 2½ scale rows wide; below whitish-cream or light yellow with coral-pink beneath tail; spotted on head and behind snout. Scale rows on mid-back 17-17-13.
Similarities
Striped Whipsnake has lateral white stripe bisected by black stripe, 15 scale rows at mid-body; Sonora Whipsnake has 2–3 light stripes anteriorly that fade in front of tail.
Habitat
In brushy areas, often among oaks, digger pines, pion-juniper areas.
Habits
A good climber, forages and suns in bushes and trees; sometimes makes escape by climbing trees; nervous, active, fast.
Range
Calif. (except nw. coast, n., Great Central Valley, e. of Sierra crest and e. of s. Calif. mts.) into Baja California.

COACHWHIP
Masticophis flagellum   **77:4**

Description
Length, 36–102 in. (91.4–259.1 cm). Scales dorsally smooth, in 17 rows at midbody and 11–13 in front of vent; lower preocular wedged between upper labial. Above variable from ash- to dark gray, yellow tones, gray-brown, pink, or reddish- to bluish-black; often faint body crossbanding and indistinct longitudinal streaking. Crossbands in neck area usually darker, sometimes rather extensive; usually no distinct longitudinal stripes. Below whitish, yellowish, brownish, to pale pinkish-orange, and dusky-spotted. Body very slender, head narrow, eyes large. Juveniles crossbarred or blotched with dark brown or black.
Habitat
Dry uplands, arid and semiarid regions, creosote bush, mesquite flats, sagebrush, short-grass prairies, pastures, fields, roadsides.
Habits
Fast; diurnal, terrestrial.

Remarks
Can give a nasty bite when handled.
Range
In West irregularly s. of 40th parallel. Absent from n. Calif. and
Sierras, ne. Nev., nw. Utah, extr. n. and cen. Colo.

STRIPED WHIPSNAKE
Masticophis taeniatus   76:9

Description
Length, 30–70 in. (76.2–177.8 cm). Scale rows at mid-body 15.
Color similar to that of Striped Racer, but dorsolateral light stripe
on each side bisected by dark line, broken to nearly continuous.
Additional black and white stripes often present on lower sides.
Ground color between dorsal stripes dark brown to black or
grayish, with lighter cast of bluish, olive, or greenish; below cream
to yellowish, lighter in front, coral posteriorly.
Similarities
Striped Racer has unbisected lateral stripe; Sonora Whipsnake has
stripes fading posteriorly.
Habitat
In warm, rocky, brushy foothills and grasslands to 6300 ft. (1920.2
m) altitude. Moves quickly; very alert.
Food
Lizards.
Range
Se.-cen. Wash., e. Oreg., s. Idaho; Calif., far ne. and far e.-cen.;
Nev., Utah, far sw. corner of Wyo., w. Colo.; ne. Ariz., sw. N.
Mex. (except extr. sw. corner); sw. Tex.

SONORA WHIPSNAKE
Masticophis bilineatus   76:11

Description
Length, 30–67 in. (76.2–170.2 cm). Above olive, bluish-gray, or
light grayish-brown, lighter on posterior ⅔ of body, including tail;
snout yellowish-brown; 2 or 3 light lateral stripes on each side that
fade in front of tail. Below white, changing to yellow toward tail.
Scales dorsally usually 17–17–12 or 13.
Habitat
In scattered oaks, yuccas, saguaro–paloverde–ocotillo associations,
chaparral, to 5500 ft. (1676.4 m) elevation.
Habits
A good climber, often found in bushes.
Food
Birds and lizards.
Range
S. Ariz., far sw. N. Mex.

WESTERN PATCH-NOSED SNAKE
Salvadora hexalepis   76:13

Description
Length, 20–45 in. (50.8–114.3 cm). Rostral scale much enlarged,
edges free; scales smooth in less than 19 rows, anal plate divided,
1–3 preoculars, loreal single or divided, posterior pair of chin
shields separated by 2–3 small scales; 9 upper labials. Above light
gray, 2 dark olive-brown to blackish-brown dorsal stripes (perhaps
only 1 stripe along coast); below whitish, unmarked. Body slender,
head moderately distinct from neck.

Similarities
Mountain Patch-nosed Snake has posterior chin shields in contact or separated by only 1 scale, 8 upper labials.
Habitat
Low, arid and semiarid regions, such as brushy desert, chaparral foothills, borders of washes.
Habits
Diurnal, active, fast; mostly ground-dwelling (occasionally climbs bushes and trees); burrows.
Range
Se. Calif.; Nev., n.-cen. to se. (absent from Carson Valley and ne.); very extr. sw. Utah; Ariz., far nw. and w., plus Grand Canyon, sw.; far nw. N. Mex., extr. w. Tex.

MOUNTAIN PATCH-NOSED SNAKE
Salvadora grahamiae   **76:15**

Description
Length, 20–30 in. (50.8–76.2 cm). Very similar to Western Patch-nosed Snake, but posterior chin shields in contact, or separated by 1 small scale; 8 instead of 9 upper labials. See coloration of Western Patch-nosed Snake; light vertebral stripe, white, grayish, or yellowish, margined by light brown to black; sides greenish, bluish, or gray; below unpatterned yellowish to bluish-white.
Habitat
In mountain woodlands from 4000 to 6500 ft. (1219.2 to 1981.2 m) altitude in w. of range, but in e. of range from prairies to sea level.
Habits
Similar to Western Patch-nosed Snake.
Range
Se. Ariz., s. N.Mex., sw. Tex.

CORN SNAKE
Elaphe guttata   **77:6**

Description
Length, 18–72 in. (45.7–182.9 cm). Above light grayish-brown with dorsal blotches, 42–55 in males, 44–55 in females (not including tail) of rich reddish-brown and saddle-shaped, 2 rows of smaller similar blotches on each side of dorsal row; usually with a pair of dark lines on neck that unite to form a spearpoint between eyes; below cream or white, usually with a few spots. Body slender, head small; 25–35 scale rows at midbody, mostly smooth, but weakly keeled on the back. Anal divided.
Habitat
In woods, woods edges, cornfields, outbuildings, roadsides, prairies, plains, towns.
Habits
Basically ground-dwelling but can climb; quite nocturnal; discharges noxious anal scent when annoyed.
Range
Island comprising far e.-cen. Utah and extr. w.-cen. Colo. on w. side of Rockies; se. Colo., Kans. (except far nw.), e. N.Mex., Okla., Tex.

TRANS-PECOS RAT SNAKE
Elaphe subocularis   **77:8**

Description
Length, 36–54 in. (91.4–137.2 cm). Above yellow with H-shaped
black or dark brown blotches with pale centers. Lateral arms of
blotches sometimes join to form longitudinal stripes. Row of
suboculars between eye and labials. Dorsal scales 31–35, midbody,
weakly keeled along middle of back. Body slender, head broad, eyes
large.
Habitat
Arid and semiarid associations of creosote bush and ocotillo; also
rocky places.
Habits
Appears to prefer rocky habitats.
Range
Extr. s.-cen. N.Mex., w. Tex.

GREEN RAT SNAKE
Elaphe triaspis   **77:10**

Description
Length, 24–50 in. (61.0–127.0 cm). Dorsal scales 25 or more,
weakly keeled along middle of back; anal divided. Above green or
olive, black marks sometimes visible at base of dorsal scales when
skin is stretched; below unmarked whitish.
Similarities
Smooth Green Snake is uniformly green, has 15–17 midbody scale
rows.
Habitat
Wooded, rocky, and canyon areas on mountains.
Habits
During daylight may frequent bushes; after dark seeks rocks and
burrows.
Range
Extr. se. Ariz. in Santa Rita, Chiricahua, and Pajaritos mts.

BAIRD'S RAT SNAKE
Elaphe obsoleta   **79:1**

Description
Length, 33–54 in. (83.8–137.2 cm). Above dark gray-brown with 4
poorly defined longitudinal dark stripes, the 2 in center darkest.
Dorsum often has orange overwash.
Habitat
Wooded uplands and rocky canyons.
Habits
Feeds on rabbits and birds.
Range
Cen. Tex., w. to Big Bend area.

GLOSSY SNAKE
Arizona elegans   **79:3**

Description
Length, 27–56 in. (68.6–142.2 cm). Above appears faded or
bleached; light or pinkish-brown, buff, cream, or yellowish-gray,
usually with median row of 55–60 small dark blotches; dark streak
from eye to mouth; sides marked with small blotches; below
unmarked pale yellowish or white. Body moderately slender, head

barely wider than neck, snout rather sharp, lower jaw deeply inset, pupil slightly vertically elliptical. Scales smooth and glossy in 27–31 rows at midbody; snout scale projects backward; anal plate single. Male has longer tail.

Similarities
Gopher Snakes have keeled dorsal scales, brighter colors; Night Snake has anal plate divided, distinctly vertically elliptical pupil.

Habitat
Sandy soil in deserts, fields, plains.

Habits
Nocturnal; not easily approached, gentle, a good burrower.

Range
Calif., San Joaquin Valley and se.; s. Nev., extr. sw. Utah; Ariz., extr. nw., sw., Mogollon Rim e.; N.Mex., cen.-s.-sw. with nw.-cen. arm to Ariz., far e.-cen.-se.; extr. ne. Colo.; extr. sw. Nebr., Kans., Okla. (except extr. w. panhandle); Tex. (except very extr. nw. panhandle). Major discontinuities in total range.

GOPHER SNAKE
Pituophis melanoleucus   77:16

Description
Length, 36–100 in. (91.4–254.0 cm). Above light brown, clay, buff, or whitish with row of large squarish to oval blotches, 48–76 on body, 13–24 on tail; sides checkered; black bar from eye to mouth, dark band across head; below white, cream, or yellowish, often spotted. Body moderately stout, head little wider than neck. Scales dorsally keeled, ventrally smooth, usually 4 prefrontals, anal plate single.

Habitat
In wide variety of habitats from seashore to 9000 ft. (2743.2 m) elevation. One of the commonest snakes in every ecological situation except swamps.

Habits
Active by day or, in hot weather, at dusk and night; hides in rodent burrows or under surface objects, can dig and is good climber, vibrates tail tip when alarmed, kills prey by constriction. Usually makes a good pet.

Voice
Very loud hiss; can be heard for 100 yds. (91.4 m).

Remarks
Also called Pine Snake and Bullsnake in various parts of range.

Range
All western states except w. Wash., extr. nw. Oreg., far n. Idaho panhandle, extr. nw. Mont., n. N.Dak., and high Rockies of Wyo. and cen. Colo.

COMMON KINGSNAKE
Lampropeltis getulus   77:15

Description
Length, 30–82 in. (76.2–208.3 cm). A highly variable species. Scales smooth, shiny. Above black or dark brown with white or yellow rings or speckling; below marked with yellow or white and black or brown in banded or checkered (or combination) arrangement. Body stout, neck constriction slight. Anal plate single, loreal scale present.

Habitat
Woods, fields, pastures, meadows, roadsides, near water, about farm buildings, to above 4800 ft. (1463.0 m) altitude.

Habits
Active by day to dusk, at night in hot desert areas; may climb into bushes or be found on or under ground surface; usually gentle and makes a good pet, but may strike when annoyed.

Range
Far sw. Oreg. (except coast); Calif. (except extr. nw. and far ne.); s. Nev., extr. sw. Utah, Ariz. (except ne.); s. N.Mex., Tex. (except w. panhandle).

CALIFORNIA MOUNTAIN KINGSNAKE
Lampropeltis zonata   77:11

Description
Length, 20–40 in. (50.8–101.6 cm). Snout usually black; body sometimes reddish-patterned with alternating white and black rings, often broken or split by various amounts of red; white rings do not broaden conspicuously on lower scale rows; underparts marked with white, black, and red, less regularly and less intense than above. Scale rows at midbody 21–23, ventrals 194–227, caudals 46–61.

Similarities
Sonora Mountain Kingsnake has white or yellow snout, white bands broader on lower scale rows; Arizona Coral Snake has red markings bordered with white or yellow rather than black.

Habitat
Coniferous forests, from sea level to 9000 ft. (2743.2 m) elevations.

Habits
Secretive, partially nocturnal.

Range
Very extr. s.-cen. Wash., far sw. Oreg. (except coast); Calif., n. of San Francisco Bay area (except coastal strip and e. of Cascade-Sierra crest); s. of San Francisco Bay area along coast and in Sierras, plus sw. mts.

SONORA MOUNTAIN KINGSNAKE
Lampropeltis pyromelana   77:12

Description
Length, 18–41 in. (45.7–104.1 cm). Color similar to that of California Mountain Kingsnake, but has light-colored snout and black bands that are narrow or disappear on sides; snout whitish, cream, to pale yellow, may show lighter spots; black rings split with coral-red; usually over 40 white to cream rings on body and tail. Scale rows dorsally at midbody 23–25, ventrals 213–235, caudals 59–79.

Habitat
Mainly in mts. in piñon-juniper belt, chaparral, or pine-fir forests.

Range
Utah, cen.-s. corridor s. of Salt Lake City, widening slightly toward far sw.; Ariz., extr. nw.-cen., broad diagonal belt from nw.-cen. through se.; far sw. N.Mex.

MILK SNAKE
Lampropeltis triangulum   77:13

Description
Length, 14–54 in. (35.6–137.2 cm). In w. range coloration very like California Mountain Kingsnake, marked with red-orange or red-brown rings or saddles bordered by black and separated by

light rings; the latter broaden on the lowermost scales. A highly variable species. Body slender, similar in structure to California Mountain Kingsnake. Scale rows at dorsal midbody 19–23, ventrals 176–231, caudals 29–59.

Habitat
In wide variety of situations; see habitat for California Mountain Kingsnake.

Habits
Largely nocturnal but some day activity; found under surface objects or beneath bark of logs and stumps.

Range
Extr. s.-cen.-se. Mont.; e. Wyo.; s. S.Dak.; Utah (except far n., far w., extr. s., and se-cen.); Colo. (except extr. nw. and cen. corridor n.-s.); Ariz., ne. corner, discontinuous; N.Mex., thence e. across U.S.

GRAY-BANDED KINGSNAKE
Lampropeltis mexicana

77:14

Description
Length, 20–36 in. (50.8–91.4 cm). Above, gray ground with widely separated crossbands of dark brown that may be split by red. Below gray with black blotches that may fuse. Head distinctly wider than neck, eyes protuberant.

Similarities
Lyre snakes have elliptical pupils.

Habitat
Desert flats and canyons.

Habits
Active mostly in morning.

Range
W. Tex.

LONG-NOSED SNAKE
Rhinocheilus lecontei

77:9

Description
Length, 20–32 in. (50.8–80 cm). Unique among U.S. harmless snakes in having most or all caudals single. Above pinkish interspaces between black saddle-shaped blotches bordered by white, cream, or yellow; snout cream, dark marks between scales; below unmarked white or cream. Body moderately slender; snout tapered, pointed. Scales smooth, anal plate single.

Similarities
The kingsnakes have divided caudals.

Habitat
Arid and semiarid regions; also fields and grasslands near rivers, valleys, foothills, plains; rare in mountains; coastal chaparral, brush, rocks.

Habits
Largely nocturnal. Good burrower, hides in rock crevices; generally docile but will bite, vibrates tail if annoyed.

Range
Sw.-cen. Idaho, as island; Calif., extr. ne., Central Valley, s.; Nev. (except far n. and deep n.-cen.); far w. Utah (except extr. nw. corner); Ariz., far w., s.; se. N.Mex.; far se. Colo.; sw.-s. Kans.; w. Okla.; w. Tex.

PLAIN-BELLIED WATER SNAKE

Nerodia erythrogaster   77:5

Description
Length, 18–67 in. (45.7–170.2 cm). Scales strongly keeled. Above
blotched with black, gray, or reddish-brown, usually unpatterned;
top of head dark; below reddish, salmon, or pink to yellowish,
unmarked. Young have brownish blotches. Body stout, head
slightly wider than neck, eyes rather large. Dorsal scale rows
usually 23 to 27, anal plate divided, 1–3 preoculars.

Similarities
Common Water Snake has underparts boldly marked.

Habitat
Near water.

Habits
Shy, wary, hard to capture; active at dusk and at night; found
under surface objects near water; occasionally basks; pugnacious,
tries to bite when caught, produces foul odor from anal glands.

Range
Extr. se. N.Mex., Tex. (except extr. nw. panhandle and far w.).

NORTHERN WATER SNAKE

Nerodia sipedon   77:7

Description
Length, 18–54 in. (45.7–137.2 cm). Scales strongly keeled. Below
white or yellowish-gray patterned with red dots, black blotches,
half-moons, or narrow crescents; above light gray, brown, tan, to
dark gray, commonly patterned with wide crossbands on anterior ⅓
of body, squarish blotches farther back, with pattern becoming
obscure in old age. Body moderately heavy, head somewhat broader
than neck. Has 19–23 dorsal rows at midbody, 2–3 postoculars,
internasal scales much narrowed in front, anal plate divided.
Young generally lighter and more contrastingly colored.

Similarities
Plain-bellied Water Snake has unmarked belly, crossbands.

Habitat
In or near rivers, streams, marshes, ponds, lakes.

Habits
Frequently congregates; gives off foul odor from anal scent glands
when handled.

Range
An eastern species that occurs as far w. as e. Colo.

RED-BELLIED SNAKE

Storeria occipitomaculata   79:5

Description
Length, 8–16 in. (20.3–40.6 cm). Above 4 narrow dark stripes on
brown background. Occasionally a broad, light-colored median
stripe. Head black above with large light blotches at rear. Below
usually bright red, sometimes orange or yellow. Dorsal scales
keeled, in 15 rows, anal divided.

Habitat
Under rocks, boards, logs, especially near human dwellings.

Habits
Feeds on soft insects and bears live young.

Range
E. U.S.; sw. S.Dak. and ne. Wyo.

GARTER SNAKES
Genus *Thamnophis*

The garter snakes are characterized by three narrow, light stripes on a darker dorsal ground color. Their heads, typically, are distinct from their necks, and the scales are keeled in less than twenty-seven rows, usually 17–23, on the upper midbody; the anal scale is not divided. Most species are diurnal and are found near water. All bear living young.

NARROW-HEADED GARTER SNAKE
Thamnophis rufipunctatus   78:7

Description
Length, 20–34 in. (50.8–86.4 cm). Above dark olive, olive-gray, or brown with many blackish-brown spots, becoming smaller and obscure on tail; faint side stripe, top of head brownish; below brownish-gray, whitish on throat, underside of tail light grayish-brown; ventral black markings often wedge-shaped. Body moderately slender; head long, narrow, eyes protuberant. Scales dorsally in 21 rows at midbody, usually 7 upper and 10 lower labials, and 2–3 preoculars.
Habitat
Near water, in streams, on rocks nearby; prefers rocky permanent streams.
Habits
When startled, commonly dives into water and swims to bottom.
Range
A broad island from ne.-cen. Ariz. into far sw. N.Mex. above Mexican border.

WESTERN TERRESTRIAL GARTER SNAKE
Thamnophis elegans   78:12

Description
Length, 18–42 in. (45.7–106.7 cm). Above, distinct middorsal light stripe; lateral stripe on scale rows 2 and 3 on each side; ground color is dark with scattered white flecks, or, less often, pale checkered with dark spots. Internasal scales broader than long and not pointed anteriorly; usually 8 upper labials, the 6th and 7th enlarged, often higher than wide; usually 10 lower labials; chin shields about equal in length.
Similarities
Northwestern Garter Snake has 17-17-15 scale rows, 7 upper labials, 8 or 9 lower labials, bright red, orange, or yellow dorsal stripe; in areas of overlap, note that Western Terrestrial Garter Snake has 19-19-17 or more scale rows, 8 upper labials, 10 lower labials, and a dull yellow, brown, or gray dorsal stripe; Common Garter Snake has large eyes, 7 upper labials, a plain blue-gray belly, and green-yellow dorsal stripe; Western Aquatic has internasals narrower than long and pointed anteriorly, 6th and 7th upper labials not enlarged.
Habitat
Grasslands, chaparral, forests, damp habitats near water.
Habits
Often seeks refuge on land rather than water.
Range
Nw. N.Mex., cen. and w. Colo., Wyo., Mont., w. to coastal Calif., Oreg., Wash.

WESTERN AQUATIC GARTER SNAKE
Thamnophis couchi   78:14

Description
Length, 18–57 in. (45.7–144.8 cm). Interanal scales are narrower
than long and pointed anteriorly; 8 upper labials, 6th or 7th not
enlarged; chin shields about equal in length. Middorsal stripe
weak, faint, or absent; lateral stripe on scale rows 2 and 3 on each
side when present; background variable, but usually light brown or
green with dark blotches.
Similarities
Common Garter Snake has 7 upper labials, larger eyes, well-
defined dorsal stripe, red blotches on sides; Western Terrestrial
Garter Snake has internasals wider than long, not pointed
anteriorly, 6th and 7th upper labials enlarged, usually higher than
wide.
Habitat
Wide variety of aquatic habitats—streams, ponds, lakes, coastal
marshes, rivers.
Habits
Takes refuge in water when alarmed.
Range
Sw. Oreg., Calif. (except desert areas and ne.), extr. w. Nev.

NORTHWESTERN GARTER SNAKE
Thamnophis ordinoides   78:16

Description
Length, 14–26 in. (35.6–66.0 cm). Above black, brown, greenish,
or bluish, scale edges black-spotted; 3 longitudinal light stripes,
medial distinct—yellow, orange, red, or blue; laterals distinct to
obscure; underparts olive, yellowish, gray, or slate, often with
bright red spots and occasionally marked with black (mostly in
north of range). Head narrow. Scales dorsally 17–17–15, upper
labials 7, lower labials 8–9, preoculars occasionally divided.
Similarities
Western Terrestrial Garter Snake has 8 upper labials, more than
17 scales at midbody; in area of overlap, note Western Aquatic
Garter Snake has 8 upper labials and dull dorsal stripe; Common
Garter Snake has 19 scale rows at midbody, no red on belly.
Habitat
Humid, dense forest meadows and clearings.
Habits
Docile, retiring.
Range
S. Vancouver Is., sw. B.C., w. Wash., w. Oreg.

BLACK-NECKED GARTER SNAKE
Thamnophis cyrtopsis   78:9

Description
Length, 16–37 in. (40.6–94.0 cm). Back of head deep gray,
yellowish on nose, cream markings on side of face, labials boldly
marked with black; whitish crescent on each side of head. Above
olive-gray to olive-brown and lacking red; 3 stripes, middle stripe
orange anteriorly, grading to yellow posteriorly; lateral stripes
cream on neck to ash-gray or whitish posteriorly, on scale rows 2
and 3; sides light olive-gray or gray. Between stripes are black
blotches flecked with light. Below cream on throat and neck, bluish
posteriorly. Scales dorsally in 19 midbody rows, 1 preocular usual.
Male tail longer.

Similarities

Checkered Garter Snake has 21 midbody dorsal scale rows, checkered pattern extends onto tail, lateral stripe confined to scale row 3 anteriorly; Mexican Garter Snake has lateral stripe on scale rows 3 and 4.

Habitat

In mountains or hilly areas, usually near water.

Habits

May seek refuge in water; when caught, may flatten and broaden head and body and release excrement and noxious scent.

Range

Far se. corner of Utah., e. Ariz., N.Mex. (except very extr. n.-cen. and far e.), e. to cen. Tex.

MEXICAN GARTER SNAKE
Thamnophis eques

78:8

Description

Length, 18–40 in. (45.7–101.6 cm). Above olive or brown; 3 longitudinal stripes, middle stripe yellow or yellowish-white; laterals on 3rd and 4th rows anteriorly; dorsal and lateral stripes separated by 2 rows of black spots; 2 black neck blotches; light crescent each side of head, white bar in front of and behind eye. Body moderately stout, tail relatively long, to ¼ body length; eyes large. Scales dorsally in 19–21 midbody rows, 8–9 upper labials.

Similarities

In Black-necked Garter Snake and Checkered Garter Snake lateral stripe anteriorly does not involve 4th row scale.

Habitat

Arid or semiarid regions in or near water.

Habits

Feeds on frogs and bears living young.

Range

Cen. to se. Ariz., extr. sw. N.Mex.

CHECKERED GARTER SNAKE
Thamnophis marcianus

78:11

Description

Length, 18–42 in. (45.7–106.7 cm). Above brownish-yellow, brown, to olive; 3 longitudinal stripes of pale yellow, borders usually irregular; lateral stripes may be obscure, but are confined to 3rd scale row anteriorly. Between median and lateral stripe, on each side, are usually 2 rows of large, squarish, black blotches in checkered pattern, sometimes brown-streaked. Sides brownish-gray, spotted below lateral stripes; head has vertical creamy marks on sides, upper labials black-barred, light crescent on each side of head; below whitish, often clouded. Scales dorsally in 21 midbody rows.

Similarities

Black-necked Garter Snake has 19 midbody dorsal scale rows, lateral stripe anteriorly on 2nd and 3rd scale rows.

Habitat

Generally in lowland ponds, streams, rivers in arid and semiarid regions.

Habits

Primarily nocturnal.

Range

Extr. se. Calif., s. Ariz.; N.Mex.; far e.-se. Colo., e. Kans.; w. Okla., w. Tex.

PLAINS GARTER SNAKE
Thamnophis radix
78:10

Description
Length, 20–42 in. (50.8–106.7 cm). Distinguished by 3 distinct yellow stripes with 2 rows of squarish black spots between them, lateral stripes on 3rd and 4th scale rows anteriorly and on 2nd and 3rd scale rows posteriorly. Above dark olive to dark brown or black, middle stripe orangish and laterals bright to pale yellow or cream with sometimes a faint purplish-gray cast; upper lip yellow, bordered with black; below greenish with black spots on edges. Body stout; head broad, distinct; tail ¼ total length. Scales dorsally usually in 21 midbody rows, upper labials usually 7.

Similarities
Mexican Garter Snake has 8 or 9 upper labials; Western Ribbon Snake has long tail, ¼–⅓ total length; all other garter snakes in the West have lateral stripes not involving row 4.

Habitat
In variety of situations including wet prairies, roadside ditches, near water, prairie-forest transition areas.

Habits
Aggressive, may bite when captured; releases excrement and scent.

Range
Great Plains region from far s. Canada, s. to Tex. panhandle and far ne. N.Mex. on e. side of Rockies.

COMMON GARTER SNAKE
Thamnophis sirtalis
78:15

Description
Length, 18–51 in. (45.7–129.5 cm). Above green to orange-brown or black with red blotches between lateral stripes that are well defined and on scale rows 2 and 3. Green-yellow dorsal stripe. Belly blue-gray posteriorly, becoming pale on throat. Eyes relatively large. Scales in 19 dorsal midbody rows, 17 on posterior ⅓ of body, upper labials usually 7, lower labials usually 10, ventrals 146–177, caudals 70–94. Male considerably smaller and has knobbed keels on dorsal scales above vent.

Similarities
Distinguishable from other garter snakes by red markings on sides, usually 7 upper labials, and relatively large eyes.

Habitat
Widely varied, including fields, meadow, marshes, roadsides, gardens; often near water.

Habits
Diurnally active; quite pugnacious, tries to bite when captured, may produce strong anal scent when handled.

Range
S. B.C. and Vancouver Is., s. Alta., sw. Sask., s. Man.; all U.S. except se. Calif., Nev., far w. and extr. s. Utah, Ariz., far w. N.Mex., and cen.-s. Tex.

WESTERN RIBBON SNAKE
Thamnophis proximus
78:13

Description
Length, 18–51 in. (45.7–129.5 cm). Body very slender, tail long, more than ¼ total length. Top of head dark, upper lip yellow, yellow spot in front of large eyes. Above velvety black or brown with yellowish ribbonlike stripe down back; brown band below

light side stripes that cover 3rd and 4th scale rows; below unmarked yellowish or greenish-white; 7–8 upper labials.
Habitat
In or near water, marshes, meadows, swamps, damp low places with ample vegetation, to water's edge.
Habits
Quick, active, seeks escape by swimming, and may try to bite on capture; releases strong-smelling fluid from anal glands.
Range
Extends w. only to extr. se. Colo., extr. e. N.Mex., and s. into Mexico.

GROUND SNAKES
Genus *Sonora*

WESTERN GROUND SNAKE
Sonora semiannulata   **78:2**
Description
Length, 8–19 in. (20.3–48.3 cm). Small with head only slightly larger than neck. Three color phases: (1) dark crossbands; (2) broad pink or orange longitudinal middorsal stripe; (3) plain yellow-brown or tan. Smooth scales, 53 or more caudals in males; 45 or more in females. Anal divided.
Similarities
Ground Snake has 52 or more caudals in male, 44 or more caudals in females.
Habitat
Arid and simiarid regions of loose, sandy soil; frequents slopes or flats with or without rocks.
Habits
Secretive.
Range
E. Calif., rather narrowly along boundaries of Nev. and Ariz.; Nev. (except nw. boundary, far n., ne. quadrant); extr. sw. Utah, w. and s.; nw., w., sw., and se. Ariz.; far sw. N.Mex.; extr. w. Tex.

GROUND SNAKE
Sonora episcopa   **78:1**
Description
Length, 9–16 in. (22.9–40.6 cm). Similar to Western Ground Snake, but with 52 or more caudals in males and 44 or more in females. Color variable; may be gray, greenish, dark brown, or red above; cream, white, or greenish below. Sometimes a black collar on neck or black oval crossbands on back. Head broad, flattened above, barely distinct from body. Scales smooth.
Habitat
Prairies, dry places, buried in soil, under stones or boards; roadsides, hillsides.
Habits
Secretive and nocturnal.
Range
Far se. Colo., s. Kans., extr. w. Okla., extr. e. N.Mex., Tex. (except extr. w.).

SHOVEL-NOSED AND SAND SNAKES
Genera *Chionactis* and *Chiloneniscus*

WESTERN SHOVEL-NOSED SNAKE
Chionactis occipitalis   **79:9**

Description
Length, 10–16 in. (25.4–40.6 cm). Head has countersunk lower
jaw, flattish snout, nasal scale undivided, nasal valves well
developed; little or no neck constriction. Above yellow or white
patterned with black or brown crossbands, usually 21 or more on
body (minus tail); characteristic dark crescent on head; dorsal
interspaces between bands distinctly marked with orange or reddish
saddles or spotted with black or brown, or interspaces suffused
with pink or red; below often similarly banded and sometimes
spotted. Scales smooth, dorsally usually in 15 rows, anal plate
divided.

Similarities
Banded Sand Snake has 13 scale rows at midbody, shorter tail;
rostral separates internasals; more rounded snout, dark spot at base
of each dorsal scale; in Arizona Coral Snake, red bands form
complete rings.

Habitat
Barren sandy places, brushy desert, rocky and grassy situations.

Habits
Completely nocturnal, prefers dark-of-the-moon periods; a rapid
burrower to depths of 2 ft. (0.6 m); may bask just beneath surface;
docile or pugnacious, may strike, but bite too small to do harm.

Range
Se. Calif., far s. Nev., along Calif. border; sw. Ariz., s. of Lake
Mead to Tucson, absent from s.-cen. Mexican border area.

Note: The **SONORA SHOVEL-NOSED SNAKE,** *Chionactis palarostris*
(79:11), has 21 or fewer black bands on its body and broad red
saddles. It occurs in extreme southern Arizona, in the Organ Pipe
Cactus National Monument.

BANDED SAND SNAKE
Chilomeniscus cinctus   **78:4**

Description
Length, 7–10 in. (17.8–25.4 cm). Similar to Western Shovel-nosed
Snake, but shorter and stouter; head more flattened, eyes and
nostrils directed more upward, snout shovel-shaped, lower jaw
countersunk. Scales smooth, glossy, in 13 longitudinal rows; nasal
valves present, rostral separates internasals. Above white,
yellowish, or reddish-orange with 19–49 crossbands on head and
body and tail; tail usually ringed with black; below whitish.

Similarities
Western Shovel-nosed Snake has 15 scale rows at midbody,
internasals not separated by rostral, and longer tail.

Habitat
In loose, windblown dune sand; less common in coarser sand of
arroyos, rarely among rocks.

Habits
Highly adapted for desert sand-burrowing with its streamlined
head; nocturnally active; travels just beneath surface.

Range
Extr. s. Ariz., except extr. se. corner.

LINED AND HOOK-NOSED SNAKES
Genera *Tropidoclonion* and *Ficimia*

LINED SNAKE
Tropidoclonion lineatum

79:7

Description
Length, 9–21 in. (22.9–53.3 cm). Similar structurally to both garter and water snakes. Head small, pointed, not distinct from neck; tail short. Above brownish with 3 yellowish stripes; 2 rows of dark dots flank dorsal stripe; light side stripe on 2nd and lower part of 3rd scale rows; underparts uniquely white, yellow, or greenish with 2 rows of dark triangular dots. Scales dorsally keeled, upper labials 5–6, anal plate single.
Habitat
Near water; old fields, rocky places, under surface litter; urban vacant lots; primarily an inhabitant of prairies and thin woods.
Habits
Secretive; seldom tries to bite, but may void excrement and anal scent when captured.
Range
In West confined to Colo., e.-cen. island, a small area w. of Denver, and 6 mi. (9.7 km) ne. of Higbee in Otero Co.

WESTERN HOOK-NOSED SNAKE
Ficimia cana

79:2

Description
Length, 7–14 in. (17.8–35.6 cm). Body stout; tail short, thick; little or no neck constriction, snout rather flattened dorsally. Rostral scale broad, flattened, upturned. Above grayish-brown or yellowish with gray tinge; 38–42 dark transverse bars, dark head band; below whitish with salmon or light reddish-orange. Scales in 17 midbody dorsal rows, smooth; no loreal; internasal scales small, anal plate divided.
Similarities
Hognosed snakes have spadelike snout, rostral with median ridge, dorsal scales keeled.
Habitat
Semiarid regions of sand, loose gravelly soil, sparse grass.
Habits
Secretive; seldom observed or captured.
Range
Far se. Ariz., far s. N.Mex., w. Tex.

Note: The **DESERT HOOK-NOSED SNAKE,** *Ficimia quadrangularis* **(79:4)**, has red body with black bands. It occurs in south-central Arizona.

BLACK-HEADED SNAKES
Genus *Tantilla*

These are small, rather slender snakes with a head only slightly broader than the neck, uniform body color, smooth scales, and a small, dark-colored head. Western representatives range from sixteen to eighteen inches (20.3–40.6 cm) in length and have fifteen rows of smooth scales. Secretive burrowers, they can be distinguished from the ring-necked snakes (*Diadophis*) by the absence of the loreal scale and a lack of dark ventral spotting.

WESTERN BLACK-HEADED SNAKE
Tantilla planiceps   79:6

Description
Length, 7–15 in. (17.8–38.1 cm). Head has black cap that extends
0–3 scale rows behind parietals at midline and is not pointed
behind; usually bordered behind by narrow white collar, sometimes
followed by a row of black dots. Above, plain brown to olive-gray,
often with a faint, narrow middorsal stripe. Below, orange or red
stripe down middle of belly. Scales, in 15 rows at midbody, smooth;
anal divided, loreal absent.

Similarities
Plains Black-headed Snake is larger, has no light collar; cap
extends 3–5 scales behind parietals.

Habitat
Under rocks and other cover.

Range
Coastal Calif., e. through deserts to s. Nev., se. Utah, w.-cen.
Colo., cen. and se. Ariz., s. N.Mex., w. Tex.

Note: The **HUACHUCA BLACK-HEADED SNAKE,** *Tantilla wilcoxi*
(79:10), has a broad white collar crossing the tips of the parietals.
It occurs in the Huachuca and Patagonia mountains of
southeastern Arizona.

PLAINS BLACK-HEADED SNAKE
Tantilla nigriceps   79:8

Description
Length, 7–18 in. (17.8–45.7 cm). The only black-headed snake
with a convex or V-shaped black cap. Head flattened, barely
distinct from neck. Above light brown; black head cap extending 3–
5 scale lengths behind parietals and not bordered by white band;
below white.

Similarities
Western Black-headed Snake is smaller, has a light collar; cap
extends 0–3 scales behind parietals.

Habitat
Under flat rocks on dry hillsides.

Habits
Nocturnal.

Range
E. Ariz., se. N.Mex., extr. e. Colo., sw. Nebr., w. Kans., w. Okla.,
w. Tex.

LYRE SNAKES
Genus *Trimorphodon*

CALIFORNIA LYRE SNAKE
Trimorphodon vandenburghi   79:12

Description
Length, 24–43 in. (61.0–109.2 cm). Body slim, head broad, neck
slender, eyes large and protuberant, pupils vertically elliptical. V-
shaped mark on head with dark crossband, like a lyre. Above light
brown, buff, or gray, dorsally blotched with row of dark brown
spots, each with a narrow, light crossbar usually present; 28–43
(average 35) blotches on body, excluding tail. Below light cream to
yellow, usually with spots and blotches. Scales smooth, in 21–24

rows; 2–3 loreals, 3–4 postoculars, no suboculars, 2–3 temporals, usually 9 upper labials, anal plate usually single.

Similarities

Sonora Lyre Snake has anal usually divided, 34 or fewer blotches on body.

Habitat

A rock dweller of slopes and canyons, found by day in crevices of massive rocks, under exfoliating rock shelves.

Habits

Nocturnal, but active in early morning in rock crevices; venomous but not dangerous to man.

Range

Far sw. Calif. w. of Imperial Valley.

SONORA LYRE SNAKE
Trimorphodon lambda   78:5

Description

Length, 24–41 in. (61.0–104.1 cm). Very similar to California Lyre Snake, but anal plate divided, 23–34 (average 28) blotches on body, tail proportionately longer.

Habitat, Habits

Same as for California Lyre Snake.

Range

Extr. se. Calif., extr. s. Nev., very extr. sw. Utah, sw. two-fifths Ariz.

Note: The **TEXAS LYRE SNAKE,** *Trimorphodon vilkinsoni* **(79:14)**, has blotches widely spaced, fewer than twenty-three on its body; and a lyre mark that is faint. It occurs in southern New Mexico and extreme western Texas.

VINE AND NIGHT SNAKES
Genera *Oxybelis* and *Hypsiglena*

VINE SNAKE
Oxybelis aeneus   79:13

Description

Length, 40–60 in. (101.6–152.4 cm). Body extremely slender; tail long, slender, to over ½ body length. Head and snout *very long* and tapered; eyes relatively small. Above ash-gray, light yellowish-brown on anterior ⅕ of body, sides darker gray, dark brown stripe on side of head, sides of jaws cream to pale yellow; below very similar. Scales smooth, in 17 dorsal midbody rows; no loreal, anal plate divided.

Habitat

In mountains with brushy slopes and oak canyons.

Habits

Active by day; moderately fast (equal to a brisk walk) with head carried well above ground; when encountered, often bluffs with open jaws.

Range

Extr. s.-se. Ariz.

NIGHT SNAKE
Hypsiglena torquata   **78:6**

Description
Length, 12–26 in. (30.5–66.0 cm). Pupils vertically elliptical.
Above gray, yellowish, or beige with dark spots, largest along
vertebrae; side spots alternate with central row; large neck blotches
(sometimes absent); below yellowish or white, unmarked. Body
moderately slender; head somewhat flattened, a bit broader than
neck; ungrooved, enlarged teeth in rear of jaw. Scales smooth
above, loreal single or double, 2 postoculars, 1 anterior temporal,
7–8 upper labials, anal plate divided.
Similarities
Juvenile Glossy Snake has single anal plate, pupil less vertically
elliptical (nearly round).
Habitat
Extremely varied in arid and semiarid regions from below sea level
to 7000 ft. (2133.6 m) elevations.
Habits
Active at dusk and night; hides under surface objects by day; will
bite but venom, toxic to its prey, is harmless to man.
Range
Se. Wash., far e. Oreg., sw. Idaho., Calif. (except coast n. of San
Francisco Bay area, n., ne., Sierras, or San Joaquin Valley); Nev.,
Utah (except n.–s. middle corridor and extr. ne.); extr. w. Colo.,
Ariz., N.Mex. (except far n.-cen. to ne.); extr. s.-cen. Kans., w.
Okla. (except very extr. w. panhandle), w. Tex.

CORAL SNAKES
Family Elapidae

Members of this venomous family have rigid, hollow fangs toward
the front of the maxillary bones, one or more solid teeth, no loreal
scale, and they lack the pits behind the nostrils of the pit vipers.

ARIZONA CORAL SNAKE
Micruroides euryxanthus   **78:3**

Description
Length, 15–21 in. (38.1–53.3 cm). Body and tail brilliantly ringed
with yellow or whitish, red, and black; broad red bands bordered
on sides with whitish; head black with yellowish band behind;
below with lighter bands. Body moderately slender, flattish; head
rather flattened, neck constriction slight; snout broad, blunt; eyes
small. Scales smooth and glossy, in 15 dorsal midbody rows; no
loreal; anal divided.
Similarities
In kingsnakes red bands are bordered by black; Banded Sand
Snake and Western Shovel-nosed Snake have pale snout.
Habitat
To 5000 ft. (1524.0 m) elevations.
Habits
Secretive, spends much time underground; flattens body to hide in
crevices.
Remarks
Venom very poisonous to man.
Range
S. Ariz. (except far sw.), far sw.-cen. N.Mex., extr. w. Tex.

PIT VIPERS
Subfamily Crotalidae

The pit vipers are poisonous, heavy-bodied snakes with a distinct pit on either side of the head between the nostrils and the eyes. In the front of the upper jaw are a pair of movable hollow fangs. The head is distinct and triangular, the eye pupils vertical, and all western forms but one have a horny, jointed rattle on the end of the tail. All species bear living young. Bites from the pit vipers may prove fatal to man.

COPPERHEAD
Agkistrodon contortrix   80:9

Description
Length, 20–36 in. (50.8–91.4 cm). Only western snake that has a facial pit but no rattles. A poisonous snake with heavy body and head distinct from neck; deep facial pit between eye and nostril. Above, hourglass-shaped, dark chestnut-colored blotches, wider on side of body, narrower at midline, separated by cream interspaces. Below rich chestnut to nearly black. Scales weakly keeled; a single row of scales under tail (at least anteriorly).
Habitat
Canyons and riparian woodland.
Habits
Quiet, almost lethargic unless aroused.
Range
W. and cen. Tex. to Atlantic Coast.

RATTLESNAKES
Genera *Sistrurus* and *Crotalus*

Members of these genera all have a series of interlocking, horny links on the tail that form a rattle. When disturbed, a rattlesnake vibrates the tail to create the characteristic buzzing noise, or rattle, for which these snakes are famed. When undisturbed, a "rattler" moves in caterpillar fashion, leaving a wide track in the sand.

MASSASAUGA
Sistrurus catenatus   80:11

Description
Length, 18–40 in. (45.7–101.6 cm). The only rattlesnake with head that has elongate dark brown markings extending onto neck and 9 large symmetrically placed plates, instead of small scales, on top. Above light gray to grayish-brown, stippled; 21–50 dark blotches in middorsal row, exclusive of tail, 3 secondary rows of blotches on each side; tail with dark bands; below white or cream (may or may not be marked) to gray, or pale yellow spotted with dark. Tail short, stout, rattle well developed. Male tail longer, with thicker base.
Habitat
Swamps, edges of streams, ponds; marshes, meadows, fence rows, fields in summer.
Habits
Not usually aggressive.
Remarks
Venom deadly.
Range
Extr. se. Ariz., se. two-fifths N.Mex.; far se. Colo., Kans. (except far nw.), w. four-fifths Okla., w. four-fifths Tex. (except far sw.).

WESTERN DIAMONDBACK RATTLESNAKE

Crotalus atrox   80:10

Description
Length, 30–89 in. (76.2–226.1 cm). Body heavy, to 15 lb. (6.8 kg).
Above usually gray, grayish-brown, cream, or buff; row of 24–45
brown diamond- or hexagon-shaped dorsal blotches; markings often
somewhat indefinite and peppered in appearance; tail, with broad
black and white rings about equal in width, contrasts with body. A
light diagonal stripe from eye intersects lip well in front of corner
of mouth.
Similarities
Mojave Rattlesnake has black tail rings narrower than white ones,
white stripe from eye passes behind corner of mouth; Red
Diamond Rattlesnake is reddish, has less definite pattern, 1st lower
labial divided transversely.
Habitat
Arid prairies, desert flats, low foothills; prefers brushy areas.
Habits
Active by day, but most active at dusk and night; bold, pugnacious,
inclined to defend self when encountered by an enemy.
Remarks
One of the most dangerous of the western rattlesnakes.
Range
S. Calif., very extr. s. tip of Nev., s. Ariz., s. N.Mex., all Tex.
(except panhandle).

RED DIAMOND RATTLESNAKE

Crotalus ruber   80:3

Description
Length, 30–65 in. (73.5–160.0 cm). Similar to Western
Diamondback, but 1st pair of lower labials usually divided
transversely. Color tones generally different, though basic colors
similar; above reddish-cinnamon, brick-red, sometimes brownish-
yellow (coastal forms more reddish-brown, desert forms lighter
red); row of diamond-shaped dorsal blotches, each usually outlined
with light color; below yellow, pink, or salmon; tail ringed with
black and ash-white. Newborn dark gray.
Habitat
Scarce at high elevations; frequents heavy brush- and rock-studded
slopes; occasionally in fields and grassland.
Habits
Disposition mild; active by day or dusk, nocturnal in hot weather.
Remarks
Quantity of its venom makes it dangerous.
Range
Far sw. Calif.

SIDEWINDER

Crotalus cerastes   80:6

Description
Length, 17–31 in. (43.2–78.7 cm). Body stout; head large, with
broad, hornlike projections over eyes. Above variable—cream, tan,
pink, light brown, gray; row of faint vertebral blotches, smaller
dark spots on sides, commonly dark stripe on each side of face; tail
conspicuously light with dark bands and whitish below. Dorsal
scales strongly keeled. Male is smaller.

Habitat

Deserts, especially in sandy flats and washes with low, sparse shrubs; occasionally on hardpan or areas of small rocks on hard soil, from below sea level to 4500 ft. (1371.6 m) elevations.

Habits

Primarily nocturnal, but active at dusk and occasionally in daytime; travels by moving body forward sidewise in looping S-series.

Remarks

Venom is deadly but small size makes it less dangerous than other western rattlesnakes.

Range

S. Calif. e. of mts.; s. Nev.; very extr. sw. Utah; Ariz., very extr. w., sw.-cen.

ROCK RATTLESNAKE

Crotalus lepidus
80:7

Description

Length, 15–30 in. (38.1–76.2 cm). Above distinctive gray, greenish-gray, or bluish-gray, darkly mottled and patterned with 14–24 regularly spaced, narrow, dark brown or black crossbands; below white, cream, pale gray, or pinkish, variously spotted but sometimes unmarked. Body rather slender, head small, rattle relatively large; distinguished by vertically divided upper preocular, prenasal curved under postnasal.

Habitat

In mountains (often to high elevations); frequents limestone areas, igneous rocks, etc.

Range

Far se. corner of Ariz., far sw.-cen. N.Mex., far w. Tex.

SPECKLED RATTLESNAKE

Crotalus mitchelli
80:4

Description

Length, 24–52 in. (61.0–132.1 cm). Prenasals usually separated from rostral by small scales or supraoculars pitted, sutured, or outer edges broken. Above variable—cream, light gray, straw, tan, pink, salmon, buff, or brown; 23–43 dorsal blotches variously shaped and colored, composed of groupings of dots, or crossbands formed from blotches; secondary blotches usual on sides. Total dorsal effect is salt-and-pepper speckling. Tail has 3–9 dark rings, below cream, buff, tan, or pink, usually speckled or blotched.

Similarities

Western Diamondback Rattlesnake has "coontail"; Mojave Rattlesnake has well defined dorsal pattern; Tiger Rattlesnake has small head, large rattle, no small scales between rostral and prenasals.

Habitat

Deserts, among rocks in buttes and mountains; occasional on level sandy plains and alluvial fans.

Habits

Commonly nocturnal; alert, nervous, often rattles when disturbed.

Range

Se. one-third Calif., sw.-s. Nev., w. (except nw.-cen. Ariz.).

BLACK-TAILED RATTLESNAKE
Crotalus molossus   80:12

Description
Length, 28–51 in. (71.1–129.5 cm). Readily distinguishable by
solid black on tail and sometimes on snout. Enlarged scales on
upper surface of snout. Above bright yellow, grayish, brownish-olive
(greenish in Texas), with dark crossbands edged with whitish and
broken by whitish center patch; below white or pale yellow, often
clouded or mottled with gray or brown. Tail solid black (sometimes
with paler areas on sides suggesting crossbands).

Habitat
Highly varied, from desert rocky areas, brushy foothills, to rock
slides at 9000 ft. (2743.2 m) altitude.

Habits
Active day or night.

Range
Ariz. (except far w., nw., far n., ne.-e.); sw. one-third N. Mex.,
w.-cen. Tex.

TWIN-SPOTTED RATTLESNAKE
Crotalus pricei   80:8

Description
Length, 12–26 in. (30.5–66.0 cm). Color pattern distinctive; above
gray, slaty, or bluish-gray with fine brown stippling; dorsal pattern
of many rounded spots and 2 rows of paired larger spots (50–55)
along spine, with 3 rows of smaller alternating lateral spots on
each side; tail with 5–10 brown bands; below with brown and gray
stippling, throat often salmon.

Habitat
On rock slides in pine-oak zone.

Habits
Generally not aggressive, rattle not loud but rather high-pitched,
like a cicada.

Range
Far se. Ariz. from Graham Mts. s.; also Chiricahua, Huachuca,
Dos Cabezas, Santa Rita Mts.

TIGER RATTLESNAKE
Crotalus tigris   Fig. 66

Description
Length, 18–36 in. (45.7–91.4 cm). Distinctively crossbanded (more
than most western species), 40–50 dark gray or brown blotches on
body, 5–10 on tail; ground color above gray, blue-gray, pinkish-
gray, lavender, or buff; below straw, yellow, pink, or whitish, often
mottled and dotted with brown. Head unusually small, body about
25 times as long as head, neck slender, rattle large; 2 internasals,
rostral contacting prenasals.

Fig. 66   Tiger Rattlesnake

Similarities
Speckled Rattlesnake has a relatively larger head and smaller rattle; small scales between rostral and prenasals, or supraoculars pitted; Western Rattlesnake has dark blotches, rather than crossbands, anteriorly.
Habitat
Canyons and rocky foothills of desert mountains; occurrence spotty.
Habits
Active day or night; mild temperament.
Range
Cen. and s. Ariz.

MOJAVE RATTLESNAKE
Crotalus scutulatus

80:1

Description
Length, 24–51 in. (31.0–129.5 cm). Usually large scales between anterior part of supraoculars. Above green, greenish-gray, olive-green, yellowish, or greenish-brown; light stripe from eye to behind corner of mouth; dorsal pattern of 27–44 brown diamond-shaped marks bordered with white scales, characteristically unmarked and uncut by edges of blotches; tail with 2–8 black rings that are narrower than separating light rings.
Similarities
Western Diamondback has light and black tail rings of equal width, stripe from eye intersecting lip well in front of corner of mouth.
Habitat
Deserts and desert grassland to 5000 ft. (1524.0 m) elevations.
Habits
Most active at night.
Remarks
Venom extremely dangerous!
Range
Calif. Mojave Desert, far se. Nev., sw. two-fifths Ariz., extr. sw. N.Mex., far sw. Tex.

WESTERN RATTLESNAKE
Crotalus viridis

80:2

Description
Length, 15–62 in. (38.1–157.5 cm). Distinguished by having more than 2 internasals in contact with rostral. Color is considerably varied; sides of head dark with 2 white diagonal lines; back has dark oval blotches edged with white and flanked by 2 rows of alternating smaller dark blotches; below yellow, mottled. Tail has light and dark rings usually not strongly contrasting.
Habitat
Varies from brushy lowlands to timberlines, from grasslands, to prairies, to forests. To 11,000 ft. (3352.8 m) elevations.
Habits
Mostly diurnal; hides in rock crevices or rodent burrows; in north of range may den up for winter.
Range
Extreme sw. Canada to c. Baja Calif. and n. Coahuila, Mexico; Pacific coast to 100th meridian.

RIDGE-NOSED RATTLESNAKE
Crotalus willardi   80:5

Description
Length, 15–24 in. (38.1–61.0 cm). Above buff or light gray with
brown blotches, vertical white stripe on snout; body brown-spotted;
tail ringed forward and striped longitudinally in center; below buff,
heavily spotted. Rattles darker and more rounded than in other
species. Scales dorsally in 25–27 midbody rows; internasals sharply
uptilted to form ridge bordering each side of snout.

Habitat
In mountains to above 9000 ft. (2743.2 m) elevations in mixed
forests.

Habits
Prone to turn and bite when captured; coils and strikes.

Range
Very extr. se. Ariz. and sw. N.Mex.

Amphibians

Consulting Editor
Robert L. Bezy
Associate Curator of Herpetology
Los Angeles County Museum of Natural History

Illustrations
Plates 81, 84–85 Sue Thompson
Plates 82–83 Jennifer Emry-Perrott
Text Illustrations John Cameron Yrizarry and
Jennifer Emry-Perrott

Amphibians

Class Amphibia

This class of vertebrates originated from the fishes in the Devonian Period about 340 million years ago. For some 140 million years it was the dominant class of land vertebrates. Some species attained a length of over six feet, but with the subsequent rise of the reptiles, the amphibians declined in size and abundance. Today, about 2500 species live widely spread throughout all land areas outside the polar regions.

Named from the fact that they live both on land and in water, the amphibians have limbs instead of fins and have no claws on their toes. They are cold-blooded vertebrates with moist skin unprotected by scales, feathers, or hair. As evolutionary descendants of fishes, they have gills at some stage in their lives, and they breathe either through these gills or through the skin or lungs, or a combination of all three. Most species must return to the water or to very damp soil to deposit their eggs, which are usually jelly-coated, and from which the young generally hatch into free-swimming gilled larvae that later transform into land-dwelling animals.

Conventionally, herpetologists study both amphibians and reptiles. However, although reptiles are descended from amphibian stock, they differ from living amphibians more than they do from either birds or mammals. Nevertheless, both classes are cold-blooded, meaning not that the blood is actually cold, but that the body temperature varies with that of the surrounding environment and cannot be maintained at a level as constant as that of most birds and mammals.

Many amphibians spend the first part of their lives in the water as larvae or tadpoles; later their bodies transform and they emerge to live most of their lives on land. However, some spend the entire life span on land; others, like the mud puppies *(Necturus)*, spend it entirely in the water. Amphibians usually do not drink, but absorb moisture directly from the water or damp earth surrounding them.

Distribution and Diversity
There are approximately 2500 species of amphibians worldwide. Amphibians occupy almost all land masses except the Arctic and Antarctic, but overall species diversity generally increases as one approaches the tropics. The factors ultimately limiting amphibian distribution are moisture and temperature.

Habits
Amphibians breathe through skin, gills, or lungs, or a combination of them. Needed moisture is also absorbed through the skin. Like fishes and reptiles they are cold-blooded, or ectothermic; that is, their body temperature is maintained at high levels by external sources and is not internally controlled, as in birds and mammals.

Reproduction
The typical pattern of reproduction is egg laying in or near water, although some species utilize moist areas on land to deposit eggs that hatch into miniature adults. The phenomenon of reproduction around water often provides a herpetologist with the best opportunity to observe and collect amphibians, which are encountered either migrating to the breeding site or about the ponds, streams, and rivers themselves. Larval salamanders have external gills that disappear during metamorphosis and mouthparts similar to those of the adult. Larval frogs, called tadpoles, lack

visible external gills and have a horny beak. Both types of larvae have a tail which is used in swimming, but in frogs this structure is later resorbed. Identification of larval forms, particularly in the early stages, is difficult and much too complex for this volume to cover adequately.

Range and Scope

This chapter includes twenty-five species of the salamanders and thirty-eight species of frogs and toads that are commonly found in western North America, west of the 100th meridian, from Alaska to the U.S.-Mexican border. Many of the species ranges are spotty and sporadic, at best, within that territory. Given are the most prominent areas, within the range, in which the species are likely to be found.

Nomenclature

The common and scientific names are in accord with either those of Robert Stebbins (1966) or Roger Conant (1975).

USEFUL REFERENCES

Bishop, S. C. 1943. *Handbook of Salamanders*. Ithaca, N.Y.: Comstock.

Cochran, D. M., and Goin, C. J. 1970. *A New Field Book of Reptiles and Amphibians*. New York: Putnam.

Conant, R. 1975. *A Field Guide to Reptiles and Amphibians of Eastern and Central North America*. 2d ed. Boston: Houghton Mifflin.

Dickerson, M. C. 1931. *The Frog Book*. New York: Doubleday, Doran.

Leviton, A. E. 1972. *Reptiles and Amphibians of North America*. New York: Doubleday.

Noble, G. K. 1931. *The Biology of Amphibia*. New York: McGraw-Hill, Dover, 1954.

Oliver, J. A. 1955. *Natural History of North American Amphibians and Reptiles*. Princeton: Van Nostrand.

Stebbins, R. C. 1951. *Amphibians of Western North America*. Berkeley: Univ. of Calif. Press.

————. 1954. *Amphibians and Reptiles of Western North America*. New York: McGraw-Hill.

————. 1966. *A Field Guide to Western Reptiles and Amphibians*. Boston: Houghton Mifflin.

Wright, A. A., and Wright, A. H. 1949. *Handbook of Frogs and Toads of the United States and Canada*. 3d ed. Ithaca, N. Y.: Comstock.

Zweifel, R. G.; Zug, G. R.; McCoy, C. J.; Rossman, D. A.; and Anderson, J. D. (eds.). 1963–Present. *Catalogue of American Amphibians and Reptiles*. Soc. for Study of Amphibians and Reptiles, New York.

GLOSSARY

Adpressed limbs Front leg held straight back against side of body; hind leg held straight forward against side of body.

Anterior Toward the front (of the body).

Cloaca Area through which internal wastes discharge.

Costal groove Vertical furrows in skin on the flanks of salamanders.

Cranial crests Raised area framing inner border of upper eyelids in toads.

Diurnal Daytime.

Dorsolateral Pertaining to the upper side of the animal.

Frontals Bony membranes that form the forehead.

Gular fold Fold of skin across posterior section of throat in salamanders.

Intercostal fold Skin fold bounded on each side by a costal groove.

Larva The earliest form that certain species take; unlike the parent (e.g., tadpole).

Lateral Of or pertaining to the side of the body.

Maxillary bone Bone on each side of head forming side border of upper jaw and bearing most of the upper teeth.

Nasolabial groove Groove from nostril to edge of upper lip.

Neotenic Retaining gills throughout entire life.

Nuptial pad A patch of darkly pigmented, roughened skin that appears on certain digits during breeding season.

Occipitals Bony membranes that form the posterior part of the skull.

Parietal bones A pair of membrane bones in the roof of the skull between the frontals and occipitals.

Parotoid Gland One of two large wartlike glands found on the rear of the head in toads.

Posterior Toward the rear (of the body).

Postocular Behind the eye.

Postorbital Behind the eye.

Reticulate Resembling the form or appearance of a net.

Riparian Relating to the bank of a river, lake, or pond.

Rugose Rough, wrinkled.

Tubercle Any of various small knoblike prominences.

Tympanum Thin round or oval external ear covering; prominent in frogs and many lizards.

Vent Opening on the surface of the cloaca.

Venter Belly.

Vertebral stripe Stripe down the midline of the back.

Vertical pupil Pupil that is elliptical; long axis is vertical.

Vocal sac Loose skin on the throat that can distend, forming a chamber which echoes the animal's vocalization.

Vomerine teeth Teeth anchored to the vomerine bones that form part of the forepart of the roof of the mouth.

Salamanders
Order Caudata

The essentially voiceless salamanders are distinguished from the other living North American amphibians, the frogs and toads, by the possession of a tail throughout their lives (most frogs and toads have a tail only during the tadpole stage). Salamanders usually have moist, relatively smooth skin, no external ear openings, no claws, and never more than four toes on their front feet. They breathe through gills, lungs, or skin. They can regenerate a lost tail or limb.

Identification

Features useful in identification include color and color pattern; the presence or absence of gular fold (skin flap across the throat) and the number of costal grooves (the vertical grooves on the sides of the body).

Fig. 67
Parts of a Typical Salamander

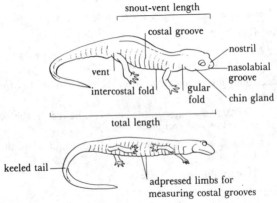

The length of salamanders, in inches and centimeters, is given for the *average snout-vent length of adults and does not include the tail.* The text provides a basic description of coloration, but individual salamanders vary a good deal in color and somewhat in pattern. For example, the color often becomes darker and the pattern more obscure with age.

The identification of salamanders provokes dissatisfaction even among herpetologists. One problem is the dearth of important distinguishing characteristics to measure and count, such as the useful scale patterns in reptiles. Many obvious differences in color or pattern are due to variations within one species, well illustrated in humans by variation in hair color. Other problems are the small size and secretiveness of salamanders, making it hard to observe all features without an examination in the hand.

Perhaps the worst problem in identification is one experienced by the salamanders themselves. At times they cannot tell who belongs to their own species when choosing a mate and they produce hybrid offspring derived from parents of two different, but usually closely related, species. Hybrids are often less likely to survive and reproduce themselves, but in certain areas they are successful.

468

Habitat and Habits

Salamanders prefer a moist habitat and avoid direct sunlight. Some spend their entire lives in water; some spend most of their lives on land. Some are arboreal, others are cliff dwellers; some, usually blind, dwell in caves or in deep wells. Salamanders live from near sea level to an altitude of over 13,000 ft. (3962.4 m). Where the winters are cold they hibernate, but they may be active underground. Unlike snakes and lizards, however, they prefer cool weather to warm, and some are active in icy water close to freezing.

Look for terrestrial salamanders in the spring and fall, particularly at or near the pools where they breed. After breeding they usually return to their hiding places under rocks, logs, bark, leaf litter, and debris left lying by human activities. Specific habitats of each species are listed in the descriptions, but the successful salamander hunter will overturn all cover, and then, of course, carefully replace it in its original position. Most species are nocturnal.

Food

Salamanders are entirely carnivorous, and among other items will feed on aquatic or terrestrial insects, or other small invertebrates and their eggs and larvae, and on small vertebrates such as salamander larvae. In captivity they will eat earthworms, beef, or other meat, and insects when these are available.

Reproduction

When male and female encounter each other during the mating season they usually go through an elaborate and species-characteristic courtship ritual accompanied by tactile and chemical sensory cues. You might observe this behavior in captive animals, given the proper sexes, the mating season, and no distractions from the observer. Fertilization is internal, except in the Hellbender, whose male fertilizes the eggs as they are released from the body of the female. In a few species the sperm is transferred to the female through the touching of vent to vent, but in most species after the courtship dance the male will deposit small spermatophores that contain sperm in a little sac on top of a jellylike pyramid. The female follows the male and picks up these sperm sacs with the sides of her vent and introduces them into her spermatheca, a specialized storage receptacle where sperm may be stored for several years. One clutch of eggs laid by the female can potentially be fertilized by sperm from two different males. It should be noted that mating and egg laying may not occur in the same season.

Eggs are deposited singly, in strings, or in masses either in the water or on land, where they are suspended from the roof of a moist retreat. The eggs may be brooded by the female. In due time, depending on the temperature, the eggs in aquatic situations hatch into larval salamanders with gills. Lower ambient temperatures will decrease the growth rate in these stages. The length of the larval period varies from a few days to over a year or two. In terrestrial species the larval period is passed within the egg, and upon hatching the salamander has only rudimentary gills, if any, which are shed shortly after birth. Another adaptation, neoteny, allows salamanders to spend their entire lives with gills and other larval features while coming to reproductive maturity. This last strategy is often found in an inhospitable or underground environment where the perils of transformation are too great.

MOLE SALAMANDERS
Family Ambystomatidae

Western ambystomids vary from 1⅔ to 6½ in. (4.23–16.5 cm) in length. The body is stout, the head broad, with relatively small eyes that are widely spaced. The tail is laterally flattened, and the sturdy limbs have four toes on the front feet and five on the hind feet. They have blunt snouts, prominent gular folds, and large fleshy tongues. They lack nasolabial grooves and have fewer costal grooves than the lungless salamanders. The adults have lungs but no gills. The larvae have bushy external gills and a keel on the back and tail. Adults are usually terrestrial except in the breeding season, but the Olympic Salamander, more aquatic than the others, inhabits mountain creeks and seepages.

Members of the Genera *Ambystoma, Dicamptodon,* and *Rhyacotriton* vary in length. They have a rounded body, well-defined costal grooves, and a broad, somewhat flattened, U-shaped head (viewed dorsally). The sturdy limbs terminate in long, tapered toes. The tail is oval in cross section at the base, but flattened toward the tip.

NORTHWESTERN SALAMANDER
Ambystoma gracile   81:8

Description
Length, 3–4½ in. (7.6–11.4 cm). Differs from other ambystomids in having parotoid glands, and a glandular ridge along upper edge of tail. No tubercles on foot; 3–4 segments (phalanges) in 4th toe of hind foot. Above nearly uniform dark brown, may be marked with cream, bronze, or yellow flecks, especially in north of range; below brown to slate. Costal grooves 11–12 (sometimes 10). Male commonly with longer tail than female; breeding male with swollen vent lined with numerous small papillae; female vent not swollen and walls pleated.
Habitat
In damp places near water beneath surface objects, as under high-water driftwood.
Range
Nw. coastal belt, s. from se. Alaska, including Vancouver Is. to n. Sonoma Co., Calif., mouth of Gualala R.

TIGER SALAMANDER
Ambystoma tigrinum   81:7a, 7b

Description
Length, 3–6½ in. (7.6–16.5 cm). Tubercles on foot well defined. Eyes relatively small, distance between 1½–2 times their width. Above variable in color and pattern geographically, but most commonly blackish or sooty with large spots ("California" form), bars ("Barred" form), or blotches of yellow, cream, or ash white. Other colorations are uniformly yellowish, vaguely dark-spotted on yellowish, or uniformly grayish with vague dark spots. Below gray to flesh, occasionally with subdued yellowish markings. Costal grooves 11–13. Breeding male has swollen vent region, papillae on walls of cloaca; tail longer than female.
Similarities
Northwestern Salamander has parotoid glands.
Habitat
Semiarid deserts to over 11,000 ft. (3352.8 m).

Habits

A burrower; in dry weather inhabits burrows of ground squirrels, marmots, prairie dogs, badgers, etc., crevices, or decayed logs and stumps; makes nocturnal migrations to temporary or permanent ponds after early spring rains.

Remarks

Large gilled larval form of this species is often sold in the southwest as fish bait under the name "waterdog."

Range

Far s.-cen. and extr. se. B.C., s.-cen. to se. Alta., s. Sask.; e. Wash., Idaho, Mont., Wyo., Utah, Colo., ne. Ariz., N.Mex., Tex.; also isolated population in w.-cen. Calif. Continues e. to Atlantic Ocean.

LONG-TOED SALAMANDER
Ambystoma macrodactylum   81:2

Description

Length, 2⅛–3¼ in. (5.3–8.1 cm). Differs from other ambystomids in having long slender toes and usually a dorsal stripe. Tubercles on feet present (sometimes ill defined); toes long, slender. Above dark brown to black with broad vertebral stripe of tan, yellow, olive-green; stripe usually with irregular edges, may be broken into variously shaped patches; flanks white-speckled; below sooty or brown with small white specks. Costal grooves 12–13 (sometimes 14). Male vent lined with papillae, female's with folds.

Habitat

Found at sea level to 9000 ft. (2743.2 m); in breeding season under logs, rocks, etc., near water; also under loose bark of down-timber and inside rotten logs.

Reproduction

Spawns right after snow melts; mature larvae reach 2½–3 in. (6.2–7.5 cm) total length.

Range

Se. Alaska, sw. B.C. (and offshore islands), extr. sw. Alta., Wash., Idaho, far w. Mont., Oreg., n. Calif. (except extr. ne. and Cascade-Sierra to just s. of Lake Tahoe; also isolated subspecies in Santa Cruz Co.

PACIFIC GIANT SALAMANDER
Dicamptodon ensatus   81:9

Description

Length, 4½–6 in. (11.4–15.2 cm). Heavy-bodied; large head; sturdy limbs; tail narrow along upper edge, rounded below, laterally flattened, especially toward tip. Color pattern distinctive; above coarsely marbled with black on brown, gray, or purplish, forming a coarse network of dark against light; below light brown to yellowish-white; sides lightly speckled with white. Foot has 3 segments in 4th toe of hind foot (instead of usual 4, as in *Ambystoma);* tubercles absent. Costal grooves 11–13 (usually 12), often indistinct.

Similarities

Large size and irregular reticulations of black on brown, gray, or purple distinguish this species from other ambystomids.

Habitat

In humid, well-forested areas; adults in water or on land usually in damp situations under logs, bark, rocks, etc., not far from water.

Habits
Sometimes seen by day; can climb well; able to vocalize, producing "a low-pitched rattling and an explosive cry suggestive of the bark of a dog" (Stebbins 1954).

Range
Nw. coastal belt of w. Cascades from extr. sw. B.C. to Monterey Bay, Calif. Also isolated population in n.-cen. Idaho, with overlap into extr. w. Mont.

OLYMPIC SALAMANDER
Rhyacotriton olympicus   **81:1**

Description
Length, 1⅔–2½ in. (4.2–6.4 cm). Smallest ambystomid, with eyes largest in proportion to body size. Body slim, limbs and toes short; tail relatively short, narrow above, oval at base, laterally flattened toward tip. Color (1) in Washington, dorsal surfaces uniformly brown finely sprinkled with ash-white flecks; below yellowish-orange, (2) in Oregon and California, above mottled and flecked with dusky on olive; white flecks less conspicuous, vent greenish-yellow. Costal grooves 14–15. Only male has prominent, squarish lobe on each side of vent.

Similarities
Larvae distinguishable from those of Pacific Giant Salamander by lack of well-formed gill rakers, presence of groove behind each nostril, and dorsal speckling.

Habitat
In and near small, cold, rapidly flowing, well-shaded, permanent creeks and seepages (splash zone); adults and larvae under mossy stones or about water edges.

Habits
Agile on land but rarely found there; a rapid swimmer.

Range
Coastal belt of w. Wash., s. to nw. Calif.

NEWTS
Family Salamandridae

The stout-bodied salamandrids of the West range from two to four inches (5.1–10.2 cm) in snout-vent length. They have sturdy limbs and a broad, somewhat flattened head. The tail is vertically oval in cross section except when flattened in the breeding male. There are four toes on the forefoot, five on the hind foot. The skin is roughened by closely set tubercles, except in breeding males, and they lack costal grooves.

ROUGH-SKINNED NEWT
Taricha granulosa   **81:5**

Description
Length, 2¼–3½ in. (6.4–8.9 cm). Lower eyelid is dark; iris yellow to pale greenish with horizontal bar. Color uniformly black to dark brown above (or tannish); dorsal color sharply distinct from ventral orange. Lips of female vent form laterally flattened cone with opening of vent at apex, dark stripe across vent more common in male.

Similarities
California Newt has larger eyes and light lower eyelid; snout not as blunt.

Habitat
Humid coastal forests and open grasslands near streams, lakes, ponds, reservoirs; found under logs, boards, rocks, etc. (in wet weather on their surfaces).

Habits
When disturbed holds head erect, flattens body, extends legs stiffly outward, elevates tail.

Range
Nw. coast from se. Alaska to Monterey Bay, Calif.; occurs e. of Cascade crest in cen. and s. Wash. and s.-cen. Oreg.; also isolated population embracing parts of Latah Co., Idaho, and Sanders Co., Mont. In Sierras of Calif. ranges to just s. of Magalia, Butte Co.

Note: The **RED-BELLIED NEWT**, *Taricha rivularis* (**81:6**), has dark eyes and a dark band across its vent. It occurs in coastal California from Sonoma to Humboldt County.

CALIFORNIA NEWT
Taricha torosa   **81:4**

Description
Length, 2¾–3¼ in. (7.0–8.3 cm). Eyes medium-size; corneal surfaces often extend to or beyond jaw outline as viewed from above; eye color yellow to greenish-silver, with horizontal black bar that includes pupil. Body tan to reddish-brown above, pale yellow to orange below; dorsal and ventral colors shade into one another on sides; light color of upper jaw extends onto lower eyelid. Sexes difficult to differentiate except during breeding season.

Similarities
Rough-skinned Newt has darker lower eyelids and smaller eyes.

Habitat.
In areas of Live Oaks, Ponderosa and Digger Pines, in and near streams.

Habits
Good swimmer (by lateral undulations of tail and body); assumes defensive pose like Rough-skinned Newt's.

Remarks
In some localities (such as Boulder Creek, San Diego Co., Calif.) extremely warty individuals occur.

Range
Calif. coastal mts. from Mendocino Co. to Baja Calif.; cen. Sierras; island at Squaw Creek headwaters of Shasta Reservoir, Shasta Co.

LUNGLESS SALAMANDERS
Family Plethodontidae

All western members of this family are terrestrial and are found in damp places under rocks, boards, logs, and bark, and inside rotten logs, in leaf litter, and in the burrows of other animals. They lack free-living larvae; their young emerge fully formed. They have a groove from nose to lip, no lungs, and the adult is without gills. There are five toes on the hind feet and four on the front feet.

WOODLAND SALAMANDERS
Genus *Plethodon*

These have long, slender bodies; bluntly rounded snouts; pronounced gular folds; long, slender, rounded, and tapering tails; and more costal grooves than any other full-limbed salamanders.

DUNN'S SALAMANDER
Plethodon dunni   82:5

Description
Length, 2–3 in. (5.1–7.6 cm). Dorsal stripe yellowish-tan to dull greenish-yellow, brighter on tail, obscured with blackish toward tip, sprinkled with flecks of dusky, sometimes so abundant as to obscure stripe; sides dark brown to black, flecked with white and spotted with tan or yellowish; limbs have upper surfaces of bases like dorsal stripe, underparts slaty with small spots of yellowish or orange. Costal grooves usually 15; 2½–4 intercostal folds between tips of toes of adpressed limbs. Male generally has broader head and longer tail than female; lower jaw more pointed.

Similarities
Western Red-backed Salamander has 16 costal grooves, light and dark reticulations on belly, stripe extending to tip of tail.

Habitat
Like that of Olympic Salamander. The most nearly aquatic of the western plethodontids, commonly found in moss-covered rock rubble in seepage areas and along permanent, well-shaded small streams.

Range
Extr. sw. Wash., w. Oreg. (absent from Willamette Valley).

Note: The **LARCH MOUNTAIN SALAMANDER,** *Plethodon larselli,* with a red or red-orange belly, occurs on the lower Columbia River, Oregon, and Archer Falls, Washington.

The **DEL NORTE SALAMANDER,** *Plethodon elongatus* **(82:1),** has 18 costal grooves and short, partially webbed toes. It occurs along the coast from Rogue River, Oregon, to Orick, California.

The **SISKIYOU MOUNTAIN SALAMANDER,** *Plethodon stormi,* is profusely speckled with white or yellow above, and occurs along the California-Oregon border near the Applegate River.

JEMEZ MOUNTAINS SALAMANDER
Plethodon neomexicanus   82:3

Description
Length, 2–3 in. (5.1–7.6 cm). Most slender western *Plethodon.* Has short toes, 5th reduced to usually 1 segment plus occasional small terminal segment. Costal grooves usually 19; 7½–8½ intercostal folds between tips of toes of adpressed limbs. Color uniformly brown above with fine pale gold stippling; juveniles have vague light gray to pale gold dorsal strip, which in adults is absent, or only edges present; below sooty; throat and underside of tail cream or beige. Sexes show no external differences.

Similarities
Del Norte Salamander has 5th toe with 2 segments.

Habitat
Coniferous forests above 8000 ft. (2438.4 m), under bark and inside rotting logs.

Habits
Little known about this species; probably most active during summer rainy season.

Range
Jemez Mts. of N.Mex.

VAN DYKE'S SALAMANDER
Plethodon vandykei

82:2

Description
Length, 2–3 in. (5.1–7.6 cm). Unique in having parotoid glands. Dorsal stripe yellowish-tan margined by black to dark brown; edges often irregular and commonly scalloped; sides of head, body, tail black to brown, lighter below; throat pale; white to light gray stippling on sides; limbs black to dark brown above and continuing down to stripe at bases; underside black to dark brown with scattered whitish flecks; gular area pale yellow. Costal grooves usually 14, and 2–3 intercostal folds between tips of toes of adpressed limbs. Toes short. Adult male has tubercular projection on each side of upper lip and fingerlike posterior projection from each side of vent.

Similarities
Dunn's and Western Red-backed Salamanders lack pale throat.

Habitat
Damp to wet places under rocks along streams, in seepages; in damp, often mossy, talus, and occasionally under bark and surface litter.

Range
W. Wash., n. Idaho, and nw. Mont.

WESTERN RED-BACKED SALAMANDER
Plethodon vehiculum

82:4

Description
Length, 1½–3¼ in. (3.8–8.3 cm). Dorsal stripe reddish-brown, tan, yellowish-tan, or yellow, usually even-edged, well defined. Occasionally unstriped individuals occur, as in (1) melanistic, or black, animals, and (2) those predominantly orange or yellowish. Sides black or dark brown suffused with whitish flecks; underparts bluish-sooty with light gray, yellowish, or orange flecks and finer white stippling. Costal grooves usually 16, intercostal folds between toe tips of adpressed limbs, 4½–5½. Male similar to Van Dyke's, except for lip tubercles.

Similarities
Dunn's Salamander has 16 costal grooves, dorsal stripe not reaching tip of tail.

Habitat
Damp to saturated situations under logs, rocks, pieces of bark, leaves, and under bark, moss, and in crevices of downed and standing dead timber in humid forests.

Range
Nw. coast from sw. B.C. and Vancouver Is. to sw. Oreg., n. of Rogue R.

ENSATINA
Genus *Ensatina*

ENSATINA
Ensatina eschscholtzi

81:3

Description
Length, 1½–3 in. (3.8–7.6 cm). Distinguished from all other western salamanders by a swollen, basally constricted tail. Four toes on forefoot with a pair of palmar tubercles on underside of each; 5 toes on hind foot. Eyes large, protuberant. Costal grooves,

12–13. Color variable among 7 subspecies, of which 3 are blotched (Yellow-blotched form), 3 uniformly colored (Monterey form), and 1 mottled. Male has longer, more slender tail and longer, broader snout than female; also prominent forking of nasolabial groove at edge of lip.

Habitat
Fairly uniform for all subspecies, i.e., they avoid steep slopes, prefer areas of surface litter such as leaves, logs, bark, boards, rocks; avoid saturated soil and prefer moderately damp situations.

Habits
Often takes refuge in other animals' burrows, in woodrat nests, or beneath bark or in rotten interiors of logs, in dry or cold weather.

Range
Coastal from sw. B.C. and e. Vancouver Is. to Baja Calif., w. from Cascade-Sierra crest.

SLENDER SALAMANDERS
Genus *Batrachoseps*

These are elongate salamanders with tiny legs, conspicuous costal and caudal grooves, four toes on all feet, and usually a tan to red dorsal stripe. They live under logs, rocks, or litter in forests and yards.

CALIFORNIA SLENDER SALAMANDER
Batrachoseps attenuatus   **82:8**

Description
Length, 1–2 in. (2.5–5.1 cm). Ground color of sides and underparts sooty, with sprinklings of fine white flecks, also along ventral midline of tail, and abundant on sides; belly with dark unbroken network. Dorsal stripe may be absent; when present, may be obliterated in varying degrees with dark markings; begins on head and extends well out onto tail, variable in color—brick, tan, brown, beige, or yellowish, commonly margined with black. Costal grooves, 18–21; vomerine teeth in clusters, sometimes in rows. Male generally with broader, blunter, snout, anteriorly raised vent margins, less elongate vent opening than female. Juveniles are more *Plethodon*-like than adults, with longer limbs, stouter body, shorter tail.

Similarities
Pacific Slender Salamander is larger, paler; dorsal stripe is faint or lacking; black network of belly broken.

Habitat
Same as for Ensatina; size differences may preclude competition for food and burrows between these two species.

Range
Ext. sw. Oreg. coastally to Baja Calif.; w. slope of Sierra Nevada; Santa Cruz Is.

PACIFIC SLENDER SALAMANDER
Batrachoseps pacificus   **82:6**

Description
Length, 1⅔–2½ in. (4.2–6.4 cm). Similar to California Slender Salamander, but more *Plethodon*-like and larger; head and body broader, limbs and toes longer, tail shorter. Brown above, with fine white stippling, often sparse and concentrated often along sides; sometimes has rust on eyelids, snout, tail; dorsal stripe seldom

present in adults; gular area and underside of tail whitish to pinkish-tan; underparts pale slate or whitish, the dark color not enough to form a continuous network. Costal grooves 18–21. Vomerine teeth usually in 2 arched rows. Sexes differ as in California Slender Salamander.

Similarities

California Slender Salamander has dark pigment of belly forming fine continuous network.

Habitat

Under rocks, logs, bark. Usually in Live Oak.

Range

Coastal s. Calif. from Pasadena to Escondido; Calif. Channel Isls.

Note: The **OREGON SLENDER SALAMANDER,** *Batrachoseps wrighti* **(82:7),** has a black belly with large white blotches, and occurs in northern Oregon along the Columbia River and the western slope of the Cascades.

CLIMBING SALAMANDERS
Genus *Aneides*

BLACK SALAMANDER
Aneides flavipunctatus

82:15

Description

Length, 2⅝–3 in. (6.7–7.6 cm). Body round; limbs, toes relatively short; toes tapered with rounded tips. Ventral color of all forms deep slate to black. Costal grooves, 14–15; intercostal folds between toe tips of adpressed limbs, 3–5. Male head broader, more triangular, upper lip more enlarged, other male characteristics similar to those of Clouded Salamander. Dorsum black; south of Golden Gate almost solid black, rarely finely spotted with white; north of Gate, along coast, with cream or pale yellow spots; farther north, in redwood belt, suffused with olive, greenish, or greenish-gray over all dorsal surfaces and spotted reduced or absent; interior, in coast ranges and east to Mt. Shasta, prominently white-spotted. Juveniles black with pale gold flecking.

Similarities

Arboreal and Clouded Salamanders have squarish toe tips and less than 2 costal folds between toe tips of adpressed limbs.

Habitat

Same as for Clouded Salamander, but more commonly found on rocks, especially those accumulated along old road cuts and in fill on downhill side of roads; also rock and soil mixtures near seepage areas and along streams.

Habits

When disturbed, holds head high and tail elevated, arched, and waves entire body from side to side.

Range

Nw. Calif. s. of Crescent City and e. to Mt. Shasta; in coast range to Monterey Bay.

CLOUDED SALAMANDER
Aneides ferreus

82:14

Description

Length, 2–3 in. (5.1–7.6 cm). Body slim, flattened; tail round, limbs long and slender, tips of toes truncate, expanded, innermost toe of both fore- and hind foot much reduced. Tips of toes of adpressed limbs separated by 1½, or less, intercostal folds, or may overlap; costal grooves, 16. Dark brown above, usually clouded or mottled with pale gold, whitish- to greenish-gray; in dark phase, light colors much reduced to nearly uniform dark brown; below whitish to slate or dark brown, speckled variously with white. Breeding male with heart-shaped mental gland on underside of lower jaw; walls of vent with small papillae; in female, smooth pleats. Newly hatched have rust to brassy dorsal stripe; older young with reddish, copper, or brass on upper snout to eyelids, on shoulders, tail, and uppersides of limb bases.

Similarities
Arboreal Salamander has 15 costal grooves; Black Salamander, 3–5 costal folds between toe tips of adpressed limbs.

Habitat
Humid coastal forests; usually under bark of Douglas Fir, Port Orford Cedar, redwood, alder, etc.; also under leaf litter on tops of stumps; occasionally under objects on ground.

Habits
Most arboreal of genus; climbs trees to 20 ft. (6.1 m); especially common in well-lighted breaks in Douglas Fir and redwood forests and in clearings with numerous downed trees and logs; not necessarily near water.

Range
Vancouver Is.; w. Oreg., except far nw. to cen. coast; nw. Calif. to cen. Mendocino Co.

SACRAMENTO MOUNTAIN SALAMANDER
Aneides hardyi   82:13

Description
Length, 1¾–2¼ in. (4.4–5.7 cm). Body slim, rounded; tail round, head moderately triangular in adults, limbs relatively short, toes lacking enlarged tips. Costal grooves, 14–15; intercostal folds between toe tips, 4–4½. Color blackish to brown above, with varying amounts of greenish-gray to bronze mottling; some large individuals may lack light color; below, light brown to purplish-brown on abdomen and chest, gular area cream, underside of tail slate. Adult male with circular mental gland and transverse furrows in margins of vent; simple slit in female vent. Juveniles may have dorsal stripe of brown to rusty-bronze, whitish throat, and darker ground color than adults.

Habitat
Confined to Douglas Fir–spruce zone at high elevations, mostly over 8000 ft. (2438.4 m). Found beneath and within rotting logs, under bark, and in rock rubble.

Habits
Surface activity limited to period of summer rains.

Range
S.-cen. N.Mex. in Sacramento, Capitan, and White Mts.

ARBOREAL SALAMANDER
Aneides lugubris   82:12

Description
Length, 2½–3¾ in. (6.4–9.5 cm). Similar to Clouded Salamander in form, but body stockier, jaw muscles of adults powerfully developed. Costal grooves, 14–16. Tips of toes blunt, broadened; adpressed limbs overlap or fail to meet up to 1 intercostal fold. Tail round, tapered, somewhat prehensile. Color uniformly brown above, usually with pale yellow spots, highly variable in size,

number, and position; below whitish to pale gray with underside of tail often dull yellowish-olive. Sexes differentiated as in Clouded Salamander. Juveniles marked like young Clouded Salamander, dark brown ground color, rust to brassy marks on snout, shoulders, limb bases, tail.

Similarities
Clouded Salamander has 16 costal grooves, mottled back, dark belly finely speckled with white.

Habitat
Under logs, bark, boards, rocks, etc., and in tree cavities, inside rotten logs, in woodrat nests, rodent burrows, mine openings.

Habits
Appears on surface of ground following first fall rains; remains active until spring, except in freezing weather. Good climbers, facilitated by expanded digits and prehensile tail; colonial, with as many as 35 individuals found in one cavity.

Range
Calif. coast, s. from Humboldt Co. to Baja Calif.; island in Sierra foothills from Calveras Co. to Madera Co.; also South Farallon and Catalina Isls.

WEB-TOED SALAMANDERS
Genus *Hydromantes*

MOUNT LYELL SALAMANDER
Hydromantes platycephalus

82:11

Description
Length, 1¾–2¾ in. (4.4–7.0 cm). Body flattened; tail round, relatively short, tip blunt; feet broad, toes short and partly webbed; tongue pediceled, edges free all around. Color brown to nearly black above, variously obscured by flecks and patches of metallic, pale gold, gray, to whitish; below dark brown to sooty, usually with whitish guanistic spots on lower sides, on underside of limbs, and in gular area. White blotches underneath usually do not extend across chest or abdomen. Costal grooves, usually 13, with ½–1½ intercostal folds between toe tips of adpressed limbs. Male larger, head generally broader than female. Juveniles black above with pale gold stippling, giving greenish cast; below sooty with white stippling.

Habitat
Commonly under granite slabs on moist to saturated subsurfaces; see habitats for Dunn's and Olympic Salamanders. Also in areas of seepages at foot of ledges and cliffs and under rocks at mouths of caves and recesses.

Range
Calif. Sierras between Sonora Pass and Twin Peaks area of Sequoia National Park.

Note: The following two species are compared with the Mount Lyell Salamander:

The **SHASTA SALAMANDER,** *Hydromantes shastae* **(82:10),** is found only in certain limestone areas near Mt. Shasta. It differs in having less flattened body, relatively larger eyes, a longer snout, longer and blunter toes, longer limbs, and less-contrasting color pattern.

The **LIMESTONE SALAMANDER,** *Hydromantes brunus* **(82:9),** of Mariposa County, California, is uniformly tan-colored dorsally in adults.

Frogs and Toads

Order Anura

As adults, frogs and toads are distinguished by their short, squat bodies, four limbs, powerful hind legs, and lack of tail. Of the world's 2000 species, in some eleven families, about sixty species in seven families inhabit North America. Of these, thirty-eight occur in the West. The terms "frog" and "toad" are more confusing than useful. In general, frogs have smooth skins and are more aquatic, whereas toads have warty skins and are more terrestrial.

Fig. 68
Parts of a Typical Toad and Frog

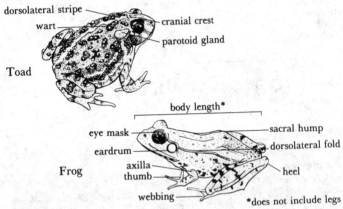

Identification

Identification can be made with certainty for some forms only with the specimen in hand. Frogs and toads of the same species often vary widely in color depending on the environment, temperature, and humidity (the cooler and moister the environment, the darker the color). Note size or shape, the presence or absence of webbing or enlarged pads on the feet, exposed and hidden markings, or the presence of glands. Females, as a rule, are larger than males and may have a greatly distended belly full of eggs in the breeding season. Males often have dark throats and special clasping pads on their thumbs, fingers, forelegs, or chest. Measurements given for anurans are the *range of snout-vent lengths for adults.*

Habitat

Look for frogs in wet areas, under cover or in water. They are easily caught by hand but a net may assist in water. Some anurans, especially toads, are found during the day in mountainous, forested, or prairie situations some distance from water.

Vocalization

Vocalizations in frogs are usually limited to males, but in some species the female may utter a nonmating call, or both sexes may be silent. In addition to other visual, olfactory, and tactile cues, the mating call is a major means of mate and species recognition. Vocalizations may be divided into the following categories: mating calls, in males, often in a chorus heard over miles; release calls, given by a male or unreceptive female when clasped by a breeding male; a territorial call, which may not necessarily be associated with breeding activity; and various calls to indicate distress or give a warning. The cessation of calling may also be a signal.

Food

Adults eat insects and other invertebrates, and bullfrogs are known to eat anything that moves and is of a suitable size, even the young bullfrogs. Tadpoles are usually vegetarian. Major changes in metamorphosis from tadpole to adult are in the shortening of the digestive tract and the disappearance of larval mouthparts.

Reproduction

Frogs are somewhat less formal than salamanders when they encounter a potential mate, and the major device that prevents a mixing of species is the mating call. At the breeding grounds the males mount and clasp the females, holding this position of amplexus from a few minutes to over a day, at the end of which the male fertilizes the eggs as they are extruded from the female. Eggs are laid under water or on the surface, depending on the species. They may be laid singly, in long strings or cables, in globular masses, or in flat films. They may float or sink or be attached to vegetation. Clusters range from a few to hundreds.

Eggs hatch into tadpoles in a short period, developmental rates often being dependent on the temperature. The tadpoles have horny beaks and often toothlike structures in rows around the mouth, and they breathe through gills covered by a flap of skin, the operculum. In general, most toad tadpoles are small and black and transform at very small sizes; at the other extreme, the true frogs have tadpoles that are larger and greenish, and which may live more than a year before transforming. At transformation they develop lungs, four limbs appear, the tail is resorbed, and structures related to feeding change. They spend less time in deeper water and more time on the edges of ponds and streams.

TAILED FROGS
Family Ascaphidae

Members of this family are small, usually two inches (5.1 cm) or less in length. They lack a true tail, but tail-wagging muscles are present and, in the male, the vent opens from a taillike prolongation of the body that serves as the copulatory organ.

TAILED FROG
Ascaphus truei   **85:1**

Description

Length, 1–2 in. (2.5–5.1 cm). Body length does not include "tail." Fifth (outermost) digit of hind foot broadest; eye has vertically oval pupil; eardrums absent. Above old-rose to brick-red, creamy white, gray or brown to almost black, usually patterned with dark streaks and blotches; eyestripe commonly present; below cream, with yellow in femoral region. Tongue large, fingers free, toes slightly webbed. Female anal tube shorter; breeding male with much-enlarged forearms and inner palmar tubercles.

Habitat

Well-forested areas, frequently permanent, relatively rapid, low-temperature streams (41–54°F, 15.0–12.2°C). Adults and larvae inhabit shallow mountain streams, beneath stones; after heavy rains or at night may leave water.

Voice

Probably voiceless.

Range

Nw. coastal forests from extreme sw. B.C., s. to Mendocino Co., Calif.; n. Idaho (except for panhandle) and adjoining w. Mont.

SPADEFOOT TOADS
Family Pelobatidae

The spadefoots are squat, toadlike anurans with fairly smooth, loose, and translucent skin, *vertical eye pupils,* and teeth in the upper jaw. Each hind foot has a sharp-edged black "spade," or metatarsal tubercle, on the inner side. They are nocturnal, terrestrial, and burrowing animals, distributed widely through the temperate and subtropical areas of the Northern Hemisphere, particularly in deserts.

PLAINS SPADEFOOT
Scaphiopus bombifrons

83:1
Fig. 69

Description
Length, 1½–2½ in. (3.8–6.4 cm). Conspicuous bony bump between eyes with elliptical pupils. Has protruding forehead. Above greenish or grayish with dark spots; below whitish. No distinct eardrums or head ridges; no neck glands. End of snout and lump between eyes sometimes covered with layer of black horn. Male with dusky throat and dark pads on fingers.
Similarities
Western Spadefoot has no lump between eyes.
Habitat
Arid regions, mixed grass prairies, sandy soil; generally same as for Western Spadefoot. Often found in flooded fields, ditches, cattle tanks.
Voice
Calls from shallow water; a short, loud quack.
Remarks
May breed in same pools as Western Spadefoot.
Range
Far s. Alta., Mont. (except ne. border and w. of Rocky Mts. e. foothills); w. N.Dak., w. S.Dak., e. Wyo., e. Colo., se. N.Mex., extr. se. Ariz., Okla., Tex. (panhandle and w.) e. into Great Plains.

Note: The **GREAT BASIN SPADEFOOT**, *Scaphiopus intermontanus,* has a bump between the eyes, and is glandular, not bony. It occurs in northwestern New Mexico, northern Arizona, western Colorado, northeastern California, north to Washington and British Columbia.

Fig. 69

Gulf Coast
Toad, p. 488

Woodhouse's
Toad, p. 488

Great Plains
Toad, p. 485

Plains
Spadefoot

Western
Spadefoot

Green Toad,
p. 486

Red-spotted
Toad, p. 487

Texas Toad,
p. 486

Western Toad,
p. 485

WESTERN SPADEFOOT

Scaphiopus hammondi

83:3
Fig. 69

Description

Length, 1½–2½ in. (3.8–6.4 cm). No bony lump between eyes. Above dusky-green, gray, or brown, with scattered splotches of darker color; on each side of dorsal midline an irregularly outlined stripe of cream or whitish that may be broken, particularly in female; tubercles of skin tipped with orange or reddish, especially in juveniles; below white. Eyes large, protuberant, with vertically elliptical pupil; tip of snout upturned. Spade single, prominent, rounded, sharp-edged. Male slightly smaller than female, with dark throat.

Similarities

Plains Spadefoot has lump between eyes, snout tip not upturned; Couch's Spadefoot has sickle-shaped spade.

Habitat

Short-grass hills and plains, alkaline flats in arid and semiarid regions; absent from extreme desert areas and from high mountain elevations.

Habits

Nocturnal, secretive, seldom seen except in breeding season mid-February to August; spends much time below ground in self-made burrows.

Voice

A single low-pitched, hoarse, rasping *"a-a-a-ah* or *w-a-a-ah"* (Stebbins 1954). A distant lively chorus may sound like a person sawing wood.

Remarks

May breed in same pools as Plains Spadefoot.

Range

Calif., except nw. coast and inland mt. systems and se. deserts; s. into n. Baja Calif.; se. B.C., e. Wash., extr. nw. Mont., e. Oreg., Idaho (except cen. and sw. Mont. boundary area), sw. Wyo.; Nev., Utah, sw. Colo., Okla. panhandle, w. Tex., Ariz., and N.Mex.

COUCH'S SPADEFOOT

Scaphiopus couchi

83:2

Description

Length, 2½–3½ in. (6.4–8.9 cm). Larger than other western species. Above dull brownish-yellow to bright greenish-yellow with coarse, irregular network of brown to blackish; below whitish. Eyes large, pupil vertically elliptical. Spade sickle-shaped. Skin of back rather uniformly tuberculate. Male smaller, more yellow-green with less-conspicuous, dark markings.

Similarities

Western Spadefoot has snout tip upturned, spade rounded; fingers and parts of hind limb from tibia to toes relatively shorter.

Habitat

Arid and semiarid regions supporting growth of yucca, cactus, mesquite, short grasses.

Voice

"A loud, resonant *ye-ow* or *wow,* with a sighing drop in pitch, each call lasting ¾–1¼ seconds" (Stebbins 1954) suggesting a human moan or bleat of a lamb.

Reproduction

Commonly breeds in rain pools.

Range

Se. Calif., Ariz., e. to ne. N.Mex., Okla. panhandle, w. Tex.

LEPTODACTYLIDS
Family Leptodactylidae

Members of this family are froglike or toadlike with prominent tubercles on undersides of toes near joints. They have teeth on the upper jaw, and the eardrum is smooth and not apparent. These species lay eggs on land. Most are tropical.

BARKING FROG
Hylactophryne augusti   **84:7**

Description
Length, 2–3¾ in. (5.1–9.5 cm). Head large, broad, with transverse fold of skin behind dark brown eyes. Large ventral disk on belly formed by circular fold of skin. Toes unwebbed, with prominent-pointed tubercles on undersides at joints. Above light purplish-gray or brown more or less clouded with cream, often abundant across head, on mid-back, and on dorsal surfaces of limbs; head, back, limbs blotched with dark brown; iris dark brown grading to dull gold in upperpart; below unmarked white with pinkish or purplish cast, especially on underside of limbs and posteriorly on body. Eardrums well defined, thin, rather transparent. Male smaller than female.
Habitat
In rocky areas of canyons in crevices, under rocks, in caves and cracks of stone walls, in rock-lined wells.
Habits
Exclusively terrestrial, permanent water not required for existence; nocturnal, walks in stilted manner with elevated body and tarsi.
Voice
A rapid yapping ending in a metallic ring, usually at night or during heavy showers.
Range
Extr. se. Ariz., and s. N.Mex., w., cen., and s. Tex. into Mexico.

Note: The **CLIFF FROG**, *Syrrhophus marnocki* (**84:8**), is small, ¾–1½ in. (1.9–3.8 cm). It has a flattened head and body and is green with dark speckling. It occurs on rocks and cliffs from Edwards Plateau to trans-Pecos, Texas.

TRUE TOADS
Family Bufonidae

These are the classical, squat, warty, terrestrial, hopping anurans. They are notable for their conspicuous neck glands. Because of their thick skins they can tolerate harsh deserts as well as high elevations. They occur in all nonpolar continental areas except Australia.

COLORADO RIVER TOAD
Bufo alvarius   **83:12**

Description
Length, 3–6 in. (7.6–15.2 cm). The largest western toad. Very large, conspicuous warts on hind legs. Skin smooth for a toad, but has many small tubercles and small, scattered warts. Above uniformly dark brown to brownish green or grayish, with some warts often pale orange or orange-brown; below light, usually unmarked; iris rusty. Head has prominent, crescent-shaped cranial crests; eardrums conspicuous; neck glands almost kidney-shaped,

divergent posteriorly, 2–3 times as long as wide, smooth; area between glands 3–4 times width of gland. Several prominent round warts, usually in a row, extend backward from angle of jaw. Breeding male has dark nuptial pads on thumb and inner fingers. In juveniles scattered warts are light-colored, set in black areas.

Habitat
In arid places.

Habits
Nocturnal.

Voice
Said to resemble a ferryboat whistle, but usually drowned out by chorusing of spadefoots.

Remarks
Skin secretions toxic to dogs and other animals, irritating to human eyes through contact.

Range
Extr. se. Calif., s. Ariz., mainly in Gila R. drainage.

WESTERN TOAD
Bufo boreas

83:9
Fig. 69

Description
Length, 2½–5 in. (6.4–12.7 cm). Distinguished by whitish vertebral stripe, often dark-bordered and sometimes broken, and by virtual absence of head ridges. Above grayish, blackish, dusky-brown, or dull greenish, with numerous pitted warts light and usually brownish; below whitish spotted with black. Eardrums small, neck glands oval and well separated, slightly larger than upper eyelid. Leg glands prominent, well-developed fold of skin on tarsus. No external vocal sac. Female larger, heavier, and stouter than male; male skin smoother, pattern more subdued, has no dark throat. Juveniles usually more spotted than adults, undersides of feet brighter yellow.

Habitat
Varied between sea level to high mountains; valleys, meadows, around water; less common in forests.

Habits
Active at night; seeks shelter under boards, logs, rocks, in rodent burrows; usually walks instead of hopping, with slow, awkward gait.

Voice
High-pitched, trembling; some notes suggest calls of a brood of goslings.

Range
S. Alaska to Baja Calif.; Rocky Mts. to Pacific coast.

Note: The **YOSEMITE TOAD**, *Bufo canorus*, has large, flat neck glands separated by less than the width of one gland. It occurs mostly above 9000 ft. (2743.2 m) in the Sierra Nevada of California, from Ebbets to Kaiser Passes.

GREAT PLAINS TOAD
Bufo cognatus

83:7
Fig. 69

Description
Length, 2–4½ in. (5.1–11.4 cm). Has conspicuous large blotches arranged symmetrically on back and sides. Head ridges well developed, converging toward front. Above brownish-yellow, greenish, or grayish, sometimes with narrow stripe down back; below light, unmarked; legs green-spotted. Eardrums distinct; neck

glands small, oval; vocal sac sausage-shaped when inflated. Male smaller than female, vocal sac dusky when uninflated and light when inflated.

Similarities
In Texas Toad spots may be paired but are smaller, less conspicuous.

Habitat
Considerably varied in different parts of range, but principally deserts and prairies, farms, irrigation ditches, long-grass areas, rain pools.

Habits
Mainly nocturnal, but occasionally active by day; constructs shallow burrows.

Voice
A shrill, harsh, vibrating trill, usually one pitch, sustained 8–17 sec.

Range
Great Plains area e. from se. Alta., extr. s. Sask., e. Mont., ne. Wyo., e., s.-cen., and extr. sw. Colo., extr. se. Calif., extr. se. Nev., Utah, Ariz., N.Mex., w. Tex.

TEXAS TOAD
Bufo speciosus

83:10
Fig. 69

Description
Length, 2–3½ in. (5.1–8.9 cm). Head ridges weak or absent. Above reddish, yellowish, or gray, with or without olive spots; no vertebral stripe; below yellowish or white, plain or lightly spotted. Body broad, snout short; eardrums distinct; neck glands small, oval, separated by 1½–2 times own width; vocal sac sausage-shaped when inflated. Foot processes or spades with cutting edges, inner one sickle-shaped. Male smaller with buff, olivaceous-centered area on throat.

Similarities
Southwestern Toad has light band across head, light patch in front of neck glands; Great Plains Toad has large symmetrical blotches on back.

Habitat
In arid, semiarid, or cultivated areas in mesquite, short-grass plains, or prairies; abundant.

Habits
Gregarious.

Voice
Explosive, loud, shrill 1-second trill; chorus can be deafening.

Range
Extr. sw. Kans., w. Okla. (except w. panhandle), w. Tex.; also se. Calif., Nev., Ariz., cen. Mont. to e. Colo., extr. se. N.Mex.

GREEN TOAD
Bufo debilis

83:6
Fig. 69

Description
Length, 1½–2 in. (3.8–5.1 cm). Head and body flattened, small. Above bright green or yellow green with many black spots, and yellow warts, capped with small brown tubercles. Neck glands large, elongate, located obliquely on shoulders; eardrums indistinct, snout pointed, narrow cranial crests near eyes, vocal sac spherical. Below pale with pinkish or bluish cast. Black bar through each eyelid and hind limbs barred with black. Male with dark throat and dark pad on thumb when breeding.

Similarities
Red-spotted Toad is larger, gray.
Habitat
In short-grass prairies, arid or semiarid plains of mesquite, creosote bush, bunch grass.
Habits
Nocturnal.
Voice
Cricketlike low, steady trill, lasting 3–7 sec., with 5–9-sec. intervals, a mixture of a buzz and a whistle; less musical and more mechanical than voice of Desert Toad.
Range
Se. corner of Colo., sw. corner of Kans., w. Okla., w. Tex., se. Ariz., N.Mex.

Note: The **SONORAN GREEN TOAD**, *Bufo retiformis,* has a strong network of black on bright yellow-green. It occurs in southern Arizona from Organ Pipe Cactus National Monument to Kitt Park.

SOUTHWESTERN TOAD
Bufo microscaphus   **83:5**

Description
Length, 2–3 in. (5.1–7.6 cm). Neck glands oval and widely separated, with a light spot on front portion. Above variable greenish-gray, gray-brown, yellow-brown, or rusty; warts usually rusty, low and variable in size, with largest ones usually in dusky blotches that may be united. Obtuse light-colored V on head, each arm of which crosses an eyelid near or at center. Underparts unmarked whitish to yellowish or light yellowish-orange. Head shows low, weak, occasionally absent cranial crests. Male throat same color as underparts, not dark. Juveniles have warts reddish-brown to yellowish-brown, closer together and relatively larger than in adults; conspicuous dorsal mark in some California individuals; and undersides of feet brighter yellow than in adults.
Similarities
Woodhouse's Toad has white stripe down back, prominent cranial crests, elongate neck glands; Texas Toad lacks light area on eyelid and front end of neck gland.
Habitat
In dry arroyos, sandy riverbanks, and washes; coastally along washes and arroyos bordered with Live Oak, willow, or cottonwood; inland along ditches, in flooded fields, along streams.
Habits
Hops, fast and high (to 18-in. [45-cm] jumps), instead of walking.
Voice
A clear, 2–14-sec. trill suggesting *wo-o-e-e-e-e-e,* beginning with a slurring rise in pitch from about middle C and ending abruptly.
Range
Discontinuous; coastal s. Calif., in disconnected islands from s. Nev., se. Utah, across cen. Ariz. to se. N.Mex.

RED-SPOTTED TOAD   **83:8**
Bufo punctatus   *Fig. 69*

Description
Length, 1½–3 in. (3.8–7.6 cm). Head flat, rough, no crests; neck glands round, smaller than eye. Above greenish, pale gray, or brown with reddish spots and rusty warts; below light, plain or

dotted. Eyes widely spaced, warts in front of eye and on eyelid, eardrums distinct. Male smaller, throat dark, dorsal coloration generally darker.

Similarities

Green Toad is smaller, green or yellow-green.

Habitat

Prairies, deserts, near water; frequents rocky canyons from below sea level to 6500 ft. (1981.2 m).

Habits

Usually nocturnal, occasionally seen by day.

Voice

Clear, high-pitched, pleasing birdlike trill about 2 octaves above middle C; calls last 4–10 sec. with nearly equal intervals.

Range

Spotty distribution coinciding with presence of water in semiarid regions embracing e. and s. Calif., s. Nev., far s. Utah, extr. sw. and se. corners of Colo., Ariz., N.Mex., sw. Kans., w. Okla., and w. Tex.

WOODHOUSE'S TOAD

Bufo woodhousei

83:11
Fig. 69

Description

Length, 2–4 in. (5.1–10.2 cm). Has prominent white vertebral stripe. Head thick, ridges prominent; bony ridges behind eyes touch neck glands, which are smooth and narrow, 1½ × length of eyelid. Above yellow-brown, greenish, grayish, or blackish; dark spots usually include several small warts; below yellowish, unspotted. Skin smoothish, snout rounded. Male has dark throat, most noticeable during breeding season.

Similarities

Southwestern Toad has no vertebral stripe and has weak head ridges; Dakota Toad has parallel-sided head bars, body heavily spotted below.

Habitat

In gardens, fields, woods, deserts, valleys; frequents sandy river banks, marshes, irrigated areas.

Habits

Active by day and night; burrows into soil.

Voice

A prolonged wheezy trill, for example, *wa-a-a-a-a-a-h,* at distance suggesting sheep's baaing; call lasts 1–2½ sec., at intervals of 5–13 sec., pitched about D to A above middle C.

Range

Far se. Wash., s. Mont., sw. N.Dak., sw. S.Dak., Nebr., Kans., Okla., Tex.; extr. n.-cen. and extr. e.-cen. Oreg., Idaho, Wyo., Utah, Colo.; extr. se. Calif, extr. se. Nev., e. Ariz.; also very extreme sw. tip of N.Mex.

Note: The **DAKOTA TOAD,** *Bufo hemiophrys,* has parallel-sided bars running from its snout between the eyes to the back of the head. It occurs in southeastern Wyoming.

GULF COAST TOAD

Bufo valliceps

83:4
Fig. 69

Description

Length, 2–4 in. (5.1–10.2 cm). Dark stripe along side of body, bordered above by light stripe, will distinguish it from all other toads. Large, somewhat flat, with strongly developed head ridges separated by a deep trough and with triangular neck glands. Light vertebral stripe. Above variable from almost black to rust-brown. Male with a clear yellow-green throat.

Habitat
Roadside and irrigation ditches, yards, coastal grasslands.

Voice
Sound of wooden rattles, short, 2–6 sec., repeated several times after 1–4-sec. pauses.

Range
S. Tex., e. to Ark.

TREEFROGS AND ALLIES
Family Hylidae

Members of this family are typically small, thin, narrow-waisted, with rounded pads at the tips of the toes and teeth in the upper jaw. They lack neck glands and cranial crests, and the eardrums are concealed or exposed. Most species are good climbers; others are terrestrial or aquatic, and some are more or less burrowers. They abound in the New World tropics, and occur on all continents except Antarctica.

CHORUS FROG
Pseudacris triseriata

84:1

Description
Length, ⅝–1½ in. (1.6–3.8 cm). Head pointed; fingers not webbed, toe webs short, toe disks small. Above variable, green, gray, or brown, usually with 3 broad dark stripes, middle stripe often broken; upper lip dark-edged, no dark triangle between eyes; below whitish, sometimes with a few black spots. Eardrum round, smallish, not touching jaws; skin smooth. Male has greenish-yellow to dark olive throat.

Similarities
Pacific Treefrog has wider head, blunter snout, toe tips more expanded, hind toes webbed.

Habitat
In damp woods, swampy, marshy places; commonly found on ground or in low bushes. In West, primarily grasslands or wooded areas.

Habits
Seen by day or night during breeding season; a poor swimmer and climber.

Voice
Only in spring, a rising series of vibrant chirps like running fingers over teeth of a comb, 30–70 calls per min., each ½-sec. long or less.

Range
Sw. District of Mackenzie, far ne. B.C., Alta., Sask., e. Mont., N.Dak., S.Dak., s. and e. Idaho, Wyo., e. Utah, w. Colo., cen.-e. Ariz. pine belt, n.-cen. N.Mex.

Note: The **SPOTTED CHORUS FROG**, *Pseudacris clarki,* has patches of bright green rimmed with black. It occurs in Kansas, Oklahoma, northwestern to southern Texas.

NORTHERN CRICKET FROG
Acris crepitans   84:2

Description
Length, ⅝–1⅜ in. (1.6–3.5 cm). Distinguished by light line down
back and light bar from eye to arm, a dark triangle between the
eyes, dark stripe on rear of thigh, and webbed hind feet. Above
variable, ground color gray, green, or various shades of brown to
nearly black; dark streak on sides; below whitish, unspotted or
dusky-spotted mainly in gular area and on chest. Skin usually
warty, but may be smooth; snout elongate, tapered; ear indistinct,
hind legs long; disks on digits very small, indistinct. Male throat
grayish to sooty, chest and throat commonly more spotted than in
female.

Habitat
Grassy borders of streams, ponds, swamps.

Habits
Active day and night, terrestrial, not a climber.

Voice
Calls from rim of pond; cricketlike; rapid, sharp, clear notes, *kick,
kick, kick,* like two stones struck together, about 1 sec. each,
accelerating to 5–6 per sec.

Remarks
Enters water when frightened; a tremendous jumper, to 3 ft.
(0.9 m) high and 4-ft. (1.2-m) distance.

Range
Nebr., e. Colo., e. N.Mex., Tex., and e. to Atlantic Ocean.

ARIZONA TREEFROG
Hyla wrightorum   84:4

Description
Length, ¾–2¼ in. (1.9–5.7 cm). Very similar to Pacific Treefrog,
but eyestripe longer, extending along sides to shoulder, where it
may be broken into segments. Back green, rear of femur and groin
orange or gold with greenish tinge; below light, generally
unmarked, may show a few dark flecks on throat. Commonly no
dark mark on head, although occasionally one on each eyelid; often
pair of longitudinal dark bars posteriorly on back, sometimes with
a pair of bars or spots farther forward; otherwise back generally
unmarked. Male throat dull greenish and tan; female throat
whitish.

Similarities
Toes less fully webbed, toe pads smaller, eyestripe longer than in
Pacific Treefrog.

Habitat
In wooded areas.

Habits
Climbs trees to considerable height.

Voice
A low-pitched, harsh, metallic clack, 2–12 notes in succession
accelerating toward end.

Reproduction
When breeding, individuals migrate toward water, favoring large,
grassy, shallow ponds.

Range
Ponderosa Pine belt of cen.-e. Ariz. and e. to n.-cen. N.Mex.

CANYON TREEFROG
Hyla arenicolor

84:3

Description
Length, 1¾–2½ in. (4.4–5.7 cm). Has no well-defined eyestripe. Above variable, ash-gray to dark brown, may be almost black; scattered splotches of darker color usually on back; below whitish or cream, yellow, or orange in femoral region, groin, and axilla. Male has dark throat.

Similarities
Resembles Pacific Treefrog, except skin generally rougher, head broader, snout blunter; toes with more expanded tips.

Habitat
Commonly in rocky canyons among boulders and scattered rock-bound pools in arid or semiarid regions.

Habits
Highly camouflaged; perches by day in crevices or niches in boulders within a jump or so of water, crouches when approached.

Voice
Single-pitch whir, 1–3 seconds.

Range
Far s. Utah, extr. sw. Colo., Ariz., N.Mex., far w. Tex.

Note: The **CALIFORNIA TREEFROG,** *Hyla californiae,* is pale gray rather than brown, with a well-developed web on the hind toes, and ducklike quack. It occurs in the mountains of southern California.

PACIFIC TREEFROG
Hyla regilla

84:6

Description
Length, ¾–2 in. (1.9–5.1 cm). Has black eyestripe and small adhesive disks at tips of toes; hind toes webbed, margins of webs between toes incurved when spread. Ground color variable—green, various shades of brown, light gray to nearly black; below usually unspotted whitish or pale yellow, more pronounced posteriorly and on undersurfaces of limbs. Conspicuous blackish eyestripe from nostril to well back of eye, broadest behind eye, narrowing toward front; triangular or Y-shaped head mark; several dusky longitudinal stripes on back that may be broken into spots, bars, or blotches. Male throat olivaceous to dusky.

Similarities
Canyon and California Treefrogs have no eyestripe; in Arizona Treefrog eyestripe extends beyond shoulder. In Canyon Treefrog margins of webs between toes more nearly straight when spread.

Habitat
Sheltered in fissures, crevices, vegetation along streams, burrows, nooks and crannies of buildings, culverts; frequents bodies of water but may occur up to ½ mi. (0.8 km) away.

Habits
Largely nocturnal, not attracted to trees (despite name), and usually terrestrial.

Voice
A startlingly loud *kreck-ek,* last syllable rising, uttered as rapid series; also single, prolonged, lower-pitched *kr-r-r-eck.*

Range
Vancouver Is., far s. B.C.; Wash., Idaho, far w. Mont.; Oreg., Calif., Nev.

BURROWING TREEFROG
Pternohyla fodiens   84:5

Description
Length, 1–2 in. (2.5-5.1 cm). The only casque-headed frog in the
U.S. Skin on top of head joined to bony plate beneath; bony ridge
from nostril to eye; fold of skin at back of head. Body squat. Toes
lacking pads, hind foot with one large tubercle. Above, green-
brown to yellow-brown with large, irregular brown blotches
rimmed with black; below white. Male with dark area on each side
of throat.
Habitat
Cattle tanks and puddles in mesquite grassland.
Voice
"Walk, walk, walk," 2–3 per sec., ⅛ –½ sec. each, all at a single
low pitch.
Range
S. Ariz. between Sells and Ajo.

TRUE FROGS
Family Ranidae

These generally long-legged, slim-waisted anurans are the only
frogs with teeth in the upper jaw, large distinct eardrums, and
broadly webbed hind feet and usually with a prominent ridge down
each side of the back. They lack head ridges, neck glands, disks on
the digits, and their skin is usually quite smooth. Worldwide there
are about 400 species.

MOUNTAIN YELLOW-LEGGED FROG
Rana muscosa   85:5

Description
Length, 2–3¼ in. (5.1-8.3 cm). Above blotched or spotted; below,
white with yellow or orange beneath hind legs and sometimes
entire belly. Toe tips dusky-colored. Dorsolateral folds present but
indistinct.
Similarities
Foothill Yellow-legged Frog has triangular buff patch on snout,
more granular skin, no dark toe tips.
Habitat
Only frog of high Sierra Nevadas where it inhabits streams, rivers,
pools, and lakes, especially along sunny, rocky, sloping banks.
Male with base of thumb swollen and darkened.
Remarks
Smells like garlic when handled.
Range
Calif.: (1) Sierra Nevadas, mostly about 6000 ft. (1829 m); (2) mts.
of s. Calif. from Pacoima R. s. to Mt. Palomar.

FOOTHILL YELLOW-LEGGED FROG
Rana boylei   85:3

Description
Length, 1¾–2¾ in. (4.4–7 cm). Above reddish-brown, gray,
olivaceous, or greenish with varying degrees of dusky spotting and
mottling; some individuals largely unspotted; triangular buff patch
on snout; below whitish to cream, becoming yellow posteriorly and
on hind limbs; dusky mottling in varying degrees on gular area,

pectoral region, sides of body, and forepart of femur. Skin variously roughened by tiny tubercles; eardrums rough or smooth, usually same color as head; dorsolateral folds obscure. Breeding male has swollen, darkened thumb base; female thumb longer, not swollen.

Similarities

Mountain Yellow-legged Frog has smoother skin, no snout patch, dark toe tips, heavier spotting and mottling above; Red-legged Frog has dark eye mask, well-defined dorsolateral folds, red on underside of legs, smooth eardrums.

Habitat

Nearly always close to water, favors creeks with rocky courses and commonly found in slow-moving water.

Habits

Active by day, often basks on shore or on rocks; when disturbed, seeks to hide beneath stones or in stream sediments.

Voice

Guttural, grating, on one pitch or with rising inflection; lower notes begin about 2 octaves below middle G.

Range

W. Oreg. s. of Salem, Calif., s. of Cascade crest to coast, and s. coastally to n. Los Angeles Co. and along w. foothills of Sierras.

Note: The **TARAHUMARA FROG**, *Rana tarahumarae*, has no mask or light-colored jaw stripe. It is dusky below, including the throat, and occurs in southern Arizona in the Pajarito and Santa Rita Mountains.

RED-LEGGED FROG
Rana aurora   85:6

Description

Length, 2–5 in. (5.1–12.7 cm). Back brownish to olive, with well-defined, or fuzzy, spots commonly having light centers; limbs blotched and crossbarred with blackish; groin mottled with black, yellow, and red. Usually blackish to dark brown eye mask from nostril to angle of jaw, but sometimes vague or absent; whitish streak above mouth extends from below eye toward shoulder. Skin either smooth or rough, dorsolateral folds present; eardrums smaller than eye opening. Male toes more webbed than females, and thumb short, swollen, basally darkened; female thumb elongate, lacks nuptial pad.

Similarities

Spotted Frog has light stripe on jaw extending onto shoulder, groin not mottled; Foothill Yellow-legged Frog has legs yellow below, pale triangle on snout, vague dorsolateral folds.

Habitat

Near bodies of water. In north of range may be considerable distance from water, but in damp, vegetated places.

Habits

Highly aquatic; can jump 3 ft. (0.9 m).

Voice

"A stuttering series of 5 or 6 low, guttural, grating sounds, *r-r-r-r-rowr*, followed by a low-pitched growl or yowl" (Stebbins 1954), lasting about 3 seconds.

Range

Coastal belt from e. Vancouver Is. and extr. sw. B.C. to Baja Calif., in Sierras as far s. as n. Butte Co.

Note: The **CASCADES FROG**, *Rana cascadae*, has light-centered ink spots on its back. It occurs in the Cascade and Olympic Mountains of Washington, south to Lassen Peak of California.

BULLFROG
Rana catesbeiana   85:9

Description
Length, 3½–8 in. (8.9–20.3 cm). The largest frog in North
America. Above green (especially forward), brownish, or sometimes
blackish with darker spots; hind legs with dark crossbars; below
whitish, often mottled. Skin fairly smooth, no ridges down sides of
back, but with ridge from back of eye around part of eardrum.
Web of 4th toe extends to tip. Male eardrum much larger than
eye, dark brown marginally, female eardrum as large as eye; male
throat yellowish. Juveniles usually show well-defined dark dorsal
markings.
Similarities
Green Frog is smaller, has well-defined dorsolateral folds; 4th toe
web does not reach tip.
Habitat
Permanent bodies of water, in varied habitats between tule ponds
or arid regions to mountains.
Habits
Aquatic, nocturnal; hibernates in mud, emerges late in spring; in
breeding season, males establish croaking posts.
Voice
Deep, hoarse, *"jug o'rum, more rum, better go around"* (Stebbins
1954).
Range
Very many discontinuous localities throughout all western states;
highly restricted localities in arid regions; artificially introduced
from the East and, proving highly successful, has spread rapidly.

Note: The **GREEN FROG,** *Rana clamitans,* has dorsolateral folds
that extend halfway down its back. It has been introduced at Toad
Lake, Washington, and Weber River, Utah.

NORTHERN LEOPARD FROG
Rana pipiens   85:4

Description
Length, 3–4 in. (7.6–10.2 cm). The only frog with irregularly
distributed round or oval dark spots above and white underthighs.
Above brown, gray, or green; below white; legs with dark bands;
light line along upper jaw toward shoulder. Skin smooth, body
slender, ridges down sides of back light; vocal sacs paired between
arms and ears. Breeding male has much enlarged thumb and
convex webs on hind feet; is usually darker than female. Juveniles
may be unspotted or show much-reduced spotting.
Similarities
Grouped under this name is a complex of 3 or more species that
are loosely related and very difficult to distinguish.
Habitat
Marshes in spring, grassy woodlands and swamps in summer,
pools or marshes in winter. Seems to prefer cattail marshes and
shallow dead streams.
Habits
Can change color rapidly; a good jumper and swimmer.
Voice
Long, low, guttural moaning, chuckling notes; long and short notes
variously interspersed; a rattle and a grunt; a deep and musical
ker-r-r-r-ock; may croak both on or under water.
Remarks
Will urinate on collector's hand.

Range

Far s.-cen. Mackenzie, se. B.C., Alta., Sask., s. Man., s. into Mexico. W. boundary of range irregular along far e. state lines of Wash. and Oreg.; extension along Columbia R. to Portland, far ne. and extr. se. Calif. Absent from extr. w. part of s. Nev., tip and extr. sw.-cen. Ariz.

SPOTTED FROG
Rana pretiosa 85:7

Description

Length, 2–4 in. (5.1–10.2 cm). Above light to dark brown, spotted with irregular, variously developed ink-black blotches lacking any light margins, but with centers sometimes light. Eyes slightly upturned; eye mask brownish, sometimes obscure, extends to angle of lower jaw; light stripe snout to shoulder. Below red, reddish-orange, or yellow; throat and sometimes entire undersides spotted and mottled; limbs blotched and spotted, markings may form bands. Skin usually smooth, dorsal surfaces in older individuals may be finely roughened; dorsolateral folds usually present, although only moderately developed. Limbs and toes relatively short; webbing rather extensive. Male usually smaller, thumb base swollen. Immatures may lack yellow, orange, or red color below.

Similarities

Red-legged Frog has longer limbs and toes, less extensive webbing, smoother skin; eyes not upturned.

Habitat

Frequents marshy places, ponds, lakes, springs, streams; ranges to 10,000 ft. (3048 m) elevation.

Habits

Highly aquatic, not a strong swimmer or jumper; hibernates in north of range.

Remarks

Easily captured.

Range

Extr. s.-cen. Yukon, se. Alaska, B.C., far sw. Alta.; Wash. (e. of Cascade crest), Idaho, w. Mont., far w. Wyo.; Oreg., n. and n.-cen. Nev., n.-cen. Utah, as far as s. Sevier Co.

WOOD FROG
Rana sylvatica 85:8

Description

Length, 1½–3 in. (3.8–7.6 cm). Well-defined blackish or dark brown eye mark, contrasting white line on upper jaw below mask. Above dark brown to reddish-brown, greenish, yellowish-gray, or gray; usually 2 light stripes on prominent dorsolateral folds, occasionally a light vertebral line from snout to vent; below whitish or cream, may show dark mottlings on throat and breast. Light-centered dark spots, occasionally elongate, may occur on back and sides. Male hind foot webs convex, thumb swollen, skin darker in breeding season.

Similarities

Spotted Frog is red, yellow, or orange below; Red-legged Frog is red beneath legs, has less distinct eye mask.

Habitat

In damp woods, sometimes far from water.

Habits

Hibernates in logs, stumps, under stones, but not in water.

Voice
Hoarse, grating, clacking, not unlike a duck's quack, resembling voice of Leopard Frog. May croak underwater.
Range
Above 8700 ft. (2652 m) elevation. Alaska; Yukon; sw. Mackenzie; B.C.: Alta.; Sask.; Jackson Co., Colo.

NARROW-MOUTHED TOADS
Family Microhylidae

These small, chunky anurans have a smooth skin, a small head with pointed snout, and short limbs. Primarily tropical, various members may be arboreal, terrestrial, or burrowing in habit, but of some twenty-seven world species, only one occurs in Western North America.

GREAT PLAINS NARROW-MOUTHED TOAD
Gastrophryne olivacea   **85:2**

Description
Length, ⅞–1⅜ in. (2.2–3.5 cm). The only anuran in the West with a narrow, pointed, turtlelike head. Body somewhat flattened, waist broad. Above gray, brown, or blackish; below lighter, unmarked or with flecks. Hind legs stout and short, commonly marked with faint to distinct transverse bars; feet like webs. Skin smooth, fold of skin across head behind eyes; eardrums, neck glands absent. Male throat dark, minute tubercles on lower jaw, chest, abdomen.
Habitat
Cattle tanks, pools along intermittent streams, spring seepages; under rocks, boards, logs.
Habits
Nocturnal; may be active by day during rains.
Voice
Calls from shallow water; loud, unmusical, rapidly vibrating, buzzing, suggesting a honeybee at close range.
Range
Extr. s. Ariz.; w. Tex., ne. to Okla. and Kans.

Fishes

Consulting Editors

Saltwater Fishes
John E. Fitch
Research Director
California State Fisheries Laboratory
Department of Fish and Game, State of California

Freshwater Fishes
Robert R. Miller
Curator of Fishes
Museum of Zoology, University of Michigan

Illustrations
Plates 86–98 Nina L. Williams
Plates 99, 100–102 Jennifer Emry-Perrott
Text Illustrations by Nina L. Williams and
Jennifer Emry-Perrott

Fishes
Superclass Pisces

Of the world's estimated 20,000 to 25,000 different species of fishes—more than all other vertebrates put together—a total of 276 species is included in this section, divided between 133 saltwater and 143 freshwater forms. These represent the species most likely to be observed in a market or a fisherman's bag, or be seen by the amateur naturalist, whether they be fishes of prey or those commonly seen by tidepool scroungers, skindivers, or aquarium habitués.

Food
Many fishes are carnivorous. Most carnivorous fishes feed on other kinds of fishes or on marine invertebrates such as jellyfish, copepods, and squid and other mollusks, and on plankton, the minute animal and plant life that floats in the top layers of the sea. Some fishes, however, are herbivorous and feed on seaweeds or on the microscopic one-celled diatoms. Other fishes are scavengers and help keep the sea and rivers clean by feeding on carrion and waste.

Reproduction
In most fishes, as in all the land vertebrates, the sexes are separate. Female fishes produce eggs, the males milt. In most species fertilization is external, the males discharging milt over the eggs as they are extruded by the female. However, in the sharks, skates, and rays, and in some killifishes and gambusia, fertilization is internal. Sharks and rays bring forth their young alive. Skates lay eggs. In these three groups the pelvic fins of the male are modified into claspers that serve as organs of intromission. In males of the viviparous killifishes and gambusia the anal fin is so modified.

In the fishes that practice external fertilization, the eggs may vary in size from 1/50 of an inch (0.05 cm) in diameter to ⅞ of an inch (2.2 cm). Eggs of different species may variously float or sink and be free in the water or become attached to weeds, snags, or rocks. They may be separate or they may adhere to each other in globular or stringy masses.

The number of eggs of any one species has little or nothing to do with the number of individuals of that species in the sea. Population sizes are controlled by the environment, in particular by the availability of food. Given ample food supplies and an increase in the amount of suitable territory, any species, no matter how few eggs it may lay, will soon have occupied the additional territory and stocked it to the limit.

Only a few saltwater fishes pay attention to their eggs after they are laid. Many freshwater species, on the other hand, build nests of sticks or stones, or scooped-out sand. The males often assist incubation by stirring the water around the eggs. They guard the eggs during incubation and sometimes the young after they are hatched.

Evolution
The earliest fishlike vertebrates, which possessed a notochord (embryological and evolutionary predecessor of the spinal column), that appear in the geologic column* were the armored, jawless ostracoderms. They arose in the late Ordovician period, flourished during the Devonian "Age of Fishes," and then died out, leaving as

*Ransom, J. E., *Fossils in America*. New York: Harper & Row, 1964, pp. 85–107.

Fig. 70

Parts of Typical Fishes

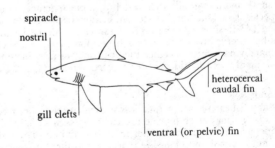

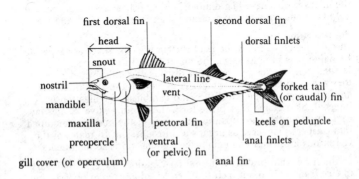

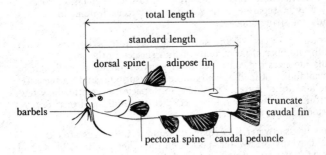

the sole relic of their class the hagfishes and lampreys. Jaws were gradually developed by the Placoderms in one of the great steps in the history of evolution. Although these early jawed fishes are all extinct now, they were ancestral to the sharks, and finally, the true bony fishes. Well-jawed, well-toothed sharks first appeared in the Devonian and for the rest of the Paleozoic era dominated the seas. They have continued, in an aggressive though subordinate capacity, to roam the salt waters of the world ever since.

The bony fishes first appeared in the Ordovician or late Silurian, but the more advanced forms did not become conspicuous until the Mesozoic era. From then on many multiplied their varieties and their numbers and successfully withstood predation and competition, including that of the marine reptiles, such as the mosasaurs and ichthyosaurs, and of marine mammals like the whales—all members of more advanced vertebrate groups that reinvaded the sea. If success were to be reckoned by variety and numbers, the bony fishes would be judged the most successful backboned animals on the globe today. They display a greater variety of species and an astronomically greater number of individuals than all other vertebrates combined.

Conservation

Fishes are of great importance to man. We eat probably several billion of them a year; we use many for oil, fertilizer, and other purposes; and we catch them for sport. Angling for sport, in its various branches, has increased considerably in recent years. Some 25,000,000 North Americans buy licenses each year to fish the fresh waters.

Fishes need conservation just as land animals and birds do. Even though commercial fisheries often take only a relatively small harvest of the great bounty of the sea, there are many problems to consider, such as the overexploitation of fishery resources by the efficient use of large commercial trawls and factory ships, and the pollution caused by industrial wastes. These and many other conservation questions engage the attention of federal and state fishery authorities.

In fresh waters the situation is quite different. Entire river systems in the past have been polluted and the fish destroyed by the dumping of toxic factory wastes. Much of this still goes on. Anglers, too, can soon overfish a stream and do so in heavily fished areas. Most states have had to institute elaborate hatchery and restocking programs in order to offset the effects of pollution and of heavy angling pressure. Such programs, if wisely conducted, are beneficial; but much must still be done to stop pollution.

The question of conservation leads inevitably to the question of the sea as a new food frontier for man, because fish are notably rich in vitamins, minerals, and proteins and make an excellent human food. The resources of the oceans and even the topography of their floors are still most inadequately understood. Scientists recognize that there is a finite limit to fishery resources and that the situation is presently at its upper limit or even has exceeded the capacity of fish populations to renew themselves.

Where to Look for Fishes

To find the best places for fishes, the angler consults the rod and gun columns in area newspapers, his local tackle shop, and his fellow anglers. To see fishes one may inspect the local and metropolitan markets after deliveries or visit fishing docks when the boats come in. The beginning ichthyologist will find museums, aquariums, and marinelands (oceanariums) both interesting and informative. Many fish watchers practice skin diving with snorkel and flippers or, if more advanced, with an Aqua-lung. For the sake of safety, enthusiasts usually go in pairs.

Range and Scope
Described in this chapter are the most readily identifiable species of saltwater and freshwater fishes in the field that (1) occur annually within a depth of 100 fathoms (182.8 m) off the west coast of North America between the Arctic Ocean and the northern border of Mexico, and (2) occur annually in the fresh waters of continental North America west of the 100th meridian between Pt. Barrow, Alaska, and the southern boundary of the United States.

This chapter will help identify fishes that have been caught as well as those that are found alive in aquariums or are displayed in fish markets. Range delineation is generalized, couched in terms of an overall area rather than in specific watersheds. Many species of freshwater fishes, for example, occur in a discontinuous distribution with a broad range; other species may be confined or restricted to a few water holes, a few isolated lakes, or the waters of a few widely separated streams. Still other species are found as transplants from distant areas, and their western distribution conforms more to the interest of sportsmen than to the ecological advantages of environment. The sizes given are the known maximum (record) sizes.

Classification
In this chapter, the classification of saltwater fishes is a compromise between the AFS Checklist, as modified by staff consultants in the Department of Zoology at the University of California in Berkeley, and the classification system of Greenwood et al. (see Useful References). The freshwater fishes classification is based on that given in *Ichthyology*.

Nomenclature
Within recent decades, numerous changes have been proposed for the scientific names of many fishes. There is as yet no firm agreement among scientists as to the proper generic placement for many fishes or even wholehearted agreement as to the relative order in which some fish families should appear. In this chapter, the scientific names and phylogenetic order of families follow the *Guide to the Coastal Marine Fishes of California* (1972). Within the family category, the order of presentation of the genus and species is alphabetical.

Ever since a broad spectrum of nationalities settled the West Coast originally, there has been exceptional confusion in the use of common names for both saltwater and freshwater fishes. Even regional fish and game department publications vary from one another in their choice of many common names. Thus, the Striped Seaperch, *Embiotoca lateralis,* bears such locally popular names between Alaska and Southern California as "Striped Surfperch," "Rainbow Perch," "Blue Perch," "Squawfish," and "Crugnoll." The common saltwater names used here follow essentially those suggested by Miller & Lee in 1972, whereas the common freshwater names follow those given in *A List of Common and Scientific Names of Fishes.*

USEFUL REFERENCES

Saltwater Fishes

Bailey, R. M., et al. 1970. A List of Common and Scientific Names of Fishes from the United States and Canada. 3rd ed. Washington, D.C.: American Fisheries Society, Special Publication 6.

Baxter, J. L. 1964. *Inshore Fishes of California.* 2d. rev. ed. Sacramento, Calif.: Fish & Game Department.

Cannon, R. 1964. *How to Fish the Pacific Coast: A Manual for Salt Water Fishermen.* 2d. ed. Menlo Park, Calif.: Lane Book Co.

Fitch, J. E., 1963. *Offshore Fishes of California.* 2d. rev. ed. Sacramento, Calif.: Fish & Game Department.

Fitch, J. E., and Lavenberg, R. J. 1968. *Deep-Water Teleostean Fishes of California.* Berkeley, Calif.: Univ. of Calif. Press.

———. 1971. *Marine Food and Game Fishes of California.* Berkeley, Calif.: Univ. of Calif. Press.

———. 1975. *Tidepool and Nearshore Fishes of California.* Berkeley, Calif.: Univ. of Calif. Press.

Gates, D. E., and Frey, H. W. 1976. Designated Common Names of Certain Marine Organisms of California. Bull. 161. Sacramento, Calif.: Dept. of Fish & Game.

Gilbert, P. W., ed. 1963. *Sharks and Survival.* Lexington, Mass.: D. C. Heath & Co.

Gosline, W. A. 1971. *Functional Morphology and Classification of Teleostean Fishes.* Honolulu, Hawaii: Univ. of Hawaii Press.

Gotshall, D. W. 1978. *Fishwatcher's Guide to the Inshore Fishes of the Pacific Coast.* Monterey, Calif.: Sea Challengers.

Greenwood, P. H.; Rosen, D. E.; Weitzman, S. H.; and Myers, G. S. Phyletic Studies of Teleostean Fishes, with a Provisional Classification of Living Forms. Bull. 131. New York: Amer. Mus. of Nat. Hist.

Hart, J. L. 1973. Pacific Fishes of Canada. Bull. 180. Ottawa: Fisheries Resources Board of Canada.

Kato, S.; Springer, S.; and Wagner, M. H. 1967. Field Guide to Eastern Pacific and Hawaiian Sharks. Circular 271. Washington, D.C.: U. S. Fish & Wildlife Service.

Miller, D. J., and Lea, R. N. 1972. Guide to the Coastal Marine Fishes of California. Fish Bulletin 157. Sacramento: Calif. Dept. of Fish & Game. Reprinted in 1976 with 13-page addendum by the Division of Agricultural Sciences, Univ. of Calif., Richmond.

Nelson, J. S. 1976. *Fishes of the World.* New York: John Wiley & Sons.

North, W. J. 1976. *Underwater California.* Berkeley, Calif.: Univ. of Calif. Press.

McClane, A. J. 1974. *McClane's Standard Fishing Encyclopedia and International Angling Guide.* New York: Holt, Rinehart & Winston.

Quast, J. C., and Hall, E. L. 1972. List of Fishes of Alaska and Adjacent Waters with a Guide to Some of Their Literature. Fisheries Series 658. Washington, D.C.: National Marine Fisheries Services, Special Science Report.

Somerton, D., and Murray, C. 1976. *Field Guide to the Fish of Puget Sound and the Northwest Coast.* Seattle, Wash.: Univ. of Wash. Press.

Turner, C. H., and Sexmith, J. C. 1967. *Marine Baits of California.* Sacramento, Calif.: Dept. of Fish & Game.

Freshwater Fishes

Bailey, R. M., and Allum, M. O. 1962. *Fishes of South Dakota.* Ann Arbor: Miscellaneous Publications, No. 119, Museum of Zoology, Univ. of Mich.

Baxter, G. T., and Simon, J. R. 1970. Wyoming Fishes. Bull. 4. Cheyenne: Wyo. Game & Fish Dept.

Beckman, W. C. 1952. Guide to the Fishes of Colorado. Leaflet No. 11. Boulder: Univ. of Colo. Museum.

Blair, W. F.; Blair, A. P.; Brodkorb, P.; Cagle, F. R.; and Moore, G. A. 1968. *Vertebrates of the United States.* Vol. II. New York: McGraw-Hill.

Brown, C. J. D. 1911. *Fishes of Montana.* Bozeman, Mont.: Big Sky Books, Mont. State Univ.

Caine, L. S. 1949. *North American Fresh Water Sport Fish.* New York: A. S. Barnes.

Carl, G. C.; Clemens, W. A.; and Lindsey, C. C. 1959. *The Fresh-Water Fishes of British Columbia.* Handbook 5. Vancouver: British Columbia Provincial Museum.

Cross, F. B. 1969. *How to Know the Freshwater Fishes.* Dubuque, Iowa: Wm. C. Brown Co.

Jordan, D. S., and Evermann, B. W. 1896–1900. The Fishes of North and Middle America. Parts I-IV. Bull. 47. Washington: Smithsonian Institution.

Koster, W. J. 1957. *Guide to the Fishes of New Mexico.* Albuquerque: Univ. of N. Mex. Press.

Lagler, K. F., et. al. 1962. *Ichthyology.* New York: John Wiley & Sons.

LaMonte, F. 1950. *North American Game Fishes.* New York: Doubleday.

LaRivers, I. 1962. *Fishes and Fisheries of Nevada.* Reno: Nev. State Fish & Game Commission.

McPhail, J. D., and Lindsey, C. C. 1970. Freshwater Fishes of Northwestern Canada. Bull. 173. Ottawa: Fisheries Research Board of Canada.

Miller, R. J., and Robison, H. W. 1973. *The Fishes of Oklahoma.* Stillwater, Okla.: Okla. State Univ. Press.

Minckley, W. L. 1973. *Fishes of Arizona.* Phoenix: Ariz. Game & Fish Dept.

Moyle, P. B. 1976. *Inland Fishes of California.* Berkeley: Univ. of Calif. Press.

Paetz, M. L., and Nelson, J. S. 1970. *The Fishes of Alberta.* Edmonton: Govt. of Alberta.

Sigler, M. F., and Miller, R. R. 1963. *Fishes of Utah.* Salt Lake City: Utah State Dept. of Fish & Game.

Walters, V. 1955. Fishes of Western Arctic America and Western Arctic Siberia: Taxonomy and Zoogeography. Bull. 106. New York: Amer. Mus. of Nat. Hist.

GLOSSARY

Abdominal ridge Belly ridge from region below gills to anus.

Adipose fin A fleshy fin, without rays, between dorsal and caudal fin.

Anadromous Ascending rivers from salt water to spawn.

Anal fin Fin situated between anus and caudal fin.

Anterior Toward the front of the body (usually toward the head or cranial end).

Barbel Fleshy appendage projecting from upper jaw, lower jaw, or chin region.

Body depth Greatest vertical distance through body, not including fins.

Branchiostegal rays Bony rays which support the membranes under the head below the opercular bones.

Caudal fin Tail fin.

Caudal peduncle Posterior end of body from last ray of anal fin to base of caudal fin.

Cephalic fins Detached part of the pectorals on the heads of certain rays.

Circuli A series of concentric ridges on the scales.

Cirrus (*pl.* cirri) Skin flap projecting outward from either head or body.

Claspers Extension from the paired pelvic fins of male sharks, rays, and chimaeras that serve as copulatory organs.

Compressed Flattened from side to side (laterally compressed) or from top to bottom (dorsoventrally compressed).

Ctenoid scales Bony fish scales with small toothlike projections from the posterior edge.

Cycloid scales Bony fish scales with a smooth posterior edge.

Dermal denticles Spinelike "scales" of a cartilaginous fish; also termed placoid scales.

Disk The flat, circular or diamond-shaped forepart of skates and rays, formed by fusion of pectoral fins to head.

Dorsal Toward the upper back region of body.

Dorsal fin Fin situated on top of the back, not including the adipose fin.

Furcate Forked.

Epipelagic The oceanic zone into which enough light penetrates for photosynthesis.

Gill arch Bony or cartilaginous structure to which the gill filaments and rakers are attached.

Gill cover Bony or cartilaginous gill cover; synonomous with *operculum*.

Gill filaments Fleshy red protuberances on the outer sides of the gill arch that serve in respiration.

Gill rakers Cartilaginous (sometimes bony) protuberances on the inner sides of the gill arch that direct food into the gullet.

Gill slits Openings between gill arches for passage of water, visible externally in cartilaginous fishes, covered by an operculum in bony fishes.

Heterocercal Unequally lobed; said of caudal fin when upper lobe is larger than lower.

Keel A longitudinal fin usually at the midline of the body near the posterior end of the fish.

Lateral line Sensory organ composed of a canal which connects a series of openings along sides of body; a posterior extension of the sensory canals on the head.

Lobule Either a small lobe or a subdivision of a lobe.

Lunate Between crescent and halfmoon in shape.

Mandible Lower jaw, composed of from four to seven bones in fishes.

Maxilla The bone lying on each side of the two halves of the upper jaw of fishes.

Medial Toward the center of the body.

Mesopelagic Relating to ocean depths from about 600 to 3000 feet.

Nostrils Paired, cuplike structures on the snout of fishes that do not connect with the mouth cavity for respiration but serve in the reception of chemical stimuli.

Notochord A flexible, rodlike structure that supports the body of vertebrate embryos and the adults of jawless fishes. In most higher vertebrates, the notochord is later replaced by the vertebral column.

Operculum Bony or cartilaginous gill cover.

Palatines Paired bones of the palate, lateral and posterior to the upper jaw bones.

Parr marks Dark vertical markings on sides of young fish, especially trout.

Pectoral fins Paired fins attached to the shoulder or pectoral girdle, located just posterior to gill openings. They may be fused with pelvic fins into a sucking disk (in the clingfish) or completely absent (as in eels).

Pelvic fins Paired fins attached to pelvic girdle, located on belly between throat and anus. Both fins may be united into a sucking disk (in gobies), be completely absent (in eels and eellike fishes), or form barbels (as in cusk-eels and brotulas).

Peritoneum The membrane lining the visceral or abdominal cavity.

Pharyngeal teeth Teeth attached to bones of the paired fifth gill (pharyngeal) arch, located immediately anterior to esophagus or gullet. Pharyngeal tooth counts are listed in order from left to right.

Placoid scales Spinelike scales of a cartilaginous fish; also termed dermal denticles.

Posterior Towards the rear of the body, usually toward caudal fin.

Precaudal pit Depression that is found immediately anterior to the base of the caudal fin in some sharks.

Premaxilla Bone at front of upper jaw, or forming entire upper jaw.

Preopercle Paired bones of the posterior cheek region, anterior to bones of gill cover (operculum).

Preorbital Region between eye and tip of snout.

Prickles Small, fine, sometimes curved spines in place of scales.

Ray Flexible (cartilaginous) supports of the soft fins; soft-fin rays.

Roe Fish eggs.

Scute An external scale usually with a sharp ridge.

Shagreen Rough, hard-scaled skin of some sharks (often used for polishing).

Spinous rays (spines) Spines present in the fins of advanced bony fishes, which are sharp, bony and usually inflexible.

Spiny Composed of sharp, inflexible spines.

Spiracle A vestigial gill slit present behind eyes of sharks, rays, and a few primitive bony fishes. In rays it is large and serves for intake of water during respiration.

Striae Small ridges or lines, usually on scales, opercula, or spines.

Tubercle Small rounded lump; a modified scale, hard or soft.

Vent Anus.

Viviparous Gives birth to living young.

Vomer Bone at the midline of the palate, immediately behind the upper jaw bone.

Saltwater Fishes

JAWLESS FISHES
Class Agnatha (Petromyzones)

Hagfishes
Order Myxiniformes (Hyperotreti)

HAGFISHES
Family Myxinidae

Members of this family are eel-shaped, jawless fishes lacking paired fins and scales. They generally inhabit cool marine waters. They are scavengers, feeding on dead or dying fish. When disturbed or handled, hagfishes secrete a great quantity of slime. Each egg is contained in a horny capsule.

PACIFIC HAGFISH
Eptatretus stoutii   *Fig. 71*

Description
Size, to 25 in. (63.5 cm). Body lacking scales and spined fins; skin thin, lax, separable from muscles; snout 1/20 total length; gill openings 10–14 on either side; teeth, 10 in each series. Color light brown to gray, paler below; edge of lower fold pale.
Habitat
Sandy or muddy bottoms.
Other name
California Hagfish.
Range
Se. Alaska to Baja Calif.; abundant in n.; occasional off San Diego.

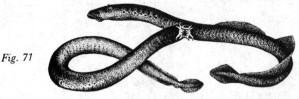

Fig. 71

Lamprey (with Pacific Hagfish attached)

CARTILAGINOUS FISHES
Class Chondrichthyes

Members of this class, which includes the sharks, skates, and rays, are distinguished from the bony fishes by an entirely cartilaginous skeleton. Most species give birth to live young but some lay eggs.

Bullhead Sharks
Order Heterodontiformes

HORN SHARKS
Family Heterodontidae

HORN SHARK
Heterodontus francisci

86:10

Description
Size, to 38¼ in. (97.2 cm). Body characterized by strong, thick spine in front of each dorsal fin and presence of an anal fin. Head with prominent forehead, steep profile; front teeth pointed, tricuspid; rear teeth blunt, molarlike. Color brown with scattered black spots.
Habitat
Bottom waters around kelp beds.
Remarks
Rather common but of little interest to anglers.
Other name
California Horn Shark.
Range
Monterey Bay to Gulf of Calif.

Cow Sharks
Order Hexanchiformes

COW SHARKS
Family Hexanchidae

Members of this family of viviparous sharks inhabit warm waters and sometimes reach a very large size. Most sharks have either six or seven gills.

SEVENGILL SHARK
Notorhynchus maculatus

86:11

Description
Size, to 8½ ft. (2.6 m). Body slender, rounded; single dorsal fin, long upper lobe to caudal fin, pelvic fin midway on belly. Head has broad, rounded snout; spiracle nearer eye than gill slits. Distinguished by 7 gill slits on each side. Color reddish-brown to sand-grayish with many black spots.
Habitat
Moderately deep water, 10–40 fathoms (18.3–73.2 m).
Remarks
Will bite. Keep fingers away from mouth until fish is dead.
Range
B.C. to n. Baja Calif.; rare s. of Pt. Conception.

Thresher, Mackerel, and Other Sharks
Order Squaliformes

These sharks have two dorsal fins, with or without spines, have no anal fins, and have five or six gill slits.

THRESHER SHARKS
Family Alopiidae

COMMON THRESHER SHARK
Alopias vulpinus   86:6

Description
Size, to 18 ft. (5.5 m) in Pacific. One of 2 eastern Pacific sharks with a tail as long as its head and body. Body stout, snout blunt; first dorsal about as high as it is long, set midway between sickle-shaped pectorals and small pelvics; dorsal and anal fins extremely small; caudal peduncle thick. Color blue-gray to purplish back fading to white belly; fins dark gray to purplish.

Habitat
Surface of temperate seas and shallow inshore waters.

Habits
Thrashes schools of forage fishes with its powerful tail.

Remarks
An excellent food fish.

Range
B.C. to tropics.

MACKEREL SHARKS
Family Lamnidae

Members of this family are distinguished by a streamlined body, slender caudal peduncle, keel on sides of peduncle, and large, sharp teeth. The last gill opening is set in front of the pectorals, which are large and half as long as they are high. The first dorsal fin is high and its base is wholly in front of the small pelvics; the second dorsal fin and the anal fin are very small. The tail is crescent-shaped and less than one-third the total length of the fish.

MAKO SHARK
Isurus oxyrhynchus   86:8

Description
Size, to 11 ft. 5 in. (348.0 cm); world record 12 ft. 6 in. (381.0 cm). Body robust; long prominent keel on sides of caudal peduncle from tail to point above pelvic fins; very small second dorsal and anal, about equally situated; first dorsal well back of base of pectoral; tail lunate. Head pointed; teeth long, sharp, knifelike edges. Color dark metallic blue back fading to white belly.

Habitat
Offshore or near islands.

Remarks
Moody, a fighter; a prime game and food fish.

Other name
Bonito Shark.

Range
Columbia R. region to Baja Calif.

WHITE SHARK
Carcharodon carcharias

Description
Size, to 21 ft. (6.4 m). Body with great depth. Small second dorsal situated in advance of anal fin; caudal fin lunate, large pectoral fins in advance of base of first dorsal; keel on each side at base of tail. Head relatively small; teeth very large; upper jaw triangular; lower jaw narrower, saw-edged. Color dark metallic gray fading to spectacularly white belly; black blotch at base of pectoral fins.

Habitat
Temperate and tropical deep water; also in shallow waters inshore.

Remarks
Extremely dangerous; most voracious fish of the open seas. Reliable reports tell of stomach contents including one or more other sharks 4–7 ft. (1.2–2.1 m) long, and 100-lb. (45.4-kg) seals and sea lions. Over 95 percent of unprovoked attacks on skindivers, surfers, and swimmers off the west coast of the U.S. are by white sharks.

Other name
Maneater Shark.

Range
Alaska to Baja Calif.

BASKING SHARKS
Family Cetorhinidae

BASKING SHARK
Cetorhinus maximus

86:13

Description
Size, to 45 ft. (13.7 m). Gill openings so large that they almost meet under the throat. Body thick, snout short, teeth very small; base of first dorsal wholly in front of small pelvics; second dorsal and anal very small, former slightly in front of latter; caudal peduncle keeled; tail crescent-shaped. Color dark metallic gray fading to light belly.

Habitat
Worldwide; surface waters inshore in summer.

Habits
Occasionally in schools of 20–30 during winter in Monterey and San Simeon bays, Calif.; basks on surface.

Remarks
Edible, liver a source of oil.

Range
Alaska to Gulf of Calif.; more common on Pacific coast of B.C. and Wash. in summer.

CAT SHARKS
Family Scyliorhinidae

SWELL SHARK
Cephaloscyllium ventriosum

86:5

Description
Size, to 3⅓ ft. (1 m). First dorsal back of mid-body, slightly behind pelvic fins; second dorsal directly above anal fin; head broad, flat. Marked with irregular brown to black spots and bars across back

and sides; below yellowish; entire surface covered with round whitish and dark spots.

Habitat
Shallow water, around kelp beds to 900 ft. (274.3 m).

Habits
For defense may inflate body with air or water to triple its circumference.

Remarks
Not a food fish; eating it may cause distress.

Range
Monterey, Calif., to Gulf of Calif.; abundant s. of San Diego.

SMOOTHHOUNDS
Family Triakididae

GRAY SMOOTHHOUND
Mustelus californicus   86:2

Description
Size, to 64¼ in. (163.2 cm). Body similar to that of Brown Smoothhound, but relatively heavier; midpoint of base of first dorsal closer to front of pelvic fins than to rear of pectoral fin base. Teeth flat, blunt. Metallic, iridescent dark gray back fading to lighter belly.

Habitat
Shallow bays to deep coastal waters; most abundant in shallow waters.

Range
Cape Mendocino, Calif. to Mazatlán; more abundant s. of Dana Point, Calif.

BROWN SMOOTHHOUND
Mustelus henlei

Description
Size, to 38 in. (96.5 cm). Body slender; first dorsal well in advance of pelvics, base of second dorsal beginning slightly in advance of anal; caudal peduncle without keels. Teeth small, flattened, with tiny pointed cusp, set in 5 or more rows. Brownish to brassy back fading to lighter sides and whitish belly.

Similarities
Soupfin, second dorsal directly above, about size of anal fin; also Gray Smoothhound, teeth flat, blunt.

Habitat
Close to shore, around piers and moored boats.

Remarks
An excellent food fish.

Range
Humboldt Bay, Calif., to Baja Calif.; most abundant shark from Humboldt Bay to San Francisco.

LEOPARD SHARK
Triakis semifasciata   86:9

Description
Size, to 6½ ft. (2 m). The only West Coast shark with large, dark gray to black bars (saddles) across back followed by spots on sides and tail. First dorsal well in advance of pelvics; insertion of second

dorsal in advance of insertion of anal; fins rounded. Color grayish back fading to lighter belly.

Habitat
Shallow surf waters to well offshore, fast-moving water in channels of bays, often near inshore colonial masses of sand crabs.

Remarks
Exceptionally fine food fish.

Range
Coos Bay, Oreg., to Mazatlán.

SOUPFIN SHARK
Galeorhinus zyopterus   86:1

Description
Size, to 6½ ft. (2 m). Body heavy; second dorsal directly above anal, forming a diamond-shaped pattern when viewed from side; pronounced angular lobule on upper lobe of tail; precaudal pits absent. Head spiracle prominent; teeth sharp, serrate on inner edges. Color dark gray above fading to lighter sides; belly whitish.

Habitat
Shallow to moderately deep water.

Habits
Feeds 8–10 ft. (2.4–3.1 m) above bottom in deep water, at varying depths in bays and sloughs.

Remarks
A fine food fish.

Range
Vancouver Is. to San Juanico Bay, Baja Calif.; also Peru and Chile.

REQUIEM SHARKS
Family Carcharhinidae

Members of this, the largest of the shark families, are distinguished by the following characteristics: teeth are sharp and saw-edged; eyes have nictitating, or winking, membranes; the last gill opening is above the base of pectorals; the first dorsal fin is entirely in front of pelvics and is much larger than the second dorsal, which lies above the similar-sized anal fin; the tail is notched, less than one-third the total length, and the upper lobe is longer.

BLUE SHARK
Prionace glauca   86:7

Description
Size, to 13 ft. (4 m). Body slender; very long pectoral fin, twice height of first dorsal; base of first dorsal about mid-body in advance of pelvics; second dorsal directly above anal; upper precaudal pit present. Head with long, pointed snout; upper teeth serrate, slightly curved. Color indigo blue on back and fins, lighter on sides; belly whitish.

Habitat
Offshore, pelagic.

Remarks
Second-rate food fish; potentially dangerous.

Range
Gulf of Alaska to Gulf of Calif.; abundant.

DOGFISH SHARKS
Family Squalidae

SPINY DOGFISH
Squalus acanthias   86:3

Description
Size, to 5½ ft. (1.7 m). Body slender, rounded; hard spine in front of each dorsal fin; no anal fin; caudal peduncle long, slender. Color gray to brownish back fading to lighter belly, white spots scattered over back of young.
Habitat
Inshore waters and to 1200 ft. (365.8 m).
Remarks
A great pest to fishermen, destroying nets and gear; a second-rate food fish; greatest use is in biology-class dissection.
Range
Japan to Aleutians to cen. Baja Calif.

PACIFIC SLEEPER SHARK
Somniosus pacificus   86:12

Description
Size, to 13 ft. (4 m). Body tapering to rear, cylindrical in front; snout rounded; pectorals small, hardly longer than pelvics, first dorsal about midway between pectoral and second dorsal; no spines on dorsals, bases of dorsal fins about equal in length; no anal fin.
Habitat
Deep waters near bottom of cold seas.
Range
Japan to Alaska, rarely to s. Calif.

Note: The **GREENLAND SHARK,** *Somniosus microcephalus,* is a close relative of this species.

ANGEL SHARKS
Family Squatinidae

PACIFIC ANGEL SHARK
Squatina californica   86:4

Description
Size, to 5 ft. (1.5 m). Body flattened, raylike; pectorals and pelvics expanded, pectorals not attached to sides of head as in the rays; no anal fin; gill openings crowded in a deep notch behind head. Color gray to dusky above, with dark spotting; underparts white.
Habitat
An inshore bottom dweller.
Remarks
Has strong jaws with sharp, upright teeth, so no attempt should be made to remove the hook until the fish is lifeless. Pectoral fins and meat along back edible.
Range
S. Alaska to Gulf of Calif.

SKATES AND RAYS
Order Rajiformes

Members of this order have large spiracles and greatly flattened bodies. They lack anal fins, and the dorsal fins, where there are any, are located on the tail. The snout and connecting pectoral and pelvic fins extend horizontally to form the more or less triangular "disk." Skates and rays are distinguishable from the sharks in the following respects: they have no free eyelids; gill openings are below only; the edges of the pectoral fins are attached to the sides of the head in front of the gill openings. Most members live and feed on the bottom.

GUITARFISHES
Family Rhinobatidae

SHOVELNOSE GUITARFISH
Rhinobatos productus 87:7

Description
Size, to 5 ft. (1.5 m). Body has small, blunt spines on middle of back, tail, shoulders, and around eyes; first dorsal situated near middle of very thick tail; caudal fin present; pectoral and pelvic fins extend horizontally. Head forms broad-based acute triangle from long, pointed snout; teeth tiny, flattened, smooth. Color back brownish-gray, underside pale.
Habitat
Shallow water.
Remarks
Often caught off piers.
Range
San Francisco Bay to Gulf of Calif.; common.

THORNBACK
Platyrhinoidis triseriata 87:8

Description
Size, to 2½ ft. (0.8 m). Distinguished by 3 rows of very strong spines along back and tail. Body disk circular, wider than long; prickles on front edge of disk, on rounded snout, and sometimes around eyes; patch of small spines on each shoulder; 2 dorsal fins back of mid-tail, caudal fin present. Color light greenish-brown to black; belly white or buff.
Habitat
Sandy shallow-water bottoms.
Remarks
Edible.
Range
San Francisco, Calif., to cen. Baja Calif.; common in s., rare in n.

SKATES
Family Rajidae

The skates have a muscular tail usually supporting two dorsal fins and sometimes a caudal fin. Their skin is more or less covered with spines. They are distinguishable from rays by their concave pelvic fins and by the absence of a tail spine. Similar to the rays, the snout and connecting pectoral and pelvic fins are extended horizontally to form the "disk," with the frontal or outer margin more or less triangular. Skates feed on the bottom and, when hooked, often cup the edges of the disk to form a partial vacuum which makes it difficult to pry them loose.

BIG SKATE
Raja binoculata   87:3

Description
Size, to 8 ft. (2.4 m). Spines along top of tail, one spine on midback behind eyes; pelvic fins shallowly notched, pectoral tips pointed. Head depressed, snout pointed; orbital spines present; frontal margin slightly concave; outline posteriorly angular, not rounded. Color brown to very dark gray above, usually with large dark spot surrounded by lighter spots at base of each pectoral fin; light spots about size of eye scattered over body; belly white.
Habitat
Moderately deep water.
Remarks
Commonly taken by trawl boats.
Range
Bering Sea to n. Baja Calif.; abundant in n. and cen. Calif., uncommon s. of Pt. Conception.

LONGNOSE SKATE
Raja rhina   87:1

Description
Size, 4–5 ft. (1.2–1.5 m). Front margin of body disk deeply concave; outer margin of pelvic fins deeply concave when held at right angles to tail; pectoral tips somewhat rounded; spines on top of tail, around eyes, and one in mid-back. Snout long, sharp, pointed, tapering. Color dark brown to black above, dusky underside.
Habitat
Moderately deep water.
Remarks
A fine food fish.
Range
Se. Alaska to cen. Baja Calif.

STINGRAYS
Family Dasyatidae

Stingrays, which resemble the edible skates, lack dorsal fins; they are distinguishable from skates by their convex (instead of concave) pelvic fins and the presence of a venomous spine on the tail. Serious injury may result from their sting. They are bottom dwellers and feeders and bear live young.

CALIFORNIA BUTTERFLY RAY
Gymnura marmorata **87:6**

Description
Size, to 5 ft. (1.5 m) wide. Width of body disk nearly twice length, tail very short; no dorsal or tail fins. Color back dark brown or grayish-olive, sometimes of varying shades with patterns of dark and light spots; disk bordered with buff; underparts white.
Habitat
Shallow water bottoms.
Remarks
Sting small or absent.
Range
Pt. Conception, Calif., to Peru; common s. of San Diego.

ROUND STINGRAY
Urolophus halleri **87:4**

Description
Size, to 22 in. (55.9 cm). Nearly circular body disk, tail shorter than disk; caudal fin present, dorsal fins absent; skin smooth, with few spines; sharp spine (sting) on top of tail. Color dark to slaty brown above, with spots or blotches; underparts yellow.
Habitat
Bottom of bays, sloughs, along sandy beaches; very abundant.
Remarks
Inflicts extremely painful wounds; serious threat to anglers and swimmers. Often caught.
Range
Humboldt Bay, Calif., to Panama.

EAGLE RAYS
Family Myliobatidae

Members of this family are distinguished as follows: the pectoral fins stop on the sides of the head; the cephalic, or head, fins are not below the level of the body and are not long and earlike; the crown is high-domed and the eyes and spiracles are on the sides of the head; the tail is long and whiplike, with the single dorsal fin near its root. Eagle rays are swimmers as well as bottom feeders.

BAT RAY
Myliobatis californica **87:2**

Description
Size, to 4 ft. (1.2 m) wide and 210 lb. (95.2 kg). Body smooth-skinned; 1 dorsal fin, followed by 1 or more long stings attached one above the other, not in series; tail whiplike, as long as or longer than width of disk. Head distinct, elevated above disk, projecting beyond angle of pectoral fins; teeth flat and hard. Color dark brown to greenish-black or entirely black; undersides white edged with gray.
Habitat
Inshore waters, bays, sloughs.
Habits
Very destructive to shellfish such as oysters, mussels, abalones, clams.
Range
Oreg. to Gulf of Calif.

ELECTRIC RAYS
Family Torpedinidae

PACIFIC ELECTRIC RAY
Torpedo californica   **87:5**

Description
Size, to 4 ft. (1.2 m) and 90 lb. (40.8 kg). Body very flabby, disk almost circular, skin smooth; caudal fin pronounced; tail short, stout; snout blunt. Color bluish-black to dark gray with black spots; white underside. Similar to Round Stingray, but has 2 dorsal fins on tail and no spine.

Habitat
Shallow to moderately deep water.

Remarks
Electric discharge can knock a man down.

Range
B.C. to cen. Baja Calif.; abundant in widely separated localities.

Chimaerae
Order Chimaeriformes

CHIMAERAS
Family Chimaeridae

Members of this group have fishlike shapes but are more closely related to the sharks and rays in having a cartilaginous skeleton.

RATFISH
Hydrolagus colliei   *Fig. 72*

Description
Size, to 38 in. (96.5 cm). Body long, tapering to a pointed tail; skin smooth, scaleless; large triangular pectoral fins; 2 dorsal fins, first preceded by large spine; caudal fin lancelike. Head markedly large, ugly; mouth small; teeth prominent, incisorlike, united into bony plate; upper lip notched; snout rounded; gill openings single on each side, gill cover of flesh or soft cartilage; eye iridescent green. Male has club-shaped appendage between eyes and long clasper behind pelvic fin. Color silvery with iridescent reflections of gold, blue, and green, or various shades of metallic brown and gray.

Habitat
Moderately shallow to 1200 ft. (365.8 m).

Reproduction
Females lay horny, dart-shaped eggs.

Remarks
Spine is venomous, can inflict very painful wound.

Range
Nw. Alaska to cen. Baja Calif. and in Gulf of Calif.; fairly common off Calif.

Fig. 72

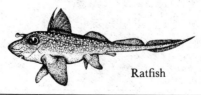

Ratfish

BONY FISHES
Class Osteichthyes

Members of this enormous class are the modern, bony fishes distinguished from the sharks, skates, and rays by the possession of true bones. Generally they have erectile fins with bony or soft rays between which stretches skin, often transparent. Normally their bodies are covered with thin overlapping scales, but in some species the body is naked. The gill opening is usually covered with a bony gill cover, or operculum. An air bladder is generally present.

Herringlike Fishes
Order Clupeiformes

These species are silvery and have the ventral fins in an abdominal position. They have branchiostegal rays, usually fewer than fifteen, and the abdomen often has keeled scutes along the ventral midline. Most are plankton feeders with numerous long gill rakers.

HERRINGS
Family Clupeidae

The herrings are compressed, silvery, fork-tailed fishes lacking a distinct lateral line. They have very small teeth or none, and many fine gill rakers. Family characteristics include abdominal pelvic fins, a single short dorsal fin near the middle of the back, no adipose fin, no scales on the head, and a relatively small terminal mouth.

PACIFIC HERRING
Clupea pallasii
88:3

Description
Size, to 18 in. (45.7 cm). Dorsal fin on midback; fleshy appendage at base of each pelvic fin. Head unscaled, lower jaw projecting, maxilla reaches to point below center of eye. Color above bluish-green to purple; sides silvery, without spots; belly silvery. Body deeper than that of Pacific Sardine; no striae on gill cover.

Similarities
Pacific Sardine, sides spotted, belly rather smooth.

Habitat
From brackish waters in bays, to open ocean; often around piers at night or dawn. Enters bays in winter and spring to spawn in shallow water.

Remarks
Commercially important in northern range.

Range
Japan to Aleutians, s. to Baja Calif.; not abundant s. of Pt. Conception, Calif.

PACIFIC SARDINE
Sardinops sagax
88:2

Description
Size, to 16 in. (40.6 cm). Body elongate; dorsal fin single, small, located slightly forward of mid-back; caudal fin deeply forked; last ray in anal fin longer than preceding rays. Head compressed; lower jaw slightly projecting or equal to upper; teeth absent. Back

greenish to dark blue, fading to silvery belly; 1 or more rows of round to oblong black spots along sides. Very similar to Pacific Herring, but has raised striae across gill covers.

Habitat
Inshore and near offshore waters, around piers.

Habits
Occurs in schools, often associating with other fishes (Pacific and Jack Mackerel, Anchovies); migratory along coast.

Range
Se. Alaska to Gulf of Calif.; spawning grounds off s. Calif.

ANCHOVIES
Family Engraulididae

The anchovies have a small, compressed body; an undershot, cleft mouth; and no lateral line. They have abdominal pelvic fins, a single short dorsal fin near the middle of the back, and no scales on the head. Members are distinguishable by an extremely large mouth with the snout projecting beyond the tip of the lower jaw, and the very long maxilla, which reaches almost to the edge of the gill cover. Anchovies occur in large schools, often close to shore; they are a vital food supply for most predatory fishes.

NORTHERN ANCHOVY
Engraulis mordax   **88:4**

Description
Size, to 9 in. (22.9 cm), but rarely exceeding 7 in. (17.8 cm). Body depth less than length of head; short single dorsal fin; base of anal fin shorter than head; pectoral lateral scale more than ½ length of pectoral fin. Eyes near tip of snout; mouth very large; gill covers not united under head. Color above opalescent or metallic blue or green with bluish or greenish reflections; sides, belly silver.

Habitat
Shallow to brackish water and to 50 miles (80.5 km) or more offshore.

Habits
A schooling fish.

Remarks
Of primary importance as bait and as prey for birds, fish, mammals. Also canned and tremendous tonnages processed into meat.

Range
B.C. to Cape San Lucas; abundant throughout range.

Trouts and Salmons
Order Salmoniformes

SMELTS
Family Osmeridae

Members of this family have small, slender bodies; moderately small scales; a well-developed lateral line, a single soft-rayed dorsal fin; and a dorsal adipose fin. The pelvic fins are abdominal, with no scaly appendage above the base. There are no scales on the head.

SURF SMELT
Hypomesus pretiosus **88:5**

Description
Size, to 10 in. (25.4 cm). Body has 66–76 small scales along lateral lines. Fins short; dorsal, 8–11 rays; anal, 12–17 rays. Very small teeth on tongue and vomer, maxilla reaches only to middle of eye. Pale green back with pronounced metallic side stripes, darkens when removed from water; belly silvery. Similar to Night Smelt, but maxilla does not extend beyond middle of eye.

Habitat
Sandy beaches on outer coast.

Range
Alaska to s. Calif. Spawns in surf during daytime.

Note: Another species, the **DELTA SMELT,** *Hypomesus transpacificus,* is similar to the surf smelt but has fifty-three to fifty-six larger scales along lateral line, larger fins, and darker color. San Joaquin and Sacramento River systems. Spawns in fresh water.

NIGHT SMELT
Spirinchus starksi **88:8**

Description
Size, to 5½ in. (14 cm). Distinguished by a number of small teeth, not caninelike, on vomer. Pectoral fin relatively short; extends 71–92% of distance from pectoral insert to pelvic insert. Eye diameter less than caudal peduncle depth; maxilla reaching past eye. Color very pale brownish-green; side and belly silvery.

Habitat
Shallow waters along sandy shores of outer coast; spawns in surf.

Range
Skelikof Bay, Alaska, to cen. Calif. (Pt. Arguello).

Note: The **LONGFIN SMELT,** *Spirinchus thaleichthys,* is similar to the Night Smelt but has slightly longer pectoral fins.

EULACHON
Thaleichthys pacificus **88:6**

Description
Size, to 12 in. (30.5 cm). Lateral line complete, with 70–78 pored scales; origin of pelvic fins in advance of mid-body and ahead of base of dorsal fin; well-defined concentric striae in opercle. Teeth caninelike, a few on vomer; gill rakers on first arch, 4–6 on upper, 13–18 on lower. Color above bluish-brown; belly silvery white.

Habitat
High-tide waters, mouth of streams and backwaters; also freshwater lakes and streams; spawns in rivers (fresh water).

Other name
Candlefish.

Range
Bering Sea to cen. Calif.; common.

PACIFIC SMOOTHTONGUE
Leuroglossus stilbius   88:1

Description
Size, to 6 in. (15.2 cm). Distinguished by silvery body, presence of adipose fin, and symphysial knob (immovable cartilage), at tip of lower jaw. Scales if present impossible to find; middle two rays of caudal fin twice as thick as rays on either side. Color silvery, dusky on dorsal surface and fins.

Habitat
Abundant in mesopelagic realm, found from near the surface to depths of several hundred feet.

Habits
Schooling fish.

Range
Various subspecies are found along entire Pacific coast.

Lanternfishes
Order Myctophiformes (Iniomi)

BLUE LANTERNFISH
Tarletonbeania crenularis

Description
Size, to 4 in. (10.2 cm). Ventral half of body covered with photophores in characteristic pattern; only one luminous spot at base of tail. Snout projects beyond mouth; caudal peduncle skinny; adipose fin present. Color dusky blue overall; photophores appear as yellowish-white dots.

Habitat
Mesopelagic; can be caught at sea surface at night; has been captured over 2500 ft. (762 m) beneath surface.

Range
Central Baja Calif. to B.C.

LIZARDFISHES
Family Synodontidae

CALIFORNIA LIZARDFISH
Synodus lucioceps   88:7

Description
Size, to 25 in. (63.5 cm). Body long, slender, lizardlike; insertion of dorsal fin slightly behind pelvic-fin base; pectorals reach base of pelvic; adipose fin above anal. Head broad, snout nearly triangular, mouth large, numerous sharp teeth in jaws; area between eyes slightly concave. Color above greenish-brown with brassy luster, abruptly ending at lateral line; sides, belly light gray; lower jaw, pelvics yellow.

Habitat
Moderately shallow water.

Remarks
Edible.

Range
San Francisco to Baja Calif. and Gulf of Calif.; common.

EELLIKE FISHES
Order Anguilliformes

MORAYS
Family Muraenidae

CALIFORNIA MORAY
Gymnothorax mordax   *Fig. 73*

Description
Body long, slender, eel-shaped; no scales; no pelvic or pectoral fins; dorsal and anal fins fleshy ridges. Head small; jaws well developed; teeth pointed, some dartlike. Color dark brown to greenish with many small, lighter spots; throat and sometimes belly marked with dark horizontal streaks.
Habitat
Rocky coastline and under rocks in kelp beds.
Range
Pt. Conception, Calif., to Magdalena Bay, Baja Calif.

Fig. 73

California Moray

Sauries, Flyingfishes, and Silversides
Order Atheriniformes

Members of this order are characterized by spineless fins, united lower pharyngeal bones, and a lateral line that forms a ridge along the lower lateral part of the body.

SAURIES
Family Scomberesocidae

PACIFIC SAURY
Cololabis saira   **90:3, 88:11**

Description
Size, to 14 in. (35.6 cm). Body long, slender. Single dorsal fin far back beyond origin of anal fin; both fins followed by 4–6 finlets. Lateral line low on belly; dorsal rays, 9–12, anal rays, 12–15. Head has sharp beaklike jaws; lower jaw flexible. Color dark green to deep blue above; sides silvery, scales tipped with blue or green; belly silvery; base of pectorals bright blue, fins colorless.
Habitat
Offshore waters, epipelagic.
Range
Japan to Alaska to Baja Calif.; abundant throughout.

FLYINGFISHES
Family Exocoetidae

CALIFORNIA FLYINGFISH
Cypselurus californicus   **88:9**

Description
Size, to 16 in. (40.6 cm). One of two West-Coast fishes which exceed 12 in. (30.5 cm) and have long, spreading pectoral fins that serve as gliding wings. Predorsal scales, 47–50; eyes large; lower jaw projecting. Color back and sides deep blue; belly silvery; pectoral fins evenly dusky.
Habitat
Open sea offshore.
Habits
Can remain gliding above water for some time.
Range
Astoria, Oreg., to Cape San Lucas, Baja Calif.; rare n. of Pt. Conception.

SILVERSIDES
Family Atherinidae

TOPSMELT
Atherinops affinis   **89:11**

Description
Size, to 14⅝ in. (36.6 cm). Distinguished by small first dorsal fin at about mid-body, a single row of forked teeth in each jaw, and 5 to 8 scales between dorsal fins. First dorsal slightly in front of anus; tip of blunt or rounded upper jaw projecting very slightly over tip of lower; eyes large. Color above blue-gray to clear green; lateral stripe thin, bright blue; below silver; yellow blotch on cheek at level of pectoral fin.
Habitat
Bays, sloughs, and kelp beds; close inshore in loose schools or aggregations.
Habits
Often associates with Jacksmelt.
Range
B.C. to Gulf of Calif.; abundant.

JACKSMELT
Atherinopsis californiensis   **89:9**

Description
Size, to 17½ in. (44.5 cm). First dorsal base well ahead of anal fin insertion; prominent snout, terminal mouth, equal jaws. Color above greenish-gray to green with bluish tinge; sides and belly silvery; lateral stripe blue with lighter blue upper border; yellow blotch on cheek at level of pectoral fin. Distinguished from Topsmelt and California Grunion by small, uniform teeth set in bands, and 10–12 scales between dorsal fins.
Habitat
About piers and kelp beds.
Range
Oreg. to Baja Calif.

CALIFORNIA GRUNION
Leuresthes tenuis

90:14

Description
Size, to 7¼ in. (18.4 cm). Body very slender; front of first dorsal above or slightly back of vent, 5 flexible spines in fin; 7–9 scales between dorsal fins. Greatly extensible upper jaw (when lower jaw is pulled down); teeth absent. Color above bluish-green; lateral stripe thin, blue and silver; belly silver, blue blotch on cheek at level of pectoral fin.

Habitat
Sandy beaches.

Remarks
Spawns by night on beach and can be caught by hand.

Range
Monterey, Calif., to Magdalena Bay, Baja Calif.; abundant in s.

Note: The **GULF GRUNION**, *Leuresthes sardina,* is very similar to the California Grunion but spawns by day or by night. It occurs in upper portions of the Gulf of California.

Codfishes
Order Gadiformes

CODFISHES
Family Gadidae

The codfishes and their allies have soft fins only. Their ventrals are under or in front of their pectorals, not behind them. The order includes some of humanity's most important food fishes. Most are bottom feeders.

PACIFIC TOMCOD
Microgadus proximus

88:12 90:9

Description
Size, to 12 in. (30.5 cm). Has 3 separate dorsal fins, vent under posterior portion of first dorsal; 2 separate anal fins; lateral line curving. Upper jaw extending beyond lower; barbel under tip of lower jaw, about ½ length of eye dia.; 22–28 gill rakers on first gill arch. Color above olive or gray with brownish tinge, becoming white or silvery on sides and belly; fins dusky-tipped (except first anal and pelvics).

Habitat
Bays, sloughs, and offshore to depths of 720 ft. (219.5 m).

Habits
Bottom feeder.

Remarks
A minor game fish esteemed by some sportsmen; a very important prey species.

Range
Alaska to cen. Calif.

525

WALLEYE POLLOCK
Theragra chalcogramma   88:10

Description
Size, to 3 ft. (0.9 m). Body elongate, 3 dorsal fins well-separated;
third dorsal and second anal form symmetrical pattern; lateral line
present. Head pointed, eyes large; lower jaw slightly protruding,
minute barbel on tip; 34–40 gill rakers on first gill arch. Color
above olive-green to brown; sides silver, belly white; dusky or black
on fins.
Habitat
Rather shallow water into depths exceeding 650 ft. (198.1 m).
Range
Japan to Bering Sea to cen. Calif.; abundant in n. range.

HAKES
Family Merlucciidae

PACIFIC HAKE
Merluccius productus   88:13

Description
Size, to 3 ft. (0.9 m). Body elongate; 2 separated soft dorsal fins
(first short; second long and deeply notched, forming symmetrical
sideview pattern with long anal fin); caudal peduncle small; scales
loosely attached, small. Head with W-shaped ridges; mouth large,
lower jaw projecting; teeth strong; eyes large, maxilla extends past
pupil. Color metallic blackish or silvery above; sides and belly
silver; inside mouth jet-black.
Habitat
Moderately shallow to deep water.
Range
Asiatic Coast to Alaska to Magdalena Bay, Baja Calif; abundant.

CUSK-EELS
Family Ophidiidae

SPOTTED CUSK-EEL
Chilara taylori   89:2

Description
Size, to 14½ in. (36.8 cm). Distinguished by spotted eellike body
and pair of feelers (modified pelvic fins) near tip of lower jaw;
despite name and appearance, is not an eel. Color yellowish-brown,
darker above; spots dark brown; lips of large, old individuals
orange-brown.
Habitat
Sandy or muddy bottom areas including among rocks; subtidal to
depths exceeding 800 ft. (243.8 m).
Habits
Burrows into the bottom tail first; nocturnal, but may be observed
standing on tail at the bottom during overcast days or when water
is turbid.
Range
San Cristobal Bay, Baja Calif., to n. Oreg.

LIVEBEARING BROTULAS
Family Bythitidae

RED BROTULA
Brosmophycis marginata **89:8**

Description
Size, to 16 in. (40.6 cm). Skin on body and fins rather loose, covered with heavy red mucus. Head large, body tapering toward tail; pelvic fins modified, filamentous, composed of two rays each. Body brownish under red mucous, fins bright red, lips pink.

Habitat
Secretive; hides in rocky crevices and under rubble and debris. Small individuals most often seen in shallow subtidal; larger individuals have been captured on the bottom in greater depths and to 840 ft. (256 m).

Range
Ensenada Bay, Baja Calif. to Petersberg, Alaska.

EELPOUTS
Family Zoarcidae

BLACKBELLY EELPOUT
Lycodopsis pacifica **89:7**

Description
Size, to 18 in. (45.7 cm). Elongate eellike body (though not an eel) with long dorsal and anal fins joined to tail fin. Pectorals large, fanlike; pelvics reduced in size, with 3 rays each. Black lining of visceral cavity shows through belly wall. Color light gray to pale reddish-brown, light spots over scales, pale vertical bars across body; elongate black spot at anterior end of dorsal fin, margins of dorsal and anal fins black.

Habitat
Sandy or sandy mud bottoms at 30–1300 ft. (9.1–396.2 m).

Range
Ensenada, Baja Calif., to Afognak Is., Alaska.

GRENADIERS
Family Macrouridae

PACIFIC GRENADIER
Coryphaenoides acrolepis **89:5**

Description
Size, to 34 in. (86.4 cm). Large head, snout projecting beyond the mouth; elongate body covered with coarse scales which have well-developed spinules on them. Highest ray of first dorsal fin with sharp spines on leading edge; short barbel at tip of lower jaw. Color grayish-brown to black.

Habitat
On or above the bottom at depths of 2000–8000 ft. (609.6–2438.4 m) or more.

Range
N. Baja Calif. to Alaska and s. to Japan.

Trumpetfishes
Order Gasterosteiformes

This order includes species with a long snout ending in a small mouth.

PIPEFISHES
Family Syngnathidae

These fishes have segmented body plates. The gill opening is a small pore, and the gills are small, rounded tufts. The male carries the eggs and young in a brood pouch formed of two folds of skin on the underside of either the tail or the belly, depending on the species. They are small, slender fishes of warm waters, feeding on crustaceans and diatoms.

KELP PIPEFISH
Syngnathus californiensis   **89:1**

Description
Size, to 19½ in. (49.5 cm). Body with 17–22 body rings and 44–50 tail rings; dorsal fin (36–47 rays) covers 9 rings; caudal pouch of males covers 21–25 rings. . Head sometimes not shorter than snout. Color olivaceous to brownish-red; yellow below; head and body marbled and speckled with whitish, anterior portion has horizontal grayish streaks.
Habitat
Common in kelp beds, rare elsewhere.
Range
San Francisco Bay to s. Baja Calif.

BAY PIPEFISH
Syngnathus leptorhynchus

Description
Size, to 13 in. (32.5 cm). Body with 17–20 body rings and 36–46 tail rings; dorsal fin (28–44 rays) shorter than head, covering 7 rings. Color varies with surroundings from brown to green.
Habitat
Eelgrass beds in bays.
Range
Alaska to cen. Baja Calif.; abundant.

Opahs and Ribbonfishes
Order Lampridiformes

OPAHS
Family Lampridae

OPAH
Lampris guttatus   **90:10**

Description
Size, to 4½ ft. (1.4 m) and 160 lb. (72.6 kg) in Pacific. Body deep, ovate, much compressed; dorsal fin high, pelvics long. Jaws without teeth. Color above bluish, shading to silver tinged with red below; prominent silver spots on body; fins bright red; flesh light salmon.

Habitat
Surface to 1680 ft. (512.1 m) in offshore waters.
Range
Worldwide open seas; at times common off Calif.

RIBBONFISHES
Family Trachipteridae

KING-OF-THE-SALMON
Trachipterus altivelis   **89:6**
Description
Size, to 6 ft. (1.8 m). Body very thin, knifelike; no anal fin; dorsal
fin long. Distinguished by overall silvery pigment with several
large black blotches; lips bright crimson.
Habitat
Offshore waters, surface to 1800 ft. (548.6 m).
Range
Alaska to Chile; fairly common.

Top Minnows
Order Microcyprini

KILLIFISHES
Family Cyprinodontidae

These are small fishes, mostly of fresh water, many of them
viviparous. The single marine species in western waters, however,
is oviparous, spawning in the summer months.

CALIFORNIA KILLIFISH
Fundulus parvipinnis

Description
Size, to 4 in. (10.2 cm). Body stout; dorsal rays 13–14, anal rays
10–11; first rays of dorsal fin near mid-back (directly above anus);
caudal peduncle deep, long. Head pointed, mouth small. Male fins
high, large; breeding male turns dark above, distinctly yellow
below. Color above dark to light green, belly yellowish.
Habitat
Bays, backwaters, sloughs; occasional in freshwater streams
tributary to salt water.
Remarks
Important forage fish; good bait.
Range
Pt. Conception, Calif., to Baja Calif.

Perchlike Fishes
Order Perciformes

This large and varied order constitutes one of the dominant types of fish life today. The dorsal and anal fins contain both spiny and soft fin rays. The pelvic fins are usually located on the throat instead of the abdomen, and they usually have one hard spine and five soft rays.

BARRACUDAS
Family Sphyraenidae

These fishes have strong jaws and teeth and a small first dorsal of a few spines, set well forward of the second.

CALIFORNIA BARRACUDA
Sphyraena argentea   90:13

Description
Size, to 4 ft. (1.2 m) and 18 lb. (8.2 kg). Distinguished by length and by sharp, pointed head. Body slender, cigar-shaped; 2 well-separated dorsal fins, first with 5 spines, second with 1 spine and 8–10 soft rays; lateral lines present, with row of raised, well-set scales. Head sharp-nosed, lower jaw pointed; teeth fanglike, strong, unequal. Color above dark gray or brownish with blue tinge; silver below lateral line; caudal fin yellow.
Habitat
Abundant near kelp beds and around offshore islands; open ocean, in schools.
Range
Alaska to Cape San Lucas, Baja Calif.; common in s.

SEA BASSES
Family Serranidae

These carnivorous fishes are highly valued for food. They have well-developed spiny and soft-rayed portions of the dorsal fin, which may be separated or divided by a deep notch. The ventral fins are under the pectorals; the anal fin is about as long as the soft part of the dorsal. There is a deep caudal peduncle and a broad tail.

KELP AND BARRED SAND BASSES
Genus *Paralabrax*

Members of this genus of fishes can be distinguished from most other West Coast basses because they have more dorsal soft rays (thirteen to fifteen) than spines (ten), and six to eight anal soft rays. They may be confused with two or three kinds of rockfishes (Genus *Sebastes*); these, however, have thirteen dorsal spines.

KELP BASS
Paralabrax clathratus   89:10

Description
Size, to 28¼ in. (72.1 cm) and 14½ lb. (6.6 kg). Single dorsal fin deeply notched between sections; third to fifth spines of about equal

length and larger than rays. Pectoral fin large; caudal fin straight; 32–36 gill rakers on first gill arch. Color above dark gray, brownish, or greenish-gray; dorsolateral area mottled brownish; fins yellowish; below tinged with yellow. Markings less distinct in older fishes; young marked with oval opalescent spots on darker background.

Habitat
In and around kelp beds.

Range
Columbia R. to Magdalena Bay, Baja Calif., rare n. of Pt. Conception; abundant in s.

BARRED SAND BASS
Paralabrax nebulifer   **89:3**

Description
Size, to 25⅝ in. (65.0 cm). Body stout; dorsal fin not deeply notched. Third dorsal spine longest, twice length of second; only third and fourth spines longer than rays. Pectorals large; caudal straight; 22–27 gill rakers on first gill arch. Color above dark greenish-gray or brownish, several broad dusky bars on back and sides; snout and cheeks speckled with golden-brown; belly gray or whitish.

Habitat
Sandy bottoms near kelp beds and among rocks.

Range
Santa Cruz, Calif., to Magdalena Bay, Baja Calif., rare n. of Pt. Conception; abundant in s.

TEMPERATE BASSES
Family Percichthyidae

GIANT SEA BASS
Stereolepis gigas   **89:4**

Description
Size, to 7 ft. (2.1 m) and 557 lb. (252.6 kg). Body heavy, robust; 2 dorsal fins contiguous, (first with 11 spines less than half as long as the 9–10 soft rays in second; pectorals comparatively small, caudal slightly furcate. Head depressed, mouth large, area between eyes broad. Color above dark brown to gray with large blackish blotches on sides (these fade when fish is taken from water); paler below. Young differ greatly in both shape and color (body brick-red, distinctly black-spotted).

Habitat
Inshore shallow waters to at least 150 ft. (45.7 m), on bottom near kelp beds.

Range
Humboldt Bay, Calif., to Gulf of Calif.; abundant in s.

JACKS
Family Carangidae

In this family the spinous first dorsal is much shorter than the soft-rayed second, and it may be reduced to a few short spines or be lacking. The anal fin usually is preceded by three short spines. The caudal peduncle is very slender and the tail deeply forked. Most species have a prominent hard keel on the rear portion of the lateral line, which serves to strengthen the peduncle. These are largely ocean fishes.

CALIFORNIA YELLOWTAIL
Seriola dorsalis   90:4

Description
Size, to 5 ft. (1.5 m) and 80 lb. (36.3 kg). Body elongate, oval; 2
dorsal fins. First dorsal low, composed of spines less than half the
height of the first soft rays; second dorsal extending nearly to
caudal fin base. Blunt low keel on sides of slender caudal peduncle;
caudal fin deeply forked; pectorals short; lateral line unshielded.
Pointed snout. Color above metallic blue to green; distinguished by
brassy to metallic, bright yellow lateral stripe from snout to caudal
fin; tail yellow; belly silvery.
Habitat
Moderately shallow near-shore waters; abundant spring and summer.
Habits
Travels in schools, often following other schooling fish close inshore.
Remarks
Excellent game, sport fish.
Range
S. Wash. to Gulf of Calif., rare n. of Pt. Conception.

JACK MACKEREL
Trachurus symmetricus   91:8

Description
Size, to 32 in. (81.3 cm) and 5 lb. (2.3 kg). Body has 2 dorsal fins,
close together and about same height; second dorsal and anal
extend almost to base of tail; caudal fin deeply notched, pectoral fin
long. Accessory lateral line runs high on back from head to
insertion of second dorsal; primary lateral line bends sharply down
above vent, bony shields form ridge posteriorly. Blunt snout,
projecting lower jaw. Color above green to bluish with iridescent
luster, sides, belly silvery.
Habitat
Inshore waters to far at sea.
Habits
Schooling fish.
Remarks
A good food fish.
Range
Se. Alaska to Magdalena Bay, Baja Calif.; abundant in s.

BUTTERFISHES
Family Stromateidae

PACIFIC BUTTERFISH
Peprilus simillimus   91:4

Description
Size, to 11 in. (27.9 cm). Body deep, compressed, oval; dorsal and
anal fins long, form symmetrical pattern; pelvic fins absent;
pectorals long, pointed; caudal fin deeply notched; caudal peduncle
short, slender; lateral line straight. Head rounded in forward
profile. Color above iridescent greenish to blue; belly silver.
Habitat
Bays, backwaters, and open ocean near shore.
Other name
Pompano.
Range
B.C. to Magdalena Bay, Baja Calif.; fairly common.

MACKERELS AND TUNAS
Family Scombridae

These are swift, powerful, spindle-shaped fishes, frequently large. They have velvety skins, small scales, no bony scales on the rear lateral line, both soft and spiny dorsal fins, and deeply forked or crescent-shaped tails. They are highly regarded as food.

PACIFIC MACKEREL
Scomber japonicus   **91:1**

Description
Size, to 25 in. (63.5 cm). Body covered with small, easily lost scales. Interspace between dorsal fins long; first dorsal high, with 8–10 long spines; dorsal and anal finlets, 4–6; caudal peduncle very slightly keeled. Color above metallic green to bluish with 25–30 dark wavy streaks and reticulations running obliquely down back to just below lateral line; belly silver.
Habitat
Shallow water to 50 miles (80.5 km) or more offshore.
Habits
Schooling fish.
Range
Gulf of Alaska to cen. Mexico; abundant in s.

PACIFIC BONITO
Sarda chiliensis   **90:1**

Description
Size, to 40 in. (101.6 cm) and 25 lb. (11.3 kg). Body covered with scales; very brief interspace between first and second dorsal fins, 6–9 finlets after second dorsal, 6–7 finlets following anal; soft keel on each side of caudal peduncle. Color above dark greenish-blue to violet with metallic luster, shading into silvery below; 8–11 narrow black or blackish stripes extending obliquely along back.
Habitat
In and around kelp beds and offshore for 35 miles (56.3 km) or more; sometimes enters coastal bays.
Range
Gulf of Alaska to s. Baja Calif., also Peru and Chile; absent in tropics; uncommon n. of Pt. Conception.

ALBACORE
Thunnus alalunga   **91:5**

Description
Size, to 5 ft. (1.5 m) and 93 lb. (42.2 kg). Distinguished by exceptionally long pectoral fins, reaching to or beyond anus. Anal round; preopercle angle square. Color above dark steel-blue; sides and underparts silvery; flesh white.
Similarities
Bluefin Tuna, pectorals shorter.
Habitat
Marginal between green inshore waters and deep, blue open sea, temperature range 58–70° F (14.4–21.1° C).
Range
Transpacific; Alaska to cen. Baja Calif.; Hawaii, Japan, and in s. hemisphere off Chile and Peru.

Note: The **BIGEYE TUNA,** *Thunnus obesus,* has pectoral fins almost as long as those of the Albacore.

BLUEFIN TUNA
Thunnus thynnus 91:7

Description
Size, to 14 ft. (4.8 m) and 1600 lb. (725.6 kg) in eastern Pacific.
Distinguished from other tunas by relatively short pectoral fins,
extending usually only to eleventh or twelfth dorsal spine and
shorter than head. Anus round; preopercle angle rounded; ventral
surface of live specimen striated with blood vessels. Color above
deep blue, iridescent; finlets yellowish, other fins dusky tinged with
yellowish-green; belly metallic silver with small silvery spots; flesh
grayish or sometimes pinkish.
Habitat
Shallow nearshore to offshore several miles and near islands to 100
ft. (30.5 m) depth.
Habits
Schooling fish.
Range
Se. Alaska to s. Baja Calif.; common s. of Pt. Conception, Calif.,
rare n. transpacific.

BILLFISHES
Family Istiophoridae

Members of this family differ from the swordfishes in having a
shorter, rounded (not flattened) bill; narrow, often embedded scales;
and a very long first dorsal fin capable of being depressed into a
groove.

STRIPED MARLIN
Tetrapturus audax 90:12

Description
Size, to 12 ft. (3.7 m) and 350 lb. (158.7 kg) in eastern Pacific.
Body very robust; first dorsal long, high-peaked on first spines,
extending almost full length of back; second dorsal short, pelvic fins
elongate, 2 small keels on either side of base of tail. Head dished;
upper jaw prolonged into a round spear; lower jaw very sharp,
extending less than ½ length of upper jaw. Color above purplish-
blue, light blue bars from dorsal line to upper belly, dorsal fin
violet with brighter blue spots; pectoral fins dark gray, black-edged;
belly silver.
Habitat
Open sea.
Habits
A furious fighter, leaps above surface, makes 1000-ft. (305-m)
runs.
Range
Pt. Conception, Calif., to Mexico and Peru and Chile.

SWORDFISHES
Family Xiphiidae

SWORDFISH
Xiphias gladius 90:6

Description
Size, to 15 ft. (4.6 m) and 1182 lb. (536.0 kg). Distinguished by
greatly prolonged upper jaw forming a sharp-edged, flat "sword."

Body scaleless; first dorsal very high, base less than ⅓ length of back, tip curved back; pelvic fins absent; single wide keel on sides of caudal peduncle. Color above purplish to almost black, paler on sides; belly silver-gray.

Habitat
Warm offshore waters and near islands.

Range
Worldwide in warm seas; Oreg. to Chile; uncommon to rare n. of Pt. Conception, Calif. in e. Pacific.

GRUNTS
Family Pomadasyidae

SARGO
Anisotremus davidsonii 91:2

Description
Size, to 17.2 in. (44.2 cm). Body deep; dorsal fin with 11–12 spines, 14–16 rays; anal fin with 3 spines, 9–11 rays. Small, thick-lipped mouth; no teeth on vomer; fine, single-pointed, unmovable teeth on jaws. Distinguished by black band (saddle) across back and sides; body entirely metallic silvery, iridescent with grayish tinge above; back, head, side sometimes vaguely blotched; caudal, soft dorsal, anal fins yellowish-tinged; edge of opercle black; dark spot at base of pectoral fin.

Habitat
Relatively shallow water; abundant in rocky habitat, around pier pilings, in bays, etc.

Range
Santa Cruz, Calif. to Gulf of Calif. and in Salton Sea.

CROAKERS
Family Sciaenidae

Members of this widespread family are typically found along sandy coasts in warm seas. They have a lateral line extending onto the caudal fin. The anal fin has only one or two spines. The spiny and soft-rayed dorsals are often continuous, but where there are two dorsals, the first is usually triangular in shape. Members of this family have air bladders which, by vibrating, can produce peculiar croaking sounds.

CALIFORNIA CORBINA
Menticirrhus undulatus 91:13

Description
Size, to 28 in. (71.1 cm) and 7 lb. (3.2 kg). Body long, slender, somewhat flattened below; spiny and soft dorsal fins connected by low membrane; 1 weak spine (sometimes 2) at front of anal fin; pectorals large, low-set, rather fan-shaped. Small eyes; snout tip projects beyond lower jaw; short, fleshy barbel on tip of lower jaw. Color above dark sooty-gray to steel-blue with metallic reflections; sides gray, fins dusky; below grayish-white.

Habitat
Low, sandy beaches and bays; rarely among rocks.

Habits
Bottom feeder in 2–20 ft. (0.6–6.1 m).

Range
Pt. Conception, Calif., to Gulf of Calif.

WHITE CROAKER
Genyonemus lineatus   91:10

Description
Size, to 16⅓ in. (41.4 cm) and 2¼ lb. (1 kg). Distinguished by
conspicuous black spot under upper base of pectoral fin. Body has
2 weak spines in front of anal fin; first dorsal with 12–15 spines;
first and second dorsals connected below very deep notch; lateral
line low, curving, distinct. Tip of snout projecting, several very
small barbels on chin. Color above brownish to yellowish with
brassy luster, fading to silvery belly; indistinct wavy lines following
scale rows backward and upward; fins pale yellowish, bases
darkish.
Habitat
Very shallow water to depths of 600 ft. (182.9 m) over sand
bottoms; also in bays, sloughs.
Habits
Often associates with Queenfish.
Remarks
Considered a good food fish but mistakenly thrown back by most
anglers.
Range
Vancouver Is. to Magdalena Bay, Baja Calif., rare n. of San
Francisco.

SPOTFIN CROAKER
Roncador stearnsii   91:11

Description
Size, to 27 in. (68.6 cm) and 9+ lb. (4.1+ kg). Distinguished by
long, pointed pectorals (may be longer than head), with black spot
at base. Dorsal fins connected below very deep notch, first dorsal
with less than 10 spines; 2 stout spines in front of anal. Short, stiff
serrations on preopercle; projecting upper jaw, no barbel on lower
jaw. Color above metallic gray or brassy or grayish-silver with
bluish luster; below silvery to white. Male during spawning season
is distinctly golden or brassy on lower sides.
Habitat
Shallow water along sandy beaches of outer coast; also in bays or
sloughs.
Range
Pt. Conception, Calif. to Magdalena Bay, Baja Calif.

QUEENFISH
Seriphus politus   91:14

Description
Size, to 13 in. (33 cm). Body compressed; first and second dorsals
widely separated, first with 7–9 spines; base of anal fin long,
nearly equal to that of second dorsal, with 2 weak spines in front.
Large eyes and mouth; mouth terminal. Color above metallic blue
to bronze, sometimes with brassy reflections; fins yellowish, base of
pectorals dusky; sides and belly silvery.
Habitat
Shallow water over sandy bottoms; also bays, sloughs.
Habits
Accompanies White Croaker, but is harder to catch.
Range
Yaquina Bay, Oreg. to s. Baja Calif.

WHITE SEABASS
Atractoscion nobilis

91:12

Description
Size, to 5 ft. (1.5 m) and 83 lb. (37.6 kg). Body large, elongated; dorsals in contact, base of second much longer than base of anal; 2 weak spines in front of anal; pectorals small; raised ridge on belly from pelvic base to vent. Head rather pointed; mouth large, lower jaw projecting, no barbel; eyes small. Color above steel-blue to gray; sides and belly silvery, dark spot on inner base of pectoral fin. Young have 3–6 dark vertical bars across back.

Habitat
Over shallow submerged banks, on edges of banks, near kelp beds.

Habits
A schooling fish.

Remarks
Important to commercial and sport fishermen.

Range
Juneau, Alaska, to Magdalena Bay, Baja Calif., and in upper Gulf of Calif.

BLANQUILLOS
Family Branchiostegidae

OCEAN WHITEFISH
Caulolatilus princeps

91:3

Description
Size, to 40 in. (101.6 cm). Distinguished by very long base of single, rather even, unnotched dorsal fin. Anal fin long, originating near center of belly; caudal fin slightly lunate; caudal peduncle rounded, slender. Color above rich brown to yellowish extending to sides; paler below; fins tinged with green or yellow; pectorals bluish, yellow streak in center; dorsals and anal with blue streak near edges.

Habitat
Shallow to moderately deep water over rocky bottoms and around kelp beds.

Habits
Feeds 4–6 ft. (1.2–1.8 m) from bottom.

Remarks
A good sport fish.

Range
Vancouver Is., B.C., to Gulf of Calif.; rare n. of Pt. Conception; abundant in s.

SURFPERCHES
Family Embiotocidae

Members of this family, generally referred to as "surfperches," "saltwater perches," or "ocean perches," are not true perches but relatively small fishes generally confined to surf along both sandy and rocky coasts. The body is deep, compressed, and elliptical in outline; the caudal fin is rather deeply notched. The male can be recognized by a modification of the forepart of the anal fin into a thickened, glandlike structure. These fishes are viviparous.

BARRED SURFPERCH
Amphistichus argenteus   93:1

Description
Size, to 16⅓ in. (41.5 cm) and 4¼ lb. (1.9 kg). Base of anal fin with a row of scales extending over base of soft rays; spinous section of dorsal low, ¾ height of rays, with 21–27 soft rays; anal fin (24–29 rays) nearly divided, 6 or more solid or broken yellowish or olive-green vertical bars on side. Head moderately large, eyes small. Color above brassy-olive, gray, or bluish between distinguishing, pronounced, greenish to brown vertical bars; sides and belly white, silvery, or light gray.

Habitat
Surf zone on sandy, wave-swept outer coast beaches.

Range
Bodega Bay, Calif. to cen. Baja Calif.; abundant s. of Pismo Beach.

SHINER PERCH
Cymatogaster aggregata   91:9

Description
Size, to 7 in. (17.8 cm). Body with three vertical yellow bars; spinous section of dorsal taller than rayed section; lateral line high; caudal peduncle short, slender. Head short, forehead depressed. Color above silvery to dusky, greenish-tinged; sides and belly silver, sides with some 8 horizontal sooty lines below lateral line. Fins and sides of males become dusky with onset of breeding season in late winter and early spring.

Habitat
Shallow water along sandy shores and in bays.

Range
Port Wrangell, Alaska, to n. Baja Calif.; common.

PILE PERCH
Damalichthys vacca   93:8

Description
Size, to 17⅖ in. (44.2 cm). First soft dorsal rays (21–25) about twice height of last dorsal spines; caudal fin deeply forked; caudal peduncle slender; space between pelvic and anal fins short, containing 25–31 rays. Color above blackish to dusky-gray to brownish with silver luster; sides and belly silvery; dusky vertical bar on body at mid-length, which fades after fish dies; fins dusky, pelvic fin yellowish tipped with black.

Habitat
Along sandy and rocky shores; especially abundant in kelp and around pilings of piers and docks.

Range
Port Wrangell, Alaska, to n. Baja Calif.

WALLEYE SURFPERCH
Hyperprosopon argenteum   **91:6**

Description
Size, to 12 in. (30.5 cm). Distinguished by very large eye and
black-tipped pelvic fin. Body very deep, greatly compressed; dorsal
fin with 25–29 rays, all shorter than longest spine; anal rays, 30–
35. Head small, snout very short, eyes prominent. Color above
steel-blue with indistinct dusky vertical bars, which fade upon
death; pelvic fins black-tipped; sides, belly white.
Habitat
Shallow waters along sandy beaches and around man-made
structures.
Range
Vancouver Is., B.C., to n. Baja Calif.; abundant in s.

BLACK PERCH
Embiotoca jacksoni   **93:9**

Description
Size, to 15⅓ in. (38.3 cm). Spinous dorsal very low; caudal
peduncle short, deep; distinctive cluster of large scales between
pectoral and pelvic fins; anal fin with row of small scales along
base. Head with rather thick yellowish-orange lips. Color
brownish-black, reddish-brown, and other shades of brown shading
to yellowish-brown on belly; several darker vertical bars on sides of
body; pelvic and anal fins often banded alternately with yellow and
blue.
Habitat
Along rocky coastlines; piers, docks, and other man-made
structures; eelgrass beds in bays.
Range
Fort Bragg, Calif., to cen. Baja Calif.; abundant in s.

REDTAIL SURFPERCH
Amphistichus rhodoterus

Description
Size, to 16 in. (40.6 cm). Distinguished by 9–10 vertical, orange to
brassy bars offset above and below lateral line; pinkish-red cast to
anal, caudal, and some other fins. Body moderately deep,
compressed; scales small; dorsal spines taller than rays. Color above
light green; fins, except dorsal and pectoral, light red; sides, belly
silver.
Habitat
Shallow water above sandy bottoms and around man-made
structures. Most common surfperch on sandy, wave-swept outer
coast beaches north of Bodega Bay, Calif.
Range
Vancouver Is., B.C. to Monterey, Calif.

RAINBOW PERCH
Hypsurus caryi   **90:8**

Description
Size, to 12 in. (30.5 cm). Body distinguished by long, straight
abdomen; vent below origin of rayed dorsal; last dorsal spines
somewhat shorter than rays; base of anal short. Head profile
straight, eyes small. Color extremely variegated; striped
horizontally with red, orange, blue; irregular streaks of orange and

sky-blue on head; fins with bright orange shades and sometimes mottled, caudal sometimes with dusky bars, blackish blotch on rayed dorsal and anal. One of most spectacularly colored West Coast fishes.

Similarities
Striped Seaperch, no black blotches on dorsal and anal fins.

Habitat
Mostly off rocky shores.

Range
N. Calif. to n. Baja Calif.; fairly common n. of Monterey, Calif.

RUBBERLIP SEAPERCH
Rhacochilus toxotes   93:12

Description
Size, to 18½ in. (47 cm). Distinguished by exceptionally thick, rubbery lips. Body and head moderately deep, heavy; dorsal rays 20–25, taller than spines; anal rays 27–30. Color usually brassy yellow, with dusky back; juveniles pinkish (fish up to 4 in. or 10.2 cm); pelvic fins dusky; lips white or light pinkish.

Habitat
Along rocky and sandy coasts, in bays, among sea growths or around pier pilings.

Habits
In small schools or aggregations.

Range
N. Calif. to cen. Baja Calif.

STRIPED PERCH
Embiotoca lateralis   93:7

Description
Size, to 15 in. (38.1 cm). Distinguished by prominent red, blue, and yellow horizontal body stripes. Body moderately deep, compressed; caudal peduncle short, deep; caudal fin lunate; spinous section of dorsal low. Color coppery, darker above and finely speckled with black; head with blue spots and streaks; dull orange and blue stripes along scale rows. Resembles Rainbow Perch, but dorsal and anal fins have no black blotches.

Habitat
Along rocky shores among sea growths and mussel beds.

Range
Port Wrangell, Alaska, to n. Baja Calif.; uncommon in s. Calif., more common in n.

DAMSELFISHES
Family Pomacentridae

BLACKSMITH
Chromis punctipinnis   92:8

Description
Size, to 12 in. (30.5 cm). Body oblong, scales large; dorsal fin long, continuous, 12–13 spines, 10–13 rays; caudal fin slightly notched, lobes pointed; lateral line ends under rear of dorsal fin. Head short, blunt; snout very short, upturned; eye interspace broad, rounded. Color above dark slate or green, fins blue-black; small dark brown or blackish spots on rear of body, dorsal rays, caudal fin; overall tinge of blue or violet.

Habitat
Shallow water around kelp beds or other sea growths and around rocks.
Range
Monterey Bay, Calif., to cen. Baja Calif., rare n. of Pt. Conception; common.

GARIBALDI
Hypsypops rubicunda   92:2

Description
Size, to 14 in. (35.6 cm). Body very deep, scales large, single dorsal fin with 11–13 spines, 15–17 rays in broad triangular section; caudal fin well notched, lobes hemispherically rounded; 2 spines in front of anal fin. Head profile steep, irregular. Distinguished from all other West Coast fishes by brilliant, overall orange colorations; newly hatched young with large spots and streaks of iridescent blue. Males very territorial.
Habitat
Swirling waters along rocky shores; abundant.
Remarks
One of the most brilliant fishes, often seen in marine gardens.
Range
Monterey Bay, Calif., to s. Baja Calif., rare n. of Pt. Conception. Protected species.

WRASSES
Family Labridae

These chiefly tropical fishes have large scales, small thick-lipped mouths, and strong canine teeth. Their dorsal fins are continuous, usually with a long spiny portion.

SEÑORITA
Oxyjulis californica   92:4

Description
Body very long, slender, cigar-shaped; scales large; dorsal fin long, continuous, 9–10 weak flexible spines, 13 rays; lateral line drops abruptly under posterior portion of dorsal. Sharp-pointed snout; teeth very small, sharp, protruding, none on vomer. Color above kelp-brown with brownish and bluish streaks on side of head, large black blotch at base of caudal fin; belly cream.
Habitat
Inshore shallow water around kelp; very abundant.
Habits
A cleaner fish (parasite picker).
Remarks
A pest to anglers; a good food fish but with strange flavor.
Range
San Francisco Bay to cen. Baja Calif.

PACIFIC SANDFISH
Trichodon trichodon   92:3

Description
Size, to 12 in. (30.5 cm). Distinguished by nearly vertical mouth and strikingly fringed lips. Body without scales; pectoral fins large, with out-turned edges. Color light brown or dusky green above, silvery-white below; back and fins with dusky streaks.

541

Habitat
Sandy or muddy bottom areas from intertidal zone to depths of 180 ft. (4.6 m), burrows into the substrate.

Range
Bering Sea to San Francisco, Calif.

CALIFORNIA SHEEPHEAD
Semicossyphus pulcher   **90:2**

Description
Size, to 3 ft. (0.9 m) and 36¼ lb. (16.4 kg). Body heavy; dorsal spines shorter than rays; rayed dorsal and anal fins pointed; caudal fin lobes pointed, particularly in males; caudal peduncle deep; scales large, heavy. Head with small, high-set eyes; chin deep; large protruding caninelike teeth in heavy jaws. Hermaphrodites; change sex from female to male when 3 to 5 years old; over 90% of fish larger than about 4 lb. (1.8 kg) are males. Males have head and posterior end of body including rayed dorsal, caudal, and anal fins reddish- to jet-black; other areas red; lower jaw and chin white. Females light to dull red, sometimes blackish; belly lighter; chin whitish.

Habitat
Shallow water along rocky shores, particularly in areas of mussel and kelp beds.

Range
Monterey Bay to Cape San Lucas, Baja Calif. and in upper Gulf of Calif.; common in s.

NIBBLERS
Family Girellidae

OPALEYE
Girella nigricans   *Fig. 74*

Description
Size, to 25⅓ in. (64.4 cm) and 13½ lb. (6.1 kg). Distinguished by bright opalescent-blue eyes and usually two white spots at base of dorsal fin. Body stout; dorsal spines 12–14, rays, 12–15; anal spines 3, rays 10–13, long; caudal fin straight, caudal peduncle deep. Head and snout profile rounded; 2 bands of tiny, tricuspid teeth on jaws, outer teeth freely movable. Color above dark greenish-blue; grayish-brown to greenish below; usually 1–2 whitish blotches on either side of back.

Habitat
Along rocky shores; common in kelp.

Range
San Francisco, Calif. to Cape San Lucas, Baja Calif.; abundant in s.

Fig. 74   Opaleye   Halfmoon

SEA CHUBS
Family Kyphosidae

HALFMOON
Medialuna californiensis

Fig. 74

Description
Size, to 19 in. (48.3 cm) and 4 lb. 12 oz. (2.2 kg). Body scaly; soft dorsal and anal fins so covered with scales that the rays are hidden. Spinous and soft portions of dorsal fin connected, spines shorter than first rays; anal fin with 3 spines and 17–21 rays; caudal fin moderately furcate. Head profile curved; eyes very small. Color above dark blue to gray-blue, below light blue.
Habitat
Shallow water along rocky shores, often abundant near kelp beds feeding near surface.
Habits
Associates with Opaleye and other dark-colored perches.
Range
Klamath R. to Gulf of Calif., rare n. of Pt. Conception.

POACHERS
Family Agonidae

Members of this group are elongate, usually quite slender, fishes with an armor of nonoverlapping bony plates. Dorsal and anal fins are short. The dorsal fin is composed of two separate fins, the first spinous and the second soft. Each pelvic fin has one spine and two soft rays. Poachers live on either rocky or muddy bottoms in cold, sometimes deep, marine waters in northern regions.

STURGEON POACHER
Podothecus acipenserinus

92:7

Description
Body elongate; head large and spinous, snout depressed (spadelike or sturgeonlike), mouth inferior and surrounded by large clusters of slender cirri. First dorsal fin with 8–10 spines, second with 7–9 soft rays; anal composed of 6–9 soft rays; body plates spinous. Body with 6 or more brownish-black bands; dorsal, pectoral and caudal fins dusky; pelvics and anal white, anal with a posterior black blotch.
Habitat
Over muddy bottoms in water between 60–180 ft. (18.3–54.4 m) deep.
Remarks
Commonly displayed in marine aquaria.
Range
Bering Sea to Eureka, Calif.; fairly common.

SCORPIONFISHES AND ROCKFISHES
Family Scorpaenidae

Members of this large family include a number of important commercial and game fishes, most of which live in relatively shallow to moderately deep water along rocky coasts. Some sixty-nine species of rockfishes (Genus *Sebastes*) occur in the eastern north Pacific, of which fifty-seven species are found in California waters. To those unfamiliar with rockfishes, they appear to be quite similar and difficult to differentiate. Those included here are among the most common sport and commercial forms. Members of the family are characterized by a bony support extending back from the lower part of the eye across the cheek just under the skin; the anal fin has three strong spines, there is no slit behind the fourth gill, and the body is covered with scales. Life histories have been worked out thus far for only a few of the species.

CALIFORNIA SCORPIONFISH
Scorpaena guttata   90:11

Description
Size, to 17 in. (43.2 cm). Distinguished from other family members by 12 dorsal-fin spines; presence of palatine teeth; and many small cirri about the body, head, and snout and over eyes. Spinous portion of dorsal fin deeply incised. Numerous spines on head and about cheek; eyes set high, snout blunt. Color variable, but generally reddish-brown above, spotted and mottled with reddish-brown, olive, gray, or purple; below light pink; pelvic fins reddish with dark spots; other fins with dark spots or bars.

Habitat
Moderately deep to very shallow water, around man-made marine structures; plentiful over rocky reefs.

Remarks
Fin spines are venomous and can inflict very painful wounds.

Other name
Sculpin.

Range
Santa Cruz, Calif., to s. Baja Calif.; rare n. of Pt. Conception; abundant in s.

PACIFIC OCEAN PERCH
Sebastes alutus   92:10

Description
Size, to 20 in. (50.8 cm). Distinctive for long, sharp, greatly projecting lower jaw with very large pointed knob at tip. Head with small, weak spines, except on opercle; mouth rather large, maxilla reaches past pupil; eyes large, interspace broad and slightly convex to flat. Color above light red, olive-brown blotches at base of dorsal fin and on upper area of caudal peduncle; lower lip black; belly lighter than back.

Habitat
Deep water.

Range
Japan to Bering Sea to La Jolla, Calif.; uncommon in s.

CHILIPEPPER
Sebastes goodei   90:5

Description
Size, to 22 in. (55.9 cm) and 5 lb. (2.3 kg). Body slender; anal rays, normally 8. Head spineless; lower jaw projecting, maxilla reaches pupil; eye interspace broad, convex. Color above pinkish-red, below pink; narrow clear-pink stripe along lateral line.

Habitat
Moderately shallow water.

Range
Eureka, Calif., to Baja Calif.; common.

COPPER ROCKFISH
Sebastes caurinus 93:11

Description
Size, to 22 in. (55.9 cm). Body elongate; pectoral fin long, blackish, reaching past anus; dorsal fins high, spinous section moderately incised. Head spines thick, prostrate; those on cheeks strong, sharp. Lower jaw projecting; eyes small, interspace narrow, smoothly concave. Color above dark greenish-brown to a coppery-pink and yellowish; greenish oblique stripes behind eyes; fins blackish; posterior ⅔ of body whitish or pinkish with lateral line; belly light; all surfaces tinged with copper.
Habitat
Moderately shallow water on rocky substrate.
Range
Se. Alaska to cen. Baja Calif.

GREENSPOTTED ROCKFISH
Sebastes chlorostictus 93:2

Description
Size, to 16 in. (40.6 cm). Very high dorsal fin, deeply incised; pectorals extend past anus; second anal spine longer than third. Head and shoulders deep, heavy; head short, profile steep; very strong, high spines on top of head, on snout, above eyes, on preopercle and opercle; jaws equal, knob on tip of lower, maxilla extends past eye; spinous ridge between eyes. Color above greenish with 3–5 pinkish spots on back; fins red, may show yellow rays; base of dorsal green.
Habitat
Rather deep water.
Range
Eureka, Calif., to cen. Baja Calif.; abundant in s.

COW ROCKFISH
Sebastes levis 93:10

Description
Size, to 37 in. (94 cm) and 28½ lb. (12.9 kg). Deep, heavy shoulders, tapering abruptly to small caudal peduncle; dorsal spines very high, deeply incised, rays low; space between eye and upper jaw as wide as eye diameter. Head with 5 or 6 distinct pairs of sharp spines; lower jaw projecting, small knob on tip. Color pink to light red, indistinctly marked with dusky or blackish crossbars (sometimes absent from older individuals).
Habitat
Deep water.
Range
N. Calif. to cen. Baja Calif.; abundant in far s.

BLACK ROCKFISH
Sebastes melanops 92:11

Description
Size, to 23¾ in. (60.3 cm). Very similar to Blue Rockfish, but has a larger eye, maxilla extends to back of eye or slightly beyond; anal fin rounded. Color above dark gray, almost black, to olive-brown; all fins dark gray to blackish, top of head blackish, distinctive array of black spots on lower part of dorsal spinous section; sides pale; lighter below.

Habitat
Shallow to moderately deep water.
Range
Se. Alaska to San Miguel Is., and Paradise Cove, Calif.; abundant
n. of Calif.

VERMILION ROCKFISH
Sebastes miniatus   92:5

Description
Size, to 30 in. (76.2 cm) and 14 lb. (6.4 kg). Dorsal fin deeply
notched between first and second sections; pelvic fins large,
reaching past anus; pectorals not reaching anus. Head and snout
spines small; lower jaw slightly projecting, with small knob on tip;
mandible rough, scaly. Color above deep vermilion (principal
distinguishing feature), mottled with gray on back; fins black-
tipped; below light red.
Similarities
The vermilion color may fade to orange in some individuals, such
as the color of the Canary Rockfish.
Habitat
Moderately deep water.
Range
Vancouver Is. to cen. Baja Calif.; common.

BLUE ROCKFISH
Sebastes mystinus   93:5

Description
Size, to 21 in. (53.3 cm). Anal fin straight-edged; head spines
obsolete; eyes small, interspace broad and convex; maxilla reaches
past pupil. Color above blue-black, paler on sides; below dirty-
white; fins dark. Abdominal cavity of large individuals black to
dusky-white.
Similarities
Black Rockfish.
Habitat
Shallow to moderately deep water.
Range
Bering Sea to n. Baja Calif.; abundant in s. half of range.

CANARY ROCKFISH
Sebastes pinniger   93:13

Description
Size, to 26½ in. (67.3 cm) and 10 lb. (4.5 kg). Body has lateral line
in a clear gray zone; posterior margin of anal fin with an anterior
slant. Head weakly spined on top; lower jaw projecting with
pronounced knob; mandible smooth; eye interspace slightly convex.
Color above orange-red or orange-yellow, gray mottlings on back;
sides paler, belly whitish; fins occasionally red-blotched but
generally bright orange; lips and lining of mouth pale red with
dusky or black mottling. Young have black blotch on dorsal
membrane between seventh and tenth spines.
Similarities
Vermilion Rockfish, scales on mandible rough to touch.
Habitat
Moderately deep water.
Range
Alaska to n. Baja Calif.; common.

BOCACCIO
Sebastes paucispinis   **92:9**

Description
Size, to 36 in. (91.4 cm) and 17 lb. (7.7 kg). Body slender; deep notch between first and second sections of dorsal fin, nearly separating them. Head long, narrow; lower jaw greatly projecting beyond profile of head, maxilla extends beyond rear margin of orbit, no spines; eye interspace broad, convex. Color above dark brown to olivaceous; sides dull orange-reddish; belly pale pink or white. Young with many black spots; colors less distinctive in larger individuals.
Habitat
Moderately deep to deep water.
Range
Alaska to n. Baja Calif.; more abundant in s.

YELLOWEYE ROCKFISH
Sebastes ruberrimus   **93:6**

Description
Size, to 36 in. (91.4 cm). Body very deep; second anal spine large, equal to third. Head and mouth large, snout blunt, lower jaw slightly projecting; small, flattened, rasplike spines around eyes and head of adults, larger spines on upper operculum angle; eye interspace flat, narrow. In individuals over 11 in. (27.9 cm), ridges on top of head between eyes become rugose, breaking into many low spines. Color above reddish-orange, sometimes with dark blotches along back; fins black-tipped on posterior margins, all black in young, faded in old. Young to 12 in. (30.5 cm) with two pinkish or white stripes on side; belly light.
Habitat
Moderately shallow to deep water.
Range
Alaska to n. Baja Calif.; abundant n. of Pt. Arena, Calif.

FLAG ROCKFISH
Sebastes rubrivinctus   **93:3**

Description
Size, to 21 in. (53.3 cm). Distinguished by 4 broad, crimson bars; bar below first dorsal insert covers operculum. Body moderately deep, compressed; second anal spine much larger than third. Head large, pointed, strongly spined above eyes, on top, on preopercle, and on upper edge of operculum; lower jaw slightly projecting. Color above pale rose, sometimes faded pink; first crimson crossbar below first dorsal insert, second behind pectoral, third encircling body at base of anal, fourth around caudal base; underparts light.
Habitat
Moderately deep water.
Range
Se. Alaska to n. Baja Calif.; common in Calif.

OLIVE ROCKFISH
Sebastes serranoides   **93:4**

Description
Size, to 24 in. (61 cm). Body long, slender; dorsal fin low, deeply notched; anal fin with 8–10 rays. Head smooth, with long pointed snout, lower jaw projecting; nasal spines concealed; preopercular

spines long, slender; eyes small. Color above olive-brown with white blotches on back; below grayish; sides finely stippled on scales; fins tinted yellowish and tipped with dark gray, caudal fin greenish-yellow.

Habitat
Fairly shallow water in or near kelp beds.

Range
Crescent City, Calif. to cen. Baja Calif.; abundant in s.

SHORTSPINE THORNYHEAD
Sebastolobus alascanus   **92:1**

Description
Size, to 30 in. (76.2 cm). Body long, slender, covered with large scales; dorsal fin deeply notched, 15–17 spines, 8–10 rays; pectoral fins with deep notch forming upper and lower lobes. Head large, conspicuously spined on snout, top, preopercle; sharp spinous ridge across cheek. Color uniformly bright red, 1 or more dark blotches on spinous dorsal, other fins with black markings.

Habitat
Flat relief substrate (mud to sand) in 84 to 5000 ft. (25.6–1524 m).

Range
Bering Sea to cen. Baja Calif.; abundant n. of Pt. Conception, Calif.

SABLEFISHES
Family Anoplopomatidae

SABLEFISH
Anopolopoma fimbria   **92:6**

Description
Size, to 40 in. (101.6 cm) and 56 lb. (25.4 kg). Body long, slender, codlike, covered with small scales; 2 well-separated dorsal fins; caudal peduncle long, slender; anal fin with 2 spines, 16–23 rays. Head broad, smooth, flat; maxilla narrow, barely reaches pupil; eyes small. Color above blackish, dark gray, or dark greenish; sides lighter below lateral line; spinous dorsal black-edged, other fins gray-tipped; throat and belly light gray. Juveniles to 6 in. (15.2 cm) are blue above, silvery below.

Habitat
Deep water in winter, shallower in summer; juveniles to 6 in. (15.2 cm) are pelagic and live in upper water layers.

Range
Japan to Bering Sea to n. Baja Calif.; abundant n. of Coos Bay, Oreg.

GREENLINGS
Family Hexagrammidae

Members of this family are commercially referred to as "sea trout." In some species males and females are markedly different in color. The dorsal and anal fins are notably long, and in *Hexagrammos*, there is only a moderate notch between the two dorsal sections.

KELP GREENLING
Hexagrammos decagrammus   94:10

Description
Size, to 21 in. (53.3 cm). Body with 5 lateral lines; pectoral fin
very large. Head usually with 2 pairs of fleshy flaps on top, first
pair ½ eye length; inside of mouth yellowish. Males brown,
reddish-brown, green, or gray, sometimes tinged with bluish or
coppery; bright blue spots on head, cheeks, forepart of dorsal fins,
body; pectoral fin with whitish spots and thin lines. Females
predominantly light brown with very small reddish or orange spots
on head, back, sides; dorsal fin with red to orange; pelvic, anal fins
with orange to yellow.
Habitat
Shallow water in rocky areas among sea growths.
Range
Aleutians to La Jolla, Calif.; very abundant n. of Pt. Arena, Calif.

PAINTED GREENLING
Oxylebius pictus   94:11

Description
Size, to 10 in. (25.4 cm). Body with single, nearly straight lateral
line; single dorsal fin moderately notched between sections; pectoral
fins reach past anus; pelvics shorter; anal fin 3–4 hard spines.
Head long, pointed, compressed; 2 pairs of fleshy flaps, 1 pair
above eyes, other on top of head. Color variable according to
habitat; gray to light brown but distinguished by 6–7 dark vertical
crossbars across body and sometimes blotched with light orange;
pectoral and pelvic fins with dark stripes; caudal and pectoral fins
sometimes orange; head brown or gray on top, orange beneath,
with dark stripes about eye (sometimes absent); males with much
black on body during breeding season.
Habitat
Shallow water along rocky shores.
Range
B.C. (Straits of Georgia) to n. Baja Calif.

LINGCOD
Ophiodon elongatus   94:9

Description
Size, 4–5 ft. (1.2–1.5 m) and to 105 lb. (47.6 kg) in British
Columbia. Body long, single lateral line; dorsal fin very long,
rather deeply notched; caudal peduncle long, slender. Head long,
mouth large with characteristically large canine teeth; single large,
thick cirrus over each eye. Color extremely variable according to
habitat; above brown, blue, or green; belly usually lighter (grayish,
cream, whitish); large golden spots over most of body.
Habitat
Adults in moderately deep to deep water (spawns in shallow
water), rocky habitat; young common in shallower waters.
Remarks
Flesh may be greenish or blue but cooks up white and is excellent
in flavor.
Range
Nw. Alaska to n. Baja Calif.

SCULPINS
Family Cottidae

Most members of this family have broad heads, slender bodies, and large fanlike pectoral fins. The top of the head is usually scaleless and covered with cirri; the skin and muscles of the cheeks are supported by a bony process. No detached rays precede the pectoral fins. Of nearly one hundred West Coast species, only three are described here; the others are mostly small fishes inhabiting a variety of marine and freshwater environments. The sculpins are chiefly cold-water bottom feeders.

STAGHORN SCULPIN
Leoptocottus armatus   94:8

Description
Size, to 12 in. (30.5 cm). Distinguished by large, black spot on spinous dorsal fin. Body elongate, scaleless; heavy, stout shoulders tapering to small caudal peduncle; fins large, pelvic with 1 spine and 4 rays (spine separable from first ray only by dissection); single lateral line. Head with gill membranes completely joined to isthmus, large antlerlike spine or preopercle. Color above dark green to brown; fins, except pelvics and anal, green with black bars; belly silver to yellowish.

Habitat
Intertidal to 300 ft. (91.4 m); also bays, backwaters.

Habits
Young enter brackish and fresh water at stream mouths.

Range
Nw. Alaska to n. Baja, Calif.; very common.

GRUNT SCULPIN
Rhamphocottus richardsonii   94:7

Description
Size, to 3⅓ in. (8.5 cm). Body short, stout; body and head covered with spinelike scales, each mounted on a fleshy papilla (vascular protuberances). First dorsal fin with 7–8 spines, second dorsal with 12–14 soft rays; anal fin with 6–8 soft rays; pectoral fins large, lower rays free, fingerlike. Head extremely large, snout pointed, jaws short. Color cream with brown streaks; streaks on head radiate from center of eye, those on sides of body run from dorsal fin downward and forward; fins orange to coral-red.

Habitat
Tidal pools, rocky shores, sandy beaches; shallow to moderately deep water (90 fathoms, or 164.6 m).

Habits
Crawls over rocks using pectoral fins; grunts when removed from water; rolls eyes independently of one another.

Range
Bering Sea to s. Calif.; rare in Calif., common elsewhere.

CABEZON
Scorpaenichthys marmoratus   90:7, 94:6

Description
Size, to 39 in. (99.1 cm) and 20–25 lb. (9.1–11.3 kg). Body scaleless; appears wrinkled, but skin is smooth and slick. Dorsal fin deeply notched; anal fin spineless, rays soft and thick. Head large, moderately broad; eyes large, highly placed, tall cirrus above each

eye; snout short with large cirrus on tip; mouth broad, with many small sharp teeth. Color extremely variable from browns and reds to tan, gray, or greenish, generally mottled or blotched; ground color in adults usually green in females, red in males; lips, mouth, and flesh green or bluish.

Habitat
Shallow inshore water over hard bottom to 30–40 fathoms (54.9–73.2 m).

Remarks
A delicious food fish, but roe poisonous.

Range
Se. Alaska to cen. Baja Calif.; abundant.

GOBIES
Family Gobiidae

Of some twelve native species of gobies in West Coast warmer waters, most occur abundantly on mud flats and in tidepools and one inhabits freshwater streams. Several are important to fishermen as bait. Gobies are distinguished by their pelvic fins that are completely joined. These fishes are usually small-sized carnivorous bottom feeders.

LONGJAW MUDSUCKER
Gillichthys mirabilis   94:3

Description
Size, to 8 in. (20.3 cm). Body long, slender; 2 separate dorsal fins; base of anal shorter than head. Head very large with huge mouth and maxilla greatly developed, reaching to base of pectorals in adults; eyes highly placed. Color above dark greenish-brown with darker mottlings, speckles, or bars; fins olive-green; below yellowish.

Habitat
Bays, sloughs, backwaters.

Remarks
Makes excellent bait as can live out of water for several days if packed in wet seaweed.

Range
Tomales Bay, Calif., to Gulf of Calif.; common in s.

BLUEBANDED GOBY
Lythrypnus dalli

Description
Size, to 2½ in. (6.4 cm). Body short, laterally compressed; first dorsal fin with 6 long, filamentous spines; second with 17 soft rays; anal fin with 14 soft rays. Head blunt, lower jaw slightly projecting, mouth directed upward. Color coral-red with 4–6 narrow, neon-blue bars on front of body, fading to orange on rear of body and caudal fin; eyes and forehead with a blue "mask"; fins colorless.

Habitat
Tidal pools to 300 ft. (91.4 cm).

Range
Morro Bay, Calif., to Mexico.

Other name
Neon Goby.

Note: The **ZEBRA GOBY,** *Lythrypnus zebra,* is similar but has fifteen blue bands over entire body.

CLINIDS
Family Clinidae

GIANT KELPFISH
Heterostichus rostratus   94:5

Description
Size, to 24 in. (61 cm). Distinguished by very sharp, pointed head
and snout. Long, continuous dorsal fin, with deep notch following
fifth spine; caudal peduncle slender, caudal fin forked; pelvic fins
small, with 1 spine, 3 rays, and ahead of pectorals; base of anal fin
extends ½ body length. Color highly variable but mostly brown,
ranging from light brown through purple, orange, yellow, coral,
reddish, and green; usually with large lighter blotches over body
and fins; fins with transparent areas; body may be barred, mottled,
or striped with silver or orange to reddish. Fish in kelp beds are
kelp-colored; those in eelgrass are bright green with brilliant
silvery stripes.
Habitat
Inshore shallow waters in kelp beds or other sea growths along
rocky coasts.
Range
B.C. to Cape San Lucas, Baja Calif.; common.

ONESPOT FRINGEHEAD
Neoclinus uninotatus

Description
Size, to 9 in. (22.9 cm). Large head and mouth, elongate body;
elongate cirrus with fringe at tip above each eye. Dorsal fin
extends from above gill opening to base of tail; tail fin distinct,
rounded on posterior margin. Color gray or gray-green to dark
brown, black speckling on much of body; a single black ocellus on
dorsal membrane between first and second spines.
Habitat
Any conveniently sized hole on ocean floor, including beer bottles
and similar containers discarded by man; depth 10–90 ft. (3.1–
27.4 m).
Habits
Lies in shelters with only head protruding; may leave briefly in
search of food.
Range
San Diego to Bodega Bay, Calif.

LUMPSUCKERS
Family Cyclopteridae

PACIFIC SPINY LUMPSUCKER
Eumicrotremus orbis   94:12

Description
Size, to 5 in. (12.7 cm). Stout, globose body; body and head covered
with scattered cone-shaped tubercles; pelvic fins modified to
support a large adhesive disk. Eyes large; mouth tiny, terminal;
lips thickened. Color usually light to dark green or brown, lighter
below; lips lavender; tubercles of male orange or reddish-brown, of
female, pale green.

Habitat
Commonest subtidally in rocky habitat, often where currents are fast moving; reported to depths of 480 ft. (146.3 m).
Range
Washington to Bering Sea and down Asian coast to Okhotsk Sea and Sakhalin.

WOLFFISHES
Family Anarhichadidae

WOLF-EEL
Anarhichthys ocellatus   94:1
Description
Size, to 8 ft. (2.4 m). Body very long, eellike, tail pointed. Dorsal fin runs full body length, composed of flexible spines; anal fin of soft rays along posterior ¾ of body; no pelvic fins or lateral line. Head thick, blocky; mouth large, pugnacious; teeth (canine, molar) strong. Color dark to grayish-green; dorsal fin and body with large, round, darker spots; smaller spots about head and on pectorals; anal fin pale.
Habitat
Shallow water, under rocks (difficult to dislodge).
Remarks
Bites viciously when caught.
Range
Kodiak Is., Alaska, to San Diego, Calif.

MONKEYFACE-EELS
Family Cebidichthyidae

MONKEYFACE-EEL
Cebidichthys violaceus   94:2
Description
Size, to 30 in. (76.2 cm) and 6 lb. (2.7 kg) or more. Body long, slender, eellike. Dorsal fin begins back of head, extends full length of body, with 22–25 sharp spines followed by 40–43 soft rays; connects with small, rounded caudal fin; anal fin with 1–2 spines and 39–42 soft rays, begins at mid-body and extends to single lateral line. Head small, adults with fleshy lumps on crest; snout blunt. Color uniform dull to brownish-green, 2 dark bars below eye.
Habitat
Intertidal zone to 80 ft. (24.4 m), under rocks.
Range
Crescent City, Calif. to n. Baja Calif.; rare s. of Pt. Conception, common elsewhere.

PRICKLEBACKS
Family Stichaeidae

ROCK PRICKLEBACK
Xiphister mucosus   94:4

Description
Size, to 23 in. (58.4 cm). Body long, slender, slippery, covered with
well-imbedded scales; 4 lateral lines, each with series of cross-
branches. Pectoral fins tiny, shorter than eye; dorsal fin with 71–78
sharp, needlelike spines; anal fin with 1 spine and 46–50 soft rays;
no pelvic fins. Head small, snout blunt, eyes very small. Color
blackish-green, sometimes with yellowish blotches in older fishes; 2
prominent olive-brown eye streaks edged with black; below pale.
Habitat
Intertidal zone and 60 ft. (18.3 m), under rocks, ledges.
Range
Alaska to Pt. Arguello, Calif.; common.

Flatfishes
Order Pleuronectiformes

Members of this order are bottom fishes and have the body
flattened so that the fish rest on one side on the ocean floor. Except
when semitransparent larvae, all flatfishes have both eyes on the
same side of the head. Normally, the eyed side of the body is
colored and the blind, or underside, is white. All are edible, and
several species are commercially fished. Although the order is
divided generally into lefteyed and righteyed types, a few species
may have the eyes on either side, and occasional reversed
individuals are found in many species.

LEFTEYED FLOUNDERS
Family Bothidae

Members of this family are sinistral, that is, they have the eyes and
coloring on the left side of the body in adults. Pelvic fins are
asymmetrically located on abdominal ridge.

PACIFIC SANDDAB
Citharichthys sordidus   95:12

Description
Size, to 16 in. (40.6 cm) and 2 lb. (0.9 kg). Asymmetrical pelvic
fins; pelvic on eyed side attached to ridge of abdomen. Lateral line
nearly straight; scales large, loosely attached; pectoral fin on eyed
side shorter than head. Lower eye longer than snout; eye interspace
concave; 12–16 gill rakers on lower limb of first arch. Color
brownish to tan to light olive, mottled with dull orange or black.
Habitat
Moderately deep water (10–80 fathoms, or 18.3–146.3 m) over
sandy bottoms.
Reproduction
Spawns in summer.
Remarks
Considered a delicacy.
Range
Bering Sea to Cape San Lucas, Baja Calif.

CALIFORNIA HALIBUT
Paralichthys californicus   **95:8**

Description
Size, to 5 ft. (1.5 m) and 72 lb. (32.7 kg). Small pectoral fin, ½ head length; lateral line about 100 scales long, forms high arch over pectoral fin. Eyes on either left or right side; small; separated by broad, flat area. Jaws large; teeth strong and sharp; maxilla reaches to or beyond hind border of lower eye. Color above greenish-black to muddy brown, sometimes mottled or blotched with lighter shades; occasionally with small, vague whitish spots.

Habitat
Shallow waters to 300 feet (91.4 m), sloughs and bays; spawns in inshore waters.

Range
B.C. to Magdalena Bay, Baja Calif., and in upper Gulf of Calif.; very abundant in s., uncommon n. of Pt. Conception.

FANTAIL SOLE
Xystreurys liolepis   **95:3**

Description
Size, to 20 in. (50.8 cm). Distinguished from other flatfishes by short, blunt snout. Long, pointed pectoral fins, longer than head; caudal fin rounded; lateral line highly arched just above pectoral fin. Jaws about equally developed on both sides, but more teeth on blind side. Maxilla extends below middle of lower eye; eyes large, separated by narrow scaly ridge; eyes may be on either right or left side. Color brownish and olive mottled darker, occasionally with many gray and reddish-brown blotches; large eyelike spot behind head and usually toward tail fin.

Habitat
Outer coast to 260 ft. (79.3 m); bays, backwaters, sloughs in summer.

Range
Monterey Bay, cen. Calif., to Gulf of Calif.

RIGHTEYED FLOUNDERS
Family Pleuronectidae

Members of this family usually have the eyes on the right side, as well as coloration. A number of exceptions do occur, however. Pelvic fins are symmetrically located, one on each side of the abdominal ridge.

REX SOLE
Glyptocephalus zachirus   **95:11**

Description
Size, to 23¼ in. (59.1 cm). Distinguished from other flatfishes by black, very long pectoral fin on eyed side, longer than head. Body slender, smooth; caudal peduncle short; lateral line nearly straight. Very small mouth; jaws and teeth better developed on blind side; maxilla to below front of lower eye. Color light brown, pectoral fins black, other fins dusky to dark brown.

Habitat
Moderately deep (60 ft., or 18.3 m) to deep (2100 ft., or 640.1 m) water.

Range
Bering Sea and Aleutians to San Diego, Calif.

ARROWTOOTH FLOUNDER
Atheresthes stomias   95:10

Description
Size, to 30 in. (76.2 cm). Distinguished by long maxilla, extending well past lower eye. Body elongate; scales large, deciduous; lateral line upcurving over pectoral fin; caudal fin lunate, incurved; caudal peduncle long, slender. Head elongate, upper eye entering margin of profile, almost on rim of head; jaws about equal; most teeth in adults with small, arrowlike tips. Color above greenish- to olive-brown; edges of scales brown-tipped; underside off-white.

Habitat
Deep water in winter, shallow in summer and fall.

Range
Aleutians to San Pedro, Calif.

Note: The similar **GREENLAND HALIBUT,** *Reinhardtius hippoglossoides,* has the teeth in one row and the blind side pigmented.

PETRALE SOLE
Eopsetta jordani   95:6

Description
Size, to 27½ in. (69.9 cm) and 6–8 lb. (2.7–3.6 kg). Body covered with small scales; about 30 rows between lateral line and dorsal fin at widest point; 88–100 scales along lateral line. Very smooth on blind side. Pectoral fins shorter than head; lateral line slightly curving over pectoral fin, no dorsal branch. Moderately large mouth; jaws and teeth about equally developed on both sides; 2 rows of small teeth on each side of upper jaw; maxilla extends to below middle of lower eye. Color brown or olive-brown, vague dusky blotches on dorsal and anal fins.

Habitat
Flat relief bottom, sandy to muddy; in 60 to 1500 ft. (18.3–457.2 m).

Range
Nw. Alaska to n. Baja Calif.

PACIFIC HALIBUT
Hippoglossus stenolepis   95:4

Description
Size, male to 4½ ft. (1.4 m) and 123 lb. (55.8 kg), females to 9 ft. (2.7 m) and 507 lb. (229.9 km). Body deep; dorsal fin with 89–109 rays; caudal fin slightly truncate; anal fin with 61–84 rays; lateral line with high arch over pectoral fin, no dorsal branch. Jaws about equally developed on both sides; teeth strong; snout pointed; maxilla reaching to anterior edge of lower eye. Color nearly uniform dark brown, often with vague paler blotches.

Habitat
Moderately deep (20 ft., or 6.1 m) to deep (3600 ft., or 1097.3 m) water. Large individuals congregate offshore; fish up to 40 lb. (18.1 kg) occur near shore during spring and summer and are caught off piers, docks, small boats.

Range
Japan to Bering Sea to Santa Rosa Is., Calif., rare s. of Eureka, Calif.

DIAMOND TURBOT
Hypsopsetta guttulata   **95:5**

Description
Size, to 18 in. (45.7 cm) and 2 lb. (0.9 kg). Body depth ½ full
length including tail; pectoral fin shorter than head; lateral line
with long branch along base of dorsal fin. Small mouth, jaws better
developed on blind side; teeth small, few if any on eyed side;
maxilla to forepart of eye; no ridge between eyes. Color dark
greenish-black to brown, mottled with bluish, lemon-yellow around
mouth on porcelain-white underside.

Habitat
Shallow water of bays, sloughs and outer coast.

Range
Cape Mendocino, Calif., to Magdalena Bay, Baja Calif., and in n.
Gulf of Calif.; uncommon n. of Pt. Conception; very abundant
elsewhere.

DOVER SOLE
Microstomus pacificus   **95:7**

Description
Size, to 2½ ft. (0.8 m) and 10½ lb. (4.8 kg). Distinguished by
slippery mucous secretion over body and by very dark brown to
black fins. Body slender; caudal peduncle short; a single, straight
lateral line; a long loop of the intestine extends back into a
"pocket" that parallels the anal fin. Lower eye slightly in advance
of upper; gill opening barely extends above pectoral fin base;
mouth small. Color greatly varied, from blackish and shades of
brown to greenish-yellow, often mottled; underside various densities
of dark colors.

Habitat
Moderately shallow (90 ft., or 27.4 m) to deep (3000 ft., or 914.4
m) water.

Range
Bering Sea to cen. Baja Calif.

ENGLISH SOLE
Parophrys vetulus   **95:9**

Description
Size, to 22½ in. (57.2 cm). Body slender; lateral line nearly
straight, long dorsal branch; smooth scales on eyed side, no scales
on fins. Head very pointed; eyes large, upper almost on rim of
head, visible from blind side; jaws rather pointed, stronger on blind
side; teeth chiefly on blind side; maxilla to below forepart of lower
eye. Color yellowish-brown above, yellowish to white below; dorsal
and anal fins dark-tipped.

Habitat
Moderately shallow (60 ft., or 18.3 m) to deep (1000 ft., or 304.8
m) water over sandy or muddy bottoms.

Range
Nw. Alaska to cen. Baja Calif., abundant in n.

STARRY FLOUNDER
Platichthys stellatus   *Fig. 75*

Description
Size, to 3 ft. (0.9 m) and 20 lb. (9.1 kg). Body with rough,
scattered, spinous plates formed of small scales; lateral line nearly

straight. Small eyes; jaws and teeth better developed on blind side; maxilla to below forepart of lower eye; about 60% have eyes on left side. Color dark brown to black with vague blotchings; distinguished by alternating orange and black stripes on fins.

Habitat
Backwaters, around stream mouths; abundant in northern bays; shallow water to about 150 fathoms (274.3 m) over all types of bottoms except rock.

Range
Japan to Arctic Ocean, to Santa Barbara, Calif.

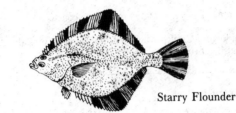

Fig. 75

Starry Flounder

HORNYHEAD TURBOT
Pleuronichthys verticalis   **95:1**

Description
Size, to 14½ in. (36.8 cm). Body smooth, scales deeply embedded; first 4–6 dorsal rays are on blind side; lateral line along base of dorsal fin. High, narrow bony ridge between eyes, with sharp, prominent spines projecting at either end; blunt spine overhanging mouth in front of eye; mouth small, no teeth on eyed side. Color above chocolate-brown, irregularly blotched and mottled darker; scattered pale spots along dorsal and anal fins and over body; underside white.

Habitat
Moderately shallow water (30–600 ft., or 9.1–182.8 m).

Range
Pt. Reyes, Calif., to Magdalena Bay, Baja Calif., and in n. Gulf of Calif., abundant in s.

TONGUEFISHES
Family Cynoglossidae

CALIFORNIA TONGUEFISH
Symphurus atricaudus   **95:2**

Description
Size to 8¼ in. (21 cm). Body tapers to a point posteriorly; dorsal and anal fins meet at posterior end of body; no lateral line. Eyes on left side, small, close-set; mouth small, curved. Color brownish with dark vertical bars from dorsal and anal fin bases toward body center.

Habitat
Shallow water.

Remarks
No commercial or sport importance.

Range
Big Lagoon, Calif., to Cape San Lucas, Baja Calif.; common.

Triggerfishes, Puffers, Molas, and Their Relatives
Order Tetraodontiformes

MOLAS
Family Molidae

COMMON MOLA
Mola mola

Description
Size, estimated to 13 ft. (4 m) and 3300 lb. (1496.6 kg). Body notable for great size, flatness, and extreme shortness; depth equal to half body length. Skin leathery; small, porelike gill opening in front of pectoral fin; teeth fused into plate; no pelvic fins. Color above dark gray to gray-blue; sides and belly silvery.

Habitat
Epipelagic, strays into shallower water in some areas.

Habits
Swims slowly, often in small schools of 5–20 individuals, or drifts just below surface; smaller molas occasionally leap clear of water; basks on side at surface.

Range
Se. Alaska to tropics, all world oceans; sporadically abundant.

Toadfishes
Order Batrachoidiformes

TOADFISHES
Family Batrachoididae

PLAINFIN MIDSHIPMAN
Porichthys notatus

Fig. 76

Description
Size, to 15 in. (38.1 cm). Body scaleless; 2 dorsal fins, first very small, with 2 spines; caudal fin rounded. Head broad, mouth large, lower jaw projecting, teeth large, maxilla reaching edge of operculum; sharp spine projecting on operculum at back of head; eyes protruding, interspace wide; second row of photophores, or "light organs," under chin form an inverted V. Color above purplish-black to grayish-brown, sides lighter, anal fin dusky, belly dusky white.

Habitat
Flat relief bottom, subtidal to 1000 ft. (304.8 m).

Range
Alaska to Cape San Lucas, Baja Calif., and Gulf of Calif.

Note: The **SPECKLEFIN MIDSHIPMAN,** *Porichthys myriaster,* is similar to the Plainfin, but all fins are speckled and the photophores under the jaw form an inverted U.

Fig. 76

Plainfin Midshipman

Freshwater Fishes

JAWLESS FISHES
Class Agnatha

Lampreys
Order Petromyzoniformes

LAMPREYS
Family Petromyzonidae

Highly specialized, greatly modified descendants of the earliest known type of vertebrates, the members of this eellike family have never possessed upper or lower jaws, true teeth, or paired fins. The skeleton is purely cartilaginous; the circular mouth, adapted for sucking blood, is covered with horny spines called "teeth." Behind the head are seven paired gill clefts. These fishes possess a long dorsal fin, sometimes divided, that is more or less continuous with the caudal fin. Lampreys vary in length from 6 inches (15.2 cm) to several feet, and most freshwater species are pale brown or fawn color.

PACIFIC LAMPREY
Lampetra tridentata

Description
Size, to 30 in. (76.2 cm). Body distinguished by lack of scales and absence of spines or rays in the fleshy dorsal fin. Mouth a ventrally placed sucking disk (no true jaws), with four pairs of tooth plates on each side and three sharp teeth above. Young lack sucking disk and eyes but are distinguished from other ammocoetes by 64–70 muscle segments (myomeres) between gills and vent. Color adults plain dark brown, rarely mottled; when fresh from sea, dark blue above, silver below; young uniform olive-green, lighter below.
Habitat
Anadromous; adults marine, young develop in fresh water.
Habits
Adults parasitic on large bony fishes; dwarf landlocked populations exist in which adults are not parasitic.
Range
Sw. Alaska to s. Calif.

ARCTIC LAMPREY
Lampetra japonica   *Fig. 77*

Description
Size, 10–14 in. (25.4–35.6 cm). Body size variable; small (under 10 in.) in landlocked populations to large (more than 10 in.) in anadromous populations. Mouth with only 2 or 3 pairs of lateral tooth plates. Color dark brown to blue-black above, light brown below.
Habitat
Anadromous and nonanadromous.

Habits
Parasitic.
Range
Yukon R. drainage, Alaska, to White Sea in Europe.

Fig. 77

Arctic Lamprey

BONY FISHES
Class Osteichthyes

These are the true fishes; all have well-developed jaws, paired fins (at least pectorals in all freshwater species) supported by rays (adipose fin excepted) which may be branched or unbranched, segmented or not. Unlike the lampreys, these fishes have only one pair of gill slits and a gill cover, or operculum. An air bladder is typically present and, in some species, can be used as a lung. Some forty-two families are represented in United States and Canadian freshwater habitats alone, including about 154 genera and 634 species. Only a relatively small, representative number are described here.

Sturgeons and Paddlefishes
Order Acipenseriformes

Members of this order have a long snout, inferior mouth, and two to four barbels, threadlike sense organs, on the undersurface of the snout. The tail lobes are unequal in size, the upper being the larger. The skeleton is largely of cartilage.

STURGEONS
Family Acipenseridae

Sturgeons and paddlefishes are living relics of early types of bony fishes, possessed of a rather primitive structure. Their skeleton is largely cartilaginous and they retain a notochord into the adult stage; typically the scales are reduced to rows of bony plates. There are two to four barbels or threadlike sense organs under the long snout in advance of the inferior mouth. The caudal fin lobes are of unequal size, the upper being the larger and bearing the end of the vertebral column.

LAKE STURGEON
Acipenser fulvescens

Description
Size, to 7 ft. (2.1 m) and 300 lb. (136.1 kg). Rounded, cone-shaped snout. Lateral plates 29–42, usually about 36; tips of pelvic fins typically fall short of front of dorsal fin. Back and sides varying from dark slate to light brown or yellow-olive; belly white.
Habitat
Fresh water, in large rivers and lakes.
Range
Nw. North America, s. through Miss. R. basin to Mo.

561

GREEN STURGEON
Acipenser medirostris   **96:6**

Description
Size, to 7 ft. (2.1 m) and 350 lb. (158.7 kg). Dorsal fin almost as
long as anal fin; dorsal rays, 33–36; anal rays, 22–28. Dorsal
plates, 8–11; lateral plates, 23–30; 4–10 rows of smaller star-
shaped plates between dorsal and lateral plate rows. Gill rakers,
18–20. Color olive-green, with olive stripe on median line of belly
and on each side above central plates. Rather similar to White
Sturgeon, but smaller and with barbels closer to mouth than to tip
of snout.
Habitat
Anadromous, but not to the extent of the White Sturgeon; may
spawn in brackish water and estuaries and are more often seen in
the ocean than the White Sturgeon.
Range
Alaska to Ensenada, Baja California.

WHITE STURGEON
Acipenser transmontanus   **96:5**

Description
Size, to 20 ft. (6.1 cm) and 1387 lb. (629 kg). Largest U.S.
sturgeon. Body elongate, subcylindrical, armed with 5 rows of bony
plates (between pelvic and anal fins in 2 rows of 4–8 each) and
38–48 lateral plates; dorsal rays, 44–48. Short, blunt snout; sharp
in young. More than 25 long gill rakers. Color above grayish-
green, below grayish-white.
Habitat
Anadromous.
Range
Alaska to Ensenada, Baja California.

SHOVELNOSE STURGEON
Scaphirhynchus platorhynchus   **96:4**

Description
Size, to 3 ft. (0.9 m) and seldom exceeding 5 lb. (2.3 kg). A small
sturgeon with flattened, shovel-shaped snout and long caudal
peduncle completely covered with bony plates; belly covered with
small plates except in young; upper lobe of caudal fin with a long
filament, sometimes broken off. Color back and sides light brown
or buff, belly white.
Habitat
In open channels of large rivers, on the bottom, often in strong
current; very tolerant of high turbidity.
Range
River basins near 100th meridian, nw. into Montana; formerly in
Rio Grande, N.Mex., and Tex.

Note: A related, larger species, the **PALLID STURGEON,**
Scaphirhynchus albus, which never has plates on the belly, is rare
in the same rivers.

PADDLEFISHES
Family Polyodontidae

Paddlefishes, like sturgeons, are an ancient group represented by
only two living species—one in North America and the other in the

valley of the great Yangtze River in China. The fins, like those of sturgeons, are archaic and sharklike, and the scales have degenerated to a patch on the upper lobe of the caudal fin. The long, paddle-shapped snout, whose precise function is still unknown, sets this fish apart from all others in North America. Although of large size, the paddlefish feeds throughout life on microscopic plants and animals.

PADDLEFISH
Polyodon spathula   *Fig. 78*

Description
Size, to 7 ft. (2.1 m) and 160 lb. (72.6 kg). Sharklike in appearance, with a much elongated, paddle-shaped snout and large mouth that lacks teeth, except in young; eyes very small, directed downward and forward, lying just above front edge of mouth. Posterior margin of operculum prolonged into a fleshy, pointed flap. Upper lobe of deeply forked caudal fin longer than lower lobe. Body scaleless. Gill rakers exceedingly numerous, usually long and slender. Color bluish-gray to nearly black on upper parts, grading to white on belly.
Habitat
Open water of large rivers and lakes, frequenting quiet water except when spawning.
Range
Miss. R. system, from e. Mont. to e. Tex.

Fig. 78

Paddlefish

Gars
Order Lepisosteiformes

GARS
Family Lepisosteidae

Gars comprise seven species confined to North and Central America that are distinguished by having an elongate, cylindrical body covered with diamond-shaped, nonoverlapping, thick ganoid scales arranged in oblique rows. (These scales are used in jewelry.) The jaws are extended forward into a beak and are armed with rows of strong, needle-sharp teeth. The dorsal and anal fins are very far back on the body and nearly opposite each other. The caudal fin is rounded. These fishes often are referred to as "living fossils" since virtually all of their relatives are extinct.

LONGNOSE GAR
Lepisosteus osseus

Description
Size, to 5 ft. (1.5 m) and average 25 lb. (11.3 kg). Snout very long and narrow, at its narrowest about $\frac{1}{15}$ or $\frac{1}{20}$ its length, except in

young; width at nostrils less than eye diameter. Large teeth in upper jaw arranged in a single row on each side. Scales in lateral line, usually 60–63; scales in diagonal row from the one at front of anal fin to that on midline of back, usually 17–19. Color on upper parts brown or dark olive, grading to white on belly; unpaired fins with numerous roundish black spots; body often spotted in individuals taken from clear water. Young with a conspicuous black stripe along mid-side.

Habitat
Typically in sluggish pools, backwaters, and oxbow lakes, along large, moderately clear streams and rivers. Thrives in man-made impoundments.

Range
E. Mont. to the 100th meridian and n., s. to ne. Mexico.

Herringlike Fishes
Order Clupeiformes

This order includes the herrings, salmons, chars, trouts, whitefishes, pikes, mooneyes, shad, tarpons, and anchovies. Characteristics include cycloid scales, reduced heterocercal caudal fins (upper lobe larger than lower), intermuscular bones, and upper jaws usually bordered by premaxillae and maxillae. These fishes have soft-rayed fins, and their pelvic fins are abdominal.

SHADS AND HERRINGS
Family Clupeidae

GIZZARD SHAD
Dorosoma cepedianum   96:2

Description
Size, 10–16 in. (25.4–40.6 cm). Body usually with more than 55 scales in lateral series; usually 29–35 anal rays; upper jaw with small notch. Color upper parts silvery blue, grading to silvery white on lower sides and belly; purplish postocular spot larger than eye; fins dusky. Similar to Threadfin Shad, but lower jaw does not project beyond tip of snout and fins lack any yellow color.

Habitat
Quiet water of rivers, lakes, and reservoirs; both clear and turbid waters.

Range
100th meridian, including the upper Mo. R. to ne. Mexico; abundant.

THREADFIN SHAD
Dorosoma petenense   96:3

Description
Size, 5–9 in. (12.7–22.9 cm). Body with 42–48 scales in lateral series; anal rays, 20–25; lower jaw projecting beyond tip of snout. Color silvery, with much yellow in all fins except dorsal; postocular spot smaller than eye.

Habitat
Bays, sloughs, and freshwater streams and lakes; also river deltas.

Range
Introduced to Calif. and Ariz.; common, particularly in Sacramento–San Joaquin delta and in Salton Sea.

MOONEYES
Family Hiodontidae

This family contains only two living species, the mooneye and goldeye. They resemble true herrings, but they lack the row of spiny scutes down the midline of the keeled belly and have unusually large eyes and prominent teeth on the jaws, roof of mouth, and tongue. They also have a lateral line, and the dorsal fin is much farther back than in herrings. The group is restricted to North America.

GOLDEYE
Hiodon alosoides   *Fig. 79*

Description
Size, average 14–16 in. (35.6–40.6 cm). A flat-sided fish with front of dorsal fin slightly behind front of anal fin, the dorsal with 9–10 rays; keel on midline of belly reaches anteriorly nearly to bases of pectoral fins. Color upper parts greenish with silvery or golden iridescence; sides and belly silvery white; iris of eye golden.
Habitat
Open waters of large rivers and streams; tolerant of high turbidity.
Range
Nw. N. America, e. of Rocky Mts., s. through Miss. R. basin.

Fig. 79

Goldeye

SALMONS, TROUTS, AND WHITEFISHES
Family Salmonidae

Members of this large family, which includes whitefishes, graylings, salmons, trouts, and chars, are slim, predatory, bony fishes. They may be exclusively fresh- or saltwater inhabitants, or divide their time between fresh water, and some in salt water. The saltwater species are anadromous, entering streams to spawn. The elongated body is covered with cycloid scales, and the dorsal fin is approximately midway between the tip of the snout and the base of the caudal fin. An adipose fin is present between the dorsal and caudal fins. The caudal fin is forked or somewhat truncate. The mouth may be small to large and the jaw teeth weakly to very strongly developed. Both the tongue and the vomer bear teeth, sometimes lost in old adults. The upper jaw may reach well beyond the eye.

TRUE WHITEFISHES
Genus *Coregonus*

Members of this genus (formerly Family Coregonidae) have an oblong or elongate, compressed body; the scales are moderately sized, thin, cycloid, and rather firm; the dorsal fin is moderately sized, the caudal fin deeply forked, the anal fin somewhat elongate, and the pelvic fins well developed. The head is more or less conic, but somewhat compressed. The snout projects somewhat beyond the lower jaw, and each nostril is divided by a double skin flap. The mouth is small; the maxillae are short, and the teeth are extremely small. There are twenty-three or more gill rakers on the first arch.

LAKE WHITEFISH
Coregonus clupeaformis

Description
Size, 12–20 in. (30.5–50.8 cm), record 30 in. (76.2 cm). Body deep, compressed; 11 dorsal rays; sometimes a fleshy hump on shoulders in adults. Snout projects well beyond tip of lower jaw; about 24–34 gill rakers on first arch; premaxilla wider than long. Color above olivaceous, below white to silvery.
Habitat
Shallow to moderate depths of lakes; entering streams in northern part of range.
Range
Yukon R. drainage of Alaska and Canada, and e.; also Cheesman Reservoir in Colo.

SALMONS
Genus *Oncorhynchus*

These carnivorous fishes have a long, stout body, small cycloid scales, a naked head, a lateral line, and an adipose fin; they have no spines. The mouth is large and has well-developed teeth. The dorsal fin is shorter than the head and has fewer than fifteen rays; the anal fin has more than twelve rays. These fish prefer water cooler than 70°F (21.1°C). All salmon are anadromous by nature, living most of their lives in salt water but entering streams to spawn, only to die shortly thereafter. A few salmon, becoming landlocked, spend their entire life cycle in fresh water.

PINK SALMON
Oncorhynchus gorbuscha   **97:10**

Description
Size, to 30 in. (76.2 cm) and 14 lb. (6.4 kg). Slender caudal peduncle; caudal fin slightly furcate; 147–198 small scales along lateral line; 28–32 gill rakers on first arch. Head small; teeth small, loosely set. Color bright gray to steel-blue; distinguished by very large black oval or irregular spots on back and caudal fin; belly silvery. Male, sides reddish; females greenish, sometimes with dusky stripes. Flesh pink.
Habitat
Anadromous.
Range
Bering Strait to La Jolla, Calif.; abundant B.C. and n.

CHUM SALMON
Oncorhynchus keta   **97:9**

Description
Size, to 40 in. (101.6 cm) and 43 lb. (19.5 kg). Body distinguished by long, slender caudal peduncle; caudal fin furcate; adipose fin small, slender; 11–17 smooth gill rakers on lower limb of first arch, 18–21 total. Head moderate-size; teeth large, conical, rigid. Color

above metallic blue, sparsely speckled; distinguished by black tips on pectoral, anal, caudal fins; no distinct spots on back and fins; sides, belly silvery; flesh pale pink.
Habitat
Anadromous.
Range
Bering Strait to Del Mar, San Diego Co., Calif.; abundant n. of Oreg.

COHO SALMON
Oncorhynchus kisutch   **97:11**

Description
Size, to 38 in. (96.5 cm) and 30 lb. (13.6 kg). Small, short caudal peduncle; adipose fin small, slender; 121–148 lateral-line scales. Head conical; teeth sharp, rigid; first-arch gill rakers, 19–25, widely spaced, rough. Color above metallic blue to greenish-blue, speckled; sides, caudal peduncle, belly silver; spawning males with brilliant red stripe; flesh pink.
Habitat
Anadromous.
Range
Alaska to San Diego, Calif.; abundant n. of Coos Bay, Oreg., rare s. of Pt. Conception.

SOCKEYE SALMON
Oncorhynchus nerka   **97:7**

Description
Size, to 33 in. (83.8 cm) and 15½ lb. (7 kg). Small, short caudal peduncle; adipose fin fleshy, slender; caudal fin moderately furcate; 125–145 lateral-line scales. Head conical; teeth small, sharp, loose-set; first-arch gill rakers, 31–43, large, slender, rough, close-set. Color above greenish to blue, finely speckled; no black spots on back or caudal fin, head brighter green; belly silver; flesh very deep red. Males flushed with reddish, females with yellowish blotches (may be dark red).
Habitat
Anadromous; in salt water around islands, stream entrances, swift currents.
Range
Aleutians to Los Angeles Harbor, Calif.; rather abundant, especially in B.C. and Wash.

CHINOOK SALMON
Oncorhynchus tshawytscha   **97:6**

Description
Size, to 58 in. (147.3 cm) and 125 lb. (56.7 kg). Body heavy, robust; 131–151 lateral-line scales; caudal peduncle short; caudal fin short, upper and lower rays stout, rigid; adipose fin rather short, fleshy. Head comparatively small, conical; teeth moderately large, pointed, loosely set; first-arch gill rakers, 19–28, rough, widely spaced. Color above greenish to dark blue or blackish, well spotted; dark spots on both lobes of caudal fin; belly silvery; flesh pink, sometimes white.
Habitat
Anadromous.
Range
Alaska to Ensenada, Baja Calif.; fairly abundant n. of cen. Calif.

WHITEFISHES
Genus *Prosopium*

This genus is most easily separated from *Coregonus* by the single, rather than double, flap of skin between the nostrils and the dark parr marks on the sides of juveniles.

BONNEVILLE CISCO
Prosopium gemmiferum   **96:1**

Description
Size, to 7½ in. (19.1 cm). Slender-bodied, with long, sharply pointed snout; lower jaw projecting beyond tip of upper jaw; 37–45 gill rakers on first arch; 70–80 lateral-line scales. Color above dark bluish, shading to silvery below.
Habitat
Deep, cold water of lakes.
Range
Bear Lake, Utah, and Idaho.

MOUNTAIN WHITEFISH
Prosopium williamsoni   **96:7**

Description
Size, 11–15 in. (27.9–38.1 cm). Body rounded in cross section; adipose fin large; 74–90 scales in lateral line; pectorals rather small; 19–26 gill rakers on first arch, 9–13 on lower limb; 24–27 scales around caudal peduncle. Head smallish, about ⅕ total length; mouth very small, maxilla not reaching anterior rim of orbit. Color above light brown to olive-green; all fins black-tipped; below silvery white.
Habitat
Upper and cooler waters in streams, lakes.
Range
W. of Rocky Mts. from B.C., to "ancient" Lake Lahontan basin in Calif. and Nev.; and upper Colo. R. in Utah, Colo., and Wyo.

ROUND WHITEFISH
Prosopium cylindraceum

Description
Size, to 20 in. (50.8 cm) and 4½ lb. (2 kg). Similar to Mountain Whitefish, but body more elongate, with smaller adipose fin, 20–23 scales around caudal peduncle, and 13–18 gill rakers on lower limb of first arch.
Habitat
Cold lakes and rivers.
Range
Alaska and nw. Canada.

TROUTS
Genus *Salmo*

The trouts are very difficult to characterize. The spots on the body are dark brown or black over a light background. The mouth is characteristically large, with teeth on the jaws, palatines, and tongue, and on the head and shaft of the vomer. The dorsal and anal fins have eight to twelve rays, and the caudal fin is forked—appearing nearly truncated, especially in old adults.

GOLDEN TROUT
Salmo aguabonita

Description
Size, 8–12 in. (20.3–30.5 cm), record 20 in. (50.8 cm); to 1 lb. (0.45 kg). Distinguished by golden-yellow color and very small scales. Body with about 180–210 lateral-line scales. Color olive above, golden-yellow below lateral line; broad rosy lateral stripe crossed by about 10 dark parr marks; belly and cheeks bright red to red-orange.

Habitat
Small mountain streams and high lakes (to 11,000 ft. or 3.4 km).

Range
Original home in cen. Sierras of upper Kern R. basin, Calif.; now widely transplanted to many w. lakes.

ARIZONA TROUT
Salmo apache

Description
Size, 12–18 in. (30.5–45.7 cm). Body deep with short peduncle, dorsal fin large, tips of dorsal, anal, and pelvic fins milky white to orange, 142–175 lateral scales. Color sides and belly yellow to golden-yellow, back and top of head rich olive-green; no red stripe on side, as in Rainbow Trout.

Habitat
Clear, cold, forested streams with rocky riffles and pools at elevations from 7800–11,000 ft. (2.4–3.4 km).

Range
Streams and lakes in the White Mts. of Ariz.

RAINBOW TROUT
Salmo gairdneri

Description
Size, to 35 in. (88.9 cm), 20–30 lb. (9.1-13.6 kg). Body rounded in cross section. Head comparatively short, maxilla reaches to scarcely beyond eye; lining of mouth white. Color in salt water, above steel-blue with small black spots, sides and belly silvery; when spawning in fresh water, a broad, lateral, red stripe appears, especially in males. Distinguished from salmon by long, deep caudal peduncle and 12 or fewer anal rays; from Cutthroat Trout by absence of "cutthroat" mark and small posterior teeth and by less slender body.

Similarities
Chinook and Coho Salmon, mouth cavities dark in adults.

Habitat
Anadromous, enters nearly all coastal streams to spawn; abundant. Also in cold, fast streams.

Other name
Steelhead Trout.

Remarks
This species displays a bewildering number of forms and, hence, goes by many different local names; generally the saltwater fish is a "steelhead," and the freshwater run a "rainbow."

Range
Sw. Alaska to n. Baja. Calif., w. of Rocky Mts.; widely introduced in islands and continents.

CUTTHROAT TROUT
Salmo clarki   97:5

Description
Size, 10–15 in. (25.4–38.1 cm), record 30 in. (76.2 cm); 6–30 lb. (2.7–13.6 kg). Bright red dash "cutthroat" mark under each side of lower jaw is distinguishing feature, but not always present. Body compressed, elongate; caudal peduncle long; adipose fin small, slender. Head relatively long, maxilla reaches to well back of eye; small teeth posterior to tongue at base of first gill arch. Color above greenish to greenish-blue, with many rather large black spots, very variable, over body, head, fins; sides may be yellowish, belly silvery.

Habitat
Small mountain streams, around rocks, in riffles and pools, under logs and overhanging banks; may also be anadromous.

Range
S. Alaska to n. Calif., Utah, and n. N.Mex., w. of Rocky Mts. In upper Rio Grande to Sask. R., e. of Rocky Mts.

Note: There are many similar subspecies. Also similar is the closely related **GILA TROUT**, *Salmo gilae*, found in headwaters of the Gila River in New Mexico, but not common.

BROWN TROUT
Salmo trutta   97:3

Description
Size, to 3 ft. (0.9 m) and 30 lb. (13.6 kg). Distinguished by reddish-orange spots on back and sides of body, ringed with lighter pigment to form halos. Large adipose fin, usually orange in young; 10–13, dorsal rays 9–10; anal rays caudal peduncle deep. Back of tongue toothless; maxilla reaches to point below back margin of eye. Color yellow-brown to brown, with many black spots on top of head, cheeks, back, and unpaired fins; sides yellow-brown, belly yellow to white.

Habitat
Anadromous, found occasionally in brackish water near mouths of streams and in warm and slow trout waters (beaver ponds, pools, lakes).

Range
A native of Europe widely introduced into most U.S. and Canadian waters; not abundant.

CHARS
Genus *Salvelinus*

ARCTIC CHAR
Salvelinus alpinus   *Fig. 80*

Description
Size, 2–3 ft. (0.6–0.9 m), 10–15 lb. (4.5–6.8 kg). Body elongate, compressed; 195–200 scales along lateral line; 10–12 dorsal rays, 8–11 anal rays, 60–71 vertebrae. Head moderate size; maxilla reaches little beyond orbit; 19–30 gill rakers on first arch, 12–19 on lower limb. Color above dark blue to olive-green or grayish; sides with large, round reddish spots, usually larger than pupil of eye; below red, especially in males; lower fins margined with white.

Habitat
Cold lakes and mountain streams.
Range
N. N. America; circumpolar. In U.S. only in Me., N.H., and Alaska.

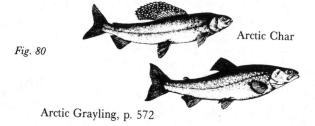

Arctic Char

Fig. 80

Arctic Grayling, p. 572

BULL TROUT
Salvelinus confluentus

Description
Size, to 37 in. (94 cm) and 40 lb. (18.3 kg). Head long and broad, small posterior (basibranchial) teeth in 1 row; gill rakers on first arch average 17; tip of lower jaw has fleshy knob fitting into notch in upper jaw. Color olive-green, with small yellowish spots on back and small but prominent red spots on sides; body lacks black spots and fins are clear except for a few yellow spots on base of caudal fin; leading edges of paired and anal fins white or cream colored.
Habitat
Pools of rivers and creeks and in cold lakes.
Range
Columbia R. basin of Idaho, Mont., and Nev., s. to n. Calif., and in upper Sask. R. system of Alta. northward to Alaska.

BROOK TROUT
Salvelinus fontinalis

97:2

Description
Size, to 34 in. (86.4 cm) and 14½ lb. (6.6 kg). Tail lunate in adult, forked only in young. Color above dark olive with many blue-bordered red spots; back and dorsal fin with dark green mottling; sides and belly lighter, reddish in males; lower fins with conspicuous white margins.
Habitat
Cold, small streams and ponds with cover.
Habits
Cautious, easily frightened away.
Range
Native to e. N. America; widely introduced into temperate areas worldwide.

DOLLY VARDEN
Salvelinus malma

97:8

Description
Size, to 36 in. (91.4 cm), and 30 lb. (13.6 kg). Body troutlike, comparatively slender, rounded. Head moderately large, somewhat

rounded; eyes large; maxilla reaches well past eye; basibranchial teeth usually in more than 1 row; fewer than 17 gill rakers on first arch. Color above light to dark olive-green, paler on sides, white on belly; yellow to orange spots on back, red spots on sides, few or no spots on fins.

Habitat
Anadromous, in salt and brackish water near streams, more abundant in n. Also occurs in nearly all lakes and streams, with dwarf populations in remote headwaters.

Range
Nw. Alaska to n. Calif., w. of Rocky Mts.

LAKE TROUT
Salvelinus namaycush   97:12

Description
Size, to 4 ft. (1.2 m) and 100 lb. (45.4 kg). Tail deeply forked, fins not emarginated. Color variable, from light gray, green, brown, to nearly black, with profuse irregular whitish (near pink) spots on back and sides, and pale spots on dorsal fin; color never bright; belly sometimes spotted.

Habitat
Deep-water lakes; shallow water in fall and winter.

Range
Alaska and Canada (except extreme se. Alta. and very extreme sw. Man.); s. to Great Lakes and St. Lawrence R. drainages of U.S., and parts of Wis., and Mont.; widely introduced w. of Rocky Mts.

GRAYLINGS
Genus *Thymallus*

ARCTIC GRAYLING
Thymallus arcticus   *Fig. 80*

Description
Size, to 24 in. (61 cm) and 5 lb. (2.3 kg). Body rather elongate, compressed; dorsal fin greatly enlarged (with 19–24 rays), saillike; scales moderate-size, 77–98 in lateral line. Head short, mouth small but wide, teeth small. Color back dark purple or blue, paling to gray on sides with scattered black spots; dusky stripe from below pectoral fins to pelvic fins, black stripe along inner edge of lower jaw; dorsal fin dark, with rows of orange spots, its margin edged with red or orange.

Habitat
Cold stream waters.

Habits
Migratory.

Range
Arctic n. Siberia and N. America (Alaska and Canada). Has been widely planted in cold streams and lakes.

SMELTS
Family Osmeridae

Members of this family are small, slender, silvery fishes inhabiting either marine, brackish, or fresh water, some species being anadromous. They bear rather small cycloid scales and have a single soft dorsal fin. Smelts can quite easily be distinguished from other small fishes by a definite lengthwise band of silver on the sides coupled with the presence of an adipose fin. Usually they have larger mouths, teeth of various types but always sharp, and a shaftless vomer. The maxilla forms the upper jaw margin. Smelts are considered excellent food fishes. The species described here are those most often caught in freshwater streams and are closely related to species described in the saltwater section.

POND SMELT
Hypomesus olidus   **100:2**

Description
Size, 6–8 in. (15.2–20.3 cm). Body small, slender; 7–10 dorsal rays, 12–18 anal rays, 51–62 lateral-line scales. Head small; mouth small, oblique; teeth small, in 2 rows on vomer and palatine; maxilla does not reach to pupil center. Color adults light brown to olive-green on back, silvery white on abdomen.

Habitat
Anadromous; enters streams, ponds to spawn.

Range
Alaska and Canada; abundant.

LONGFIN SMELT
Spirinchus thaleichthys   **100:3**

Description
Size, to 6 in. (15.2 cm). Body slender, all fins very large and greatly expanded. Mouth large; teeth large, but on single row on vomer and palatine; 39–44 gill rakers on first arch. Breeding males have greatly dilated scales along lateral line. Color silvery, with dusky back.

Habitat
Anadromous; enters streams to spawn.

Range
Alaska to Sacramento–San Joaquin estuary, Calif.

MUDMINNOWS
Family Umbridae

These are small reddish-brown or dark brown fishes with no lateral line or adipose fin. They live in soft-bottomed, sluggish, or stagnant water, burrow into the mud when alarmed, and are extremely resistant to adverse conditions.

ALASKA BLACKFISH
Dallia pectoralis   *Fig. 81*

Description
Size, to 8 in. (20.3 cm). Body slender, ovate, with 40–42 vertebrae, skeleton delicate; dorsal and anal fins far back on body and

Fig. 81

Alaska Blackfish

573

opposite each other; pectorals with 32–36 rays, rounded; caudal fin rounded; pelvics with 3 rays; lateral line rudimentary. Head short, with blunt snout; lower jaw projecting. Color adults dark brown, with 4–6 bars on sides; underside pale, with dark brown speckling.

Habitat
Brooks, lakes, weed-choked swamps and ponds.

Habits
Spends winter in silt; very docile; does not migrate far; can survive freezing unless liquids in body cavity freeze.

Remarks
Important for human and dog food.

Range
Alaska and Siberia.

PIKES
Family Esocidae

These carnivorous fishes, represented by four species in North America, are distinguished by a duck-billed snout; a large mouth with sharp teeth; a long, cylindrical body; and dorsal and anal fins that lie far back and are opposite each other. The tail is forked. The body is covered with cycloid scales that are deeply scalloped on their front margins. The larger species are important game fishes.

NORTHERN PIKE
Esox lucius 101:1

Description
Size, to 55 in. (139.7 cm) and 46 lb. (20.9 kg). Fully scaled cheeks but no scales on lower half of the gill cover. Body elongate; dorsal fin large, set far back; lateral-line scales, 119–128. Head with 5 or fewer pores on each side of ventral surface of lower jaw; branchiostegals, usually 14–16. Color back and sides dark green to brown, the sides with irregular light yellow spots roughly arranged in vertical rows; dark spots on fins.

Habitat
Summers in shallows, winters in deep water.

Range
Alaska, n. and nw. Canada, s. to Nebr.

Note: A variant form, the **SILVER PIKE,** has lost all body spots.

Minnowlike Fishes
Order Cypriniformes

This is the largest order of freshwater fishes, comprising an estimated 4500 to 5000 species, some native to all continents except Australia. Four families are described here: characins, minnows, suckers, and catfishes. The head is always, and the body sometimes, without scales; spines may or may not be present.

CHARACINS
Family Characidae

This is a large family of fishes essentially confined to the New World tropics and Africa. Only one species reaches the United States.

MEXICAN TETRA
Astyanax mexicanus

101:8

Description
Size, to 4 in. (10.2 cm). Body rather short and deep, with adipose fin; dorsal rays, 10–11, anal rays, 18–24. Teeth very sharp, in 2 rows on premaxillae. Conspicuous black lateral band extending to end of caudal fin, intensified on caudal peduncle; lateral band overlaid by a broad silvery band.
Habitat
Coastal streams.
Habits
Pugnacious.
Range
N.Mexico, sw. Tex., s. N.Mex.; introduced into lower Colo. R. drainage of sw. Ariz. and se. Calif.

MINNOWS
Family Cyprinidae

The minnows constitute the largest group of freshwater fishes, with more species and more individuals than any other family. More than 250 species live in North America alone. Most minnows are small and, in general, have a naked head and a scaly body; teeth in the throat or pharynx, not in the jaws; a forked tail; and a single dorsal fin, with eleven or fewer rays, in the middle of the back. In the United States only the introduced Carp and Goldfish and three native genera—*Lepidomeda, Meda,* and *Plagopterus*— have any spines in the fins.

Many species are difficult to identify. Determination is often based on the number of pharyngeal teeth. These teeth are borne on the lower pharyngeal bones located immediately posterior to the gills and covered by skin and muscle. A circular incision carefully made with a sharp-pointed knife anterior to the pectoral or shoulder girdle will allow removal of the pharyngeal bones. Careful cleaning and subsequent examination of the bones with a hand lens will reveal the pharyngeal tooth formula. In the following species descriptions, a formula of 5/5, for example, indicates a single row of five teeth on the left pharyngeal bone and a row of five on the right. Tooth counts of additional rows, if present, are listed from left to right and separated by commas; for example, "2,5/4,2" indicates rows of two and five on the left and four and two on the right, with the rows of two on the outside.

CHISELMOUTH
Acrocheilus alutaceus

98:9

Description
Size, to 12 in. (30.5 cm). Body slender, fine-scaled; about 85 scales in lateral line. Dorsal rays, 10; anal rays, 9. Mouth wide; lower jaw bearing a sharp horny sheath; pharyngeal teeth 5/4 or 5/5. Color very dark, belly somewhat lighter, most body parts studded with minute dark points.
Habitat
Lakes, rivers.
Remarks
Rarely used for food.
Range
Columbia R. drainage in Wash., Oreg., and n. Nev.

LONGFIN DACE
Agosia chrysogaster **98:7**

Description
Size, to 4 in. (10.2 cm). Similar to the daces of Genus *Rhinichthys*, but having a slight frenum hidden in the groove of the premaxilla (easily overlooked), in the elongated anal fin in the adult female, and pharyngeal teeth arranged 4/4. Color dark above, silvery below, with dark lateral band; male may show yellow-orange sides.
Habitat
Sandy streams in low desert regions.
Range
Lower Colo. R. drainage, Ariz. and N.Mex., s. into Mexico.

CENTRAL STONEROLLER
Campostoma anomalum *Fig. 82*

Description
Size, 3–7 in. (7.6–17.8 cm). Intestine very long, encircling the swim bladder with many loops; lining of body cavity black; jaws of adults with thin cartilaginous sheaths; fins short and rounded; eyes small; anal fin with 7 rays. Color back and upper sides tan or light brown, lower sides and belly silvery white; spawning males with orange-tinted sides and much orange and black in fins.
Habitat
Streams of moderate or high gradients, with rocky riffles and permanent flow; generally on riffles or in pools.
Range
S. of Canada from the Rocky Mts. to the 100th meridian.

Fig. 82

Central Stoneroller   Goldfish

GOLDFISH
Carassius auratus *Fig. 82*

Description
Size, 12–16 in. (30.5–40.6 cm), to 2 lb. (0.9 kg). Large, without barbels and with a strong saw-toothed spine at front of dorsal and anal fins. Mouth oblique. First-arch gill rakers, 37–43. Color metallic blue to gray in wild populations, no spot on base of scales.
Habitat
Warm, often very shallow water, especially in lakes.
Range
Widely introduced throughout w.

LAKE CHUB
Couesius plumbeus **99:6**

Description
Size, to 4–6 in. (10.2–15.2 cm). Body slender, head short; mouth slightly oblique, with distinct barbel near end of upper jaw; origin of dorsal fin slightly behind origin of pelvic fins; dorsal and anal rays, 8; pharyngeal teeth, 2,4/4,2; scales in lateral line about 55–

70. Color dark brown or green above, pale below, with weak mid-lateral stripe on posterior half of body, sometimes extending forward onto head in young.

Habitat

Lakes and rivers, in both clear and muddy waters; also outlets of hot springs.

Range

Nw. N. America from Yukon River, Alaska and s. (on Pacific slope) to B.C., and as isolated populations to Iowa, S.Dak., Nebr., Mont., Wyo., and Colo.

NORTHERN REDBELLY DACE
Chrosomus eos

Description

Size, to 3 in. (7.6 cm). Body with about 85–90 scales in lateral line, snout rather short; mouth small and oblique; dorsal and anal rays, 8; pharyngeal teeth, 5/5 or 5/4. Color 3 dark lateral bands in adults; upper band beginning at edge of operculum and extending to base of upper caudal-fin lobe, ending in a number of spots; lower band beginning on snout and ending at caudal-fin base. Males brilliant red or yellow in spring.

Range

N. B.C., and s. to n. Mont., S.Dak., and Nebr.

CARP
Cyprinus carpio   *Fig. 83*

Description

Size, to 3½ ft. (1.1 m) and 55 lb. (24.9 kg). Large, with strong saw-toothed spines at base of dorsal and anal fins and with 2 barbels on each side of upper jaw. Body with large scales, 35–38 in lateral line; long dorsal fin, 19–22 rays; 5–6 anal rays. Mouth straight. Gill rakers, 21–27 on first arch. Color sides brassy yellow, back dark, belly lighter; dark spot on base of each scale.

Habitat

Warm, often very shallow water, especially in lakes.

Range

Originally native to Asia but established early in Europe; widely introduced into U.S.; abundant.

Fig. 83

Carp

ROUNDNOSE MINNOW
Dionda episcopa   **99:7**

Description

Size, to 2½ in. (6.4 cm). An elongate minnow, round in cross section; very small mouth, end of maxilla not reaching beyond nostrils; blunt snout; conspicuous, rounded, black spot at base of caudal fin; pharyngeal teeth, 4/4; dorsal rays, 8, anal rays, 7–8. Color silvery or brassy, with a prominent dark stripe on midside, from snout to caudal-fin base, ending in a black spot; fins of spawning males yellow.

Habitat
Current of usually clear creeks and spring outflows over gravel bottom.

Range
S. N.Mex. and Tex., s. into Mexico.

DESERT DACE
Eremichthys acros

Description
Size, to 10 in. (25.4 cm). Body with low, rounded fins; anal rays, 7–8; caudal fin shallowly emarginated; lateral line, with 68–75 scales, almost complete. Mouth ridges inside jaws covered by easily removed horny sheaths; pharyngeal teeth, 5/4. Differs strikingly from Chiselmouth in having horny sheaths on both upper and lower jaws.

Habitat
Desert springs.

Range
Nev. (w. Humboldt Co., Soldier Meadows; completely isolated); rare.

CALIFORNIA ROACH
Hesperoleucus symmetricus   **98:6**

Description
Size, to 5 in. (12.7 cm). Body with rather large scales, 47–61 along lateral line; dorsal fin has 8–9 rays, originates behind pelvic insertion; anal rays, 7–9. Head relatively short; mouth small, slightly inferior; pharyngeal teeth, 5/4. Color variable, but usually with dark lateral stripe passing from tip of snout to base of caudal fin; breeding individuals have a second black stripe extending from operculum to above anus, and red-orange pigment around jaws, above operculum, and at bases of lower fins.

Habitat
Clear creeks and small rivers of foothills.

Range
Calif. in watersheds of Sacramento, Russian, San Joaquin, Salinas, and adjacent river systems.

WESTERN SILVERY MINNOW
Hybognathus argyritis   **99:2**

Description
Size, 3–5 in. (7.6–12.7 cm). Very similar to Silvery Minnow, but with smaller eye; eye smaller than or same size as mouth opening.

Habitat
Parts of large rivers, over silt or sand bottoms with little current, tolerating high turbidity.

Range
Mo. R. basin, Mo. to Mont.

BRASSY MINNOW
Hybognathus hankinsoni   **99:1**
Description
Size, 3–4 in. (7.6–10.2 cm). Similar to Silvery Minnow, but body
scales with about 20 faint radii. Head blunt. Color yellowish.
Habitat
In creeks and lakes; most frequently in bog waters.
Range
B.C., e. to Hudson R., s. to Mont., Kan., and Colo.

SILVERY MINNOW
Hybognathus nuchalis

Description
Size, to 6 in. (15.2 cm). Body scales with about 10 radii; first
dorsal ray thin, small, attached to first full ray; dorsal fin begins
forward of pelvic fin. Color back yellowish-olive with emerald
reflections; sides silvery. Similar to Plains Minnow but eyes larger.
Habitat
Clear, low-gradient, moderately large streams.
Range
Miss. R. basin, from Wis. and Ohio s. to La. and Ala.

PLAINS MINNOW
Hybognathus placitus   **99:8**
Description
Size, to 6 in. (15.2 cm). Body scales large, with few strong radii;
dorsal fin over or in front of pelvic insertion. Head elongate, eye
small. Color back yellowish-olive with green reflections; sides
silvery.
Habitat
River channels of the Great Plains in current over sandy bottom.
Range
Lower Miss. and upper Mo. R. drainage from e. Mont. s. and e.
to La.

GILA CHUBS
Genus *Gila*

Considerable variation occurs among the member species. One
group, formerly placed in *Siphateles,* has one row of teeth on each
pharyngeal bone; all others have two rows of pharyngeal teeth.
Scale size is very variable, with about forty-eight to ninety-six
scales in the lateral line. The scales are uniformly distributed on
the body. The body varies from relatively short and stout to long
and slender, and the fins vary greatly in size and shape. The dorsal
fin is placed over or slightly behind the pelvic insertion. There are
eight species in the waters of the Pacific Slope, the Great Basin,
and the Rio Grande that are briefly described here. Most are
shown on Plate 98.

UTAH CHUB, *Gila atraria,* **98:10,** 12–15 in. (30.5–38.1 cm).
Dorsal rays, 9; anal rays, 8; lateral-line scales, 51–63; dorsal origin
over pelvics. Widespread in "ancient" Lake Bonneville basin and
in upper Snake R. in e. Nev., Utah, Idaho, Wyo.; introduced as
bait fish in other parts of the west, including Mont. where it is
established as far s. as the mouth of the Madison R.

TUI CHUB, *Gila bicolor,* **98:13,** 12 in. (30.5 cm). Very similar to Utah chub but pharyngeal teeth in only 1 row. Widespread in "ancient" Lake Lahontan basin, Nev., nw. into ne. Calif., s. and e. Oreg., adjacent Idaho and Wash.; also in Owens R. and Mohave R., s. Calif.

BLUE CHUB, *Gila coerulea,* 12 in. (30.5 cm). Pharyngeal teeth, 2,5/5,2. Color, bluish above, silvery below. Klamath Lake drainage of se. Oreg., ne. Calif.

LEATHERSIDE CHUB, *Gila copei,* **98:11,** 6 in. (15.2 cm). Scales small, about 80 in lateral line; pharyngeal teeth, 2,4/4,2 or 1,4/4,1. Color bluish above, silvery below, with dusky lateral stripe. "Ancient" Bonneville R. and upper Snake R. drainages, Nev., Utah, and Wyo.

ARROYO CHUB, *Gila orcutti,* **98:12,** 3–5 in., (7.6–2.7 cm) to 10–12 in. (25.4–30.5 cm) in lakes. Silvery or gray to olive-green on back, white on belly, usually with graying lateral stripe. Coastal streams of s. Calif. from Santa Ynez to San Luis Rey; introduced in Mohave R. in San Bernardino Co.

RIO GRANDE CHUB, *Gila pandora,* 6–12 in. (15.2–30.5 cm). Color dusky above, silvery below, often with 1 or 2 lateral stripes. Rio Grande R. drainage of Colo., N.Mex., Tex.

ROUNDTAIL CHUB, *Gila robusta,* **98:5,** 12–15 in. (30.5–38.1 cm). Dorsal and anal rays, typically 9. Body completely scaled to naked on breast and back. Color dusky above, pale below; several subspecies in range. Colo. R. drainage in Calif., Nev., Utah, Ariz., Wyo., Colo., N.Mex., s. to nw. Mexico.

HUMPBACK CHUB, *Gila cypha,* 12–15 in. (30.5–38.1 cm). Bizarre abrupt hump on back behind head; some individuals with almost no scales; dorsal rays, 9; anal rays, 10; fins large and falcate. Colorado R. drainage in Ariz., Utah, Colo.; now rare.

CHUBS
Genus *Hybopsis*

These slender-bodied fishes vary in color from dull to silvery. The eye size varies from small to large; the mouth is rather small and usually horizontal, with the upper jaw protractile. There is a slender barbel at the rear of the maxilla.

SPECKLED CHUB
Hybopsis aestivalis   **99:5**

Description
Size, 2–2½ in. (5.1–6.4 cm). Eye small, diameter much less than snout length; mouth small and horizontal. Barbel well-developed, sometimes very long. Color back and upper sides pale yellow, with silvery reflections and scattered black spots; lower sides and belly silvery white.

Habitat
Open channels of rivers and prairie streams, and in lowland ditches, most commonly in current over bottom of sand or gravel; will tolerate high turbidity.

Range
Miss.–Mo. R., including s. Great Plains from Ill. and Ohio to the Rio Grande, s. into Mexico.

Note: The **FLATHEAD CHUB**, *Hybopsis gracilis*, **99:11**, 10 in. (25.4 cm), has a dorsoventrally flattened head and falcate pectoral fins. It is olive above and silvery below. It occurs east of the Rockies from Canada to N.Mex.

HITCH
Lavinia exilicauda   **99:10**

Description
Size, to 12 in. (30.5 cm). Body deep at front, compressed; anal rays, 10–15; lateral line deeply decurved. Mouth short, not extending much behind nostrils, and oblique; pharyngeal teeth, 5/4 or 5/5; gill rakers, 17–32. Color dark above, light below. Closely resembles Golden Shiner but lacks the distinctive belly keel and has 10–13 dorsal rays.

Habitat
Inland streams and lakes.

Remarks
There are several similar species in the range.

Range
Cen. Calif.

MOAPA DACE
Moapa coriacea

Description
Size, to 3 in. (7.6 cm). An interesting relict of Ice Age waters. Body with tiny, deeply embedded scales, 70–80 in complete, slightly decurved lateral line; skin leathery; dorsal and anal rays, 7–8. Pharyngeal teeth, 5/4. Color, deep olive above, blotched on sides, white on belly; sides marked with golden-brown lateral stripe bordered above by light streak.

Habitat
Pools and currents of warm spring outflows.

Range
Restricted to Moapa R., Nev.

HARDHEAD
Mylopharodon conocephalus

Description
Size, to 3 ft. (0.9 m). Body long, slender; dorsal fin originates slightly behind pelvic insertion; scales small, 70–80 in complete, anteriorly decurved lateral line; 8 dorsal rays, 8–9 anal rays. Snout long, pointed; mouth rather large, maxilla reaching orbit, premaxillae not protractile; anterior 3–4 pharyngeal teeth in main row heavy, molarlike, without hooks; posterior pharyngeal teeth tending to be slender, hooked; gill rakers short, 10–14 on first arch.

Habitat
Warm, clear streams with large, deep pools of sand or rock bottom.

Range
Sacramento R. system, Calif.

VIRGIN SPINEDACE
Lepidomeda mollispinis   **99:3**

Description
Size, to 3 in. (7.6 cm). Body with minute scales, about 75–86 in lateral line; 2 dorsal spines, anterior one grooved posteriorly to receive the second; 9 anal rays. Head small, eye comparatively large; pharyngeal teeth, 2,5/4,2. Color olivaceous above, silvery below with lateral band.

Habitat
Desert streams.

Range
Virgin R. system of Utah, Nev., Ariz.

Note: Similar species are the **WHITE RIVER SPINEDACE,** *Lepidomeda albivallis,* found in the White River valley, Nevada; the **PAHRANAGAT SPINEDACE,** *Lepidomeda altivelis,* in Pahranagat Valley, Nevada; and the **LITTLE COLORADO SPINEDACE,** *Lepidomeda vittata,* in the Little Colorado River system, Arizona.

SPIKEDACE
Meda fulgida

Description
Size, to 3 in. (7.6 cm). Easily recognized by absence of scales and presence of 2 sharp dorsal spines. Dorsal rays, 7; anal rays, 9; pharyngeal teeth, 1,4/4,1. Color above dusky, silvery on sides and below; somewhat speckled.

Habitat
Riffles and pool heads of flowing streams.

Range
Ariz., N.Mex., in Gila R. drainage.

Note: A similar species, the **WOUNDFIN,** *Plagopterus argentissimus,* has a barbel at end of maxilla, 8 to 9 dorsal and 10 anal rays, and pharyngeal teeth arranged 1,5/4,1. It inhabits the Virgin River in Utah, Arizona, and Nevada.

PEAMOUTH
Mylocheilus caurinus   **99:12**

Description
Size, to 14 in. (35.6 cm). Similar to Hardhead, but smaller; mouth smaller; maxilla not reaching orbit, with terminal barbel; pharyngeal teeth, 1 or 2,5/5,1 or 2, hooked in young, larger teeth becoming molarlike with age; premaxillae protractile. Color above dark brownish or greenish grading to silvery below, with two dark lateral stripes, the upper one extending to tail, lower ending opposite anus; spawning fish with reddish across cheek and on sides below the lower dark stripe.

Habitat
Deep to shallow parts of cool lakes and rivers.

Habits
Known to enter the sea.

Range
Nass R., B.C. to Mont., Idaho; Wash. and Oreg. (lower Columbia R. drainage).

GOLDEN SHINER
Notemigonus crysoleucas

99:13

Description

Size, to 10 in. (25.4 cm). Body deep, distinguished by a sharp
naked keel on belly behind pelvic fins; lateral line deeply decurved,
dorsal rays 8, anal fin relatively long, 11–15 rays. Mouth very
oblique, pharyngeal teeth 5–5. Color gold.

Similarities

Hitch lacks belly keel.

Habitat

Quiet, heavily vegetated sloughs, ponds, lakes, and impoundments.

Range

Native to most e. N. America, introduced into Ariz. and s. Calif.

SHINERS
Genus *Notropis*

This is the largest genus of American minnows and is very difficult
to characterize. The scales are relatively large and often deciduous.
There are usually eight dorsal rays, but anal rays vary from seven
to thirteen. Barbels are almost never present, but the mouth is
highly variable, horizontal to oblique; eye size also varies. The
body is often silvery, with or without conspicuous lateral bands and
caudal spots; some species are brilliantly colored, usually with
bright yellows or reds, others with iridescent greens and blues; the
peritoneum may be spotless silvery, silvery with scattered cells
containing melanin (melanophores), or inky black. Originally
Notropis was confined to North America (north of Mexico) east of
the Rocky Mountains, with *Notropis formosus* being the only
native to the southwest Pacific drainage. Others have been
introduced on the West Coast. There are more than one hundred
North American species, but only a handful occur west of the
100th meridian. Not all of these are included here. Most are
shown on Plate 98.

EMERALD SHINER, *Notropis atherinoides,* 3½ in. (8.9 cm).
Slender, well streamlined, with dorsal fin originating behind pelvic
fins; sicklelike anal fin of 10–20 rays; large eyes; terminal oblique
mouth; pharyngeal teeth, 2,4/4,2. Color yellowish-olive on back,
sides silvery, with narrow iridescent emerald stripe. E. N. America
into n. Great Plains and w. Canada s. to Tex.

RIVER SHINER, *Notropis blennius,* 3½ in. (8.9 cm). Dorsal fin
origin equidistant between tip of snout and base of caudal fin;
mouth terminal, oblique, length of upper jaw greater than eye
diameter; mouth horizontal, pharyngeal teeth 2,4/4,2. Color
silvery, dark stripe along midline of back well-defined and of
uniform width. Alta. and s. to e. Wyo. and Okla.

BIGMOUTH SHINER, *Notropis dorsalis,* **98:4,** 3 in. (7.6 cm).
Eyes directed upward, lower margins of pupils usually visible
when fish viewed directly from above; mouth nearly horizontal;
head long, lower surface broad, flat; pharyngeal teeth, 1,4/4,1.
Color back olive-yellow, with narrow dusky stripe; sides silvery
often with faint dusky stripe. N.-cen. U.S., w. from N.Dak. to
Wyo.

SPOTTAIL SHINER, *Notropis hudsonius,* **98:3,** 5 in. (12.7 cm).

Conspicuous round, black spot at base of caudal fin; eye large; dorsal fin originates much nearer tip of snout than base of caudal; head bluntly rounded. Color back olive-yellow, with dusky stripe along midline, sides silvery, belly silvery white. W. Canada to N.Dak., e. and s.

RED SHINER, *Notropis lutrensis,* **98:2,** 3 in. (7.6 cm).

Adult deep-bodied; dorsal fin origin over pelvic insertion; anal rays, usually 9; lateral-line scales, 32–36; pharyngeal teeth, 4/4 or 1,4/4,1. Color steel-blue above, silvery below, lower fins red; breeding males with purple shoulder crescent and nonbreeding male with orange on fins, belly, and behind shoulders. E. of Rocky Mts. from Wyo. to Mexico e. to Miss. R.; also established in Colo. R. of Utah, Ariz., Calif.

SAND SHINER, *Notropis stramineus,* **98:1,** 2½ in. (6.4 cm).

Anal rays, 7; dark stripe along midline of back forms wedge-shaped spot at front of dorsal fin; mouth small, slightly oblique, length of upper jaw not greater than eye diameter; pharyngeal teeth 4/4. Color back olive-yellow; scales prominently dark-edged; sides silvery, with pores of lateral line marked by dark pigment. E. of Rocky Mts. along Great Plains from s. Canada to Mexico.

SACRAMENTO BLACKFISH
Orthodon microlepidotus   **98:8**

Description
Size, 12–16 in. (30.5–40.6 cm). Body elongate; scales very small, about 100 in decurved lateral line. Pharyngeal teeth knifelike, elevated in 1 row of 5/5 or 6/6; dorsal rays, 10, anal rays, 8; gill rakers, about 30, brushlike at tip. Color dark olivaceous above, lighter below.
Range
Sacramento R. system, cen. Calif.

FATHEAD MINNOW
Pimephales promelas

Description
Size, 2–3 in. (5.1–7.6 cm). Body robust, with blunt, rounded snout and short, rounded fins; mouth small, oblique; anal fin with 7 rays; predorsal region broad, flat, with scales smaller than those on sides; pharyngeal teeth, 4/4. Color back tan or yellowish-olive, with dark stripe along midline; sides silvery, often with dusky stripe; belly silvery white. Breeding males blue-black.
Range
E. North America, w. to base of Rocky Mts. from s. Canada to Mexico; introduced into Colo. R. basin.

SQUAWFISHES
Genus *Ptychocheilus*

This genus includes the largest members of American Cyprinidae. The body is slender and pikelike, the snout long and pointed, . with protractile premaxillae. The scales are small, with seventy-three to ninety-five in the complete and decurved lateral line. The dorsal fin with eight to ten rays is placed well back; there are eight or nine anal rays. The gill rakers are very short and the pharyngeal teeth (2,5/4,2) are pointed and lack grinding surfaces. Three species are known; all are carnivorous.

SACRAMENTO SQUAWFISH, *Ptychocheilus grandis,* 3–4 ft. (0.9–1.2 m). Dark above, light below. Sacramento R. system, Calif.

COLORADO SQUAWFISH, *Ptychocheilus lucius,* 3–5 ft. (0.9–1.5 m), 80 lb. (36.3 kg). Largest of American minnows. Dark above, light below. Lower Colo. R. drainage in Ariz., Utah, Colo., Wyo. An endangered species.

NORTHERN SQUAWFISH, *Ptychocheilus oregonensis,* **99:14,** 3–4 ft. (0.9–1.2 m). Columbia R. system, n. to Nass R. basin of B.C.; also in disconnected basin of Malheur Lake in e. Oreg.

DACES
Genus *Rhinichthys*

Members of this genus are difficult to define because of the variability of included forms. The pharyngeal teeth are always in two rows (with either one or two teeth in outer rows and always four in the main rows). There are seven to nine dorsal rays and almost invariably seven anal rays. A barbel is usually present, and there are thirty-five to ninety scales in the lateral line; these typically have radii on all fields (like the spokes of a wheel).

LONGNOSE DACE, *Rhinichthys cataractae, Fig. 84,* 4–6 in. (10.2–15.2 cm). Dorsal rays, 8, lateral-line scales, 58–68; snout long, prominently pointed, overhanging the inferior mouth; premaxillae not protractile, bound to snout by a broad frenum. Back olive to dark green or black, light below; a weak mid-lateral stripe. Widely distributed in N. America.

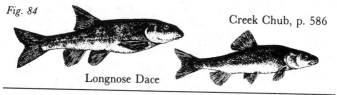

Fig. 84

Creek Chub, p. 586

Longnose Dace

LEOPARD DACE, *Rhinichthys falcatus,* 2–3 in. (5.1–7.6 cm). Dorsal rays, 9–10; lateral-line scales, 52–57; premaxillae protractile. Mottled or blotched above. Columbia R. basin e. of Cascades.

SPECKLED DACE, *Rhinichthys osculus,* **99:4,** 1½–3½ in. (3.8–8.9 cm). Dorsal rays, 7–9; anal rays, 6–7, lateral-line scales, 55–85; pharyngeal teeth, 1,4/4,1 or 2,4/4,2. Sides nearly plain, with poorly defined blotches, speckled, or with a mid-lateral stripe; found in creeks, rivers, springs. Wash., Idaho, Oreg., to s. Calif.; also e. of Sierras in "ancient" Lake Lahontan and "ancient" Lake Bonneville basins and Colo. R. drainage.

REDSIDES
Genus *Richardsonius*

Members of this group constitute a genus of moderately scaled (fifty-two to sixty-three scales in lateral line) cyprinids closely related to the Genus *Gila*. The pharyngeal teeth are arranged 2,5/ 4,2; there are eight to eleven dorsal-fin rays and eight to twenty-two anal-fin rays. Barbels are absent. Breeding individuals have on their sides either a bright orange or red stripe, which may be visible in nonbreeding fishes as a dark lateral band.

REDSIDE SHINER, *Richardsonius balteatus*, **99:9,** 6 in. (15.2 cm). Dark olive or brownish above with silvery sides and belly. Nass R. system of B.C. and Columbia R. drainage of Wash. and Puget Sound, e. to Idaho; also "ancient" Lake Bonneville basins, Utah.

LAHONTAN REDSIDE, *Richardsonius egregius*, 4 in. (10.2 cm). Very dark on back, belly golden; sides marked with 2 dark lateral bands separated by golden streak. "Ancient" Lake Lahontan basin and related waters of w. cen. Nev. and ne. Calif.

CREEK CHUB
Semotilus atromaculatus   *Fig. 84*

Description
Size, to 12 in. (30.5 cm). Body slender, cylindrical, with dark blotch at front of dorsal-fin base and small dark spot at base of caudal fin; scales in lateral line, 51–64; a small, flaplike barbel in groove of upper lip near corner of mouth; mouth large; pharyngeal teeth, 2,5/4,2. Color dark olive on back, with broad dusky stripe along midline; sides silvery, with greenish or purplish reflections; juveniles with dusky mid-lateral stripe.

Habitat
Small creeks, often with flow reduced to pools in dry season; spawns over gravel bottom.

Range
Widespread in e. N. America, w. to Mont., Wyo., Colo., and ne. N.Mex.

LOACH MINNOW
Tiaroga cobitis

Description
Size, to 2½ in. (6.4 cm). Body slender, flattened ventrally. Mouth small and strongly oblique; lower lip very thick, lateral creases suggest side lobes; upper lip attached to snout by broad frenum; pharyngeal teeth, 1,4/4,1. Color olivaceous, strongly blotched with darker pigment, with small caudal spot; males spotted with bright red; pair of yellowish-white spots at base of caudal fin.

Habitat
Flowing currents of rocky riffles with good growth of green algae.

Range
Gila R. drainage of Ariz., N.Mex.

SUCKERS
Family Catostomidae

Members of this family are closely allied to the minnows. They are soft-rayed fishes possessing toothless jaws and a more or less sucking, protractile mouth, typically opening downward, with thick lips, located usually behind the point of the snout. Pharyngeal teeth lie in the throat in a single comblike row, distinguishing suckers from minnows which have either more than one row of teeth or one row with only a few (6) teeth. All fins lack spines. There usually are more than 10 dorsal rays. The anal fin lies far back on the body and has 7 to 9 rays. An adipose fin is absent. Suckers are bottom feeders and often move in large schools. Their young provide an important food for game fishes.

RIVER CARPSUCKER
Carpiodes carpio

Description
Size, 15–18 in. (38.1–45.7 cm), record 30 in. (76.2 cm); 2–3 lb. (0.9–1.4 kg), record 10 lb. (4.5 kg). Body moderately deep, back not strongly arched; dorsal fin low, sickle-shaped anteriorly, rays 23–30; lateral-line scales, 34–36. Mouth small, horizontal; small knob at tip of mandible. Color silvery-gray.
Habitat
Large silty streams and rivers.
Range
E. U.S., w. to Rocky Mts. from Mont. to Mexico; abundant.

UTAH SUCKER
Catostomus ardens   100:7

Description
Size, to 18 in. (45.7 cm.) Body with 60–70 lateral-line scales; dorsal rays, 11–13; caudal peduncle relatively short and deep. Color above dark, fins dark; below whitish.
Habitat
Lakes, rivers, creeks from warm (80°F, or 26.7°C) to cold water with current rapid or absent, clear or silty, over varying bottoms; a bottom dweller.
Range
Basin of "ancient" Lake Bonneville in Idaho, Nev., Utah, Wyo.; also above Shoshone Falls in Idaho and Wyo.; common.

LONGNOSE SUCKER
Catostomus catostomus   101:10

Description
Size, 2–2½ ft. (0.6–0.8 m). Body elongate; dorsal rays, 9–11; lateral-line scales, 90–120. Head with long snout (longer than in White Sucker); eyes small, behind middle of head. In spring males have head and anal fin profusely tuberculate, sides with broad rosy band.
Habitat
Cold-water streams, lakes.
Range
Alaska, s. to Columbia R. basin; common.

WHITE SUCKER
Catostomus commersoni   **100:5**

Description
Size, to 2 ft. (0.6 m) and 5 lb. (2.3 kg). Body heavier-set than in
Longnose Sucker; dorsal rays, 10–13; lateral line scales, 55–75;
diagonal rows of scales between front of dorsal fin and lateral line,
8–10. Mouth with upper lip thick, papillose, protractile; lower lip
large, almost divided to base. Color back and sides greenish with
brassy or silvery luster, belly white. Spring males somewhat rosy;
young brownish with series of blotches along mid-side.

Habitat
Adaptable to various conditions in small to large streams, bottoms
of lakes.

Remarks
Common and easily caught.

Range
N. North America along Great Plains e. of Rocky Mts. from
Mont. to n. N.Mex.; Pacific slope of B.C.; introduced into upper
Colo. R. basin.

Note: There are more than fifteen closely related American species,
including some western forms having restricted distribution, such as
the **SACRAMENTO SUCKER,** *Catostomus occidentalis,* in central
California; the **LARGESCALE SUCKER,** *Catostomus macrocheilus,*
found from the Columbia River basin to Idaho and western
Montana; the **BRIDGELIP SUCKER,** *Catostomus columbianus,* in the
middle and lower Columbia River; and many others.

SONORA SUCKER
Catostomus insignis   **100:6**

Description
Size, to 2 ft. (0.6 m) and 5 lb. (2.3 kg). A rather chubby, coarse-
scaled sucker; typically fewer than 60 scales in lateral line. Large
head and rather enlarged lower lips; dorsal fin usually square
along posterior margin, typically with 11 rays. Sharply bicolored,
brownish above, yellow below; scales on upper parts sharply
outlined to produce a variably distinct spot on each.

Habitat
Deep, quieter parts of rivers and creeks, around gravelly or rocky
pools.

Range
Gila R. and Bill Williams R., Ariz., N.Mex.

FLANNELMOUTH SUCKER
Catostomus latipinnis   **100:11**

Description
Size, 1½–2 ft. (0.5–0.6 m). Body elongate, with narrow caudal
peduncle; dorsal fin large, sickle-shaped, with 10–14 rays, usually
12–13; lateral-line scales small, 90–115. Mouth large, with large
fleshy lobes on lower lips, very prominent in large adults. Color
above typically light gray or tan, sometimes greenish, scales with
dusky outline; lower sides yellowish, abdomen pale; underside of
head pinkish.

Habitat
Pools of streams and large rivers, usually unvegetated, clear to
murky, in strong current.

Range
Confined to drainage systems of Colo. R. basin; common.

LOST RIVER SUCKER
Catostomus luxatus

100:9

Description
Size, to 3 ft. (0.9 m), 8–10 lb. (3.6–4.5 kg). Body robust; mouth subterminal to nearly terminal, lower jaw slightly oblique; premaxillae projecting to form hump on top of snout; head long; lips rather thin, with weak papillae; gill rakers short, triangular, 24–33 on first arch; scales small, 82–88 in lateral line. Color back and sides dark, fading to white or yellow on belly.

Habitat
Lakes, including reservoirs, and rivers, ascending tributaries to spawn.

Remarks
Largest of the Klamath Lake suckers, and once placed in a separate genus, *Deltistes*. Formerly an important food fish, now rare.

Range
Upper Klamath R. basin, Oreg. and Calif.

TAHOE SUCKER
Catostomus tahoensis

100:8

Description
Size, to 2 ft. (0.6 m). Body elongate; caudal peduncle thick, snout long; fine-scaled, 82–95 scales in lateral line; dorsal rays, 10–11; mouth large, with lower lips so deeply incised that only 1 row of papillae crosses completely. Color dark above, olive, yellow, or whitish on lower sides and belly; breeding males with bright red lateral stripe.

Habitat
Cold mountain lakes.

Range
Lakes and rivers of the "ancient" Lahontan drainage system of w. cen. Nev. and adjacent Calif.; Lake Tahoe and "ancient" Lake Lahontan basin, Calif. and Nev.

MOUNTAIN SUCKERS
Subgenus *Pantosteus*

This group of suckers, formerly separated from *Catostomus* in the Genus *Pantosteus*, is treated here separately from *Catostomus* to emphasize the distinctiveness of these essentially western fishes. Typically they occur in creeks and rivers of strong gradient at higher elevations. They do not attain a length much greater than 1 foot (30.5 cm). In contrast to *Catostomus*, the upper and lower lips are separated by a lateral notch, with a shallow, median incision on the lower lip, and the jaws have cartilaginous scraping edges. As in *Catostomus*, breeding males have an orange or reddish lateral stripe, often bordered below by a black stripe, with the lower sides and belly yellowish to whitish. The most widespread species is treated first, with mention of other species.

MOUNTAIN SUCKER
Catostomus platyrhynchus

Description
Size, to 8½ in. (21.6 cm). Body elongate, round in cross-section; head cone-shaped, terminating in a long, blunt snout that

overhangs the mouth; usually 75–90 lateral-line scales; dorsal rays, typically 10; lower lip with large papillae, typically absent from outer edge of upper lip. Color back and sides dusky brown to greenish, usually with dark lateral stripe or series of blotches; belly white to light golden yellow.

Habitat
Small, clear mountain streams, with rubble, sand, or boulder bottoms; usually in pools or behind submerged rocks in swift water. May occur in large rivers, turbid streams, or lakes.

Remarks
An important forage fish, particularly for trout.

Range
Widespread in Great Basin of Utah, Wyo., Nev., Calif.; Fraser R. (B.C.) and upper Columbia R. drainages; Green R. in Utah, Wyo.; upper Mo. R. drainage and upper Sask. R. drainage.

DESERT SUCKER, *Catostomus clarki*, 8–12 in. (20.3–30.5 cm). Gila R. and Bill Williams R. systems of N.Mex. and Ariz.; Virgin R. in Nev., Ariz., Utah; drainages in sw. Nev.

BLUEHEAD SUCKER, *Catostomus discobolus*, 12 in. (30.5 cm). Colo. R. system of Ariz., N. Mex., Colo., Utah, Wyo.

RIO GRANDE SUCKER, *Catostomus plebeius*, 12 in. (30.5 cm). Rio Grande system from Colo. s. to n. Mexico.

CUI-UI
Chasmistes cujus   100:10

Description
Size, to 25 in. (63.5 cm) and 7 lb. (3.2 kg). Body plump, robust, coarse-scaled, about 60–65 scales in lateral line; caudal peduncle thick. Head very large, blunt; eyes proportionately very small, in anterior part of head; mouth ventroterminal, oblique, unsuckerlike; lips thin, the papillae weak or nearly absent. Color above pale olive to blackish-brown, broken laterally, fading to flat-white on belly; breeding males reddish on sides.

Habitat
Lakes; normally spawning in river mouth.

Range
Pyramid Lake, Nev., formerly spawning in mouth of Truckee River; formerly abundant.

Note: Members of this genus occur also in Klamath Lakes (and tributaries), in Oregon and California; and in Utah Lake, Utah. Once very abundant, they have declined drastically in recent decades and are now extinct, rare, or endangered species. Attempts are being made to culture the Cui-ui to restore the fishery in Pyramid Lake.

SMALLMOUTH BUFFALO
Ictiobus bubalus

Description
Size, 15–30 in. (38.1–76.2 cm), 2–15 lb. (0.9–6.8 kg). Similar to Bigmouth Buffalo, but with small, nearly horizontal mouth; thicker, more strongly grooved, lips; front of upper lip well below level of lower margin of eye; forward part of back usually strongly

keeled. Color back and sides slate-gray or pale brownish to silvery, belly whitish or pale yellow.

Habitat

Large rivers and major tributaries; prefers clearer water than Bigmouth Buffalo.

Range

Cen. U.S. and s. Canada, e. of Rocky Mts. from Mont. to Tex. and ne. Mexico. Introduced into some reservoirs in Ariz. and n. Calif.

BIGMOUTH BUFFALO

Ictiobus cyprinellus

101:4

Description

Size, 1½–3 ft. (0.5–0.9 m), 5–30 lb. (2.3–13.6 kg), record 50 lb. (22.7 kg). Body large, heavy, deep, carplike; dorsal fin long, with 27–29 rays; anal rays, 9; lateral-line scales, 35–40. Large oblique mouth, terminal, thin-lipped, protractile forward; front of upper lip about level with lower margin of eye. Color dull brownish-olive, with coppery and greenish reflections; belly whitish or pale yellow; all fins dusky. In breeding season, head becomes slate-gray with greenish tinge, sides of head olive-green, general dorsal surface coppery.

Habitat

Large rivers, oxbow sloughs and lakes; browses in soft mud.

Range

Cen. U.S. and s. Canada, from Sask. s. and e. to Tex. Introduced into some Ariz. impoundments.

SHORTHEAD REDHORSE

Moxostoma macrolepidotum

Description

Size, to 18 in. (45.7 cm) and 2 lb. (0.9 kg). Body slender; head short. Scales large, not small and crowded anteriorly as in *Catostomus*; 12 scales around caudal peduncle; dorsal rays, 12–13. Posterior margin of lower lips forms almost straight line; lips without papillae. Color back and upper sides olive-brown with golden reflections, scales notably dark-edged; rest of sides rich golden yellow; belly white; caudal fin bright red.

Habitat

Streams and rivers with predominance of gravelly or rocky bottoms and a permanent, strong flow.

Range

N. North America from Sask. R. to James Bay, B.C., s. to Colo. and Tenn. and to N.C. on Atlantic Coast.

RAZORBACK SUCKER

Xyrauchen texanus

101:3

Description

Size, to 2 ft. (0.6 m), 8–10 lb. (3.6–4.5 kg). Easily recognized (except when young) by razorlike keel on anterior part of back; lower lip with deep median cleft completely separating the two halves; dorsal rays, 13–16. Color dusky to olivaceous on back, grading to yellow-orange on belly; spawning males become nearly black on back and sides, brilliant orange on belly and anal fin.

Habitat

Large rivers and reservoirs, tolerating high turbidity.

Remarks

Regarded to be a threatened species. Once a major food source for Indians and early settlers, and taken commercially in Ariz. reservoirs 25 to 30 years ago.

Range

Colo. R. basin, Wyo. to Ariz.; rare.

CATFISHES
Family Ictaluridae

Members of this family have scaleless skin; a broad, flat head; a lateral line; a single, strong spine in both dorsal and pectoral fins; an adipose fin; eight barbels, two on the snout, two on the jaw, and four on the chin; and bristlelike teeth in bands in the upper jaw. Catfishes are virtually omnivorous and are principally active after dark, when they feed on the bottom. Caution should be exercised in handling them, as a poison gland at the base of the pectoral spines in some species can cause a painful wound; however, the poison is no more dangerous to man than a wasp sting and does not affect the edibility of the fish. There are some five genera containing thirty-seven species in the United States, of which six species are described here.

WHITE CATFISH
Ictalurus catus

Description

Size, to 18 in. (45.7 cm). Moderately forked tail; 18–24 anal rays, with tips forming an arc. Lower jaw shorter than upper; maxillary barbels long and dark-colored; chin barbels white. Color olive-blue above, silvery to white below; unspotted, sometimes mottled.

Habitat

Fresh to brackish streams, reservoirs, ponds, sloughs. Often introduced into private lakes, ponds, streams.

Range

Atlantic Coast native, introduced into Calif. and some other w. states.

YELLOW BULLHEAD
Ictalurus natalis   **101:9**

Description

Size, to 18 in. (45.7 cm). Body chunky, caudal fin rounded; anal rays, 24–27, most about equal length; rear edge of pectoral spines sharply barbed. Color yellow-brown to blackish, belly white, chin barbels white.

Habitat

Shallow waters of large ponds, lakes, streams.

Range

U.S., e. of Rocky Mts. Introduced into w. U.S.

BLACK BULLHEAD
Ictalurus melas   **100:1**

Description

Size, to 18 in. (45.7 cm) and 8 lb. (3.6 kg). Body rather slender, caudal fin very slightly emarginate; anal rays, 17–24, with the membranes darkly pigmented; anal fin rounded; rear edges of pectoral spines smooth, never toothed; base of caudal fin with a pale, vertical bar. Color brownish-yellow to black, belly yellow to milky-white, chin barbels black.

Habitat

Mud-bottomed lakes, ponds, oxbows, large rivers.

Range

E. North America, e. of Rocky Mts., s. to Tex. Widely introduced into w. U.S.

BROWN BULLHEAD
Ictalurus nebulosus

100:4

Description

Size, to 18 in. (45.7 cm). Rays of anal fin unicolored, numbering 21–24; pectoral spines with rear edges sharply barbed. Color back dark brownish, belly gray to yellowish, sides and back often mottled; chin barbels black.

Habitat

Quiet, weedy mud-bottomed lakes, ponds; also large rivers.

Range

U.S., e. of Rocky Mts.; introduced into Calif. and other w. states.

CHANNEL CATFISH
Ictalurus punctatus

101:13

Description

Size, 20–55 lb. (9.1–24.9 kg). Body slender, caudal fin deeply forked; anal fin, 24–29 rays, slightly convex, anterior rays longer. Color bluish or silvery, often with black spots.

Habitat

Chiefly large waters.

Range

U.S., e. of Rocky Mts.; widely introduced in w.; abundant.

STONECAT
Noturus flavus

Fig. 85

Description

Size, to 12 in. (30.5 cm) and 1 lb. (0.45 kg). Body slender; back edges of pectoral spines smooth; adipose fin entirely attached to back, separated from caudal fin only by a notch; tail rectangular. Lower jaw shorter than upper; teeth in upper jaw in bands with backward lateral extensions. Color back and sides yellowish-brown; underside of head and belly white or pale yellow; fins yellow-edged, caudal fin with light border.

Habitat

Fast-water streams and riffles; also weedy lake-shore waters.

Range

N. half of U.S. e. of Rocky Mts., Mont. to Colo., s. to Okla.

Fig. 85

Stonecat

Troutperch, p. 594

Burbot, p. 594

Percopsiform Fishes
Order Percopsiformes

TROUTPERCHES
Family Percopsidae

TROUTPERCH
Percopsis omiscomaycus   *Fig. 85*

Description
Size, 6–8 in. (15.2–20.3 cm). Noted for peculiar translucence of the body; adipose fin; fine saw-toothed edges of scales; long pectoral fins, reaching well beyond origin of pelvics. Body elongate, tapering; 2 thin, weak spines in dorsal fin; 1 very weak spine in anal fin; soft anal rays, 6–7; lateral line complete, about 45–50 scales, caudal peduncle long, slender. Head naked, mouth small and straight, overhung by conical snout. Color back and sides greenish-yellow or straw-colored, with silvery reflections; mottled with row of spots on lateral line, above which is another row of spots.
Habitat
Lakes and slow streams; nocturnal.
Range
N. North America, from upper Miss. R. basin to Yukon and Mackenzie R. systems.

SAND ROLLER
Percopsis transmontana

Description
Size, to 6 in. (15.2 cm). Body moderately deep, compressed; 2 stout spines in dorsal fin, 2 very stout spines in anal fin; 44–46 scales in the incomplete lateral line. Color greenish-yellow, sides mottled with numerous spots above and on lateral line.
Habitat
Quiet backwaters or sluggish parts of cool streams over sandy or silty bottoms.
Range
Columbia R. basin of Wash., Oreg., w. Idaho (Snake R.).

Codlike Fishes
Order Gadiformes

CODFISHES
Family Gadidae

BURBOT
Lota lota   *Fig. 85*

Description
Size, to 38 in. (96.5 cm) and 60 lb. (27.2 kg). The only strictly freshwater species of cod. Body elongate, covered with tiny embedded scales; no spines in fins; pelvic fins beneath pectorals; only 1 chin barbel. Color back and sides dark olive or brown, marbled with darker brown or black.

Habitat
Deep, cool lakes and rivers.
Range
N. North America, in w. from Alaska and nw. Canada s. to
Columbia R.

Cyprinodont Fishes
Order Cyprinodontiformes

Members of this order lack spines in their fins. The head as well
as the body is scaled. The pelvic fins, when present, are small and
positioned near the middle of the belly. There is no lateral line and
the caudal fin is straight or rounded (never forked). Cyprinodont
fishes are represented on all continents except Australia, with
many species inhabiting salt and brackish water. Included are
many popular aquarium fishes, such as guppies, swordtails, and
mollies.

KILLIFISHES
Family Cyprinodontidae

The small fishes of this shallow-water family are all egg-layers.
Many species show strong differences between the sexes in
markings and color. Many are surface feeders.

WHITE RIVER SPRINGFISH
Crenichthys baileyi
102:12

Description
Size, to 3 in. (7.6 cm). Body robust; dorsal and anal fins located far
back; no pelvic fins; jaw teeth with two cusps; a double row of
spots on the sides.
Habitat
Warm and cool springs and their outflows.
Range
Basin of Pleistocene White R. and Pahranagat valleys in e. Nev.;
also Moapa R., s. Nev.

Note: A related species, the **RAILROAD VALLEY SPRINGFISH,**
Crenichthys nevadae, lives in Railroad Valley, Nye County,
Nevada.

PUPFISHES
Genus *Cyprinodon*

These fishes have a typically short, stout body. The teeth are
tricuspid and incisorlike; there is a single series in each jaw. The
scales are large, with twenty to thirty-four (usually twenty-five or
twenty-six) from opercular angle to caudal base. The dorsal rays
typically number 10 or 11 and anal rays 8 to 12. Pupfishes average
about 3 inches (7.6 cm) in length and are typical inhabitants of
desert springs and creeks, some very saline, in the American
southwest.

DESERT PUPFISH
Cyprinodon macularius   **101:12**

Description
Size, to 2½ in. (6.4 cm). Circuli of scales with marked spinelike projections, interspaces clear; front of dorsal fin about midway between caudal base and snout; pelvic-fin rays, typically 7. Females and juveniles have silvery sides with dark, vertical bars usually interrupted to form a disjunct lateral stripe; males bright blue on body with posterior part of caudal peduncle and entire caudal fin yellow to orange.
Habitat
Marshes, springs, slow-moving parts of creeks.
Range
Lower Colo. R. basin and Rio Sonoyta basin in Ariz., Calif., and n. Mexico. Now rare, but can be seen in a refuge in Anza-Borrego State Park, San Diego Co., Calif.; protected in Organ Pipe Cactus National Monument, Ariz.

DEVILS HOLE PUPFISH, *Cyprinodon diabolis,* ½–1¼ in. (1.3–3.3 cm). No pelvic fins, mature at less than 1 in. (2.5 cm). Restricted to Devils Hole, Ash Meadows, Nye Co., Nev. An endangered species.

AMARGOSA PUPFISH, *Cyprinodon nevadensis,* 1½–2½ in. (3.8–6.4 cm). Interspaces between scale circuli with strong reticulations; no yellow-orange on males; pelvic rays 6 or fewer; dorsal fin more posterior than in Desert Pupfish. Amargosa R. basin, from Death Valley, Calif. to Nev.

RED RIVER PUPFISH, *Cyprinodon rubrofluviatilis,* 1½–2½ in. (3.8–6.4 cm). Distinguished by a scaleless abdomen; a thin, dark, vertical bar near base of caudal fin, in front of which is a broad yellowish area; and anal fin of breeding males with reddish border. Brazos R. and Red R. drainages of Tex. and sw. Okla.

SALT CREEK PUPFISH, *Cyprinodon salinus,* 1–2 in. (2.5–5.1 cm). Similar to Amargosa Pupfish, but scales much smaller. Restricted to Salt Creek in Death Valley, Calif.

KILLIFISHES
Genus *Fundulus*

This is a genus of slender fishes with pelvic fins, three to six pores on the lower jaw, and no lateral line.

PLAINS KILLIFISH
Fundulus zebrinus   *Fig. 86*

Description
Size, to 3 in. (7.6 cm). Body elongated, with 12–16 dark, vertical bars on side; front of dorsal fin before front of anal fin, dorsal with 14–15 rays, anal with 13–14 rays; scales along side, about 40–60. Color olive-brown to greenish above, fading to silvery on sides and belly; breeding males with orange or red on lower fins.

Habitat
Alkaline to saline streams of the Great Plains.
Range
W. of Miss. R., Wyo. to Tex.; introduced into Colo. R. drainage,
Ariz. and Utah.

Fig. 86

Plains Killifish

Plains Topminnow

PLAINS TOPMINNOW
Fundulus sciadicus   *Fig. 86*
Description
Size, to 2½ in. (6.4 cm). Body chunky, without any markings on
side; dorsal rays, 10–11; anal rays, 12–14. Color uniformly olive-
brown punctulated with fine dots; belly pale; fins yellowish or
plain in females and juveniles, orange-red in breeding males.
Habitat
Quiet pools of small creeks, backwaters, and overflow pools of
larger streams.
Range
Missouri R. system of Mont., Wyo., S.D., Colo.

LIVEBEARERS
Family Poeciliidae

Members of this family are similar to the killifishes, but with the
anal fin of the male placed farther forward, its anterior rays
modified to function as an intromittent organ. The livebearers and
the viviparous perches (Family Embiotocidae), with one exception,
are the only viviparous freshwater fishes in the United States.

MOSQUITOFISH
Gambusia affinis   **102:3**
Description
Size, to 2 in. (5.1 cm). Body with origin of dorsal fin behind that
of anal fin; dorsal fin with 6–8 rays, the distance from its origin to
caudal fin ½ distance to snout; anal rays, 9–10; 29–32 scales along
side; Color light olive, each scale dark-edged; vertical rows of spots
on caudal fin.
Habitat
Clear vegetated water in ponds, pools, ditches, marshes.
Range
Miss. R. basin of cen. U.S. Introduced widely into w. waters.

Note: Closely allied to the Mosquitofish is the **TEXAS GAMBUSIA,**
Gambusia nobilis, in the Pecos River system of New Mexico and
western Texas. The **GILA TOPMINNOW,** *Poeciliopsis occidentalis,* is
similar to the Mosquitofish, but it has a dark (often black in male)
lateral stripe, and no row of spots on the caudal fin. It inhabits the
Gila River system in Arizona and is an endangered species.

SAILFIN MOLLY
Poecilia latipinna   **102:5**

Description
Size, 3–4 in. (7.6–10.2 cm). Body short and thick, especially in
female; more compressed in male. Dorsal rays 14–16, dorsal fin
greatly expanded in large, breeding males. Color olivaceous to
brownish-gray; sides with longitudinal rows of dark dashes. Large
males with 6–10 deep-lying vertically elongated spots on dorsal fin;
margin of caudal fin blackened; parr marks on lower sides; blue
and orange on caudal, sometimes on dorsal.

Habitat
Quiet lowland streams, marshes, and canals; tolerate high salinity.

Range
Coastal e. U.S. to ne. Mexico, invading fresh water; introduced
into Ariz., Calif. (Salton Sea), Nev.

Sticklebacks
Order Gasterosteiformes

STICKLEBACKS
Family Gasterosteidae

These are all small scaleless fishes with slender, streamlined
bodies. They have a series of free dorsal spines in front of the soft
dorsal fin, and the pelvic fins are reduced to heavy spines. The
sides may have bony plates. Although several species are
anadromous, only one species is restricted to fresh water.
Sticklebacks are noted for their pugnacious habits. The males build
elaborate nests of plants and sticks and guard the eggs and young.

THREESPINE STICKLEBACK
Gasterosteus aculeatus   *Fig. 87*

Description
Size, to 4 in. (10.2 cm). Body rather stout, sides either naked or
usually partially or completely covered by bony plates; 3 sharp
dorsal spines; pelvic bones joined. Gill membranes not free;
attached to the isthmus. Color olive-green to dark brown above and
on sides; light below; underside yellow, white, or silvery. Breeding
males red below, blue along sides.

Habitat
Freshwater streams, lakes, rivers; also marine waters; extremely
variable.

Range
Alaska, Canada, and s. throughout U.S. to Baja Calif.

Fig. 87   northern / southern   Threespine Stickleback

Perchlike Fishes
Order Perciformes

Members of this large order have the head scaled. The pelvic fins are forward and typically have a single spine and five soft rays. There are spines in both dorsal and anal fins which may be united or separate.

TEMPERATE BASSES
Family Percichthyidae

Members of the temperate bass family, some of which are anadromous, have pointed gill covers, two dorsal fins, three anal spines, and a lateral line which does not extend onto the slightly forked tail.

WHITE BASS
Morone chrysops   *Fig. 88*

Description
Size, 12–18 in. (30.5–45.7 cm) and to 3 lb. (1.4 kg). Body with separated dorsal fins, with the spines graduated; anal soft rays, 11–12; second anal spine ⅓ head length; body depth usually more than ⅓ standard length. Head with projecting lower jaw. Color back blue-gray, sides silvery, with 7 longitudinal stripes (may be broken) on each side.

Habitat
Lakes, large rivers.

Habits
Usually travels in schools.

Range
Primarily e. U.S.; introduced into w. U.S.

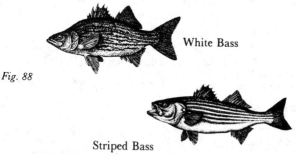

White Bass

Fig. 88

Striped Bass

STRIPED BASS
Morone saxatilis   *Fig. 88*

Description
Size, 4–5 ft. (1.2–1.5 m) and to 100 lb. (453.5 kg). Body streamlined, with 6–9 horizontal black stripes. Dorsal fins sharply separated, anterior fin with 9 spines, second with 1–2 spines followed by 12 soft rays. Mouth large; lower jaw projecting, but maxilla does not reach beyond hind margin of eye. Color dark olive-green to bluish-black above, silvery white on lower sides and belly.

599

Habitat
Unusually adaptable, migrating from the ocean in the spring into larger rivers to spawn in moderate to strong current; also inhabits bays.

Range
Introduced on Pacific Coast; s. B.C. to Mexico; abundant in Sacramento–San Joaquin estuary and river system, Calif., and some landlocked lakes on Colorado R. system.

SUNFISHES
Family Centrarchidae

This family contains some of the most important game fishes in North America, including the crappies and black basses. They are carnivorous fishes with oblong or circular bodies that may be thin to deep, dorsal fins that are completely joined (except in Genus *Micropterus*), a lateral line, and three or more anal spines. The males make a shallow depression for a nest (the Sacramento Perch is an exception) and guard the eggs and fry. Except for the Sacramento Perch, this family is native only east of the Rocky Mountains.

SACRAMENTO PERCH
Archoplites interruptus   101:11

Description
Size, to 20 in. (50.8 cm). Body oblong, fairly deep, compressed; scales strongly ctenoid; 12–13 spines in dorsal fin; longest anal spine about equal to spinous part of anal fin. Large, oblique mouth; maxilla broad, reaching to about middle of eye; teeth on vomer, palatines, tongue, and upper jaw; tongue teeth in 2 patches; gill rakers long, 25–30. Color back dark, below silvery with 6–7 dark vertical bars on each side, somewhat interrupted and irregular.

Habitat
Streams, lakes.

Range
Cen. Calif., in Sacramento, San Joaquin, and Salinas river systems (rare); successfully introduced into alkaline lakes of Nev., Utah, Colo., Nebr., N.Dak. and S.Dak.

SUNFISHES
Genus *Lepomis*

The sunfishes have deep, thin bodies; ten dorsal and three anal spines; a notched tail; and no teeth on the tongue. They are colorful and popular game fishes, and some species have been widely introduced.

GREEN SUNFISH
Lepomis cyanellus   102:11

Description
Size, 6–8 in. (15.2–20.3 cm). Short, rounded, pectoral fin; lateral line complete; large mouth extending past front edge of eye; gill cover stiff to the smooth rear edge. Green; dark spot on gill cover and dark on bases of dorsal and anal fins.

Habitat
Warm streams and ponds.

Habits
Hybridizes with Bluegill.
Remarks
A good pan fish.
Range
Widely introduced and common throughout w. U.S.

PUMPKINSEED
Lepomis gibbosus

102:10

Description
Size, 6–8 in. (15.2–20.3 cm) and 8–10 oz. (226.8–283.5 g).
Distinguished by long, pointed pectoral fins and a small red spot
near edge of gill cover. Body very deep. Mouth small, not reaching
eye; no teeth on roof of mouth; gill cover stiff to smooth rear edge.
Color orange to olive, sometimes with blue markings on cheek; rear
of dorsal fin mottled.
Habitat
Clear, still water with many submerged plants.
Range
Introduced throughout w. U.S.

BLUEGILL
Lepomis macrochirus

101:6

Description
Size, 10–15 in. (25.4–38.1 cm) and to 4¾ lb. (2.2 kg). Body with
long, pointed pectoral fins; anal soft rays, 10–12. Mouth small,
rarely reaching front edge of eye; no teeth on roof of mouth; rear
edge of gill cover thin and flexible. Olive to bronze on back, with
blue and orange on sides; 2 bluish bars extend back from mouth
and chin; lower edge of gill cover bluish; dusky spot on last 3
dorsal rays.
Habitat
Warm, reedy waters of bays, ponds, lakes.
Habits
Hybridizes with Green Sunfish.
Range
Introduced into various w. lakes, particularly in Calif.

BLACK BASSES
Genus *Micropterus*

The basses have an elongate body, and a large mouth with a
prominent lower jaw. They are the only sunfishes in which the
dorsal fins are separated by a moderate to deep notch.

SMALLMOUTH BASS
Micropterus dolomieui

102:9

Description
Size, to 16 in. (40.6 cm) and 12 lb. (5.4 kg). Notch in dorsal fin
shallow; shortest dorsal spine more than ½ length of longest; dorsal
rays, 13–15; pectoral rays, 16–18; anal rays, 11. Head with upper
jaw rarely reaching to rear of eye. Back olive, belly dusky silver; no
side stripe, but young have vertical bar on sides.
Habitat
Clear, rocky lakes and rivers.

Habits
Male fans out nest in gravelly bottom, guards eggs and fry.
Range
Widely introduced throughout w. U.S.

LARGEMOUTH BASS
Micropterus salmoides   **101:2**

Description
Size, to 20 in. (50.8 cm) and 22¼ lb. (10.1 kg). Distinguished by
dorsal notch almost dividing the fins, and upper jaw extending
beyond the eye. Shortest dorsal spine less than ½ longest; dorsal
rays, 12–13; pectoral rays, 13–17; anal rays, 11. Color, back olive,
sides and belly silver; dark band on side, becoming broken with
age.
Habitat
Ponds, small lakes, oxbows; fairly tolerant of turbidity.
Range
Widely introduced throughout U.S.

CRAPPIES
Genus *Pomoxis*

WHITE CRAPPIE
Pomoxis annularis   **101:5**

Description
Size, to 12 in. (30.5 cm) and 5 lb. (2.3 kg). Base of dorsal fin
shorter than distance from eye to dorsal fin; dorsal spines 5–7,
usually 6. Head with large, oblique mouth; lower jaw heavy,
protruding; gill cover notched. Iridescent olive-green on back,
silvery on sides, with indistinct vertical bars.
Habitat
Ponds, lakes, oxbows, sluggish pools, preferably with vegetation;
very tolerant of turbidity.
Range
Widely introduced throughout U.S.

BLACK CRAPPIE
Pomoxis nigromaculatus   **102:7**

Description
Size, to 12 in. (30.5 cm); 2–3 lb. (0.9–1.4 kg), record 5 lb. (2.3 kg).
Base of dorsal fin equal to distance from eye to dorsal fin; dorsal
spines, usually 7–8. Mouth smaller than in White Crappie. Olive-
green to silvery, mottled with blue or black, lacking any evidence of
vertical bars in adults.
Habitat
Ponds, lakes, impoundments, slow-moving waters; not tolerant of
turbidity.
Range
Widely introduced throughout U.S.

PERCHES
Family Percidae

These predaceous fishes have ctenoid scales, two completely separate dorsal fins, and one or two anal spines. They are native only east of the Rocky Mountains and include three popular game fishes, described below.

YELLOW PERCH
Perca flavescens

102:4

Description
Size, to 12 in. (30.5 cm); 1–2 lb. (0.45–0.91 kg). Body moderately deep, compressed; first dorsal with 13–15 spines, second dorsal with 1 spine and 13–15 soft rays; anal fin with 2 spines, 6–8 rays. Yellowish with black bars, fins dusky; 6–8 pronounced vertical bars on sides.
Habitat
Lakes, ponds, sluggish streams.
Range
N. North America; widely introduced in w. U.S.

SAUGER
Stizostedion canadense

Description
Size, to 15 in. (38.1 cm) and 8 lb. (3.6 kg). Large with many round black spots, but no single large black blotch, on the dorsal fin. Body elongated; dorsal spines 12–13, soft rays 17–20. Brownish above, usually with about 4 dark bars extending obliquely forward onto sides; belly white.
Habitat
Shallow, turbid lakes and large rivers.
Range
Great Plains region of s. Can. as far w. in U.S. as e. Mont., e. Wyo., e. Colo., s. through lower Miss. R. Valley.

WALLEYE
Stizostedion vitreum vitreum

101:7

Description
Size, to 36 in. (91.4 cm) and 22½ lb. (10.2 kg). Body elongate, cylindrical in cross-section; dorsal spines 12–15, soft rays 19–23. Dark olive with irregular dark bars across sides in fish under 12 in.; black spot at base of last dorsal spine; lower lobe of caudal fin with white tip; no pronounced vertical bars on sides.
Habitat
Cold lakes, rivers.
Range
Nw. Canada to Great Slave Lake, s. to lower Miss. R. Valley, e. to Atlantic. Introduced into lower Colo. R. drainage, Ariz. and Nev.; also in Utah.

IOWA DARTER
Etheostoma exile

102:1

Description
Size, to 2½ in. (6.4 cm). Body elongate, slender; caudal peduncle long; snout rounded; lateral line incomplete, not extending beyond

base of second dorsal fin; 46–60 scales along side; dorsal spines 8–10, rays 10–11; anal fin with 2 spines (rarely 1), 7–8 rays.
Breeding males, olivaceous on back, with about 8 dark saddles, the sides with about 10 vertically elongate dark blotches overlaid with bluish-green, the interspaces red; lower sides dull red to orange, belly white or creamy. Females similar, but sides lack bright colors, belly silvery.

Habitat
Moderately cool lakes and sluggish streams on sandy to muddy bottoms about vegetation.

Range
N. North America, Mont. and Alta., e. to Lake Champlain, s. to Colo. and s. Ill.

LOGPERCH
Percina caprodes   102:2

Description
Size, to 6 in. (15.2 cm). Body elongate but rather chunky; snout long, pointed, overhanging mouth; 75–90 scales in lateral line; 1 or more enlarged scales between pelvic fins, with breast and rest of midline of belly otherwise naked or with modified scales. Back and sides pale yellowish-olive, with 15–20 narrow, vertical dark bars alternating in length, every other bar extending down sides only about to lateral line, the others nearly to belly.

Habitat
Streams with rocky riffles and pools; gravelly waveswept shores of lakes and reservoirs.

Range
E. North America from s. Canada through Miss. R. Valley to Fla. and Tex.

DRUMS
Family Sciaenidae

Most of the fishes of this family are marine and are highly valued as food and game fishes. They are readily distinguished from other perciform fishes by the extension of the lateral line onto the caudal fin.

FRESHWATER DRUM
Aplodinotus grunniens   102:13

Description
Size, to 20 in. (50.8 cm) and 5 lb. (2.3 kg). Body deep at front, sloping steeply upward from snout to dorsal fin, tapering backward to slender caudal peduncle; mouth low, horizontal; dorsal fin long, distinctly divided into spiny (usually 10 stiff spines) and soft (29–32 rays) parts; anal fin short, with 2 spines; lateral line extending onto tail. Back gray, with blue and purple reflections; sides silvery; ventral surface milky white.

Habitat
Large rivers, lakes, and impoundments; avoids strong current but tolerant of high turbidity; prefers pools and depths of 30 ft. (9.1 m) or more in reservoirs.

Range
E. North America between Rocky Mts. and Appalachians, from s. Canada to n. Guatemala.

SURFPERCHES
Family Embiotocidae

TULE PERCH
Hysterocarpus traski

Description
Size, to 8 in. (20.3 cm). The only freshwater member of this viviparous family. Body deep; complete lateral line; cycloid scales. Single dorsal fin with 15–19 spines, 9–15 rays; anal fin rather long, with 3 spines, 20–26 rays. Head with scaly cheeks and opercula; mouth small, terminal; premaxillae protractile; maxilla short, lying under preorbital. Color variable; back dark, often bluish or purplish, sides olivaceous to silvery, throat and belly usually yellow. Two color phases: (1) barred, marked on sides with irregular blotches forming diagonal bars; (2) unbarred, silvery to dusky.

Habitat
Lowland streams, tule lakes.

Range
Cen. Calif. in Sacramento R. system and in other tributaries of San Francisco Bay; also Clear Lake, Russian R., and Pajaro–Salinas basin.

SCULPINS
Family Cottidae

The sculpins are primarily marine but a number of species occur in fresh water. They have broad flattened heads, large pectoral fins, lack scales, and have one or more preopercular spines on the side of the head. Also, the gill membranes are attached to the isthmus. They are bottom dwellers.

SCULPINS
Genus *Cottus*

MOTTLED SCULPIN
Cottus bairdi

Fig. 89

Description
Size, 4 in. (10.2 cm). Dorsal spines 6–8, rays 16–18; anal rays 11–13; pelvic fins with 1 spine and 4 rays; preopercular spines, usually 3. Lateral line incomplete, ending below base of soft dorsal; prickles variable; palatine teeth present.

Habitat
On rocky riffles and in pools of cool, clear streams.

Range
Columbia R. system and adjacent drainages of B.C., Wash., Oreg., Idaho, Mont., to upper Colo. R. system of Wyo., Colo., N.Mex., and e. and s. U.S.

Fig. 89

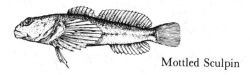

Mottled Sculpin

PRICKLY SCULPIN
Cottus asper

Description
Size, 8–10 in. (20.3–25.4 cm). Dorsal spines 8–10, rays 19–23;
pelvic fins with 1 spine, 4 preopercular spines. Palatine teeth and
prickles well developed; a single chin pore. Mottled dark brown to
gray above and on sides, white to yellow-white below; usually 3
dark, oblique bars under second dorsal fin, black spot on posterior
part of spiny dorsal fin.
Habitat
Fresh and brackish waters; coastal streams, rivers, lakes.
Range
Alaska to Ventura R., s. Calif.

PIUTE SCULPIN
Cottus beldingi   102:6

Description
Size, 4–6 in. (10.2–15.2 cm). Dorsal spines 6–8, rays 15–18;
palatine teeth few or none; prickles absent; only 1 preopercular
spine. Lateral line almost complete.
Habitat
On rocky riffles and in pools of cool, clear streams.
Range
Wash., Idaho, Oreg., Nev., Utah, Colo., Wyo., in Colo. R.
"ancient" Lake Lahontan, "ancient" Lake Bonneville, Columbia
R. systems and drainages.

RIFFLE SCULPIN
Cottus gulosus   102:8

Description
Size, 3–5 in. (7.6–12.7 cm). Dorsal spines 7–8, rays 16–19; anal
rays, 14–16; preopercular spines, 2–3. Lateral line complete or
incomplete; prickles present; palatine teeth usually present.
Habitat
Coastal streams.
Range
Wash., Oreg., s. to Morro Bay, Calif., with gap in range in se.
Oreg. and nw. Calif.

DEEPWATER SCULPIN
Myoxocephalus quadricornis

Description
Size, 6–8 in. (15.2–20.3 cm). Body naked; gill membranes free
from isthmus. Dorsal fins distinctly separated; dorsal spines 8–9,
rays 11–16; rayed portion much higher than spinous; male soft
dorsal very large, usually reaching base of caudal when depressed.
Anal rays, 12–15; pelvic fins with 1 spine, 3 rays; lateral line
incomplete, extending to below soft dorsal; 4 unbranched
preopercular spines, second conspicuous. Pale brown or cream.
Habitat
Streams and deep lakes.
Range
Arctic regions of Alaska., nw. and n. Canada s. to Great Lakes;
circumpolar.

Mollusks

Consulting Editor
William K. Emerson
Curator
Department of Invertebrates
American Museum of Natural History

Illustrations
Plates 115, 117–119, 121–124 Walter Hortens
Plates 103–112, 113, 114, 116, 120 William Downey

Mollusks
Phylum Mollusca

The enormously successful phylum of mollusks, sometimes termed the "soft ones" from translation of the Latin name, constitutes one of the most advanced groups of animals in the invertebrate world. To be sure, an oyster may not seem very advanced, but the squid that can remove the lid from a box with its tentacles to get the food within is certainly more intelligent than some of the lower vertebrates. Moreover, its eye is the most highly organized visual organ known among the invertebrates.

Nature of Mollusks
The phylum Mollusca forms one of the major branches of the animal kingdom, including the clams, oysters, scallops, snails, slugs, chitons, squids, octopi, and a few others, and totaling between 50,000 and 80,000 living species. To these can be added thousands of extinct ancestors, preserved as fossils in strata of sedimentary rocks going back to the early Cambrian period of the Paleozoic era. About three-fourths of all the living mollusks are gastropods (the coiled snails), distributed through more than 1650 genera; second are the bivalves or pelecypods (clams), with about 420 genera; third in numbers but largest in size are the cephalopods, with some 150 genera.

Habitat
Mollusks have penetrated all habitats except the air in their evolutionary development. The limpets generally cling to tidewater rocks, although there are deep-water forms; snails crawl, dig, or swim; bivalves anchor themselves, burrowing in mud, wood, stone, even transoceanic lead cables; cephalopods jet through the water or move on their arms among bottom rocks. Some mollusks are parasitic in the bodies of other creatures. From the polar seas to the tropics, mollusks occur on sandy, muddy, pebbly, or rocky coasts; in clear, muddy, brackish, or polluted water; in rocky crevices and tide pools, clinging to eelgrass, underneath driftwood and rocks, on pilings and the undersides of boats, even adrift on the open ocean. Some mollusks not included in this chapter inhabit freshwater lakes and streams, live underground, or are found on trees.

Classification
In general terms, mollusks are soft-bodied, unsegmented, and usually hard-shelled invertebrate animals. The living mollusks comprise seven classes; the five major ones are discussed in this chapter. The five discussed are:

Class Polyplacophora
The chitons, also known as coat-of-mail shells, are primitive, ancestral-type mollusks of very sluggish habits. Elongated and flattened, the chitons bear a shelly armor of eight saddle-shaped plates (imbricating valves) arranged in an overlapping series along the back and held together by a surrounding girdle often bearing spiny hairs. On the underside is the broad, flat foot. Although most chitons inhabit shallow water close to shore, others live in depths as great as 2000 fathoms (3.7 km).

Class Gastropoda
The gastropods are the snails, most of which have shells. Most gastropods are univalves; that is, they have a single shell, or valve, which is usually spiral. A few exhibit two valves. Gastropods have a low order of sight, hearing, and possibly taste but well-developed

senses of smell and touch. Most are carnivorous and predatory. They move over the ocean floor, and some are lifelong wanderers on the open sea.

Class Scaphopoda

The scaphopods comprise a small class of burrowing univalve mollusks having a tapering conical shell open at both ends and slightly arched, like a miniature elephant's tusk. The members of this class are commonly known as tooth or tusk shells, after their shape. From the larger end of the shell projects the foot and several slender tentacles. The animals live partially buried in sand in clear water at various depths, feeding on minute marine organisms.

Class Bivalvia

The bivalves, also known as pelecypods, are the clams, which have two valves joined along a hinge line by a tough ligament and held together by one or two strong muscles. Except in a short larval stage, most clams are sedentary throughout their lives.

Fig. 90

Parts of a Typical Snail and Clam

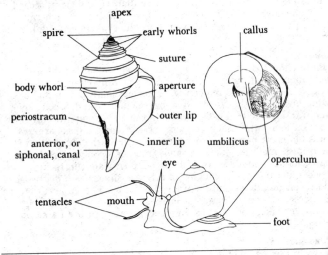

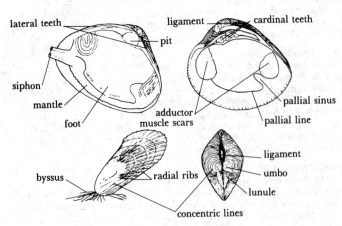

Class Cephalopoda

The cephalopods are highly specialized, carnivorous mollusks with relatively keen sight and a well-developed nervous system. The shell, if present, is generally internal and lengthened. The head supports two large eyes and is armed with a powerful parrotlike beak; it is surrounded by long flexible tentacles, or arms, studded with powerful sucking disks. Most cephalopods can discharge a caustic, inky fluid in defense. They move rapidly through the water with a jet-propelled action, the tentacles streaming behind.

Evolution

The earliest reliably datable shell heaps appear to be Aurignacian, discarded by Late Paleolithic Cro-Magnon men along the Mediterranean shores of Europe, Africa, and Asia Minor, prior to 50,000 B.C. Both Cro-Magnon and later Bronze and Iron Age men used shellfish for food and the shells for ornaments, implements, money, and even clothing. Much later, the Romans began farming oysters and edible snails as a major industry.

Seashells

The seashells are the species most usually associated with the phylum Mollusca, for they number more than two-thirds of all the mollusk species. Below we discuss some important points for the collector of seashells to know.

Shell Collecting Techniques

The most productive place to collect shells is the seashore at low tide. Strolling along a sandy stretch of beach can prove fruitful in finding shells that have been washed up by a late tide. Much of such shelly material, however, will be wave-eroded, broken, or damaged. The valves of a bivalve shell are commonly found separated, and the lips of a univalve usually show broken edges and worn shoulders.

When no longer content with beach-worn specimens, one should learn to search for living invertebrates in their natural habitats. This opens the door to unlimited possibilities for acquiring a worthwhile shell collection. It is thus important to pay attention to tides, seasonal moods of the ocean, and the effects of wind and current.

Some states, particularly California, may require possession of a regular angling license before collecting of living seashells is permitted. Additionally, for certain species, there are stringent protective regulations. These laws should be understood before collecting living mollusks.*

The best time to look for and collect living shellfish is at low tide, when exploration of the exposed tidal zone is easiest. Because many marine creatures hide under objects to await the return of the tide, any stone, plank, or piece of driftwood should be turned over, and then replaced as originally found. Bivalves can often be detected in sand or on mud flats by small holes which reveal the siphon; and weight above may cause the animal to withdraw its siphon with a sharp squirt of water. Snails, which usually bury themselves just beneath the surface, reveal their locations by small telltale mounds.

Equipment for Collecting

In a specimen bag, it is advisable to carry a few small vials or empty match boxes for tiny shells, and one or more plastic jars or

---

*At the time of this writing, California had declared a temporary moratorium on the collecting of living mollusks and other invertebrate animals, though dead specimens can be collected.

paraffin-treated milk cartons, to be filled with sea water, for very fragile specimens. A prying tool of some sort is useful for probing and overturning heavy stones, and a shovel or trowel should be available for digging. A sieve or wire screen can be very handy for working over sandy deposits. Because iron-mesh sieves corrode rapidly in sea water, it is more desirable to use one that is plastic-coated. Copper mesh should be avoided because copper is extremely toxic to all marine invertebrates. Also handy to carry are a magnifying glass and a pair of tweezers. If the main interest is in microspecimens, taking home a gunnysack of bottom sand or beach litter and screening it leisurely brings richer rewards than on-the-spot hunting for individual small specimens.

For shallow underwater exploration, or diving depths of less than 20 feet (6 m), a diving mask and water goggles are indispensable. A glass-bottomed box can perform the same function. For deep-water dredging, in waters up to 100 feet (30.5 m) or deeper, a boat with small motor, 300 feet of ½″ (1.3 cm) Manila line, periodic weights, and a dredging frame enclosed with fishnet are all necessities.

Preparation and Display
For all live specimens, removing every vestige of the animal's soft parts is the first step. The shellfish should be placed in a pan of water, the water brought to a boil, and the shell kept in the water for a few minutes. Brightly colored, glossy, or enameled shells should never be placed directly into hot water, since abrupt expansion may crack the polished surface. After heating, the soft material can be easily scraped from the interior. The shell then should be set to dry in a cool place. Never should formaldehyde or formalin—not even the buffered product—be used to preserve. Both chemicals eventually break down to formic acid, which dissolves calcareous material, causing the shell to lose color and eventually crumble away. Tiny shells should be soaked for several days in 70-percent alcohol and then dried. It is important to be careful when using chemicals for cleaning, as most will inevitably cause some damage to the shell. Shells are usually displayed phylogenically in boxes or drawers. Accurate labeling is a must.

Nomenclature
In this chapter every effort has been made to use the latest acceptable scientific nomenclature. The scientific name (genus and species) tells much that is known about a particular mollusk from its most distant ancestry to the present. This double name is always italicized. Common, or popular, names have been included as an aid to the beginning malacologist. To scientists the translations may seem unnecessary. It is suggested that the student concentrate on mastering the scientific name.

Range and Scope
About 6000 species of mollusks are known from North American waters. Owing to the limitations of space, this chapter describes 228 species of the more commonly found mollusks of the Pacific Coast, from the Arctic Ocean to the Gulf of California, in habitats from just above the high-tide mark to depths of 100 fathoms (183 m). Land and freshwater mollusks and the smaller marine specimens, those less than ¼ inch (0.64 cm) in height, are excluded.

Illustrations
Virtually every species discussed is illustrated. The illustrations are captioned with the common names of the individual species, and each name is followed by a number indicating the size of the illustration relative to the actual life size of the species.

USEFUL REFERENCES

Abbott, R. T. 1974. *American Seashells*. 2nd ed. New York: Van Nostrand Reinhold Co. 664 pp., 24 plates, text figs.

Jacobson, M. K., and Emerson, W. K. 1971. *Wonders of the World of Shells: Sea, Land and Fresh-water*. New York: Dodd, Mead & Co. 80 pp., illus.

Johnson, M. E., and Snook, H. J. 1927. *Seashore Animals of the Pacific Coast*. New York: Macmillan. 659 pp., 11 color plates. Reprinted by Dover Publications, New York, 1967.

Keen, A. M., and Coan, E. 1974. *Marine Molluscan Genera of Western North America: An Illustrated Key*. Palo Alto, Calif.: Stanford Univ. Press. 208 pp., illus.

Keep, J., and Baily, J. L., Jr. 1935. *West Coast Shells*. Palo Alto, Calif.: Stanford Univ. Press. 350 pp., 334 text figs.

MacGinitie, G. E., and MacGinitie, N. 1968. *Natural History of Marine Animals*. 2nd ed, pp. 327–401, 478–484. New York: McGraw-Hill.

McLean, J. H. 1969. *Marine Shells of Southern California*. Science Series 24, Zoology no. 11. Los Angeles, Calif.: County Museum of Natural History. 104 pp., illus.

Morris, P. A. 1952. *A Field Guide to Shells of the Pacific Coast and Hawaii*. Boston: Houghton Mifflin. 220 pp., 50 plates.

Morton, J. E. 1967. *Molluscs*. 4th ed. London: Hutchinson & Co.

Solem, A. 1974. *The Shell Makers: Introducing Mollusks*. New York: John Wiley & Sons. 289 pp., illus.

Yonge, C. M., and Thompson, T. E. 1976. *Living Marine Molluscs*. London: Collins. 288 pp., illus.

GLOSSARY

Animal   The fleshy part of a shellfish.

Annulated   Marked with rings.

Anterior   The forward end of a bivalve shell.

Aperture   The entrance or opening into a shell.

Apex   The tip of the spire in snail shells.

Apical   At the tip of the growth point.

Auricle   Either of the wings at the hinged border of bivalve mollusks.

Bathymetry   The measurement of depths of water in oceans, lakes, etc.

Beak   Noselike angle near hinge of bivalve shell, where growth started.

Bivalve   A shell with two halves, or valves.

Bridle   Retractable edges.

Byssus   Strong threads that attach some clams to a support.

Calcareous   Composed of lime (calcium carbonate, $CaCO_3$).

Callus   Shelly deposit sometimes covering the umbilicus.

Canal   An extension of gastropod aperture; siphonal canal (posterior); anal canal (anterior).

Cancellate   Marked with numerous crisscrossed lines or ridges.

Carina   Any of various keel-shaped anatomical structures, processes, or ridges.

Carinate   Shaped like the keel of a ship.

Chondrophore   A cavity or process that supports the internal hinge cartilage of the shell of a bivalve mollusk.

Columella   Central pillar around which the spiral whorls form.

Cords   Thickened, rounded raised lines.

Costa (pl. costae)   Rib.

Costate   Ribbed.

Crenate   Having the margin cut into rounded scallops; of gastropods.

Dorsal   Relating to the back.

Eaves   The sites where the insertion plate and the sutural laminae extend from under the tegmentum.

Escutcheon   Depression behind the beak of certain bivalves.

Fimbriation   A fringelike process.

Foot   Muscular extension of animal's body, used in locomotion.

Fusiform   Spindle-shaped.

Globose   Rounded like a globe or ball.

Hinge   Place where the valves of a bivalve join.

Lamella (pl. lamellae)   Thin plates.

Lips   Margins of the aperture, inner and outer.

Lunule   Impressed or modified area in front of the beak on the outside of many bivalve shells.

Mantle   A membranous flap or outer covering of the soft parts of a mollusk; it secretes the shelly material.

Mouth   Opening or aperture of a snail shell.

Nacre   Iridescent inner layer of various mollusk shells.

Nacreous   Pearly.

Nympha (pl. nymphae)   One of the thickened marginal processes behind the beak of many bivalves where the ligament attaches.

Operculum   In some snails, a plate or "trap door" that closes the aperture when the foot is withdrawn.

Ossicle   A small nodular structure.

Ovate   Egg-shaped.

Pallial line   A mark on the inner surface of a bivalve shell more or less parallel with the margin, caused by attachment of the mantle.

Parietal wall   The broader, upper portion of the inner lip.

Periostracum   Thin, noncalcareous outer layer covering many shells, often miscalled epidermis; usually eroded by beach wear.

Plait   One of the flattened folds on the inner wall of some gastropod shells.

Prodissoconch   Rudimentary or embryonic shell of a bivalve mollusk.

Quadrate   Squarish.

Radula   Ribbonlike, rasplike "tongue," or dental apparatus, of most snails and in other classes of mollusks, except bivalves.

Resilium   Internal part of the hinge ligament of a bivalve shell, chitinous or horny in nature.

Rhinophore   One of two tentacles that are chemosensory organs on the back of the head or neck of a mollusk of the order Nudibranchia.

Rostrate   Having a rostrum.

Rostrum   Snout of gastropod when nonretractile; grooved extension of many gastropod shells protecting the siphon.

Sinus   A deep cut.

Siphon   Organ for admitting and expelling water from body cavity of bivalve mollusks.

Spire   Upper whorls from the apex to the body whorl.

Suture   Spiral line of spire, where one whorl touches another.

Teeth   Pointed protuberances at hinge of a bivalve shell; in snails, the toothlike structures are in the aperture.

Test   The external shell or other hard covering of many invertebrates.

Trigonal   Symmetry of three parts; triangular.

Turbinate   Top-shaped; spiral with whorls decreasing rapidly from base to apex.

Turreted   The flattened tops of whorls.

Umbilicus   Hole in center of snail shell at base of body whorl.

Umbo (pl. umbones)   One of the lateral prominences just above the hinge of a bivalve shell.

Varix (pl. varices)   One of the prominent ridges across each whorl of certain gastropods showing a former position of the outer lip of the aperture; often called a "rib."

Volution   A whorl.

Whorl   One of the distinct turns of a snail shell.

Wing   A more or less triangular projection or expansion of the shell of a bivalve, either in the plane of the hinge or extending above it; also called "ear."

CHITONS
Class Polyplacophora

Chitons, or coat-of-mail shells, are sluggish, primitive saltwater mollusks. Typically they are flat and oblong, with eight armorlike overlapping plates held together by a surrounding girdle often bearing spiny hairs. The animal uses the foot beneath to move about on and to hold fast to a rock or other substrate. Most chitons live in shallow water and avoid exposure to sunlight. They are often found on the underside of rocks, where they feed commonly on small algae or less commonly on small invertebrates. There are some 600 species distributed in all but polar seas.

True Chitons
Order Chitonida

SCALY CHITONS
Family Lepidochitonidae

LINED CHITON
Tonicella lineata   103:1

Description
Size, 1 in. (2.5 cm). Shell oval to oblong, elevated, rather acutely angular; upper surfaces smooth; valves glossy. Color on intermediate valves bright reddish black, in oblique lines bordered with white; on end valves the same colors are concentrically arranged.
Habitat
Rocky shores (in Alaska) to onshore waters.
Range
Japan; Aleutian Is. to San Diego, Calif.; uncommon s. of Monterey, common elsewhere.

RED CHITON
Tonicella ruber   104:7

Description
Size, ½–1 in. (1.3–2.5 cm). Shell oblong, moderately elevated, valves rather rounded; girdle covered with minute, elongated scales, not overlapping; 15–18 gill lamellae. Anterior valve with convex front slope; posterior valve with 7–11 slits. Sculptured above with smoothish growth lines. Color yellowish brown, heavily suffused with red all over, or with orange-red marblings; valve interiors bright pink.
Similarities
White Chiton is slightly smaller, with front slope of anterior valve straight to slightly concave.
Habitat
1–80 fathoms (1.8–146.3 m); common.
Range
Arctic Ocean to Monterey, Calif.

Note: The **MOTTLED CHITON**, *Tonicella marmorea,* is similar to the Red Chiton but with the girdle naked, and occurs in Japan and from Alaska to Washington.

GOULD'S BABY CHITON
Lepidochitona dentiens   **103:2**

Description
Size, ½ in. (1.3 cm). Shell oval, slightly elevated; upper surface covered with sharp, microscopic, unaligned granulations; lateral areas sometimes slightly raised, may be bounded by very low frontal rib; girdle very narrow, with minute granules; insertion teeth prominently developed with widely V-shaped bounding slits. Posterior valve with raised apex at center, valve concave behind apex, teeth very sharply angled on side. Color tawny, olivaceous, slaty, or brownish, usually speckled darker.
Habitat
Intertidal rocks; common.
Range
Alaska to s. Calif.

Note: A common intertidal form south of Monterey, California is *Lepidochitona keepiana,* similar to Gould's Baby Chiton with numerous short, narrowly slitted teeth on terminal valves, and with the eaves (projecting edges) narrower, extremely thin, and less open.

HARTWEG'S CHITON
Cyanoplax hartwegii   **103:3**

Description
Size, 1–1½ in. (2.5–3.8 cm). Shell oval, rather flattened; girdle narrow, finely granulated. Anterior valve with 10–11 slits; intermediate valves with lateral areas with distinct, very tiny warts, as on end valves; posterior valve with 9–12 slits, central raised apex followed by convex area. Color same as in Gould's Baby Chiton, with interior intense blue-green.
Similarities
Gould's Baby Chiton has area behind central raised apex concave.
Habitat
Intertidal zone; moderately common.
Range
Wash. to Baja Calif.

CALIFORNIA NUTTALL CHITON
Nuttallina californica   **103:5**

Description
Size, 1 in. (2.5 cm). Shell width about ⅓ length; upper surface finely granulated and with shallow furrow on each side of smooth dorsal ridge; head valve with many radiating ribs; girdle with short, rigid spinelets, appearing mossy. Posterior valve about as wide as long; 8–9 slits; apex far back, extending beyond posterior margin of eaves. Color dark brown to olive-brown with whitish streaks; girdle spinelets chiefly brown, but some intermingled white spines; valve interiors blue-green; animal's foot reddish.
Habitat
Intertidal zone; moderately common. Lives higher up on rocks and more exposed to light than some other species.
Range
Vancouver, B.C., to n. Baja Calif.

MOPALIIDS
Family Mopaliidae

MOSSY CHITON
Mopalia muscosa   103:6

Description
Size, 1–2 in. (2.5–5.1 cm). Shell oval to oblong; valves strong,
surface lusterless; "central areas have close, fine longitudinal riblets
with crenulated or latticed interstices which may diverge"
(Johnson); girdle fringed with mosslike hairs. Very similar to
Hairy Chiton but with posterior tip having a very small, shallow
notch. Color usually dull brown, but may be grayish or blackish-
olive; valve interiors blue-green, rarely pink-stained.
Habitat
Intertidal zone; common.
Range
Aleutian Is. to Baja Calif.

WOODY CHITON
Mopalia lignosa   103:8

Description
Size, 1–2½ in. (2.5–6.4 cm). Shell oblong; girdle solid, straplike
hairs not numerous. Valves sculptured with numerous small
pittings near center; concentric growth lines distinct in smoother
specimens. Color grayish- or blackish-green, rarely marked with
feathery whitish cream and brown; girdle spotted with yellows and
browns; hairs brown; valve interiors white to greenish white.
Habitat
Intertidal zone: moderately common.
Range
Alaska to Baja Calif.

HINDS' CHITON
Mopalia hindsii   103:7

Description
Size, 2–4 in. (5.1–10.2 cm). Shell oblong, flattened, similar to
Hairy Chiton but smoother; girdle fairly wide, rather thin, and
almost naked, with a few short hairs only. Colors as in Hairy
Chiton but girdle brown, valve interiors white, short crimson rays
under beaks.
Habitat
Intertidal zone, often on pilings in estuaries; moderately common.
Range
Alaska to Gulf of Calif.

HAIRY CHITON
Mopalia ciliata   103:4

Description
Size, 1–1½ in. (2.5–3.8 cm). Shell oblong, rather depressed, valves
slightly beaked; surface finely sculptured, lusterless; central areas
with many coarse, wavy, longitudinal riblets and sometimes pitted
between; lateral area coarsely granulated or wrinkled, separated
from central area by prominent raised row of beads; girdle fairly
wide, generally notched at posterior end and covered with curly

brown and white hairs. Anterior valve granulated, 8–9 coarse, raised rays of beads; posterior valve small, with deep slit on each side, broad deep notch at tip end. Color sometimes grayish green mottled with white or grayish black, sometimes black and emerald-green splotches; girdle yellowish- to blackish-brown; valve interiors greenish-white.

Habitat
Intertidal zone; common.

Range
Alaska to Monterey, Calif.

VEILED CHITON
Placiphorella velata   **103:11**

Description
Size, 1–2 in. (2.5–5.1 cm). Shell flat, oval, sometimes almost circular; girdle wide, very broad in front, a few hairs microscopically coated with diamond-shaped scales. Posterior valve with 1–2 slits, sutural plates very large. Color dull olivaceous brown streaked with buff, blue, pink, or chestnut; valve interiors very faintly bluish.

Habitat
Intertidal zone; fairly common.

Range
Monterey, Calif., to Baja Calif.

BLACK KATY CHITON
Katharina tunicata   **103:9**

Description
Size, 2–3 in. (5.1–7.6 cm). Shell distinguished by black, leathery girdle nearly covering valves, with about ⅓ valve width exposed; oval, elevated; valves usually eroded; girdle shiny, naked. Anterior valve densely covered with very small holes. Girdle black, valves gray, valve interiors white; foot reddish to salmon.

Habitat
Intertidal zone.

Range
Aleutian Is. to s. Calif.; very common, especially in n.

CRYPTOPLACIDS
Family Cryptoplacidae

GIANT CHITON
Cryptochiton stelleri   **103:10**

Description
Size, 5–6 in. (12.7–15.2 cm) wide, up to 12 in. (30.5 cm) long. Largest chiton in world. Shell oblong, flattened; valves large, butterfly-shaped; girdle firm, leathery, gritty from minute spicules, entirely covering valves. Anterior valve with 4–7 slits; posterior valve with 1 slit on each side of notch. Girdle reddish- to yellowish-brown, spinelets red; valves white.

Habitat
Intertidal zone, but generally just below low-tide mark.

Range
Japan; Alaska to s. Calif.; common in n.

ISCHNOCHITONS
Family Ischnochitonidae

HEATH'S CHITON
Stenoplax heathiana   **104:1**

Description
Size, 2–3 in. (5.1–7.6 cm). Shell long, narrow; central areas with
irregular fine, longitudinal cuts and diamond-shaped pits near
prominently raised lateral areas; girdle rather narrow, with
sandpaper texture from tiny round scales. Anterior valve with
straight front slope, 10–13 slits; posterior valve with 10–12 slits.
Lateral areas sculptured with 10–12 coarse radial ribs of low,
rough beads. Color drab-greenish, commonly eroded to gray-white;
girdle alternately barred with each valve whitish.
Habitat
Intertidal zone; common.
Range
Coos Bay, Oreg., to Baja Calif.

Note: The **MAGDALENA CHITON,** *Stenoplax magdalenensis,* is
similar to Heath's Chiton, but its size is 2 to 6 inches (5.1–15.2
cm), the sides of the central areas have wavy ribs, the girdle scales
are round and much larger, the anterior valve has a concave front
slope, and the color is drab-greenish. It occurs from Baja
California to the Gulf of California.

CONSPICUOUS CHITON
Stenoplax conspicua   **104:2**

Description
Size, 2–6 in. (5.1–15.2 cm). Shell very similar to Heath's Chiton,
but central areas practically smooth; girdle scales elongate, hard,
densely packed, velvety to touch; anterior valve with very concave
front slope. Color drab-greenish; central areas green-flecked; valve
interiors pinkish and blue.
Habitat
Intertidal zone; moderately common. Found concealed under rocks
and sand by day.
Range
Santa Barbara, Calif., to Gulf of Calif.

REGULAR CHITON
Ischnochiton regularis   **104:3**

Description
Size, 1–1½ in. (2.5–3.8 cm). Shell oblong, smoothish, width ½
length; valves slightly carinate, lateral areas slightly raised; girdle
covered with microscopic, closely packed, low, round scales.
Sculptured on central areas with fine, even radial striations and
very fine longitudinal threads parallel to dorsal ridge; radial
threads on lateral areas. Color uniform slate- or olive-blue; valve
interiors gray-blue.
Habitat
Intertidal zone; moderately common.
Range
Mendocino Co. to s. Calif.

WHITE CHITON
Ischnochiton albus   **104:10**

Description
Size, ½ in. (1.3 cm). Shell oblong, moderately elevated; girdle sandpapery with tiny, closely packed, gravelly scales; 17–19 gill lamellae on each side, beginning about halfway along foot. Anterior valve with front slope straight to slightly concave; posterior valve with 12–13 weak slits. Sculptured above with irregular, concentric growth lines and slight sandpapery effect. Color whitish, cream, or light orange, rarely marked with brown; valve interiors white.
Similarities
Red Chiton is slightly larger and has a convex anterior-valve front slope.
Habitat
Shore to several fathoms, in cold water; common.
Range
Arctic Ocean to San Diego, Calif.

TRELLISED CHITON
Lepidozona pectinulata   **104:5**

Description
Size, 1–1½ in. (2.5–3.8 cm). Shell oval to oblong; girdle closely packed with tiny, convex, split-pea scales. Sculpture heavy; central area with longitudinal and cross ribs, giving netted appearance; raised lateral areas with 4 rows of conspicuous beads; posterior edge of valves serrated with about 20 small, toothlike beads; anterior valve with 20–27 strongly granular ribs. Color dull greenish with yellowish blotches; dark brown area on top of each valve.
Habitat
Intertidal zone to offshore waters; moderately common.
Remarks
Formerly known as *L. californiensis.*
Range
Cen. Calif. to Baja Calif.

MERTEN'S CHITON
Lepidozona mertensi   **104:6**

Description
Size, 1–1½ in. (2.5–3.8 cm). Shell rather oval with angular dorsal ridges and straight sides; girdle covered with tiny, low, smooth, split-pea scales. Sculptured on central areas with strong longitudinal ribs and smaller, lower cross ridges, giving a netted appearance; V-shaped ridged area with 5–6 smooth longitudinal ribs; raised lateral areas smoothish but with a few conspicuous warts; anterior valves with 30 or more radial rows of warts, largest near girdle. Color variable, commonly yellowish with dark reddish brown streaks and blotches; girdle alternately banded yellowish and reddish; valve interiors whitish, rarely pink-tinged.
Habitat
Just off shore.
Range
Aleutian Is. to Baja Calif.; very abundant, especially in n.

SNAILS
Class Gastropoda

Worldwide there are some 40,000 species of gastropods. A large and varied class of mollusks, most have a single, usually spiral, shell. In the limpets the shell is more apparent in the early stages. The nudibranchs have the shell greatly reduced or entirely lacking.

As with all mollusks, the soft mantle secretes the shell, which grows around the outer edge of the mantle and increases in size as the animal grows. Truly marine snails breathe with gills and move by means of an extensible foot. Most snails are quite mobile, traveling about at will over the ocean floor, and are carnivorous, although some snails are scavengers and others are vegetarians.

Various collections of ganglia control the movements of the animal. In attacking their prey most gastropods use a long ribbonlike "tongue," or radula, well-toothed and remotely resembling a rasp. This is protruded through the opening and is worked back and forth like a file over the shell of the snail's prey, such as a clam. Finally, a hole is bored into the interior of the prey, and the snail sucks out the internal soft parts. Most snails are nocturnal.

Ancient Stomach-footed Gastropods
Order Archaeogastropoda

ABALONES
Family Haliotidae

The ovate, greatly flattened shells of this family are cup-shaped and depressed and have a small spire, so that the body whorl constitutes most of the shell. Along the left margin is a row of round to oval perforations, and those holes near the edge are open. The interior is nacreous and often varicolored. Popularly called abalones, the animals live attached to rocks by a broad foot, very much like greatly enlarged limpets. All species are edible; in California they constitute a dietary staple on seafood menus. However, strongly enforced laws protect some species from the collector.

BLACK ABALONE
Haliotis cracherodii   **105:1**

Description
Size, 5–6 in. (12.7–15.2 cm). Shell bluntly oval, rather plump, fairly deep; sculptured with coarse growth lines and only faint reflections of spiral sutures, otherwise smoothish; 5–9 holes, usually open. Some pathological shells may lack holes. Color bluish black to deep greenish black; interior pearly white or silvery with green and pink reflections.
Habitat
Near high-tide mark to 20-ft. (6.1-m) depth, clinging to rocks or in rock crevices close to shore between tidemarks; most abundant of West Coast abalones but not commercially fished in California.
Range
Coos Bay, Oreg., to Baja Calif.

GREEN ABALONE
Haliotis fulgens   105:4

Description
Size, 7–8 in. (17.8–20.3 cm). Shell almost circular, flatly coiled,
moderately deep, aperture wide and flaring; sculptured with 30–40
raised, coarse, spiral, rounded ribs; 5–7 open holes near one
margin, notch at edge. Interior muscle scar near center. Color dull
reddish brown; interior iridescent blue and green, highly polished;
area of muscle scar sparkling blue, with a prismatic luster.
Habitat
Low tide to 60 ft. (18.3 m), on rocky shores.
Remarks
Legal minimum 6¼ in. (15.9 cm); fished commercially in southern
California.
Range
Pt. Conception, Calif., to Baja Calif.

PINK ABALONE
Haliotis corrugata   105:7

Description
Size, 5–7 in. (12.7–17.8 cm). Shell almost circular, fairly deep,
rather highly arched; sculptured with strong wavy corrugations,
edge scalloped; 2–4 tubular holes, open, large, rims sharply
elevated. Color dull green to reddish brown; interior brilliantly
iridescent.
Habitat
Intertidal zone to 180 ft. (54.9 m) along rocky shores.
Remarks
Legal minimum size 6 in. (15.2 cm).
Other name
Corrugated Abalone.
Range
Monterey, Calif., to Baja Calif., abundant in s.

RED ABALONE
Haliotis rufescens   105:8

Description
Size, 7–12 in. (17.8–30.5 cm). Shell oval, flattish; large, thick,
heavy; sculptured with rounded spiraling ribs and low radiating
waves; surface generally rough; 3–4 holes, usually open. Outer rim
projects over interior to form a distinct edge; interior shows large
central muscle scar. Surface dull brick-red, edge of shell bordered
narrowly with red; interior iridescent, highly polished variegated
hues of bluish, greenish (predominating), copper.
Habitat
Near high-tide mark to 540 ft. (164.6 m) on rocky shores.
Remarks
Commercially fished from 20–50 ft. (6.1–15.2 m), especially
between Monterey and Pt. Conception, Calif. A very popular food;
legal minimum size for sportsmen 7 in. (17.5 cm). Polished shells
valued for decorative purposes.
Range
Sunset Bay, Oreg., to Baja Calif.

WHITE ABALONE
Haliotis sorenseni   105:6

Description
Size, 5–8 in. (12.7–20.3 cm), record 10 in. (25.4 cm). Shell thin,
light, oval, highly arched; sculptured regularly with low spiral ribs,
sometimes covered by lush marine growth, especially in tube-
dwelling mollusks; 3–5 holes, highly elevated; interior muscle scar
typically absent or poorly differentiated. Color reddish brown;
interior striking pearl-white with iridescent tints, mainly pink;
outer edge of lip with narrow red border; body yellow or orange;
tentacles green and yellowish, extending beyond shell margin.
Habitat
Deep water, from 15 ft. (rare) to 150 ft. (4.6–45.7 m); most
abundant between 80–100 ft. (24.4–30.5 m).
Range
Pt. Conception, Calif., to Baja Calif.

FLAT ABALONE
Haliotis walallensis   105:2

Description
Size, 4–5 in. (10.2–12.7 cm). Shell elongate, flattened; sculptured
with numerous spiral threads; 4–8 holes, open, edges not elevated.
Color dark brick-red, pale bluish green mottlings.
Habitat
Shallow water near shore to 70 ft. (21.3 m); not plentiful, but
locally abundant.
Remarks
Cannot legally be collected.
Range
B.C. to La Jolla, Calif.

THREADED ABALONE
Haliotis assimilis   105:5

Description
Size, 4–6 in. (10.2–15.2 cm). Shell rather elongate-oval, fairly
deep, moderately thin but strong; noticeable spire; sculptured with
weak, rough corrugations and weak to strong spiral threadlike
ridges; 3–6 holes, open, tubular; no obvious muscle scar inside.
Color mottled brick-red, greenish blue, and gray; interior silvery,
hued with pink and green.
Habitat
Rocky offshore bottoms, 10–120 ft. (3.1–36.6 m) deep.
Range
Pt. Conception, Calif., to Baja Calif.

PINTO ABALONE
Haliotis kamtschatkana   105:3

Description
Size, 4–6 in. (10.2–15.2 cm). Shell elongate (more oval south of Pt.
Sur, California), moderately thin, spire fairly high; sculptured with
rough corrugations, some shells may show weak spiral cords; 3-6
holes, open, edges raised. Color mottled brown; interior iridescent
pale greens and blues.
Habitat
Shallow water.

Remarks
Illegal to collect.
Range
Widely distributed. Japan; Sitka, Alaska, to Pt. Conception, Calif.; abundant in n., uncommon in Calif.

KEYHOLE LIMPETS
Family Fissurellidae

Members of this family have conical shells, oval at the base, and are the only limpets with a hole through the apex or a slit or notch in the shell margin. The shells are sculptured with many radiating lines or ribs. The animals inhabit all except the coldest seas.

ROUGH KEYHOLE LIMPET
Diodora aspera   106:3

Description
Size, 1½–2½ in. (3.8–6.4 cm). Shell base oval, margins finely crenulate, apex elevated slightly less than ⅓ maximum dia.; apical perforation small, oval to nearly circular, flat-sided; sculptured by numerous coarse, radial ribs and weaker concentric threads. Color grayish white to gray, with 12–18 irregularly sized rays of purplish brown or bluish; interior white.
Habitat
Clinging to low-tide rocks; in south to 20 fathoms (36.6 m), often attached to kelp stalks.
Range
Afognak Is., Alaska, to Baja Calif.

GIANT KEYHOLE LIMPET
Megathura crenulata   106:2

Description
Size, 2½–4 in. (6.4–10.2 cm). Shell very large, elongate-oval; rather flat, height about ⅙ length; basal margin finely crenulate, heavy enamel rim around orifice. Sculptured with numerous strong, finely beaded radiating lines and several well-separated concentric growth lines. Color grayish white to light mauve-brown; interior smoothly white. The living animal is larger than shell and can wholly conceal shell beneath an enveloping brownish to black mantle. Has a massive yellow foot.
Habitat
Low-tide rocky areas, breakwaters, to fairly deep offshore water; common.
Range
Monterey Bay, Calif., to Baja Calif.

VOLCANO LIMPET
Fissurella volcano   107:1

Description
Size, ¾–1 in. (1.9–2.5 cm). Shell base oval, somewhat narrower anteriorly, slightly crenulate; conical, well-elevated, height about ⅓–½ length. Apical opening very slightly off center, anteriorly elongate; inner sides deep, flat. Sculptured with numerous rather large, but low and rounded, radial ribs of varying sizes. Color grayish white to dark slate, with distinct radial rays of mauve-pink; interior glossy white, often with callus at apex circumscribed by a

fine pink line. The living animal is brightly colored; mantle with red stripes, foot yellow.

Habitat
Low-tide rocky rubble; very common.

Range
Monterey Bay, Calif., to Baja Calif.

TWO-SPOTTED KEYHOLE LIMPET
Megatebennus bimaculatus   **106:1**

Description
Size, ½–⅝ in. (1.3–1.6 cm). Shell ovate, ends rounded, nearly same size; sides comparatively straight, ends turned slightly upward; apical hole central, elongate-oval, length about ⅓ shell length. Sculpture cancellate, with numerous radial and concentric threads. Color dark gray to light brown with a wide darker ray on each side of hole, also occasionally at each end; interior white to grayish. Living animal much larger than shell, variously colored in reds, yellows, or white.

Habitat
Under stones in low-tide zone; common.

Range
Along entire coast, Alaska to Mexico.

LIMPETS
Family Acmaeidae

The shells of this family are conical, usually somewhat depressed, oval and open at the base, and have no apical opening. At only the earliest growth stage is there a spire, and the interior of most species lacks iridescence. The limpets are shore creatures, commonly clinging to tidal rocks and seaweeds.

GREAT OWL LIMPET
Lottia gigantea   **107:2**

Description
Size, 3–4 in. (7.6–10.2 cm). The largest West Coast limpet. Shell oval, barely arched, apex close to anterior end. Sculptured only by roughened surface; can be highly polished. Color dirty brown to mottled gray and black, often stained with algal green; interior glossy brownish black, with wide dark brown border and a bluish center mark.

Habitat
At or above high-tide line but in range of sea spray; very common.

Range
Neah Bay, Wash., to Baja Calif.; largest specimens occur in s.

WHITE-CAPPED LIMPET
Acmaea mitra   **107:3**

Description
Size, 1 in. (2.5 cm). Shell conical, thick; base circular; shell height slightly less than dia.; apex barely off center anteriorly, bluntly pointed. Often coated with small, knobby algal growths. Color dull white; growths may lend a pinkish to pale greenish overlay.

Habitat
Cold water beyond low-tide line; commonly washed ashore.

Range
Aleutian Is., Alaska to Baja Calif.

SHIELD LIMPET
Collisella pelta   107:4

Description
Size, 1–1½ in. (2.5–3.8 cm). Shell elliptical in outline, edge slightly
wavy; strong, rugged; apex central, to ⅙ shell-length off center
toward anterior end, moderately high, pointed. Sculptured with
some 25 axial, weakly developed, blunt, radial ribs. Color whitish
cream to gray background bearing strong black, often intertwining,
stripes; interior faint bluish white, sometimes with small, central
brown spot; inner border edged with alternating black and cream
bars; juveniles shiny black.
Habitat
Intertidal rock dweller; common.
Range
Aleutian Is., Alaska, to Mexico.

TEST'S LIMPET
Collisella conus   107:6

Description
Size, ¾ in. (1.9 cm). Shell elliptical in outline, low; apex off center
anteriorly; interior center smooth, evenly glossed; sculptured with
distinct, widely spaced, radial ribs. Color dirty greenish gray;
interior highly iridescent, often showing evenly colored brown
center.
Similarities
Rough Limpet is larger; interior has rough center and smeary
brown center stain.
Habitat
Clinging to rocks above high-tide line but in range of spray;
usually found with Rough Limpet and Fingered Limpet.
Remarks
Often confused with Rough Limpet.
Range
Pt. Conception, Calif., to Baja Calif., abundant s. of La Jolla,
Calif.

FINGERED LIMPET
Collisella digitalis   107:5

Description
Size, 1¼ in. (3.2 cm). Shell elliptical; apex moderately high, ⅓ in.
from anterior end, slightly hooked forward, surface between hook
and anterior end concave. Sculptured with 15–25 moderately
developed, coarse, radiating ridges on posterior ⅔, giving shell a
slightly wavy margin; smooth on anterior ⅓. Color gray, mottled
with stripes and blotches of brownish black and white; interior
faintly bluish white, margined with a solid or broken narrow band
of brownish black; large central dark brown spot, usually even.
Similarities
Rough Limpet lacks forward apical hook and has dull brown
internal spot.
Habitat
Rocks close to or above high-tide mark; the most abundant western
limpet.
Range
Aleutian Is. to Baja Calif.

FILE LIMPET
Collisella limatula   107:7

Description
Size, 1–1¾ in. (2.5–4.5 cm). Shell oval to nearly round in outline,
moderately arched to quite flat; apex off center anteriorly.
Sculptured with numerous closely set, radiating, sharply rough
riblets, actually rows of tiny beads. Color light brown to greenish
black or almost black; interior glossy bluish white with solid
brownish black narrow band rimming shell; central spot brown,
usually weak or absent. Juveniles show blue tint inside; also,
occasional albinos may be creamy-brown to tan outside.
Habitat
Low-tide rocks; abundant.
Range
Newport, Oreg., to Baja Calif.

ROUGH LIMPET
Collisella scabra   107:8

Description
Size, 1¼ in. (3.2 cm). Shell elliptical, somewhat variable; apex in
most specimens moderately elevated, ⅓ in from anterior end, not
hooked, peak sharp. Sculptured with 15–25 strong, coarse,
radiating, rounded ridges, giving margin a strong crenulation.
Color light gray; interior whitish, irregularly stained blackish
brown in center; margin between serrations blackish to purplish
brown.
Similarities
Test's Limpet is smaller; interior has glossy center, stained evenly
with brown.
Habitat
Clinging to rocks close to or above high-tide mark, but spray-
swept; quite abundant.
Range
Cape Arago, Oreg., to Baja Calif.

FENESTRATE LIMPET
Notoacmea fenestrata   107:9

Description
Size, 1–1½ in. (2.5–3.8 cm). Shell nearly circular in outline,
narrowing slightly anteriorly; rather high; smoothish. A northern
subspecies, *N. f. cribraria,* Alaska to northern California, has
exterior plain dark gray, interior variously hued with glossy
chocolate-brown with narrow solid black border. A southern
subspecies, *N. f. fenestrata,* Pt. Conception, California, and south,
has exterior patterned with regular dottings of cream on a gray-
green background, interior with a small brown apical spot in a
bluish area and margin bordered with brown.
Habitat
The only western limpet living on smooth boulders set in loose
sand; common.
Range
Alaska to Baja Calif.

UNSTABLE LIMPET
Notoacmea instabilis   **107:11**

Description
Size, 1–1¼ in. (2.5–3.2 cm). Shell elongate-oval; rims crescent-shaped in side view, shell rocks forward and back on a flat surface; apex off center anteriorly, rather high. Sculptured with fine radiating lines. Color dull brown, commonly nearly black; interior bluish white, narrowly but solidly margined with brown, center faintly brown-stained.
Habitat
Lives on the stems of large seaweed.
Range
Alaska to San Diego, Calif.

SEAWEED LIMPET
Notoacmea insessa   **107:10**

Description
Size, ½–¾ in. (1.3–1.9 cm). Shell caplike, narrowly elliptical; smooth, shiny; apex high, bluntly crested, parallel sides quite steep. Color uniform light brown inside and out.
Habitat
Clinging to stalks or holdfasts of large seaweeds, and on algae; abundant.
Range
Alaska to Baja Calif.

TOP SHELLS
Family Trochidae

The shells of these herbivorous snails are generally cone-shaped; they have a pearly interior. The snails have a thin, corneous operculum. They are widely distributed in warm seas, generally in shallow water among seaweeds.

RINGED TOP SHELL
Calliostoma annulatum   **106:4**

Description
Size, 1–1¼ in. (2.5–3.2 cm). Shell lightweight, sharply conical, not quite as wide as long; 8 whorls, rather flattened. Sculptured by numerous spiral ridges of tiny distinct beads, 5–9 rows on each spire whorl. Color yellowish brown dotted with darker brown on sculpturing; sutures set off by narrow zone of deep purple; base of columella pink to purple.
Habitat
Moderately shallow water, on seaweeds; dredged offshore; occasionally washed ashore.
Range
Alaska to Baja Calif.

CHANNELED TOP SHELL
Calliostoma canaliculatum   **106:5**

Description
Size, 1–1½ in. (2.5–3.8 cm). Shell thin, strong; base almost flat; about 7 whorls, sides flat, periphery of body whorl sharp; apex sharply pointed, no umbilicus. Sculptured by sharp, prominent

spiral cords, slightly beaded, especially on upper volutions. Color light tan to deep brown; nuclear whorls white; interior pearly; columella shows small blue patch.

Habitat
Clinging to floating offshore seaweeds; moderately common.

Range
Alaska to San Diego, Calif.

NORRIS TOP SHELL
Norrisia norrisi   106:6

Description
Size, 1½–2 in. (3.8–5.1 cm). Shell slightly wider than long, rather heavy and solid, smooth, outline orbicular; 3–4 whorls, last constituting most of shell; spire very low, flat; aperture thickened within; outer lip thin, sharp; umbilicus ovate, well-defined, very deep; operculum multispiral, horny, shaggy with spiral rows of dense bristles. Color glossy yellowish brown; aperture pearly, lip edge black; umbilicus greenish blue on columella side, bordered on other side by dark brown; animal tinged with pale red.

Habitat
Close to shore in kelp beds; moderately common.

Range
Monterey, Calif., to Baja Calif.

BLACK TOP SHELL
Tegula funebralis   106:7

Description
Size, 1–1½ in. (2.5–3.8 cm). Shell heavy, robust, top-shaped; base rounded; 4 whorls, faint puckered wrinkle at base of suture; umbilicus closed, or a slight dimple; 2 small nodules at columella. Sculpture smoothish, rarely with weak spiral cords; coarse growth lines in larger specimens; early whorls usually eroded. Periostracum strong. Operculum small, thin, circular, corneous. Color deep purplish black, substratum brilliantly pearly; periostracum black, columella pearly.

Similarities
Speckled Top Shell is lighter, grayish green with purplish zigzag axial stripes, and has coarser surface.

Habitat
Intertidal zone, on or among rocks; extremely abundant.

Range
Vancouver, B.C., to Baja Calif.

SPECKLED TOP SHELL
Tegula gallina   106:8

Description
Size, 1–1½ in. (2.5–3.8 cm). Shell solid, top-shaped; apex blunt, commonly eroded; 5–6 whorls; coarse surface roughening. Operculum small, thin, corneous, circular. Color dark purplish black longitudinally striped with gray; apex orangey; outer lip edged in black; interior pearly.

Similarities
Black Top Shell.

Habitat
Among littoral zone rocks; common.

Range
Santa Barbara, Calif., to Gulf of Calif.

BANDED TOP SHELL
Tegula eiseni   **106:9**

Description
Size, ¾–1 in. (1.9–2.5 cm). Shell solid, heavy, variably shaped but usually about as broad as it is high; 4–5 rounded whorls; both whorls and spire convex; outer lip sharp, thickened within, lower part of lip shows about 8 tiny nodules; umbilicus round, fairly narrow, very deep; numerous spinal cords, often broken into elongate beads. Color brownish tan or rusty brown, with black flecks; interior pearly.

Habitat
Littoral zone among rocks; moderately common.

Range
Santa Barbara, Calif., to Baja Calif.; Mexico.

TURBAN SHELLS
Family Turbinidae

Shells of this herbivorous family are turbinate, or top-shaped, and generally solid and heavy, with the aperture usually closed by a thick, calcareous operculum. There is no umbilicus. The exterior may be smooth, ridged, or spiny; the interior is nacreous. Commonly used for ornamental purposes, the turbans inhabit all warm-water seas.

WAVY TURBAN
Astraea undosa   **106:12**

Description
Size, 2–4 in. (5.1–10.2 cm). Shell dia. same as height, base concave and rather flat, apex sharp; 6–8 whorls, steeply sloping; periphery strong, wavy, overhanging. Each spiral twist sculptured with a series of wavy vertical ridges; suture followed by prominent knobby band, like a twisted cord; base bears 3 small, indistinct, beaded spiral lines. Periostracum fuzzy. Operculum calcareous, thickened by 3 strong, prickly ridges with concentric grooves. Color pale brown; periostracum dark brown; substratum and interior brilliantly pearly.

Habitat
Shallow water; common.

Range
Ventura, Calif., to Baja Calif.

Note: The **RED TURBAN,** *Astraea gibberosa,* is similar to the Wavy Turban but is smaller, has more vertical folds, and is red in color. It occurs from Dellwynn, British Columbia, to Baja California.

PHEASANT SHELLS
Family Phasianellidae

BANDED PHEASANT SHELL
Tricolia compta   **106:10**

Description
Size, ¼–⅓ in. (0.64–0.85 cm). Shell moderately high-spired, smooth. Sculptured with numerous spiral downward-slanting lines, not parallel with the suture; also wider zigzag axial bands. Periostracum thin, translucent. Operculum calcareous. Spiral lines

blackish green, red, brown, or purplish; periostracum grayish green.
Habitat
Clinging to eelgrasses in shallow bays; very abundant; often washed ashore.
Range
San Francisco, Calif., to Baja Calif.

Middle Stomach-footed Gastropods
Order Mesogastropoda

PERIWINKLES
Family Littorinidae

The periwinkles are a large family of shore-dwelling snails remarkably resistant to drying out. While some inhabit brackish water and many salt water, several species are amphibious in that they can survive drought and the hazards of land life at the very edge of the spring-tide saltwater splash line. Eggs are hatched inside or outside the body. The shell is usually sturdy, spiral, turbinate or globular, and it has few whorls and no umbilicus. Worldwide in distribution, periwinkles are found clinging to rocks between the tide marks and occasionally well above high-tide limits.

CHECKERED PERIWINKLE
Littorina scutulata   **108:1**

Description
Size, ½ in. (1.3 cm). Shell moderately tall, slender, apex sharp; 4 whorls; smooth, semiglossy. Color varies, commonly greenish gray checked and spotted with white; aperture interior whitish brown; columella white.
Similarities
Flat Periwinkle has eroded flattened area on body whorl beside columella.
Habitat
Littoral zone, among rocks; common.
Range
Alaska to Baja Calif.

FLAT PERIWINKLE
Littorina planaxis   **108:2**

Description
Size, ½–¾ in. (1.3–1.9 cm). Shell moderately stout, apex bluntly pointed; 3–4 whorls; aperture large, outer lip sharp; columella flattened on body whorl. Sculptured with smoothish spiral lines; surface usually badly eroded. Color pale chocolate to grayish brown, often with spots and flecks of bluish white; aperture interior chocolate-brown with white spiral band at bottom. Juveniles may be banded with white.
Similarities
Checkered Periwinkle.
Habitat
Littoral zone, on rocks near high-tide line; common.
Range
Puget Sound, Wash., to Baja Calif.

CHINK SHELLS
Family Lacunidae

The stoutly conical chink shells are smooth, rather fragile periwinkles with thin shells, a half-moon aperture, and a shelflike columella alongside which is the characteristic lengthened groove, or chinklike umbilicus. The periostracum is smooth, fairly thin, and light brown; the operculum is horny. These usually cold-water animals are frequently dredged in areas of kelp growth.

CARINATE CHINK SHELL
Lacuna carinata

106:11

Description
Size, ⅜–½ in. (0.95–1.27 cm). Shell moderately fragile, rather squat; 3–5 whorls, body whorl sloping and expanding to large semilunar aperture with thin outer lip; columellar chink wide, long, forming groove part way up columella. Sculpture smooth. Periostracum thin, always covers shell, smooth. Color chalky white; periostracum yellowish brown; chink white.

Habitat
Shallow water, clinging to kelp weeds; common.

Range
Alaska, to Monterey, Calif.

SCREW SHELLS
Family Turritellidae

The turritellas are greatly elongated, many-whorled shells resembling a turret in shape. This is a large family living generally in cool to tropical waters. The shells are very commonly found as fossils.

COOPER'S TURRET
Turritella cooperi

108:4

Description
Size, 1–2 in. (2.5–5.1 cm). Shell spinelike, handsome; 17–20 whorls, slightly convex (rather flattish); base concave; aperture round, outer lip thin and sharp; columella fairly fragile. Whorls sculptured with 2–3 small, spiral ridges and usually with several much finer variously sized spiral threads. Color yellowish to orange, often longitudinally streaked with dark brown or chocolate-brown.

Habitat
Sandy bottoms just offshore; moderately common.

Range
Monterey, Calif., to Baja Calif.

WORM SHELLS
Family Vermetidae

This small family comprises mollusks which when young have regularly spiraled shells and can creep about freely. Later they attach themselves firmly to some object and develop separate whorls that are often irregularly bent, like a worm tube.

SCALED WORM SHELL
Serpulorbis squamigerus   108:3

Description
Size, ¼–½ in. (0.6–1.3 cm dia.). Shells in large compact masses; tubes circular, last part usually erect for ½ in. (1.2 cm) and smoothish; aperture large, roundish. Sculptured with numerous rough longitudinal ridges, minutely scaled. Color gray to pinkish gray.
Habitat
Colonial masses on wharf pilings, attached to rocks below low-tide line; very common.
Range
Forrester Is., Alaska, to Peru.

HORN SHELLS
Family Potamididae

The horn shells are intertidal mud-lovers distinguished by their long shape, numerous whorls, and oblique opening. Axial rib sculpturing is most prominent on the early whorls. The aperture, with its flaring outer lip, is closed by a horny, thin, lightly spired operculum having a central nucleus. Found on mud flats.

CALIFORNIA HORN SHELL
Cerithidea californica   108:6

Description
Size, 1–1¼ in. (2.5–3.2 cm). Shell stoutly spikelike; 10–11 rounded whorls; aperture circular. Sculptured with 12–18 strong axial ribs per whorl plus weak spiral threads. Operculum round, chitinous. Color dark brown, sometimes almost black, with some individuals showing a narrow paler band; also 1–2 yellowish white, swollen varices near the base.
Habitat
Colonial on mud flats when tide is out; very common.
Range
Bolinas Bay, Calif., to Baja Calif.

WENTLETRAPS
Family Epitoniidae

These easily recognized snails have high spires consisting of many ribbed whorls increasing in size from top to bottom, usually white and polished. Known also as "staircase shells."

WROBLEWSKI'S WENTLETRAP
Opalia wroblewskii   108:5

Description
Size, 1–1¼ in. (2.5–3.2 cm). Shell solid but slender; 7–8 whorls, slightly rounded; apex blunt; aperture round, outer lip thickened.

Each whorl sculptured with 6–8 low axial ribs or varices, blunt, often indistinct, tending to fade out at center of whorl; base bounded by a strong, smooth, low spiral cord. Color grayish- to yellowish-white; usually appears beach-worn.

Habitat
Shallow to moderately deep water; fairly common.

Range
Forrester Is., Alaska, to San Diego, Calif.

BOAT SHELLS
Family Calyptraeidae

These stationary, limpetlike snails are distinguished by their oval shape and horizontal cuplike platform inside, which helps to protect the animal in its shell. They occur in all seas, living attached to hard objects such as rocks and other shells in water of shallow to moderate depth.

STRIATE CUP-AND-SAUCER SHELL
Crucibulum spinosum
                                                                        108:7

Description
Size, ¾–1½ in. (1.9–3.8 cm). Size differs by sex; males ¾–1 in. (1.8–2.5 cm), females 1½ in. (3.75 cm). Shell height variable, ⅓–¾ as high as it is broad; base nearly circular. Some specimens well arched, others flattish. Apex low, nearly central; inner platform rather large, cuplike, laterally compressed, delicate, attached to one side. Sculptured, except for a smoothish apical area, with radiating, fine, somewhat wrinkled ridges studded with short spines, sometimes erect, tubular. Color yellowish white, often flecked with brown; interior glossy chestnut-brown, sometimes with light radial rays; cup white. Occasional albino shells are found.

Habitat
Low water to 15 fathoms (27.4 m); very common in south, uncommon in north.

Range
Monterey, Calif., to Chile.

HALF SLIPPER SHELL
Crepipatella lingulata
                                                                        108:8

Description
Size, ½–¾ in. (1.3–1.9 cm). Shell thin, nearly circular, low; apex near margin. Interior characterized by shallow deck attached along only one side of shell; middle of deck may show a weakly raised ridge. Sculptured with wrinkles. Color brownish; interior tannish to mauve-white, glossy.

Habitat
Low-water rocks or on living shells; very common.

Range
Bering Sea to Panama.

ONYX SLIPPER SHELL
Crepidula onyx
                                                                        108:10

Description
Size, 1–2 in. (2.5–5.1 cm). Shell fairly thick, cuplike, well arched; oval margin; apex small, close to margin, turned to one side; interior platform large, slightly concave, free edge sinuate (wavy).

Sculpture smooth, glossy. Color grayish brown, often with dull reddish rays; interior rich chocolate-brown, highly polished; deck white.
Habitat
Shallow estuaries to 50 fathoms (91.4 m) attached to rocks or other shells or stacked one on the other; very common.
Range
Monterey, Calif., to Peru.

HOOKED SLIPPER SHELL
Crepidula adunca   108:9

Description
Size, ¾ in. (1.9 cm). Shell well arched, smooth; apex pronounced, sharp, curved, often hooked; interior platform large, edge smoothly rounded. Color brown; inner deck white.
Habitat
Just beyond low-tide mark attached to dead shells, stones; not very common.
Range
B.C. to Baja Calif.

MINIATURE COWRIES
Family Triviidae

Within this family, the subfamily Triviinae resembles miniature cowries. The shells are sculptured by stout riblets, or wrinkles, which circumscribe the shell from the slitlike aperture to the mid-back.

APPLE SEED
Erato vitellina   108:11

Description
Size, ½ in. (1.3 cm). Shell cowrylike, somewhat pear-shaped, not unlike a beach-worn dove shell; aperture long, narrow; outer lip thickened, lower ¾ curled inward slightly, bearing 7–10 tiny teeth; inner lip bearing distinct plaits; columella arched, bearing 5–8 small teeth. Sculpture smooth. Color dark reddish brown; aperture white, teeth whitish.
Habitat
Fairly shallow water, under stones and in crevices; moderately common but nowhere abundant.
Range
Bodega Bay, Calif., to Baja Calif.

CALIFORNIA COFFEE BEAN
Trivia californiana   108:13

Description
Size, ⅓–½ in. (0.9–1.3 cm). Shell rotund, broad; slightly depressed groove on dorsal midline. Sculptured with fairly coarse riblets over entire shell; outer lip has about 15. Color mauve; groove white; animal body bright red.
Habitat
Littoral zone to 40 fathoms (73.2 m), often washed ashore with seaweeds; common.
Range
Crescent City, Calif., to Acapulco, Mexico.

COWRIES
Family Cypraeidae

The cowries comprise a large family of brilliantly polished, brightly colored shells, primarily tropical, which have always been collectors' favorites.

CHESTNUT COWRIE
Cypraea spadicea

108:12

Description
Size, 1–2 in. (2.5–5.1 cm). Shell oval or egg-shaped, height ½ length; spire covered by body whorl in adults. Aperture long, narrow, running length of underside; notched at each end; lined on each side with 20–23 teeth. Sculpture smooth, hard, glossy, enamellike. Base white; sides bluish to mauve-gray.
Habitat
Low tide to 25 fathoms (45.7 m), among seaweeds; moderately common during certain seasons.
Range
Monterey, Calif., to Baja Calif.

MOON SHELLS
Family Naticidae

These active, carnivorous snails, found in all seas, have globular, sometimes depressed, smooth, and frequently polished shells. The aperture is sharp-edged, and there is a very large foot which, in some species, covers the whole shell when extended.

BABY'S EAR SHELL
Sinum scopulosum

108:14

Description
Size, 1–1 ¼ in. (2.5–3.2 cm). Shell quite flat, spire obsolete; 2–4 whorls, last extremely large, very large aperture. Sculptured with numerous finely incised spiral lines and grooves. Periostracum thin, translucent. Color chalky white to yellowish white; periostracum yellowish.
Habitat
Shallow water, sandy beaches; moderately common.
Range
Monterey, Calif., to Baja Calif.

Note: A regional associate, the **SMALL BABY'S EAR SHELL,** *Sinum debile,* occurring from Catalina Island to the Gulf of California, is similar to the Baby's Ear Shell but has a flatter spire and less inflated whorls. The **STEARN'S EAR SHELL,** *Lamellaria stearnsii,* which occurs from Puget Sound, Washington, to San Diego, California, is smaller and smoother than the Baby's Ear Shell and is globular in shape.

RECLUZ'S MOON SHELL
Polinices reclusianus

108:15

Description
Size, 1½–2½ in. (3.8–6.4 cm). Shell strong, solid, very heavy for size; 4–5 whorls, spire moderately to quite well elevated to a blunt point; umbilicus may or may not be covered by large tonguelike shelly callus. Shape of shell and callus development quite variable.

Aperture large; inner top has reinforcing callus. Sculpture smooth, except for shoulder wrinkles. Operculum horny, translucent. Color semiglossy grayish white, with brownish or greenish stains; juveniles pale blue; aperture callus white or brownish, inner callus white.

Habitat
Shallow water to 25 fathoms (45.7 m); common.

Range
Mugu Lagoon, Calif., to Mazatlán, Mexico.

FROG SHELLS
Family Bursidae

The frog shells are ovate, laterally somewhat compressed, and bear two rows of continuous varices, one row on each side.

CALIFORNIA FROG SHELL
Bursa californica   109:1

Description
Size, 3–5 in. (7.6–12.7 cm). Shell stout, rugged, moderately heavy, flattish in appearance; about 6 whorls, last having 4–5 large nodules; spire with only 1 nodule, giving angular, bulgy look to shell; aperture rather large, lip crenulate, posterior canal about size of anterior (siphonal) canal. Each whorl sculptured with 2 knobby varices opposite each other, between which are 2 stout spines. Color yellowish brown; aperture white.

Habitat
Offshore waters, occasionally washed ashore; common, especially in south.

Range
Channel Is., Calif., to Gulf of Calif.

New Stomach-footed Gastropods
Order Neogastropoda

ROCK SHELLS
Family Muricidae

The shells of these active, carnivorous snails are strong and thick, and usually spiny. They live in moderately shallow water in all seas, though most abundant in the tropics, inhabiting rocky or pebbly bottoms.

BELCHER'S CHORUS SHELL
Forreria belcheri   109:2

Description
Size, 3–6 in. (7.6–15.2 cm). One of the largest univalves. Shell pear-shaped, solid; 6–7 rather squarish whorls, body whorl very large; spire turreted, rather short. Aperture large; outer lip thin and sharp with toothlike projection near bottom; inner lip strongly reflected. Canal moderately long, curved, open; prominent to left of narrow, rather shallow umbilicus. Sculptured with 8–10 prominent, pointed, scalelike spines on shoulder of each whorl,

although general surface is smoothish; spines are the tops of varices which, flattened, continue down the sides of the whorls. Color yellowish white, brown-streaked; aperture snow-white.

Habitat
Intertidal areas near oyster beds; common.

Range
Ventura Co., Calif., to Baja Calif.

THREE-WINGED ROCK SHELL
Pteropurpura trialata

109:3

Description
Size, 2–3 in. (5.1–7.6 cm). Shell highly ornate; 5–6 turreted whorls, sutures indistinct, tops of shoulders slightly excavated; aperture small, canal rather short and tightly closed. Shell sculptured by 3 large bladelike varices; whorl surface between varices smoothish, with or without a single rounded tubercle; sometimes bearing 2–5 weak spiral threads. Anterior face of each varix crowded with fine, axial fimbriations. Color grayish, light brown, or dark brown, or with white spiral bands.

Similarities
Frill-winged Rock Shell.

Habitat
Among rocks at moderate depths; common.

Range
Bodega Bay, Calif., to Baja Calif.

FRILL-WINGED ROCK SHELL
Pteropurpura macroptera

109:4

Description
Size, 2–3 in. (5.1–7.6 cm). Similar to Three-winged Rock Shell, but with 3 thick, frondlike, frilly varices at shoulder; also with fine horizontal ribs and a smaller varix in each space between main varices. Inner lip of aperture partially reflected; canal curved to right and tightly closed. Color light yellowish brown.

Habitat
Moderately shallow water, among rocks; common.

Remarks
In 1964 Emerson showed that *Murex petri* and *Murex carpenteri* are identical and should be given the earlier name used above.

Range
Pt. Conception, Calif., to Baja Calif.

FESTIVE ROCK SHELL
Pteropurpura festiva

109:5

Description
Size, 1½–2 in. (3.8–5.1 cm). Shell stout, decorative; spire high, 6–7 whorls; aperture largish, outer lip wide, inner lip reflected; canal sharp, partially closed. Sculptured with 3 prominent varices per whorl, each thin, curled backward, surface fimbriated; between varices 1 very large, rounded nodule. Color light yellowish brown, with numerous fine, dark, spiral lines.

Habitat
Shore rocks, mud flats, to 75 fathoms (137.2 m); very common.

Range
Morro Bay, Calif., to Baja Calif.

NUTTALL'S HORNMOUTH

109:14

Ceratostoma nuttalli

Description
Size, 1½–2 in. (3.8–5.1 cm). Shell rugged, about 5 whorls, sutures indistinct; aperture large, slender spine at base projecting outward (lacking in juveniles), inner lip reflected on body whorl. Sculptured with 3 major large, thin, bladelike varices and 1 prominent, noduled rib in each space between varices. Color mottled yellowish gray or brownish, sometimes spirally banded.

Similarities
Poulson's Rock Shell has elongated aperture, siphonal canal twisted, 3–4 teeth inside outer lip.

Habitat
Littoral zone to rather deep water; common.

Range
Monterey, Calif., to Baja Calif.

Note: The **LEAFY HORNMOUTH**, *Ceratostoma foliatum,* is similar to Nuttall's Hornmouth but has more whorls, lacks the spikelike horn at the base, and has variously sized spiral cords in the spaces between the varices. It occurs from Alaska to Point Conception, Calif.

CIRCLED ROCK SHELL

109:9

Ocenebra circumtexta

Description
Size, ¾–1 in. (1.9–2.5 cm). Shell stout, spire ⅓ shell length; 4–5 rounded whorls; aperture rather large, outer lip thickened and feebly toothed inside, canal short and open. Sculptured with deeply impressed, strong, rough spiral ridges, 15 on body whorl, 6 on spire whorls; 7–9 wide, low, rounded axial ribs on each whorl. Ridges viewed under magnification present arched, crowded, raised, axial lamellae. Color grayish white, strongly banded with chocolate-brown.

Habitat
On rocks from low-tide mark to 30 fathoms (54.9 m); very abundant.

Range
Moss Beach and Monterey, Calif., to Baja Calif.

POULSON'S ROCK SHELL

109:10

Roperia poulsoni

Description
Size, 1½–2 in. (3.8–5.1 cm). Shell rugged, sturdy, elongate, high-spired; about 6 whorls; aperture elongated into narrowly open, short, twisted siphonal canal; inner lip reflected on last whorl, 3–4 teeth well inside outer lip. Sculptured with 8–9 varices; 4 or 5 raised spiral cords form nodules on the ribs. Periostracum thin, smoothish. Color grayish to glossy white, with numerous fine dark brown to yellow-brown spiral lines; periostracum grayish or brownish; aperture white.

Habitat
On intertidal to low-tide rocks, wharf pilings; extremely common.

Range
Santa Barbara, Calif., to Baja Calif.

GEM ROCK SHELL
Maxwellia gemma

Description
Size, 1–1¼ in. (2.5–3.2 cm). Shell heavy, moderately high-spired, pocked with crude squarish pits; 6–7 whorls, sutures indistinct; aperture small, canal tightly closed. Sculptured on each whorl with 6 varices, swollen, rounded, smooth, interconnected in middle area; in area of suture and near base of shell, varix is thin, elevated, curled back, and may bear 1 or more small spines. There are also several low spiral cords which become somewhat more prominent on the smoother midsection of each whorl. Coloı white, with dark brown cross-stripes.

Habitat
Shallow water, under protective rubble and worm-tube masses in rocky areas; fairly common.

Range
Santa Barbara, Calif., to Baja Calif.

Note: The **SANTA ROSA ROCK**, *Maxwellia santarosana*, size, 1½ inch (3.8 cm), is similar to the Gem Rock Shell but is low-spired and has smooth intervarical spaces. It is found in the same range but is uncommon.

ANGULAR UNICORN
Acanthina spirata

Description
Size, 1–1½ in. (2.5–3.8 cm). Shell somewhat elongate, solid, rather low-spired; 5–6 whorls, sharp-edged at the slight, sloping shoulders, producing a more or less turreted spire; sutures indistinct; aperture rather large; canal short, broadly open, 1 hornlike tooth nearby. Sculptured with numerous poorly developed spiral threads, otherwise smoothish. Color dark bluish gray with numerous rows of small reddish brown dots; aperture bluish white.

Habitat
Above high-tide mark and on mussel beds; common.

Range
Puget Sound, Wash., to Baja Calif.

CHECKERED UNICORN
Acanthina paucilirata

Description
Size, ⅓–½ in. (0.85–1.27 cm). Shell stoutly fusiform; about 4 whorls, tops slightly concave, shoulders sloping; aperture dentate, teeth quite prominent on outer lip; spine at base of outer lip small, needlelike; canal short. Sculptured with 3–5 weak spiral ribs, upper rib marking edge of shoulder; ribs smooth, raised on later whorls, early whorls cancellate. Color brownish gray or cream-white with about 6 spiral rows of small squares of blackish brown; aperture purplish.

Habitat
Above high-tide mark; common.

Range
Monterey Bay, Calif., to Baja Calif.

FRILLED DOGWINKLE
Nucella lamellosa   109:11

Description
Size, 1½–5 in. (3.8–12.7 cm). Shell solid, rugged, sturdy, quite
variable in form; spire usually fairly high, pointed; about 7 whorls;
columella nearly vertical and straight, umbilicus small, canal short.
Sculpture variable; in north with up to 15 strong frilly varices; in
south fairly smooth. The smoothish or variously developed foliated
axial ribs are sometimes spinose. Color variable; white, grayish,
cream, orange; sometimes spirally banded.
Habitat
Shallow water, among rocks; very common.
Range
Bering Strait to Santa Cruz, Calif.

CHANNELED DOGWINKLE
Nucella canaliculata   109:12

Description
Size, 1 in. (2.5 cm). Shell moderately globose, spire well elevated,
apex sharp; about 5 whorls, sutures rather distinct; aperture rather
large, inner lip flattened, slightly twisted at base; canal short;
columella arched, flattened below. Sculptured with 14–16 low, flat-
topped, closely spaced, spiral cords on body whorl, interstices deep
and vertically checked. Color white or yellowish brown, often
spirally banded; aperture stained bright yellow.
Habitat
On rocks and mussel beds; moderately common.
Range
Aleutian Is. to Cayucos, Calif.

EMARGINATE DOGWINKLE
Nucella emarginata   109:13

Description
Size, 1 in. (2.5 cm). Shell extremely variable, some short-spired
and squat, others more elongate and higher-spired; usually 5
whorls, body whorl constituting most of shell; aperture large;
columella strongly arched, flattened, slightly concave below.
Sculpture variable, but usually of coarse spiral ridges, generally
alternately large and small and often scaled or strongly noduled.
Color yellow-gray to rusty brown, often with narrow darker spiral
bands; aperture reddish- to chestnut-brown; columella light- to
chestnut-brown.
Habitat
Among rocky shores; very common.
Range
Bering Sea to Baja Calif., Mexico.

DOVE SHELLS
Family Columbellidae

These small, fusiform shells have an outer lip commonly thickened in the middle area. They are glossy and often colorful, primarily inhabiting warm seas.

KEELED DOVE SHELL
Alia carinata   **110:1**

Description
Size, ¼ in. (0.6 cm). Shell strongly keeled at shoulder, 5–7 whorls; body whorl bearing keel, i.e., shoulder usually strongly swollen; spire flat-sided, apex very sharp; aperture elongate, outer lip thickened, crooked. Sculpture smooth, but about 12 incised spiral lines on exterior of canal; on inside of outer lip about 12 small spiral threads or toothlike structures. Color pale brown or variegated with white, yellow, orange, and brown.
Habitat
Shallow water, 7–30 fathoms (12.8–54.9 m), among rocks, stems of seaweeds; abundant.
Range
Alaska to Baja Calif.

JOSEPH'S COAT AMPHISSA
Amphissa versicolor   **110:3**

Description
Size, ¼–½ in. (0.6–1.3 cm). Shell rather stout, thin; 5–7 whorls, suture well-impressed, spire rather short but sharp; aperture long, outer lip thickened by about 12 internal teeth, lower columella area with small shield. Sculptured on spire and upper ⅓ of body whorl with about 15 obliquely slanting, strong, rounded, axial ribs; on base of body whorl numerous spiral incised lines strongest. Color variable, range including pale yellow, pinkish gray with indistinct orange-brown mottlings, and reddish brown.
Habitat
Littoral zone to shallow depths; common.
Range
B.C. to Baja Calif.

COLUMBIAN AMPHISSA
Amphissa columbiana   **110:2**

Description
Size, ¾–1 in. (1.9–2.5 cm). Shell stout, 5–7 whorls, spire moderately long; aperture elongate, wider at bottom, inner lip slightly reflected. Sculptured with numerous large, weak, vertical, axial ribs, 20–24 on next-to-last whorl, absent from last part of last whorl; also numerous weaker spiral lines; lower half of body whorl shows only spiral lines, and these sharply. Periostracum thin. Color yellowish brown with indistinct mauve mottlings; periostracum yellowish brown.
Habitat
Shallow water; moderately common.
Range
Alaska to San Pedro, Calif.

WHELKS
Family Buccinidae

The shells of these carnivorous snails are usually large and thick and have few whorls. They are generally pear-shaped and have a pointed spire. The aperture is large and notched at the base.

RIDGED WHELK
Neptunea lyrata **110:9**

Description
Size, 4–5 in. (10.2–12.7 cm). Shell large, fusiform, solid, fairly heavy, width ¾ length; 5–6 robust convex whorls; spire partially turreted, aperture large, outer lip sharp, made wavy by ends of cords; canal moderately long, open. Sculptured with about 8 evenly spaced, strong to poorly developed, raised spiral cords, each spire whorl showing only 2; also smallish, faint spiral threads. Cord interspaces deeply concave. Color dull whitish- to reddish-brown, commonly darker on cords; aperture enamel-white with hint of tan.

Habitat
Shore to 50 fathoms (91.4 m); fairly common.

Range
Arctic Ocean to Puget Sound, Wash.

PHOENICEAN WHELK
Neptunea phoenicea **110:6**

Description
Size, 4 in. (10.2 cm). Shell resembling a small, delicately sculptured Ridged Whelk, but with more robust whorls and many more spiral cords; up to 20 cords on the body whorl.

Habitat
Shore to 50 fathoms (91.5 m).

Range
Alaska to Oreg.

TABLED WHELK
Neptunea tabulata **110:7**

Description
Size, 3–4 in. (7.6–10.2 cm). Shell moderately solid, sturdy, 6–8 whorls; spire well-elevated, strongly turreted; aperture moderate; canal open, curved; umbilicus tiny. Sculptured characteristically with wide, flat channel next to suture (top of each whorl flat or concave); numerous sandpapery spiral threads cover rest of each whorl. Periostracum thin. Color yellowish white; periostracum brown.

Habitat
30–200 fathoms (54.9–365.8 m); not uncommonly dredged.

Remarks
A choice collector's item.

Range
B.C. to San Diego, Calif.

KELLET'S WHELK
Kelletia kelletii

110:8

Description
Size, 4–6 in. (10.2–15.2 cm). One of largest of western gastropods. Shell fusiform, very heavy and rugged; 6–7 whorls, sutures rather indistinct, shoulders sloping. Aperture large, oval, pointed top and bottom; outer lip sharp, crenulated; canal moderately long, open. Sculptured with 8–10 very strong axial folds which form strong, rounded knobs along the periphery of each whorl; base with 6–10 incised spiral lines. Color white to yellowish; aperture glossy white.

Habitat
10–35 fathoms (18.3–64 m).

Remarks
Very commonly caught in traps.

Range
Santa Barbara, Calif., to Baja Calif.

DIRE WHELK
Searlesia dira

110:5

Description
Size, 1–1½ in. (2.5–3.8 cm). Shell fusiform, width ½ length, solid; 5–6 whorls, spire moderately tall; aperture smallish; outer lip thin, strong, finely serrated; canal short, twisted slightly to left, columella arched, slight fold at base. Sculptured on spire whorls with 9–11 low, rounded, axial ribs and on entire exterior surface with numerous fine, sharp, narrow, crowded, unequal-sized spiral threads. Color dull dark gray to purplish brown; aperture brown; columella glossy chocolate-brown.

Habitat
Shallow water; common.

Range
Alaska to Moss Beach, Calif.

LIVID MACRON
Macron lividus

110:4

Description
Size, ¾ –1 in. (1.9–2.5 cm). Shell small, fusiform; spire short, bluntly pointed; about 5 whorls; aperture elongate, upper end narrow, canal short and with a white, toothlike callus on the parietal wall; outer lip sharp, strong; siphonal canal short, slightly twisted; columella strongly concave. Sculptured only by closely spaced growth lines; small spiral thread near base of outer lip and 5–7 incised spiral lines at base of shell; upper part of columella with strong spiral ridge. Periostracum thick, feltlike. Operculum thick, oval, chitinous, nucleus at one end. Color yellowish white to bluish white; periostracum dark brown; operculum brown.

Habitat
Under stones at low tide; very common.

Range
Santa Barbara, Calif., to Baja Calif.

DOG WHELKS
Family Nassariidae

These generally small, scavenging snails have rather strong, stout shells with pointed spires, an oval aperture, a short notchlike canal, and usually a distinct columellar callus. They inhabit all seas.

FAT DOG WHELK
Nassarius perpinguis   110:10

Description
Size, ¾–1 in. (1.9–2.5 cm). Shell fairly thin, relatively stout; about 7 whorls, spire well-elevated, apex sharp; aperture moderately large, outer lip rather fragile. Sculptured with usually fine cancellate or minutely beaded spiral threads and axial riblets; variable, but spiral threads often predominant. Color grayish yellow to yellowish white, with 2 or 3 spiral bands of brown-orange.
Similarities
Channeled Dog Whelk is larger; inner lip reflected to form callus.
Habitat
Intertidal flats to 50 fathoms (91.4 m); very abundant.
Range
Puget Sound, Wash., to Baja Calif.

Note: The **CALIFORNIA DOG WHELK,** *Nassarius rhinetes,* occurring from Oregon to Baja California, is similar to the Fat Dog Whelk but is white and has coarser sculpture.

LEAN DOG WHELK
Nassarius mendicus   110:11

Description
Size, ½–¾ in. (1.3–1.9 cm). Shell rather slender, spire moderately high, about 8 whorls; aperture rather small, inner lip well-reflected, outer lip not thickened. Sculptured with numerous small beads formed by crossing of about 12 prominent axial ribs and smaller spiral threads. Color yellowish gray to brownish; aperture bluish white.
Habitat
Shallow water in north, deeper in south; common.
Range
Alaska to Baja Calif.

Note: The **COOPER'S DOG WHELK,** *Nassarius cooperi,* occurring from Puget Sound south, is similar to the Lean Dog Whelk but is shorter and stouter; has 7 to 9 strong, smoother axial ribs that persist to last of body whorl; and is grayish yellow to whitish, often spirally lined in brown or mauve.

CHANNELED DOG WHELK
Nassarius fossatus   110:12

Description
Size, 1½–2 in. (3.8–5.1 cm). Largest of genus on West Coast; a fine, showy species. Shell sharply pointed, about 7 whorls, spire well-elevated; aperture large; outer lip constricted at top, jagged-edged; inner lip reflected to form callus on columella; canal a deep notch at base of aperture. Sculptured on early whorls with coarse beads; last whorl with spiral threads and with short axial ribs.

Color shiny orange-brown to yellowish tan; callus and interior bright orange.
Habitat
Intertidal zone; common.
Range
Vancouver, B.C., to Baja Calif.

OLIVE SHELLS
Family Olividae

The shells of this family are more or less cylindrical in shape, since the much-inflated body whorl tends to conceal all the earlier whorls. Widely distributed in warm seas, the shells are often brightly colored and usually smoothly polished.

DWARF OLIVE
Olivella baetica   **111:1**

Description
Size, ½ –¾ in. (1.3–1.9 cm). Shell moderately elongate, rather lightweight, 4–5 whorls, spire somewhat prominent; aperture ½ shell length and narrow; columellar callus poorly developed. Sculpture smooth and highly polished; double-ridged spiral fold at lower end of columella. Color drab-tan to brownish, with weak bluish purple blotches usually more visible near suture. Fasciole white, often brown-stained; early whorls may be bluish purple.
Habitat
1–15 fathoms (1.8–27.4 m); moderately common.
Range
Kodiak Is., Alaska, to Baja Calif.

PURPLE OLIVE
Olivella biplicata   **111:2**

Description
Size, 1–1¼ in. (2.5–3.2 cm). Shell stoutly globular to elongate, quite heavy; about 4 whorls, spire short, body whorl much enlarged; aperture long, narrow at top, wide at bottom with distinct notch; columella heavily enameled, with 2 small plaits, upper wall with heavy callus. Sculpture smooth, polished, except for raised spiral fold at base of columella crossed by 1 to 3 spiral incised lines. Color variable, but usually bluish gray or whitish brown; violet stains about aperture base. Some specimens may be nearly white, others very dark, some brown.
Habitat
Shallow water to 25 fathoms (45.7 m); abundant in summer in sandy bays and beaches, generally colonial.
Habits
Quickly burrows into sand when tide goes out.
Range
Vancouver, B.C. to Baja Calif.

CONE SHELLS
Family Conidae

The cone shells are a large family of many-whorled shells of the tropics, variously patterned in a wide variety of colors. Some species are distinctly poisonous with each tooth in the radula bearing a venomous barb. The only representative on the northern West Coast is small and apparently harmless to humans.

CALIFORNIA CONE
Conus californicus   111:3

Description
Size, ¾–1 in. (1.9–2.5 cm). Shell shaped like an inverted cone, 6–7 whorls; smooth, glossy spire moderately elevated, slightly concave; shoulders rounded, sides very slightly rounded; aperture long, narrow. Periostracum, in life only, velvety, rather thick, hairy. Color grayish white to yellowish brown; top whitish, more or less purple-stained; aperture interior purplish; periostracum dull brown.
Habitat
Shallow water; rather common in some localities.
Range
Farallon Is., Calif., to Baja Calif.

AUGER SHELLS
Family Terebridae

The slender, elongate, many-whorled auger shells bear no plaits on the columella. They chiefly inhabit tropical seas.

SAN PEDRO AUGER
Terebra pedroana   111:4

Description
Size, 1–1¼ in. (2.5–3.2 cm). Shell tall, slender, spikelike, 10–12 whorls; aperture small, inner lip twisted at the notchlike canal. Sculptured on each whorl with numerous weak axial ridges and numerous fine, incised spiral lines; canal bounded on exterior by a sharp spiral line; also between sutures of first whorl a row of about 15–18 poorly developed nodules followed by weakly wrinkled flat area. Color grayish to bluish white, whitish yellow, or brownish in irregular splotches. Some individuals may be all brown.
Habitat
Shallow water; fairly common.
Range
San Pedro, Calif., to Baja Calif.

TURRET SHELLS
Family Turridae

This very large family of approximately 500 genera and subgenera, including several thousand species, is difficult to classify. Many shells are highly ornate, have a generally fusiform shape, and have a slit or notch in the outer lip.

SMOOTH TOWER
Ophiodermella inermis

111:5

Description
Size, 1½ in. (3.8 cm). Shell spikelike, 7–8 whorls, sutures slightly constricted; aperture large, outer lip thin, canal elongate. Sculptured with weak axial and spiral sculpture. Color light brown with darker growth lines.
Habitat
In bays on sandbars; offshore on sandy bottoms; fairly common.
Remarks
Formerly known as *O. ophioderma*.
Range
Santa Barbara, Calif., to Baja Calif.

DOLEFUL TOWER
Pseudomelatoma moesta

111:6

Description
Size, ¾ in. (1.9 cm). Shell strong, fusiform; 7–8 whorls, strong folds at shoulders, apex sharp; aperture rather long, narrow; outer lip thin, canal short. Sculptured with 9–10 slightly curved, axial ribs on each whorl, interstices smoothly polished; just below suture line is a faint row of beads. Color greenish brown.
Habitat
Intertidal zone, under rocks.
Remarks
P. torosa and *P. penicillata* are apparently synonymous.
Range
Monterey, Calif., to Baja Calif.

Bubble Shells and Sea Hares
Order Cephalaspidea

In this order the shell may or may not be present. If present, it is in a reduced condition and becomes thinner as the size of the snail increases and envelops more of the shell.

BABY BUBBLE SHELLS
Family Acteonidae

These small, solid, cylindrical shells have a sharp, short but prominent spire and a single plait on the inner lip of a long, narrow aperture. The cephalic disk (head region) is divided, and there is a thin, corneous operculum. The sculpturing consists usually of spiral grooves.

BARREL SHELL
Rictaxis punctocaelatus

111:7

Description
Size, ¾ in. (1.9 cm). Shell fusiform, solid, oblong, 3–5 whorls; body whorl most of shell; spire short, pointed; aperture long, narrow; outer lip crenulate inside, inner lip with 1 plait or fold, columella obliquely truncated at base. Sculptured with about 26 sharp and distinct spiral grooves on body whorl; 1 spiral fold on columella; surface well-polished. Color white, with 2 broad ashy or brown to black spiral bands; base orange-stained.

649

Habitat
Shallow water, in sand; common.
Range
B.C. to Baja Calif.

BUBBLE SHELLS
Family Bullidae

These carnivorous snails have small to fairly large shells that are thin and lightweight and usually rolled up like a scroll. They burrow in muddy and sandy bottoms of shallow water.

GOULD'S BUBBLE
Bulla gouldiana   111:8

Description
Size, 1½–2 in. (3.8–5.1 cm). Shell rotund, very thin, delicate, fragile, smooth; greatly enlarged body whorl completely engulfs earlier whorls; in place of a spire is a pit or depressed area. Aperture flaring, longer than shell itself, narrowing at top and wide at base; inner lip spread over body whorl like a thin enamel layer. Periostracum microscopically crinkled. Color pale grayish brown to dark brown, most specimens considerably mottled in darker brown and posteriorly bordered with cream; aperture white, but so thin that mottlings show through; periostracum dark brown.
Habitat
Shallow water.
Remarks
Abundantly collectible at night.
Range
Santa Barbara, Calif., to Gulf of Calif.

SMALL BUBBLE SHELLS
Family Haminoeidae

These snails have partly internal, fragile, glassy shells that are nearly cylindrical, with the enlarged body whorl almost engulfing the spire. The spiral line is channeled and forms a continuous groove. Because the shell appears to have been turned on a lathe, it is popularly called a "lathe shell." The animals inhabit muddy and brackish waters in warmer seas.

GREEN PAPER BUBBLE
Haminoea virescens   111:9

Description
Size, ½ in. (1.3 cm). Shell very fragile, semitransparent, quite globular, smooth; body whorl conceals earlier whorls; aperture very large, narrow at top, broad at bottom; outer lip thin, upper part high and narrowly winged, extending above top of shell; no apical hole. Color pale greenish yellow.
Habitat
Littoral zone of open coast; common.
Range
Puget Sound, Wash., to Baja Calif.

Note: Similar, less common species include the **GOULD'S PAPER BUBBLE,** *Haminoea vesicula,* ¾ inch (1.9 cm) in size, which has an apical hole; and the **OLGA'S PAPER BUBBLE,** *Hamimoea olgae,* 1 inch (2.5 cm), in which the outer lip rises well above depressed top of shell and the inner lip has a very thick enamel coating.

BARREL BUBBLE SHELLS
Family Scaphandridae

BARREL BUBBLE
Acteocina culcitella

Description
Size, ½–¾ in. (1.3–1.9 cm). Shell oblong, moderately solid, somewhat constricted in upper portions; 5 whorls; spire elevated, pointed, with minute pimplelike nucleus, often eroded in north; suture narrowly or deeply channeled; body whorl swollen in lower half; columella a single, raised, spiral cord. Sculptured with numerous spiral, wavy, incised lines. Color yellowish; sometimes with numerous fine, spiral, golden-yellow lines.
Habitat
Shallow water; common.
Range
Kodiak Is., Alaska, to Baja Calif.

Notch-banded Gastropods
Order Pyramidellida

PYRAMS
Family Pyramidellidae

This well-known family of tiny gastropods is characterized by a conical shell, usually polished-white in color and with many whorls. They hold with suction, pierce the host's body with a tiny drill, and suck its juices.

ADAMS' PYRAMIDELLA
Pyramidella adamsi   **111:10**

Description
Size, ⅝ in. (1.6 cm). Shell elongate; smooth; base fairly long, well-rounded; about 10 moderately rounded whorls, sutures deep; aperture oval, outer lip fairly thin, inner lip with strong plication. Color white to dark brown, spotted or banded.
Habitat
Shallow water, in sand; uncommon.
Range
Monterey, Calif., to Mexico.

FINE-SCULPTURED TURBONILLA
Turbonilla tenuicula   **111:13**

Description
Size, ¼ in. (0.6 cm). Shell somewhat cylindrical, high-spired, apex blunt; about 10 whorls, slightly shouldered, sutures deep; aperture

relatively tiny, outer lip thin. Each whorl sculptured with 18–25 axial grooves. Color white to dark brown.
Habitat
Shallow water, in sand; fairly common.
Range
Monterey, Calif., to Baja Calif.

Sea Butterflies or Pteropods
Order Pteropoda

Members of this order are small, pelagic gastropods very abundant in all the world's seas. Although occasionally washed ashore, the shells are usually found in dredge hauls.

CAVOLINID PTEROPODS
Family Cavolinidae

The symmetrical uncoiled shells of members of this family are fragile, white to brown, and of various configurations.

PYRAMID CLIO
Clio pyramidata   **111:11**

Description
Size, ⅝–⅞ in. (1.6–2.2 cm). Shell somewhat angular, compressed dorsoventrally, no lateral keels or spines. Lateral margins divergent. Sculptured with undivided dorsal ribs. Color like frosted glass.
Habitat
Floating in the open sea.
Range
Worldwide pelagic.

GIBBOSE CAVOLINE
Cavolina gibbosa   **111:12**

Description
Size, ⅜ in. (0.95 cm). Shell lacking lateral points, dorsal lip thin-margined, ventral lip no more developed than dorsal; keeled transversely on anterior ventral surface.
Habitat
Floating in open sea.
Range
Worldwide pelagic, between latitudes 43°N and 38°S; common.

Note: The **THREE-SPINED CAVOLINE,** *Cavolina tridentata,* another worldwide species, is similar to the Gibbose Cavoline, but it is larger and the ends of the lips are broader.

Nudibranchs and Sea Slugs
Order Nudibranchia

These shell-less gastropods bear an arc or circle of branchial plumes (gills) usually joined together at their bases and retractile into a cavity. The rhinophores invariably have a perfoliate club, appearing leaflike; and the pharyngeal bulb, the tubercle of the underside of the throat, is never suctorial. Below is a glossary of terms specific to this order:

GLOSSARY

Branchiae Respiratory organs.

Clavus An extension of rhinophore.

Club A clublike projection of the body.

Denticle Minute teeth or projecting points.

Perfoliate With leaflike projections.

Pharyngeal bulb An expansion in the region of the pharynx.

Pinnate Feathery, as in construction or arrangement.

Pleural teeth Side teeth.

Uncinal Hooked or barbed at the end.

DORISES
Family Doridae

The nudibranchs of the subfamily Cadlininae have a lamelliform labial armature that is almost annulate and that bears extremely small hooks. The middle of the radula bears a denticulated tooth, and the external margin of the pleural teeth is serrate.

Formerly Genus *Chromodoris*, the subfamily Glossodoridinae contains nudibranchs having a brilliantly blue, smooth-backed, elongate body. The minutely hooked labial armature is strong; the center of the radula is very narrow and often bears minute, compressed spurious teeth.

MONTEREY DORIS
Archidoris montereyensis

112:1

Description
Size, 1–2 in. (2.5–5.1 cm). Body relatively soft, dorsum granular or tubercular; tentacles short, thick, with external longitudinal groove; labial armature lacking. Rhinophore stalks conical; clavus slightly dilated, conical, perfoliate with 24–30 leaves on each side. Branchial plumes, 7, large, spreading, featherlike. Radula with 33 rows, center naked; 42–49 strongly hooked, denticulate, pleural teeth. Color light yellow, sprinkled on back with brown, greenish, or black dots; patches of darker color toward middle of back; branchial plumes dusty.
Habitat
Tidal pools; moderately common.
Range
Alaska to San Diego, Calif.

NOBLE PACIFIC DORIS
Montereina nobilis   **112:2**

Description
Size, 4 in. (10.2 cm). Rhinophore stalks stout, conical; clavus
perfoliate, about 24 leaves, stalks deeply retractile in low sheaths,
margins tuberculate. Branchial plumes, 6, large, spreading,
featherlike; with plumes joined by thin, membranelike expansion.
Radula with 26 rows, center naked, 55–62 strongly hooked pleural
teeth. Color variable, from rich orange-yellow to light yellow,
mottled with patches of dark brown between tubercles; branchial
plumes pinkish tipped with white.

Similarities
Monterey Doris is smaller; dark brown patches on tubercles.

Habitat
Tidal pools; moderately common.

Range
B.C. to Baja Calif.

YELLOW-SPOTTED DORIS
Cadlina flavomaculata   **112:4**

Description
Size, ¾ in. (1.9 cm). Rhinophores with 10–12 leaves in club.
Branchial plumes, 10–11, small, either simple pinnate or bipinnate.
Radula with about 77 rows; center bears tooth with 4–6 equal-
sized denticles; 23 pleural teeth on each side of central tooth.
Characterized by 2 rows of lemon-yellow spots borne upon low
tubercles; mantle yellowish white; rhinophores darker than mantle,
sometimes brown or black; branchial plumes white.

Habitat
Rocky tidal pools, all seasons; moderately common.

Range
B.C. to Baja Calif.

YELLOW-RIMMED DORIS
Cadlina marginata   **112:5**

Description
Size, 1½ in. (3.8 cm). In form similar to Noble Pacific Doris.
Rhinophores with 16–18 clavus leaves. Branchial plumes, 6,
bipinnate, retractile into sheath with yellow-tipped marginal
tubercles. Radula with 90 rows, central tooth with 4–6 even-sized
denticles, about 47 teeth on each side of center tooth. Color
translucent yellowish white, covered all over with low, yellow-
tipped tubercles surrounded by a narrow ring of white; around
margins of mantle and lateral and posterior edges of foot is a
distinct narrow lemon-yellow band.

Habitat
Rocky pools; not uncommon.

Range
B.C. to Baja Calif.

SAN DIEGO DORIS
Diaulula sandiegensis   **112:3**

Description
Size, 2–3 in. (5.1–7.6 cm). Body fairly soft; silky or velvety texture
to dorsal surface. Rhinophores conical; clavus with 20–30 leaves,
retractile into conspicuous sheath with crenulate margin. Branchial
plumes, 6, tripinnate; branchial aperture round, crenulate. Radula

broad, with 19–22 rows, each with 26–30 sickle-shaped teeth on each side of the naked center. Color pale yellow or brownish, easily distinguished by row of 2–30 dark brown or black rings varied in size and position along the back.
Habitat
Tidal seaweed zone, in rock pools; moderately common at all seasons.
Reproduction
From June to August the animal lays its broad, white, spiral egg bands.
Range
Japan; Alaska to Gulf of Calif.

PORTER'S BLUE DORIS
Glossodoris porterae   **112:6**

Description
Size, ½ in. (1.3 cm). Body narrow. Color deep ultramarine blue with 2 orange stripes; after death, the blue fades out. Light blue stripe along median line of mantle; mantle margin edged with white.
Similarities
California Blue Doris is larger; possibly adult form of Porter's Blue Doris.
Habitat
Rocky tidal pools; fairly common.
Range
Monterey, Calif., to Baja Calif.

CALIFORNIA BLUE DORIS
Glossodoris californiensis   **112:7**

Description
Size, 2 in. (5.1 cm). Body narrow, mantle aligned at sides; mantle projects beyond oral tentacles, but foot extends well behind mantle when animal crawls. Color similar to Porter's Blue Doris, but with numerous bright orange oblong spots (2 rows on mantle, 1 row down each side of foot), and a group of round spots on anterior end.
Habitat
Tidal pools; common.
Range
Monterey, Calif., to Gulf of Calif.

DENDRODORIDS
Family Dendrodorididae

The soft bodies of these nudibranchs are doris-shaped. The pharyngeal bulb and elongated sucking tube lack mandibles and radulas.

COMMON YELLOW DORIS
Doriopsilla albopunctata   **113:1**

Description
Size, 2 in. (5.1 cm). Back soft; papillalike protuberances low, white-tipped. Rhinophores with 18–20 clavus leaves, ⅔ rhinophore length, completely retractile. Branchial plumes, 5, tripinnate.
Habitat
Tidal pools at all seasons; very common, especially in summer.
Range
Bolinas Bay, Calif., to Gulf of Calif.

POLYCERIDS
Family Polyceridae

The bodies of these nudibranchs are limaciform (sluglike), and the branchial plumes cannot be retracted.

CARPENTER'S DORIS
Triopha carpenteri　　　　　　　　　　　　　　　**113:3**

Description
Size, 1 in. (2.5 cm). Rhinophores with 20–30 leaves in club. Branchial plumes, 5, large, tripinnate. Radula with 30–33 rows, center part with 4 teeth, 9–18 pleural teeth strongly hooked, 9–18 uncinal teeth quadrangular in outline. Ground color white, sometimes yellowish, often with white spots on very small tubercles; sides with irregularly arranged orange spots.
Habitat
Tidal rock pools, on kelps; very common and conspicuous.
Range
B.C. to San Diego, Calif.

MACULATED DORIS
Triopha maculata　　　　　　　　　　　　　　　**113:2**

Description
Size, 1 in. (2.5 cm). A very sluglike animal with a broad, flattened frontal margin. Rhinophore stalk and club same length; club with 18 leaves. Branchial plumes, 5, tripinnate. Radula with 14 rows, each with 4 flattened plates, 4–5 pleural teeth, 7–8 uncinal teeth. Glans penis blunt, armed with tiny hooks. Ground color yellowish brown, varying in hue in different individuals; surface dotted with bluish white, round, or oval spots, inconspicuous in young; frontal margin branching processes and branchial plumes bright orange or vermilion, shading to dark brown; rhinophore stalks yellowish, leaves and border of sheath bright orange-red.
Habitat
Tidal rock pools; abundant in summer, uncommon in winter.
Range
Bodega Bay to Baja Calif.

ORANGE-SPIKED DORIS
Polycera atra　　　　　　　　　　　　　　　**113:5**

Description
Size, ½–1 in. (1.3–2.5 cm). Body sluglike, highest in front of branchiae; frontal margin with 6 slender, pointed processes. Branchial plumes bordered with fingerlike processes. Gill plumes (branchiae), 8, anterior ones longest. Radula with 9–10 rows; center naked, flanked by 2 lateral teeth and 3–4 uncinal teeth. Ground color light, with surface striped longitudinally with blue-black lines separated by lighter bands, almost white with numerous orange-yellow spots; frontal margin processes yellow; rhinophores with yellow band near tip; gill plumes tipped and spotted with orange.
Habitat
Tidal pools, attached to brown algae; common.
Range
San Francisco, Calif., to Mexico.

OKENIIDS
Family Okeniidae

HOPKINS' DORIS
Hopkinsia rosacea   **113:4**

Description
Size, 1 in. (2.5 cm). Body flattened, firm, fragile; dorsal part sloping to margin of foot, no ridge separating back from sides; foot with broad, short tail and deep triangular notch in front. Rhinophores long, tapering, anterior side smooth, posterior ¾ with about 20 pairs of oblique plates. Branchial plumes, 7–14, narrow, naked. Radula with 1 large pleural tooth on each side flanked by a tiny, triangular pleural tooth. Color bright rose-pink; spiral egg ribbon rosy.

Habitat
Intertidal zone, under shelving rocks; moderately common at all seasons.

Range
Coos Bay, Oreg., to San Diego, Calif.

AEOLIDRIDS
Family Aeolidiidae

The body of these nudibranchs is rather broad and depressed. The branchiae are somewhat flattened and set in numerous close, transverse rows. The four tentacles are simple, the foot broad with acute anterior angles. The radula consists of a single broad, pectinate (comblike) plate.

PAPILLOSE EOLIS
Aeolidia papillosa   **113:6**

Description
Size, 1–3 in. (2.5–7.6 cm). Radula with 30 rows of a single broad, arched tooth bearing 46 denticles. Color variable; gray, brown, or yellowish always more or less spotted with lilac, gray or brown, and opaque white. Juveniles exhibit fewer papillae.

Habitat
Shoreline to moderately shallow water.

Range
Arctic seas to Santa Barbara, Calif.

LONG-HORNED HERMISSENDA
Hermissenda crassicornis   **113:7**

Description
Size, 1–2 in. (2.5–5.1 cm). Back covered with plumed gills. Color opalescent yellow-green; some variation, as cerata range from light yellow to red-brown.

Habitat
Mud flats and tidal pools.

Range
Alaska to Baja Calif.; very abundant, especially in Elkhorn Slough.

TUSK SHELLS
Class Scaphopoda

This small class of univalves is distinguished by a single, hollow, tusklike shell with an opening at each end. The animals are largely restricted to subtidal waters and the living examples are found mostly by dredging. Like most other mollusks the scaphopods possess a radula, a foot, and a mantle which secretes the shell. Unlike most other mollusks, however, they lack a heart, gills, or eyes, and they breathe through the mantle. The sexes are separate. They occur most often subtidally, with the tubelike shell partly buried in sand. They feed on minute marine organisms caught by threadlike tentacles covered by cilia that protrude from the larger opening; waste is expelled through the smaller end.

TUSK SHELLS
Family Dentaliidae

The shells of this family are elongate, curved somewhat like a miniature elephant's tusk, open at both ends with the greatest diameter at the aperture. Known also as "money shells" and "wampum," dentalia were highly valued by the American Indians. A four-inch (10.2-cm) shell carried an approximate value of five dollars by present U.S. currency standards.

INDIAN MONEY TUSK
Dentalium pretiosum   104:8

Description
Size, 2 in. (5.1 cm). Shell moderately curved, solid; apex with short notch on convex side; foot conical, bearing lobes along edges. Radula with a median tooth twice as wide as long. Color opaque white, like ivory, commonly transversely ringed with faint dirty-buff growth lines.
Habitat
Offshore shallow water, sandy bottoms; common.
Range
Alaska to Baja Calif.

SIX-SIDED TUSK
Dentalium neohexagonum   104:4

Description
Size, 1¼–2 in. (3.2–5.1 cm). Shell long, thin, tusklike; cross-section hexagonal. Color white.
Habitat
Subtidal, partially buried in sandy bottoms.
Range
Monterey, Calif., to Cen. America.

CLAMS
Class Bivalvia

Of the world's estimated 12,000 species of marine clams, oysters, and scallops, more than 400 occur along the California coast alone, of which some 60 species are commercially important. This class was formerly known as Pelecypoda, the name Bivalvia, or bivalves, coming from the animals' two valves, or shell parts. The two halves are joined at the hinge by a ligament and held together by one or two strong muscles. Except for a short larval stage, many species are sedentary throughout their lives. Some, however, have a fleshy and extensible foot and can move about, and the scallops can swim by clapping their valves together. The bivalves have adopted many ways of life. Some burrow into sand, mud, rocks, or wood; some become attached to rocks or other solid objects; others are free-living and able to travel short distances.

Bivalves breathe through both their gills and the mantle, which also, as in the other mollusks, secretes the calcium carbonate which makes up the shell. In identifying bivalves, various characteristics of the shell are usually used; some species exhibit variability in color and markings. Growth lines show seasons of relative quiet, such as winter.

The soft body of the bivalve is enclosed in and protected by the two chalky valves of the shell. The shelly material is formed in three layers: an outside layer, or periostracum, of a horny composition that is often so thin that in many species it is worn off except around the outer margins, where new growth takes place; a middle layer, called the prismatic layer, which makes up most of the shell thickness; and an inner layer, often nacreous, that is very hard and in some species very shiny.

In general the clams possess no senses of sight and sound. Their presumed senses of taste and smell are seemingly limited to identifying edible bits of food and to closing their valves if threatened by predators. The animal has no head, but three somewhat enlarged sets of ganglia function in place of a brain. Some species of bivalves, such as the scallops, have little eyespots and are able to distinguish between dark and light objects. Most clams are vegetarians, either extracting minute food particles from sea water passing over their gills or sucking up food from the mud with their siphons. A few clams and oysters form pearls around a foreign nucleus within the mantle, but in general few pearls of any commercial value are produced by species along the West Coast north of Baja California.

Nut Clams
Order Nuculoida

NUT SHELLS
Family Nuculidae

These are small, three-cornered or ovate shells with pearly interior, finely denticulated ventral margins, and a row of teeth on each side of the beak cavity but no ligamental pit between them.

SMOOTH NUT SHELL
Nucula tenuis   **114:2**

Description
Size, ⅛–³⁄₁₆ in. (0.48–0.95 cm). Shell small, ovate, rather plump and nutlike, ventral edge smooth. Sculptured by irregular growth lines; no radial lines; internal margins finely crenulate. Beaks small, near anterior end; hinge shows double row of prominent teeth, 6 fore, 9 aft. Color shiny olive-green; may show darker growth lines; interior white, often polished.

Habitat
Offshore muds.

Range
Arctic Ocean to Baja Calif.

YOLDIAS
Family Nuculanidae

Formerly classed with the Nuculidae, the yoldias have oblong shells, usually rounded in front but angled behind, and crenulated margins. A double row of teeth is separated by an oblique ligamental pit. They are widely distributed in cool seas.

TAPHRIA NUT
Nuculana taphria   **114:1**

Description
Size, ⅓–¾ in. (0.85–1.9 cm). Shell rather plump, anterior end rounded, posterior end pointed. Sculptured with numerous fine, concentric ribs. Beaks nearly central, low; row of teeth on either side of ligamental pit. Color shiny greenish brown.

Habitat
Shallow-water muds; commonly dredged; also rather common in stomachs of bottom-feeding fish.

Range
Bodega Bay, Calif., to Baja Calif.

ALMOND YOLDIA
Yoldia amygdalea   **114:3**

Description
Size, 1–2½ in. (2.5–6.4 cm). Shell thin, elongate, length twice height; narrowing posteriorly; anterior and basal margins regularly rounded; posterior tip pointed, recurved, snoutlike. Sculptured with faint, concentric growth lines. Beaks near center, small; 20–22 prominent teeth in filelike order on each side of central cartilage pit. Color shiny greenish tan to light chestnut-brown; interior glossy bluish white.

Habitat
Mud at moderate depths; rather common.

Range
Bering Sea to n. Calif.

COMB YOLDIA
Yoldia myalis   114:4

Description
Size, ½–1 in. (1.3–2.5 cm). Shell thin-walled, smooth, less elongate than most yoldias; anterior end rounded, posterior end bluntly pointed, valves only slightly inflated. Beaks nearly central, low, about 12 teeth on either side. Color yellowish green; periostracum dark greenish olive; interior yellowish white.
Habitat
7–100 fathoms (12.8–182.8 m); fairly common.
Range
Arctic Ocean to Puget Sound, Wash.

COOPER'S YOLDIA
Yoldia cooperi   114:5

Description
Size, 3 in. (7.6 cm). Shell quite thin; anterior end large, broadly rounded; posterior end small, hooked; basal margin a smooth curve, concave between hooklike posterior tip and beaks. Sculptured with distinct concentric lines. Beaks near posterior end small; about 12 V-shaped teeth in front, 40+ behind. Color shiny green.
Habitat
Offshore waters; fairly common.
Range
Cen. Calif. to Mexico.

Arks and Bittersweets
Order Arcoida

ARK SHELLS
Family Arcidae

The rather boxlike ark shells are strong, heavily ribbed or cancellate, with a narrow hinge line bearing numerous comblike teeth arranged in a straight line on both valves. The umbones are toward the posterior end. There is usually a heavy, often bristly periostracum but no siphon. Some arks move about in mud or sand; others cling to rocks by means of a silky byssus.

BAILY'S MINIATURE ARK
Barbatia bailyi   114:7

Description
Size, ¼+ in. (0.64+ cm). Shell oblong, squarish, fat. Sculpture cancellate, with beads foliating at posterior end. Ligament small, narrow, well behind beaks. Teeth, about 15. Color white to light tan.
Habitat
Underside of rocks at low tide; common.
Range
Santa Monica, Calif., to Panama.

MANY-RIBBED ARK

Anadara multicostata   114:8

Description
Size, 3–4 in. (7.6–10.2 cm). Shell strong, solid, thick, squarish; left valve slightly overlapping right valve. Sculptured with 31–36 prominent radial ribs. Teeth robust, along hinge line. Color ivory-white; periostracum velvety brown.
Habitat
Sands at 12 ft. (3.7 m) or deeper, or under stones at low tide in some localities; very common.
Range
Newport Bay, Calif., to Panama.

BITTERSWEET CLAMS

Family Glycymerididae

This family consists of a small group of heavy, usually orbicular, equivalve, porcellaneous shells, generally with a soft velvety periostracum; the beaks are incurved, the hinge heavy and with many small teeth, and the ligament external with grooves diverging from the area. The largest muscle scar is at the anterior end.

BITTERSWEET

Glycymeris subobsoleta   114:6

Description
Size, ¾–1 in. (1.9–2.5 cm). Shell subtrigonal (nearly round), slightly inflated, fairly solid. Sculptured with flat radial ribs and narrow interstices; inner shell margins strongly crenulate. Periostracum heavy, velvety, usually well-worn. Beaks central, rather prominent; ligamental area short, 2 curving inside rows of compressed hinge teeth. Color white to yellowish gray, blotched with reddish brown.
Habitat
Shallow to moderately deep water; rather common. Single valves commonly found washed ashore.
Range
Aleutian Is. to Baja Calif.

Mussels

Order Mytiloida

MUSSELS

Family Mytildae

Mussels are relatively common bivalves ranging from small to rather large, and many species are used extensively for food. The shells are oval to oblong and have a long, thin, finely dentate hinge line; a heavy, dark brownish, often hairy periostracum; fine radial ribs in succeeding pairs which cross at right angles; sharp umbones; and a weak internal ligament. The valves are dark blue to black, often with a pearly lining; they are equal in size and shape, with the umbones close to the front and bent backward. The mantle is open in front but folded at the posterior end into a stationary excurrent siphon. The worm-shaped foot is disk-shaped at the terminus.

Mussels have worldwide distribution. Some species burrow, but most attach themselves to rocks or pilings by means of a byssus, or set of threads, which they spin from a gland in the foot.

Numerous cases of poisoning from eating mussels have been reported in the past, but California Fish and Game Department investigations have shown that such cases were a result of poor judgment in gathering damaged mussels.

HORSE MUSSEL
Modiolus modiolus   **115:1**

Description
Size, 2–6 in. (5.1–15.2 cm). One of the commonest and largest cold-water mussels. Shell thick, coarse, oblong, heavy, especially in older specimens; basal margin concave, with fissure for byssus. Sculptured with coarse concentric ribs. Periostracum coarse, thick, leathery; flakes off in dried specimens; roughly bearded near shell margin. Dried shell chalky mauve-white; periostracum deep purplish- to brown-black; interior pearly white.
Habitat
Deep water below low-tide limit; very common. Empty or single valves often found on beaches of Northwest Coast.
Range
Arctic seas to Baja Calif.

CARPENTER'S HORSE MUSSEL
Modiolus carpenteri   **115:3**

Description
Size, 1 in. (2.5 cm). Shell short, swollen, "somewhat wedge-shaped, having a breadth more than half its length" (Keep); Sculpture smoothish. Beaks not quite terminal (marginal), strongly curved forward. Color white; periostracum light brown, whitish at beak end; interior dull white. Formerly known as *M. fornicatus.*
Habitat
Moderately deep water; cast ashore rarely.
Range
Monterey Bay, Calif., to San Pedro, Calif.

FAT HORSE MUSSEL
Modiolus capax   **115:4**

Description
Size, 2–6 in. (5.1–15.2 cm). Shell elongate, considerably inflated; top bluntly rounded, basal margin broadly rounded. Periostracum thickish, often bearded with coarse hairs; glossy. Color chestnut-brown; worn shell brick-red with bluish mottlings; interior bluish white, ventral half yellowish to brownish purple.
Habitat
Moderately deep water, usually solitary; not common except as small specimens.
Range
Santa Cruz, Calif., to Peru.

STRAIGHT HORSE MUSSEL
Modiolus rectus   115:6

Description
Size, 8 in. (20.3 cm). Shell large, thin; ventral margin concave;
posterior end much broadened; smoothish. Periostracum heavy,
glossy, lightly bearded on posterior ¼ of shell. Color dark;
periostracum brown; interior white; animal slightly yellowish.
Similarities
Fat Horse Mussel is smaller; ventral margin nearly straight.
Habitat
Muddy places in bays, lagoons, quiet offshore waters; often washed
ashore.
Habits
Lives embedded vertically in mud with just posterior shell tip
protruding; solitary.
Remarks
M. flabellatus is a synonym.
Range
Vancouver, B.C., to Baja Calif.

PLATFORM MUSSEL
Septifer bifurcatus   115:5

Description
Size, 1–2 in. (2.5–5.1 cm). Shell subtriangular, inflated; anterior
margin flattened, posterior margin curved. Sculptured with 20–24
narrow, prominent, wavy, bifurcating (branching), radial ribs;
inner margin crenulate. Periostracum often eroded between ribs.
Beaks pointed, at anterior tip; inside each valve under beaks in a
small, transverse, shelly platform (diaphragm). Color dark purple;
periostracum black; interior pearly white, often stained bluish
brown.
Habitat
Onshore waters, in rock crevices.
Range
Crescent City, Calif., to Gulf of Calif.

Note: Similar to the Platform Mussel are a subspecies, *Septifer
bifurcatus obsoletus,* found south from San Diego, which is much
more elongate, with the interior mostly black; and an associate, the
STEARNS' MUSSEL, *Brachidontes adamsianus,* sized ½ to 1 inch
(1.3–2.5 cm) and having the shell obtusely carinate.

CALIFORNIA MUSSEL
Mytilus californianus   115:11

Description
Size, 2–6 in. (5.1–15.2 cm), record 10 in. (25.4 cm). Shell
considerably elongated, thick, inflated; anterior or ventral margin
straight, posterior margin curved. Sculptured with weak radial ribs
not numerous but fairly prominent near basal margins; growth
lines very coarse. Beaks at apex of long triangle. Adults purplish
black; juveniles showing paler streaks of brown and white.
Habitat
Clustered colonially, often in great beds, on surf-beaten rocks; very
common.
Range
Aleutian Is. to Socorro Is., Mexico.

BLUE MUSSEL
Mytilus edulis

115:2

Description
Size, 1–3 in. (2.5–7.6 cm). The common, edible, nearly worldwide blue mussel. Shell a long, flattened, wedge-shaped oval (elongate-triangular). Sculptured with fine growth lines, no ribs. Periostracum heavy, smooth, varnishlike, thin. Beaks at apex, barely noticeable; 4 teeth. Color blue-black, eroded areas chalky purple; adults deep-toned; juveniles varying in gray, green, brown shades, may show colored rays; periostracum satiny black; interior slightly pearly- to bluish-white with deep purple-blue border. Some specimens show radial yellowish brown rays.

Habitat
Rocky shores, pilings, in cool-sea areas, in south, occasionally on driftwoods; common.

Range
Arctic Ocean to Baja Calif. and to S. Carolina.

Note: A variety, *Mytilus edulis diegensis,* occurring from northern California to Baja California, is similar to the Blue Mussel and identical to Alaska forms. It may be an ecological or physiological variant.

LITTLE BLACK MUSSEL
Musculus niger

115:7

Description
Size, 2–3 in. (5.1–7.6 cm). Shell plumply oval, slightly protruding at posterior end. Sculptured by network of radiating lines; ribs axial, crossed at right angles; growth lines faint. Periostracum rusty brown; interior pearly, often pinkish.

Habitat
Rock crevices in moderately deep water; common.

Range
Alaska to Oreg.

PEA-POD SHELL
Adula falcata

115:9

Description
Size, 2–4 in. (5.1–10.2 cm). Shell very elongate, slightly curved, extremely thin, fragile; anterior end rounded, dilated; posterior end extended to lengthy blunt point. Sculptured with vertical, wavy ribs. Beaks rounded, about ⅛ in from anterior end; strong angle between beaks and base of posterior tip. Periostracum thick, wrinkled, shiny. Periostracum chestnut-brown; interior white, more or less pearly.

Habitat
Low-water rocks; bores cylindrical hole in hard rock, lives inside attached by a silky byssus; not common.

Range
Coos Bay, Oreg., to Baja Calif.

CALIFORNIA PEA-POD SHELL
Adula californiensis

115:8

Description
Size, 1–1¼ in. (2.5–3.2 cm). Shell elongate, somewhat cylindrical, rounded at both ends, posterior end a bit broader; smooth.

Periostracum heavy, velvety, hairy over posterior end. Beaks low, close to anterior end. Color deep, shiny brownish black; interior bluish gray.
Habitat
Excavated burrows in low-water stiff clay or softened rock; moderately common.
Range
B.C. to San Diego, Calif.

ROCK BORER MUSSEL
Lithophaga plumula   **115:10**

Description
Size, 3¼ in. (8.3 cm). Shell elongate, cylindrical, rounded posteriorly, gracefully tapering anteriorly. Sculptured with 2 radial grooves back from beaks; interspace often filled with plumelike encrustation. Periostracum chestnut; interior somewhat iridescent metallic.
Habitat
Burrows into rocks, sometimes into living abalone and rock scallop shells; quite common, especially on rocky reefs.
Range
Mendocino Co., Calif., to Peru.

Scallops and Oysters
Order Pterioida

SCALLOPS
Family Pectinidae

The scallops are enormously diversified, with great numbers of fossil and living species, so nomenclature of genera tends to vary greatly among specialists. Instead of the fifty or more genera and subgenera proposed by various writers, this section is limited to the Genera *Chlamys, Leptopecten, Argopecten,* and *Hinnites.*

The valves of scallops may be, though commonly are not, equal. The right, or lower, valve is usually smaller or raised, and it may be strongly convex; the left, or upper, valve is flat or concave. Shells are usually ribbed radially, have scalloped edges, and lack teeth. There is an ear-shaped projection, or "wing," on each side of the umbones. Most interestingly, there is a row of tiny eyes along the edge of the mantle. Scallops occur worldwide at all depths.

PINK SCALLOP
Chlamys hericia   **116:2**

Description
Size, 2–2¾ in. (5.1–7 cm). Shell varicolored, obliquely ¾ circular; wings unequal. Right valve with byssal notch, sculptured with 18–21 broad, radiating, moderately scaled, primary ribs separated by 5–7 much weaker-spined secondary ribs; left valve with 10–11 more closely spaced primary ribs separated by a single, rounded secondary nearly as large; between these large ribs are 15–18 very small spined ribs, 3 being on the large secondary rib itself. Lower valve white to light yellowish; upper valve varied with broad rays of pink and lavender; color blends common.

Habitat
Deep water in north, shallower in south; common.
Remarks
Long considered a variety of the spear scallop, *Chlamys hastata*, a more northern form.
Range
Alaska to San Diego, Calif., dredged in large number in Puget Sound, Wash.

HINDS' SCALLOP
Chlamys rubida

116:1

Description
Size, 1–1½ in. (2.5–3.8 cm). Shell with microscopic reticulations between ribs either near beaks or near valve margins. Sculptured with some 25 closely set ribs with very narrow interspaces; left valve, without byssal notch, with each primary rib bearing 3 rows of spines and with a secondary rib between primaries; right valve flatter with ribs fewer, smoothish, rounded often in pairs; reticulated sculpturing is more pronounced on right valve. Wings unequal, posterior wing greatly expanded. Color variable through shades of red, light rose, mauve, pink, lemon-yellow, pale orange, to white; left valve shades darker; color blends common.
Habitat
Shallow water to 800+ fathoms (1.5+ km); rather common.
Remarks
C. hindsii is a synonym.
Range
Bering Sea to San Diego, Calif.

ICELAND SCALLOP
Chlamys islandica

116:3

Description
Size, 3–4 in. (7.6–10.2 cm). Shell valves moderately convex to flattish, upper valve more so; wings unequal, posterior shorter. Sculptured with about 50 coarse, crowded, irregular, radial ribs which divide into 2 toward margin; ribs set with tiny erect scales; rarely, ribs are more or less in groups of 2, 3, or 4. Color dirty gray or cream, sometimes tinged with yellow, peach, or purplish, inside and outside; also occasionally pale orange to dark reddish brown.
Habitat
Continental shelf; very common.
Range
Arctic Ocean to Puget Sound, Wash.

SPECKLED SCALLOP
Argopecten circularis aequisulcatus

116:4

Description
Size, 3½ in. (8.9 cm). Shell globose; valves rounded; winglike lateral projections on each side of umbos nearly equal. Sculptured with 19–22 flat-topped radiating ribs, which interlock at edges with those of opposite valve. Color variable from gray to orange or reddish with numerous dark spots and blotches; left valve usually darker. Flesh yellowish, tinged with orange or red.
Habitat
Surface of sandy or muddy bottoms just below low-tide mark, usually inside sheltered bays or in quiet coastal water.

Remarks
Currently protected in California with heavy fines for collecting, even for scientific purposes.
Range
Monterey Bay, Calif., to Baja Calif., generally rare, but in some areas fairly common; most numerous at Alamitos, Newport, Mission Bay.

KELP-WEED SCALLOP
Leptopecten latiauratus   **116:5**

Description
Size, 1 in. (2.5 cm). Shell thin, lightweight; wings about equal, strongly pointed at tips. Sculptured with 12–16 rounded to squarish, nonprominent ribs, with central rib angled about 70° to the straight hinge line. Color translucent yellowish to chestnut-brown or orange-brown, commonly with zigzag mottlings in white.
Habitat
Inshore, attached to rocks, kelps, bottoms of boats; common.
Range
Pt. Reyes, Calif., to Gulf of Calif.

Note: A form of Kelp-weed Scallop, *Leptopecten latiauratus monotimeris,* is slightly smaller, with the wings less prominent and less acutely pointed, and the ribs rounded and forming broad corrugations on the shell.

GIANT ROCK SCALLOP
Hinnites giganteus   **116:6**

Description
Size, 3–6 in. (7.6–15.2 cm), record 10 in. (25.4 cm). Shell nearly symmetrical, heavy, massive, usually spherical; adults irregularly oblong; young resemble *Chlamys.* Sculptured with many crowded wrinkled lines, coarsening with age. Adult exterior reddish to white, interior white stained with rich purple near hinge; some young bright orange; colors tend to fade with increasing age.
Habitat
Rocks beyond low-tide mark to 100 ft. (30.5 m); common.
Habits
A free-swimming scallop when young; when about 1 in. (2.5 cm) long, attaches to some object permanently.
Remarks
Formerly known as *H. multirugosus,* an unnecessary replacement name.
Range
Queen Charlotte Is., B.C., to Baja Calif.

FILE SHELLS
Family Limidae

Shells of this family are obliquely oval, gaping at both ends and winged only on one side. Popularly termed either file or scoop shells, they have a toothless hinge with a triangular pit for the ligament. They are expert swimmers, darting about with the hinge foremost and trailing a long sheaf of filaments.

HEMPHILL'S FILE
Lima hemphilli   114:9

Description
Size, 1 in. (2.5 cm). Shell obliquely elliptical, somewhat
compressed, wings very small; anterior end fairly straight, gaping;
posterior end rounded. Sculptured with fine, irregular, narrow,
sharp radiating ribs crossed by very fine, rough threads, like the
teeth of a file; all margins smooth. Color white.
Habitat
Shallow water; fairly common.
Range
Monterey, Calif., to Acapulco, Mexico.

JINGLE SHELLS
Family Anomiidae

This family is characterized by the hole in the lower valve through
which the byssal threads pass to attach the shell to some solid
support. These permanently fixed bivalves are fragile, roundish,
and waxy in luster and have unequal-sized valves. The left
(upper), valve is dome-shaped; the right (lower), one is smaller and
concave. The upper valve has one large and two small muscle scars
(except for Genus *Pododesmus*, with only two scars) and is the
"jingle shell" most commonly found washed up on shore. The
valves are rather translucent and pearly inside.

JINGLE SHELL
Anomia peruviana   114:10

Description
Size, 1–2 in. (2.5–5.1 cm). Shell variable in shape but usually
irregularly circular; rather thin, partially translucent; smooth; right
valve, on which it rests, with hole near beak for byssus. Color pale
yellowish green to orange; luster waxy.
Habitat
Intertidal zone, attached to rocks, other shells, waterlogged
driftwoods; common.
Range
Monterey Bay, Calif., to Peru.

PEARLY MONIA
Pododesmus macroschisma   114:11

Description
Size, 2½–5 in. (6.4–12.7 cm). Shell shape varied according to host;
attached by teardrop byssus through large notch in lower, or right,
valve, which is thin, fragile; left valve heavier. Left valve
sculptured with numerous fine, branching, radiating ridges; interior
with 1 large and 1 smaller muscle scar. Color yellowish or greenish
white, interior pearly gray or gray-green.
Similarities
Jingle Shell is smaller, smooth, with 3 muscle scars.
Habitat
Lives singly attached to rocks and shells, chiefly on living abalones,
occasionally on rocks of breakwaters, road fills; quite common.
Remarks
Edible.
Range
Alaska to Baja Calif.

OYSTERS
Family Ostreidae

The often large, heavy shells of this family are asymmetrical and greatly varied in shape, generally attached to some solid object by the lower valve, which is usually larger than the upper valve. The prodissoconch hinge is long; in the adult, the muscle scar is nearly central and uncolored. Species of the Genus *Ostrea* characteristically have an attached left valve, which is larger than the right valve.

CALIFORNIA OYSTER
Ostrea lurida

116:7

Description
Size, 2–3 in. (5.1–7.6 cm). The common native oyster of the West Coast. Shell variously shaped; valves not very thick or heavy; left valve attached and larger than right. Sculpture generally rough; growth lines coarse, concentric, sometimes smoothish. Color purplish black or brown, occasionally purplish brown to brown axial color bands; interior often stained with various shades of olive-green, may show slight metallic sheen.

Habitat
Intertidal zones; common.

Range
Sitka, Alaska, to Baja Calif.

Note: Two forms or ecological variants are similar to the California Oyster: *Ostrea lurida expansa*, which is roundish and often fluted at the margins; and *Ostrea lurida laticaudata*, more slender and elongate and often reddish.

JAPANESE OYSTER
Crassostrea gigas

116:8

Description
Size, 3–12 in. (7.6–30.5 cm). Shell shape varied, but typically long and straplike and usually very large; upper valve flattish, lower valve deeply cupped. Sculptured by widely spaced, coarse, concentric lamellae or thick, heavy, longitudinal flutings; inner margin smooth. Prodissoconch hinge is short, valves asymmetrical. Color dingy gray with many purple streaks and blotches radiating away from umbones; interior enamel-white; muscle scar or near shell edges faintly purplish, rarely greenish.

Habitat
Intertidal zone; common.

Remarks
This species was formerly placed in the genus *Ostrea*.

Range
B.C. to Morro Bay, Calif.

Note: A round form of this introduced Japanese species was named *C. laperousii*.

Lucines, Clams, and Razors
Order Veneroida

CARDITAS
Family Carditidae

The generally solid, equivalve shells of this family are small, thick, radially ribbed, and quadrate (squarish). They have a slight ventral gape, a byssus, and two robust teeth under the beaks.

CARPENTER'S CARDITA
Cardita subquadrata

117:3

Description
Size, ¼–½ in. (0.6–1.3 cm). Shell elongate; anterior end short, nearly straight; posterior end lengthened, rounded. Sculptured with strong, radiating ridges. Beaks well anterior, very small. Color brownish gray; interior purplish.
Habitat
Shallow to deep water, also under stones at water line; very common.
Remarks
Formerly called *C. carpenteri.*
Range
B.C. to Baja Calif.

DIPLODONS
Family Ungulinidae

Formerly called Diplodontidae, the Ungulinidae shells are thin, orbicular, and strongly inflated. The valves are split (bifid) at the left anterior and right posterior ends. Each valve shows two cardinal teeth.

ORB DIPLODONTA
Diplodonta orbellus

117:8

Description
Size, ¾–1 in. (1.9–2.5 cm). Shell thin, almost globular, outline circular; commonly hidden. This species forms a protective outer coating of sand grains cemented by mucus, with the siphons concealed in long tubelike extensions. Sculpture smoothish; growth lines, some more prominent than others, make surface somewhat uneven. Beaks small, forward-pointed; aft ligament long, raised, conspicuous; 2 prominent teeth on each valve below beaks, left anterior and right posterior teeth divided. Color grayish white.
Habitat
Shallow water; rather common.
Range
Bering Sea to Gulf of Calif.

LUCINES
Family Lucinidae

The equivalve shells of this family are orbicular, strong, and laterally compressed and have small but definite beaks. Most species are white.

FINE-LINED LUCINE
Parvilucina tenuisculpta   117:6

Description
Size, ½ in. (1.3 cm). Shell oval, slightly less high than long.
Sculptured with many small, radial, weakly raised threads; growth
lines concentric, fine, irregularly spaced; inner margin of valves
finely toothed. Periostracum thin. Beaks prominent, close together;
ligament behind beaks depressed, narrow, externally visible; lunule
in front of beaks small, depressed, heart-shaped. Color chalky
white; periostracum grayish to light olive-green.
Habitat
Just offshore; common.
Range
Bering Sea to Monterey, Calif.

Note: The **APPROXIMATE LUCINE**, *Parvilucina approximata*, is a
common species similar to the Fine-lined Lucine but more globose,
sized ⅛–¼ inch (0.3–0.6 cm), and having fewer radial riblets. It
occurs from Monterey, California, to the Gulf of California.

NUTTALL'S LUCINE
Lucinisca nuttalli   117:2

Description
Size, 1 in. (2.5 cm). Shell outline nearly circular; valves thin,
rather stout, moderately inflated. Sculptured with both concentric
and radiating fine, sharp lines, less concentrically in a somewhat
more compressed region at anterior or upper section. Beaks central;
lunule short, deep, larger in left valve. Color white.
Habitat
Sand just offshore; not uncommon.
Range
Monterey, Calif., to Gulf of Mexico.

CALIFORNIA LUCINE
Codakia californica   117:5

Description
Size, 1–1½ in. (2.5–3.8 cm). Shell oval to circular, moderately
inflated. Sculptured with many rather distinct but small, crowded,
concentric lines. Beaks central; lunule small, deep-set, wholly in
right valve. Color dull white.
Habitat
Littoral zone to 78 fathoms (142.7 m); common.
Range
Crescent City, Calif., to Baja Calif.

JEWEL BOXES
Family Chamidae

The shells of these warm-water, attached animals are thick, heavy,
irregular, and inequivalve. The fixed left valve, with which they
attach themselves to some solid object (usually to the underside of
rocks), is larger and more convex than the right valve. In the
Genus *Chama* the umbones turn from right to left; in *Pseudochama*
they turn from left to right, with attachment by the right valve.

AGATE JEWEL BOX
Chama arcana   **117:7**

Description
Size, 1½–2½ in. (3.8–6.4 cm). Shell generally circular, very strong and robust. Sculpture very rough, from concentric, frondlike frills (wrinkles or bladelike projections) extending irregularly but well beyond shell margins; interior margins finely toothed or crenulate. Color opaque white or cream, often with rosy rays; interior white. Entire shell often has a curious agatelike translucence.

Habitat
Beyond low-tide mark, attached commonly to rocks, dead shells, pilings, breakwaters, driftwoods; also may be dredged down to 25 fathoms (45.7 m).

Remarks
The free valve is often washed ashore, but to obtain a complete specimen requires underwater hammer-and-chisel work. Species formerly misidentified as *Chama pellucida,* a different species.

Range
Oreg. to Chile.

Other name
Rock Oyster.

REVERSED JEWEL BOX
Pseudochama exogyra   **117:4**

Description
Size, 1½–3 in. (3.8–7.6 cm). Shell very similar to Agate Jewel Box but reversed, i.e., attached by right valve and, when viewed from the inside, arched counterclockwise (beaks with a sinistral twist); valves thick, solid. Sculptured with fewer, somewhat less spiny irregular concentric frills; interior not bordered by crenulations. Color dull white, sometimes greenish-tinged; interior opaque white.

Habitat
Intertidal.

Range
Oreg. to Baja Calif.

ASTARTES
Family Astartidae

The shells of this family of small, brownish bivalves are thick and solid, with prominent, nearly central umbones and well-developed teeth at the hinge. Triangular in outline, they all have conspicuous concentric grooves and growth lines. The soft parts are commonly brightly colored. They inhabit chiefly cold seas.

ALASKA ASTARTE
Astarte alaskensis   **117:1**

Description
Size, 1 in. (2.5 cm). Shell obliquely triangular, both ends well-rounded, anterior somewhat extended. Sculptured with 12–14 broad and evenly spaced concentric ridges and deep furrows, covering shell from beaks to margins; inner margin smooth, untoothed. Periostracum dark brown to nearly black; interior chalky white.

Habitat
Fairly shallow water; commonly dredged, especially in Puget Sound.

Range
Bering Sea to Puget Sound, Wash.

COCKLES
Family Cardiidae

The cockles are a family of equivalved, heart-shaped (end view) clams which often gape behind, with the beaks almost central. The margins are toothed or scalloped, the shell ends gently rounded. The valves are thin but quite well inflated; the hinge teeth are arched; the pallial line is wavy behind. The animals are mobile and have no byssus; they are a dietary staple in Europe but are little used for food in America.

GIANT PACIFIC COCKLE
Trachycardium quadragenarium   118:1

Description
Size, 3–6 in. (7.6–15.2 cm). Shell large, somewhat higher than long; well-inflated with both ends rounded, posterior end flattened a bit. Sculptured with 40–44 strong, closely set, squarish radial ribs which form a scalloped margin; ribs studded with small, upright, triangular spines, especially on anterior, posterior, ventral sections. Periostracum thin, opaque. Beaks moderately large with smoothish ribs. Color whitish tan; periostracum yellowish brown; interior dull white to orange-brown.
Habitat
Tidewater to 75 fathoms (137.2 m); not common.
Remarks
Edible.
Range
Monterey Bay, Calif., to Baja Calif.

STRAWBERRY COCKLE
Americardia biangulata   118:5

Description
Size, 1½ in. (3.8 cm). Shell roundly angular; anterior end regularly rounded, posterior concavely sloping. Sculptured with about 30 strong, radiating ribs, narrowest on posterior slope. Color yellowish white; interior reddish purple.
Habitat
Shallow to moderately deep water; fairly common.
Range
Redondo Beach, Calif., to Ecuador.

HUNDRED-LINED COCKLE
Nemocardium centifilosum   118:4

Description
Size, ½–¾ in. (1.3–1.9 cm). Shell nearly circular, not quite as high as long, rather plump. Sculptured by very fine, sharp, numerous ribs; posterior ⅓ cancellated by threadlike concentric lines, separated from anterior ⅔ by a single raised rib; margins finely serrated. Periostracum fuzzy. Beaks central, prominent. Color white; periostracum gray, greenish- or brownish-gray; interior dull white.
Habitat
Moderately shallow water; fairly common.
Range
Alaska to Baja Calif.

LITTLE EGG COCKLE
Laevicardium substriatum   **118:7**

Description
Size, 1 in. (2.5 cm). Shell obliquely ovate, slightly compressed;
valves thin. Sculpture quite smooth, with obscure, close-set,
narrow, radial lines commonly interrupted. Beaks at triangle apex.
Color mottled yellowish brown or tan; interior cream-yellow with
purple-brown mottlings; radial lines reddish brown.
Habitat
Shallow water.
Range
Ventura Co., Calif., to Baja Calif.

GIANT EGG COCKLE
Laevicardium elatum   **118:2**

Description
Size, 3–7 in. (7.6–17.8 cm). Largest of the cockles. Shell oval,
slightly oblique, well inflated, higher than long. Sculptured by
some 40 quite low, radiating ribs and shallow grooves that leave
the total area relatively smoothish; end regions smooth. Color
mottled yellowish brown to orange-yellow; interior porcelain-white.
Habitat
Sandy mud in relatively shallow water.
Range
Rare to moderately common in Calif.; abundant in Gulf of Calif.

NUTTALL'S COCKLE
Clinocardium nuttallii   **118:9**

Description
Size, 2–6 in. (5.1–15.2 cm). The common West Coast cockle. Shell,
in adults, higher than long, stout, thick, moderately compressed,
somewhat brittle; in juveniles, almost circular. Sculptured with 33–
37 strong, squarish, radial ribs crisscrossed by half-moon-shaped
wavy lines near margins; in small specimens first 2 ribs behind
ligament are large rounded. Older specimens worn relatively
smooth. Periostracum thin. Beaks near center, prominent, high.
Color drab grayish white; periostracum brownish yellow.
Habitat
Mud or muddy sand in bays, sloughs, estuaries, or quiet offshore
waters; common. More abundant on tide flats of northwest coast.
Range
Bering Sea to San Diego, Calif.

VENUS CLAMS
Family Veneridae

This is the largest bivalve family and the most widely distributed
in range and depth. Its members are distinguished by their
beautiful symmetry and arresting color and sculpturing. The valves
are equal, oval-oblong generally, and porcellaneous; the teeth
interlock and the lunule is clear and deep. The shell is thick and
strong, the pallial line wavy, and the muscle scars oval. The inside
edge is often ridged or scalloped. Native to all seas, these active,
burrowing mollusks have served mankind as both food and
ornament since prehistoric times.

NORTHERN QUAHOG
Mercenaria mercenaria   **118:3**

Description
Size, 3–5 in. (7.6–12.7 cm). The common edible hard-shell clam.
Shell thick, rounded ovate-trigonal; heavy, moderately inflated.
Sculptured with numerous prominent concentric growth lines or
riblets, fairly well spaced at beaks, becoming close-set toward
margins; valves smooth and glossy in central area; inside margins
lightly ridged. Beaks toward the shorter anterior end; lunule ¾ as
wide as long. Color dingy white to dirty gray; interior chalk-white
with violet muscle scars, purple stains common.
Habitat
Low-tide mud and sand; abundant.
Range
Humboldt Bay, Calif., introduced.

CALIFORNIA VENUS
Chione californiensis   **119:1**

Description
Size, 2–2½ in. (5.1–6.4 cm). Shell robust, subtrigonal (roughly
oval), moderately inflated, margins crenulate. Sculptured with
several stout raised concentric ribs, frilly at edges; and numerous
low, rounded, rather wide radial riblets. Ribbing quite distinct
centrally, weakening toward edges. Dorsal posterior end of right
valve slightly rough, overlaps left valve. Lunule heart-shaped,
striated; escutcheon long, smooth, V-shaped in cross section. Color
creamy white to dull yellow; escutcheon striped with mauve;
interior white, often showing purple at posterior end.
Similarities
Frilled California Venus is more inflated; ribs are more numerous,
closer-spaced, thinner; has mauve-brown splotches.
Habitat
Shoreline sands; common.
Range
Ventura Co., Calif., to Panama.

FRILLED CALIFORNIA VENUS
Chione undatella   **119:5**

Description
Size, 2–2½ in. (5.1–6.4 cm). Shell very similar to California Venus
but sculptured with more numerous and more prominent, closer-
spaced, thinner, wavy concentric ridges and less distinct radial
riblets; lunule conspicuous. Color grayish white, adults retaining
violet-brown blotches.
Habitat
Low-tide sands and mud flats.
Remarks
Edible.
Range
Goleta, Calif., to Peru; most common Calif. *Chione.*

SMOOTH PACIFIC VENUS
Chione fluctifraga   **119:6**

Description
Size, 1–2½ in. (2.5–6.4 cm). Shell roundish-oval (subtrigonal),
stout, solid, moderately compressed; valves heavy, compact.

Sculptured by both radial and concentric ribs, strong over umbones and at posterior ⅓ and anterior ¼ of shell, weaker at margins; central region marked by rather wide, low, concentric ribs bearing coarse half-moon-shaped beads; inner shell margins crenulate. Beaks nearer anterior end. Color creamy white, commonly with darker blue-gray bands, semiglossy; interior white, blotched with purple near muscle scars or on teeth.

Habitat
Mud flats and sand at low tide; not uncommon.

Range
Ventura Co., Calif., to Gulf of Calif.

THIN-SHELLED LITTLENECK
Prototheca tenerrima   **119:10**

Description
Size, 2¾–4 in. (7.0–10.2 cm). Shell oval, thin, highly compressed, chalky; valves rather thin; anterior end short, rounded; posterior end, about 4/5 of shell, deep, rounded. Sculptured with several prominent, evenly spaced, raised concentric ridges and numerous tiny radial threads. Lunule fairly distinct. Color light grayish brown; interior chalky white.

Habitat
Fairly shallow water; not common, but often washed ashore in California.

Range
Vancouver, B.C., to Baja Calif.

PACIFIC LITTLENECK
Prototheca staminea   **119:8**

Description
Size, 1½–2½ in. (3.8–6.4 cm). Shell roundly oval (subovate), slightly longer than high, both ends rounded, valves thickish. Sculptured with many fine concentric and radial ribs, forming crosshatch of tiny beads; especially distinct anteriorly. Beaks almost smooth, nearer anterior end. Color creamy white to rusty brown with purplish cast, sometimes mottled or showing chevronlike chestnut markings.

Similarities
Rough-sided Littleneck is larger, coarsely cancellate and beaded, rusty brown to grayish. Philippine Littleneck is more elongate and compressed, smoother, with lunule and escutcheon more distinct.

Habitat
Low-tide coarse sands and sandy mud to moderately deep water.

Remarks
Edible.

Range
Aleutian Is. to Baja Calif.; one of commonest clams on Calif. coast; most abundant n. of San Francisco.

Note: A number of varieties occur: *Prototheca staminea petiti*, abundant north of Columbia River, is larger than Pacific Littleneck, colored yellowish, chalky white, or dull gray, without color spots; *Prototheca staminea ruderata*, a northern form, has the concentric ridges more prominent than the radiating ribs; *Prototheca staminea orbella* includes specimens misshapen from nestling in the borings of pholads.

ROUGH-SIDED LITTLENECK
Protothaca laciniata   **119:7**

Description
Size, 2–3 in. (5.1–7.6 cm). Very similar to Pacific Littleneck, but sculpture strongly reticulated, with a great many prominent radiating ribs crossed by concentric ridges; many of the ribs have sharp spines. Color rusty brown to grayish.
Habitat
Shallow to fairly deep water; not uncommon.
Range
Monterey Bay to Baja Calif.

PHILIPPINE LITTLENECK
Tapes philippinarum   **119:2**

Description
Size, 3 in. (7.6 cm). Shell oval, inflated; pallial sinus extending less than ½ way to anterior muscle scar; hinge ligament external, prominent. Sculptured by well-defined radiating ribs and less prominent concentric ridges, ribs particularly heavy and conspicuous at posterior end; inside ventral margins smooth. Color very variable, mostly yellowish or buff with geometric patterns of wavy brown or black lines and blotches on sides. Distinguished from other littleneck clams by its short pallial sinus, and from Genus *Chione* by its very prominent radiating ribs and rounded pallial sinus.
Habitat
Coarse, sandy mud of bays, sloughs, estuaries; lives about 1 in. (2.5 cm) beneath surface.
Remarks
Formerly identified as *T. semidecussata.*
Range
B.C. to Elkhorn Slough, Monterey Co., Calif.; accidentally introduced into San Francisco Bay about 1930.

PISMO CLAM
Tivela stultorum   **119:3**

Description
Size, 3–7 in. (7.6–17.8 cm), 4 lb. (1.8 kg). Shell ovate or triangular, thick, solid, moderately inflated, both ends roundly pointed. Sculpture glossy-smooth except for weak growth lines. Periostracum thin, varnishlike. Beaks centralized at triangle apex; hinge rugged, ligament large and strong, lunule broad at base and tapers to a point, with vertical scratches. Color brownish cream or grayish, often with distinct, wide, mauve radial bands; interior porcellaneous.
Habitat
Intertidal sands of open coasts; abundant.
Remarks
An important commercial edible clam with excellent flavor much sought at certain seasons.
Range
Half Moon Bay, Calif., to Baja Calif.

WASHINGTON CLAM
Saxidomus nuttalli

119:4

Description
Size, 3–5 in. (7.6–12.7 cm). Shell roughly oval, sturdy, valves slightly gaping posteriorly. Sculptured with coarse, crowded, sharp concentric ribs. Specimens under 2 in. (5.1 cm) with thin, somewhat glossy shells radially streaked with mauve both fore and aft the beaks on the dorsal edge. Beaks near anterior end, large ligament, no lunule. Color dull grayish white or reddish brown, usually with a few stains or scrawls near beaks; interior glossy white, often purple-stained at posterior margins.

Habitat
Low-tide mud and sandy muds, 12–18 in. (30.5–45.7 cm) deep; very common.

Remarks
Edible.

Range
Humboldt Bay, Calif., to Baja Calif.

SMOOTH WASHINGTON CLAM
Saxidomus gigantea

119:9

Description
Size, 5–6 in. (12.7–15.2 cm). Very similar to Washington Clam and may be only an ecological variant. Shell larger, surface smoother, usually without stains; growth lines finer, less marked; interior white, without purple tinge.

Habitat
Muddy shallow waters 10–14 in. (25.4–35.6 cm) deep; very common.

Remarks
Alaska's most desirable edible shellfish.

Range
Aleutian Is. to San Francisco, Calif.; not common in s. range.

ROCK DWELLERS
Family Petricolidae

The shells of these bivalves are elongate, rounded in front, narrowing behind, and have a weak, almost toothless hinge. They excavate burrows in clay, limestone, coral, and other surfaces, enlarging them as they grow.

HEART ROCK DWELLER
Petricola carditoides

118:8

Description
Size, 1–2 in. (2.5–5.1 cm). Shell oblong; anterior end short, rounded; posterior end elongate, sloping, partially truncated. Shape varies somewhat according to medium into which the animal burrows; normally plump, some may be distorted or slender. Sculptured only by minute radial wrinkles. Beaks low. Color dingy white.

Habitat
Tidewater burrows in limestone, soft rocks, hard clay, etc.

Range
Vancouver, B.C., to Baja Calif.

FALSE ANGEL WING
Petricola pholadiformis   118:6

Description
Size, 2–2¼ in. (5.1–5.7 cm). Shell elongate, fragile, rectangular;
anterior end short, sharply rounded. Sculptured with numerous
strong radial ribs; anterior 10–12 larger, with prominent scales;
posterior ribs crowded and weak; growth lines emphasized at
intervals. Beaks raised, close to anterior end; ligament external, just
posterior to beaks; 2 rather long, pointed cardinal teeth. Siphons
large, tubular, separated almost to base. Color chalky white;
siphons translucent gray.
Habitat
Clay and peat-moss burrower.
Range
B.C. to Baja Calif.; introduced from Atlantic Coast.

RAZOR CLAMS
Family Solenidae

The very long and narrow equivalve shells of this family, often
gaping at the ends, are the true razor clams, so called wherever
they occur. Commonly 6 inches (15.2 cm) long, oval, and laterally
compressed, they have a rather straight, raised, and ventrally
directed internal rib. Distributed worldwide in the sandy bottoms
of shallow coastal waters, all species are considered edible.

PACIFIC RAZOR CLAM
Siliqua patula   121:11

Description
Size, 5–6 in. (12.7–15.2 cm). The commercially valuable razor
clam of Washington and Oregon. Shell oval-oblong, flatly
compressed, moderately thin, both ends evenly rounded. Sculptured
with low concentric lines. Periostracum thin, varnishlike, entirely
concealing surface. Beaks nearly central; internal rib under teeth
stout, transverse, descends obliquely towards anterior end,
extending ½ way across shell to become lost near margin. Color
white; periostracum olive-green or yellowish brown; interior glossy
white, with pinkish to purplish flush.
Similarities
Transparent Razor Clam is smaller, internal rib narrower; found
more to south.
Habitat
Exposed intertidal muds, sands; abundant.
Remarks
Commercially sold fresh or canned.
Range
Aleutian Is. to Pismo Beach, Calif.

TRANSPARENT RAZOR CLAM
Siliqua lucida   121:7

Description
Size, 1–2 in. (2.5–5.1 cm). Shell very oval, flatly compressed; valves
thin, quite fragile, often translucent, smooth; ventral margin
somewhat arcuate. Periostracum thin, varnishlike, polished. Beaks
well off center; internal, narrow, fairly high rib crosses valves at
nearly right angle from beaks. Color polished whitish tan, bluish

white in some specimens, with broad, indistinct, darker radial rays; periostracum olive-green.

Similarities

Resembles young Pacific Razor Clam, which has internal rib broader, lower, descending obliquely toward anterior end, and ventral margin less arcuate.

Habitat

In sand, low tide to 25 fathoms (45.7 m); moderately common.

Range

Bolinas Bay, Calif., to Baja Calif.

MYRA'S RAZOR CLAM

Ensis myrae   **121:10**

Description

Size, 2 in. (5.1 cm). Shell elongate, slender, slightly curving; sides nearly parallel, both ends squarish; valves fragile. Periostracum thin, varnishlike. Beaks nearer anterior end. Color white; periostracum brownish green.

Similarities

Rosy Razor Clam has proportions similar, but not curving.

Habitat

Low-tide sand and mud flats; uncommon.

Range

Monterey, Calif., to Baja Calif.

BLUNT RAZOR CLAM

Solen sicarius   **121:8**

Description

Size, 2–4 in. (5.1–10.2 cm). Shell elongate, length 4 times width; anterior end abruptly truncated, posterior end bluntly rounded; valves thin, smooth, well inflated, single tooth at either end. Periostracum glossy, varnishlike. Beaks set at extreme anterior end Color white; periostracum yellowish green.

Similarities

Sometime mistaken for Transparent Razor Clam, which does not have beaks at extreme end.

Habitat

Low-tide sandy mud flats to 25 fathoms (45.7 m); also said to inhabit burrows similar to those of Jackknife Clam; moderately common in dredgings.

Range

Vancouver, B.C., to Baja Calif.

ROSY RAZOR CLAM

Solen rosaceus   **121:9**

Description

Size, 1–3 in. (2.5–7.6 cm). Shell thin, flat, fragile, length about 5 times width; rather cylindrical, anterior end more rounded and narrower than posterior end. Periostracum thin, smooth, glossy, transparent. Beaks at extreme anterior end; siphons united. Beach-worn shells whitish to pinkish white; periostracum olive to yellowish gray; interior stained with rose.

Habitat

Bayshore sands and sandy mud to 25 fathoms (45.7 m); abundant.

Range

Humboldt Bay, Calif., to Mexico.

TELLINS
Family Tellinidae

A colorful, attractive, worldwide family of graceful bivalves that are usually equivalve, rather compressed, and often somewhat curved. The valves are rounded anteriorly but sharp, slightly folded, and fairly flat behind, and the edges close evenly. The animals have an extraordinarily long siphon and a large pallial sinus.

IDA'S TELLIN
Tellina idae   120:3

Description
Size, 2–2½ in. (5.1–6.4 cm). Shell elongate, compressed, anterior end gracefully rounded, posterior end rather sharply pointed and slightly twisted. Usually either right valve has a rounded, radial ridge (near the dorsal margin) or the left valve has a rounded radial ridge posteriorly (at the dorsal margin), with a furrow below it. Sculptured with numerous fine, strong, rather sharp, evenly spaced concentric lines. Beaks central; ligament elongate, deeply contained within the long dorsal margin furrow. Color grayish white.

Habitat
Low-tide sands to shallow depths; quite common.

Range
Santa Barbara to San Diego, Calif.

MODEST TELLIN
Tellina modesta   120:2

Description
Size, ¾–1 in. (1.9–2.5 cm). Shell elongate-oval, posterior lower corner somewhat pointed; valves thin, compressed. Sculpture smooth, polished, with fine concentric lines, coarsest on extreme posterior slope, fading out at posterior ¼ of shell; sharply defined radial rib just back of anterior muscle scar. Color white, with iridescent sheen.

Habitat
Sandy areas from shoreline to 25 fathoms (45.7 m); common in some localities.

Range
Se. Alaska to Baja Calif.

CARPENTER'S TELLIN
Tellina carpenteri   120:1

Description
Size, ⅓–¾ in. (0.9–1.9 cm). Shell oval, moderately elongate, glossy smooth; anterior end rounded; posterior end shorter, rather truncate; valves thin, compressed. Color, inside and outside, ranging from creamy, whitish, or pinkish white to deep pink.

Habitat
Mud and sand from shore to 370 fathoms (676.7 m); very abundant in some areas.

Range
Alaska to Panama.

BENT-NOSED MACOMA
Macoma nasuta   **120:4**

Description
Size, 2–2½ in. (5.1–6.4 cm). Shell light, thin-edged; posterior end upturned to right, elongated siphons separate. Periostracum thin. Color white; periostracum gray.

Similarities
White Sand Macoma has ligament more conspicuous, posterior end short and more truncated.

Habitat
Muddy bays or sandy beaches; almost every mud flat on West Coast. Withstands stale, dirty water; lives in soft mud, rests on left side at depth of 6–8 in. (15.2–20.3 cm) with siphonal end upward; very common.

Remarks
Edible after thorough cleaning via several changes of water.

Range
Kodiak, Alaska, to Baja Calif.

WHITE SAND MACOMA
Macoma secta   **120:6**

Description
Size, 2–4 in. (5.1–10.2 cm). Largest western macoma. Shell generally oval; valves thin, smooth, glossy; right valve rather well inflated, left valve almost flat; anterior end long, rounded; posterior end short, partially truncated. Distinctly sharp angle (about 120°) at beak and extending toward posterior tip. Periostracum fringed at edges. Beaks central; ligament relatively short, somewhat depressed; inside each valve just behind hinge is a large, oblique, riblike extension. Color creamy to white.

Similarities
Bent-nosed Macoma is smaller; has ligament less conspicuous, posterior end elongate and upturned. Indented Macoma is shorter, slightly more elongate; posterior end more pointed, with slight indentation on posterior ventral margin.

Habitat
Clean sand or mud to 30 ft. (9.1 m); not common.

Range
B.C. to Baja Calif.

INDENTED MACOMA
Macoma indentata   **120:5**

Description
Size, 1½ in. (3.8 cm). Shell oval-subtrigonal, valves compressed; anterior end broadly rounded; posterior end sloping, quite bluntly pointed. Sculpture smooth; indented in ventral line near posterior end. Periostracum thin. Beaks somewhat closer to posterior end. Color grayish white.

Habitat
Offshore cold water.

Range
Puget Sound, Wash., to Baja Calif.

GROOVED MACOMA
Florimetis obesa   **120:7**

Description
Size, 2–3½ in. (5.1–8.9 cm). Shell strong, oval, moderately compressed. Left valve sculptured with shallow radial groove near posterior end, matched on right valve with corresponding ridge; at marginal termination of ridge is a shallow notch. Color dull grayish white; interior glossy white, with central area tinted pastel peach.
Habitat
Offshore, in shallow water; common.
Remarks
F. biangulata is a synonym.
Range
Santa Barbara, Calif., to n. Baja Calif.

WEDGE SHELLS
Family Donacidae

These small, wedge-shaped clams are long and rounded in front, short and straight in back; the valve edges are usually ridged. Each valve contains two cardinal teeth and an anterior and a posterior lateral tooth. The pallial sinus is deep.

LITTLE BEAN CLAM
Donax gouldii   **121:1**

Description
Size, ½–¾ in. (1.3–1.9 cm). Shell in 2 forms. Larger size strong, wedge-shaped, quite fat; anterior end short, sharply rounded; posterior end considerably lengthened, bluntly pointed. Sculpture glossy smooth, except for numerous microscopic axial lines anteriorly. Beaks near posterior end. Color in concentric patterns of purple, brown, or gray-green often rayed with lilac, rose, or light tan; shell margins commonly purplish; interior blotched with bluish brown or purple. Smaller form slightly fatter, usually no color rays, commoner in south.
Similarities
California Bean Clam has beaks nearly central.
Habitat
On surface of inshore sands, particularly hard, smooth sands between mid-tide and low water; common.
Range
San Luis Obispo, Calif., to Baja Calif.

CALIFORNIA BEAN CLAM
Donax californicus   **121:6**

Description
Size, ½–1 in. (1.3–2.5 cm). Similar to Little Bean Clam, but narrowly pointed at each end, with posterior end much lengthened, valves thinner. Sculpture glossy smooth except for faint radiating grooves. Periostracum heavy, varnishlike. Beaks nearly central. Color yellowish white, sometimes with pale brown rays; periostracum greenish tan; interior white or purplish white, dark blotch of purple at each end of dorsal margin.
Habitat
Shallow waters of coves, bays; common.
Range
Goleta, Calif., to Baja Calif.

SEMELES
Family Semilidae

Chiefly warm-seas animals, the semeles have rounded-oval shells that are little inflated, with the posterior end characterized by relatively obscure folds. There are two cardinal teeth in each valve; the right valve has two distinct lateral teeth, the left valve practically none. The ligament is external.

ROCK-DWELLING SEMELE
Semele rupicola   **120:9**

Description
Size, 1–1½ in. (2.5–3.8 cm). Shell rather thick, nearly circular; often irregularly oval, oval-elongate, or obliquely oval; shape corresponds somewhat to hole in which animal lives; both ends rounded, anterior short, posterior a bit longer; basal margin flattish. Sculptured with numerous concentric ridges roughened by a few weak radial cross lines. Beaks off center anteriorly; lateral teeth prominent, cardinal teeth small. Color white to dull cream; interior with glossy white center to deep rose or almost purple, especially marked around margins and at hinge.
Habitat
Rocky, creviced bottoms; in holes of boring mollusks; common in beds of *Mytilus* and *Chama*.
Range
Monterey, Calif., to Gulf of Calif.

BARK SEMELE
Semele decisa   **120:8**

Description
Size, 2–4 in. (5.1–10.2 cm). Shell stout, heavy; nearly circular, except for short slope from beaks to anterior end and abruptly truncated posterior. Sculptured with coarse, wide, irregular, concentric folds and furrows, like coarse bark. Beaks central; pallial sinus deep, prominent; cardinal teeth obsolete. Color whitish gray with purple tinges in grooves; interior china-white, tinged around hinge and margins with rosy purple.
Habitat
Rocky bottoms, coarse sand, gravel in shallow water; not common.
Range
Santa Barbara, Calif., to Baja Calif.

CALIFORNIA CUMINGIA
Cumingia californica   **120:10**

Description
Size, 1–1⅓ in. (2.5–3.4 cm). Shell somewhat variable in shape, valves distorted by hole or crevice in which animal grows, but normally elongate-oval to subtriangular, moderately compressed; anterior end rounded, posterior end pointed, sometimes with slight twist to one side. Sculptured with numerous wavy, rather sharp, largish concentric ridges. Beaks almost central, rolled inward; ligament short, small, lying just posterior to and partly under beaks; behind ligament is a wide-flaring furrow; pallial sinus very long. Color grayish white.
Habitat
Rock crevices, pilings; abundant.
Range
Crescent City, Calif., to Baja Calif.

LONG SIPHON CLAMS
Family Solecurtidae

Members of this family somewhat resemble the tellins but are larger. The sides of the shell are more nearly parallel and are often marked with fine concentric lines.

PURPLE CLAM
Nuttallia nuttallii   121:2

Description
Size, 2½–4¾ in. (6.4–12.1 cm). Shell oval, both ends rounded, posterior end a bit longer and narrower; valves thin, quite compressed, right valve almost flat, left inflated. Sculpture smooth except for fine concentric lines. Periostracum glossy all over in fresh specimens. Beaks off center anteriorly, small; ligament external, like a leather button. Color whitish, often rayed with purple; periostracum rich nut-brown; interior whitish, commonly flushed with rose or purplish.

Habitat
Bays, estuaries, in 6–8 in. (15.2–20.3 cm) of mud; common.

Range
Tomales Bay, Calif., to Baja Calif.

JACKKNIFE CLAM
Tagelus californianus   121:4

Description
Size, 2–4 in. (5.1–10.2 cm). Shell elongate, both ends bluntly rounded; valves rather thin, long margins nearly parallel; length 3 times height or longer. Periostracum partially covers valves, radially striated on posterior slope. Beaks central; pallial sinus extends only to a line vertical to beaks. Color yellowish white, periostracum dark brown.

Habitat
Intertidal sandy muds, 15–20 in. (38.1–50.8 cm) deep in smooth-lined burrow; fairly common on muddy sand flats near salt marshes.

Range
Humboldt Bay, Calif., to Panama.

FALSE DONAX
Heterodonax pacificus   121:3

Description
Size, ½–1 in. (1.3–2.5 cm). Shell oval (like a strong oval tellin), moderately inflated; anterior end truncated, posterior end more rounded. Sculptured with numerous fine growth lines, otherwise smoothish. Beaks anterior of center, forward-pointing; hinge, anterior to beaks, thick at first, then thinner and concave; 2 cardinal teeth in each valve; pallial sinus extends ⅗ shell length. Color quite variable, commonly bluish white with most valves bearing 2 crimson or purplish spots, may also have radial streaks of purplish or violet; some shells pink, yellow, mauve; others speckled with brown or black.

Habitat
Sloping sandy beaches; common, often occurring with bean clams.

Remarks
Formerly confused with *H. bimaculatus*, Atlantic Coast species.

Range
Monterey, Calif., to Peru.

SUNSET SHELL
Gari californica   **121:5**

Description
Size, 2–4 in. (5.1–10.2 cm). Shell elongate-oval, quite thick, valves compressed; both ends rounded, anterior end partially truncated above. Sculptured with irregular, strong concentric growth lines. Periostracum fairly thin, irregularly wrinkled, otherwise quite smooth. Beaks very small, low, just off center anteriorly. Color creamy to dirty white, may show faint, narrow, radial, pinkish to purplish rays extending from beaks; periostracum brownish gray; interior snow-white.

Habitat
Onshore sands to 25 fathoms (45.7 m); common.

Remarks
Edible. Often washed ashore after storms.

Range
Alaska to Baja Calif.

Note: The **DEEPWATER GARI**, *Gari edentula*, occurring from Santa Barbara to San Diego, is similar to the Sunset Shell but has the beaks toward the anterior end and no pinkish rays.

SURF CLAMS
Family Mactridae

The shells of the surf clams are equal, usually tightly closed but sometimes gaping slightly at the ends. The hinge is characterized by a large, spoon-shaped cavity to accommodate the internal ligament and two cardinal teeth. Worldwide in distribution, they are common in sandy bottoms, expecially where pounded by the surf.

CALIFORNIA SURF CLAM
Mactra californica   **122:1**

Description
Size, 1½–2½ in. (3.8–6.4 cm). Shell thin, moderately elongate, strong rather than solid; both ends rounded, anterior a little more acutely. Periostracum with heavy fold along posterodorsal margin, concentrically wrinkled on ventral ⅓ of shell; glossy. Beaks near center, prominent, with concentric undulations; hinge strong, ligamental pit large and triangular, separated from ligament by shelly plate. Color near-white; periostracum grayish.

Habitat
Shallow waters of bays, lagoons, buried 3–6 in. (7.6–15.2 cm).

Remarks
Edible.

Range
B.C. to Costa Rica; not common, but more common in s.

Note: The **PACIFIC SURF CLAM**, *Mactra nasuta*, is similar to the California Surf Clam, but it is sized 3½ inches (8.9 cm); is more oval at the ventral margin and dips down; has a wide posterior gape; and has 2 sharp, raised radial ridges on the posterior dorsal margin. It occurs from San Pedro, California, to Mexico.

HOOKED SURF CLAM
Spisula falcata   122:5

Description
Size, 2–3 in. (5.1–7.6 cm). Shell quite elongated anteriorly, end
narrower; anterior upper margin slightly concave; ventral margin
very convex to hinge below chrondrophore. Periostracum shiny,
usually partially eroded. Color chalky white; periostracum light
brown.

Habitat
Sand, just beyond low-tide mark; moderately common.

Range
Puget Sound, Wash., to Baja Calif.

HEMPHILL'S SURF CLAM
Spisula hemphilli   122:8

Description
Size, 6 in. (15.2 cm). Shell height about ¾ length; rather well
inflated, valves thin; anterior end rounded, compressed; posterior
end sloping, fatter, bluntly pointed. Sculptured by single raised rib
parallel to posterior margin; also numerous fine, concentric growth
lines; otherwise smooth. Anterodorsal slope distinctly concave; all
margins close tightly in adults. Periostracum thin, dull, quite
concentrically wrinkled on posterior slope, usually eroded. Beaks
central, prominent; pallial sinus inclined upward, moderately deep.
Color yellowish white; periostracum grayish brown.

Habitat
Low-tide sands on firm sandy mud in bays, sloughs, estuaries, to
moderately deep, quiet offshore waters. Larger animals inhabit
deeper water, lie 6–8 in. (15.2–20.3 cm) beneath surface of bottom.

Range
Santa Barbara, Calif., to Baja Calif.

PACIFIC GAPER
Tresus nuttallii   122:11

Description
Size, 6–8 in. (15.2–20.3 cm), 4 lb. (1.8 kg). Shell large, strong,
roughly oval to oblongish, well inflated; anterior end rounded,
slightly gaping; posterior end sloping, truncated, prominent gape at
end. Sculpture smoothish. Periostracum thick, often badly eroded
on large specimens. Beaks ¼–⅓ from anterior end, neatly formed;
deep triangular pit under beaks inside; pallial sinus very large,
deep. Siphons extremely long, united, nonretractile into shell,
covered with heavy, dark epidermis; tips with thick cutaneous
flaps. Color white to gray; periostracum grayish to brownish,
stained black if animal lives in mud.

Similarities
Northern Gaper is larger, broader; has more northern range.

Habitat
Buried 3+ ft. (0.9+ m) in fine sand or mud from high-tide mark
to 100 ft. (30.5 m), in bays, sloughs, estuaries, quiet offshore
waters of coves; extremely common.

Habits
Siphons form tube reaching surface; when disturbed, animal squirts
water several feet into air.

Remarks
Formerly Genus *Schizothaerus*.

Range
Puget Sound, Wash., to Baja Calif.

NORTHERN GAPER
Tresus capax

122:2

Description
Size 8–10 in. (20.3–25.4 cm). Shell similar to Pacific Gaper but larger, more oval (broader), more inflated; ventral margin well-rounded, dipping more steeply from beaks.

Habitat
Intertidal sand and mud beaches.

Range
Kodiak Is., Alaska, to Monterey, Calif.; very common, especially in Puget Sound, Wash.

Clams, Rock Borers, Piddocks, and Shipworms
Order Myoida

SOFT-SHELLED CLAMS
Family Myidae

These "steamer" clams of the menu have gaping, usually unequal valves and an internal resilium behind the beaks attached in the left valve to a horizontally projecting chondrophore, a spoon-shaped tooth which fits into a corresponding pit in the right valve. They are found worldwide.

COMMON SOFT-SHELLED CLAM
Mya arenaria

122:10

Description
Size, 1–6 in. (2.5–15.2 cm). Shell elliptical, moderately thick, lightweight, brittle, both ends gaping; anterior end rounded, posterior end slightly pointed. Sculptured by roughened, somewhat wrinkled growth lines. Periostracum very thin. Beaks central; chondrophore long, spoon-shaped, shallow; pallial sinus somewhat V-shaped. Color chalky white to gray; periostracum gray to straw.

Habitat
Intertidal heavy black mud flats, buried with just tip of siphon exposed; very common.

Habits
Reveals presence to intruder by abrupt vertical squirt of water from suddenly indrawn siphon.

Remarks
A dietary staple.

Range
Introduced accidentally from the East Coast in 1874 to San Francisco Bay; now Alaska to Elkhorn Slough, Calif.

FRAGILE SPHENIA
Sphenia fragilis

122:4

Description
Size, ¾ in. (1.9 cm). Shell rather elongate but often irregular, fragile; anterior half fat, rotund; posterior end extended into long, narrow, compressed, commonly twisted snout. Sculptured with fine concentric threads. Periostracum dull, usually eroded near beaks. Beaks near center; chondrophore in left valve large, elongate,

flattened, 2-lobed, jutting obliquely near hinge margin of right valve to large round socket. Color chalky white; periostracum yellowish gray.

Habitat
Low-tide muds to 46 fathoms (84.1 m); common.

Range
Oreg. to Peru.

Note: The Fragile Sphenia could be confused with the **FAT SPHENIA**, *Sphenia ovoidea,* which is half as large, smoother, and more ovoid; lacks a snout; is uncommon; and occurs from Alaska to Panama.

CALIFORNIA SOFT-SHELLED CLAM
Cryptomya californica   **122:9**

Description
Size, 1–1½ in. (2.5–3.8 cm). Shell oval, moderately fat; valves thin, fragile, right valve fatter; slight gape at posterior end, noticeably curved. Sculptured by growth lines only; otherwise smooth. Periostracum at posterior end faintly and radially striped. Beaks slightly off center, right beak crowding slightly over left; chondrophore in left valve large, fits closely against small, concave shelf under right beak. Color chalky to ash-gray; periostracum dull gray; interior slightly nacreous when fresh.

Habitat
Intertidal zone in sand, muds, gravels, buried up to 20 in. (50.8 cm) deep; quite common.

Habits
The short siphon may penetrate burrows of other marine animals.

Range
Alaska to Peru.

BASKET CLAMS
Family Corbulidae

Formerly the Family Aloididae, the shells are small, solid, inequivalve with one valve usually overlapping the other. Commonly ribbed centrally, the valves may gape slightly at the anterior end. Each valve has one upright conical tooth. They occur worldwide in temperate waters.

BASKET CLAM
Corbula luteola   **122:6**

Description
Size, ⅓ in. (0.85 cm). Shell sturdy, somewhat squarish, slightly obese; anterior end elliptical; posterior end sloping, rather bluntly pointed, bearing distinct line from beaks to tip; right valve fatter, overlaps left valve on ventral margin. Sculptured with weak concentric growth lines, weakest toward beaks. Beaks off center posteriorly, strong, close together. Color porcellaneous, whitish gray or yellow, occasionally pinkish to purplish; interior whitish, though commonly yellowish and reddish purple stains.

Habitat
Onshore sands, rocky bottoms, rubbly beaches to 25 fathoms (45.7 m); common in some localities.

Range
Monterey, Calif., to Gulf of Calif.

ROCK BORERS
Family Hiatellidae

The nestling and burrowing habits of these clams cause irregularities and considerable variation in their shells. The texture is chalky, they are toothless, and the pallial line is interrupted; the naked siphons are separated at the tips.

ARCTIC ROCK BORER
Hiatella arctica   **122:3**

Description
Size, 1–3 in. (2.5–7.6 cm). Shell variable, young rather evenly oblong; adults oval, oblong, twisted, or misshapen; generally elongate with dorsal and ventral margins parallel; each valve has strongish radial rib, sometimes scaled, at posterior end; may show gape at posterior end. Sculptured with coarse, irregular growth lines. Periostracum thin, weak, flakes off when dry. Beaks about ⅓ back from anterior end, close together; just posterior is a conspicuous bean-shaped external ligament. Color chalky white; periostracum gray.

Habitat
Cold water; common.

Range
Arctic Ocean to deep water off Panama.

GEODUCK
Panopea generosa   **122:7**

Description
Size, 7–9 in. (17.8–22.9 cm), 8 lb. (3.6 kg). Shell very large (but cannot nearly contain the 2 huge, joined siphons up to 3 ft., or 0.9 m, long and equal to ½ weight of entire clam), slightly elongate, inflated; valves rather thick, bluntly oval, gaping at both ends; anterior end rounded, posterior end truncate. Sculptured with coarse, concentric, wavy ridging, especially apparent near beaks. Periostracum thin. Beaks small, central, depressed; hinge with 1 large horizontal thickening. Color dull grayish white; interior may be pearly.

Habitat
Buried 4–6 ft. (1.2–1.8 m) in mud, extending long siphons to surface; common, but large specimens rare.

Remarks
Edible but tough.

Range
Forrester Is., Alaska, to Baja Calif.

PIDDOCKS
Family Pholadidae

These boring clams can penetrate hard rock, coral, and wood. The long, narrowed shells are white, thin, brittle, and have sharp toothlike ridges in front for abrading purposes. They gape at both ends and lack both ligament and hinge teeth. They are distributed worldwide in all seas.

PACIFIC PIDDOCK
Barnea subtruncata   123:4

Description
Size, 2–2½ in. (5.1–6.4 cm). Shell very elongate, rather cylindrical; valves thin, brittle, both ends moderately gaping; anterior end pointed, reflected; posterior end elongated, bluntly rounded. Anterior end sculptured with prickly, concentric lamellae, sharp and distinct; on posterior ⅓ faint or lacking; no distinct radial grooves on sides of valves. Periostracum heavy, covering siphon. Beaks ⅓ in from anterior tip; just anterior, the top shell margin folds back on itself to form a long, triangular shelly "third valve" extending above the ligament; beneath hinge is a delicate spoonlike appendage; internal rib short, curved, flattened. Color snow-white; periostracum brown.
Habitat
Low-tide hard clay or soft shale banks, in burrows 10+ in. (25.4+ cm) deep; moderately common in some localities.
Remarks
B. pacifica is a synonym.
Range
Newport, Oreg., to Chile.

COMMON PIDDOCK
Penitella penita   123:2

Description
Size, 3–4 in. (7.6–10.2 cm). Shell thin, globular at anterior end, tapering rapidly to narrow, rather compressed posterior end, characterized by extended leathery membrane protecting siphon; anterior end gaping (foot protrudes during burrowing stage, shelled over in adult life); mid-dorsal area crested by short, triangular plate. Color brownish white.
Similarities
California Piddock is larger.
Habitat
Burrows into clay, shale, sandstone, or other soft rock on open coast, seldom over 5 in. (12.7 cm) deep; common.
Habits
A hammer blow causes animal to squirt water from its siphon hole.
Remarks
Mildly flavored, good in chowder.
Range
Bering Sea to Baja Calif.

Note: Another form in the same range, *Penitella sagitta,* lacks calcareous closure over anterior gape.

PILSBRY'S PIDDOCK
Zirfaea pilsbryi   **123:1**

Description
Size, 2–4 in. (5.1–10.2 cm). Shell widely gaping at both ends; valves thin, zoned by an oblique fold into anterior section, sculptured with concentric growth lines, and posterior section, sculptured with growth lines plus sharp scales. These filelike scales constitute the boring rasp for excavating burrows in hard clay. Periostracum thin; in fresh shell covering folded-back dorsal margin at and before beaks and posterior (siphonal) end. Beaks ⅓ in from sharp posterior tip; no accessory plate; large "spoon" beneath. Siphon granulated with chitinous spots. Color snow-white; periostracum yellowish.
Habitat
Burrows to 10–14 in. (25.4–35.6 cm) in hard-packed clay at low-tide level; commonly washed ashore dead.
Range
Bering Sea to Baja Calif.

OVAL PIDDOCK
Chaceia ovoidea   **123:7**

Description
Size, 2–5 in. (5.1–12.7 cm). Shell short, stubby, oval from side view; anterior or foot end gapes, half-covered by smoothish calcareous eggshell-like material; posterior end blunt. Anterior end sculptured with sharp radiating lines set with scales (the burrowing rasp); posterior end decorated with growth lines. Periostracum often forms tubelike protection for siphon beyond posterior end, covered at base, remainder wartlike with small upraised chitinous bars. Beaks well off center anteriorly, valves widely gaping behind. Color whitish; periostracum brownish; siphon cream.
Habitat
Burrows to 20 in. (50.8 cm) deep in hard clay, soft rock; moderately common.
Remarks
Edible.
Range
Santa Cruz, Calif., to Baja Calif.

CALIFORNIA PIDDOCK
Parapholas californica   **123:9**

Description
Size, 3–6 in. (7.6–15.2 cm). Shell oval-oblong, cylindrical; wedge-shaped, almost pear-shaped; anterior end rotund; posterior end elongated, circular in cross section, pointed, tip truncated and gaping; dorsal margin with 2 long, complicated accessory plates. Not less than 8 accessory plates can be counted, 2 along basal margin extending most of shell length, ventral edge covered by 1 elongate plate; anterior gape closed by 2 thin, calcareous extensions. Exterior sculptured with long radial grooves pushed in at the middle, anterior section with sharp, wavy ridges, posterior section with concentric growth lines; interior with 2 long, descending, shelly rods under hinge center. Periostracum over sides of snout, foliated. Beaks near anterior end; diagonal line from beaks divides shell into 2 parts. Color grayish- to brownish-white; periostracum brownish; siphon tip reddish purple.

693

Habitat
Bores to 8 in. (20.3 cm) into hard clay, shale, etc., on open coast;
common.
Remarks
Edible.
Range
Santa Cruz, Calif., to Baja Calif.

SHIPWORMS
Family Teredinidae

Shipworms are an aberrant group of economically important clams
in which the shell is used for boring into saltwater-soaked wood
and the body, which maintains connection with the outside world
of sea water, becomes greatly lengthened, sometimes to forty times
the length of the shell. The two highly specialized valves at the
anterior end cannot be used in identification of species; therefore,
differentiation is based on the two featherlike pallets at the
posterior end, which the animal uses to close off the end of its
burrow. The shipworms are worldwide in distribution, the actual
range being indeterminate for any given species inasmuch as they
inhabit drifting wood.

FEATHERY SHIPWORM
Bankia setacea   **123:8**

Description
Size, shell ½ in. (1.3 cm). Shell vestigial; shelly tube of burrow,
secreted by mantle as added body protection, to several inches long.
Pallets with 15–30 parts, to 2 in. (5.1 cm) long and ¼ in. (0.6 cm)
wide, plumelike. Color white, sometimes tinged with pale rose.
Habitat
Ocean-soaked wood, in burrows to 3 ft. (0.9 m) deep and ⅞ in.
(2.2 cm) dia. Frequently so many individuals are present that the
timber becomes a mass of intertwining galleries.
Range
Bering Sea to San Diego, Calif.

NAVAL SHIPWORM
Teredo navalis   **123:3**

Description
Size, shell ½ in. (1.3 cm). Valves thin, sharp, globular; wings
subtriangular. Pallets 2 in. (5.1 cm) long; pointed, calcareous,
slightly compressed, symmetrical; with shallow cones, urn-shaped,
widening from a slender stalk and tapering to a somewhat hollow
tip. Periostracum over extreme distal ⅓ of pallet; yellowish brown.
Habitat
Ocean-soaked wood, in temperate seas; can live in brackish water;
very common.
Food
Plankton from sea water; cellulose from the wood it excavates.
Remarks
Destructive; particularly destructive to untreated piling in brackish
waters, such as those surrounding San Francisco Bay, Calif.
Range
Worldwide, arctic to tropic seas.

Paper Shells, Slender Clams, and Spoon Shells
Order Pholadomyoida

PAPER SHELLS
Family Lyonsiidae

Members of this family have shells with fragile valves which may be unequal in size and somewhat rectangular. The ligament attaches to a narrow shelly ledge along the hinge line. The various species glue sand grains to the shell surfaces.

CALIFORNIA LYONSIA
Lyonsia californica   **123:5**

Description
Size, 1 in. (2.5 cm). Shell quite elongate, moderately fat, very thin and fragile, almost transparent, anterior end somewhat inflated, rounded; posterior end tapering, laterally compressed, often crooked, tip truncated; valves equal. Periostracum thin; often eroded, revealing pearly shell surface; commonly remaining as numerous weak dark radial lines. Beaks well off center anteriorly, area swollen; ossicle under hinge inside. Color whitish, opalescent in worn specimens; periostracum olive-gray; interior pearly, ossicle opaque white.
Habitat
Bay and slough mud and sand bottom, nestled in small pits, usually close inshore but to 40 fathoms (73.2 m); common.
Range
Alaska to Baja Calif.

ROCK ENTODESMA
Entodesma saxicola   **123:10**

Description
Size, 2–5 in. (5.1–12.7 cm). Shell misshaped to conform to abode in rock holes, crevices, burrows; generally oblong to pear-shaped; anterior end narrow, short; posterior end swollen, gaping, abruptly truncate; valves weak, brittle. Periostracum thick, rough, strong, partially flakes off when dry. Beaks off center anteriorly; no teeth on hinge. Periostracum brown; interior brownish tan to whitish, with slight opalescence.
Habitat
Shoreside rocks, in holes, crevices, burrows; moderately common.
Range
Aleutian Is. to Baja Calif.

SEA BOTTLE SHELL
Mytilimeria nuttallii   **123:6**

Description
Size, 1–2 in. (2.5–5.1 cm). Shell obliquely oval, bulging in hinge region; valves equal, thin, fragile, convex. Sculptured with fine concentric and radiating lines. Periostracum very thin. Beaks small, spiral, central; hinge weak, no teeth, small calcareous ossicle present. Color white, underlayered by slightly pearly material; periostracum yellowish brown.

Habitat
Under low-tide rocks to 10 fathoms (18.3 m), usually embedded among bottlelike compound ascidians or sea squirts; common.
Range
Alaska to Baja Calif.

SLENDER CLAMS
Family Pandoridae

These small, inequivalve shells are very thin and flat. The valves are white outside, pearly inside. The inconspicuous beaks are barely noticeable, and there are two teeth at the hinge which fit into matching grooves. These animals generally live on stony or pebbly bottoms in all seas.

WESTERN PANDORA
Pandora filosa   **124:1**

Description
Size, 1 in. (2.5 cm). Shell thin, rather strong; valve outline semicircular; right valve almost flat, with 1 largish lateral tooth; left valve moderately convex; dorsal margin almost straight, basal margin gently rounded; anterior end deeply rounded, posterior end extended somewhat into a rostrum with tip slightly upturned and squared. Sculptured by fine growth lines; right valve with fine radiating grooves, 2 ossicles present. Periostracum on border. Beaks minute, about ¼ in. from anterior end. Color opalescent white; periostracum brownish; interior very pearly and opalescent.
Habitat
From 10–75 fathoms (18.3–137.1 m); moderately common.
Range
Alaska to Baja Calif.

DOTTED PANDORA
Pandora punctata   **124:2**

Description
Size, 2 in. (5.1 cm). Shell shape unusual; very thin, much compressed, somewhat sickle-shaped, area between tip and beaks deeply concave; anterior end regularly rounded; posterior end expanded, acutely rostrate and truncate, tip upturned. Color snow-white; interior vividly pearly, with tiny raised dots.
Habitat
Low-tide line to 20 fathoms (36.6 m); rare.
Range
Vancouver, B.C., to Baja Calif.

PERFORATED CLAMS
Family Thraciidae

The valves of these rather large clams are quite unequal, the right being fatter than the left, and both ends gape slightly. The shell is commonly somewhat rostrate posteriorly, with prominent beaks set so closely together that the left beak punctures the right. The ligament is external. The animals live in moderately deep water.

PACIFIC THRACIA
Thracia trapezoides   **124:4**

Description
Size, 2 in. (5.1 cm). Shell sturdy, thin, chalky, both ends gently
rounded; posterior end somewhat flattened, partially truncate, the
broadly rostrated part set off by a furrow bordered by radial ridge.
Sculptured by faint, concentric lines. Beaks near center, rather
prominent, right beak perforated. Color drab grayish white to
brown.

Similarities
Short Thracia is shorter, suboval, lacks prominent rostrum.

Habitat
Sand, mud; commonly dredged.

Range
Alaska to Redondo Beach, Calif.

SHORT THRACIA
Thracia curta   **124:6**

Description
Size, 1–1½ in. (2.5–3.8 cm). Shell very similar to Pacific Thracia
but suboval in outline, no prominent rostrum, valves rather fat;
anterior and regularly rounded, posterior end very slightly rostrate
and rather bluntly rounded. Sculptured by weak radial ridge;
ventral margin irregularly undulating. Beaks central, prominent;
pallial sinus shallow, U-shaped; hinge thickened behind beak and
below the large, wide, external ligament; no teeth. Color chalky
white.

Habitat
Extreme low tide to 20 fathoms (36.6 m), often on wharf pilings;
moderately common, frequently dredged.

Range
Alaska to Baja Calif.

SPOON SHELLS
Family Periplomatidae

These small, fragile shells are oval, with the right valve fatter than
the left; they gape slightly and have a faint pearly sheen. They are
called spoon shells because of a spoon-shaped tooth at the hinge
which is supported by a small triangular process (lithodesma),
often lost when the animal is removed from the shell. The ligament
is absent; the anterior muscle scar is long and narrow, contrasting
with the small, ovate posterior scar.

WESTERN SPOON CLAM
Periploma planiusculum   **124:3**

Description
Size, 1–1¾ in. (2.5–4.5 cm). Shell ovate (rounded rectangular),
thin, right valve fatter than left; anterior end short, rounded;
posterior end long, evenly rounded. Sculptured with weak,
concentric growth lines. Beaks close to anterior end; inside, each
valve has prominent spoonlike process projecting into beak cavity.
Color snow-white.

Habitat
Muds just beyond low-tide mark; very common, often washed
ashore.

Range
Pt. Conception, Calif., to Peru.

DIPPER SHELLS
Family Cuspidariidae

The shells of this family are pear-shaped (globose in front, rostrate behind), thin, and commonly ribbed; most are very small. The hinge shows a posterior lateral tooth in the right valve, and the ligament is external and elongated. Dipper shells are inhabitants of deep water.

OLDROYD'S DIPPER
Cardiomya oldroydi   **124:5**

Description
Size, ¼ in. (0.6 cm). Shell tiny, short, inflated; anterior end swollen, rounded; posterior end extended and narrowed to produce a small tube. Sculptured with sharp, coarse, radiating ribs, absent from posterior extension. Color rather dirty white.
Habitat
Deep water.
Range
Puget Sound, Wash., to Catalina Is., Calif.

CALIFORNIA DIPPER
Cardiomya californica   **124:7**

Description
Size, ¼ in. (0.6 cm). Shell tiny, short, inflated; anterior end swollen, rounded; posterior end extended and narrowed to produce a small tube. Sculptured with up to 20 sharp, coarse, radiating ridges. Periostracum very thin, delicate. Color creamy white.
Habitat
Deep water; not uncommonly dredged.
Range
Puget Sound, Wash., to San Diego, Calif.

CEPHALOPODS
Class Cephalopoda

Cephalopods are carnivorous mollusks with the shell, when present, lengthened and usually inside, not outside, the animal. The region of the head is surrounded by tentacles or arms, which have prehensile suckers. The mouth is round with a parrotlike beak and there are two eyes, often large. The body is either cylindrical or rounded and frequently has terminal fins. The outside has colored spots which, when contracted or expanded, change the color of the surface of the animal. Most cephalopods can discharge an inky fluid as a means of defense.

This class includes some of the most highly organized, swift-moving, and intelligent of the invertebrates. Some octopi in tests have shown an intelligence greater than that of many vertebrates.

Spirula and Squids
Order Decapoda

The members of this group all have ten arms, of which two are long and tentacular. The body is cylindrical in outline, usually with an internal shell which may be either calcareous (the cuttlebone) or thin and horny (squid pen). The arms bear small suckers, usually set on peduncles, or small stalks. Around each sucking aperture is a horny ring or a ring of hooks.

COMMON SQUIDS
Family Loliginidae

These animals have a long, tapering, cylindrical body with ten arms and large triangular terminal fins. The arms bear two rows of suckers; each sucker is surrounded by a horny, dentated ring. The tentacular arms bear four rows of suckers on small clubs. The horny internal pen is slightly broader at its base and tapers to a point, with a keel on the underside.

COMMON PACIFIC SQUID
Loligo opalescens

104:9

Description
Size, body 6–8 in. (15.2–20.3 cm). Body slightly swollen near middle; fins about ½ as long as mantle, very slightly lobed in front; siphon large, broad, with dorsal muscular bridles. Pen broad.
Habitat
Offshore waters.
Remarks
The common edible squid of the Pacific Coast, at some seasons in schools of thousands; commercially fished.
Range
Puget Sound, Wash., to San Diego, Calif.

Octopods
Order Octopoda

The octopods are characterized by having only eight arms, lacking the tentacular arms of the squids. The arms bear suckers which, unlike those of the squids, are not on stalks and lack horny rings. There is no internal pen or shell.

OCTOPI
Family Octopodidae

COMMON PACIFIC OCTOPUS
Octopus dolfleini

104:11

Description
Size, 6 in.–3 ft. (15.2–91.4 cm). Length includes body and longest arm; radial spread of Alaskan forms may reach 28 ft. (8.5 m). Skin of preserved specimens covered all over by numerous small, pimplelike tubercles with star-shaped bases, as well as with many heavy longitudinal but interrupted wrinkles; web between second and third arms (eyes pointed away from observer) usually extends out ¼ arm length. Above each eye is a rather small, conical wart behind which stands a very large, pinnaclelike protuberance.
Habitat
Shore to 100 fathoms (182.8 m); abundant.
Range
Alaska to Baja Calif.

Other Marine Invertebrates

Consulting Editor
James W. Nybakken
Professor and Staff Member
Moss Landing Marine Laboratory

Illustrations
Plates 125–126 Klarie Phipps
Plate 127 John Hamberger
Plates 128–129 Nancy Lou Gahan
Text Illustrations by Pamela Carroll

Other Marine Invertebrates

In addition to the mollusks, a large number of species of other invertebrates inhabit the shore waters of the Pacific Coast—with a few in the freshwater streams and lakes of the Pacific Northwest—for a total in excess of those that can be described in this chapter. The marine invertebrate fauna of the Pacific Coast is one of the richest in the world.

Habitat

After more than 525 million years of evolution, since Precambrian times, the marine invertebrate fauna has become adapted to every possible habitat between the high-tide mark and the abyssal depths. Accessible habitats for observation include tide pools, sea caverns, reefs, mud and sand flats, and the open sea, as well as the surf on sand and solid rock. In this area of the upper watery layers, sea life is subject to very much less seasonal change of temperature than are land animals. Along the Pacific Coast at any one locality, the maximum temperature range over the year in marine waters is low, from 48°F (9°C) to 63°F (17°C) in Monterey Bay, California, for example, compared to a hundred degree differential in the Great Central Valley of California, where temperatures vary between 20°F (7°C) and 120°F (49°C), according to the season. Because of the narrow seasonal range in seawater temperatures, many of the ocean animals have limited ability to adapt to extreme temperatures. Moreover, since cold water holds more oxygen in solution than warm water, and since Pacific Coast waters are always cool, most animals are easily stressed or killed when kept at elevated temperatures.

The best place to seek invertebrate animal life is in tidal pools, which occur in suitable areas along nearly all stretches of the western North American coastline. The best time in the spring and summer months is usually very early morning at times of the lowest tides. In fall and winter, the lowest tides occur in the afternoon and evening hours. Local newspapers usually supply exact hours.

Conservation

Many tide pools are protected because they are in state parks, and collecting or damaging specimens is prohibited by law. Furthermore, all California marine life is protected and may not be collected without a permit. An unfortunate number of tide pools that once teemed with plant and animal life have been stripped almost bare by vandals, commercial collectors, or biology students, and when taken out of their natural element, these colorful creatures soon die.

In other areas, overfishing has reduced the number of invertebrates (drastically in some places), such as lobsters, crabs, and in freshwater streams, crayfish. Pollutants have killed great numbers of shallow water inhabitants, especially in areas where sewage and industrial wastes have polluted harbors and inlets. Real estate development all along the West Coast and the sprawl from urban centers have destroyed much ocean-front, bay-shore, and salt-marsh habitat. Trawlers operating just offshore have all too frequently ravaged many submarine fields and pastures.

MARINE INVERTEBRATES

Range and Scope
This chapter discusses 153 species, covering the coastal areas of western North America from the Straits of Juan de Fuca to the border of Mexico. All those described here are marine, constituting the most common, typical, or striking forms that occur between these boundaries, to a depth of about 25 fathoms (45 m). The size given is the species' average size; occasionally the record size is also included.

Illustrations
Each species illustrated is shown as it would be seen by the naked eye, without the use of a high-powered microscope, to aid field identification. Common names, for those species for which they are accepted, have been included. When there is no accepted common name, the common name of the group to which the species belongs has been captioned in parentheses on the illustrated plates, numbers 125 through 129, and in the text illustrations to facilitate identification.

Nomenclature
The scientific names and the classification of the invertebrates described conform to the latest accepted usage as set forth in the third edition of *Light's Manual,* by Smith and Carlton, 1975.

USEFUL REFERENCES

Guberlet, M. L. 1936. *Animals of the Seashore.* Portland, Oreg.: Metropolitan Press.

Hinton, S. 1969. *Seashore Life of Southern California.* California Natural History Guides No. 26. Berkeley, Calif.: Univ. of Calif. Press.

Kozloff, E. 1973. *Seashore Life of Puget Sound, the Strait of Georgia, and the San Juan Archipelago.* Seattle, Wash.: Univ. of Wash. Press.

Johnson, M. E., and Snook, H. J. 1927. *Seashore Animals of the Pacific Coast.* New York: Macmillan.

MacGinitie, G. E., and MacGinitie, N. 1968. *Natural History of Marine Animals.* 2nd ed. New York: McGraw-Hill.

Ricketts, E. F., and Calvin, J. 1968. *Between Pacific Tides.* 4th ed., rev. by Joel Hedgepeth. Stanford, Calif.: Stanford Univ. Press.

Smith, R. I., and Carlton, J., eds. 1975. *Light's Manual: Intertidal Invertebrates of the Central California Coast.* Berkeley, Calif.: Univ. of Calif. Press.

GLOSSARY

Aboral The side away from the mouth.

Ampulla (*pl.* ampullae) A membranous sac.

Anterior Front; forepart.

Branchiae Gills.

Capitulum Barnacle body at end of stalk.

Carapace Horny covering or shell.

Chelipeds Claw-bearing appendages.

Cilia Tiny hairs.

Cirri Tendrils or fleshy appendages.

Cloaca Common duct into which digestive, excretory, and reproductive systems empty.

Detritus Organic waste.

Dorsal Pertaining to the back.

Flagellum Long, movable hairlike process of cell surface.

Ganglion (*pl.* ganglia) A mass of nerve cell bodies.

Gnathopods Crustacean appendages which manipulate food.

Gonads Reproductive glands.

Hermaphroditic Containing both male and female gonads.

Hydranth Oral end of polyp in a hydrozoan colony.

Interradius Space between radii.

Interray Space between rays.

Lobule Part of a lobe.

Madreporic plate, madreporite Water intake disk.

Mantle An enveloping external covering of the body.

Medusa Jellyfish.

Mesenteries Folds of tissue in the body cavity.

Nephridia Excretory and internal water regulatory systems.

Palp A small, often fleshy, feeler.

Papilla A small, nipplelike projection.

Parapodia Side feet.

Peduncle Stalk.

Peristome Region around mouth.

Polyp Individual animal in a colony. Synonym for hydranth above.

Prehensile Grasping; capable of suspending body.

Proboscis Nose; extension in front of face.

Radius (pl. radii) An imaginary radial plane dividing the body of a radially symmetrical animal into similar parts.

Ray Any of the radiating divisions of the body of an echinoderm with all its included parts.

Rostrum Beak or beaklike part of an animal.

Scutum Barnacle plate.

Sessile Attached directly by its base.

Setae Bristles; stiff hairlike projections.

Spicule Skeletal unit of sponges.

Spongin A horny substance.

Stolon Extension of body wall from which buds are developed.

Telson Posterior projection of last body segment.

Test Hard internal skeleton.

Thoracic Pertaining to the thorax, or middle division of the body.

Umbo Prominence above hinge in bivalves.

Ventral Pertaining to the belly or underside.

Zooecia Individual units containing zooids in a colony.

Zooid Individual animal in a colony.

SPONGES
Phylum Porifera

The sponges are the most primitive of all multicellular animals. They are characterized by a low order of organization, porosity, and permanent attachment. They display greater plasticity of form than any other animal and show almost no other biologic relationships, unless it be to a subgroup of the flagellate protozoans termed the Choanoflagellata, considered by some specialists to be ancestral. In quiet waters sponges grow upright or branch out; in surf they become flattened like a crust. A few inhabit fresh water, the majority salt water, from the surface to the greatest depths. The most evident animal motion is the outward flow of water from large openings on the body of the sponge.

Fig. 91 Typical Sponge

The structure of the sponge skeleton is the basis for higher classification. In the Calcarea the spicules, or tiny skeletal units, are of calcite. In the Demospongiae the skeleton, if there is one, may be spongin (a horny substance) alone or spongin strengthened with four-rayed siliceous spicules. Since the greatly varied species of sponges can usually be positively identified only by dissolving the organic matter in alkali (cooking in potassium or sodium hydroxide) and examining the spicules left behind under a compound microscope, there is little value in describing here more than the more readily identifiable species. The sponges which follow, like all the invertebrates in this unit, are shown as they would be seen by the naked eye.

CALCAREOUS SPONGES
Class Calcarea

LEUCOSOLENIA ELEANOR Fig. 92

Description
Size, 2–3 in. (5.1–7.6 cm). Consists of an open network of opaque white tubes with oscula (mouthlike openings) at the free ends; spicules on both one and three axes.
Habitat
Low intertidal zone of open coast, loosely attached to rocks.
Range
Calif.

Fig. 92
Sponges

Leucosolenia eleanor   *Lissodendoryx firma*, p. 708   *Leucilla nuttingi*, p. 708

LEUCILLA NUTTINGI
Fig. 92

Description
Size, 2–3 in. (5.1–7.6 cm), usually ⅕–⅖ in. (0.5–1 cm) dia. Vase-shaped, with osculum slightly narrower than body; body borne on a narrow stalk. Several sponges arise from what appears to be a common base, and thus the species occurs in tight clumps.
Habitat
Attached to rocks.
Range
Calif.

SILICEOUS AND HORNY SPONGES
Class Demospongiae

LISSODENDORYX FIRMA
Fig. 92

Description
Size, 6 in. (15.2 cm) dia. Body to 3 in. (7.6 cm) thick, irregular and lumpy (like a dried bath sponge), semicavernous. Color yellow. Has strong offensive odor.
Habitat
Crevices or cracks between rocks, underside of rocks, in contact with substratum of open coast; abundant in clusters and masses.
Range
Cen. Calif.; especially common at Pacific Grove.

OPHLITASPONGIA PENNATA
Fig. 93

Description
Size, 8 in. (20.3 cm) dia. Body irregular; thin encrusting, firm, rough; many oscula, spicules bow-shaped or long and single-pointed. Color bright red or red-orange.
Habitat
Under rocks on open coast; common.
Range
Calif. to B.C.

Fig. 93

Halichondria panicea
Breadcrumb Sponge

Ophlitaspongia pennata
(Sponge)

Haliclona permollis
Purple Encrusting Sponge

BORING SPONGE
Cliona celata

Description
Size, 8 in. (20.3 cm) dia. Body irregular, encrusting, firm, warty; spicules pointed at one end, rounded at the other. Color bright yellow.

Habitat
On old mollusk or barnacle shells, low tide to 100 ft. (30.5 m).
Habits
Free-swimming larva attached to a shell bores a series of canals throughout the shell, finally consumes it, and emerges as the actual sponge.
Range
Arctic Ocean to s. Calif.

BREADCRUMB SPONGE
Halichondria panicea   *Fig. 93*
Description
Size, 1¼ in. (3.2 cm) dia. Body encrusting, stiffened by mass of needlelike spicules with a few horny fibers, fragile; surface tuberculate to superficially smooth. Color orange to green, grayish, or yellowish.
Habitat
Beach to low tide; relatively common.
Range
Alaska to s. Calif.

PURPLE ENCRUSTING SPONGE
Haliclona permollis   *Fig. 93*
Description
Size, varies, encrusting. Body in compact, encrusting colonies, fairly soft but not slimy; oscula (volcanolike "craters") regularly spaced. Color vivid purple.
Habitat
Low tide to several hundred feet; common.
Range
Arctic Ocean to s. Calif.

CNIDARIANS
Phylum Cnidaria

Forming one of the major groups of the animal kingdom, the cnidarians are a varied, numerous, and largely marine phylum of animals with a body structure based on two cell layers, compared to three cell layers in the higher animals. They get their name from the special stinging cells, the nematocysts, contained in cells called cnidoblasts and unique to this phylum. All members have within the body only a single hollow tubular cavity—the coelenteron—which functions as the digestive sac. Around its single opening is an array of tentacles armed with nematocysts, used to kill and capture food and protect against predators. All cnidarians are carnivorous. Various members of the phylum are colonial, the

Fig. 94

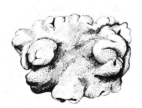

Typical Cnidaria

individual animals being known as polyps (hydranths) or medusae. Many cnidarians, particularly corals and even many anemones and hydroids, are vividly colored in greens, oranges, and brilliant reds. Luminescence is common, the light coming from granules that, in some cases, may be rubbed off in slime.

HYDROIDS
Class Hydrozoa

The Hydromedusae, or medusoids, can be distinguished from the true jellyfishes (Class Scyphomedusae) by the velum, a flap of thin tissue that forms a shelf on the underside of the bell. When swimming normally, the hydromedusae have a jump-and-glide progression, whereas the Scyphomedusae pulsate more to give a more determined but apparently less controlled movement. The hydroids, plantlike in form, are often colonial, and are easier to find than the less conspicuous polyp forms of the Scyphomedusae.

Uncupped Hydroids
Suborder Gymnoblastea

In this group the polyps do not form a protective cup (hydrotheca) around themselves, as do those of suborder Calyptoblastea. However, the stems of the colony may have a firm protective covering.

TUBULARIA CROCEA   125:10
Description
Size, 4–5 in. (10.2–12.7 cm). Body on an unbranched stem attached to a basal stolon, generally matted together and bushy; 2 sets of tentacles (shorter set 20–25 in number in circle close to mouth; longer set in single whorl at base of head), fixed medusoids borne in clusters on peduncles attached between sets of tentacles. Stems brownish or yellowish; tentacles transparent; hydranth coral-pink.
Habitat
In bays or harbors on floats, piles, wharves, in large clumps; colonial. Most common where surface is submerged.
Range
Arctic Ocean to Baja Calif.; introduced from Atlantic.

Note: The similar *Tubularia marina*, occurring north to central California, is 1 in. (2.5 cm) long; it is a large, dainty hydranth, solitary, vividly orange, with medusoids red with pink centers.

SYNCORYNE MIRABILIS   127:6
Description
Size, ⅝ in. (1.6 cm) high. Colony profusely branching; attached to support by a creeping stolon; polyps terminating main stem and branches, developing medusa buds on sides just below tentacles.
Habitat
Low-tide zone.
Range
Alaska to Chile; fairly common, especially at Friday Harbor in Puget Sound and in San Francisco Bay. Supposed to have been introduced from E. Coast.

GARVEIA ANNULATA 125:7

Description
Size, 3 in. (7.6 cm) high. Body with hydranths having conical or dome-shaped proboscis surrounded with about 16 threadlike tentacles; colony arborescent. Body uniformly brilliant orange; tentacles lighter, stalks darker.

Habitat
Commonly growing through or on sponges, on coralline algae, or at base of other hydroids; colonial.

Remarks
May also be overgrown with still other hydroids.

Range
Alaska to s. Calif.

HYDRACTINIA (Species) 127:13

Description
Size, 1 in. (2.5 cm) dia. Colony commonly observed as a flat, pink-colored "fuzzy" patch on undersides of rocks; polyps small and of 5 different types, but feeding hydranths most common.

Habitat
Attached to the undersides of rocks.

Range
Calif. to B.C.

EUDENDRIUM CALIFORNICUM 125:11

Description
Size, 6½ in. (16.5 cm) tall. Body most conspicuous of hydroids; bushy, with flowerlike hydranths at tips of secondary branches; stem with many annulations (rings), stiff, stout, brown, and hard. Reproductive zooids differ in size and shape, occur scattered over the colony. Color brown; zooids pink, tentacles white.

Habitat
Firmly attached to rock surfaces, low-tide zone; common.

Range
Puget Sound, Wash., to Pacific Grove, Calif.

Note: The similar species *Eudendrium rameum*, occurring from Puget Sound, Wash., to San Diego, Calif., has the main stem thick, more irregularly branched, and with 2 to 5 rings at bases; and has 24–27 tentacles in one whorl.

Cupped Hydroids
Suborder Calyptoblastea

In this group of hydroids the transparent protective covering not only occurs on the stems but also extends up around the polyps to form a protective cup, or hydrotheca, into which the polyp can withdraw.

OBELIA LONGISSIMA 127:10

Description
Size, 1+ ft. (30.5+ cm). Forming dense fuzzy growth as colonies often dichotomously branched; 5–8 annulations (rings) above base of each branch; reproductive zooids urn-shaped, attached by short-ringed stalk. Color brownish-white.

Habitat
Cosmopolitan; very common; preferred habitat on pilings and floats.
Remarks
At times, under favorable conditions, the medusa stage may be observed separating from the zooids, as when small groups of colonies placed in about one gallon of fresh sea water are exposed to mild sunlight.
Range
Puget Sound, Wash., to San Diego, Calif.

Note: The *Obelia geniculata*, a widely distributed species, is similar to *Obelia longissima* but has the stems usually unbranched and zigzag in outline. It occurs in shallow waters of Alaska to Puget Sound and is very abundant.

OSTRICH-PLUME HYDROID
Aglaophenia struthionides

Description
Size, 6 in. (15.2 cm). Stem central, branched regularly on each side, branches graded in length; hydrothecae in single series on one side of branch, each with 11 prominent teeth; 3 nematophores, middle one larger; large triangular nematophore at base of each branch. Closely resembles bunches of seaweed.
Habitat
Intertidal zone to deep water, larger specimens in violently surf-swept rocky clefts; common, often cast up on beach with seaweed.
Range
Vancouver Is., B.C., to San Diego, Calif.

ABIETINARIA (Species)   127:11

Description
Central stem, branched regularly in one plane; hydrothecae in two rows on opposite sides of branches with smooth margin; operculum in one piece; delicate.
Habitat
Rocky intertidal zone on open coasts; common.
Range
Alaska to Calif.

POLYORCHIS PENICILLATUS   127:4

Description
Size, 2¼ in. (5.7 cm). Body bell-shaped, higher than wide; gonads fingerlike and hanging down alongside manubrium (kneelike veil inside bell), ring canal unbranched, radial canals with 15–25 pairs of short branches. Tentacles, 40–150, arranged in single whorl; can contract to short, thick stubs or extend to twice bell length; eyespot at base of each tentacle. Bell transparent-white; stomach, gonads, tentacle bulbs, radial canals reddish-brown to purple.
Habitat
Coastal shallow-water bays; commonest Dec. to Apr., often stranded ashore.
Habits
An excellent swimmer; moves by kicking the manubrium.
Range
Vancouver Is., B.C., to s. Calif.

HYDROCORAL
Stylantheca porphyra   125:12

Description
Size, varies, encrusting. Crust corallike (but unrelated), encrusting many square feet of rocky, low-tide ledges; surface covered with tiny star-shaped pits, each with a point at the bottom and containing feeding and several protective polyps. Deep-water forms are more erect masses. Color vivid purple to orange.

Habitat
Low-tide areas of rock drenched with fairly powerful surf, to deep water.

Range
B.C. to s. Calif.

Note: The similar species *Allopora petrograpta* occurs in Alaska on exposed shores.

BY-THE-WIND-SAILOR
Velella velella   127:2

Description
Size, 3–4 in. (7.6–10.2 cm). Body (raftlike portion) flat, elliptical, with air chambers; half-circular sail, or keel, projecting above surface; individuals of the colony attached to underside of float; long contractile tentacles form a double outer row around polyps. In center of colony is a single large feeding polyp (gastrozooid) and "between it and the tentacles are the reproductive individuals which also have mouths and are able to take in food" (Johnson). Mantle covers raft above and below and on both sides of keel, forming flaps along margin. Keel runs diagonally across float (large spongy mass which serves as a float); both float and keel contain canals surrounding a central point as concentric ellipses. Float and keel nearly colorless; mantle deep-blue, tinged here and there with green; polyps and tentacles lighter blue.

Habitat
Open sea in warmer waters, usually in schools. Storms drive great numbers onto coastlines.

Range
Puget Sound to equator.

JELLYFISHES
Class Scyphozoa

The usually large jellyfishes generally observed floating on the surface of the sea or washed ashore as firm gelatinous masses constitute the Scyphomedusae, distinguishable from the Hydromedusae by the absence of a velum, or flaplike shelf within the bell. They inhabit all seas and vary in diameter from about one inch (2.5 cm) to seven and one-half feet (2.3 m). The inconspicuous hydroid form, the scyphistoma, attaches to seaweeds or to hard bottoms; it is not colonial and is usually less than one-half inch (1.3 cm) high, appearing like a stack of tiny platelets. The top platelet, showing tentacles, detaches by budding at regular intervals; the tiny platelet that is detached grows to full medusa size in several weeks. The medusa, in turn, sexually reproduces the hydroid form. The sexual jellyfish generation is the larger form, the attached asexual plantlike generation the smaller form.

HALICLYSTUS AURICULA 127:12

Description
Size, 1 in. (2.5 cm) dia. A small, fixed jellyfish. Top of bell
attached to a substrate, usually red algae or eelgrass, by a
cylindrical stalk, which can contract somewhat; mouth disk studded
with tentacles. Colors subtle, matching algal substrate.

Habitat
On algae in the rocky intertidal zone.

Range
Aleutians to Monterey Bay, Calif.; more abundant to n.

CHRYSAORA MELANASTER 127:5

Description
Size, 10–12 in. (25.4–30.5 cm) dia. Disk a flattened hemisphere
about 4–6 in. (10.2–15.2 cm) high; 8 marginal sense organs; 24
tentacles, 3 between each pair of organs; 32 marginal lappets
(lobes) of equal size, slightly narrower at base; curtainlike tapering
lips, margins well-folded. Color light bluish with 32 brown rays on
umbrella, 16 dark brownish to black radial streaks on underside of
umbrella; gonads reddish brown, tentacle tips red.

Habitat
Open sea, occasionally washed ashore; common.

Range
Aleutians to cen. Calif.

PURPLE-STRIPED JELLYFISH
Pelagia noctiluca 125:4

Description
Size, 1–1½ ft. (30.5–45.7 cm) dia. Bell with 8 marginal tentacles
alternating with 8 sense organs between which are 16 divided
lappets (lobes) (Johnson); 16 radiating stomach pouches, no ring
canal. Tentacles hang 6–8 in. (15.2–20.3 cm) below umbrella
margin and can be greatly extended. Color purple and red.

Habitat
Bays and open sea; often cast ashore.

Remarks
Nettle cells produce painful but not dangerous sting; effect of sting
lasts an hour or more.

Range
Calif. coast; very abundant in cen. Calif.

MOON JELLY
Aurelia aurita 127:1

Description
Size, 10 in. (25.4 cm) dia. Body jellylike, disk-shaped, fringed with
small tentacles; 8 marginal indentations containing club-shaped
organs bearing eyespots; mouth surrounded by 4 long, narrow
lobes; gonads horseshoe-shaped, arranged radially around center of
animal. Color bluish-white, gonads yellow in male, pink in female;
has four-leaf-clover pattern when viewed through bell.

Habitat
Open seas on surface, bays, sounds; often in large numbers.

Range
Worldwide; very abundant Alaska to Puget Sound, Wash.; common
Alaska to cen. Calif.

SEA BLUBBER
Cyanea capillata

Description
Size, 3 ft. (91.4 cm) dia. Can be largest of the West Coast
jellyfishes. Disk rather flat; margin divided by 8 deep clefts into
lobes, each with median cleft and 2 short notches; at bottom of each
median cleft is a club-shaped sense organ. Bell thick toward disk
center, thin at edges. Tentacles number several hundred in each of
8 groups; can be extended to 25 times disc diameter or to 35 ft.
(10.7 m) in larger specimens; bear powerful nettle cells. Mouth
arms, 4, long; margins greatly folded (length of bell diameter); no
ring canal. Color yellow-brown or orange; stomach, radial pouches
brownish; some tentacles reddish, some nearly white.
Habitat
Open sea in colder waters; occasionally washed ashore.
Range
N. Pacific, Aleutians to n. Calif.

SEA ANEMONES AND CORALS
Class Anthozoa

The members of this class do not have a free-swimming medusa
stage. Whether solitary or colonial, these animals have a central
mouth opening on an oval disk margined with tentacles, and a
short gullet leading to the internal cavity. Vertical partitions in the
cavity support the gullet and have a free edge, on which the gonads
are located. The hollow tentacles possess nettle cells. Most of the
Anthozoa are fixed forms of the shore or shallow waters.

SLENDER SEA PEN
Stylatula elongata   *Fig. 95*

Description
Size, colony 10–20 in. (25.4–50.8 cm). Colony supported by
slender, horny axis (when contracted, only tip shows), bulbous
lower ends buried in mud. Polyps in whorls; actually on paired
pinnules around stem, about 10 pairs per inch. Color white or
gray, phosphorescent at night, when touched.
Habitat
Muds and sands of bays, low tide to 35 fathoms (64 m); common.
Range
San Francisco to San Diego, Calif.; especially abundant in
Tomales, Mission, and Balboa bays.

Fig. 95

Renilla köllikeri
Sea Pansy, p. 716

Stylatula elongata
Slender Sea Pen

SEA PANSY
Renilla köllikeri   *Fig. 95*

Description
Size, colony 1½–3 in. (3.8–7.6 cm). Disk heart-shaped, stalks buried; polyps hand-shaped, transparent. Color purple, polyps water-white, tip of peduncle white.
Habitat
Shallow-water bay sands; muds.
Range
Wilmington, Calif., to Cedros Is.; very abundant, especially in Newport, Mission, and San Diego bays.

Note: The similar *Renilla amethystina* is violet or amethyst with heart-shaped stalk and 200+ independent individuals. Its range extends south to Panama.

EPIACTIS PROLIFERA   127:9

Description
Size, 1 in. (2.5 cm) dia. Body small, with up to 90 tentacles in a large specimen; mesenteries in sixes with 4–5 cycles, margins often fluted. Particularly characteristic are the young anemones held in pits on outside of body. Color red or reddish-brown, sometimes greenish; column marked by vertical lines of lighter color.
Habitat
Low-tide zone; abundant, especially on eelgrass at Friday Harbor, Wash.
Range
Puget Sound, Wash., to s. Calif.

TEALIA CRASSICORNIS   125:8

Description
Size, 6 in. (15.2 cm) dia. Tubercles reduced or absent. Disk bearing long, thick tentacles, usually grayish-green in color but with light cross-banding. Column mottled red and green.
Habitat
Intertidal zone, on open rocky coasts, usually on undersides of rocks or on overhangs.
Range
B.C. to s. Calif.

GREAT GREEN ANEMONE
Anthopleura xanthogrammica   125:5

Description
Size, 6–10 in. (15.2–25.4 cm) dia. Most common West Coast sea anemone. Disk to 6 in. dia., column twice as long. Column greatly extensible; has longitudinal rows of tubercles, largest near tentacle bases, where there are several lobes and numerous contained nematocysts. About 24 pairs of mesenteries. Color uniformly brilliant green in sunlight, paler in shaded areas; when shaded, often white or faintly hued with pink or lavender. Tentacles uniform in color, not pink-tipped like those of *Anthopleura elegantissima*.
Habitat
Tide pools and heavy surf entirely free from contamination; very common.
Range
Middle Aleutians and Sitka, Alaska, to Panama.

AGGREGATING ANEMONE
Anthopleura elegantissima 125:6

Description
Size, 2–3 in. (5.1–7.6 cm) dia. Body small with tentacles tipped in pink or violet. Column with longitudinal rows of tubercles, white to light green in color. Disk often with contrasting radial lines.
Habitat
Rocks, wharf pilings, at base of rocks in sand, open coast and bays; very common.
Habits
This species usually occurs in close aggregations which are clones derived by asexual budding.
Range
B.C. to s. Calif.

BROWN SEA ANEMONE
Metridium senile 127:8

Description
Size, 2–6 in. (5.1–15.2 cm) dia. One of the largest, most common anemones. Body with up to 1000 short, tapered tentacles (many fewer in young) extending over most of a greatly expanded, frilled oral disk; column smooth. (A giant may expand to fill a 10-gallon jar.) Color brown, orange, salmon, or white; young dull pink or greenish.
Habitat
Intertidal zone on bare piling or dead shells, to deep water (300 ft., or 90 m); common. Largest specimens found to 60 fathoms (110 m).
Habits
Specimens of one color usually cluster together because of common method of asexual reproduction by nasal fragmentation.
Remarks
Body wall may release stinging threads through pores.
Range
Sitka, Alaska, to Santa Barbara, Calif.

CORYNACTIS CALIFORNICA
125:9

Description
Size, ½–⅝ in. (1.3–1.6 cm) dia. A small, lovely anemone. Tentacles markedly enlarged and rounded at tips, with particularly large nematocysts. Color numerous shades from red or orange to pale pink and almost white; may fluoresce in deep water; tentacles white.
Habitat
Crevices between rocks or under overhanging ledges in intertidal and subtidal zones; very common.
Range
Sonoma Co. (especially Salt Point) to Santa Barbara, Calif., intertidally; subtidally s. to Coronados Is. and San Diego.

BALANOPHYLLIA ELEGANS
125:13

Description
Size, ½ in. (1.3 cm) dia. Cup with characteristic star pattern; base comprises about ⅓ animal's bulk. Polyps cadmium-yellow or orange.
Habitat
Mid-tide zone, well sheltered from dehydration; abundant.
Range
Puget Sound, Wash., to Monterey, Calif.

COMB JELLIES
Phylum Ctenophora

The ctenophores are characterized by their spherical, lobed, thimble-shaped or bandlike, extremely transparent bodies, often gelatinous, typically found floating at the surface of the sea. They have eight bands of comb plates of fused cilia (tiny hairlike processes) used as locomotor organs. The symmetry is biradial rather than radial as in the coelenterates. Some comb jellies have retractile tentacles with characteristic adhesive cells. The digestive system starts with a slitlike mouth that leads to a tubular stomach; this branches into a canal system which eventually leads to tubes which lie directly beneath the comb plates. In the walls of these tubes the gonads develop, both male and female in each individual. The development of the embryo is complex, but there is no alternation of generations. Almost all comb jellies are free-swimming or -floating and occur in vast numbers under favorable conditions.

SEA GOOSEBERRIES
Pleurobrachia bachei   **127:7**

Description
Size, ½ in. (1.3 cm) dia. Sphere remarkably transparent; tentacles extend more than 5 times body diameter, fringed with adhesive cells to capture prey. No color, except tentacles and esophagus may be tinted reddish; combs often somewhat iridescent.
Habitat
Floating on surface of sea, often close to shore; abundant.
Range
Puget Sound, Wash., to San Diego, Calif.

BEROE FORSKALI   **127:3**

Description
Size, 2–3 in. (5.1–7.6 cm) long. Body saclike in shape, translucent (like ground glass); network of side branches from tubes beneath comb rows; 8 rows of ciliated combs from apex of body to ⅔ distance to mouth; no tentacles, very large mouth. Very young colorless, half-grown individuals rosy, with brilliantly iridescent rows of combs.
Habitat
Surface waters; occasionally abundant.
Habits
Carnivorous, exceedingly voracious.
Range
Calif. to Antarctic.

FLATWORMS
Phylum Platyhelminthes

The bodies of these worms are flattened and well adapted to hiding under stones, in crevices, and among algae. Some are broad and leaflike, others long and ribbonlike. They are variously colored. The body is unsegmented, with no definite head and no paired appendages.

POLYCHOERUS CARMELENSIS 125:16

Description
Body bluntly pointed at anterior end, broader at posterior end, which terminates in 2 caudal lobes and a peculiar "tail." Color orange to orange-red.
Habitat
Among the green alga *Ulva* or on rocks in very high-tide pools.
Range
Monterey Bay, Calif.

ALLOIOPLANA CALIFORNICA *Fig. 96*

Description
Size, 1½ in. (3.8 cm) long. Body thick, firm, almost round in outline; nuchal (sensory) tentacles in mouth area, with minute eye-spots clustered at bases; mouth mid-body underneath. Color marks of blue-green, black, white.
Habitat
Underside of large rocks on damp gravel; fairly abundant, especially near Monterey, Calif.
Range
Monterey, Calif., to Baja Calif.

Fig. 96

Alloioplana californica
(Flatworm)

KABURAKIA EXCELSA 128:1

Description
Size, 4 in. (10.2 cm). Body large, thick, and broadly oval with conspicuous tentacles. Color brown or tan with small black spots, which are eyes, on the margin.
Habitat
On floating docks, pilings, mussel clumps, and under rocks.
Range
B.C. to s. Calif.

RIBBON WORMS

Phylum Nemertea

This phylum contains long, flattened worms differing from the true flatworms by having both a mouth and an anus. They also have an extremely specialized and extensible proboscis which is independent of the mouth. The body is soft, highly contractile, and unsegmented; there may be light-sensitive organs on the anterior end.

AMPHIPORUS BIMACULATUS   125:15

Description
Size, 1½–6 in. (3.8–15.2 cm). Body short, flattened, broad; 12–30 eyes on each side of head in irregular rows outside 2 triangular dark spots. Proboscis large; central stylet; long, slender, with short basis; usually 4 pouches. Dorsal surface of head light-colored with 2 deep red or black spots in center; pale red or orange or pinkish beneath.

Habitat
On wharf piles, among algae on low-tide rocks; common.

Range
Alaska to cen. Calif.

AMPHIPORUS IMPARISPINOSUS   128:2

Description
Size, 2½ in. (6.4 cm). Body slender, flat; eyes numerous in two patches on either side of head; accessory stylets in 2–3 pouches. Color uniformly opaque white both dorsally and ventrally.

Habitat
Among algae or beneath rocks on exposed rocky shores; common.

Range
Commander Is. (U.S.S.R.), Alaska to s. Calif.

EMPLECTONEMA GRACILE   128:3

Description
Size, 8–12 in. (20.3–30.5 cm). Body extremely narrow (about 2 mm), slightly flattened; proboscis bears long, curved stylets; eyespots in 2 groups on each side of head (8–10 anteriorly, 10–20 farther back). Color above green or yellowish green, below yellowish or white.

Habitat
Coiled in masses among mussels, in seaweeds; very abundant.

Range
Alaska to Baja Calif.

PARANEMERTES PEREGRINA   128:5

Description
Size, 6 in. (15.2 cm). Body long, slender; head variably shaped (somewhat flattened); stylet (sharp projection at tip of nose) as long as basis, braided appearance; reverse stylets in 2–4 pouches. Color variable, from dark brown to orange- or purple-brown above; white or yellowish beneath.

Habitat
Low-tide zone under rocks; common.

Range
Commander Is. (U.S.S.R.) to Baja Calif.

ENTOPROCTS
Phylum Entoprocta

PEDICELLINA CERNUA   128:4

Description
Size, ½–1 in. (1.3–2.5 cm). Body margined by tentacles, often folded; borne on a stalk rising upright from a creeping stolon. Color white.
Habitat
Furring stems of plants, hydroids, bryozoans; often on shells and among hydroid colonies.
Habits
Tentacles roll up instead of retracting.
Range
Calif. coast.

MOSS ANIMALS
Phylum Bryozoa

This phylum contains many species of very small colonial animals, each of which possesses a circle of tentacles that can be drawn into a tiny chitinous or calcareous chamber serving as a protective outside skeleton. The bryozoans—a term referring to the mosslike appearance of many species—represent one of the most ancient groups of animals with origins in the Precambrian Era more than 525 million years ago. Calcified colonies are abundant as fossils, and geologists use them to date various earth formations.

The body wall has an aperture, often with an operculum, or trapdoor, through which the lophophore is extruded to feed on plankton and detritus. The lophophore is a crown of hollow, ciliated tentacles carrying the mouth in the middle. The body is well organized, with a U-shaped digestive tract divided into pharynx, stomach, and intestine; there are muscles to contract within the calcified chamber (eversion is by hydraulic pressure) and a simple nervous system. Reproduction is both sexual and asexual; sexual reproduction produces free-swimming larvae which form new colonies.

Bryozoans occur in all the oceans of the world, from the tropics to the polar seas, and a few species inhabit fresh water. While many species occur in shallow waters, others inhabit depths as great as 3000 fathoms (5.5 km). Too many to describe here, there are at least 150 species in Puget Sound alone, and "most of us can scarcely hope to recognize more than a few" (Ricketts).

BUGULA NERITINA

Description
Size, 4 in. (10.2 cm). Zooecia in tufts, elongated, somewhat rectangular; aperture occupies nearly entire front; ovicells spherical, attached by short, broad stalk. Color brown, red, or yellow-purple. Commonly mistaken for seaweed.
Habitat
Characteristic of bare spots on pilings, boat bottoms (sparse), palmate clusters, also on submerged wood but never on rock; fairly common.
Range
Monterey, Calif., to Baja Calif.

BUGULA CALIFORNICA 128:8

Description
Size, 1–3 in. (2.5–7.6 cm). One of the most common bryozoa.
Colony spiral in form. Zooecia with 3 spines at tip; avicularia
(beak) large, attached by stalk; ovicells completely enclosed. Largest
specimens in north; southern forms rarely more than 1 inch in
height. Color, in north, purple or yellowish to greenish; in south,
straw-colored to almost white.
Habitat
Rocky intertidal zone.
Range
B.C. to Monterey Bay, Calif.; abundant, particularly in n. of
range.

CRISIA OCCIDENTALIS 128:6

Description
Size, ½–¾ in. (1.3–1.9 cm). Zooecia cylindrical, curving slightly
outward, arranged in 2 alternating rows; ovicell opens by small
pore at end of short, inconspicuous tube, directed upward near
summit but on dorsal side.
Habitat
Low tide to 100 fathoms (183 m), as growths on firm support
below water; abundant.
Range
Arctic Ocean to San Pedro, Calif.

Note: Considered by some to be identical with *Crisia eburnea,*
common moss animal of Atlantic Coast.

MEMBRANIPORA MEMBRANACEA Fig. 97

Description
Size, varies, encrusting. Colonies thin, flat, from very small to
several square inches. Zooecia outward radiating, ridged by the
calcareous walls; front membranous, transparent; at each anterior
angle a short spine, blunt, hollow, chitinous; no ovicells. Color
white.
Habitat
On fronds and floats of kelp; common.
Range
Entire Pacific Coast; also worldwide.

Note: A very similar but much more spiny form in similar habitats
is *Membranipora villosa,* occurring from Puget Sound to San
Diego, Calif.

Fig. 97

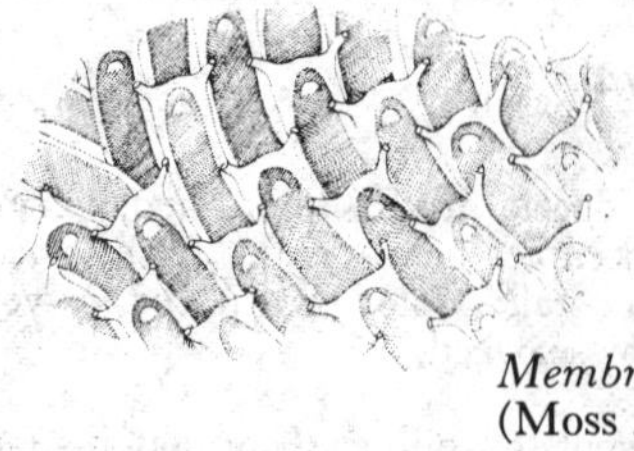

Membranipora membranacea
(Moss Animal)

LACE CORAL
Phidolopora pacifica   **128:7**

Description
Size, 1–1½ in. (2.5–3.8 cm). A small, latticed, calcareous bryozoan,
with an encrusting base above which grow thin, convoluted,
perforated orange sheets.

Habitat
Under rocks and at base of algae in south, to 35 fathoms (64 m) in
north; rather common, widely distributed between tide marks.

Range
Puget Sound, Wash., to San Diego, Calif.

LAMP SHELLS
Phylum Brachiopoda

Brachiopod shells so closely resemble certain clams that originally
they were classed with the phylum Mollusca. However, the
brachiopods, or lamp shells, are more closely related to the moss
animals (bryozoans) and phoronid worms. The upper, or dorsal,
valve is usually small and flat. The ventral, or lower, valve extends
behind the dorsal valve and has a gap through which the peduncle,
or stalk, passes. Asymmetrical in side view, both valves are
symmetrical about the longitudinal axis. The stalk, by which the
brachiopod attaches itself to the substrate, distinguishes these
animals from any mollusk. In addition, brachiopods are very
different anatomically inside the shell. Of the approximately 120
brachiopods, the two most usually found in shallower waters are
described below.

STALKED BRACHIOPOD
Glottidia albida   *Fig. 98*

Description
Size, 1 in. (2.5 cm). Fragile, elongate shell, attached to a long and
muscular pedicel (stalk), which is used for burrowing. Valves of
equal size, pointed behind, truncated in front; dorsal valve has 2
sharp internal ridges, extending about ⅓ shell length. Peduncle
fleshy, 2–3 times shell length; long, stiff setae on edge of mantle,
projecting beyond shell margin. Color whitish to brownish.

Habitat
Fine, sandy, or muddy areas, from low-water mark to 60 fathoms
(109.7 m); locally abundant. Usually found with stalk buried in
sediment and shell partly protruding.

Range
Tomales Bay to San Diego, Calif.

Fig. 98

Terebratalia transversa
Lamp Shell, p. 724

Glottidia albida
Stalked Brachiopod

LAMP SHELL
Terebratalia transversa   *Fig. 98*

Description
Size, 1½ in. (3.8 cm). Shell large, wider than long, broadest near hinge line; calcareous loop within dorsal valve long, doubly attached. Sculptured with numerous radiating ribs; edges scalloped. Color reddish.
Habitat
Low-water mark to 100 fathoms (183 m); most common brachiopod on West Coast.
Range
Alaska to s. Calif.

TUBE-DWELLING WORMS
Phylum Phoronida

The tube-dwelling, wormlike phoronids are distant relatives of the bryozoans, and all species are marine. Like the bryozoans, the phoronids possess tentacles, but instead of forming a circle or a simple horseshoe, the tentacles are attached in the form of a horseshoe with the two open ends curled inward into spirals. These whorls are called lophophores. Scientific classification depends in part on the number of tentacles and the number of spirals in the lophophore.

Phoronids are usually hermaphroditic. Some species are minute, but the majority reach eight inches (20.3 cm) long, and one West Coast species, *Phoronopsis viridis*, attains a length of about twelve inches (30.5 cm). Phoronids may be found on mud and sand flats between tide marks or in clusters attached to pilings, rocks, and floats.

PHORONIS VANCOUVERENSIS

Description
Body curiously wormlike, gelatinous.
Habitat
Intertidal zone, as great masses.
Range
Straits of Juan de Fuca (common) to Monterey Bay, Calif. (rare).

PHORONOPSIS VIRIDIS   128:9

Description
Size, 12–18 in. (30.5–45.7 cm). Body soft, wormlike, found extending vertically down into the substratum of sandy mud in quiet bays; very difficult to extract whole. Animals dioecious. Tubes sand-impregnated, mucus-lined. Body reddish with bright green tentacles often extruded from burrow to lie on the sand.
Habitat
Mud flats of estuaries, bays; as large masses or beds of animals.
Range
Calif. coast; particularly common in Elkhorn Slough and Bodega Bay.

Note: *Phoronopsis harmeri*, occurring in Washington and Oregon, is similar to *Phoronopsis viridis* but lacks green pigmentation.

ECHINODERMS

Phylum Echinodermata

This exclusively marine phylum, characterized by its radial symmetry, is divided into five living classes: Asteroidea, or starfish; Ophiuroidea, serpent stars or brittle stars; Echinoidea, or sea urchins and sand dollars; Holothuroidea, or sea cucumbers; and Crinoidea, feather stars or sea lilies. The adult body is usually made up of ten parts which radiate from a central axis: the five rays (e.g., arms of the starfish) and the five interrays, or spaces between the rays. In most forms the central axis is perpendicular to the surface over which the animal moves, but in the sea cucumbers the axis is lengthened and is parallel to the underlying surface. Most echinoderms creep or crawl on the bottom, or burrow slowly through mud or sand.

Fig. 99

Typical Echinodermata

Characteristic of the phylum is the presence of an internal calcareous skeleton, composed either of definitely shaped plates more or less rigidly joined together or scattered spicules or plates. Except in the sea cucumbers, the outer skin is ciliated.

STARFISH

Class Asteroidea

Starfish are found in all seas from between tide marks to great depths. They are bottom dwellers, creeping about with a slow, gliding motion. They derive their name from the characteristic five-ray pattern of the phylum (even though some species may show more than twenty rays, or arms). There is a disk, or central portion, from which the rays extend; the mouth is directed downward and the anus is a little off center on the dorsal surface. On the underside of each ray is a deep groove from which two to four rows of small tube feet extend. A tiny red eyespot at the tip of each arm is sensitive to light.

The starfish has a certain degree of rigidity because its skeleton of calcareous plates or rods is bound together by connective tissue. The frame is generally a multitude of spines. Although the animals have no hard jaws, most are carnivorous and voracious feeders, living largely on shellfish and barnacles; they are very destructive to oysters and mussels. Oystermen, who once tried to kill starfish by chopping them into pieces, only compounded their destructiveness: most starfish have remarkable powers of regeneration, and each of the major body parts soon grew to become a whole new starfish again.

LEATHER STAR
Dermasterias imbricata   **125:1**

Description
Size, 10 in. (25.4 cm) dia. Arms broadly attached to disk, giving
webbed effect. Body covered with thick, slippery, smooth skin
concealing spines, except along grooves; tips of the 5 rays turned
up. Color delicately purple with red markings to orange or lead-
blue mottled with dull red.

Habitat
Low-tide pools; nowhere very common, but fairly abundant in
some localities.

Range
Sitka, Alaska, to Monterey, Calif.

HENRICIA LEVIUSCULA   **125:2**

Description
Size, 5–6 in. (12.7–15.2 cm) dia. A neat 5-rayed starfish. Dorsal
surface covered with pseudopaxillae (groups of short spinelets)
arranged in a fine meshed network interspersed by small spaces;
disk small; rays slender, cylindrical, gradually tapering, giving
definite shape. Color vivid blood-red or orange, but dorsal parts
may be tan, orange, orange-red, occasionally purple, often mottled
or banded in darker shades; lower surface lighter, usually yellowish
or orange.

Habitat
Open rocky shores, low-tide pools; brooding mothers usually
hidden under or between rocks in darkness (brood eggs in
January); moderately common.

Range
Aleutians to San Diego, Calif.

Note: Because of great variation in form, color, and shape of
spines, a number of subspecies and varieties may be recognized.

DAWSON'S SUN STAR
Solaster dawsoni

Description
Size, 14 in. (35.6 cm) dia. Body with relatively broad disk
surrounded by 8–15 rigid rays, usually 11–12; surface rough to
touch and rather hard. Color purple-gray to orange.

Similarities
Solaster stimpsoni is smaller, usually with 10 rays that are more
slender, spinelets shorter than fringing spines; more in-shore
habitat.

Habitat
Low-tide zone (not common) to deep water in southern range.
Rare in intertidal zone of Calif., but more common in north.

Range
Aleutians to Monterey, Calif.

STIMPSON'S SUN STAR
Solaster stimpsoni   *Fig. 100*

Description
Size, 7–8 in. (17.8–20.3 cm) dia. Body usually with 10 rays, disk
relatively small, rays slender. Color bluish-gray in center of disk
with broad band extending almost to tip of each ray, bands
bordered by yellow-ochre or orange; lower surface light, with
narrow blue-gray stripe on each side of groove.
Similarities
Solaster dawsoni is larger, with more arms, flat-topped
pseudopaxillae, longer spinelets along grooves; marginal plates
more conspicuous; rather narrower, less common at tide level.
Habitat
Low intertidal zone; not infrequent.
Range
Bering Sea to Oreg.

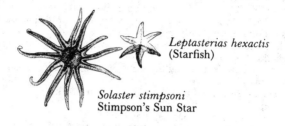

Fig. 100

Leptasterias hexactis
(Starfish)

Solaster stimpsoni
Stimpson's Sun Star

BAT STAR
Patiria miniata   **125:3**

Description
Size, 4 in. (10.2 cm) dia. One of commonest starfish of the
California shore. Body with thick inflated disk; usually 5 short,
triangular webbed rays; dorsal plates crescent-shaped, plates and
interstices roughly granulated, concave side of crescent faces disk.
Color red, yellow, brown, gray, or purple; commonly bright red or
scarlet above and yellowish below; sometimes greenish.
Habitat
Rocky intertidal zone to deep water; common.
Remarks
Because of ready availability and early summer breeding habits,
this species is extensively used for embryological experimentation.
Range
Sitka, Alaska, to s. Baja Calif.; s. of Pt. Conception, uncommon,
small, stunted; rare or absent on coasts of Oreg. and Wash.

LEPTASTERIAS HEXACTIS   *Fig. 100*

Description
Size, 3 in. (7.6 cm) dia. A 6-rayed starfish. Body with granular
surface having basketwork appearance, rays slightly swollen at
base. Color variable, but usually light brown, mottled with pink or
orange; often gray, green, or olive.
Habitat
Under rocks in the intertidal zone.
Range
Vancouver Is. to San Diego, Calif.; common from Puget Sound to
Monterey, Calif., area.

PISASTER BREVISPINUS

Description
Size, 24–28 in. (70–71.1 cm) dia. A pink-skinned, massively proportioned edition of *Leptasterias hexactis*. Body covered with short spines on upper surface, blunt, usually not describing a reticulated pattern on disk; spines may be obliterated in very large specimens. Color pink.

Habitat
Soft sand bottoms, usually just below low spring tide line and, in south, to 60 fathoms (109.1 m); rather common.

Range
B.C. to Monterey Bay, Calif.

PISASTER GIGANTEUS
126:1

Description
Size, 20–22 in. (50.8–55.9 cm) dia. Body diameter normally smaller than *Pisaster ochraceus*, symmetrical, rather delicate; spines are blunt, do not form network, and are margined by a ring of blue flesh. Color beautifully contrasting on and surrounding spines.

Habitat
Low-tide pools and shallow water among rocks; common.

Remarks
Very tough animal and difficult to pry off rocks.

Range
Vancouver Is. to Monterey Bay, Calif.

PURPLE STARFISH
Pisaster ochraceus
126:2

Description
Size, 6–14 in. (15.2–35.6 cm) dia., record 20 in. (50.8 cm). Body with thick, broad disk; 5, occasionally 6, stout tapering rays. Dorsal spines short, arranged in close-set rows forming distinct network; makes well-marked pentagon on disk. Color brown, purple, yellow, in 3 phases.

Habitat
Shallow water and on rocks at low tide, chiefly where wave-washed; most abundant starfish.

Range
Sitka, Alaska, to Pt. Conception. Range extended to n. Baja Calif. by subspecies *segnis*.

Note: A quiet water form, *Pisaster ochraceus f. confertus*, occurs in mussel beds of lower and middle tidal zones. It is vivid violet; its range is limited to British Columbia and Puget Sound, Wash.

SUNFLOWER STARFISH
Pycnopodia helianthoides
Fig. 101

Description
Size, 24–30 in. (61–76.2 cm) dia., record 48 in. (121.9 cm). Possibly the largest known starfish. Body surrounded with 20–24 rays; begins life with only 6 rays, but as age advances adds new rays in pairs between older ones; skin soft, delicate, upper surface sparsely set with spines and bearing bunches of pedicellariae (minute pincers); disk broad, soft, rays gracefully tapering. Color variable in lively pinks and purples, yellow and orange, bright red, occasionally gray.

Habitat
Low intertidal zone; common, conspicuous by size.
Range
Unalaska (Aleutians) to Monterey Bay, Calif.

Fig. 101

Ophioplocus esmarki
(Starfish)

Pycnopodia helianthoides
Sunflower Starfish

BRITTLE STARS
Class Ophiuroidea

The brittle stars, or serpent stars, are less starlike than the sea stars. The central disk is more compact; the snaky arms are spiny, thin, and fragile. Brittle stars are rarely found on the beach. The disk is sharply differentiated from the arms, and the lower surface lacks longitudinal furrows or grooves. The tube feet, which lack suckers, have functions of food capture, respiration, and excretion. The water-intake, or madreporic, plate, which is on the upper surface of sea stars, is here on the underside. Peculiar to this group of echinoderms are buccal (cheek) plates.

The members of this class are more active than starfish, moving about by wriggling their arms, and are found under stones and in protected low-tide areas. They inhabit regions from the tide pools to the deepest abyss. They feed on detritus and small invertebrates and are, in turn, fed upon by larger invertebrates and fishes. Fossil series of Silurian and Devonian ages show an intermediate structure between brittle stars and starfish, probably representing ancestral forms to both. The Ophiuroidea is perhaps the largest class of echinoderms with more than 2000 species.

OPHIOPLOCUS ESMARKI

Fig. 101

Description
Size, radial spread 3–3½ in. (7.6–8.9 cm), disk ½–¾ in. (1.3–1.9 cm) dia. Body covered with irregular swollen scales, imparting a pebbled appearance; scales nearly hide radial shields. Disk and arms flattened; arms rather short; arm spines, in groups of three, blunt and short; genital openings are short slits. Color various shades of brown.
Habitat
Sandy mud substratum under flat rocks, especially disintegrating granite; common in some localities.
Range
Pacific Grove to San Diego, Calif.

AMPHIODIA OCCIDENTALIS

Fig. 102

Description
Size, radial arms 4–5 in. (10.2–12.7 cm) long, disk less than ½ in. (1.3 cm). An exceptionally "brittle" brittle star. Body covered with overlapping scales almost microscopic in size; radial shields narrow,

joined near edge of disk; notch in disk margin between shields and base of arms; arm spines blunt, in groups of 3, at right angles; arms distinctly flat, easily shed, quickly regenerated. Color yellowish or whitish.

Habitat
Low-tide pools on the underside of rocks, particularly where embedded in sand and detritus, sometimes in substratum; also in mud or sand at base of eelgrass roots. Abundant, usually in aggregates of several to several dozen, often with arms intertwined.

Habits
At night often crawl about on top of rocks at ebb tide.

Range
Kodiak, Alaska, to cen. Calif.

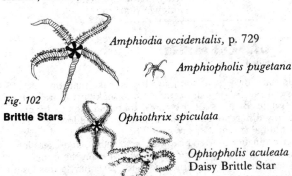

Amphiodia occidentalis, p. 729

Amphiopholis pugetana

Fig. 102

Brittle Stars

Ophiothrix spiculata

Ophiopholis aculeata
Daisy Brittle Star

DAISY BRITTLE STAR
Ophiopholis aculeata   *Fig. 102*

Description
Size, $\frac{1}{10}$–$\frac{2}{5}$ in. (3–9 mm) disk dia. A dainty, nicely proportioned brittle star. Small disk; 5 conspicuous lobes between rays, a bit inflated on lower surface in larger specimens; surface covered with small conical spines, more distinct on outer parts of lobes; 3 or more rounded plates in a row from center of disk to base of each ray. Color rusty red, curiously streaked and mottled with lighter colors; rays may be red-banded; sometimes dull greenish in part of pattern.

Habitat
Enormously varied from shoal water to 600 fathoms (1.1 km); common.

Range
Alaska; B.C.; Puget Sound s. to Pt. Sur, Calif.

AMPHIOPHOLIS PUGETANA   *Fig. 102*

Description
Size, ¾ in. (1.9 cm) spread. Smallest West Coast ophiuran. Body dainty. Color black and white.

Similarities
Larger relative *Amphiodia*.

Habitat
Shallow on-shore waters, in great beds, so common as to cover underside of all rocks. Never found with intertwined arms.

Range
Alaska to Baja Calif.

OPHIOTHRIX SPICULATA

Fig. 102

Description

Size, radial spread several inches, disk ⅔ in. (1.7 cm) dia. Disk thickly set with spicules, so appears fuzzy; egg sacs conspicuous, bulging out between arms; mouth papillae absent, tooth papillae present. Arm spines in 7 rows, longer, thinner, more numerous than in *Ophiopholis,* its common associate. Color very variable but often greenish-brown, with orange bands on arms, orange specks on disk; mouth area whitish.

Habitat

Low-tide zone, under rocks.

Range

Monterey Bay, Calif., to Central America.; rather common, particularly in n.

SEA URCHINS AND SAND DOLLARS

Class Echinoidea

The sea urchins are spiny, globular (or disk-shaped) echinoderms lacking arms. The radii and interradii (spaces between the radii), conspicuous in other echinoderms, are not obvious on living specimens. Close examination, however, reveals the five-ray pattern in the doubled rows of tube feet and in the five teeth. The spines are joined to the skeleton by a ball-and-socket joint. The skeleton itself comprises closely joined calcareous plates forming a test, or case, about the internal organs.

Although relatively few species live within tidal limits, there are great numbers of individuals; empty tests without spines are usually abundant along West Coast shorelines. Most commonly recognized is the sand dollar, a form in which the living animal is flattened both above and below. All the Echinoidea except the heart urchins possess a remarkable system of hard jaws and teeth, used in feeding.

GIANT RED URCHIN

Strongylocentrotus franciscanus

126:4

Description

Size, 5–7 in. (12.7–17.8 cm) dia. Largest West Coast sea urchin. Body globular, very similar to *Strongylocentrotus purpuratus,* but test somewhat lighter, spines relatively longer. Color usually dark red, but may be whitish to dark purple.

Habitat

Outer coasts, occasionally in tide pools; more abundant subtidally.

Range

Alaska to Cedros Is., Baja Calif.

GREEN SEA URCHIN

Strongylocentrotus droebachiensis

Description

Size, 2–3½ in. (5.1–8.9 cm) dia. Body globular; spines moderately long, pointed. Color mildly greenish, sometimes violet tinged; spines often deep green, test violet; occasionally entirely dull brown.

ECHINODERMS:
SEA URCHINS AND SAND DOLLARS

Similarities
Strongylocentrotus purpuratus is smaller; spines thick, rather short, fluted, blunt; color purple.
Habitat
Both open and protected shores; abundant.
Habits
Passes through several free-swimming stages not even remotely resembling the adult; omnivorous feeder, principally on algae.
Range
Arctic Ocean to Wash. coast and Puget Sound.

PURPLE SEA URCHIN
Strongylocentrotus purpuratus

Description
Size, 1½–2 in. (3.8–5.1 cm) dia. Body globular; spines thick, rather short, fluted, blunt. Color decidedly purple; very young specimens whitish or greenish.
Habitat
Open surf-swept coast, also in tide pools; usually found half-burrowed into rock, occasionally imprisoned; most abundant of sea urchins.
Range
Alaska to Cedros Is., Baja Calif.

SAND DOLLAR
Dendraster excentricus   *Fig. 103*

Description
Size, 3–4 in. (7.6–10.2 cm) dia. Body very flat, somewhat circular in outline, thickly covered with minute spines to ¹⁄₁₆ in. (0.2 cm) long, giving a velvety appearance. Ambulacral (tube feet) areas form flowerlike figure with 5 petals on dorsal surface; 2 zones are shorter than the other 3, thus off center (hence the name). Color light brown to purple when alive; tests, as found, white.
Habitat
On top of sand or partially buried in sand in lagoons, at the lowest tides on sandy beaches and often subtidally; very abundant, sometimes in great beds (Puget Sound, Newport and San Diego bays, Calif.).
Range
Alaska to Baja Calif.

*Eupentacta
quinquesemita*
(Sea Cucumber)

Fig. 103

Cucumaria lubrica
(Sea Cucumber)

Dendraster excentricus
Sand Dollar

SEA CUCUMBERS
Class Holothurioidea

Sea cucumbers are softer bodied than other echinoderms, and the body axis lies parallel to the surface over which it crawls, rather than perpendicular to it; thus, the animal is elongated and cylindrical, with a tough, leathery, often warty skin. The skeleton is almost nonexistent, revealed under magnification only as tiny calcareous granules in the shape of plates, buried in the skin. The radial symmetry is shown in the oral tentacles and rows of tube feet. At the end away from the mouth is the cloaca, a common chamber into which the digestive tract empties and through which water is pumped to the respiratory system.

Holothurians live in all seas, from shallow water down to abyssal depths, and range in length from a few millimeters to three feet. In China, the dried body wall of a large species—called *trepang*—is an important food. The shore collector can usually find specimens in tidal pools along rocky beaches, and occasionally on kelp holdfasts recently washed ashore. Wormlike species can be found under rocks in sand or buried in mud. When roughly handled, most sea cucumbers eject a large proportion of their internal organs; under normal conditions, the part of the animal retaining the head will regenerate the lost parts and survive.

CUCUMARIA LUBRICA
Fig. 103

Description
Size, 2–4 in. (5.1–10.2 cm) long. Body long, cylindrical; calcareous deposits in body wall form irregular, thick, knobbed plates or buttons; 10 large tentacles; tube feet in 2 rows in each radius. Color white to black, with ventral surface yellowish; northern specimens lighter. Tentacles, 8 large, 2 small.
Habitat
Intertidal rocks.
Range
Puget Sound to Monterey Bay, Calif.; common in n.

CUCUMARIA MINIATA

Description
Size, 10 in. (25.4 cm) long. Body wall with irregularly perforated, knobbed plates, margins smooth, dorsal side may show papillae or pedicels (stalks) between radii; tentacles, 10, surrounding mouth, many-branched, retracted when animal is disturbed or left stranded by receding low tide; tube feet large, in rows parallel to body axis. Color salmon-red to dark brownish purple; tentacles bright orange.
Habitat
On-shore shallow to deep waters, under or between layers of rocks; abundant.
Range
Circumpolar, cosmopolitan; especially common in Puget Sound, Wash.

EUPENTACTA QUINQUESEMITA
Fig. 103

Description
Size, to 4 in. (10.2 cm). Body covered with more or less rigid, nonretractile tube feet, giving a spiny appearance. Color white, tentacles yellowish.

Habitat
On rocks, in tide pools, on floats and pilings in marinas; common.
Range
Puget Sound; to cen. Calif.

SEA SLUG
Stichopus californicus   **126:7**

Description
Size, 18 in. (45.7 cm) long. Body cylindrical, highly contractile,
covered dorsally with elongated warts and ventrally with tube feet.
Body shape variable; flaccid when undisturbed, stiff and turgid if
annoyed. Color usually red-brown, or yellowish to chestnut-brown,
paler below; tentacles yellow-tipped; tube feet black-tipped.
Habitat
Usually subtidal; commonly noticeable along shore.
Range
Puget Sound, Wash., to Monterey, Calif.

Note: A similar, more southern form, *Stichopus parvimensis*,
occurring from Laguna Beach to Baja California, is up to 3½
inches (8.9 cm) long and has a definitely intertidal habitat.

SEGMENTED WORMS
Phylum Annelida

These animals are called annelids because of the segmented body;
that is, the body is ringed externally, or annular. The common
earthworm probably is the most familiar member of this phylum.
The peristomium, the first true segment behind the extreme
anterior end, bears the mouth, but behind it each segment has
much the same basic structure: external bristles, or setae; part of
the digestive system; a nerve ganglion; and usually a pair of
nephridia, or excretory organs. Many species have light-sensitive
organs of sufficient organization to be almost eyes. They have a
sense of touch and probably several chemical senses. Reproduction
is sexual.

Three classes of these elongated, segmented worms are generally
recognized: Polychaeta, or marine worms; Oligochaeta, including
both terrestrial and freshwater worms; and Hirudinea, or leeches.
Of these, the polychaete worms constitute by far the largest class,
with a vast literature.

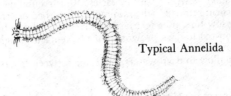

Fig. 104   Typical Annelida

MARINE SEGMENTED WORMS
Class Polychaeta

The polychaete worms possess paired paddlelike or bristlelike appendages on each segment along the length of the body, which is generally somewhat flattened rather than tubular. The head is quite pronounced and the setae are quite conspicuous. The sexes are usually separate.

Johnson and Snook observe, "The polychaetes have parapodia, leglike appendages, that are made up of dorsal and ventral branches, both of which may bear setae and cirri. In some forms the parapodia are modified to form 'fins' for swimming. The cirri, which are sensory, may in some cases have an extra blood supply, be much enlarged, and serve as gills. The bristles are useful in locomotion and probably for defense also."

Polychaetes are extremely abundant in terms of species and individuals and are common in all marine habitats. The few species described here constitute only the most common, largest, or most spectacular of the hundreds of species.

CLAM WORM
Nereis vexillosa

129:7

Description
Size, 2–12 in. (5.1–30.5 cm). One of the most common annelids on the West Coast. Body with about 118 segments, prostomium longer than broad, tentacles shorter than prostomium and well separated at bases, palps large and reaching beyond tentacles; jaws chitinous, terminating in protrusible pharynx (have severe biting capability, but rarely bite). Color dark brown to blue-green, often iridescent.
Habitat
Nearly all marine habitats, but principally among barnacles and mussels on wharf piles, rocky shores, gravel beaches, surf-pounded shores; very abundant.
Remarks
Valued as bait, much sought by fishermen.
Range
Alaska to San Diego, Calif.

LUGWORM
Abarenicola pacifica

129:2

Description
Size, 6–12 in. (15.2–30.5 cm). Body with 3 distinct regions: anterior end thick, blunt; middle more slender, with branching gills above rudimentary parapodia; tail slightly thickened. Skin rough. Color yellow, green, brown, or black.
Habitat
In U-shaped burrows in low-tide mud flats; also in barren sand of unprotected outer beaches in areas of silt deposition; very common.
Habits
Leaves characteristic coiled fecal castings.
Remarks
Popular with fishermen as bait.
Range
Bay waters of B.C. to Humboldt Bay, Calif.

EUDISTYLIA POLYMORPHA

125:14

Description
Size, 18 in. (45.7 cm). A spectacularly vivid, large-tubed worm. Body has thoracic segments distinct from the tapering abdomen; head with a crown of feathery tentacles about 30 on each side in spirals of 2–3 turns each, each tentacle with 2–10 conspicuous extremely sensitive eyespots. When the animal is undisturbed, its colorful tentacles spread out into the water above the tube, retract abruptly if touched or shadowed. Tube parchmentlike, tough, extending deeply into rock crevices. Tube dull yellow or gray; tentacles variable from purple or wine color to whitish or tawny. Tentacle crown may be of a single color or banded with 2 hues.
Habitat
Quiet tide pools and crevices; abundant.
Range
Alaska to San Pedro, Calif. (uncommon south of Pacific Grove, Calif.).

SERPULA VERMICULARIS

125:17

Description
Size, 2¼ in. (5.7 cm). Recognized by its great twisted masses of limy, white, more or less coiled tubes. Operculum (closing tube opening) funnel-shaped, on right side; border notched, made up of 150 or more ribs. Branchiae, 54; abdominal segments 250 or more. Collar, gills, operculum vividly scarlet and variously banded.
Habitat
Covering rocky reefs, in tide pools, under stones near low-water mark; abundant.
Habits
Gills snap back into tube at the least disturbance.
Range
Cosmopolitan.

HALOSYDNA BREVISETOSA

129:1

Description
Size, 2 in. (5.1 cm). Body short, fairly thick; dorsal surface covered by two rows of overlapping "scales" which may detach on capture. Color dull brownish.
Habitat
Common among byssal threads in mussel beds, in kelp holdfasts, in commensal relationships with other invertebrates or free-living in rocky areas.
Range
Alaska to s. Calif.

HEMIPODUS BOREALIS

129:8

Description
Size, to 5 in. (12.7 cm) long. Body thick with very short parapodia; head small, tapering to a point bearing 4 small tentacles; large eversible proboscis armed with 4 jaws. Color flesh.
Habitat
Active burrower in sand and mud of semienclosed bays.
Range
Calif.

EULALIA AVICULISETA

Fig. 105

Description
Size, to 2 in. (5.1 cm). Body margined with rows of leaflike cirri borne on the parapodia; head with eyes and 5 tentacles. Color bright green with black lines between body segments.
Habitat
Among byssal threads of mussel beds, algae holdfasts, under rocks.
Range
Calif.

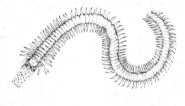

Fig. 105

Eulalia aviculiseta
(Marine Segmented
Worm)

NEPHTYS CALIFORNIENSIS

129:5

Description
Size, to 5 in. (12.7 cm) long. Body more or less rectangular in cross section; small square or pentagonal head bearing small tentacles; eversible proboscis. Color dark pinkish.
Habitat
Burrowing in open sand beaches.
Range
Calif. to Mexico.

ARABELLA IRICOLOR

129:6

Description
Size, to 7 in. (17.8 cm) long. Body very long, appearing much like an earthworm because of greatly reduced parapodia; very iridescent outer skin; head without appendages of any kind but with eyes. Color dark reddish to greenish.
Habitat
In kelp holdfasts, among mussels, on rocky shores.
Range
Cosmopolitan.

CIRRATULUS CIRRATUS

129:3

Description
Size, to 4 in. (10.2 cm). Entire length of body margined with rows of very long filamentous cirri giving worm a hairy appearance. Cirri very mobile but extremely susceptible to breakage; many fall off during collection. Color dark brownish black.
Habitat
Lies buried in sand or mud with only the cirri visible on surface, where they appear as a mass of free-moving small worms.
Range
Cosmopolitan.

THELEPUS CRISPUS 129:9

Description
Size, to 4 in. (10.2 cm). Body thick and rather stubby, without
parapodia; head obscured by a mass of long tentacles. Color orange
to yellowish.
Habitat
Rocky shore where animal either lives buried just below the
surface in sand or forms a sand-grain tube around itself; tentacles
extend out on surface for long distances away from animal.
Range
Alaska to s. Calif.

NOTOMASTUS TENUIS

Description
Size, to 7 in. (17.8 cm) long. Body extremely long and thin without
parapodia, appearing as a very thin earthworm; head small
without appendages. Color usually red.
Habitat
Live in vertical position in huge numbers in the mud or muddy
sand of protected bays and harbors.
Remarks
Difficult to remove without breaking.
Range
B.C. to s. Calif.

PHRAGMATOPOMA CALIFORNICA 129:4

Description
Size, to 2 in. (5.1 cm). Animal with cone-shaped operculum on the
head. Color black.
Habitat
Lives in a tube constructed of sand grains and forms colonies of
these tubes cemented to rocks intertidally; in certain favored
localities may form massive reefs covering the substrate; animal
never leaves tube.
Range
Calif.

SIPUNCULID WORMS
Phylum Sipuncula

The sipunculans have an elongated, cylindrical body covered by a
tough epidermis, without cilia or bristles or appendages of any
kind. The body is divided into two parts; the anterior introvert may
be invaginated into the posterior trunk. The sipunculans, often
called "peanut worms," are unsegmented, and the introvert is
usually much more slender than the trunk. The head end has a
ring or rings of tentacles surrounding the mouth, which in some
species are elaborately branched.

All sipunculans are marine. There are 320 known species. The
majority prefer shallow water; some live in the sand and others
beneath rocks; a few species live among mussels; still others inhabit
holes in rocks or live in empty shells or tubes. Sizes range from
four to eight inches (10.2–20.3 cm) in length. Some species are
very slender and threadlike. The sexes are nearly always separate.

Worms in this group are primarily burrowing or nesting forms that live in sand or mud, among roots of eelgrass, or on wharf pilings. When the worms are placed in seawater and allowed to completely extend themselves, a circle of rather prominent tentacles can be seen at the extreme anterior end.

PHASCOLOSOMA AGASSIZII   128:10

Description
Size, to 3 in. (7.6 cm). The commonest West Coast sipunculan. Body to 5 inches (12.7 cm) if fully extended, with thick, tough skin with prominent papillae; introvert thinner than trunk; tentacles small and unbranched. Color varies from shades of gray to reddish-brown; introvert often irregularly banded with dark brown; body often with blotches of dark brown or black.
Habitat
In mud, kelp holdfasts, crevices in rock, shells, holes of boring clams, and *Mytilus* beds.
Range
Alaska to s. Calif., cosmopolitan.

SIPUNCULUS NUDUS   128:11

Description
Size, 5–8 in. (12.7–20.3 cm) long. The largest West Coast sipunculan. Body to ½ in. (1.3 cm) dia., with inconspicuous tentacles and 30–32 longitudinal muscle bands; external anal opening prominent. Skin shining, iridescent, reveals muscles in small rectangular patches. Color white to flesh.
Habitat
In mud or sand at bottoms of shallow bays, rather deeply embedded; particularly common in areas of the burrowing anemone *Edwardsiella,* the brachiopod *Glottidia,* and sea pens.
Range
San Pedro, Calif., to Baja Calif., especially abundant in Newport and Mission bays.

ECHIURANS
Phylum Echiura

Once classed with the sipunculans, the echiurans are unsegmented, sausage-shaped worms with a rather shovel-shaped proboscis which, unlike that of the sipunculans, is not eversible but can be greatly shortened or lengthened. It serves as a remarkably mobile appendage and, as a tactile organ, replaces the eyes or other distinct sense organs.

The echiuran worms are quite abundant, living in burrows in mud flats or in the finer detritus beneath rocks, with some species existing at depths of more than 130 fathoms (237.8 m). Echiurans range in size from 2½ in. (6.4 cm) to a maximum of twenty inches (50.8 cm) long and two inches (5.1 cm) in diameter.

FAT INNKEEPER
Urechis caupo   126:8
Description
Size, 8–19½ in. (20.3–49.5 cm). Body large, cigar-shaped, 2 bristles under mouth and circlet of bristles around anus. The active animal

continually changes shape because of peristaltic movements.
Proboscis much reduced. Color pinkish; mouth and anal bristles
golden.

Habitat
U-shaped burrow with entrances 16–38 in. (40.6–96.5 cm) apart in
sand and mud flats; entrances constricted.

Habits
Spins slime net to catch microscopic food, long, transparent (but
gray and visible when filled with detritus), then eats net and
contents. Usually accompanied by commensal guests such as the
reddish scale worm, *Hesperonoë adventor* (½–2 in., or 1.3–5.1 cm,
long), 1 or 2 pea crabs, *Scleroplax granulata* ($\frac{5}{16}$ in., or 0.79 cm,
across carapace), and 1 or more gobies, *Clevelandia ios.*

Range
Humboldt Bay to Newport Bay, Calif.; especially common in
Elkhorn Slough.

Note: The similar *Urechis echiurus alaskensis,* very abundant in
southeastern Alaska, is smaller, has a scoop-shovel proboscis, lacks
a food net, and lives embedded in gravelly substratum as well as
mud flats.

ARTHROPODS
Phylum Arthropoda

Considered the most successful group in the animal world in point
of numbers and species, the arthropods are easily recognized
because all possess a hard, chitinous, outer body covering, divided
externally into segments to which are attached jointed appendages.
The outer covering is periodically molted. The body segments are
unequal in size and are usually organized into three definite body
regions: head, thorax and abdomen.

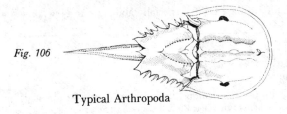

Fig. 106

Typical Arthropoda

Well over 900,000 species (800,000 insects alone) have been
identified among the barnacles, crabs, sand hoppers, insects,
waterfleas, spiders, myriapods, and others. Moreover, their
fossilized hard parts are found in all rock strata extending back
into Precambrian times. Arthropods are found in every
environment, and some are parasitic on man and other animals as
well as on plants. Most arthropods are small; the largest are the
giant lobsters and spider crabs. Some are so small that dozens of
them can parasitize another insect's egg.

CRUSTACEANS
Class Crustacea

Most marine arthropods are crustaceans, ranging in size from almost microscopic forms to an Eastern lobster three feet long, weighing thirty-five pounds (15.9 kg), and a giant Japanese spider crab with five-foot (1.5-m) legs. All have a hard shell and two pairs of antennae. Crustaceans molt by digesting away the inner parts of the old shell with enzymes, then secreting a new shell under the old, and finally breaking the old shell along zones of weakness. The soft animal then emerges. Molting is a laborious process, since every protuberance and hair must be pulled out of the old chitinous cover. When, having shed its old shell, the animal is free, it takes in more water to increase its size so that the newly forming shell when hard again will be large enough to accommodate its owner. Then the internal pressure is relaxed and the animal has room in which to grow. Among the oldest (Cambrian through Permian geologic eras) and most sought after fossil arthropods are the trilobites, whose molted dorsal shields, calcified, constitute the bulk of the fossilized remains.

BARNACLES
Subclass Cirripedia

The most aberrant form of crustacean, and quite surprising to the layman, are the barnacles. These sessile animals begin life as remotely shrimplike, free-swimming larvae which soon attach themselves to almost any firm object by means of a cement gland at the base of the antennae. They then form joined limy plates, which in some species cover the entire body. The plates, which close the opening of the shell proper, can be opened on the midline and the legs, or cirri, extended. The cirri, which look like miniature feathers, sweep through the water, trapping small organisms for food. These particles are passed to the animal's mouth within the shell.

Barnacle shells have very definite shapes. Two types commonly seen along the shores are gooseneck barnacles and acorn barnacles. Generally, barnacles occur in zones or strips, according to tide levels, from the upper part of the highest zone to somewhat below the low-tide mark. They attach themselves to rocks, pilings, and the undersides of boats, sometimes so crowded together that their shells become very much elongated. Some goosenecks float in clusters on the ocean surface; some acorn barnacles grow attached to the skin of whales. Some species are parasitic; almost all are hermaphroditic.

Stalked Barnacles
Suborder Lepadomorpha

DRIFTWOOD GOOSE BARNACLE
Lepas anatifera   *Fig. 107*

Description
Size, to 32 in. (81 cm). Most abundant goose barnacle. Capitulum ¾ in. (1.9 cm) long; radial striations on valves, internal tooth on umbo of right scutum and none on left, carina often toothed.

Habitat
Attached to floating debris.
Range
Worldwide.

Pollicipes polymerus
Pacific Goose Barnacle

Fig. 107

Lepas anatifera
Driftwood Goose Barnacle, p. 741

PACIFIC GOOSE BARNACLE
Pollicipes polymerus   *Fig. 107*

Description
Size, 3–4 in. (7.6–10.2 cm) tall. Capitulum to 1 in. (2.5 cm) dia.,
with 18 or more plates and irregularly arranged scales at base.
Peduncle finely scaled. Color red or reddish brown to yellowish
brown; plates white.
Habitat
Upper ⅔ of intertidal zone; very common, particularly in
association with *Mytilus*, on rocky coasts exposed to open sea.
Range
Bering Strait to cen. Baja Calif.

Note: A vividly colored subspecies, *Pollicipes polymerus echinata*,
inhabits the darkness of caves and under rocks and is more colorful
in darker zones.

Acorn Barnacles
Suborder Balanomorpha

Species in this group have the rostrum provided with alae, or
wings, which overlap the next lateral plate or compartment.

CHTHAMALUS DALLI

Description
Size, basal dia. ¼ in. (0.6 cm), height ⅛ in. (0.3 cm). Form
definite, clean-cut, never crowded or piled in clusters; shell conical,
opening oval. Color pale to dark gray, interior whitish with
pinkish tints; common.
Habitat
High intertidal, above other barnacles.
Range
Alaska to cen. Calif.

BALANUS CARIOSUS

Description
Size, 2 in. (5.1 cm) basal dia., height slightly less than basal dia.
Crowded specimens may be 4 in. (10.2 cm) high and ½ in. (1.3
cm) in dia. Shell wall thick, porous, conical, usually characterized

by shinglelike thatches of downward-pointing spines; many deeply
cut, irregular ribs overlapping each other; alae narrow,
indistinguishable except in cylindrical forms; back narrow, apex
beaked, spur sharply pointed. Young starlike in form. Color whitish.
Habitat
Steep shorelines with strong currents and wave action; also in quiet
waters in deep crevices and under overhanging ledges in low-tide
zone; very abundant.
Range
Alaska to Monterey Bay, Calif.; largest forms n. of Puget Sound.

BALANUS GLANDULA
Fig. 108

Description
Size, to ½ in. (1.3 cm) dia. Shell usually conical, deeply ribbed;
when crowded, often elongated; most easily identified by small pit
near center of inside of scutum (anterior valve). Color dirty white.
Habitat
High-tide rocks; abundant.
Remarks
Look like dead encrustations, but when placed in sea water or
during extreme high tides make feeding movements with appendages.
Range
Aleutian Is. to Mexico.

Fig. 108
Acorn Barnacles

Balanus glandula

Tetraclita squamosa rubescens
Thatched Barnacle, p. 744

Balanus nubilus

BALANUS NUBILUS
Fig. 108

Description
Size, 2½–3 in. (6.4–7.6 cm) high. Largest acorn barnacle in the
world. Body with basal diameter exceeding height; shell often coated
with fouling organisms. Color of mantle rich reds and purples.
Habitat
Low tide and below on pilings; often in clusters to 12 in. thick,
growing one on top of another; abundant.
Range
S. Alaska to cen. Baja Calif.

BALANUS TINTINNABULUM CALIFORNICUS

Description
Size, 2¼ in. (5.7 cm) dia., height to 1⅜ in. (3.5 cm). One of largest
and most conspicuous of barnacles. Shell nearly cylindrical, walls
finely striated, alae indistinctly striated transversely. Color pinkish-
red with white lines; mantle lips vividly red with blue and light
spots and edged with white, beneath blue edged with red.
Habitat
Wharf pilings, boat bottoms, shore rocks; common.
Range
Monterey Bay to s. Calif.

THATCHED BARNACLE
Tetraclita squamosa rubescens

Fig. 108

Description
Size, 1½ in. (3.8 cm) dia. Shell roughly conical, surface usually
eroded and roughened like thatch; 4 compartments, not always well
marked; walls permeated by pores, forming several irregular rows
in large specimens. Color usually dull red.
Habitat
On rocks exposed at low tide; rather solitary, rarely bunched;
common.
Range
Farallon Is. to Baja Calif.

Isopods
Order Isopoda

Members of this order, most commonly known as sow bugs, pill
bugs, or wood lice, generally have flattened bodies. Most isopods
are marine, living either free or as parasites, and are very
abundant both as species and individuals. Of the very large number
of species, only a few of the largest and most common are
mentioned here.

CIROLANA HARFORDI

Fig. 109

Description
Size, ⅜–⅝ in. (0.95–1.6 cm) long. Abdomen has 6 segments
bearing about 26 spines along posterior border. Head with
rostrum. Color drab, not constant; ground color generally light and
patterned by minute black dots; some specimens dark brown to
almost black, others nearly colorless; often tawny or orange-yellow
marked above with gray or brown.
Habitat
High intertidal zone under rocks or among beds of mussels; very
common.
Range
B.C. to n. Baja Calif.

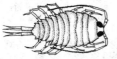

Ligia occidentalis, p. 746 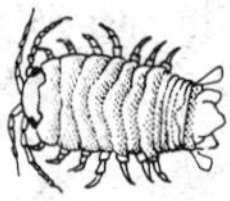   *Cirolana harfordi*

Fig. 109
Isopods

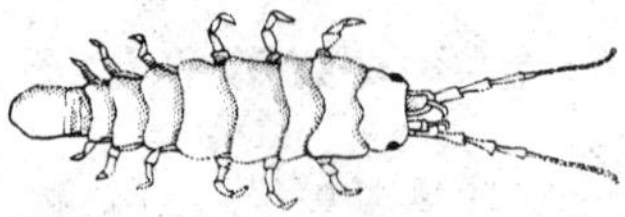

Idotea resecata    *Gnorimosphaeroma
oregonensis*

GRIBBLES
Family Limnoriidae

126:5

These wood-boring pests are all very small and destroy wooden pilings by boring holes with their mandibles, eating wood along grain and even across at the rate of half to one inch (1.3–2.5 cm) per year per animal. The two most common are *Limnoria lignorum,* occurring from Alaska to Point Arena, Calif., and *Limnoria tripunctata,* from San Francisco Bay to central Baja California.

GNORIMOSPHAEROMA OREGONENSIS

Fig. 109

Description
Size, ¼ in. (0.6 cm) long. A most ubiquitous pill bug. Body squat, flat; head set well back into first thoracic segment; eyes widely separated; last abdominal segment short, flat, lacking tubercles.
Habitat
Beach at low tide; undercrust population may run many dozen to the square foot in Puget Sound; one of commonest pill bugs in San Francisco Bay.
Range
Alaska to Baja Calif.

IDOTEA RESECATA

Fig. 109

Description
Size, 2 in. (5.1 cm) long. Body long, narrow, terminates in 2 spinelike points; abdomen equals ⅓ total length. Color yellowish brown.
Habitat
Clinging tenaciously to kelp and surf grasses; quite common.
Range
B.C. to s. Calif.

IDOTEA UROTOMA

Description
Size, ¾ in. (1.9 cm) long. Abdomen relatively short, broad; tail paddle-shaped, no spinelike points, somewhat irregular. Color brown.
Habitat
Under rocks of intertidal zone; common.
Range
Puget Sound, Wash., to cen. Calif.

IDOTEA WOSNESENSKII

Description
Size, 1½ in. (3.8 cm) long. Body similar to *Idotea resecata,* but terminal segment rounded and without 2 spines on lateral border. Male larger than female.
Habitat
Along rocky shores, also on kelps at low tide, under rocks, and in clusters of mussels; abundant; commonest small crustacean at Santa Cruz, Calif.
Range
Alaska to cen. Calif.

LIGIA OCCIDENTALIS

Fig. 109

Description
Size, 1½ in. (3.8 cm) long. Dorsoventrally flattened, oval in outline. Body of conspicuous segments. Most conspicuous isopod along southern California coast. Color dull gray or brown, pattern variable; legs tipped with orange.

Habitat
Often seen scurrying among or upon rocks well above high-tide mark (never wets its feet); common.

Habits
Fast-moving in highest part of intertidal; almost terrestrial.

Range
Sacramento R. to San Francisco Bay and s. along coast to Gulf of California.

Note: A northern relative, *Ligia pallasii,* occurring along the open coast from Alaska to San Francisco Bay, is ¾ in. (1.9 cm) long and has a flat, broad body with a granular surface.

Amphipods
Order Amphipoda

The amphipods usually have the body laterally compressed and the body axis strongly curved. They may be distinguished from the small shrimps and prawns by the absence of a carapace and by the body that is jointed and flexible along its entire length. More than 3000 species have been described, ranging in size from microscopic specimens to deep-sea giants more than twelve inches (30.5 cm) in length; most are marine. Many species commonly are to be found in beach wrack or climbing among the hydroids and ascidians on rocks and piling. These animals are extremely abundant in all marine habitats. There are a very large number of species, but most species are so similar that it is difficult for anyone but an expert to tell them apart. The few mentioned here are representative.

SAND FLEA
Orchestia traskiana

Fig. 110

Description
Size, ½ in. (1.3 cm). Similar to, but much smaller than, *Orchestoidea californiana.* Color dull-green or gray-brown; legs slightly blue.

Habitat
Among decaying algae on rocks, and under debris, at high-water mark and to 20 feet (6.1 m) above, or in mud sloughs; common.

Habits
A scavenger.

Range
Calif.

Fig. 110

Amphipods

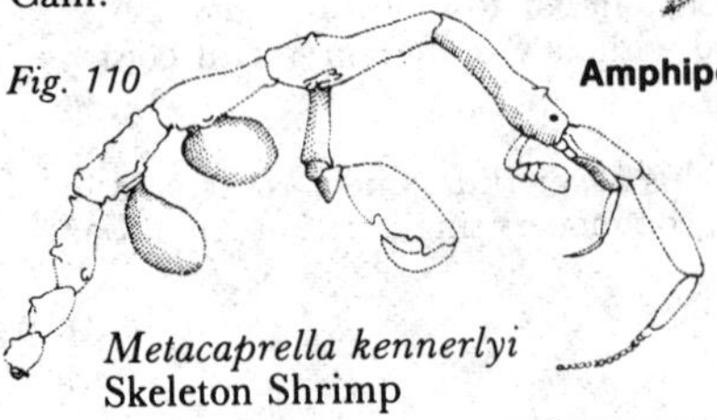

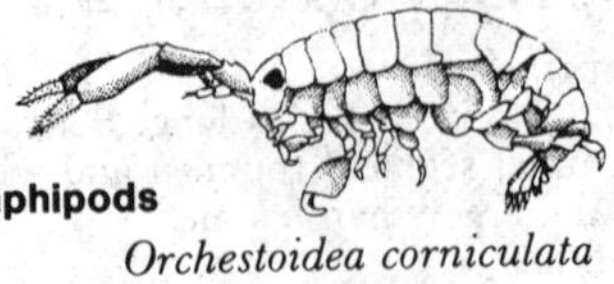

Orchestoidea corniculata

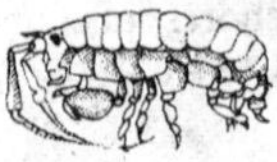

Metacaprella kennerlyi
Skeleton Shrimp

Orchestia traskiana
Sand Flea

BEACH HOPPER
Orchestoidea californiana

Description
Size, 2½ in. (6.4 cm) long. Body more than 1 in. (2.5 cm) long; total length includes antennae, first one short, second longer than body. Color white or dingy ivory; antennae and part of dorsal surface bright red-orange.

Habitat
Above tide line, usually feed on kelp in great numbers.

Habits
These animals avoid getting wet, retreating in windrows from advancing waves. They spend days buried in damp sand, but on quiet evenings leap vigorously about and retreat from an intruder like waves of grasshoppers; when chased, they rapidly dig into sand headfirst.

Range
Puget Sound to Calif.

ORCHESTOIDEA CORNICULATA   *Fig. 110*

Description
Size, body ⅝ in. (1.6 cm). Body similar to *Orchestoidea californiana* but smaller; second antenna of male about ½ body length, of female about ¼ body length. Color pink or mottled with red and brown.

Habitat
One of chief scavengers of semiprotected beaches, found by day in sand burrows or under piles of wood, by night feeding on decaying seaweeds and storm wrecks; very common.

Range
Calif.

SKELETON SHRIMP
Metacaprella kennerlyi   *Fig. 110*

Description
Size, males 1¼ in. (3.2 cm), females ½ male size. Body of male with several dorsal and ventrolateral tubercles or spines; female very rough. Head bears pair of small forward-pointing spines. Legs with prominent prongs on basal joints. Color pink-banded.

Habitat
Rock pools, in hydroid colonies; often in vast multitudes.

Range
Alaska to Santa Barbara, Calif.

Shrimps, Lobsters, and Crabs
Order Decapoda

The decapods are the most highly organized crustaceans, possessing ten walking legs, from which the name of the order is derived. The head and thorax are rigidly fused to each other to form the cephalothorax, which is covered by the carapace. To this order belong all the shrimps, lobsters, crayfishes (freshwater), hermit crabs, and crabs. With the exception of the smaller shrimps, almost all decapods deposit calcium carbonate within the chitinous skeleton. The majority of the group are marine, but the crayfishes have invaded fresh water and a few species (not described here) have adapted themselves to land habitats.

Swimming Decapods
Suborder Natantia

COON-STRIPED SHRIMP
Pandalus danae

Description
Size, 8¼ in. (21 cm). Commercially important as food. Rostrum less than 1½ times carapace length; dorsal spines extend slightly more than ½ way back on carapace; third abdominal segment not compressed or keeled, lacks median lobe or spine in front of margin.
Habitat
Usually deep water, but occasionally found in tidal pools.
Range
Alaska to San Francisco Bay, Calif.

HIPPOLYTE CALIFORNIENSIS

Description
Size, 1½ in. (3.8 cm). A graceful, excessively slender shrimp. Rostrum base rounded, not continued on carapace, 3–5 teeth on both upper and lower margin. Walking legs (first pair) short, hand broad and thick at base; second pair with wrist of 3 segments, more slender and longer. Abdomen not crested or keeled; telson (terminal abdominal segment) truncated and spinulous at tip.
Habitat
Quiet, shallow waters, usually hiding in crevices, under rocks, or in eelgrass; common.
Habits
At night, swims among eelgrass blades in large numbers.
Range
Alaska to San Diego, Calif.; most commonly between Bodega Bay, Calif., and Gulf of Calif., but common only as far s. as Elkhorn Slough.

HEPTACARPUS PALUDICOLA

Description
Size, 1¼ in. (3.2 cm). A transparent species scarcely distinguishable from *Heptacarpus pictus;* in fact, there is no way to easily distinguish among the several *Heptacarpus* species commonly found. Rostrum about as long as, or longer than, rest of carapace; sixth abdominal segment shorter than seventh; third, fourth, fifth segments unkeeled, also lack spine. Color green.
Habitat
Among eelgrass; common.
Range
B.C. to Calif.

HEPTACARPUS PICTUS   *Fig. 111*

Description
Size, ¾ in. (1.9 cm). A fairylike, semitransparent shrimp smaller than *Heptacarpus paludicola.* Rostrum shorter than rest of carapace but reaches beyond middle of antennal scale. Beating heart and all organs plainly visible within body. Color pale green, often with red bands; legs barred with crimson.

Habitat
Tidal pools, darting about or in crevices; very abundant.
Range
San Francisco to San Diego, Calif.

Fig. 111

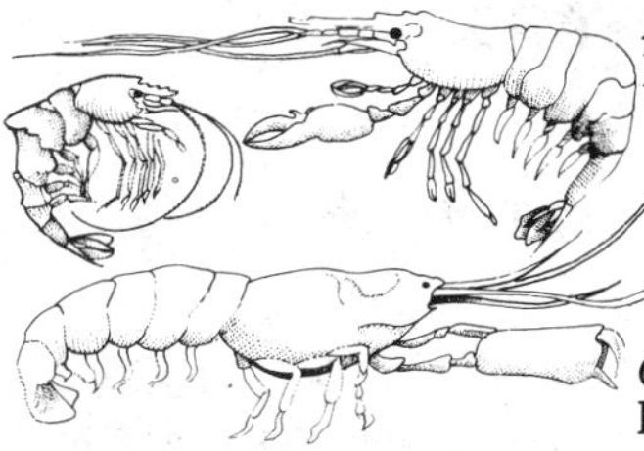

Heptacarpus pictus

Swimming Decapods

Alpheus dentipes
Pistol Shrimp

Crangon nigricauda
Black-tailed Shrimp

PISTOL SHRIMP
Alpheus dentipes   *Fig. 111*

Description
Size, to 2 in. (5.1 cm). Makes itself known by loud snapping (metallic clicking) of its large claw. Rostrum with 3 frontal spines of nearly equal size; eye stalks short, concealed under carapace margin. Claws of first pair of legs very different in size; second pair of legs slender, claws small. Color greenish, sometimes slightly tinged with blue, orange border on telson; claws mottled with brown, or green and orange; tip of palm darker, rounded tip of finger white; antennae and antennules yellow-brown; legs whitish, first pair yellowish on last few joints.
Habitat
Tide pools; quite common.
Other name:
Snapping Shrimp.
Range
Farallon Is. to Baja Calif.

Note: A similar species, *Alpheus bellimanus,* has a slightly smaller snapping claw.

BLACK-TAILED SHRIMP
Crangon nigrocauda   *Fig. 111*

Description
Size, 2¾ in. (7 cm). A commercial species. Carapace occasionally swollen out on one side into a blister from a parasitic isopod; dorsal profile nearly straight, not depressed. Rostrum characteristically very short, distinguished by one median spine with a smaller spine just back of it. Color (alive) dark gray; tail blackish; covered with salt-and-pepper markings, making it nearly invisible over sand.
Habitat
Usually in pools in sand or buried in sand, occasionally exposed; abundant.
Habits
Very agile; cannot be captured without a dip net.
Remarks
Edible.
Range
Alaska to Baja Calif.

Note: Occurring with this species is often *Crangon nigromaculata,* which differs in having a large circular spot on each side of the sixth segment toward back margin, and no dorsal median keel on the fifth abdominal segment.

Creeping or Walking Decapods
Suborder Reptantia

SPINY LOBSTERS
Section Palinuira

CALIFORNIA SPINY LOBSTER
Panulirus interruptus   *Fig. 112*

Description
Size, 10–17 lb. (3.7–6.34 kg). Carapace subcylindrical, abdomen depressed. Has enlarged, spiny antennae and lacks ponderous crushing claws of the true Atlantic lobster. All 5 pairs of legs used for walking. Mouth parts, 6; largest resemble walking legs, help hold food. Gill chamber, containing 21 plumelike gills, at each side of body between carapace and thorax wall. Color varies with surroundings.

Habitat
Subtidal waters, but young in tide pools; formerly abundant, but in danger of commercial extermination.

Remarks
A valuable commercial lobster, most usually seen in California seafood markets; females under 10½ in. (26.7 cm) protected by law.

Range
Pt. Conception, Calif., to Mexico.

Fig. 112

Panulirus interruptus
California Spiny Lobster

HERMIT CRABS AND MOLE CRABS
Section Anomura

Members of this section are greatly diversified in structure, but are distinguished by their upward-turned fifth walking legs and eyes lying inside the antenna.

PINK OR GHOST SHRIMP
Callianassa californiensis   *Fig. 113*

Description
Size, 2½ in. (6.4 cm). Distinguishable by unflexed abdomen and large tail fan. First walking legs clawlike and unequal in size; males with strikingly unequal claws. Appendages under abdomen used for swimming. Color pinkish orange, with some internal organs visible.

Habitat
Burrows into mud flats; abundant.

Remarks
Dug for bait when tide is out.
Range
Alaska to San Diego, Calif.

Fig. 113

Upogebia pugettensis
Blue Mud Shrimp

Pagurus samuelis
(Hermit Crab)

Pagurus granosimanus
(Hermit Crab)

Callianassa californiensis
Ghost Shrimp

BLUE MUD SHRIMP
Upogebia pugettensis   Fig. 113

Description
Size, 4½ in. (11.4 cm). Rostrum narrow, anteriorly roughened.
First pair of thoracic legs with imperfect claws and equal in size;
hand with 2 parallel hairy lines on upper edge, thumb bent
downward, tooth near middle; very hairy. Color dirty white
(bluish).
Habitat
Burrows into lowest areas of mud bared by tide; common.
Range
Se. Alaska to cen. Baja Calif.

PAGURUS SAMUELIS   Fig. 113

Description
Size, 1½ in. (3.8 cm). Bright red antennae, brilliant blue bands
around tips of feet, walking legs banded with blue, claws bluish
with red stripes; hairy.
Habitat
Commonest hermit crab in upper tide pools of California coast.
Range
Calif. coast predominantly.

PAGURUS GRANOSIMANUS   Fig. 113

Description
Size, 1⅛ in. (2.9 cm). Similar to *Pagurus samuelis,* but smaller,
less hairy, and without blue bands on legs. Color extremely
variable.
Habitat
Intertidal zone pools.
Range
Unalaska (Aleutians) to Baja Calif.; extremely abundant in Puget
Sound.

PAGURUS HEMPHILLI

Description
Size, 2 in. (5.1 cm). A large form. Carapace with sharp median tooth on anterior margin. Big claw has wrist laterally compressed, subtends sharp angle with upper surface. Color almost uniformly straw-tan, with characteristic yellow eye ring.
Similarities
Pagurus granosimanus is smaller, has no tooth on carapace margin.
Habitat
Upper tide-pool region; occupies *Tegula* shells almost exclusively; abundant.
Range
B.C. to Monterey, Calif.

HAIRY HERMIT
Pagurus hirsutiusculus

Description
Size, 2–3 in. (5.1–7.6 cm). Body soft, very hairy. Northern specimens larger, usually more hairy than southern. Walking legs with white or pinkish bands, antennae with white bands or rings.
Habitat
Tidal pools; one of most abundant forms.
Range
Alaska to Baja Calif.

UMBRELLA CRAB
Cryptolithodes sitchensis   *Fig. 114*

Description
Size, carapace 2 in. (5.1 cm). Carapace so large that, from above, it hides all appendages; high in middle with 2 shieldlike extensions projecting from sides; surface smooth with an occasional rounded tubercle. Rostrum narrow at body, wider at tip, 2 deep notches between it and the lateral expansions. Color very variable, from red to brown; some may be spotted with different colors.
Habitat
Low-tide pools, among rocks in shallow water; uncommon.
Range
Sitka, Alaska, to Monterey, Calif.

Fig. 114

Cryptolithodes sitchensis
Umbrella Crab

Hapalogaster cavicauda
(Hermit Crab)

HAPALOGASTER CAVICAUDA Fig. 114

Description
Size, carapace ¾ in. (1.9 cm). Furriest and fuzziest of anomuran crabs; rostrum shorter; chelipeds without rows of spines. Color brown.
Habitat
Common under rocks at low tide.
Remarks
Looks like a furry pebble and difficult to distinguish.
Range
Mendocino Co. to s. Calif.

SAND CRAB
Emerita analoga Fig. 115

Description
Size, ¾–1 in. (1.9–2.5 cm). Carapace ovate, somewhat elongated, very convex dorsally; marked with ripplelike transverse lines, less evident toward sides and posterior area. Eye stalks long, slender, jointed near base; eyes small, pigmented. Antennae plumelike, carried coiled, each antennule with 2 flagella (whips). Color gray.
Habitat
Half-tide line generally, in sand; so abundant in places that a spadeful of sand may turn over a large number.
Remarks
Favored as bait by fishermen.
Other name:
Mole Crab.
Range
Oreg. to South America.

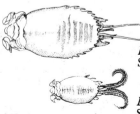

Fig. 115

Blepharipoda occidentalis
Spiny Sand Crab

Emerita analoga
Sand Crab

SPINY SAND CRAB
Blepharipoda occidentalis Fig. 115

Description
Size, carapace 2 in. (5.1 cm). Largest of sand crabs. Carapace oblong, roughened in front, marked with transverse grooves; anterior and lateral margins armed with long, sharp, pointed spines curving forward. Legs, first pair with imperfect claws, others (anterior) end in sharp, toothed claws, flattened laterally, with curved spines on margins and outer surfaces, crested by light-colored hairs; walking legs crested with hairs, terminal segments bladelike, sickle-shaped, second pair very broad and deeply notched. Eye stalks slender, jointed midway to tip; eyes small.
Habitat
Very low-tide sand flats, often found lower on beach than *Emerita*.
Range
Marin Co., Calif., and s.; quite uncommon outside Morro Bay.

THICK-CLAWED CRAB
Pachycheles rudis   *Fig. 116*

Description
Size, carapace ⅝ in. (1.6 cm). Carapace longitudinally convex,
quite smooth, slightly broader than long, finely striated in gill area.
Chelipeds large and unequal, upper surface roughly tuberculate;
prominences on claws give a rough, sandy appearance; walking legs
covered with light-colored hair. Color light brown with a few small
whitish streaks.

Habitat
Low tide under rocks; common.

Range
Alaska to Baja Calif.

Note: *Pachycheles pubescens,* occurring from Puget Sound to San
Francisco, is similar but has a flatter, rougher carapace, small
anterior transverse holes, a shallowed indentation at the rear, and
claws that are granulated rather than tuberculate.

Pachycheles rudis
Thick-clawed Crab

Fig. 116

Petrolisthes cinctipes
Porcelain Crab

PORCELAIN CRAB
Petrolisthes cinctipes   *Fig. 116*

Description
Size, carapace 9/16 in. (1.4 cm). An extremely flat crab, with equal-
sized smooth chelipeds. Carapace and abdomen greatly compressed,
abdomen folded under thorax. Eye stalks do not fit into orbits;
antennae long, whiplike, join body outside eye area. Color
brownish with transverse flecks of lighter color; chelipeds brownish
or bluish, walking legs marked with grayish bands.

Habitat
Rocky coasts, under rocks of middle zone; frequenter of crevices;
very common.

Habits
Scurries about feverishly for cover when a stone is lifted.

Range
B.C. to Gulf of Calif.

Note: The **FLAT-TOPPED CRAB,** *Petrolisthes eriomerus,* is similar,
but the palp of the maxilliped is blue, and the wrist has parallel
sides.

TRUE CRABS
Section Brachyura

These crabs are the short-tailed decapods, a number of species of
which are edible. The carapace is characteristically wide,
frequently broader than long; the rostrum may be very small or
lacking; the abdomen much reduced and practically invisible from
above. The eyes are internal to the antennae and often extend from
sockets in the carapace, into which they can be retracted. These

crabs have four walking legs on each side used in continuous and uninterrupted movement. Within this group are the largest of the crustacea, the huge spider crabs measuring up to 12 feet (3.7 m) or more across the extended legs.

MASKING CRAB
Loxorhynchus crispatus   *Fig. 117*

Description
Size, 4½ in. (11.4 cm). Usually entirely covered with hydroids, bryozoans, sponges, seaweed, so as not even to look like a crab. Carapace somewhat flattened, narrowly triangular, 9–12 tubercles; coated with hair, short, thick, plushlike. Walking legs short.
Habitat
Tide line to 50-ft. (15.2-m) depths; not uncommon.
Habits
One of the most inactive crabs.
Other name:
Moss Crab.
Range
Pt. Reyes to San Diego, Calif.

Fig. 117   *Oregonia gracilis*   *Loxorhynchus crispatus*
               Decorator Crab   Masking Crab

               Pugettia producta   *Mimulus foliatus*
               Kelp Crab, p. 756   (True Crab)

MIMULUS FOLIATUS   *Fig. 117*

Description
Size, 1¼ in. (3.2 cm). Rostrum forked in two equal parts; horns short, never more than ¼ carapace length. Carapace broad anteriorly, surface smooth, undulated, thin margin produced by leaflike expansion. Color variable, light red, tan, or purplish; legs crossed by light bands. Adults often coated with bryozoans or sponge growths.
Habitat
Among rocks at low tide; not uncommon.
Range
Unalaska (Aleutians) to cen. Mexico.

DECORATOR CRAB
Oregonia gracilis   *Fig. 117*

Description
Size, 1¾ in. (4.5 cm). Carapace triangular, with apex formed by 2 spines of the slender rostrum; covered with stiff recurved hairs and

minute tubercles (Johnson); a sharp spine curving forward just behind each orbit. Chelipeds long, roughened with fine tubercles and tiny hairs, ending in long, inward-curving, smooth fingers; walking legs slender, with a few hairs, terminal segments strongly curved. Color grayish or tan, red-spotted. Carapace usually hidden by attached seaweed, sponges, or bryozoans.

Habitat
Low-tide mark to deep water, occasionally on wharf pilings and in eelgrass beds; abundant.

Range
Bering Sea to Monterey, Calif.

KELP CRAB
Pugettia producta   *Fig. 117*

Description
Size, 3¾–4¼ in. (9.5–11.4 cm). Carapace smooth with nearly parallel sides adorned with a few sharp spines. Color dark olive-green, reddish, or olive-brown, mottled with small darker spots.

Habitat
Among or on kelp or seaweed strands; common.

Remarks
Cling viciously, so must be handled cautiously.

Range
Alaska to Baja Calif.

GRACEFUL CRAB
Cancer gracilis   *Fig. 118*

Description
Size, 2½–3 in. (6.4–7.6 cm). Very similar in appearance and range to *Cancer magister*, but carapace widest at eighth or ninth anterolateral tooth, and fingers of chelipeds light-colored; carapace more rounded, smoother; ventral surface less hairy. Walking legs slender, graceful. Color grayish or tan, with many small red spots.

Habitat
Along shore at low tide in north, deep water in south; intertidal north of Dillon Beach, Calif.

Range
Same as *C. magister*.

Note: Members of the Genus *Cancer* include the large edible crabs commonly found in the markets. Although some smaller species are also edible, their importance is limited by size or lack of numbers. Although nine species occur on the West Coast, only five are found in any abundance.

Fig. 118

Cancer gracilis
Graceful Crab

Lophopanopeus leucomanus
(True Crab)

DUNGENESS CRAB
Cancer magister

Description
Size, 7 in. (17.8 cm). The common commercial edible crab found in the markets. Carapace granulated, hairy. Anterolateral margins widest at tenth anterolateral tooth; fingers of chelipeds dark in color. Color light reddish-brown above, more yellowish below.

Habitat
On sand in fairly deep water, but the smaller, younger individuals frequent tidal pools with sandy bottoms. Large fisheries operate in 5–10 fathoms (9.1–18.3 m) of water. Abundant, but in danger of being overfished.

Range
Aleutians to cen. Baja Calif.; not commercial s. of Monterey Bay, Calif.

RED CRAB
Cancer productus

Description
Size, 7 in. (17.8 cm). Distinguished from other *Cancer* by having the front markedly pronounced beyond orbits and made up of 5 nearly equal teeth. Carapace smooth, not marked with granulations; margin serrated, cheliped fingers dark-colored. Color of dorsal surface dark red; lower parts much lighter. Young very variable, often mottled or streaked.

Habitat
Half buried in substratum under rocks; at night stalks about in tidal pools; common.

Range
Kodiak, Alaska, to Baja Calif.

LOPHOPANOPEUS BELLUS

Description
Size, 1⅜ in. (3.5 cm) wide. Lobe absent on upper margin of hand; hand and carpus smooth. Color variable; red-brown or purplish, through grays, to nearly white; variously patterned, black band always across finger and thumb.

Habitat
Rocky shores.

Range
Puget Sound, Wash. (only member of genus in this region), to Monterey, Calif.; uncommon s. of Wash.

LOPHOPANOPEUS LEUCOMANUS   *Fig. 118*

Description
Size, ¾ in. (1.9 cm) wide. Carapace front notched, dark band across thumb and finger; hands with lobes on upper margin.

Habitat
Rocky shores.

Range
Monterey, Calif., to Baja Calif.

SCLEROPLAX GRANULATA   *Fig. 119*

Description
Size, ⁵⁄₁₆ in. (0.79 cm). Carapace hard, strongly convex, center
smooth but finely granulated toward margins. Color gray-white.
Habitat
Usually commensal with *Urechis*, *Upogebia,* and *Callianassa*.
Range
B.C. to Baja Calif.

Note: A dozen or more members of this family occur along the
West Coast, all of them very small. The carapace is short and wide
and the third pair of walking legs relatively large and long. These
animals live in the tubes of annelid worms and in cavities of
holothurians and bivalve mollusks; hence, they are easily missed by
any but skilled collectors. They are extraordinarily difficult to
identify by species; furthermore, their taxonomy is not well worked
out.

Fig. 119

Hemigrapsus oregonensis
Yellow Shore Crab

Scleroplax granulata
(True Crab)

PURPLE SHORE CRAB
Hemigrapsus nudus   **126:6**

Description
Size, 1½ in. (3.8 cm). Carapace almost square, smooth, convex
forward, anterior corners with conspicuous spines pointing
forward, all corners rounded. Walking legs hairless. Color and
markings variable; greenish-yellow, reddish-brown, purplish.
Purple tinge prevalent. Red spots constant on claws; distinguish
this from *Hemigrapsus oregonensis*.
Habitat
Middle tide-pool region; uncommon in south, more common in
north.
Range
Sitka, Alaska, to Gulf of Calif.

YELLOW SHORE CRAB
Hemigrapsus oregonensis   *Fig. 119*

Description
Size, 1⅛ in. (2.9 cm). Somewhat like a small *Hemigrapsus nudus*,
but walking legs markedly hairy. Color yellow or gray, carapace
and legs mottled with brownish-purple or black spots, claw tips
light yellow or white; white beneath.
Habitat
Mud flats; enormous numbers (swarms) in some locations.
Habits
Very active in scampering about; aggressive, invites combat.
Other names:
Hairy Shore Crab, Pebble Crab.
Range
Alaska to Mexico; especially common in San Francisco Bay.

ROCK CRAB
Pachygrapsus crassipes   **126:3**

Description
Size, 1½ in. (3.8 cm). Carapace squarish, sides converging posteriorly, dorsal surface transversely striated anteriorly. Color dark-red or green with variable stripes and markings of red, green, and dark purple; claws with network of fine purple lines.

Habitat
Rocky beaches, on mud flats exposed by tide and in the winding channels; very abundant.

Habits
Very active; runs rapidly sideways or congregates about a dead fish; excellent scavenger.

Other name:
Striped Shore Crab.

Range
Oreg. to Gulf of Calif.

CHORDATES
Phylum Chordata

To this phylum belong the largest and most highly organized animals, beginning with the lowly tunicates and ending with man. All members are characterized by a notochord, a cylindrical stiff rod of tissue dorsal to the digestive system and the forerunner of the vertebrate spinal column, present at some stage of their lives; a dorsal tubular nervous system; paired gill slits in the pharynx, at least in the embryonic stages; and a peculiar structure which is apparent but often ignored—the tail. These structures, though present in the embryos, often tend to disappear in the adult stages of growth in the higher animals.

TUNICATES OR SEA SQUIRTS
Class Ascidiacea

Ascidians, popularly called sea squirts, are a very specialized group of marine chordates with a habit of ejecting water from siphons. In general they are baglike, and most are sessile whether solitary or colonial. In colonies, the zooids (individuals) are usually very small and are embedded in a common tunic which may form a tough, flexible mass on firm surfaces. The incurrent siphons of colonial forms are separate, but the excurrent siphons may open into a common cloaca. The outer coating or tunic is composed of tunicin, chemically quite similar to cellulose of the plant kingdom.

SEA PORK
Aplidium californicum   **126:10**

Description
Size, 7–8 in. (17.8–20.3 cm) dia. An irregular, variable, thick encrusting tunicate made up of tiny individuals enclosed in a gelatinous tunic with color and texture resembling that of raw salt pork, hence the name. Body often lobed, sometimes pedunculated, never encrusted with sand; zooids long, pin-headed, clearly visible in more or less distinct groupings. Color opalescent white to reddish-brown.

Habitat
Encrusting most intertidal objects, to 10–20 ft. (3–6.1 m) deep; usually abundant.
Remarks
Base may be sandy, but surface is slippery, flabby; colony is very soft to mushy.
Range
Alaska to San Diego, Calif.

POLYCLINUM PLANUM 126:11

Description
Size, 2–3 in. (5.1–7.6 cm). Body bulbous; peduncles thick, cylindrical; usually forms large regular colonies, spherical or flattened; surface smooth; zooids plainly visible in regular groups, like flowery wallpaper.
Habitat
Surfaces of intertidal solid objects; common, especially s. of Pt. Conception, Calif.
Range
Cen. Calif. and s.

ASCIDIA CERATODES 126:9

Description
Size, 1–1½ in. (2.5–3.8 cm). Test thick, gelatinous, translucent, elliptical, somewhat depressed. Mantle thin, transparent on underside. Siphons, incurrent lobes 8, excurrent lobes 6.
Habitat
On pilings.
Range
Calif. coast.

SEA VASE
Ciona intestinalis 126:14

Description
Size, 4–5 in. (10.2–12.7 cm). Body elongate, slender, cylindrical, attaches at rear; outer skin translucent; strong longitudinal muscle bands visible through skin. Siphons long, glassy, basal parts often debris-coated; incurrent siphon 8-lobed, excurrent 6-lobed. Color usually yellowish-green.
Habitat
Buoys, timbers, floats, wharf piles.
Range
N. Hemisphere, generally.

CNEMIDOCARPA FINMARKIENSIS 126:13

Description
Size, 1–1½ in. (2.5–3.8 cm). Body oval, openings 4-lobed; test thin, smooth; craterlike projections (incurrent, excurrent siphons) prominent. Color bright orange-red.
Habitat
Low-tide to intertidal zone, under rocks; abundant.
Remarks
One of most conspicuous animals in Puget Sound region.
Range
Puget Sound, Wash.

STYELA MONTEREYENSIS

Description
Size, 11–13 in. (27.9–33 cm). A stalked, simple tunicate. Body club-shaped, merges gradually into peduncle often twice as long as body. Orifices 4-lobed; siphons distinct, branchial siphon with downward curve; tentacles around branchial opening, 40–130. Color dark red, lighter on lower part of body.

Habitat
Rocky shores, wharf pilings where water is not contaminated; rather common.

Remarks
Often festooned with growths of ostrich-plumed hydroids, other tunicates, even with anemones.

Range
B.C. to San Diego, Calif.

Note: *Styela gibbsii*, occurring in the same range, usually under rocks, is similar but has a short peduncle.

Life Lists

BIRDS

_____ Abert's Towhee
_____ Acorn Woodpecker
_____ Alder Flycatcher
_____ Aleutian Tern
_____ Allen's Hummingbird
_____ American Avocet
_____ American Bittern
_____ American Coot
_____ American Golden Plover
_____ American Goldfinch
_____ American Kestrel
_____ American Redstart
_____ American Robin
_____ American White Pelican
_____ American Wigeon
_____ Ancient Murrelet
_____ Anhinga
_____ Anna's Hummingbird
_____ Arctic Loon
_____ Arctic Tern
_____ Ash-throated Flycatcher
_____ Ashy Storm-Petrel

_____ Baird's Sandpiper
_____ Baird's Sparrow
_____ Bald Eagle
_____ Band-tailed Pigeon
_____ Bank Swallow
_____ Barn Owl
_____ Barn Swallow
_____ Barred Owl
_____ Barrow's Goldeneye
_____ Bar-tailed Godwit
_____ Bay-breasted Warbler
_____ Bell's Vireo
_____ Belted Kingfisher
_____ Bendire's Thrasher
_____ Bewick's Wren
_____ Black-and-White Warbler
_____ Black-backed Woodpecker
_____ Black-bellied Plover
_____ Black-bellied Whistling-Duck
_____ Black-billed Cuckoo
_____ Black-billed Magpie
_____ Black Brant
_____ Blackburnian Warbler
_____ Black-capped Chickadee
_____ Black-capped Vireo
_____ Black-chinned Hummingbird
_____ Black-chinned Sparrow
_____ Black-crowned Night Heron
_____ Black Duck
_____ Black-footed Albatross
_____ Black-headed Grosbeak
_____ Black-headed Oriole
_____ Black-legged Kittiwake
_____ Black-necked Stilt
_____ Black Oystercatcher
_____ Black Phoebe
_____ Blackpoll Warbler
_____ Black Rail
_____ Black Rosy Finch
_____ Black Scoter
_____ Black Storm-Petrel
_____ Black Swift
_____ Black-tailed Gnatcatcher
_____ Black Tern
_____ Black-throated Blue Warbler
_____ Black-throated Gray Warbler
_____ Black-throated Green Warbler
_____ Black-throated Sparrow
_____ Black Turnstone
_____ Black Vulture
_____ Blue-footed Booby
_____ Blue-gray Gnatcatcher
_____ Blue Grosbeak
_____ Blue Grouse
_____ Blue Jay
_____ Blue-throated Hummingbird
_____ Blue-winged Teal
_____ Bobolink
_____ Bobwhite
_____ Bohemian Waxwing
_____ Bonaparte's Gull
_____ Boreal Chickadee
_____ Boreal Owl
_____ Botteri's Sparrow
_____ Brandt's Cormorant
_____ Brant
_____ Brewer's Blackbird
_____ Brewer's Sparrow
_____ Bridled Titmouse
_____ Broad-billed Hummingbird
_____ Broad-tailed Hummingbird
_____ Broad-winged Hawk
_____ Bronzed Cowbird
_____ Brown Creeper
_____ Brown-headed Cowbird
_____ Brown Pelican
_____ Brown Rosy Finch
_____ Brown Thrasher

_____ Brown Towhee
_____ Buff-breasted Flycatcher
_____ Buff-breasted Sandpiper
_____ Bufflehead
_____ Burrowing Owl
_____ Bushtit

_____ Cactus Wren
_____ California Condor
_____ California Gull
_____ California Quail
_____ California Thrasher
_____ Calliope Hummingbird
_____ Canada Goose
_____ Canada Warbler
_____ Canvasback
_____ Canyon Wren
_____ Cape May Warbler
_____ Cardinal
_____ Carolina Wren
_____ Caspian Tern
_____ Cassin's Auklet
_____ Cassin's Finch
_____ Cassin's Kingbird
_____ Cassin's Sparrow
_____ Cattle Egret
_____ Cave Swallow
_____ Cedar Waxwing
_____ Chestnut-backed Chickadee
_____ Chestnut-collared Longspur
_____ Chestnut-sided Warbler
_____ Chimney Swift
_____ Chipping Sparrow
_____ Chukar
_____ Cinnamon Teal
_____ Clapper Rail
_____ Clark's Nutcracker
_____ Clay-colored Sparrow
_____ Cliff Swallow
_____ Colima Warbler
_____ Common Black Hawk
_____ Common Crow
_____ Common Flicker
_____ Common Gallinule
_____ Common Goldeneye
_____ Common Grackle
_____ Common Ground Dove
_____ Common Loon
_____ Common Merganser
_____ Common Murre
_____ Common Nighthawk
_____ Common Raven
_____ Common Redpoll
_____ Common Snipe
_____ Common Teal

_____ Common Tern
_____ Common Wheatear
_____ Common Yellowthroat
_____ Connecticut Warbler
_____ Cooper's Hawk
_____ Costa's Hummingbird
_____ Coues' Pewee
_____ Crested Auklet
_____ Crested Caracara
_____ Crested Myna
_____ Crissal Thrasher
_____ Curve-billed Thrasher

_____ Dark-eyed Junco
_____ Dickcissel
_____ Dipper
_____ Double-crested Cormorant
_____ Downy Woodpecker
_____ Dunlin
_____ Dusky Flycatcher

_____ Eared Grebe
_____ Eastern Bluebird
_____ Eastern Kingbird
_____ Eastern Meadowlark
_____ Eastern Phoebe
_____ Eastern Wood Pewee
_____ Elegant Tern
_____ Elegant Trogon
_____ Elf Owl
_____ Emperor Goose
_____ Eskimo Curlew
_____ Eurasian Bluethroat
_____ Eurasian Wigeon
_____ Eurasian Yellow Wagtail
_____ Evening Grosbeak

_____ Ferruginous Hawk
_____ Ferruginous Owl
_____ Field Sparrow
_____ Flammulated Owl
_____ Fork-tailed Storm-Petrel
_____ Forster's Tern
_____ Fox Sparrow
_____ Franklin's Gull
_____ Fulvous Whistling-Duck

_____ Gadwall
_____ Gambel's Quail
_____ Gila Woodpecker
_____ Glaucous Gull
_____ Glaucous-winged Gull
_____ Golden-cheeked Warbler
_____ Golden-crowned Kinglet
_____ Golden-crowned Sparrow

_____ Golden Eagle
_____ Golden-fronted Woodpecker
_____ Grace's Warbler
_____ Grasshopper Sparrow
_____ Gray Catbird
_____ Gray-cheeked Thrush
_____ Gray-crowned Rosy Finch
_____ Gray Flycatcher
_____ Gray Hawk
_____ Gray-headed Chickadee
_____ Gray Jay
_____ Gray Partridge
_____ Gray Vireo
_____ Great Blue Heron
_____ Great Crested Flycatcher
_____ Great Egret
_____ Greater Prairie Chicken
_____ Greater Scaup
_____ Greater Yellowlegs
_____ Great Gray Owl
_____ Great Horned Owl
_____ Great-tailed Grackle
_____ Green Heron
_____ Green Jay
_____ Green Kingfisher
_____ Green-tailed Towhee
_____ Groove-billed Ani
_____ Gull-billed Tern
_____ Gyrfalcon

_____ Hairy Woodpecker
_____ Hammond's Flycatcher
_____ Harlequin Duck
_____ Harris' Hawk
_____ Harris' Sparrow
_____ Hawk Owl
_____ Heermann's Gull
_____ Hepatic Tanager
_____ Hermit Thrush
_____ Hermit Warbler
_____ Herring Gull
_____ Hoary Redpoll
_____ Hooded Merganser
_____ Hooded Oriole
_____ Horned Grebe
_____ Horned Lark
_____ Horned Puffin
_____ House Finch
_____ House Sparrow
_____ House Wren
_____ Hudsonian Godwit
_____ Hutton's Vireo

_____ Inca Dove
_____ Indigo Bunting

_____ Killdeer

_____ King Eider
_____ King Rail
_____ Kiskadee Flycatcher
_____ Kittlitz's Murrelet

_____ Ladder-backed Woodpecker
_____ Lapland Longspur
_____ Lark Bunting
_____ Lark Sparrow
_____ Laughing Gull
_____ Lawrence's Goldfinch
_____ Lazuli Bunting
_____ Leach's Storm-Petrel
_____ Least Auklet
_____ Least Bittern
_____ Least Flycatcher
_____ Least Grebe
_____ Least Sandpiper
_____ Least Tern
_____ Le Conte's Sparrow
_____ Le Conte's Thrasher
_____ Lesser Goldfinch
_____ Lesser Nighthawk
_____ Lesser Prairie Chicken
_____ Lesser Scaup
_____ Lesser Yellowlegs
_____ Lewis' Woodpecker
_____ Lincoln's Sparrow
_____ Little Blue Heron
_____ Loggerhead Shrike
_____ Long-billed Curlew
_____ Long-billed Dowitcher
_____ Long-billed Marsh Wren
_____ Long-billed Thrasher
_____ Long-eared Owl
_____ Long-tailed Jaeger
_____ Lucifer Hummingbird
_____ Lucy's Warbler

_____ McCown's Longspur
_____ MacGillivray's Warbler
_____ Magnificent Frigatebird
_____ Mallard
_____ Manx Shearwater
_____ Marbled Godwit
_____ Marbled Murrelet
_____ Masked Duck
_____ Merlin
_____ Mew Gull
_____ Mexican Chickadee
_____ Mexican Duck
_____ Mexican Jay
_____ Mississippi Kite
_____ Montezuma Quail
_____ Mountain Bluebird
_____ Mountain Chickadee
_____ Mountain Plover
_____ Mountain Quail

_____ Mourning Dove
_____ Mourning Warbler

_____ Nashville Warbler
_____ New Zealand
Shearwater
_____ Northern Beardless
Flycatcher
_____ Northern Fulmar
_____ Northern Goshawk
_____ Northern Harrier
_____ Northern Mockingbird
_____ Northern Oriole
_____ Northern Parula
_____ Northern Phalarope
_____ Northern Pintail
_____ Northern Shoveler
_____ Northern Shrike
_____ Northern Skua
_____ Northern Waterthrush
_____ Nuttall's Woodpecker

_____ Oldsquaw
_____ Olivaceous Flycatcher
_____ Olive-sided Flycatcher
_____ Olive Sparrow
_____ Olive Warbler
_____ Orange-crowned
Warbler
_____ Orchard Oriole
_____ Osprey
_____ Ovenbird

_____ Painted Bunting
_____ Painted Redstart
_____ Palm Warbler
_____ Parakeet Auklet
_____ Parasitic Jaeger
_____ Pectoral Sandpiper
_____ Pelagic Cormorant
_____ Peregrine Falcon
_____ Phainopepla
_____ Philadelphia Vireo
_____ Pied-billed Grebe
_____ Pigeon Guillemot
_____ Pileated Woodpecker
_____ Pine Grosbeak
_____ Pine Siskin
_____ Pink-footed Shearwater
_____ Pinyon Jay
_____ Piping Plover
_____ Plain Chachalaca
_____ Plain Titmouse
_____ Pomarine Jaeger
_____ Poor-will
_____ Prairie Falcon
_____ Purple Finch
_____ Purple Gallinule
_____ Purple Martin
_____ Pygmy Nuthatch
_____ Pygmy Owl

_____ Pyrrhuloxia

_____ Red-bellied
Woodpecker
_____ Red-breasted
Merganser
_____ Red-breasted Nuthatch
_____ Red-breasted
Sapsucker
_____ Red Crossbill
_____ Reddish Egret
_____ Red-eyed Vireo
_____ Red-faced Cormorant
_____ Red-faced Warbler
_____ Redhead
_____ Red-headed
Woodpecker
_____ Red Knot
_____ Red-legged Kittiwake
_____ Red-naped Sapsucker
_____ Red-necked Grebe
_____ Red Phalarope
_____ Red-shouldered Hawk
_____ Red-tailed Hawk
_____ Red-throated Loon
_____ Red-winged Blackbird
_____ Rhinoceros Auklet
_____ Ring-billed Gull
_____ Ring-necked Duck
_____ Ring-necked Pheasant
_____ Rivoli's Hummingbird
_____ Roadrunner
_____ Rock Pigeon
_____ Rock Ptarmigan
_____ Rock Sandpiper
_____ Rock Wren
_____ Roseate Spoonbill
_____ Rose-breasted
Grosbeak
_____ Rose-throated Becard
_____ Ross' Goose
_____ Rough-legged Hawk
_____ Rough-winged Swallow
_____ Royal Tern
_____ Ruby-crowned Kinglet
_____ Ruby-throated
Hummingbird
_____ Ruddy Duck
_____ Ruddy Turnstone
_____ Ruffed Grouse
_____ Rufous-crowned
Sparrow
_____ Rufous Hummingbird
_____ Rufous-sided Towhee
_____ Rufous-winged
Sparrow
_____ Rusty Blackbird

_____ Sabine's Gull
_____ Sage Grouse
_____ Sage Sparrow

_____ Sage Thrasher
_____ Sanderling
_____ Sandhill Crane
_____ Savannah Sparrow
_____ Saw-whet Owl
_____ Say's Phoebe
_____ Scaled Quail
_____ Scarlet Tanager
_____ Scissor-tailed
Flycatcher
_____ Scott's Oriole
_____ Screech-Owl
_____ Scrub Jay
_____ Semipalmated Plover
_____ Semipalmated
Sandpiper
_____ Sharp-shinned Hawk
_____ Sharp-tailed Grouse
_____ Sharp-tailed Sandpiper
_____ Sharp-tailed Sparrow
_____ Short-billed Dowitcher
_____ Short-billed Marsh
Wren
_____ Short-eared Owl
_____ Slaty-backed Gull
_____ Smith's Longspur
_____ Snow Bunting
_____ Snow Goose
_____ Snowy Egret
_____ Snowy Owl
_____ Snowy Plover
_____ Solitary Sandpiper
_____ Solitary Vireo
_____ Song Sparrow
_____ Sooty Shearwater
_____ Sora
_____ Spotted Dove
_____ Spotted Owl
_____ Spotted Sandpiper
_____ Sprague's Pipit
_____ Spruce Grouse
_____ Starling
_____ Steller's Jay
_____ Stilt Sandpiper
_____ Strickland's (Arizona)
Woodpecker
_____ Sulphur-bellied
Flycatcher
_____ Summer Tanager
_____ Surfbird
_____ Surf Scoter
_____ Swainson's Hawk
_____ Swainson's Thrush
_____ Swamp Sparrow

_____ Tennessee Warbler
_____ Thayer's Gull
_____ Thick-billed Kingbird
_____ Thick-billed Murre
_____ Three-toed
Woodpecker

_____ Townsend's Solitaire
_____ Townsend's Warbler
_____ Tree Sparrow
_____ Tree Swallow
_____ Tricolored Blackbird
_____ Tropical Kingbird
_____ Tropical Parula
_____ Trumpeter Swan
_____ Tufted Puffin
_____ Tufted Titmouse
_____ Turkey
_____ Turkey Vulture

_____ Upland Sandpiper

_____ Varied Bunting
_____ Varied Thrush
_____ Vaux's Swift
_____ Veery
_____ Verdin
_____ Vermilion Flycatcher
_____ Vesper Sparrow
_____ Violet-crowned
Hummingbird
_____ Violet-green Swallow
_____ Virginia Rail
_____ Virginia's Warbler

_____ Wandering Tattler
_____ Warbling Vireo
_____ Water Pipit
_____ Western Bluebird
_____ Western Flycatcher
_____ Western Grebe
_____ Western Gull
_____ Western Kingbird
_____ Western Meadowlark
_____ Western Sandpiper
_____ Western Tanager
_____ Western Wood Pewee
_____ Whimbrel
_____ Whip-poor-will
_____ Whiskered Owl
_____ Whistling Swan
_____ White-breasted
Nuthatch
_____ White-crowned
Sparrow
_____ White-faced Ibis
_____ White-fronted Dove
_____ White-fronted Goose
_____ White-headed
Woodpecker
_____ White-necked Raven
_____ White-rumped
Sandpiper
_____ White-tailed Hawk
_____ White-tailed Kite
_____ White-tailed
Ptarmigan
_____ White-throated
Sparrow

_____ White-throated Swift
_____ White-winged Crossbill
_____ White-winged Dove
_____ White-winged Scoter
_____ Whooping Crane
_____ Wied's Crested
Flycatcher
_____ Willet
_____ Williamson's Sapsucker
_____ Willow Flycatcher
_____ Willow Ptarmigan
_____ Wilson's Phalarope
_____ Wilson's Plover
_____ Wilson's Warbler
_____ Winter Wren
_____ Wood Duck
_____ Wood Stork (Wood
Ibis)
_____ Wrentit

_____ Xantus' Murrelet

_____ Yellow-billed Cuckoo
_____ Yellow-billed Magpie
_____ Yellow-breasted Chat
_____ Yellow-crowned Night
Heron
_____ Yellow-eyed Junco
_____ Yellow-headed
Blackbird
_____ Yellow Rail
_____ Yellow-rumped
Warbler
_____ Yellow-throated Vireo
_____ Yellow Warbler

_____ Zone-tailed Hawk

MAMMALS

_____ Abert's Squirrel
_____ Agile Kangaroo Rat
_____ Allen's Chipmunk
_____ Antelope Jackrabbit
_____ Apache Pocket Mouse
_____ Arctic Fox
_____ Arctic Ground Squirrel
_____ Arctic Hare
_____ Arctic Shrew
_____ Arizona Cotton Rat
_____ Arizona Pocket Mouse

_____ Badger
_____ Bailey's Pocket Mouse
_____ Banner-tailed
Kangaroo Rat
_____ Bearded Seal
_____ Beaver
_____ Belding's Ground
Squirrel

_____ Big Brown Bat
_____ Big-eared Kangaroo
Rat
_____ Big Free-tailed Bat
_____ Bison
_____ Black Bear
_____ Black-footed Ferret
_____ Black Rat
_____ Black Right Whale
_____ Black-tailed Jackrabbit
_____ Black-tailed Prairie
Dog
_____ Blue Whale
_____ Bobcat
_____ Botta's Pocket Gopher
_____ Bowhead Whale
_____ Brazilian Free-tailed
Bat
_____ Broad-footed Mole
_____ Brown Lemming
_____ Brush Mouse
_____ Brush Rabbit
_____ Bushy-tailed Woodrat

_____ Cactus Mouse
_____ California Ground
Squirrel
_____ California Leaf-nosed
Bat
_____ California Mouse
_____ California Myotis
_____ California Pocket
Mouse
_____ California Sea Lion
_____ California Vole
_____ Camas Pocket Gopher
_____ Canyon Mouse
_____ Caribou
_____ Cascade Golden-
mantled Ground
Squirrel
_____ Cave Myotis
_____ Chisel-toothed
Kangaroo Rat
_____ Cliff Chipmunk
_____ Coast Mole
_____ Coati
_____ Collared Lemming
_____ Collared Peccary
_____ Colorado Chipmunk
_____ Columbian Ground
Squirrel
_____ Common Dolphin
_____ Common Pilot Whale
_____ Coronation Island Vole
_____ Coyote
_____ Creeping Vole

_____ Dall's Porpoise
_____ Dall's Sheep
_____ Dark Kangaroo Mouse

_____ Deer Mouse
_____ Desert Cottontail
_____ Desert Kangaroo Rat
_____ Desert Pocket Gopher
_____ Desert Pocket Mouse
_____ Desert Shrew
_____ Desert Woodrat
_____ Douglas' Squirrel
_____ Dusky-footed Woodrat
_____ Dusky Shrew
_____ Dusky Tree Vole
_____ Dwarf Shrew
_____ Dwarf Sperm Whale

_____ Eastern Cottontail
_____ Eastern Woodrat
_____ Elk
_____ Ermine
_____ False Killer Whale
_____ Fin Whale
_____ Fisher
_____ Franklin's Ground Squirrel
_____ Fresno Kangaroo Rat
_____ Fringed Myotis
_____ Fulvous Harvest Mouse

_____ Giant Kangaroo Rat
_____ Golden-mantled Ground Squirrel
_____ Goose-beaked Whale
_____ Grampus
_____ Gray-collared Chipmunk
_____ Gray-footed Chipmunk
_____ Gray Fox
_____ Gray Squirrel
_____ Gray Whale
_____ Gray Wolf
_____ Great Basin Pocket Mouse
_____ Grizzly Bear
_____ Gunnison's Prairie Dog

_____ Harbor Porpoise
_____ Harbor Seal
_____ Harris' Antelope Squirrel
_____ Heather Vole
_____ Heermann's Kangaroo Rat
_____ Hispid Cotton Rat
_____ Hispid Pocket Mouse
_____ Hoary Bat
_____ Hoary Marmot
_____ Hog-nosed Skunk
_____ Hooded Skunk
_____ Horse
_____ House Mouse
_____ Hump-backed Whale

_____ Idaho Ground Squirrel
_____ Idaho Pocket Gopher
_____ Insular Vole

_____ Keen's Myotis
_____ Killer Whale
_____ Kit Fox

_____ Least Chipmunk
_____ Least Weasel
_____ Little Brown Myotis
_____ Little Pocket Mouse
_____ Lodgepole Chipmunk
_____ Long-eared Chipmunk
_____ Long-eared Myotis
_____ Long-legged Myotis
_____ Long-tailed Pocket Mouse
_____ Long-tailed Vole
_____ Long-tailed Weasel
_____ Long-tongued Bat
_____ Lynx

_____ Marten
_____ Masked Shrew
_____ Meadow Jumping Mouse
_____ Meadow Vole
_____ Merriam's Chipmunk
_____ Merriam's Kangaroo Rat
_____ Merriam's Mouse
_____ Merriam's Shrew
_____ Mexican Ground Squirrel
_____ Mexican Long-nosed Bat
_____ Mexican Woodrat
_____ Mink
_____ Minke Whale
_____ Mohave Ground Squirrel
_____ Moose
_____ Mountain Beaver
_____ Mountain Goat
_____ Mountain Lion
_____ Mountain Pocket Gopher
_____ Mountain Sheep
_____ Mountain Vole
_____ Mule Deer
_____ Muskox
_____ Muskrat

_____ Narrow-faced Kangaroo Rat
_____ Narwhal
_____ Nayarit Squirrel
_____ Nelson's Antelope Squirrel
_____ Nelson's Pocket Mouse

MAMMALS

_____ Nine-banded Armadillo
_____ North Atlantic Bottle-nosed Whale
_____ Northern Bog Lemming
_____ Northern Elephant Seal
_____ Northern Flying Squirrel
_____ Northern Fur Seal
_____ Northern Grasshopper Mouse
_____ Northern Pocket Gopher
_____ Northern Pygmy Mouse
_____ Northern Red-backed Vole
_____ Northern Right-whale Dolphin
_____ Northern Sea Lion
_____ North Pacific Beaked Whale
_____ North Pacific Bottle-nosed Whale
_____ Norway Rat
_____ Nutria
_____ Nuttall's Cottontail

_____ Olive-backed Pocket Mouse
_____ Ord's Kangaroo Rat
_____ Ornate Shrew

_____ Pacific Bottle-nosed Dolphin
_____ Pacific Coast Black-tailed Deer
_____ Pacific Dolphin
_____ Pacific Jumping Mouse
_____ Pacific Shrew
_____ Pacific Water Shrew
_____ Pacific White-Sided Dolphin
_____ Pale Kangaroo Mouse
_____ Pallid Bat
_____ Palmer's Chipmunk
_____ Palo Duro Mouse
_____ Panamint Chipmunk
_____ Panamint Kangaroo Rat
_____ Pika
_____ Pinyon Mouse
_____ Plains Harvest Mouse
_____ Plains Pocket Gopher
_____ Plains Pocket Mouse
_____ Polar Bear
_____ Porcupine
_____ Prairie Vole
_____ Pronghorn

_____ Pygmy Rabbit
_____ Pygmy Shrew
_____ Pygmy Sperm Whale

_____ Raccoon
_____ Red Bat
_____ Red Fox
_____ Red Squirrel
_____ Red-tailed Chipmunk
_____ Red Tree Vole
_____ Richardson's Ground Squirrel
_____ Ringtail Cat
_____ River Otter
_____ Rock Mouse
_____ Rock Pocket Mouse
_____ Rock Squirrel
_____ Rough-toothed Porpoise
_____ Round-tailed Ground Squirrel

_____ Sagebrush Vole
_____ Salt-marsh Harvest Mouse
_____ San Diego Pocket Mouse
_____ San Joaquin Pocket Mouse
_____ Sea Otter
_____ Shrew-mole
_____ Silky Pocket Mouse
_____ Silver-haired Bat
_____ Singing Vole
_____ Siskiyou Chipmunk
_____ Sitka Mouse
_____ Small-footed Myotis
_____ Snowshoe Hare
_____ Sonoma Chipmunk
_____ Southern Grasshopper Mouse
_____ Southern Plains Woodrat
_____ Southern Pocket Gopher
_____ Southern Red-backed Vole
_____ Southern Yellow Bat
_____ Sperm Whale
_____ Spiny Pocket Mouse
_____ Spotted Bat
_____ Spotted Ground Squirrel
_____ Stephen's Kangaroo Rat
_____ Stephens' Woodrat
_____ Striped Skunk
_____ Swift Fox

_____ Texas Antelope Squirrel

772

_____ Texas Kangaroo Rat
_____ Texas Mouse
_____ Texas Pocket Gopher
_____ Thirteen-lined Ground
Squirrel
_____ Townsend's Big-eared
Bat
_____ Townsend's Chipmunk
_____ Townsend's Ground
Squirrel
_____ Townsend's Mole
_____ Townsend's Pocket
Gopher
_____ Townsend's Vole
_____ Trowbridge's Shrew
_____ Tundra Vole

_____ Uinta Chipmunk
_____ Uinta Ground Squirrel
_____ Utah Prairie Dog

_____ Vagrant Shrew
_____ Virginia Opossum

_____ Walrus
_____ Washington Ground
Squirrel
_____ Water Shrew
_____ Water Vole
_____ Western Gray Squirrel
_____ Western Harvest
Mouse
_____ Western Jumping
Mouse
_____ Western Mastiff Bat
_____ Western Pipistrelle
_____ Western Pocket
Gopher
_____ Western Red-backed
Vole
_____ Western Spotted Skunk
_____ White-ankled Mouse
_____ White-eared Pocket
Mouse
_____ White-footed Mouse
_____ White-footed Vole
_____ White-tailed Antelope
Squirrel
_____ White-tailed Deer
_____ White-tailed Jackrabbit
_____ White-tailed Prairie
Dog
_____ White-throated
Woodrat
_____ White Whale
_____ Wild Boar
_____ Wild Burro
_____ Wolverine
_____ Woodchuck

_____ Yellow-bellied Marmot

_____ Yellow-cheeked
Chipmunk
_____ Yellow-cheeked Vole
_____ Yellow-eared Pocket
Mouse
_____ Yellow-nosed Cotton
Rat
_____ Yellow Pine Chipmunk
_____ Yuma Myotis

REPTILES

_____ Arizona Alligator
Lizard
_____ Arizona Coral Snake

_____ Baird's Rat Snake
_____ Banded Gecko
_____ Banded Rock Lizard
_____ Banded Sand Snake
_____ Black-necked Garter
Snake
_____ Black-tailed
Rattlesnake
_____ Blunt-nosed Leopard
Lizard
_____ Bunch Grass Lizard

_____ California Legless
Lizard
_____ California Lyre Snake
_____ California Mountain
Kingsnake
_____ Canyon Lizard
_____ Checkered Garter
Snake
_____ Checkered Whiptail
_____ Chihuahua Whiptail
_____ Chuckwalla
_____ Clark's Spiny Lizard
_____ Coachella Valley
Fringe-toed Lizard
_____ Coachwhip
_____ Coast Horned Lizard
_____ Collared Lizard
_____ Colorado Desert
Fringe-toed Lizard
_____ Common Garter Snake
_____ Common Kingsnake
_____ Copperhead
_____ Corn Snake
_____ Crevice Spiny Lizard

_____ Desert-grassland
Whiptail
_____ Desert Hook-nosed
Snake
_____ Desert Horned Lizard
_____ Desert Iguana
_____ Desert Night Lizard
_____ Desert Spiny Lizard

REPTILES

_____ Desert Tortoise

_____ Eastern Fence Lizard
_____ Eastern Hognose Snake

_____ Flat-tailed Horned Lizard
_____ Four-lined Skink

_____ Giant Spotted Whiptail
_____ Gila Monster
_____ Gilbert's Skink
_____ Glossy Snake
_____ Gopher Snake
_____ Granite Night Lizard
_____ Granite Spiny Lizard
_____ Gray-banded Kingsnake
_____ Great Plains Skink
_____ Greater Earless Lizard
_____ Green Rat Snake
_____ Green Turtle
_____ Ground Snake

_____ Huachuca Black-headed Snake

_____ Island Night Lizard

_____ Leaf-toed Gecko
_____ Leatherback
_____ Leopard Lizard
_____ Lesser Earless Lizard
_____ Lined Snake
_____ Little Striped Whiptail
_____ Loggerhead
_____ Long-nosed Snake
_____ Long-tailed Brush Lizard

_____ Many-lined Skink
_____ Massasauga
_____ Mexican Garter Snake
_____ Milk Snake
_____ Mojave Fringe-toed Lizard
_____ Mojave Rattlesnake
_____ Mountain Patch-nosed Snake
_____ Mountain Skink

_____ Narrow-headed Garter Snake
_____ New Mexican Whiptail
_____ Night Snake
_____ Northern Alligator Lizard
_____ Northern Water Snake
_____ Northwestern Garter Snake

_____ Orange-throated Whiptail

_____ Pacific Ridley
_____ Painted Turtle
_____ Panamint Alligator Lizard
_____ Plain-bellied Water Snake
_____ Plains Black-headed Snake
_____ Plains Garter Snake
_____ Plateau Whiptail

_____ Racer
_____ Red-bellied Snake
_____ Red Diamond Rattlesnake
_____ Regal Horned Lizard
_____ Ridge-nosed Rattlesnake
_____ Ringneck Snake
_____ River Cooter
_____ Rock Rattlesnake
_____ Rosy Boa
_____ Rough Green Snake
_____ Round-tailed Horned Lizard
_____ Rubber Boa
_____ Rusty-rumped Whiptail

_____ Saddled Leaf-nosed Snake
_____ Sagebrush Lizard
_____ Sharp-tailed Snake
_____ Short-horned Lizard
_____ Side-blotched Lizard
_____ Sidewinder
_____ Six-lined Racerunner
_____ Slider
_____ Small-scaled Lizard
_____ Smooth Green Snake
_____ Smooth Softshell
_____ Snapping Turtle
_____ Sonora Lyre Snake
_____ Sonora Mountain Kingsnake
_____ Sonoran Mud Turtle
_____ Sonora Shovel-nosed Snake
_____ Sonora Whipsnake
_____ Southern Alligator Lizard
_____ Speckled Rattlesnake
_____ Spiny Softshell
_____ Spotted Leaf-nosed Snake
_____ Striped Plateau Lizard
_____ Striped Racer
_____ Striped Whipsnake

_____ Texas Alligator Lizard
_____ Texas Banded Gecko
_____ Texas Blind Snake
_____ Texas Horned Lizard
_____ Texas Lyre Snake
_____ Texas Spotted Whiptail
_____ Tiger Rattlesnake
_____ Trans-Pecos Rat Snake
_____ Tree Lizard
_____ Twin-spotted
Rattlesnake

_____ Vine Snake

_____ Western Aquatic
Garter Snake
_____ Western Black-headed
Snake
_____ Western Blind Snake
_____ Western Box Turtle
_____ Western Diamondback
Rattlesnake
_____ Western Fence Lizard
_____ Western Ground Snake
_____ Western Hognose
Snake
_____ Western Hook-nosed
Snake
_____ Western Patch-nosed
Snake
_____ Western Pond Turtle
_____ Western Rattlesnake
_____ Western Ribbon Snake
_____ Western Shovel-nosed
Snake
_____ Western Skink
_____ Western Terrestrial
Garter Snake
_____ Western Whiptail

_____ Yarrow's Spiny Lizard
_____ Yellow Mud Turtle

_____ Zebra-tailed Lizard

AMPHIBIANS

_____ Arboreal Salamander
_____ Arizona Treefrog

_____ Barking Frog
_____ Black Salamander
_____ Bullfrog
_____ Burrowing Treefrog

_____ California Newt
_____ California Slender
Salamander
_____ California Treefrog
_____ Canyon Treefrog
_____ Cascades Frog

_____ Chorus Frog
_____ Cliff Frog
_____ Clouded Salamander
_____ Colorado River Toad
_____ Couch's Spadefoot

_____ Dakota Toad
_____ Del Norte Salamander
_____ Dunn's Salamander

_____ Ensatina

_____ Foothill Yellow-legged
Frog

_____ Great Basin Spadefoot
_____ Great Plains Narrow-
mouthed Toad
_____ Great Plains Toad
_____ Green Frog
_____ Green Toad
_____ Gulf Coast Toad

_____ Jemez Mountains
Salamander

_____ Larch Mountain
Salamander
_____ Limestone Salamander
_____ Long-toed Salamander

_____ Mountain Yellow-
legged Frog
_____ Mount Lyell
Salamander

_____ Northern Cricket Frog
_____ Northern Leopard
Frog
_____ Northwestern
Salamander

_____ Olympic Salamander
_____ Oregon Slender
Salamander

_____ Pacific Giant
Salamander
_____ Pacific Slender
Salamander
_____ Pacific Treefrog
_____ Plains Spadefoot

_____ Red-bellied Newt
_____ Red-legged Frog
_____ Red-spotted Toad
_____ Rough-skinned Newt

_____ Sacramento Mountain
Salamander
_____ Shasta Salamander

FISHES

_____ Siskiyou Mountain Salamander
_____ Sonoran Green Toad
_____ Southwestern Toad
_____ Spotted Chorus Frog
_____ Spotted Frog

_____ Tailed Frog
_____ Tarahumara Frog
_____ Texas Toad
_____ Tiger Salamander

_____ Van Dyke's Salamander

_____ Western Red-backed Salamander
_____ Western Spadefoot
_____ Western Toad
_____ Wood Frog
_____ Woodhouse's Toad

_____ Yosemite Toad

FISHES

_____ Alaska Blackfish
_____ Albacore
_____ Amargosa Pupfish
_____ Arctic Char
_____ Arctic Grayling
_____ Arctic Lamprey
_____ Arizona Trout
_____ Arrowtooth Flounder
_____ Arroyo Chub

_____ Barred Sand Bass
_____ Barred Surfperch
_____ Basking Shark
_____ Bat Ray
_____ Bay Pipefish
_____ Bigeye Tuna
_____ Bigmouth Buffalo
_____ Bigmouth Shiner
_____ Big Skate
_____ Blackbelly Eelpout
_____ Black Bullhead
_____ Black Crappie
_____ Black Perch
_____ Black Rockfish
_____ Blacksmith
_____ Bluebanded Goby
_____ Blue Chub
_____ Bluefin Tuna
_____ Bluegill
_____ Bluehead Sucker
_____ Blue Lanternfish
_____ Blue Rockfish
_____ Blue Shark
_____ Bocaccio
_____ Bonneville Cisco

_____ Brassy Minnow
_____ Bridgelip Sucker
_____ Brook Trout
_____ Brown Bullhead
_____ Brown Smoothhound
_____ Brown Trout
_____ Bull Trout
_____ Burbot

_____ Cabezon
_____ California Barracuda
_____ California Butterfly Ray
_____ California Corbina
_____ California Flyingfish
_____ California Grunion
_____ California Halibut
_____ California Killifish
_____ California Lizardfish
_____ California Moray
_____ California Roach
_____ California Scorpionfish
_____ California Sheephead
_____ California Tonguefish
_____ California Yellowtail
_____ Canary Rockfish
_____ Carp
_____ Central Stoneroller
_____ Channel Catfish
_____ Chilipepper
_____ Chinook Salmon
_____ Chiselmouth
_____ Chum Salmon
_____ Coho Salmon
_____ Colorado Squawfish
_____ Common Mola
_____ Common Thresher Shark
_____ Copper Rockfish
_____ Cow Rockfish
_____ Creek Chub
_____ Cui-ui
_____ Cutthroat Trout

_____ Deepwater Sculpin
_____ Delta Smelt
_____ Desert Dace
_____ Desert Pupfish
_____ Desert Sucker
_____ Devils Hole Pupfish
_____ Diamond Turbot
_____ Dolly Varden
_____ Dover Sole

_____ Emerald Shiner
_____ English Sole
_____ Eulachon

_____ Fantail Sole
_____ Fathead Minnow
_____ Flag Rockfish

_____ Flannelmouth Sucker
_____ Flathead Chub
_____ Freshwater Drum

_____ Garibaldi
_____ Giant Kelpfish
_____ Giant Sea Bass
_____ Gila Topminnow
_____ Gila Trout
_____ Gizzard Shad
_____ Golden Shiner
_____ Golden Trout
_____ Goldeye
_____ Goldfish
_____ Gray Smoothhound
_____ Greenland Halibut
_____ Greenland Shark
_____ Greenspotted Rockfish
_____ Green Sturgeon
_____ Green Sunfish
_____ Grunt Sculpin
_____ Gulf Grunion

_____ Halfmoon
_____ Hardhead
_____ Hitch
_____ Horn Shark
_____ Hornyhead Turbot
_____ Humpback Chub

_____ Iowa Darter

_____ Jack Mackerel
_____ Jacksmelt

_____ Kelp Bass
_____ Kelp Greenling
_____ Kelp Pipefish
_____ King-of-the-Salmon

_____ Lahontan Redside
_____ Lake Chub
_____ Lake Sturgeon
_____ Lake Trout
_____ Lake Whitefish
_____ Largemouth Bass
_____ Largescale Sucker
_____ Leatherside Chub
_____ Leopard Dace
_____ Leopard Shark
_____ Lingcod
_____ Little Colorado
  Spinedace
_____ Loach Minnow
_____ Logperch
_____ Longfin Dace
_____ Longfin Smelt
_____ Longjaw Mudsucker
_____ Longnose Dace
_____ Longnose Gar
_____ Longnose Skate

_____ Longnose Sucker
_____ Lost River Sucker

_____ Mako Shark
_____ Mexican Tetra
_____ Moapa Dace
_____ Monkeyface-eel
_____ Mosquitofish
_____ Mottled Sculpin
_____ Mountain Sucker
_____ Mountain Whitefish

_____ Night Smelt
_____ Northern Anchovy
_____ Northern Pike
_____ Northern Redbelly
  Dace
_____ Northern Squawfish

_____ Ocean Whitefish
_____ Olive Rockfish
_____ Onespot Fringehead
_____ Opah
_____ Opaleye

_____ Pacific Angel Shark
_____ Pacific Bonito
_____ Pacific Butterfish
_____ Pacific Electric Ray
_____ Pacific Grenadier
_____ Pacific Hagfish
_____ Pacific Hake
_____ Pacific Halibut
_____ Pacific Herring
_____ Pacific Lamprey
_____ Pacific Mackerel
_____ Pacific Ocean Perch
_____ Pacific Sanddab
_____ Pacific Sandfish
_____ Pacific Sardine
_____ Pacific Saury
_____ Pacific Sleeper Shark
_____ Pacific Smoothtongue
_____ Pacific Spiny
  Lumpsucker
_____ Pacific Tomcod
_____ Paddlefish
_____ Pahranagat Spinedace
_____ Painted Greenling
_____ Pallid Sturgeon
_____ Peamouth
_____ Petrale Sole
_____ Pile Perch
_____ Pink Salmon
_____ Piute Sculpin
_____ Plainfin Midshipman
_____ Plains Killifish
_____ Plains Minnow
_____ Plains Topminnow
_____ Pond Smelt
_____ Prickly Sculpin

_____ Pumpkinseed

_____ Queenfish

_____ Railroad Valley
Springfish
_____ Rainbow Perch
_____ Rainbow Trout
_____ Ratfish
_____ Razorback Sucker
_____ Red Brotula
_____ Red River Pupfish
_____ Red Shiner
_____ Redside Shiner
_____ Redtail Surfperch
_____ Rex Sole
_____ Riffle Sculpin
_____ Rio Grande Sucker
_____ Rio Grande Chub
_____ River Carpsucker
_____ River Shiner
_____ Rock Prickleback
_____ Roundnose Minnow
_____ Round Stingray
_____ Roundtail Chub
_____ Round Whitefish
_____ Rubberlip Seaperch

_____ Sablefish
_____ Sacramento Blackfish
_____ Sacramento Perch
_____ Sacramento Squawfish
_____ Sacramento Sucker
_____ Sailfin Molly
_____ Salt Creek Pupfish
_____ Sand Roller
_____ Sand Shiner
_____ Sargo
_____ Sauger
_____ Señorita
_____ Sevengill Shark
_____ Shiner Perch
_____ Shorthead Redhorse
_____ Shortspine Thronyhead
_____ Shovelnose Guitarfish
_____ Shovelnose Sturgeon
_____ Silver Pike
_____ Silvery Minnow
_____ Smallmouth Bass
_____ Smallmouth Buffalo
_____ Sockeye Salmon
_____ Sonora Sucker
_____ Soupfin Shark
_____ Speckled Chub
_____ Speckled Dace
_____ Specklefin Midshipman
_____ Spikedace
_____ Spiny Dogfish
_____ Spotfin Croaker
_____ Spottail Shiner
_____ Spotted Cusk-eel

_____ Staghorn Sculpin
_____ Starry Flounder
_____ Stonecat
_____ Striped Bass
_____ Striped Marlin
_____ Striped Perch
_____ Sturgeon Poacher
_____ Surf Smelt
_____ Swell Shark
_____ Swordfish

_____ Tahoe Sucker
_____ Texas Gambusia
_____ Thornback
_____ Threadfin Shad
_____ Threespine Stickleback
_____ Topsmelt
_____ Troutperch
_____ Tui Chub
_____ Tule Perch

_____ Utah Chub
_____ Utah Sucker

_____ Vermilion Rockfish
_____ Virgin Spinedace

_____ Walleye
_____ Walleye Pollock
_____ Walleye Surfperch
_____ Western Silvery
Minnow
_____ White Bass
_____ White Catfish
_____ White Crappie
_____ White Croaker
_____ White River Spinedace
_____ White River Springfish
_____ White Seabass
_____ White Shark
_____ White Sturgeon
_____ White Sucker
_____ Wolf-eel
_____ Woundfin

_____ Yellow Bullhead
_____ Yelloweye Rockfish
_____ Yellow Perch

_____ Zebra Goby

MOLLUSKS

_____ Adams' Pyramidella
_____ Agate Jewel Box
_____ Alaska Astarte
_____ Almond Yoldia
_____ Angular Unicorn
_____ Apple Seed
_____ Approximate Lucine
_____ Arctic Rock Borer

_____ Baby's Ear Shell
_____ Baily's Miniature Ark
_____ Banded Pheasant Shell
_____ Banded Top Shell
_____ Bark Semele
_____ Barrel Bubble
_____ Barrel Shell
_____ Basket Clam
_____ Belcher's Chorus Shell
_____ Bent-nosed Macoma
_____ Bittersweet
_____ Black Abalone
_____ Black Katy Chiton
_____ Black Top Shell
_____ Blue Mussel
_____ Blunt Razor Clam

_____ California Bean Clam
_____ California Blue Doris
_____ California Coffee Bean
_____ California Cone
_____ California Cumingia
_____ California Dipper
_____ California Dog Whelk
_____ California Frog Shell
_____ California Horn Shell
_____ California Lucine
_____ California Lyonsia
_____ California Mussel
_____ California Nuttall Chiton
_____ California Oyster
_____ California Pea-pod Shell
_____ California Piddock
_____ California Soft-shelled Clam
_____ California Surf Clam
_____ California Venus
_____ Carinate Chink Shell
_____ Carpenter's Cardita
_____ Carpenter's Doris
_____ Carpenter's Horse Mussel
_____ Carpenter's Tellin
_____ Channeled Dog Whelk
_____ Channeled Dogwinkle
_____ Channeled Top Shell
_____ Checkered Periwinkle
_____ Checkered Unicorn
_____ Chestnut Cowrie
_____ Circled Rock Shell
_____ Columbian Amphissa
_____ Comb Yoldia
_____ Common Pacific Octopus
_____ Common Pacific Squid
_____ Common Piddock
_____ Common Soft-shelled Clam
_____ Common Yellow Doris

_____ Conspicuous Chiton
_____ Cooper's Dog Whelk
_____ Coopers Turret
_____ Cooper's Yoldia
_____ _Crassostrea laperousii_

_____ Deepwater Gari
_____ Dire Whelk
_____ Doleful Tower
_____ Dotted Pandora
_____ Dwarf Olive

_____ Emarginate Dogwinkle

_____ False Angel Wing
_____ False Donax
_____ Fat Dog Whelk
_____ Fat Horse Mussel
_____ Fat Sphenia
_____ Feathery Shipworm
_____ Fenestrate Limpet
_____ Festive Rock Shell
_____ File Limpet
_____ Fine-lined Lucine
_____ Fine-sculptured Turbonilla
_____ Fingered Limpet
_____ Flat Abalone
_____ Flat Periwinkle
_____ Fragile Sphenia
_____ Frilled California Venus
_____ Frilled Dogwinkle
_____ Frill-winged Rock Shell

_____ Gem Rock Shell
_____ Geoduck
_____ Giant Chiton
_____ Giant Egg Cockle
_____ Giant Keyhole Limpet
_____ Giant Pacific Cockle
_____ Giant Rock Scallop
_____ Gibbose Cavoline
_____ Gould's Baby Chiton
_____ Gould's Bubble
_____ Gould's Paper Bubble
_____ Great Owl Limpet
_____ Green Abalone
_____ Green Paper Bubble
_____ Grooved Macoma

_____ Hairy Chiton
_____ Half Slipper Shell
_____ Hartweg's Chiton
_____ Heart Rock Dweller
_____ Heath's Chiton
_____ Hemphill's Surf Clam
_____ Hemphill's File
_____ Hinds' Chiton
_____ Hinds' Scallop
_____ Hooked Slipper Shell

MOLLUSKS

_____ Hooked Surf Clam
_____ Hopkins' Doris
_____ Horse Mussel
_____ Hundred-lined Cockle

_____ Iceland Scallop
_____ Ida's Tellin
_____ Indented Macoma
_____ Indian Money Tusk

_____ Jackknife Clam
_____ Japanese Oyster
_____ Jingle Shell
_____ Joseph's Coat Amphissa

_____ Keeled Dove Shell
_____ Kellet's Whelk
_____ Kelp-weed Scallop

_____ Leafy Hornmouth
_____ Lean Dog Whelk
_____ _Lepidochitona keepiana_
_____ Lined Chiton
_____ Little Bean Clam
_____ Little Black Mussel
_____ Little Egg Cockle
_____ Livid Macron
_____ Long-horned
        Hermissenda

_____ Maculated Doris
_____ Magdalena Chiton
_____ Many-ribbed Ark
_____ Merten's Chiton
_____ Modest Tellin
_____ Monterey Doris
_____ Mossy Chiton
_____ Mottled Chiton
_____ Myra's Razor Clam

_____ Naval Shipworm
_____ Noble Pacific Doris
_____ Norris Top Shell
_____ Northern Gaper
_____ Northern Quahog
_____ Nuttall's Cockle
_____ Nuttall's Hornmouth
_____ Nuttall's Lucine

_____ Oldroyd's Dipper
_____ Olga's Paper Bubble
_____ Onyx Slipper Shell
_____ Orange-spiked Doris
_____ Orb Diplodonta
_____ Oval Piddock

_____ Pacific Gaper
_____ Pacific Littleneck
_____ Pacific Piddock
_____ Pacific Razor Clam
_____ Pacific Surf Clam

_____ Pacific Thracia
_____ Papillose Eolis
_____ Pea-pod Shell
_____ Pearly Monia
_____ Philippine Littleneck
_____ Phoenicean Whelk
_____ Pilsbry's Piddock
_____ Pink Abalone
_____ Pink Scallop
_____ Pinto Abalone
_____ Pismo Clam
_____ Platform Mussel
_____ Porter's Blue Doris
_____ Poulson's Rock Shell
_____ Purple Clam
_____ Purple Olive
_____ Pyramid Clio

_____ Recluz's Moon Shell
_____ Red Abalone
_____ Red Chiton
_____ Red Turban
_____ Regular Chiton
_____ Reversed Jewel Box
_____ Ridged Whelk
_____ Ringed Top Shell
_____ Rock Borer Mussel
_____ Rock-dwelling Semele
_____ Rock Entodesma
_____ Rosy Razor Clam
_____ Rough Keyhole Limpet
_____ Rough Limpet
_____ Rough-sided Littleneck

_____ San Diego Doris
_____ San Pedro Auger
_____ Santa Rosa Rock
_____ Scaled Worm Shell
_____ Sea Bottle Shell
_____ Seaweed Limpet
_____ Shield Limpet
_____ Short Thracia
_____ Six-sided Tusk
_____ Small Baby's Ear Shell
_____ Smooth Nut Shell
_____ Smooth Pacific Venus
_____ Smooth Tower
_____ Smooth Washington
        Clam
_____ Speckled Scallop
_____ Speckled Top Shell
_____ Stearn's Ear Shell
_____ Stearn's Mussel
_____ Straight Horse Mussel
_____ Strawberry Cockle
_____ Striate Cup-and-Saucer
        Shell
_____ Sunset Shell

_____ Tabled Whelk
_____ Taphria Nut

_____ Test's Limpet
_____ Thin-shelled Littleneck
_____ Threaded Abalone
_____ Three-spined Cavoline
_____ Three-winged Rock Shell
_____ Transparent Razor Clam
_____ Trellised Chiton
_____ Two-spotted Keyhole Limpet

_____ Unstable Limpet

_____ Veiled Chiton
_____ Volcano Limpet

_____ Washington Clam
_____ Wavy Turban
_____ Western Pandora
_____ Western Spoon Clam
_____ White Abalone
_____ White-capped Limpet
_____ White Chiton
_____ White Sand Macoma
_____ Woody Chiton
_____ Wroblewski's Wentletrap

_____ Yellow-rimmed Doris
_____ Yellow-spotted Doris

OTHER MARINE INVERTEBRATES

_____ *Abietinaria* (species)
_____ Aggregating Anemone
_____ *Alloioplana californica*
_____ *Allopora petrograpta*
_____ *Alpheus bellimanus*
_____ *Amphiodia occidentalis*
_____ *Amphiopholis pugetana*
_____ *Amphiporus bimaculatus*
_____ *Amphiporus imparispinosus*
_____ *Arabella iricolor*
_____ *Ascidia ceratodes*

_____ *Balanophyllia elegans*
_____ *Balanus cariosus*
_____ *Balanus glandula*
_____ *Balanus nubilus*
_____ *Balanus tintinnabulum californicus*
_____ Bat Star
_____ Beach Hopper
_____ *Beroe forskali*
_____ Black-tailed Shrimp
_____ Blue Mud Shrimp
_____ Boring Sponge

_____ Breadcrumb Sponge
_____ Brown Sea Anemone
_____ *Bugula californica*
_____ *Bugula neritina*
_____ By-the-Wind Sailor

_____ California Spiny Lobster
_____ *Chrysaora melanaster*
_____ *Chthamalus dalli*
_____ *Cirolana harfordi*
_____ *Cirratulus cirratus*
_____ Clam Worm
_____ *Cnemidocarpa finmarkiensis*
_____ Coon-striped Shrimp
_____ *Corynactis californica*
_____ *Crangon nigromaculata*
_____ *Crisis eburnea*
_____ *Crisia occidentalis*
_____ *Cucumaria lubrica*
_____ *Cucumaria miniata*

_____ Daisy Brittle Star
_____ Dawson's Sun Star
_____ Decorator Crab
_____ Driftwood Goose Barnacle
_____ Dungeness Crab

_____ *Emplectonema gracile*
_____ *Epiactis prolifera*
_____ *Eudendrium californicum*
_____ *Eudendrium rameum*
_____ *Eudistylia polymorpha*
_____ *Eulalia aviculiseta*
_____ *Eupentacta quinquesemita*

_____ Fat Innkeeper
_____ Flat-topped Crab

_____ *Garveia annulata*
_____ Ghost Shrimp
_____ Giant Red Urchin
_____ *Gnorimosphaeroma oregonensis*
_____ Graceful Crab
_____ Great Green Anemone
_____ Green Sea Urchin
_____ Gribbles

_____ Hairy Hermit
_____ Hairy Shore Crab
_____ *Haliclystus auricula*
_____ *Halosydna brevisetosa*
_____ *Hapalogaster cavicauda*
_____ *Hemipodus borealis*
_____ *Henricia leviuscula*
_____ *Heptacarpus paludicola*

OTHER MARINE INVERTEBRATES

_____ *Heptacarpus pictus*
_____ *Hippolyte californiensis*
_____ *Hydractinia* (species)
_____ Hydrocoral

_____ *Idotea resecata*
_____ *Idotea urotoma*
_____ *Idotea wosnesenskii*

_____ *Kaburakia excelsa*
_____ Kelp Crab

_____ Lace Coral
_____ Lamp Shell
_____ Leather Star
_____ *Leptasterias hexactis*
_____ *Leucilla nuttingi*
_____ *Leucosolenia eleanor*
_____ *Ligia occidentalis*
_____ *Ligia pallasii*
_____ *Limnoria lignorum*
_____ *Limnoria tripunctata*
_____ *Lissodendoryx firma*
_____ *Lophopanopeus bellus*
_____ *Lophopanopeus leucomanus*
_____ Lugworm

_____ Masking Crab
_____ *Membranipora membranacea*
_____ *Membranipora villosa*
_____ *Mimulus foliatus*
_____ Mole Crab
_____ Moon Jelly
_____ Moss Crab

_____ *Nephtys californiensis*
_____ *Notomastus tenuis*

_____ *Obelia geniculata*
_____ *Obelia longissima*
_____ *Ophioplocus esmarki*
_____ *Ophiothrix spiculata*
_____ *Ophlitaspongia pennata*
_____ *Orchestoidea corniculata*
_____ Ostrich-plume Hydroid

_____ *Pachycheles pubescens*
_____ Pacific Goose Barnacle
_____ *Pagurus granosimanus*
_____ *Pagurus hemphilli*
_____ *Pagurus samuelis*
_____ *Paranemertes peregrina*
_____ Pebble Crab
_____ *Pedicellina cernua*
_____ *Phascolosoma agassizii*
_____ *Phoronis vancouverensis*
_____ *Phoronopsis harmeri*

_____ *Phoronopsis viridis*
_____ *Phragmatopoma californica*
_____ Pink Shrimp
_____ *Pisaster brevispinus*
_____ *Pisaster giganteus*
_____ Pistol Shrimp
_____ *Polychoerus carmelensis*
_____ *Polyclinum planum*
_____ *Polyorchis penicillatus*
_____ Porcelain Crab
_____ Purple Encrusting Sponge
_____ Purple Sea Urchin
_____ Purple Shore Crab
_____ Purple Starfish
_____ Purple-striped Jellyfish

_____ Red Crab
_____ *Renilla amethystina*
_____ Rock Crab

_____ Sand Crab
_____ Sand Dollar
_____ Sand Flea
_____ *Scleroplax granulata*
_____ Sea Blubber
_____ Sea Gooseberries
_____ Sea Pansy
_____ Sea Pork
_____ Sea Slug
_____ Sea Vase
_____ *Serpula vermicularis*
_____ *Sipunculus nudus*
_____ Skeleton Shrimp
_____ Slender Sea Pen
_____ Snapping Shrimp
_____ Spiny Sand Crab
_____ Stalked Brachiopod
_____ *Stichopus parvimensis*
_____ Stimpson's Sun Star
_____ Striped Shore Crab
_____ *Styela gibbsii*
_____ *Styela montereyensis*
_____ Sunflower Starfish
_____ *Syncoryne mirabilis*

_____ *Tealia crassicornis*
_____ Thatched Barnacle
_____ *Thelepus crispus*
_____ Thick-clawed Crab
_____ *Tubularia crocea*
_____ *Tubularia marina*

_____ Umbrella Crab
_____ *Urechis echiurus alaskensis*

_____ Yellow Shore Crab

Index

Animal species are indexed in general by common name. A number of species, however, especially the marine invertebrates, do not have common names, and these species are indexed by their Latin (genus-species) nomenclature. Higher taxa, such as classes, orders, and families, are indexed by both common and scientific names. A number in *italics* indicates a text page on which the species is illustrated; a reference in **bold face** indicates the number of the color plate on which the species is illustrated and the position of the species on the color plate.